Fundamentals of Air Pollution

FOURTH EDITION

Authors of Third Edition

RICHARD W. BOUBEL

Department of Mechanical Engineering
Oregon State University
Corvallis, Oregon

DONALD L. FOX

Department of Environmental Science
School of Public Health
University of North Carolina
Chapel Hill, North Carolina

D. BRUCE TURNER

Trinity Consultants, Inc.
Chapel Hill, North Carolina

ARTHUR C. STERN

(14 March 1909–17 April 1992)

Fundamentals of Air Pollution

FOURTH EDITION

DANIEL A. VALLERO
Civil and Environmental Engineering Department
Pratt School of Engineering
Duke University
Durham, North Carolina

AMSTERDAM • BOSTON • HEIDELBERG • LONDON • NEW YORK • OXFORD
PARIS • SAN DIEGO • SAN FRANCISCO • SINGAPORE • SYDNEY • TOKYO
Academic Press is an imprint of Elsevier

Academic Press is an imprint of Elsevier
30 Corporate Drive, Suite 400, Burlington, MA 01803, USA
525 B Street, Suite 1900, San Diego, California 92101-4495, USA
84 Theobald's Road, London WC1X 8RR, UK

∞ This book is printed on acid-free paper.

First Edition 1973
Second Edition 1984
Third Edition 1994
Fourth Edition 2008

Library of Congress Cataloging-in-Publication Data
Vallero, Daniel A.
Fundamentals of air pollution / Daniel A. Vallero — 4th ed.
p. cm.
Includes index.
ISBN 978-0-12-373615-4 (alk. paper)
1. Air—Pollution. I. Title.
TD883.V25 2007
628.5'3—dc22 2007028062

British Library Cataloguing-in-Publication Data
A catalogue record for this book is available from the British Library

ISBN: 978-0-12-373615-4

For information on all Academic Press publications
visit our website at www.books.elsevier.com

Typeset by Charon Tec Ltd (A Macmillan Company), Chennai, India
www.charontec.com
Printed and bound in the USA

07 08 09 10 9 8 7 6 5 4 3 2 1

Dedicated to the four authors of the previous edition.

I am standing on the shoulders of giants.

Contents

2 *The Earth's Atmosphere*

3 *Scales of the Air Pollution Problem*

Part II

The Physics and Chemistry of Air Pollution

4 *Air Pollution Physics*

5 *The Physics of the Atmosphere*

6 *Air Pollution Systems and Processes*

Part III

Risks from Air Pollution

Part VII

Preventing and Controlling Air Pollution

Preface to the Third Edition

The authors of this book include a chemist (Donald L. Fox), a meteorologist (D. Bruce Turner), and a mechanical engineer (Richard W. Boubel). This 1:1:1 ratio has some relevance in that it approximates the ratio of those professionally involved in the field of air pollution. In the environmental protection and management field, the experience of the recent past has been that physicists and electrical engineers have been most attracted to the radiation, nuclear, and noise areas; biologists and civil engineers to the aquatic and solid waste areas; chemists, meteorologists, and chemical and mechanical engineers to the area of air pollution and its control. These remarks are not intended to exclude all others from the party (or from this course). The control of air pollution requires the combined efforts of all the professions mentioned, in addition to the input of physicians, lawyers, and social scientists. However, the professional mix of the authors, and their expectation of a not-too-dissimilar mix of students using this book, forewarns the tenor of its contents and presentation.

Although this book consists of six parts and three authors, it is not to be considered six short books put together back-to-back to make one large one. By and large, the several parts are the work of more than one author. Obviously, the meteorologist member of the author team is principally responsible for the part of the book concerned with the meteorology of air pollution, the chemist author for the chapters on chemistry, and the engineer author for those on engineering. However, as you will see, no chapters are signed, and all authors accept responsibility for the strengths and weaknesses of the chapters and for the book as a whole.

In the 20 years since publication of the first edition of *Fundamentals of Air Pollution* (1973), and the 9 years since the second edition (1984), the fundamentals have not changed. The basic physics, chemistry, and engineering are still the same, but there is now a greater in-depth understanding of their application to air pollution. This edition has been edited, revised, and updated to include the new technology available to air pollution practitioners. Its contents are also influenced to a great extent by the passage of the US Clean Air Act Amendments of 1990 (CAAA90). These amendments have changed the health and risk-based regulations of the US Clean Air Act to technology-driven regulations with extensive penalty provisions for noncompliance.

We have added more detailed discussion of areas that have been under intensive study during the past decade. There has been a similar need to add discussion of CAAA90 and its regulatory concepts, such as control of air toxics, indoor air pollution, pollution prevention, and trading and banking of emission rights. Ten more years of new data on air quality have required the updating of the tables and figures presenting these data.

We have expanded some subject areas, which previously were of concern to only a few scientists, but which have been popularized by the media to the point where they are common discussion subjects. These include "Global Warming," "The Ozone Hole," "Energy Conservation," "Renewable Resources," and "Quality of Life."

With each passing decade, more and more pollution sources of earlier decades become obsolete and are replaced by processes and equipment that produce less pollution. At the same time, population and the demand for products and services increase. Students must keep these concepts in mind as they study from this text, knowing that the world in which they will practice their profession will be different from the world today.

The viewpoint of this book is first that most of the students who elect to receive some training in air pollution will have previously taken courses in chemistry at the high school or university level, and that those few who have not would be well advised to defer the study of air pollution until they catch up on their chemistry.

The second point of view is that the engineering design of control systems for stationary and mobile sources requires a command of the principles of chemical and mechanical engineering beyond that which can be included in a one-volume textbook on air pollution. Before venturing into the field of engineering control of air pollution, a student should, as a minimum, master courses in internal combustion engines, power plant engineering, the unit processes of chemical engineering, engineering thermodynamics, and kinetics. However, this does not have to be accomplished before taking a course based on this book but can well be done simultaneously with or after doing so.

The third point of view is that *no one*, regardless of their professional background, should be in the field of air pollution control unless they sufficiently understand the behavior of the atmosphere, which is the feature that differentiates *air* pollution from the other aspects of environmental protection and management. This requires a knowledge of some basic atmospheric chemistry in addition to some rather specialized air pollution meteorology. The viewpoint presented in the textbook is that very few students using it will have previously studied basic meteorology. It is hoped that exposure to air pollution meteorology at this stage will excite a handful of students to delve deeper into the subject. Therefore, a relatively large proportion of this book has been devoted to meteorology because of its projected importance to the student.

The authors have tried to maintain a universal point of view so that the material presented would be equally applicable in all the countries of the world. Although a deliberate attempt has been made to keep American provincialism out of the book, it has inevitably crept in through the exclusive use of English language references and suggested reading lists, and the preponderant use of American data for the examples, tables, and figures. The saving grace in this respect is that the principles of chemistry, meteorology, and engineering are universal.

As persons who have dedicated all or significant parts of their professional careers to the field of air pollution, the authors believe in its importance and relevance. We believe that as the world's population increases, it will become increasingly important to have an adequate number of well-trained professions engaged in air pollution control. If we did not believe this, it would have been pointless for us to have written this textbook.

We recognize that, in terms of short-term urgency, many nations and communities may rightly assign a lower priority to air pollution control than to problems of population, poverty, nutrition, housing, education, water supply, communicable disease control, civil rights, mental health, aging, or crime. Air pollution control is more likely to have a higher priority for a person or a community already reaping the benefits of society in the form of adequate income, food, housing, education, and health care than for persons who have not and may never reap these benefits.

However, in terms of long-term needs, nations and communities can ignore air pollution control only at their peril. A population can subsist, albeit poorly, with inadequate housing, schools, police, and care of the ill, insane, and aged; it can also subsist with a primitive water supply. The ultimate determinants for survival are its food and air supplies. Conversely, even were society to succeed in providing in a completely adequate manner all of its other needs, it would be of no avail if the result were an atmosphere so befouled as not to sustain life. The long-term objective of air pollution control is to allow the world's population to meet all its needs for energy, goods, and services without sullying its air supply.

Preface to the Fourth Edition

In the Preface to the Third Edition of this book, Donald L. Fox, D. Bruce Turner, and Richard W. Boubel expressed the importance of a multidisciplinary approach to air pollution. I wholeheartedly agree. Nothing has changed in this regard, making it a daunting challenge to update the impressive work of these renowned experts (as well as the late Arthur C. Stern in previous editions). It was easier to add new material than to remove old material. A new edition is an optimization exercise. The book must not change so much that professors using it have to change the course structure so severely that it constitutes a completely new text. On the other hand, a text must be up to date in terms of current technologies and programs, as well as in addressing threats on the horizon.

Over a decade has passed since the publication of previous version. From a regulatory perspective, this is a very long time. By conventional measures, such as the National Ambient Air Quality Standards, the past decade has been very successful. But, science marches on. I recall that in the 1970s, detection in the parts per million (ppm) was impossible for most compounds. During the 1980s detection limits continued to decrease. Now, detections have improved to allow for measurements below parts per billion for many compounds. We have also witnessed sea changes in risk assessment and management. For example, the US Environmental Protection Agency laboratories were realigned to address risks, with separate laboratories to conduct research exposure, effects, risk characterization, and risk reduction.

Indeed, the previous authors were quite prescient in predicting the effects of the then newly amended Clean Air Act. The major changes started to kick in as the focus moved from technology-based approaches (best available and maximum achievable control technologies) to risk-based decision-making (residual risks remaining even after the required control technologies).

The fundamentals of the science underlying air pollution have not changed, but their applications and the appreciation of their impacts have. For example, I have endeavored to enhance the discussion and explanation of the physical and chemical processes at work, particularly those related to air toxics. This has been a tendency through all four editions. New technologies must be explained, better models and computational methods have become available, analytical procedures have evolved and improved, and acute and chronic effects have become better understood. All of these have enhanced the science and engineering knowledge available to practitioners, teachers, and students. And, the savvy of the lay public about air pollution has grown substantially during the previous decade.

I am indebted to my fellow scientists and engineers for their insights and comments on how to incorporate the new trends. I particularly want to note Alan Huber, who shared his work in atmospheric dispersion modeling, especially computational fluid dynamics. Others include Russ Bullock (mercury fate and transport), Paul Lioy and Panos Georgopoulos (modeling), Mark Wiesner (nanotechnology), John Kominsky and Mike Beard (asbestos), and Aarne Vesilind (history).

As in previous editions, my expectation is that the reader has received some formal background in chemistry. I agree with the previous authors that anyone interested in air pollution must have a solid grounding in chemistry and the physical sciences. Without it, there is no way of knowing whether a rule or policy is plausible. I have seen too many instances of "junk science" in environmental decision-making. Often, these are underlain with good intentions. But, so-called "advocacy" does not obviate the need for sound science. That said, with a bit of effort, much of this edition can be a useful tool to any audience who is motivated to understand the what, how and why of air pollution.

Another trend that I have hoped to capture is the comprehensiveness needed to address air quality. A problem need not occur if the processes leading to air pollution are approached from a life cycle or "green" perspective. This goes beyond pollution prevention and calls for an integrated and sustainable view. I have dedicated an entire chapter to this emergent environmental expectation.

The authors of the previous edition introduced discussions about some emerging continental and global threats to the atmosphere. Since then, the urgency of some has abated (e.g. acid rain and some threats to the ozone layer), some have increased in concern (e.g. global warming), and others have continued but the contaminants of concern have varied (long-range transport of persistent chemicals). The scientific credibility of arguments for and against regulatory and other actions has been uneven. The best defense against bad policy decisions is a strong foundation in the physical sciences.

Let me rephrase that a bit more proactively and optimistically:

> My overall objective of this book is to give you, the reader, the ability to design and apply the tools needed to improve and sustain the quality of the air we breathe for many decades. These tools can only be trusted if they are thoroughly grounded in the *Fundamentals of Air Pollution*.

DAV

Part I

Air Pollution Essentials

1

The Changing Face of Air Pollution

I. DEFINING AIR POLLUTION

Air pollution: The presence of contaminants or pollutant substances in the air that interfere with human health or welfare, or produce other harmful environmental effects.

United States Environmental Protection Agency (2007)
"Terms of Environment: Glossary, Abbreviations and Acronyms"

The US Environmental Protection Agency's (EPA) definition is a good place to start thinking about what makes something an air pollutant. The key verb in the definition is "interfere." Thus, we obviously have a desired state, but these substances are keeping us from achieving that state. So, then, what is that state? The second part of the definition provides some clues; namely, air must be of a certain quality to support human and other life.

Some decades ago, few people were familiar with the term pollution. Of course, most people knew that something was amiss when their air was filled with smoke or when they smell an unpleasant odor. But, for most pollutants, those that were not readily sensed, a baseline had to be set to begin to take action. Environmental professionals had to reach agreement on what is and what is not "pollution." We now have the opposite problem, nearly everyone has heard about pollution and many may have their own working definitions. So, once again, we have to try to reach consensus on what the word actually means.

Another way to look at the interferences mentioned in the EPA definition is to put them in the context of "harm." The objects of the harm have received varying levels of interests. In the 1960s, harm to ecosystems, including threats to the very survival of certain biological species was paramount. This concern was coupled with harm to humans, especially in terms of diseases directly associated with obvious episodes, such as respiratory diseases and even death associated with combinations of weather and pollutant releases.

Other emerging concerns were also becoming apparent, including anxiety about nuclear power plants, particularly the possibilities of meltdown and the generation of cancer-causing nuclear wastes, petrochemical concerns, such as the increasing production and release of ominous-sounding chemicals like Dichloro-Diphenyl-Trichloroethane (DDT) and other pesticides, as well as spills of oil and other chemicals. These apprehensions would increase in the next decade, with the public's growing wariness about "toxic" chemicals added to the more familiar "conventional" pollutants like soot, carbon monoxide, and oxides of nitrogen and sulfur. The major new concern about toxics was cancer. The next decades kept these concerns, but added new ones, including threats to hormonal systems in humans and wildlife, neurotoxicity (especially in children), and immune system disorders.

Growing numbers of studies in the last quarter of the twentieth century provided evidence linking disease and adverse effects to extremely low levels of certain particularly toxic substances. For example, exposure to dioxin at almost any level above what science could detect could be associated with numerous adverse effects in humans. During this time, other objects of pollution were identified, including loss of aquatic diversity in lakes due to deposition of acid rain. Acid deposition was also being associated with the corrosion of materials, including some of the most important human-made structures, such as the pyramids in Egypt and monuments to democracy in Washington, DC. Somewhat later, global pollutants became the source of public concern, such as those that seemed to be destroying the stratospheric ozone layer or those that appeared to be affecting the global climate.

This escalation of awareness of the multitude of pollutants complicated matters. For example, many pollutants under other circumstances would be "resources," such as compounds of nitrogen. In the air, these compounds can cause respiratory problems directly or, in combination with hydrocarbons and sunlight indirectly can form ozone and smog. But, in the soil, nitrogen compounds are essential nutrients. So, it is not simply a matter of "removing" pollutants, but one of managing systems to ensure that optimal conditions for health and environmental quality exist. When does something in our air change from being harmless or even beneficial to become harmful? Impurities are common, but in excessive quantities and in the wrong places they become harmful. One of the most interesting definitional quandaries about pollution has come out of the water pollution literature, especially by the language in the Federal Water Pollution Control Act Amendments of 1972 (Public Law 92-500). The objective of this law is to restore and maintain the chemical, physical, and biological integrity of the nation's waters. To achieve this objective, the law set two goals: the elimination of the discharge of all pollutants into the navigable waters of the United States by 1985; and to provide an interim level of water quality to protect fish, shellfish, and wildlife and recreation by 1983.[1] Was Congress serious? Could they really mean that they had expected all sources that drained into US lakes and rivers to be completely free of pollutants in 13 years? Or did this goal hinge upon the definition of pollutant? In other words, even toxic substances are not necessarily "pollutants" if they exist below a threshold of harm. In light of the fact that this same law established so-called "effluent limitations," there is a strong likelihood that the definition called for in this goal was concentration-based.[2]

This paradigm spilled over into air pollution circles. More recently, the term "zero emission" has been applied to vehicles, as the logical next step following low emission vehicles (LEVs) and ultra-low emission vehicles (ULEVs) in recent years. However, zero emissions of pollutants will not be likely for the foreseeable future, especially if one considers that even electric cars are not emission free, but actually emission trading, since the electricity is generated at a power plant that is emitting pollutants as it burns fossil fuels or has the problem of radioactive wastes if it is a nuclear power plant. Even hydrogen, solar and wind systems are not completely pollution free since the parts and assemblages require energy and materials that may even include hazardous substances.

These definitional uncertainties beg the question, then, of when does an impurity become a pollutant? Renaissance thinking may help us here. Paracelsus, the sixteenth century scientist is famous for his contention that "dose alone makes a poison… . All substances are poisons; there is none

[1] 33 USC 1251.

[2] In fact, my own environmental career began shortly after the passage of this law, when it, along with the National Environmental Policy Act and the Clean Air Act of 1970, was establishing a new environmental policy benchmark for the United States. At the time environmentalists recited an axiom frequently: "Dilution is not the solution to pollution!" I recall using it on a regular basis myself. However, looking back over those three decades, it seems the adage was not completely true. Cleanup levels and other thresholds are concentration based, so if one does an adequate job in diluting the concentrations (e.g. dioxin concentrations below 1 part per billion, ppb), one has at least in part solved that particular pollution problem. Also, when it came to metal pollution, dilution was a preferred solution, since a metal is an element and cannot be destroyed. A sufficient amount of the metal wastes are removed from water or soil and moved to a permanent storage site. The only other engineering solution to metal pollution was to change its oxidation state and chemical species, which is not often preferable because when environmental conditions change, so often do the oxidation states of the metals, allowing them to again become toxic and bioavailable.

which is not a poison. The right dose differentiates a poison and a remedy."[3] Paracelsus' quote illuminates a number of physical, chemical, and biological concepts important to understanding air pollution. Let us consider two. First, the poisonous nature, i.e. the toxicology, of a substance must be related to the circumstances of exposure. In other words, to understand a pollutant, one must appreciate its context. Air pollutants become a problem when they come into contact with the receptor. This leads to some important questions that must be answered if we are to address air pollution:

1. What is the physical, chemical, and biological nature of the agent to which the receptor (e.g. a person, an endangered species, or an entire population or ecosystem) is exposed?
2. What is that person's existing health status?
3. What is the condition of the ecosystem?
4. What are the chemical composition and physical form of the contaminant?
5. Is the agent part of a mixture, or is it a pure substance?
6. How was the person or organism exposed; from food, drink, air, through the skin?

These and other characterizations of a contaminant must be known to determine the extent and degree of harm.

The second concept highlighted by Paracelsus is that dose is related to response. This is what scientists refer to as a biological gradient or a *dose–response* relationship. Under most conditions, the more poison to which one is exposed the greater the harm.

The classification of harm is an expression of *hazard*, which is a component of risk. The terms hazard and risk are frequently used interchangeably in everyday parlance, but hazard is actually a component of risk. As we will see throughout this text, hazard is not synonymous with risk. A hazard is expressed as the potential of unacceptable outcome, while risk is the likelihood (i.e. probability) that such an adverse outcome will occur. A hazard can be expressed in numerous ways. For chemical or biological agents, the most important hazard is the potential for disease or death (referred to in medical literature as "morbidity" and "mortality," respectively). So, the hazards to human health are referred to collectively in the medical and environmental sciences as "toxicity." Toxicology is chiefly concerned with these health outcomes and their potential causes.

To scientists and engineers, risk is a straightforward mathematical and quantifiable concept. Risk equals the probability of some adverse outcome. Any risk is a function of probability and consequence.[4] The consequence can take many forms. In environmental sciences, a consequence is called a "hazard."

[3] Kreigher, W. C., Paracelsus: dose response in *Handbook of Pesticide Toxicology*, (San Diego, C. A, Kreiger, R., Doull, J., and Ecobichon, D., eds.), 2nd ed. Elsevier Academic Press, 2001. New York, NY.

[4] Lewis, H.W., *Technological Risk*, Chapter 5: *The Assessment of Risk*. W.W. Norton & Company, Inc., New York, 1990.

Risk, then, is a function of the particular hazard and the chances of person (or neighborhood or workplace or population) being exposed to the hazard. For air pollution, this hazard often takes the form of toxicity, although other public health and welfare hazards abound.

II. THE EMERGENCE OF AIR POLLUTION SCIENCE, ENGINEERING, AND TECHNOLOGY

Environmental science and engineering are young professions compared to many other disciplines in the physical and natural sciences and engineering. In a span of just a few decades, advances and new environmental applications of science, engineering, and their associated technologies have coalesced into a whole new way to see the world. Science is the explanation of the physical world, while engineering encompasses applications of science to achieve results. Thus, what we have learned about the environment by trial and error has incrementally grown into what is now standard practice of environmental science and engineering. This heuristically attained knowledge has come at a great cost in terms of the loss of lives and diseases associated with mistakes, poor decisions (at least in retrospect), and the lack of appreciation of environmental effects.

Environmental awareness is certainly more "mainstream" and less a polarizing issue than it was in the 1970s, when key legislation reflected the new environmental ethos (see Fig. 1.1). There has been a steady march of advances in environmental science and engineering for several decades, as evidenced by the increasing number of Ph.D. dissertations and credible scientific journal articles addressing a myriad of environmental issues. Corporations and government agencies, even those whose missions are not considered to be "environmental," have established environmental programs.

Arguably, our understanding of atmospheric processes is one of the more emergent areas of environmental science and technology; growing from the increasing awareness of air pollution and advances of control technologies in the twentieth century. However, the roots of the science of air pollution can be traced to the Ancients.

The environmental sciences, including its subdisciplines specializing in air pollution, apply the fundamentals of chemistry, physics, and biology, and their derivative sciences such as meteorology, to understand these abiotic[5]

[5] The term "abiotic" includes all elements of the environment that are non-living. What is living and non-living may appear to be a straightforward dichotomy, but so much of what we call "ecosystems" is a mixture. For example, some soils are completely abiotic (e.g. clean sands), but others are rich in biotic components, such as soil microbes. Vegetation, such as roots and rhizomes are part of the soil column (especially in the "A" horizon or topsoil). Formerly living substances, such as detritus exist as lignin and cellulose in the soil organic matter (SOM). In fact, one of the problems with toxic chemicals is that the biocidal properties kill living organisms, reducing or eliminating the soil's productivity.

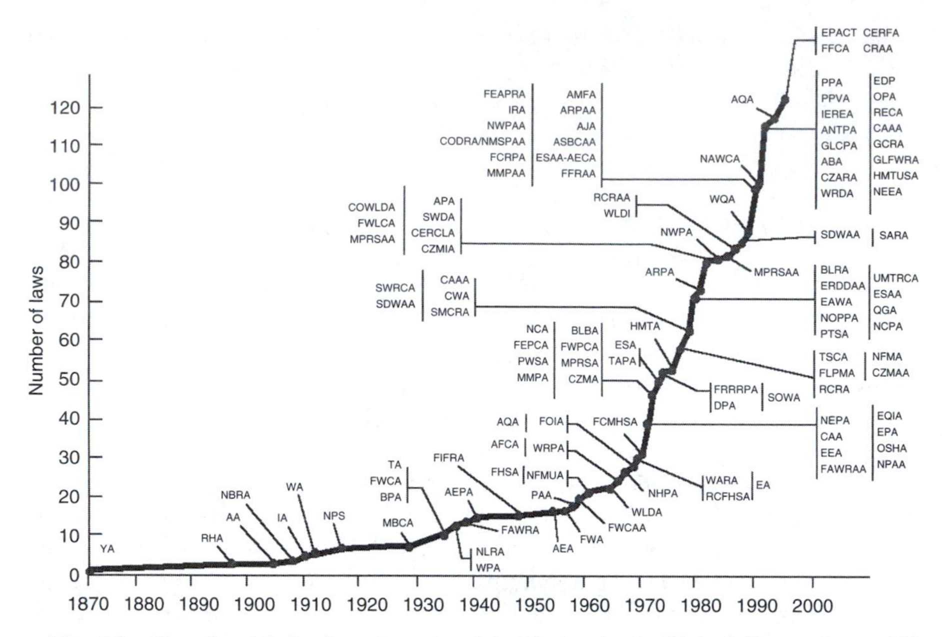

Fig. 1.1. Growth of federal environmental legislation in the United States. *Source*: US Environmental Protection Agency, 2004, Shonnard, D. R., Chapter 3, http://www.epa.gov/oppt/greenengineering/images/regulation_big.jpg; and Allen, D. T., and Shonnard, D. R., *Green Engineering: Environmentally Conscious Design of Chemical Processes*. Prentice Hall, Upper Saddle River, NJ, 2002.

and biotic relationships. Expanding these observations to begin to control outcomes is the province of environmental engineering.

As scientists often do, systematic and specific explanations must be applied to practical knowledge. So, biologists and their subdisciplines began to specialize in what came to be known as the environmental sciences. Health scientists, like Paracelsus and William Harvey, provided insights into how the human body interacts with and reacts to environmental stimuli. In fact, Paracelsus' studies of metal contamination and exposure to miners may well be among the earliest examples of *environmental epidemiology*.

Not only are the environmental disciplines young, but also many of the environmental problems faced today differ from most of the earth's history. The difference is in both kind and degree. For example, the synthesis of chemicals, especially organic compounds has grown exponentially since the mid-1900s. Most organisms had no mechanisms to metabolize and eliminate these new compounds. Also, stresses put on only small parts of ecosystems prior to the Industrial Revolution were small in extent of damage. For example, pollutants have been emitted into the atmosphere throughout human history, but only recently were such emissions so large and long lasting, or of pollutants with such high toxicity, that they have diminished the quality of entire *airsheds*.[6]

DISCUSSION: WHAT IS THE DIFFERENCE BETWEEN AN ENVIRONMENTAL ENGINEER AND A SANITARY ENGINEER?

When air pollution engineering first became recognized as a unique discipline, most of the engineers involved called themselves *sanitary engineers*, but now they are usually considered to be *environmental engineers*. Why has "environmental engineering" for the most part replaced "sanitary engineering" in the United States?

Discussion

There were many reasons for the name change. One certainly is the greater appreciation for the interconnections among abiotic and biotic systems in the protection of ecosystems and human health. Starting with the New Deal in second quarter of the twentieth century, engineers engaged in "public works" projects, which in the second half of the century evolved to include sanitary engineering projects, especially wastewater treatment plants, water supplies, and sanitary landfills. Meanwhile, sanitary engineers were designing cyclones, electrostatic precipitators, scrubbers, and other devices to remove air pollutants from industrial emissions.

The realization that there was much more that engineers design beyond these structures and devices has led to comprehensive solutions to environmental problems. Certainly, structural and mechanical solutions are still the foundation of air pollution engineering, but these are now seen as a part of an overall set of solutions. Thus, systems engineering, optimizations, and the application of more than physical principles (adding chemical and biological foundations) are better reflected in "environmental engineering" than in sanitary engineering. As mentioned by Vesilind, *et al*,[7] "everything seems to matter in environmental engineering."

[6] An airshed is analogous to a watershed, but applies to the atmosphere. For example, the California Air Resources Board defines an airshed as a subset of an "air basin, the term denotes a geographical area that shares the same air because of topography, meteorology, and climate." An air basin is defined as "land area with generally similar meteorological and geographical conditions throughout. To the extent possible, air basin boundaries are defined along political boundary lines and include both the source and receptor areas. California is currently divided into 15 air basins." *Source*: California Air Resources Board, "Glossary of Air Pollution Terms," http://www.arb.ca.gov/html/gloss.htm, update on May 3, 2006.

[7] This quote comes from Aarne Vesilind, P., Jeffrey Peirce, J., and Weiner, Ruth F., "Environmental Engineering," 4th ed. Butterworth-Heinemann: Boston, MA, 2003. The text is an excellent introduction to the field of environmental engineering and one of the sources of inspiration for this book.

Another possible reason for the name change is that "sanitary" implies human health, while "environmental" brings to mind ecological and welfare as well as human health as primary objectives of the profession. Sanitation is the province of industrial hygienists and public health professionals. The protection of the environment is a broader mandate for engineers.

THIS LEADS US TO ANOTHER QUESTION

Why is environmental engineering often a field in the general area of civil engineering, and not chemical engineering?

Discussion

The historical "inertia" may help to explain why environmental engineering is a discipline of civil rather than chemical engineering. As mentioned, environmental protection grew out of civil engineering projects of the New Deal and beyond. Chemical engineering is most concerned with the design and building of systems (e.g. "reactors") that convert raw materials into useful products. So, in a way, chemical engineering is the mirror image of environmental engineering, which often strives to return complex chemicals to simpler compounds (ultimately CO_2, CH_4, and H_2O). So, one could view the two fields as a chemical equilibrium where the reactions in each direction are equal! Most importantly, both fields are crucial in addressing air pollution, and contribute in unique ways.

A. What is a Contaminant?

The definition of air pollution considered at the beginning of this chapter included the term "contaminants." This is arguably even more daunting than "pollutant." If you were told your yard, your home, your neighborhood, or your air is contaminated, it is very likely that you would be greatly troubled. You would probably want to know the extent of the contamination, its source, what harm you may have already suffered from it, and what you can do to reduce it. Contamination is also a term that is applied differently by scientists and the general public, as well as among scientists from different disciplines.

So, then, what is *contamination*? The dictionary[8] definition of the verb "*contaminate*" reads something like "to corrupt by contact or association," or "to make inferior, impure, or unfit." These are fairly good descriptions of what

[8] Webster's Ninth New Collegiate Dictionary, Merriam-Webster, Inc., Springfield, MA, 1990.

environmental contaminants do. When they come into contact with people, ecosystems, crops, materials, or anything that society values, they cause harm. They make resources less valuable or less fit to perform their useful purposes. From an air quality perspective, contamination is usually meant to be "chemical contamination" and this most often is within the context of human health. However, air pollution abatement laws and programs have recognized that effects beyond health are also important, especially *welfare* protection. Thus, public health is usually the principal driver for assessing and controlling environmental contaminants, but ecosystems are also important *receptors* of contaminants. Contaminants also impact structures and other engineered systems, including historically and culturally important monuments and icons, such as the contaminants in rainfall (e.g. nitrates and sulfates) that render it more corrosive than would normally be expected (i.e. *acid rain*).

Contaminants may also be physical, such as the energy from ultraviolet (UV) light. Often, even though our exposure is to the physical contamination, this exposure was brought about by chemical contamination. For example, the release of chemicals into the atmosphere, in turn react with ozone in the stratosphere, decreasing the ozone concentration and increasing the amount of UV radiation at the earth's surface. This has meant that the mean UV dose in the temperate zones of the world has increased. This has been associated with an increase in the incidence of skin cancer, especially the most virulent form, melanoma.

One of the advantages of working in an environmental profession is that it is so diverse. Many aspects have to be considered in any environmental decision. From a scientific perspective, this means consideration must be given to the characteristics of the pollutant *and* the characteristics of the place where the chemical is found. This place is known as the "environmental medium." The major environmental media are air, water, soil, sediment, and even biota. This book is principally concerned with the air medium, but every medium affects or is affected by air pollution actions and inactions, as demonstrated by the fuel additive MTBE:

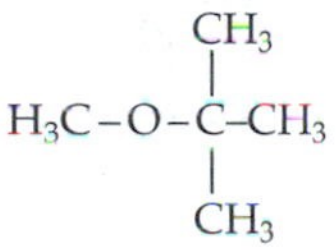

$$
\begin{array}{c}
\qquad\qquad\quad CH_3 \\
\qquad\qquad\quad | \\
H_3C-O-C-CH_3 \\
\qquad\qquad\quad | \\
\qquad\qquad\quad CH_3
\end{array}
$$

Methyl tertiary-butyl ether (MTBE)

Automobiles generally rely on the internal combustion engine to supply power to the wheels.[9] Gasoline is the principal fuel source for most cars. The

[9] The exception is electric cars, which represent a very small fraction of motorized vehicles; although a growing number of hybrid power supplies (i.e. electric systems charged by internal combustion engines) are becoming available.

exhaust from automobiles is a large source of air pollution, especially in densely populated urban areas. To improve fuel efficiency and to provide a higher octane rating (for anti-knocking), most gasoline formulations have relied on additives. Up to relatively recently, the most common fuel additive to gasoline was tetraethyl-lead. But with the growing awareness of lead's neurotoxicity and other health effects, tetraethyl-lead has been banned in most parts of the world, so suitable substitutes were needed.

Methyl tertiary-butyl ether (MTBE) was one of the first replacement additives, first used to replace the lead additives in 1979. It is manufactured by reacting methanol and isobutylene, and his been produced in very large quantities (more than 200 000 barrels per day in the US in 1999). MTBE is a member of the chemical class of oxygenates. MTBE is a quite volatile (vapor pressure = 27 kPa at 20°C), so that it is likely to evaporate readily. It also readily dissolves in water (aqueous solubility at 20°C = $42\,g\,L^{-1}$) and is very flammable (flash point = −30°C). In 1992, MTBE began to be used at higher concentrations in some gasoline to fulfill the oxygenate requirements set the 1990 Clean Air Act Amendments. In addition, some cities, notably Denver, used MTBE at higher concentrations during the wintertime in the late 1980s.

The Clean Air Act called for greater use of oxygenates in an attempt to help to reduce the emissions of carbon monoxide (CO), one of the most important air pollutants. CO toxicity results by interfering with the protein hemoglobin's ability to carry oxygen. Hemoglobin absorbs CO about 200 times faster than its absorption rate for oxygen. The CO-carrying protein is known as carboxyhemoglobin and when sufficiently high it can lead to acute and chronic effects. This is why smoking cigarettes leads to cardiovascular problems, i.e. the body has to work much harder because the normal concentration of oxygen in hemoglobin has been displaced by CO. CO is also a contributing factor in the photochemistry that leads to elevated levels of ozone (O_3) in the troposphere. In addition, oxygenates decrease the emissions of volatile organic compounds (VOCs), which along with oxides of nitrogen are major precursors to the formation of tropospheric O_3. This is one of the most important roles of oxygenates, since unburned hydrocarbons can largely be emitted before catalytic converters start to work.

Looking at it from one perspective, the use of MTBE was a success by providing oxygen and helping gasoline burn more completely, resulting in less harmful exhaust from motor vehicles. The oxygen also dilutes or displaces compounds such as benzene and its derivatives (e.g. toluene, ethylbenzene, and xylene), as well as sulfur. The oxygen in the MTBE molecule also enhances combustion (recall that combustion is oxidation in the presence of heat). MTBE was not the only oxygenate, but it has very attractive blending characteristics and is relatively cheap compared to other available compounds. Another widely used oxygenate is ethanol.

The problem with MTBE is its suspected links to certain health effects, including cancer in some animal studies. In addition, MTBE has subsequently been found to pollute water, especially ground water in aquifers. Some of the pollution comes from unburned MTBE emitted from tailpipes, some from fueling, but a large source is underground storage tanks (USTs) at gasoline stations or other fuel operations (see Fig. 1.2). A number of these tanks have leaked into the surrounding soil and unconsolidated media and have allowed the MTBE to migrate into the ground water. Since it has such a high aqueous solubility, the MTBE is easily dissolved in the water.

When a pollutant moves from one environmental compartment (e.g. air) to another (e.g. water) as it has for MTBE, this is known as cross-media transfer. The problem has not really been eliminated, just relocated. It is also an example of a risk trade-off. The risks posed by the air pollution have been traded for the new risks from exposure to MTBE-contaminated waters.

The names that we give things, the ways we describe them, and how we classify them for better understanding is uniquely important to each discipline. Although the various environmental fields use some common language, they each have their own lexicons and systems of taxonomy. Sometimes the

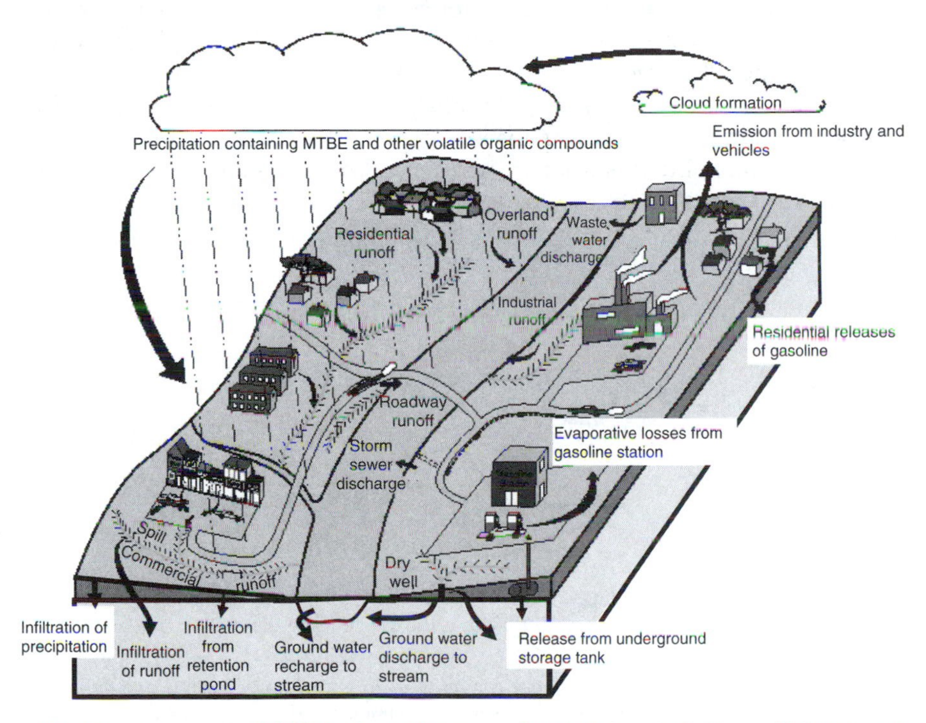

Fig. 1.2. Migration of MTBE in the environment. *Source*: Delzer, G. C., Zogorski, J. S., Lopes, T. J., and Bosshart, R. L., US Geological Survey, *Occurrence of the Gasoline Oxygenate MTBE and BTEX in Urban Stormwater in the United States, 1991–95*. Water Resources Investigation Report 96-4145, Washington, DC, 1996.

difference is subtle, such as different conventions in nomenclature and symbols.[10] This is more akin to different slang in the same language.

Sometimes the differences are profound, such as the use of the term "particle." In atmospheric dispersion modeling, a particle is the theoretical point that is followed in a fluid (see Fig. 1.3). The point represents the path that the pollutant in the air stream is expected to take. Particle is also used interchangeably with the term "aerosol" in atmospheric sciences and exposure studies (see Fig. 1.4). Particle is also commonly used to describe one part of an unconsolidated material, such as soil or sediment particle (see Fig. 1.5). In addition, engineering mechanics defines a particle as it applies to kinematics; i.e., a body in motion that is not rotating is called a particle. At an even more basic level, particle is half of the particle-wave dichotomy of physics, so the quantum mechanical properties of a particle (e.g. a photon) are fundamental to detecting chemicals using chromatography. Different members of the science community who contribute to our understanding of air pollution use the term "particle" in these different ways.

Let us consider a realistic example of the challenge of science communications related to the environment. There is concern that particles emitted from a power plant are increasing aerosol concentrations in your town. To determine if this is this case, the state authorizes the use of a Lagrangian (particle) dispersion model to see if the particles are moving from the source to the town. The aerosols are carrying pollutants that are deposited onto soil particles, so the state asks for a soil analysis to be run. One of the steps in this study is to extract the pollutants from individual soil particles before analysis. The soil scientist turns his extract into an analytical chemist who uses chromatography (which is based on quantum physics and particle-wave dichotomy).

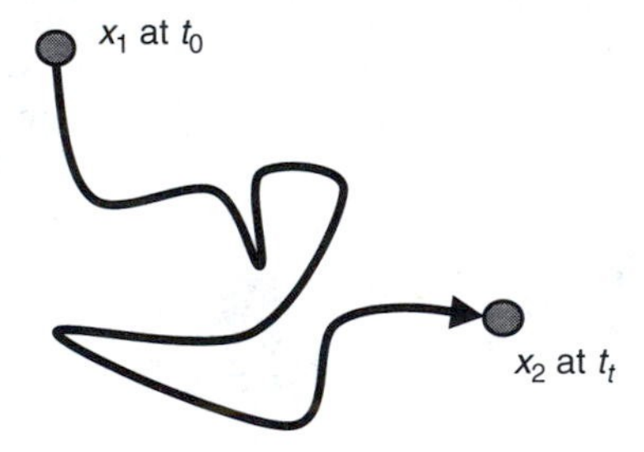

Fig. 1.3. Atmospheric modeling definition of a particle; i.e. a hypothetical point that is moving in a random path during time interval (t_0–t_t). This is the theoretical basis of a Lagrangian model.

[10] This text intentionally uses different conventions in explaining concepts and providing examples. One reason is that is how the information is presented. Environmental information comes in many forms. A telling example is the convention the use of K. In hydrogeology, this means hydraulic conductivity, in chemistry it is an equilibrium constant, and in engineering it can be a number of coefficients. Likewise, units other than metric will be used on occasion, because that is the convention in some areas, and because it demonstrates the need to apply proper dimensional analysis and conversions. Many mistakes have been made in these two areas!

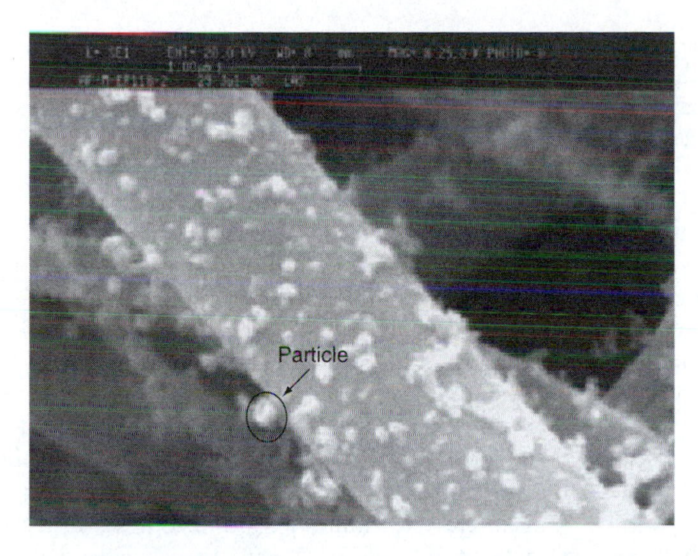

Fig. 1.4. Electron micrograph (>45 000× enlargement) showing an example of a particle type of air pollutant. These particles were collected from the exhaust of an F-118 aircraft under high throttle (military) conditions. The particles were collected with glass fiber filters (the 1 μm width tubular structures in the micrograph). Such particles are also referred to as PM or aerosols. Size of the particle is important, since the small particles are able to infiltrate the lungs and penetrate more deeply into tissues, which increases the likelihood of pulmonary and cardiovascular health problems. *Source*: Shumway, L., *Characterization of Jet Engine Exhaust Particulates for the F404, F118, T64, and T58 Aircraft Engines*. US Navy, Technical Report 1881, San Diego, CA, 2002.

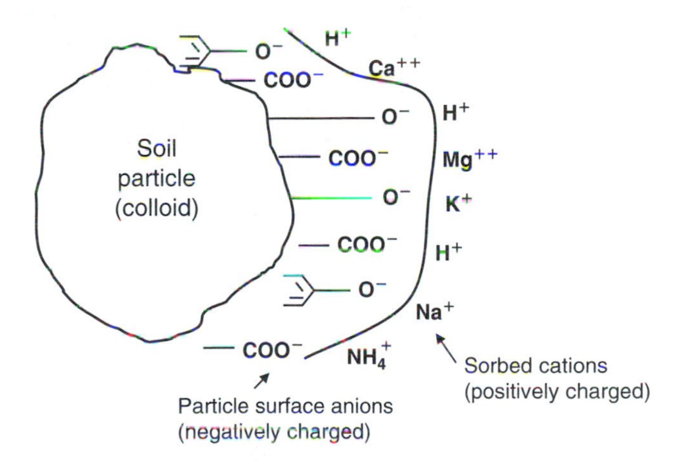

Fig. 1.5. Particle of soil (or sediment) material; in this instance, humic matter with a negative surface that sorbs cations. The outer layer's extent depends on the size of the cations (e.g. a layer of larger sodium (Na^+) cations will lead a large zone of influence than will a layer of smaller magnesium (Mg^{++}) cations.

You invite the dispersion modeler, the soil scientist, the chromatographer, as well as an exposure scientist to explain the meaning of their findings. In the process of each explanation, the scientists keep referring to particles. They are all correct within their specific discipline, but together they leave you confused. This is akin to homonyms in the same language or different languages altogether.

Under most environmental conditions, air pollution represents are solutions or suspensions of minute amounts of harmful compounds in air. For example, air's solutes represent small percentages of the solution at the highest (e.g. water vapor) and most other solutes represent parts per million (ppm). For example, on average carbon dioxide concentrations in the troposphere are a bit more than 300 ppm. Most "contaminants" in air and water, thankfully, are found in the parts per billion (ppb) range, if found at all. We usually include only the air in the atmosphere, but air exists in all media. It is dissolved in surface and ground water. In addition, soil and sediment are conglomerations of all states of matter. Soil is predominantly solid, but frequently has large fractions of liquid (soil water) and gas (soil air, methane, carbon dioxide) that make up the matrix. The composition of each fraction is highly variable. For example, soil–gas concentrations are different from those in the atmosphere and change profoundly with depth from the surface (Table 1.1 shows the inverse relationship between carbon dioxide and oxygen).

So, another way to think about these environmental media is that they are compartments, each with boundary conditions, kinetics and partitioning relationships within a compartment or among other compartments. Chemicals, whether nutrients or contaminants, change as a result of the time spent in each compartment. The environmental professional's challenge is to describe, characterize, and predict the behaviors of various chemical species as they

TABLE 1.1

Composition of Two Important Gases in Soil Air[a]

	Silty clay		Silty clay loam		Sandy loam	
Depth from surface (cm)	O_2 (% volume of air)	CO_2 (% volume of air)	O_2 (% volume of air)	CO_2 (% volume of air)	O_2 (% volume of air)	CO_2 (% volume of air)
30	18.2	1.7	19.8	1.0	19.9	0.8
61	16.7	2.8	17.9	3.2	19.4	1.3
91	15.6	3.7	16.8	4.6	19.1	1.5
122	12.3	7.9	16.0	6.2	18.3	2.1
152	8.8	10.6	15.3	7.1	17.9	2.7
183	4.6	10.3	14.8	7.0	17.5	3.0

[a] Evangelou, V.P., *Environmental Soil and Water Chemistry: Principles and Applications*, John Wiley and Sons, Inc., New York, 1998.

move through the media. When something is amiss, the cause and cure lie within the physics, chemistry, and biology of the system. It is up to the professional to properly apply the principles.

1. *Social Aspects of Air Pollution*

Environmental quality is important to everyone. We will be stressing the importance of sound science, but it should always be understood that when it comes to the environment and public health, the social sciences can never be ignored. Certain social values have been passed down for generations, while others have been recently adapted to changes. For example, if a person or neighborhood has just experienced a nasty battle with an industry or governmental organization regarding an environmental decision, such as the location of landfill or prohibition of a certain type of use of their own land, they may be reticent to trust the environmental professional who promises that it will be different this time. The second half of the twentieth century, particularly the time from the mid-1960s to through the 1970s ushered in a new environmental ethos, or at least memorialized the fact that most, if not all, people expect a certain quality of the environment. The laws and policies stemming from these years are still evolving.

Environmental awareness grew in the second half of the twentieth century. With this awareness the public demand for environmental safeguards and remedies to environmental problems was an expectation of a greater role for government. A number of laws were on the books prior to the 1960s, such as early versions of federal legislation to address limited types of water and air pollution, and some solid waste issues, such as the need to eliminate open dumping. The real growth, however, followed the tumultuous decade of the 1960s. Environment had become a social cause, akin to the civil rights and anti-war movements. Major public demonstrations on the need to protect "spaceship earth" encouraged elected officials to address environmental problems, exemplified by air pollution "inversions" that capped polluted air in urban valleys, leading to acute diseases and increased mortality from inhalation hazards, the "death" of Erie Canal and rivers catching on fire in Ohio and Oregon.

2. *The National Environmental Policy Act*

The movement was institutionalized in the United States by a series of new laws and legislative amendments. The National Environmental Policy Act (NEPA) was in many ways symbolic of the new federal commitment to environmental stewardship. It was signed into law in 1970 after contentious hearings in the US Congress. NEPA was not really a technical law. It did two main things. It created the Environmental Impact Statement (EIS) and established the Council on Environmental Quality (CEQ) in the Office of the President. Of the two, the EIS represented a sea change in how the federal government was to conduct business. Agencies were required to prepare EISs on any major action that they were considering that could "significantly" affect the quality of the environment. From the outset, the agencies had to reconcile often-competing values; i.e. their mission and the protection of the environment.

The CEQ was charged with developing guidance for all federal agencies on NEPA compliance, especially when and how to prepare an EIS. The EIS process combines scientific assessment with public review. The process is similar for most federal agencies. Agencies often strive to receive a so-called "FONSI"[11] or the finding of no significant impact, so that they may proceed unencumbered on a mission-oriented project.[12] Whether a project either leads to a full EIS or a waiver through the FONSI process, it will have to undergo an evaluation. This step is referred to as an "environmental assessment." An incomplete or inadequate assessment will lead to delays and increases the chance of an unsuccessful project, so sound science is needed from the outset of the project design. The final step is the record of decision (ROD). The ROD describes the alternatives and the rationale for final selection of the best alternative. It also summarizes the comments received during the public reviews and how the comments were addressed. Many states have adopted similar requirements for their RODs.

The EIS documents were supposed to be a type of "full disclosure" of actual or possible problems if a federal project is carried out. This was accomplished by looking at all of the potential impacts to the environment from any of the proposed alternatives, and comparing those outcomes to a "no action" alternative. At first, many tried to demonstrate that their "business as usual" was in fact very environmentally sound. In other words, the environment would be better off with the project than without it (action is better than no action). Too often, however, an EIS was written to justify the agency's mission-oriented project. One of the key advocates for the need for a national environmental policy, Lynton Caldwell, is said to have referred to this as the federal agencies using an EIS to "make an environmental silk purse from a mission-oriented sow's ear!"[13] The courts adjudicated some very important laws along the way, requiring federal agencies to take NEPA seriously. Some of the aspects of the "give and take" and evolution of federal agencies' growing commitment to environmental protection was the acceptance of the need for sound science in assessing environmental conditions and

[11] Pronounced "Fonzie" like that of the nickname for character Arthur Fonzarelli portrayed by Henry Winkler in the television show, *Happy Days*.

[12] This is understandable if the agency is in the business of something not directly related to environmental work, but even the natural resources and environmental agencies have asserted that there is no significant impact to their projects. It causes the cynic to ask, then, why are they engaged in any project that has no significant impact? The answer is that the term "significant impact" is really understood to mean "significant adverse impact" to the human environment.

[13] I attribute this quote to Timothy Kubiak, one of Professor Caldwell's former graduate students in Indiana University's Environmental Policy Program. Kubiak has since gone on to become a successful environmental policy maker in his own right, first at EPA and then at the US Fish and Wildlife Service.

possible impacts, and the very large role of the public in deciding on the environmental worth of a highway, airport, dam, waterworks, treatment plant, or any other major project sponsored by or regulated by the federal government. This was a major impetus in the growth of the environmental disciplines since the 1970s. We needed experts who could not only "do the science" but who could communicate what their science means to the public.

All federal agencies must follow the CEQ regulations[14] to "adopt procedures to ensure that decisions are made in accordance with the policies and purposes of the Act." Agencies are required to identify the major decisions called for by their principal programs and make certain that the NEPA process addresses them. This process must be set up in advance, early in agency's planning stages. For example, if waste remediation or reclamation is a possible action, the NEPA process must be woven into the remedial action planning processes from beginning with the identification of the need for and possible kinds of actions being considered. Noncompliance or inadequate compliance with NEPA rules regulations can lead to severe consequences, including lawsuits, increased project costs, delays, and the loss of the public's loss of trust and confidence, even if the project is designed to improve the environment, and even if the compliance problems seem to be only "procedural."

The US EPA is responsible for reviewing the environmental effects of all federal agencies' actions. This authority was written as Section 309 of the Clean Air Act. The review must be followed with the EPA's public comments concerning the environmental impacts of any matter related to the duties, responsibilities, and authorities of EPA's administrator, including EISs. The EPA's rating system (see Table 1.2) is designed to determine whether a proposed action by a federal agency is unsatisfactory from the standpoint of public health, environmental quality, or public welfare. This determination is published in the *Federal Register* (for significant projects) and referred to the CEQ.

3. *Clean Air Legislation*

The year 1970 was a watershed[15] year in environmental awareness. The 1970 amendments to the Clean Air Act arguably ushered in the era of environmental legislation with enforceable rules. The 1970 version of the Clean Air Act was enacted to provide a comprehensive set of regulations to control air emissions from area, stationary, and mobile sources. This law authorized the EPA to establish National Ambient Air Quality Standards (NAAQS) to protect public health and the environment from the "conventional" (as opposed to "toxic") pollutants: carbon monoxide, particulate matter (PM), oxides of nitrogen, oxides of sulfur, and photochemical oxidant smog or ozone (see Discussion Box: Evolution of Environmental Indicators for Air). The metal lead (Pb) was later added as the sixth NAAQS pollutant.

[14] 40 CFR 1507.3

[15] Pun intended! Or should it be an "airshed" year?

TABLE 1.2

Summary of the CEQ Guidance for Compliance with the NEPA of 1969[a]

Title of guidance	Summary of guidance	Citation	Relevant regulation/Documentation
Forty most often asked questions concerning CEQ's NEPA regulations	Provides answers to 40 questions most frequently asked concerning implementation of NEPA.	46 FR 18026, dated March 23, 1981	40 CFR Parts 1500–1508
Implementing and explanatory documents for Executive Order (EO) 12114, Environmental effects abroad of major federal actions	Provides implementing and explanatory information for EO 12114. Establishes categories of federal activities or programs as those that significantly harm the natural and physical environment. Defines which actions are excluded from the order and those that are not.	44 FR 18672, dated March 29, 1979	EO 12114, Environmental effects abroad of major federal actions
Publishing of three memoranda for heads of agencies on: • Analysis of impacts on prime or unique agricultural lands (Memoranda 1 and 2) • Interagency consultation to avoid or mitigate adverse effects on rivers in the Nationwide Inventory (Memorandum 3)	1 or 2 Discusses the irreversible conversion of unique agricultural lands by Federal Agency action (e.g. construction activities, developmental grants, and federal land management). Requires identification of and cooperation in retention of important agricultural lands in areas of impact of a proposed agency action. The agency must identify and summarize existing or proposed agency policies, to preserve or mitigate the effects of agency action on agricultural lands. 3 "Each Federal Agency shall, as part of its normal planning and environmental review process, take care to avoid or mitigate adverse effects on rivers identified in the Nationwide Inventory prepared by the Heritage Conservation and Recreation Service in the Department of the Interior." Implementing regulations includes	[45 FR 59189, dated September 8, 1980]	1/2 Farmland Protection Policy Act (7 USC §4201 et seq.) 3 The Wild and Scenic Rivers Act of 1965 (16 USC §1271 et seq.)

	determining whether the proposed action: affects an Inventory River; adversely affects the natural, cultural and recreation values of the Inventory river segment; forecloses options to classify any portion of the Inventory segment as a wild, scenic or recreational river area, and incorporates avoidance/mitigation measures into the proposed action to maximum extent feasible within the agency's authority.		
Memorandum for heads of agencies for guidance on applying Section 404(r) of the Clean Water Act at Federal projects which involve the discharge of dredged or fill materials into waters of the US including wetlands	Requires timely agency consultation with US Army Corps of Engineers (COE) and the US EPA before a Federal project involves the discharge of dredged or fill material into US waters, including wetlands. Proposing agency must ensure, when required, that the EIS includes written conclusions of EPA and COE (generally found in Appendix).	CEQ, dated November 17, 1980	Clean Water Act (33 USC §1251 et seq.) EO 12088, Federal compliance with pollution control standards
Scoping guidance	Provides a series of recommendations distilled from agency research regarding the scoping process. Requires public notice; identification of significant and insignificant issues; allocation of EIS preparation assignments; identification of related analysis requirements in order to avoid duplication of work; and the planning of a schedule for EIS preparation that meshes with the agency's decision-making schedule.	46 FR 25461, dated May 7, 1981	40 CFR Parts 1500–1508
Guidance regarding NEPA regulations	Provides written guidance on scoping, CatEx's, adoption regulations, contracting provisions, selecting alternatives in licensing and permitting situations, and tiering.	48 FR 34263, dated July 28, 1983	40 CFR Parts 1501, 1502, and 1508

(*Continued*)

TABLE 1.2 (*Continued*)

Title of guidance	Summary of guidance	Citation	Relevant regulation/Documentation
NEPA implementation regulations, Appendices I, II, and III	Provides guidance on improving public participation, facilitating agency compliance with NEPA and CEQ implementing regulations. Appendix I updates required NEPA contacts, Appendix II compiles a list of Federal and Federal-State Agency Offices with jurisdiction by law or special expertise in environmental quality issues; and Appendix III lists the Federal and Federal-State Offices for receiving and commenting on other agencies' environmental documents.	49 FR 49750, dated December 21, 1984	40 CFR Part 1500
Incorporating biodiversity considerations into environmental impact analysis under the NEPA	Provides for "acknowledging the conservation of biodiversity as national policy and incorporates its consideration in the NEPA process"; encourages seeking out opportunities to participate in efforts to develop regional ecosystem plans; actively seeks relevant information from sources both within and outside government agencies; encourages participating in efforts to improve communication, cooperation, and collaboration between and among governmental and nongovernmental entities; improves the availability of information on the status and distribution of biodiversity, and on techniques for managing and restoring it; and expands the information base on which biodiversity analyses and management decisions are based.	CEQ, Washington, DC, dated January 1993	Not applicable

Pollution prevention and NEPA	Pollution-prevention techniques seek to reduce the amount and/or toxicity of pollutants being generated, promote increased efficiency of raw materials and conservation of natural resources and can be cost-effective. Directs Federal agencies that to the extent practicable, pollution prevention considerations should be included in the proposed action and in the reasonable alternatives to the proposal, and to address these considerations in the environmental consequences section of an EIS and EA (when appropriate).	58 FR 6478, dated January 29, 1993	EO 12088, Federal Compliance with Pollution Control Standards
Considering cumulative effects under NEPA	Provides a "framework for advancing environmental cumulative impacts analysis by addressing cumulative effects in either an environmental assessment (EA) or an environmental impact statement". Also provides practical methods for addressing coincident effects (adverse or beneficial) on specific resources, ecosystems, and human communities of all related activities, not just the proposed project or alternatives that initiate the assessment process.	January 1997	40 CFR §1508.7
Environmental justice guidance under NEPA	Provides guidance and general direction on EO 12898 which requires each agency to identify and address, as appropriate, "disproportionately high and adverse human health or environmental effects of its programs, policies, and activities on minority populations and low-income populations."	CEQ, Washington, DC, dated December 10, 1997	EO 12898, Federal actions to address environmental justice in a minority populations and low-income populations

[a] *Source*: National Aeronautics and Space Administration, 2001, Implementing The National Environmental Policy Act And Executive Order 12114, Chapter 2.

DISCUSSION: EVOLUTION OF ENVIRONMENTAL INDICATORS FOR AIR

The term *smog* is a shorthand combination of "smoke–fog." However, it is really the code word for photochemical oxidant smog, the brown haze that can be seen when flying into Los Angeles, St. Louis, Denver, and other metropolitan areas around the world (see Fig. 1.6). Point-of-fact

Fig. 1.6. Photo of smog episode in Los Angeles, CA taken in May of 1972. *Source*: Documerica, US Environmental Protection Agency's Photo Gallery; Photographer: Gene Daniels.

is that to make smog, at least three ingredients are needed: light, hydrocarbons, and radical sources, such the oxides of nitrogen. Therefore, smog is found most often in the warmer months of the year, not because of temperature, but because these are the months with greater amounts of sunlight. More sunlight is available for two reasons, both attributed to the earth's tilt on its axis. In the summer, the earth is tilted toward the sun, so the angle of inclination of sunlight is greater than when the sun tipped away from the earth leading to more intensity of light per earth surface area. Also, the days are longer in the summer, so these two factors increase the light budget.

Hydrocarbons come from many sources, but the fact that internal combustion engines burn gasoline, diesel fuel, and other mixtures of hydrocarbons makes them a ready source. Complete combustion results in carbon dioxide and water, but anything short of complete combustion will be a source of hydrocarbons, including some of the original ones found in the fuels, as well as new ones formed during combustion. The compounds that become free radicals, like the oxides of nitrogen, are also readily available from internal combustion engines, since the three quarters of the troposphere is made up of molecular nitrogen (N_2). Although N_2 is relatively not chemically reactive, with the high temperature and pressure conditions in the engine, it does combine with the O_2 from the fuel/air mix and generates oxides that can provide electrons to the photochemical reactions.

The pollutant most closely associated with smog is ozone (O_3), which forms from the photochemical reactions mentioned above. In the early days of smog control efforts, O_3 was used more as a surrogate or *marker* for smog, since one could not really take a sample of smog. Later, O_3 became recognized as a pollutant in its own right since it was increasingly linked to respiratory diseases.

The original goal was to set and to achieve NAAQS in every state by 1975. These new standards were combined with charging the 50 states to develop state implementation plans (SIPs) to address industrial sources in the state. The ambient atmospheric concentrations are measured at over 4000 monitoring sites across the US. The ambient levels have continuously decreased, as shown in Table 1.3.

The Clean Air Act Amendments of 1977 mandated new dates to achieve attainment of NAAQS (many areas of the country had not met the prescribed dates set in 1970). Other amendments were targeted at insufficiently addressed types of air pollution, including acidic deposition (so-called "acid rain"), tropospheric ozone pollution, depletion of the stratospheric ozone layer, and a new program for air toxics, the National Emission Standards for Hazardous Air Pollutants (NESHAPs).

TABLE 1.3

Percentage Decrease in Ambient Concentrations of National Ambient Air Quality Standard Pollutants from 1985 through 1994

Pollutant	Decrease in Concentration
CO	28%
Lead	86%
NO_2	9%
Ozone	12%
PM_{10}	20%
SO_2	25%

Source: US Environmental Protection Agency

The 1990 Amendments to the Clean Air Act profoundly changed the law, by adding new initiatives and imposing dates to meet the law's new requirements. Here are some of the major provisions.

Urban Air Pollution Cities that failed to achieve human health standards as required by NAAQS were required to reach attainment within 6 years of passage, although Los Angeles was given 20 years, since it was dealing with major challenges in reducing ozone concentrations.

Almost 100 cities failed to achieve ozone standards, and were ranked from *marginal* to *extreme*. The more severe the pollution, the more rigorous controls required, although additional time was given to those extreme cities to achieve the standard. Measures included new or enhanced inspection/maintenance (I/M) programs for autos; installation of vapor recovery systems at gas stations and other controls of hydrocarbon emissions from small sources; and new transportation controls to offset increases in the number of miles traveled by vehicles. Major stationary sources of nitrogen oxides will have to reduce emissions.

The 41 cities failing to meet carbon monoxide standards were ranked *moderate* or *serious*; states may have to initiate or upgrade inspection and maintenance programs and adopt transportation controls. The 72 urban areas that did not meet PM_{10} standards were ranked *moderate*; states will have to implement Reasonably Available Control Technology (RACT); use of wood stoves and fireplaces may have to be curtailed.

The standards promulgated from the Clean Air Act Amendments are provided in Table 1.4. Note that the new particulate standard addresses smaller particles; i.e., particles with diameters $\leq 2.5\,\mu m$ ($PM_{2.5}$). Research has shown that exposure to these smaller particles is more likely to lead to health problems than do exposures to larger particles. Smaller particles are able to penetrate further into the lungs and likely are more bioavailable than the larger PM_{10}. Recently, however, concerns about larger particles (e.g. PM_{10}) have increased, particularly as research is beginning to link coarse particles to chronic diseases like asthma.

TABLE 1.4

National Ambient Air Quality Standards

Pollutant	Averaging period[a]	Standard	Primary standards[b]	Secondary standards[c]
Ozone	1 h	Cannot be at or above this level on more than 3 days over 3 years	125 ppb	125 ppb
	8 h	The average of the annual 4th highest daily 8 h maximum over a 3-year period cannot be at or above this level	85 ppb	85 ppb
Carbon monoxide	1 h	Cannot be at or above this level more than once per calendar year	35.5 ppm	35.5 ppm
	8 h	Cannot be at or above this level more than once per calendar year	9.5 ppm	9.5 ppm
Sulfur dioxide	3 h	Cannot be at or above this level more than once per calendar year	—	550 ppb
	24 h	Cannot be at or above this level more than once per calendar year	145 ppb	—
	Annual	Cannot be at or above this level	35 ppb	—
Nitrogen dioxide	Annual	Cannot be at or above this level	54 ppb	54 ppb
Respirable PM (aerodynamic diameter $\leq 10\,\mu m = PM_{10}$)	24 h	The 3-year average of the annual 99th percentile for each monitor within an area cannot be at or above this level	$155\,\mu g\,m^{-3}$	$155\,\mu g\,m^{-3}$
	Annual	The 3-year average of annual arithmetic mean concentrations at each monitor within an area cannot be at or above this level	$51\,\mu g\,m^{-3}$	$51\,\mu g\,m^{-3}$
Respirable PM (aerodynamic diameter $\leq 2.5\,\mu m = PM_{2.5}$)	24 hr	The 3-year average of the annual 98th percentile for each population-oriented monitor within an area cannot be at or above this level	$66\,\mu g\,m^{-3}$	$66\,\mu g\,m^{-3}$
	Annual	The 3-year average of annual arithmetic mean concentrationsfrom single or multiplecommunity-oriented monitors cannot be at or above this level	$15.1\,\mu g\,m^{-3}$	$15.1\,\mu g\,m^{-3}$

(Continued)

TABLE 1.4 (Continued)

Pollutant	Averaging period[a]	Standard	Primary standards[b]	Secondary standards[c]
Lead	Quarter	Cannot be at or above this level	$1.55\,\mu g\,m^{-3}$	$1.55\,\mu g\,m^{-3}$

[a] Integrated time used to calculate the standard. For example, for particulates, the filter will collect material for 24 h and then analyzed. The annual integration will be an integration of the daily values.
[b] Primary NAAQS are the levels of air quality that the EPA considers to be needed, with an adequate margin of safety, to protect the public health.
[c] Secondary NAAQS are the levels of air quality that the EPA judges necessary to protect the public welfare from any known or anticipated adverse effects.

Mobile Sources Vehicular tailpipe emissions of hydrocarbons, carbon monoxide, and oxides of nitrogen were to be reduced with the 1994 models. Standards now have to be maintained over a longer vehicle life. Evaporative emission controls were mentioned as a means for reducing hydrocarbons. Beginning in 1992, "oxyfuel" gasolines blended with alcohol began to be sold during winter months in cities with severe carbon monoxide problems. In 1995, reformulated gasolines with aromatic compounds will were introduced in the nine cities with the worst ozone problems; but other cities were allowed to participate. Later, a pilot program introduced 150 000 low emitting vehicles to California that meet tighter emission limits through a combination of vehicle technology and substitutes for gasoline or blends of substitutes with gasoline. Other states are also participating in this initiative.

Toxic Air Pollutants The number of toxic air pollutants covered by the Clean Air Act was increased to 189 compounds in 1990. Most of these are carcinogenic, mutagenic, and/or toxic to neurological, endocrine, reproductive, and developmental systems. All 189 compound emissions were to be reduced within 10 years. The EPA published a list of source categories issued Maximum Achievable Control Technology (MACT) standards for each category over a specified timetable.

The next step beyond MACT standards is to begin to address chronic health risks that would still be expected if the sources meet these standards. This is known as residual risk reduction. The first step was to assess the health risks from air toxics emitted by stationary sources that emit air toxics after technology-based (MACT) standards are in place. The residual risk provision sets additional standards if MACT does not protect public health with an "ample margin of safety," as well as additional standards if they are needed to prevent adverse environmental effects.

What an "ample margin of safety" means is still up for debate, but one proposal for airborne carcinogens is shown in Fig. 1.7. That is, if a source can demonstrate that it will not contribute to greater than 10^{-6} cancer risk, then

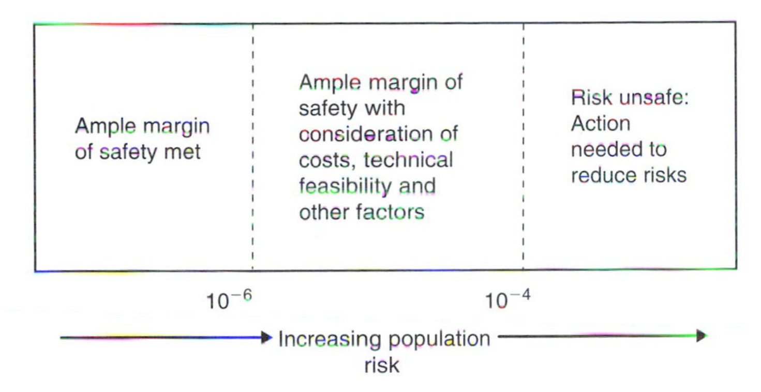

Fig. 1.7. Ample margin of safety based on airborne contaminant's cancer risk.

it meets the ample margin of safety requirements for air toxics. The ample margin needed to protect populations from non-cancer toxins, such as neurotoxins, is being debated, but it will involve the application of some type of hazard quotient (HQ). The HQ is the ratio of the potential exposure to the substance and the level at which no adverse effects are expected. An HQ < 1 means that the exposure levels to a chemical should not lead to adverse health effects. Conversely, an HQ < 1 means that adverse health effects are possible. Due to uncertainties and the feedback that is coming from the business and scientific communities, the ample margin of safety threshold is presently ranging from HQ $= 0.2$ to 1.0. So, if a source can demonstrate that it will not contribute to greater than the threshold (whether it is 0.2, 1.0, or some other level established by the federal government) for non-cancer risk, it meets the ample margin of safety requirements for air toxics.

Acid Deposition The introduction of acidic substances to flora, soil, and surface waters has been collectively called "acid rain" or "acid deposition." Acid rain is generally limited to the so-called "wet" deposition (low pH precipitation), but acidic materials can also reach the earth's surface by dry deposition ("acid aerosols") and acid fog (airborne droplets of water that contains sulfuric acid or nitric acid). Generally, acid deposition is not simply materials of pH < 7, but usually of pH < 5.7, since "normal" rainfall has a pH of about 5.7, due to the ionization of absorbed carbon dioxide (see discussion in Chapter 3). There is much concern about acid deposition because aquatic biota can be significantly harmed by only slight changes in pH. The problem of acidified soils and surface waters is a function of both the increase in acidity of the precipitation and the ability of the receiving waters and soil to resist the change in soil pH. So, it is the contribution of the human generated acidic materials, especially the oxides of sulfur and the oxides of nitrogen, that are considered to be the sources of "acid rain."

The 1990 amendments introduced a two-phase, market-based system to reduce sulfur dioxide emissions from power plants by more than half. Total annual emissions were to be capped at 8.9 million tons, a reduction of 10 million tons from the 1980 baseline levels. Power facilities were issued allowances based on fixed emission rates set in the law, as well as their previous fossil-fuel use. Penalties were issued for exceedances, although the allowances could be banked or traded within the fuel burning industry. In Phase I, large, high-emission plants in eastern and Midwestern US were required to reduce emissions by 1995. Phase II began in 2000 to set emission limits on smaller, cleaner plants and further tightening of the Phase I plants' emissions. All sources were required to install continuous emission monitors to assure compliance. Reductions in the oxides of nitrogen were also to be reduced; however, the approach differed from the oxides of sulfur, using EPA performance standards, instead of the two-phase system.

Protecting the Ozone Layer[16] The ozone layer filters out significant amounts of UV radiation from the sun. This UV radiation can cause skin damage and lead to certain forms of cancer, including the most fatal form, melanoma. Therefore, the international scientific and policy communities have been concerned about the release of chemicals that find their way to stratosphere and accelerate the breakdown of the ozone (see Discussion box: Ozone: Location, Location, and Location).

OZONE: LOCATION, LOCATION, AND LOCATION

Like the three rules of real estate, the location of ozone (O_3) determines whether it is essential or harmful. As shown in Fig. 1.8, O_3 concentrations are small, but increase in the stratosphere (about 90% of the atmosphere's O_3 lies in the layer between 10 and 17 km above the earth's surface up to an altitude of about 50 km). This is commonly known as the ozone layer. Most of the remaining ozone is in the lower part of the atmosphere, the troposphere. The stratospheric O_3 concentrations must be protected, while the tropospheric O_3 concentrations must be reduced.

Stratospheric ozone (the "good ozone") absorbs the most of the biologically damaging UV sunlight (UV-B), allowing only a small amount to reach the earth's surface. The absorption of UV radiation by ozone

[16] Sources for this discussion are: "Frequently Asked Questions" of the World Meteorological Organization/United Nations Environment Programme report, Scientific Assessment of Ozone Depletion: 1998 (WMO Global Ozone Research and Monitoring Project-Report No. 44, Geneva, 1999); and the Center for International Earth Science Information Network (CIESIN) website: http://www.ciesin.org/index.html.

generates heat, which is why Fig. 1.8 shows an increase in temperature in the stratosphere. Without the filtering action of the ozone layer, greater amounts of the sun's UV-B radiation would penetrate the atmosphere and would reach the earth's surface. Many experimental studies of plants and animals and clinical studies of humans have shown the harmful effects of excessive exposure to UV-B radiation.

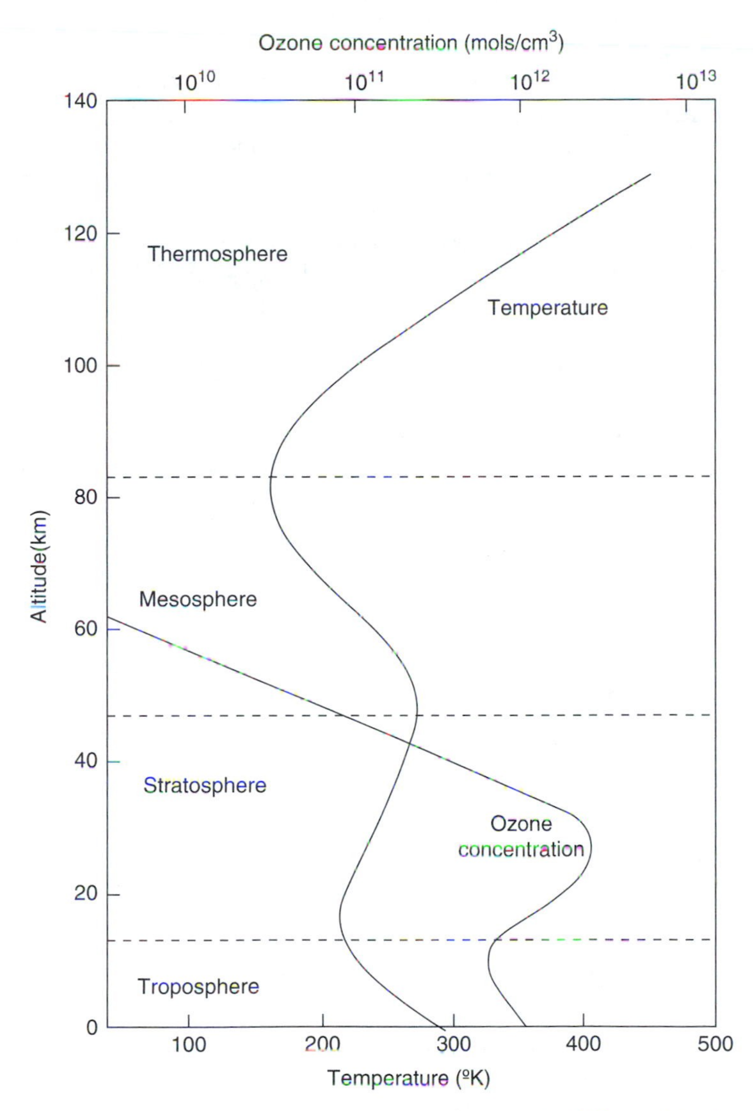

Fig. 1.8. Ozone concentrations and temperature profile of the earth's atmosphere. *Source*: Watson, R. T., Geller, M. A., Stolarski, R. S., and Hampson, R. F., *Present State of Knowledge of the Upper Atmosphere: An Assessment Report*, NASA Reference Publication, 1986.

In the troposphere, O_3 exposure is destructive ("bad ozone"), because it is highly reactive with tissues, leading to ecological and welfare effects, such as forest damage and reduced crop production, and human health effects, especially cardiopulmonary diseases.

Chlorofluorocarbons (CFCs), along with other chlorine- and bromine-containing compounds, can accelerate the depletion of stratospheric O_3 layer. CFCs were first developed in the early 1930s for many industrial, commercial, and household products. They are generally compressible, non-flammable, and nonreactive, which led to many CFC uses, including as coolants for commercial and home refrigeration units, and aerosol propellants. In 1973, chlorine was found to catalyze ozone destruction. Catalytic destruction of ozone removes the odd-numbered oxygen species [atomic oxygen (O) and ozone (O_3)], but leaves the chlorine unaffected. A complex scenario involving atmospheric dynamics, solar radiation, and chemical reactions accounts for spring thinning of the ozone layer at the earth's poles. Global monitoring of ozone levels from space using National Aeronautics and Space Administration's (NASA) Total Ozone Mapping Spectrometer (TOMS) instrument has shown significant downward trends in ozone concentrations at all of the earth's latitudes, except the tropics. Even with the international bans and actions, stratospheric ozone levels are expected to be lower than pre-depletion levels for many years because CFCs are persistent in the troposphere, from where they are transported to the stratosphere. In the high-energy stratosphere, the compounds undergo hundreds of catalytic cycles involving O_3 before the CFCs are scavenged by other chemicals.

A search for more environmentally benign substances has been underway for some time. One set of potential substitutes is the hydrochlorofluorocarbons (HCFCs). Obviously, the HCFC molecules contain Cl atoms, but the hydrogen increases the reactivity of the HCFCs with other tropospheric chemical species. These low altitude reactions decrease the probability of a Cl atom finding its way to the stratosphere. Hydrofluorocarbons (HFCs) are potential substitutes that lack chlorine.

The new act built upon the Montreal Protocol on substances that deplete the ozone layer, the international treaty where nations agreed to reduce or eliminate ozone-destroying gas production and uses of chemicals that pose a threat to the ozone layer. The amendments further restricted the use, emissions, and disposal of these chemicals, including the phasing out of the production of CFCs, as well as other chemicals that lead to ozone attacking halogens, such as tetrachloromethane (commonly called carbon tetrachloride) and methyl chloride by the year 2000, and methyl chloroform by 2002. The act also will freeze the production of

CFCs in 2015 and requires that CFCs be phased out completely by 2030. Companies servicing air conditioning for vehicles are now required to purchase certified recycling equipment and train employees. EPA regulations require reduced emissions from all other refrigeration sectors to lowest achievable levels. "Nonessential" CFC applications are prohibited. The act increases the labeling requirements of the Toxic Substances Control Act (TSCA) by mandating the placement of warning labels on all containers and products (such as cooling equipment, refrigerators, and insulation) that contain CFCs and other ozone-depleting chemicals.

4. *Solid and Hazardous Wastes Laws*

Although the Clean Air Act and its amendments comprise the most notable example of US legislation passed to protect air quality, other laws address air pollution more indirectly. For example, air pollution around solid and hazardous waste facilities can be problematic. The two principal US laws governing solid wastes are the Resource Conservation and Recovery Act (RCRA) and Superfund. The RCRA law covers both hazardous and solid wastes, while Superfund and its amendments generally address abandoned hazardous waste sites. RCRA addresses active hazardous waste sites.

Management of Active Hazardous Waste Facilities With RCRA, the US EPA received the authority to control hazardous waste throughout the waste's entire life cycle, known as the "cradle-to-grave." This means that manifests must be prepared to keep track of the waste, including its generation, transportation, treatment, storage, and disposal. RCRA also set forth a framework for the management of non-hazardous wastes in Subtitle D.

The Federal Hazardous and Solid Waste Amendments (HSWA) to RCRA required the phase out of land disposal of hazardous waste. HSWA also increased the federal enforcement authority related to hazardous waste actions, set more stringent hazardous waste management standards, and provided for a comprehensive UST program.

The 1986 amendments to RCRA allowed the federal government to address potential environmental problems from USTs for petroleum and other hazardous substances.

Addressing Abandoned Hazardous Wastes The Comprehensive Environmental Response, Compensation, and Liability Act (CERCLA) is commonly known as Superfund. Congress enacted it in 1980 to create a tax on the chemical and petroleum industries and to provide extensive federal authority for responding directly to releases or threatened releases of hazardous substances that may endanger public health or the environment.

The Superfund law established prohibitions and requirements concerning closed and abandoned hazardous waste sites; established provisions for the liability of persons responsible for releases of hazardous waste at these sites; and established a trust fund to provide for cleanup when no responsible party could be identified.

The CERCLA response actions include:

- Short-term removals, where actions may be taken to address releases or threatened releases requiring prompt response. This is intended to eliminate or reduce exposures to possible contaminants.
- Long-term remedial response actions to reduce or eliminate the hazards and risks associated with releases or threats of releases of hazardous substances that are serious, but not immediately life threatening. These actions can be conducted only at sites listed on EPA's National Priorities List (NPL).

Superfund also revised the National Contingency Plan (NCP), which sets guidelines and procedures required when responding to releases and threatened releases of hazardous substances.

CERCLA was amended by the Superfund Amendments and Reauthorization Act (SARA) in 1986. These amendments stressed the importance of permanent remedies and innovative treatment technologies in cleaning up hazardous waste sites. SARA required that Superfund actions consider the standards and requirements found in other state and federal environmental laws and regulations and provided revised enforcement authorities and new settlement tools. The amendments also increased state involvement in every aspect of the Superfund program, increased the focus on human health problems posed by hazardous waste sites, encouraged more extensive citizen participation in site cleanup decisions, and increased the size of the Superfund trust fund.

SARA also mandate that the Hazard Ranking System (HRS) be revised to make sure of the adequacy of the assessment of the relative degree of risk to human health and the environment posed by uncontrolled hazardous waste sites that may be placed on the NPL.

Environmental Product and Consumer Protection Laws Although most of the authorizing legislation targeted at protecting and improving the environment is based on actions needed in specific media (i.e. air, water, soil, and sediment), some law has been written in an attempt to prevent environmental and public health problems while products are being developed and before their usage. In this way, air pollution is prevented to some extent at the potential source, such as products that can lead to indoor air pollution.

The predominant product laws designed to protect the environment are the Federal Food, Drug, and Cosmetics Act (FFDCA); the Federal Insecticide, Fungicide, and Rodenticide Act (FIFRA); and the Toxic Substances Control

Act (TSCA). These three laws look at products in terms of potential risks for yet-to-be-released products and estimated risks for products already in use. If the risks are unacceptable, new products may not be released as formulated or the uses will be strictly limited to applications that meet minimum risk standards. For products already in the marketplace, the risks are periodically reviewed. For example, pesticides have to be periodically re-registered with the government. This re-registration process consists of reviews of new research and information regarding health and environmental impacts discovered since the product's last registration.

FIFRA's major mandate is to control the distribution, sale, and applications of pesticides. This not only includes studying the health and environmental consequences of pesticide usage, but also to require that those applying the pesticides register when they purchase the products. Commercial applicators must be certified by successfully passing exams on the safe use of pesticides. FIFRA requires that the EPA license any pesticide used in the US. The licensing and registration makes sure that pesticide is properly labeled and will not cause unreasonable environmental harm.

An important, recent product production law is the Food Quality Protection Act (FQPA), including new provisions to protect children and limit their risks to carcinogens and other toxic substances. The law is actually an amendment to FIFRA and FFDCA that includes new requirements for safety standard—reasonable certainty of no harm—that must be applied to all pesticides used on foods. FQPA mandates a single, health-based standard for all pesticides in all foods; gives special protections for infants and children; expedites approval of pesticides likely to be safer than those in use; provides incentives for effective crop protection tools for farmers; and requires regular re-evaluation of pesticide registrations and tolerances so that the scientific data supporting pesticide registrations includes current findings.

There is some ongoing debate about the actual routes and pathways that contribute the most to pesticide exposure. For example, certain pesticides seem to be most likely to be present in dietary pathways (e.g. the purchased food contains the pesticide), nondietary ingestion (e.g. the food is contaminated by airborne pesticides in the home or the people living in the home touch contaminated surfaces and then touch the food), and dermal exposure (people come into contact with the pesticide and it infiltrates the skin). This seems to vary considerably by the type of pesticide, but it suffices to say that the air pathways are important.

Another product-related development in recent years is the screening program for endocrine disrupting substances. Research suggests a link between exposure to certain chemicals and damage to the endocrine system in humans and wildlife. Because of the potentially serious consequences of human exposure to endocrine disrupting chemicals, Congress added specific language on endocrine disruption in the FQPA and recent amendments to the Safe Drinking Water Act (SDWA). The FQPA mandated that the EPA develop an endocrine disruptor screening program, and the SDWA authorizes EPA to screen endocrine disruptors found in drinking water systems. The newly developed Endocrine Disruptor Screening Program focuses on methods and procedures to detect and to characterize the endocrine activity of pesticides and other chemicals (see Fig. 1.9). The scientific data needed for the estimated 87 000 chemicals in commerce does not exist to conduct adequate assessments of potential risks. The screening program is being used by EPA to collect this information for endocrine disruptors and to decide appropriate regulatory action by first assigning each chemical to an endocrine disruption category.

The chemicals undergo sorting into four categories according to the available existing, scientifically relevant information:

1. Category 1 chemicals have sufficient, scientifically relevant information to determine that they are not likely to interact with the estrogen, androgen, or thyroid systems. This category includes some polymers and certain exempted chemicals.
2. Category 2 chemicals have insufficient information to determine whether they are likely to interact with the estrogen, androgen, or thyroid systems, thus will need screening data.
3. Category 3 chemicals have sufficient screening data to indicate endocrine activity, but data to characterize actual effects are inadequate and will need testing.
4. Category 4 chemicals already have sufficient data for the EPA to perform a hazard assessment.

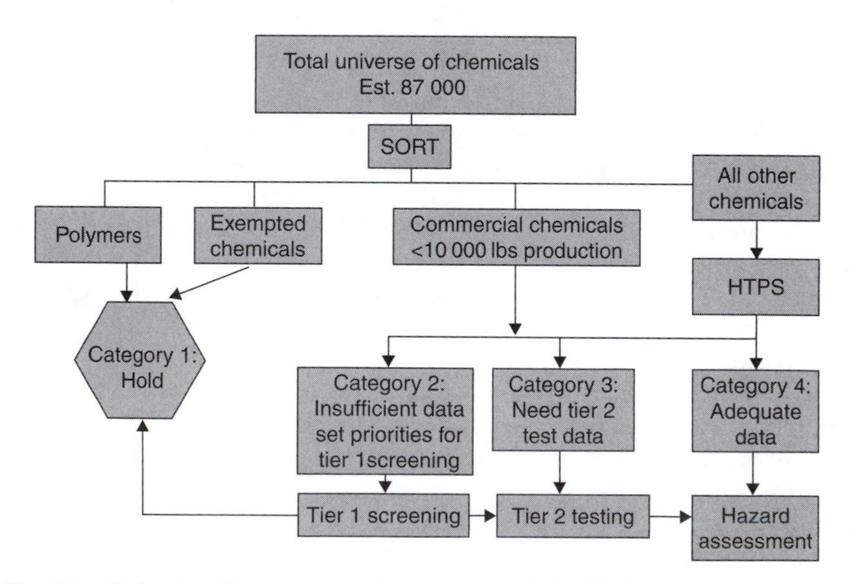

Fig. 1.9. Endocrine disruptor screening program of the US Environmental Protection Agency. *Source*: US EPA, *Report to Congress: Endocrine Disruptor Screening Program*, 2000.

TSCA gives the EPA the authority to track 75 000 industrial chemicals currently produced or imported to the US. This is accomplished through screening the chemicals and requiring that reporting and testing be done for any substance that presents a hazard to human health or the environment. If chemical poses a potential or actual risk that is unreasonable, the EPA may ban the manufacture and import of that chemical.

The EPA has tracked thousands of new chemicals being developed by industries each year, if those chemicals have either unknown or dangerous characteristics. This information is used to determine the type of control that these chemicals would need to protect human health and the environment. Manufacturers and importers of chemical substances first submitted information about chemical substances already on the market during an initial inventory. Since the initial inventory was published, commercial manufacturers or importers of substances not on the inventory have been subsequently required to submit notices to the EPA, which has developed guidance about how to identify chemical substances to assign a unique and unambiguous description of each substance for the inventory. The categories include:

- polymeric substances;
- certain chemical substances containing varying carbon chains;
- products containing two or more substances, formulated, and statutory mixtures; and
- chemical substances of unknown or variable composition, complex reaction products and biological materials (UVCB Substance).

Environmental policy and science set a framework for addressing air pollution. While the programs to address air pollution are recent, controlling air pollution is rooted in Antiquity.

III. AIR POLLUTION BEFORE THE INDUSTRIAL REVOLUTION

One of the reasons the tribes of early history were nomadic was to move periodically away from the stench of the animal, vegetable, and human wastes they generated. When the tribesmen learned to use fire, they used it for millennia in a way that filled the air inside their living quarters with the products of incomplete combustion. Examples of this can still be seen today in some of the more primitive parts of the world. After its invention, the chimney removed the combustion products and cooking smells from the living quarters, but for centuries the open fire in the fireplace caused its emission to be smoky. In AD 61 the Roman philosopher Seneca reported thus on conditions in Rome:

> As soon as I had gotten out of the heavy air of Rome and from the stink of the smoky chimneys thereof, which, being stirred, poured forth whatever pestilential vapors and soot they had enclosed in them, I felt an alteration of my disposition.

Air pollution, associated with burning wood in Tutbury Castle in Nottingham, was considered "unendurable" by Eleanor of Aquitaine, the wife of King Henry II of England, and caused her to move in the year 1157. One hundred sixteen years later, coal burning was prohibited in London; and in 1306, Edward I issued a royal proclamation enjoining the use of *sea coal* in furnaces. Elizabeth I barred the burning of coal in London when Parliament was in session. The repeated necessity for such royal action would seem to indicate that coal continued to be burned despite these edicts. By 1661, the pollution of London had become bad enough to prompt John Evelyn to submit a brochure "Fumifugium, or the Inconvenience of the Aer, and Smoake of London Dissipated (together with some remedies humbly proposed)" to King Charles II and Parliament. This brochure has been reprinted and is recommended to students of air pollution [1].

Fig. 1.10. Lead smelting furnace. *Source*: Agricola, G., *De Re Metallica*, Book X, p. 481, Basel, Switzerland, 1556. Translated by Hoover, H. C. and Hoover, L. H. *Mining Magazine*, London, 1912. Reprinted by Dover Publications, New York, 1950.

It proposes means of air pollution control that are still viable in the 21st century.

The principal industries associated with the production of air pollution in the centuries preceding the Industrial Revolution were metallurgy, ceramics, and preservation of animal products. In the bronze and iron ages, villages were exposed to dust and fumes from many sources. Native copper and gold were forged, and clay was baked and glazed to form pottery and bricks before 4000 BC. Iron was in common use and leather was tanned before 1000 BC. Most of the methods of modern metallurgy were known before AD 1. They relied on charcoal rather than coal or coke. However, coal was mined and used for fuel before AD 1000, although it was not made into coke until about 1600; and coke did not enter metallurgical practice significantly until about 1700. These industries and their effluents as they existed before 1556 are best described in the book "De Re Metallica" published in that year by Georg Bauer, known as Georgius Agricola (Fig. 1.10). This book was translated into English and published in 1912 by Herbert Clark Hoover and his wife [2].

Examples of the air pollution associated with the ceramic and animal product preservation industries are shown in Figs. 1.11 and 1.12, respectively.

Fig. 1.11. A pottery kiln. *Source*: Cipriano Piccolpasso, *The Three Books of the Potters's Art*, fol. 35C, 1550. Translated by Rackham, B., and Van de Put, A. Victoria and Albert Museum, London, 1934.

Fig. 1.12. A kiln for smoking red herring. *Source*: Duhamel due Monceau, H. L., Traité général des pêches, Vol. 2, Section III, Plate XV, Fig. 1, Paris, 1772.

IV. AIR POLLUTION AND THE INDUSTRIAL REVOLUTION

The Industrial Revolution was the consequence of the harnessing of steam to provide power to pump water and move machinery. This began in the early years of the eighteenth century, when Savery, Papin, and Newcomen designed their pumping engines, and culminated in 1784 in Watt's reciprocating engine. The reciprocating steam engine reigned supreme until it was displaced by the steam turbine in the twentieth century.

Steam engines and steam turbines require steam boilers, which, until the advent of the nuclear reactor, were fired by vegetable or fossil fuels. During most of the nineteenth century, coal was the principal fuel, although some oil was used for steam generation late in the century.

The predominant air pollution problem of the nineteenth century was smoke and ash from the burning of coal or oil in the boiler furnaces of stationary power plants, locomotives, and marine vessels, and in home heating fireplaces and furnaces. Great Britain took the lead in addressing this problem, and, in the words of Sir Hugh Beaver [3]:

> By 1819, there was sufficient pressure for Parliament to appoint the first of a whole dynasty of committees "to consider how far persons using steam engines and furnaces could work them in a manner less prejudicial to public health and comfort." This committee confirmed the practicability of smoke prevention, as so many succeeding committees were to do, but as was often again to be experienced, nothing was done.

> In 1843, there was another Parliamentary Select Committee, and in 1845, a third. In that same year, during the height of the great railway boom, an act of Parliament disposed of trouble from locomotives once and for all (!) by laying down the dictum that they must consume their own smoke. The Town Improvement Clauses Act 2 years later applied the same panacea to factory furnaces. Then 1853 and 1856 witnessed two acts of Parliament dealing specifically with London and empowering the police to enforce provisions against smoke from furnaces, public baths, and washhouses and furnaces used in the working of steam vessels on the Thames.

Smoke and ash abatement in Great Britain was considered to be a health agency responsibility and was so confirmed by the first Public Health Act of 1848 and the later ones of 1866 and 1875. Air pollution from the emerging chemical industry was considered a separate matter and was made the responsibility of the Alkali Inspectorate created by the Alkali Act of 1863.

In the United States, smoke abatement (as air pollution control was then known) was considered a municipal responsibility. There were no federal or state smoke abatement laws or regulations. The first municipal ordinances and regulations limiting the emission of black smoke and ash appeared in the 1880s and were directed toward industrial, locomotive, and marine rather than domestic sources. As the nineteenth century drew to a close, the pollution of the air of mill towns the world over had risen to a peak (Fig. 1.13); damage to vegetation from the smelting of sulfide ores was recognized as a problem everywhere it was practiced.

The principal technological developments in the control of air pollution by engineering during the nineteenth century were the stoker for mechanical firing of coal, the scrubber for removing acid gases from effluent gas streams,

Fig. 1.13. Engraving (1876) of a metal foundry refining department in the industrial Saar region of West Germany. *Source*: The Bettmann Archive, Inc.

cyclone and bag house dust collectors, and the introduction of physical and chemical principles into process design.

V. RECENT AIR POLLUTION

A. 1900–1925

During the period 1900–1925 there were great changes in the technology of both the production of air pollution and its engineering control, but no significant changes in legislation, regulations, understanding of the problem, or public attitudes toward the problem. As cities and factories grew in size, the severity of the pollution problem increased.

One of the principal technological changes in the production of pollution was the replacement of the steam engine by the electric motor as the means of operating machinery and pumping water. This change transferred the smoke and ash emission from the boiler house of the factory to the boiler house of the electric generating station. At the start of this period, coal was hand-fired in the boiler house; by the middle of the period, it was mechanically fired by stokers; by the end of the period, pulverized coal, oil, and gas firing had begun to take over. Each form of firing produced its own characteristic emissions to the atmosphere.

At the start of this period, steam locomotives came into the heart of the larger cities. By the end of the period, the urban terminals of many railroads had been electrified, thereby transferring much air pollution from the railroad right-of-way to the electric generating station. The replacement of coal by oil in many applications decreased ash emissions from those sources. There was rapid technological change in industry. However, the most significant change was the rapid increase in the number of automobiles from almost none at the turn of the century to millions by 1925 (Table 1.5).

TABLE 1.5

Annual Motor Vehicle Sales in the United States[a]

Year	Total	Year	Total
1900	4192	1945	725 215
1905	25 000	1950	8 003 056
1910	187 000	1955	9 169 292
1915	969 930	1960	7 869 221
1920	2 227 347	1965	11 057 366
1925	4 265 830	1970	8 239 257
1930	3 362 820	1975	8 985 012
1935	3 971 241	1980	8 067 309
1940	4 472 286	1985	11 045 784
		1990	9 295 732

[a] Data include foreign and domestic sales for trucks, buses, and automobiles.

The principal technological changes in the engineering control of air pollution were the perfection of the motor-driven fan, which allowed large-scale gas-treating systems to be built; the invention of the electrostatic precipitator, which made particulate control in many processes feasible; and the development of a chemical engineering capability for the design of process equipment, which made the control of gas and vapor effluents feasible.

B. 1925–1950

In this period, present-day air pollution problems and solutions emerged. The Meuse Valley, Belgium, episode [4] occurred in 1930; the Donora, Pennsylvania, episode [5] occurred in 1948; and the Poza Rica, Mexico, episode [6] in 1950. Smog first appeared in Los Angeles in the 1940s (Fig. 1.14). The Trail, British Columbia, smelter arbitration [7] was completed in 1941. The first National Air Pollution Symposium in the United States was held in Pasadena, California, in 1949 [8], and the first United States Technical Conference on Air Pollution was held in Washington, DC, in 1950 [9]. The first large-scale surveys of air pollution were undertaken—Salt Lake City, Utah (1926) [10]; New York, (1937) [11]; and Leicester, England (1939) [12].

Air pollution research got a start in California. The Technical Foundation for Air Pollution Meteorology was established in the search for means of disseminating and protecting against chemical, biological, and nuclear warfare agents. Toxicology came of age. The stage was set for the air pollution scientific and technological explosion of the second half of the twentieth century.

Fig. 1.14. Los Angeles smog in the 1940s. *Source*: Los Angeles County, California.

(a)

Fig. 1.15. (a) Pittsburgh after the decrease in black smoke. *Source*: Allegheny County, Pennsylvania. (b) Pittsburgh before the decrease in black smoke.

A major technological change was the building of natural gas pipelines, and where this occurred, there was rapid displacement of coal and oil as home heating fuels with dramatic improvement in air quality; witness the much publicized decrease in black smoke in Pittsburgh (Fig. 1.15) and St. Louis. The diesel locomotive began to displace the steam locomotive, thereby slowing the pace of railroad electrification. The internal combustion engine bus started its displacement of the electrified streetcar. The automobile continued to proliferate (Table 1.5).

During this period, no significant national air pollution legislation or regulations were adopted anywhere in the world. However, the first state air pollution law in the United States was adopted by California in 1947.

(b)

Fig. 1.15. (*Continued*)

C. 1950–1980

In Great Britain, a major air pollution disaster hit London in 1952 [13], resulting in the passage of the Clean Air Act in 1956 and an expansion of the authority of the Alkali Inspectorate. The principal changes that resulted were in the means of heating homes. Previously, most heating was done by burning soft coal on grates in separate fireplaces in each room. A successful effort was made to substitute smokeless fuels for the soft coal used in this manner, and central or electrical heating for fireplace heating. The outcome was a decrease in "smoke" concentration, as measured by the blackness of paper filters through which British air was passed from $175\,\mu g\,m^{-3}$ in 1958 to $75\,\mu g\,m^{-3}$ in 1968 [14].

During these two decades, almost every country in Europe, as well as Japan, Australia, and New Zealand, experienced serious air pollution in its larger cities. As a result, these countries were the first to enact national air pollution control legislation. By 1980, major national air pollution research centers had been set up at the Warren Springs Laboratory, Stevenage, England; the Institut National de la Santé et de las Recherche Medicale at Le Visinet, France; the Rijksinstituut Voor de Volksgezondheid, Bilthoven and the Instituut voor Gezondheidstechniek-TNO, Delft, The Netherlands; the Statens Naturvardsverk, Solna, Sweden; the Institut für Wasser-Bodenund Luft-hygiene, Berlin; and the Landensanstalt für Immissions und Bodennutzungsshutz, Essen, Germany. The important air pollution research centers in Japan are too numerous to mention.

In the United States, the smog problem continued to worsen in Los Angeles and appeared in large cities throughout the nation (Fig. 1.16). In 1955 the first federal air pollution legislation was enacted, providing federal support for air pollution research, training, and technical assistance. Responsibility for the administration of the federal program was given to the Public Health Service (PHS) of the United States Department of Health, Education, and Welfare, and remained there until 1970, when it was transferred to the new US EPA. The initial federal legislation was amended and extended several times between 1955 and 1980, greatly increasing federal

Fig. 1.16. Smog in New York City in the 1950s. *Source*: Wide World Photos.

authority, particularly in the area of control [15]. The automobile continued to proliferate (Table 1.5).

As in Europe, air pollution research activity expanded tremendously in the United States during these three decades. The headquarters of federal research activity was at the Robert A. Taft Sanitary Engineering Center of the PHS in Cincinnati, Ohio, during the early years of the period and at the National Environmental Research Center in Triangle Park, North Carolina, at the end of the period.

An International Air Pollution Congress was held in New York City in 1955 [16]. Three National Air Pollution Conferences were held in Washington, DC, in 1958 [17], 1962 [18], and 1966 [19]. In 1959, an International Clean Air Conference was held in London [20].

In 1964, the International Union of Air Pollution Prevention Associations (IUAPPA) was formed. IUAPPA has held International Clean Air Congresses in London in 1966 [21]; Washington, DC, in 1970 [22]; Dusseldorf in 1973 [23]; Tokyo in 1977 [24]; Buenos Aires in 1980 [25]; Paris in 1983 [26]; Sydney in 1986 [27]; The Hague in 1989 [28]; and Montreal in 1992 [29].

Technological interest during these 30 years has focused on automotive air pollution and its control, on sulfur oxide pollution and its control by sulfur oxide removal from flue gases and fuel desulfurization, and on control of nitrogen oxides produced in combustion processes.

Air pollution meteorology came of age and, by 1980, mathematical models of the pollution of the atmosphere were being energetically developed. A start had been made in elucidating the photochemistry of air pollution. Air quality monitoring systems became operational throughout the world. A wide variety of measuring instruments became available.

VI. THE 1980s

The highlight of the 1970s and 1980s was the emergence of the ecological, or total environmental, approach. Organizationally, this has taken the form of departments or ministries of the environment in governments at all levels throughout the world. In the United States there is a federal EPA, and in most states and populous counties and cities, there are counterpart organizations charged with responsibility for air and water quality, sold waste sanitation, noise abatement, and control of the hazards associated with radiation and the use of pesticides. This is paralleled in industry, where formerly diffuse responsibility for these areas is increasingly the responsibility of an environmental protection coordinator. Similar changes are evident in research and education.

Pollution controls were being built into pollution sources—automobiles, power plants, factories—at the time of original construction rather than later on. Also, for the first time, serious attention was directed to the problems caused by the "greenhouse" effect of carbon dioxide and other gases building up in the atmosphere, possible depletion of the stratospheric ozone layer by fluorocarbons, long-range transport of pollution, prevention of significant deterioration (PSD), and acidic deposition.

VII. RECENT HISTORY

The most sweeping change, in the United States at least, in the decade of the 1990s was the passage of the Clean Air Act Amendments on November 15, 1990 [29]. This was the only change in the Clean Air Act since 1977, even though the US Congress had mandated that the Act be amended much earlier. Michigan Representative John Dingell referred to the amendments as "the most complex, comprehensive, and far-reaching environmental law any Congress has ever considered." John-Mark Stenvaag has stated in his book, "Clean Air Act 1990 Amendments, Law and Practice" [30], "The enormity of the 1990 amendments begs description. The prior Act, consisting of approximately 70 000 words, was widely recognized to be a remarkably complicated, unapproachable piece of legislation. If environmental attorneys, government officials, and regulated entities were awed by the prior Act, they will be astonished, even stupefied, by the 1990 amendments. In approximately 145 000 new words, Congress has essentially tripled the length of the prior Act and geometrically increased its complexity."

The 1990s saw the emergence, in the popular media, of two distinct but closely related global environmental crises, uncontrolled global climate changes and stratospheric ozone depletion. The climate changes of concern were both the warming trends caused by the buildup of greenhouse gases in the atmosphere and cooling trends caused by PM and sulfates in the same atmosphere. Some researchers have suggested that these two trends will cancel each other. Other authors have written [31] that global warming may not be all bad. It is going to be an interesting decade as many theories are developed and tested during the 1990s. The "Earth Summit," really the UN Conference of Environment and Development, in Rio de Janeiro during June 1992 did little to resolve the problems, but it did indicate the magnitude of the concern and the differences expressed by the nations of the world.

The other global environmental problem, stratospheric ozone depletion, was less controversial and more imminent. The US Senate Committee Report supporting the Clean Air Act Amendments of 1990 states, "Destruction of the ozone layer is caused primarily by the release into the atmosphere of chlorofluorocarbons (CFCs) and similar manufactured substances—persistent chemicals that rise into the stratosphere where they catalyze the destruction of stratospheric ozone. A decrease in stratospheric ozone will allow more UV radiation to reach Earth, resulting in increased rates of disease in humans, including increased incidence of skin cancer, cataracts, and, potentially, suppression of the immune system. Increased UV radiation has also been shown to damage crops and marine resources."

The Montreal Protocol of July 1987 resulted in an international treaty in which the industrialized nations agreed to halt the production of most ozone-destroying CFCs by the year 2000. This deadline was hastily changed to 1996, in February 1992, after a US NASA satellite and high-altitude sampling aircraft found levels of chlorine monoxide over North America that were 50% greater than that measured over Antarctica.

Global problems continue to hold sway. In the early years of the 21st Century, global air pollution has taken on a greater urgency, both within the scientific community and the general public. Courts in the United States have mandated that carbon dioxide be considered and regulated as an air pollutant due to its radiant properties and link to global warming. The long-range atmospheric transport of persistent, bioaccumulative toxic substances (PBTs) has led to elevated concentrations in the tissues of marine and arctic mammals, as well as increased PBT levels in the mother's milk of indigenous people, such as the Inuit. Stratospheric ozone depletion is strongly suspected to have played a role in diminished biodiversity around the world.

VIII. THE FUTURE

The air pollution problems of the future are predicated on the use of more and more fossil and nuclear fuel as the population of the world increases. During the lifetime of the students using this book, partial respite may be offered by solar, photovoltaic, geothermal, wind, non-fossil fuel (hydrogen and biomass), and oceanic (thermal gradient, tidal, and wave) sources of energy. Still, many of the agonizing environmental decisions of the next decades will involve a choice between fossil fuel and nuclear power sources and the depletion of future fuel reserves for present needs. Serious questions will arise regarding whether to conserve or to use these reserves—whether to allow unlimited growth or to curb it.

Other problems concerning transportation systems, waste processing and recycling systems, national priorities, international economics, employment versus environmental quality, and personal freedoms will continue to surface. The choices will have to be made, ideally by educated citizens and enlightened leaders.

FURTHER READING

1. The Smoake of London—Two Prophecies [Selected by Lodge Jr., J. P.]. Maxwell Reprint, Elmsford, NY, 1969.
2. Agricola, G., *De Re Metallica*, Basel, 1556 [English translation and commentary by Hoover, H. C., and Hoover, L. H., *Mining Magazine*, London, 1912]. Dover, New York, 1950.
3. Beaver, Sir Hugh E. C., The growth of public opinion, in *Problems and Control of Air Pollution* (Mallette, F. S., ed.). Reinhold, New York, 1955.

4. Firket, J., *Bull. Acad. R. Med. Belg.* **11**, 683 (1931).
5. Schrenk, H. H., Heimann, H., Clayton, G. D., Gafefer, W. M., and Wexler, H., *US Pub. Health Serv. Bull.* **306**, 173 (1949).
6. McCabe, L. C., and Clayton, G. D., *Arch. Ind. Hyg. Occup. Med.* **6**, 199 (1952).
7. Dean, R. S., and Swain, R. E., Report submitted to the Trail Smelter Arbitral Tribunal, *US Bur. Mines Bull.* **453**. US Government Printing Office, Washington, DC, 1944.
8. *Proceedings of the First National Air Pollution Symposium*, Pasadena, CA, 1949. Sponsored by the Stanford Research Institute in cooperation with the California Institute of Technology, University of California, and University of Southern California..
9. McCabe, L. C. (ed.), *Air Pollution* Proceedings of the United States Technical Conference on Air Pollution, 1950. McGraw-Hill, New York, 1952.
10. Monett, O., Perrott, G. St. J., and Clark, H. W., *US Bur. Mines Bull.* **254** (1926).
11. Stern, A. C., Buchbinder, L., and Siegel, J., *Heat. Piping Air Cond.* **17**, 7–10 (1945).
12. *Atmospheric Pollution in Leicester—A Scientific Study*, D.S.I.R. Atmospheric Research Technical Paper No. 1. H.M. Stationery Office, London, 1945.
13. *Ministry of Health, Mortality and Morbidity during the London Fog of December 1952*, Report of Public Health and Related Subject No. 95, H.M. Stationery Office, London, 1954.
14. *Royal Commission on Environmental Pollution*, First Report, H.M. Stationery Office, London, 1971.
15. Stern, A. C., *Air Pollut. Control Assoc.* **32**, 44 (1982).
16. Mallette, F. S. (ed.), *Problems and Control of Air Pollution*. American Society of Mechanical Engineers, New York, 1955.
17. *Proceedings of the National Conference on Air Pollution*, Washington, DC, 1958. US Public Health Service Publication No. 654, 1959.
18. *Proceedings of the National Conference on Air Pollution*, Washington, DC, 1962. US Public Health Service Publication No. 1022, 1963.
19. *Proceedings of the National Conference on Air Pollution*, Washington, DC, 1966. US Public Health Service Publication No. 1699, 1967.
20. *Proceedings of the Diamond Jubilee International Clean Air Conference 1959*. National Society for Clean Air, London, 1960.
21. *Proceedings of the International Clean Air Congress*, London, October 1966, Part I. National Society for Clean Air, London, 1966.
22. Englund, H., and Beery, W. T. (eds.), *Proceedings of the Second International Clean Air Congress*, Washington, DC, 1970. Academic Press, New York, 1971.
23. *Proceedings of the Third International Clean Air Congress*, Dusseldorf, September 1973. Verein Deutcher Inginieure, Dusseldorf, 1974.
24. *Proceedings of the Fourth International Clean Air Congress*, Tokyo, 1977. Japanese Union of Air Pollution Prevention Associations, Tokyo.
25. *Proceedings of the Fifth International Clean Air Congress*, Buenos Aires, 1981, Vol. 2 Associon Argentina Contra La Contaminacion del Aire, Buenos Aires, 1983.
26. *Proceedings of the Sixth International Clean Air Congress*, Paris, 1983. Association Pour La Prevention de la Pollution Atmospherique, Paris.
27. *Proceedings of the Seventh International Clean Air Congress*, International Union of Air Pollution Prevention Associations, Sydney, 1986.
28. *Proceedings of the Eighth International Clean Air Congress*, International Union of Air Pollution Prevention Associations, The Hague, 1989.
29. Public Law 101–549, 101st Congress. November 15, 1990, an Act, to amend the Clean Air Act to provide for attainment and maintenance of health protective national ambient air quality standards, and for other purposes.

30. Stenvaag, J. M., *Clean Air Act 1990 Amendments, Law and Practice*. Wiley, New York, 1991.

31. Ausubel, J. H., A second look at the impacts of climate change. *Am. Sci.* **79** (May–June, 1991).

SUGGESTED READING

Daumas, M. (ed.), *The History of Technology and Invention*, Vol. 3. Crown, New York, 1978.

Fishman, J., *Global Alert: The Ozone Pollution Crisis*. Plenum, New York, 1990.

Gore, A., *Earth in Balance*. Houghton Mifflin, New York, 1992.

Singer, C., Holmyard, E. K., Hall, A. R., and Williams, T. I. (eds.), *A History of Technology*, Vol. 5. Oxford University Press (Clarendon), New York, 1954–1958.

Williams, T. (ed.), *A History of Technology—Twentieth Century, 1900–1950*, Part I (Vol. 6), Part II (Vol. 7). Oxford University Press (Clarendon), New York, 1978.

Wolf, A., *A History of Science, Technology and Philosophy in the Sixteenth and Seventeenth Centuries*. George Allen & Unwin, London, 1935.

Wolf, A., *A History of Science, Technology and Philosophy in the Eighteenth Century*. George Allen & Unwin, London, 1961.

QUESTIONS

1. Discuss the development of the use of enclosed space for human occupancy over the period of recorded history.
2. Discuss the development of the heating and cooling of enclosed space for human occupancy over the period of recorded history.
3. Discuss the development of the lighting of enclosed space for human occupancy over the period of recorded history.
4. Discuss the development of means to supplement human muscular power over the period of recorded history.
5. Discuss the development of transportation over the period of recorded history.
6. Discuss the development of agriculture over the period of recorded history.
7. Discuss the future alternative sources of energy for light, heat, and power.
8. Compare the so-called soft (i.e. widely distributed small sources) and hard (i.e. fewer very large sources) paths for the future provision of energy for light, heat, and power.
9. What have been the most important developments in the history of the air pollution problem since the publication of this edition of this book?
10. Explain the science behind the phrase, "Think globally, but act locally."
11. What are some of the major differences between how the Clean Air Act (NESHAPs) addresses hazardous chemicals and how they are addressed by the hazardous waste laws (RCRA and Superfund)?
12. Choose two chemicals from the list of 189 "air toxics" (visit: http://www.epa.gov/ttn/atw/orig189.html). Give at least three steps that a business can take to approach air pollution from these substances.

2

The Earth's Atmosphere

I. THE ATMOSPHERE

Before we can determine the extent of air pollution, we need a baseline to which present conditions can be compared. While dynamic, the atmosphere can be characterized quantitatively.

On a macroscale (Fig. 2.1) as the earth's atmospheric temperature varies with altitude, as does the density of the substances comprising the atmosphere.[1] In general, the air grows progressively less dense with increasing altitude moving upward from the troposphere through the stratosphere and the chemosphere to the ionosphere. In the upper reaches of the ionosphere, the gaseous molecules are few and far between as compared with the troposphere.

The ionosphere and chemosphere are of interest to space scientists because they must be traversed by space vehicles en route to or from the moon or the planets, and they are also regions in which satellites travel in the earth's orbit. These regions are also of interest to communications scientists because of their influence on radio communications. However, these layers are of interest to air pollution scientists primarily because of their absorption and scattering of solar energy, which influence the amount and spectral distribution of solar energy and cosmic rays reaching the stratosphere and troposphere.

The stratosphere is of interest to aeronautical scientists because it is traversed by airplanes; to communications scientists because of radio and television communications; and to air pollution scientists because global transport of pollution, particularly the debris of aboveground atomic bomb tests and volcanic eruptions, occurs in this region and because absorption

[1] U.S. Environmental Protection Agency, SI:409 Basic Air Pollution Meteorology, 2007: Accessed at: http://yosemite.epa.gov/oaqps/eogtrain.nsf/DisplayView/SI_409_0-5?OpenDocument

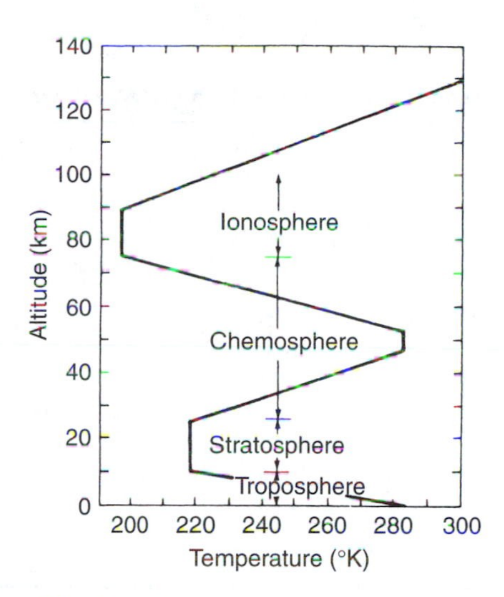

Fig. 2.1. The regions of the atmosphere.

and scattering of solar energy also occur there. The lower portion of this region contains the stratospheric ozone layer, which absorbs harmful ultraviolet (UV) solar radiation. Global change scientists are interested in modifications of this layer by long-term accumulation of chlorofluorocarbons (CFCs) and other gases released at the earth's surface or by high-altitude aircraft.

The troposphere is the region in which we live and is the primary focus of this book; however, we will discuss stratospheric ozone formation and destruction in some detail in Chapter 15.

II. BASELINE CONDITIONS: UNPOLLUTED AIR

The gaseous composition of unpolluted tropospheric air is given in Table 2.1. Unpolluted air is a concept, i.e., what the composition of the air would be if humans and their works were not on earth. We will never know the precise composition of unpolluted air because by the time we had the means and the desire to determine its composition, humans had been polluting the air for thousands of years. Now even at the most remote locations at sea, at the poles, and in the deserts and mountains, the air may be best described as dilute polluted air. It closely approximates unpolluted air, but differs from it to the extent that it contains vestiges of diffused and aged human-made pollution.

The real atmosphere is more than a dry mixture of permanent gases. It has other constituents—vapor of both water and organic liquids, and particulate matter (PM) held in suspension. Above their temperature of condensation, vapor molecules act just like permanent gas molecules in the air. The predominant vapor in the air is water vapor. Below its condensation temperature,

TABLE 2.1

The Gaseous Composition of Unpolluted Air (Dry Basis)

	ppm (vol.)	$\mu g\, m^{-3}$
Nitrogen	780 000	8.95×10^8
Oxygen	209 400	2.74×10^8
Water	—	—
Argon	9300	1.52×10^7
Carbon dioxide	315	5.67×10^5
Neon	18	1.49×10^4
Helium	5.2	8.50×10^2
Methane	1.0–1.2	$6.56–7.87 \times 10^2$
Krypton	1.0	3.43×10^3
Nitrous oxide	0.5	9.00×10^2
Hydrogen	0.5	4.13×10^1
Xenon	0.08	4.29×10^2
Organic vapors	*ca.* 0.02	—

if the air is saturated, water changes from vapor to liquid. This phenomenon gives rise to fog or mist in the air and condensed liquid water on cool surfaces exposed to air. The quantity of water vapor in the air varies greatly from almost complete dryness to supersaturation, i.e., between 0% and 4% by weight (see Fig. 2.2). If Table 2.2 is compiled on a wet air basis at a time when the water vapor concentration is 31 200 parts by volume per million parts by volume of wet air (Table 2.3), the concentration of condensable organic vapors is seen to be so low compared to that of water vapor that for all practical purposes the difference between wet air and dry air is its water vapor content.

Gaseous composition in Tables 2.1 and 2.2 is expressed as parts per million by volume—ppm (vol). (When a concentration is expressed simply as ppm, it is unclear whether a volume or weight basis is intended.) To avoid confusion caused by different units, air pollutant concentrations in this book are generally expressed as micrograms per cubic meter of air ($\mu g\, m^{-3}$) at 25°C and 760 mmHg, i.e., in metric units. To convert from units of ppm (vol.) to $\mu g\, m^{-3}$, it is assumed that the ideal gas law is accurate under ambient conditions. A generalized formula for the conversion at 25°C and 760 mmHg is

$$
\begin{aligned}
1 \text{ ppm(vol) pollutant} &= \frac{1\text{ L pollutant}}{10^6 \text{ L air}} \\
&= \frac{(1\text{ L}/22.4) \times \text{MW} \times 10^6 \mu\text{g} \times \text{gm}^{-1}}{10^6 \text{L} \times 298°\text{K}/273°\text{K} \times 10^{-3}\ \text{m}^3 \times \text{L}^{-1}} \\
&= 40.9 \times \text{MW}\, \mu\text{g}/\text{m}^{-3}
\end{aligned}
\qquad (2.1)
$$

where MW equals molecular weight. For convenience, conversion units for common pollutants are shown in Table 2.3.

TABLE 2.2

The Gaseous Composition of Unpolluted Air (Wet Basis)

	ppm (vol.)	$\mu g\, m^{-3}$
Nitrogen	756 500	8.67×10^8
Oxygen	202 900	2.65×10^8
Water	31 200	2.30×10^7
Argon	9000	1.47×10^7
Carbon dioxide	305	5.49×10^5
Neon	17.4	1.44×10^4
Helium	5.0	8.25×10^2
Methane	0.97–1.16	$6.35–7.63 \times 10^2$
Krypton	0.97	3.32×10^3
Nitrous oxide	0.49	8.73×10^2
Hydrogen	0.49	4.00×10^1
Xenon	0.08	4.17×10^2
Organic vapors	*ca.* 0.02	—

TABLE 2.3

Conversion Factors Between Volume and Mass Units of Concentration (25°C, 760 mmHg)

	To convert from	
Pollutant	ppm (vol.) to $\mu g\, m^{-3}$, multiply by	$\mu g\, m^{-3}$ to ppm (vol.), multiply by ($\times 10^{-3}$)
Ammonia (NH_3)	695	1.44
Carbon dioxide	1800	0.56
Carbon monoxide	1150	0.87
Chlorine	2900	0.34
Ethylene	1150	0.87
Hydrogen chloride	1490	0.67
Hydrogen fluoride	820	1.22
Hydrogen sulfide	1390	0.72
Methane (carbon)	655	1.53
Nitrogen dioxide	1880	0.53
Nitric oxide	1230	0.81
Ozone	1960	0.51
Peroxyacetylnitrate	4950	0.20
Sulfur dioxide	2620	0.38

A minor problem arises in regard to nitrogen oxides. It is common practice to add concentrations of nitrogen dioxide and nitric oxide in ppm (vol.) and express the sum as "oxides of nitrogen." In metric units, conversion from ppm (vol.) to $\mu g\, m^{-3}$ must be done separately for nitrogen dioxide and nitric oxide prior to addition.

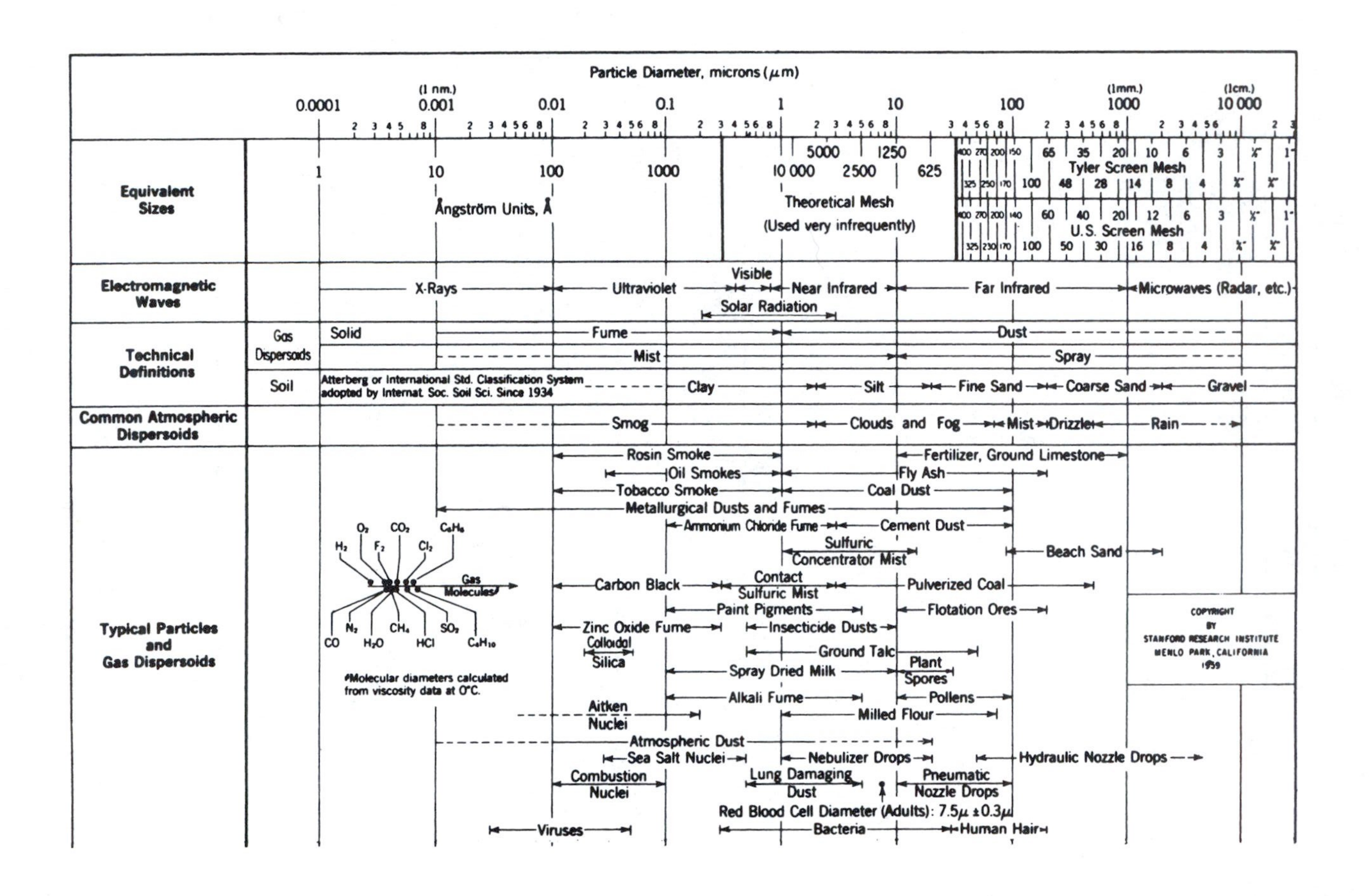

Particle Diameter, microns (μm)
0.0001
(1 nm.) 0.001
0.01
0.1
1
10
100
(1mm.) 1000
(1cm.) 10 000
Equivalent Sizes
Ångström Units, Å
Theoretical Mesh (Used very infrequently)
Tyler Screen Mesh
U.S. Screen Mesh
Electromagnetic Waves
X-Rays
Ultraviolet
Visible
Near Infrared
Far Infrared
Microwaves (Radar, etc.)
Solar Radiation
Technical Definitions
Gas Dispersoids
Solid
Fume
Dust
Mist
Spray
Soil
Atterberg or International Std. Classification System adopted by Internat. Soc. Soil Sci. Since 1934
Clay
Silt
Fine Sand
Coarse Sand
Gravel
Common Atmospheric Dispersoids
Smog
Clouds and Fog
Mist
Drizzle
Rain
Typical Particles and Gas Dispersoids
Rosin Smoke
Oil Smokes
Tobacco Smoke
Metallurgical Dusts and Fumes
Ammonium Chloride Fume
Fertilizer, Ground Limestone
Fly Ash
Coal Dust
Cement Dust
Sulfuric Concentrator Mist
Beach Sand
Gas Molecules*
Carbon Black
Contact Sulfuric Mist
Pulverized Coal
Paint Pigments
Flotation Ores
Zinc Oxide Fume
Colloidal Silica
Insecticide Dusts
Ground Talc
Spray Dried Milk
Plant Spores
Alkali Fume
Pollens
Aitken Nuclei
Milled Flour
Atmospheric Dust
Sea Salt Nuclei
Nebulizer Drops
Hydraulic Nozzle Drops
Combustion Nuclei
Lung Damaging Dust
Pneumatic Nozzle Drops
Red Blood Cell Diameter (Adults): 7.5μ ± 0.3μ
Viruses
Bacteria
Human Hair
*Molecular diameters calculated from viscosity data at 0°C.
COPYRIGHT BY STANFORD RESEARCH INSTITUTE MENLO PARK, CALIFORNIA 1959

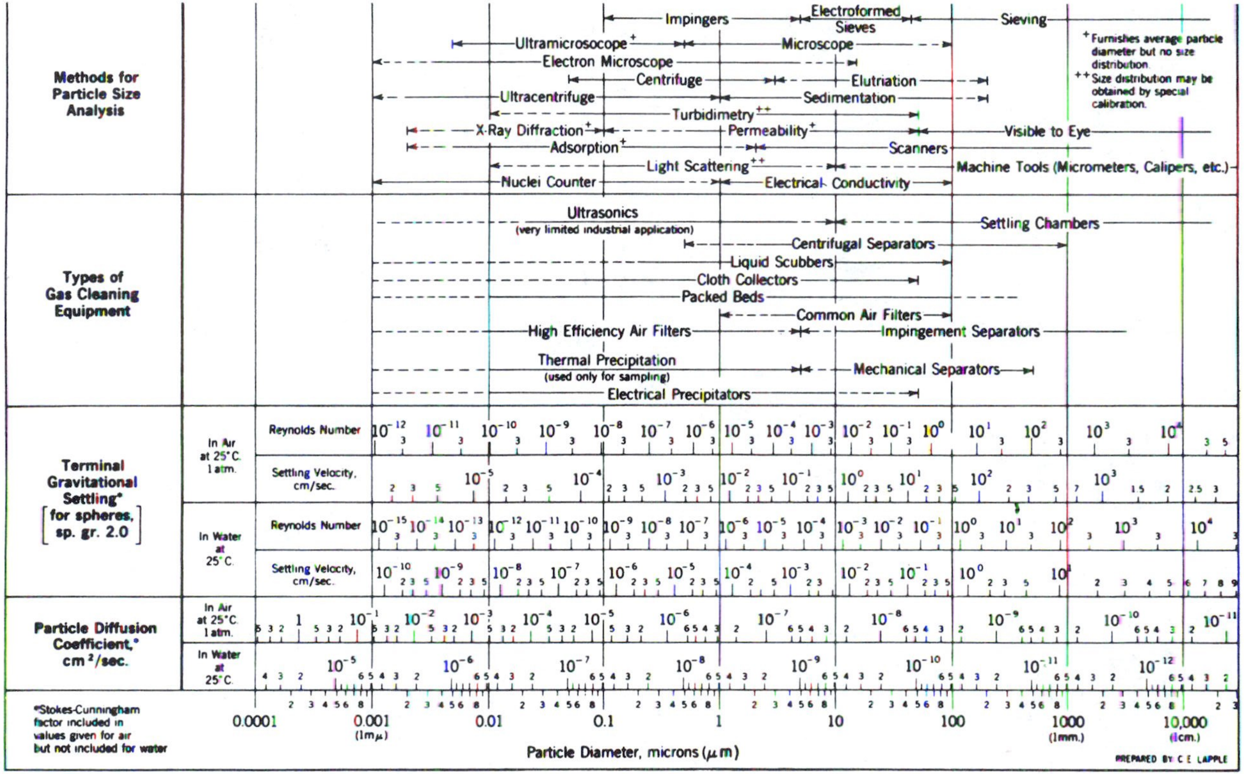

Methods for Particle Size Analysis

Impingers — Electroformed Sieves — Sieving
Ultramicroscope+ — Microscope
Electron Microscope
Centrifuge — Elutriation
Ultracentrifuge — Sedimentation
Turbidimetry++
X-Ray Diffraction+ — Permeability+ — Visible to Eye
Adsorption+ — Scanners
Light Scattering++ — Machine Tools (Micrometers, Calipers, etc.)
Nuclei Counter — Electrical Conductivity

+Furnishes average particle diameter but no size distribution.
++Size distribution may be obtained by special calibration.

Types of Gas Cleaning Equipment

Ultrasonics (very limited industrial application) — Settling Chambers
Centrifugal Separators
Liquid Scubbers
Cloth Collectors
Packed Beds
Common Air Filters
High Efficiency Air Filters — Impingement Separators
Thermal Precipitation (used only for sampling) — Mechanical Separators
Electrical Precipitators

Terminal Gravitational Settling* [for spheres, sp. gr. 2.0]

In Air at 25°C, 1 atm.: Reynolds Number; Settling Velocity, cm/sec.
In Water at 25°C.: Reynolds Number; Settling Velocity, cm/sec.

Particle Diffusion Coefficient,* $cm^2/sec.$

In Air at 25°C, 1 atm.
In Water at 25°C.

*Stokes-Cunningham factor included in values given for air but not included for water

Particle Diameter, microns (μm): 0.0001, 0.001 ($1 m\mu$), 0.01, 0.1, 1, 10, 100, 1000 (1mm.), 10,000 (1cm.)

PREPARED BY C. E. LAPPLE

Fig. 2.2. Characteristics of particles and particle dispersoids. Adapted from figure reproduced by permission of SRI International, Menlo Park, CA, 1959.

III. WHAT IS AIR POLLUTION?

In Chapter 1, we attempted to define air pollution. Let us now consider what distinguishes polluted air from acceptable air quality. Over the past few decades the central feature of air pollution has been its association with harm, especially harm to humans in terms of diseases, such as respiratory diseases associated with air pollutants. Harm implies a value; i.e., something that society values is lost or diminished. In the United States, a handful of commonly found air pollutants are known to cause three specific types of harm. They impair health, destroy and adversely affect environmental resources, and damage property. To address these harms, as mentioned in Chapter 1, the Clean Air Act of 1970 established the National Ambient Air Quality Standards to address six so-called "criteria air pollutants":

1. particulate matter (PM),
2. ozone (O_3),
3. carbon monoxide (CO),
4. sulfur dioxide (SO_2),
5. nitrogen dioxide (NO_2),
6. lead (Pb).

They are called criteria air pollutants because the US Environmental Protection Agency (EPA) regulates them by using two sets of criteria for pollutant standards. The first set of standards is designed to protect public health based on sound science. This set of limits (known as *primary standards*) protects health. A second set of limits (known as *secondary standards*) aims to prevent environmental and property damage. When an urban area or other geographic area has concentrations of a criteria pollutant below the standard it is said to be "in attainment" and the area is declared to be an "attainment area." Conversely, any area that has concentrations of a criteria pollutant above the standard is called a "nonattainment area." Such a designation is not only problematic because of the potential health and environmental effects, but also because it means that the local and state governments will have to take actions to bring the area into attainment. The Clean Air Act gives the federal government a range of possible sanctions to encourage these actions, including withholding certain federal funds (e.g. road-building and other transportation projects). At present, many urban areas are classified as nonattainment for at least one criteria air pollutant. In fact about 90 million Americans live in nonattainment areas.[2]

The public's apprehensions, however, extend beyond the criteria pollutants and increase with the growing wariness about "toxic" chemicals added to the more familiar "conventional" pollutants. The clearest association of toxic air pollutants in recent decades has been with cancer, although neurotoxicity (especially in children) from lead and mercury grew in importance in the 1970s and 1980s. By the end of the twentieth century, new toxic pollutants also

[2] US Environmental Protection Agency, *Plain English Guide to the Clean Air Act*, http://www.epa.gov/oar/oaqps/peg_caa/pegcaa03.html#topic3a; accessed on October 6, 2006.

competed for the public's attention, including air pollutants that threaten hormonal systems in humans and wildlife, as well as those associated with immune system disorders.

IV. PARTICULATE MATTER

Neither Table 2.1 nor Table 2.2 lists among the constituents of the air the suspended PM that it always contains. The gases and vapors exist as individual molecules in random motion. Each gas or vapor exerts its proportionate partial pressure. The particles are aggregates of many molecules, sometimes of similar molecules, often of dissimilar ones. They age in the air by several processes. Some particles serve as nuclei upon which vapors condense. Some particles react chemically with atmospheric gases or vapors to form different compounds. When two particles collide in the air, they tend to adhere to each other because of attractive surface forces, thereby forming progressively larger and larger particles by agglomeration. The larger a particle becomes, the greater its weight and the greater its likelihood of falling to the ground rather than remaining airborne. The process by which particles fall out of the air to the ground is called *sedimentation*. Washout of particles by snowflakes, rain, hail, sleet, mist, or fog is a common form of agglomeration and sedimentation. Still other particles leave the air by impaction onto and retention by the solid surfaces of vegetation, soil, and buildings. The particulate mix in the atmosphere is dynamic, with continual injection into the air from sources of small particles; creation of particles in the air by vapor condensation or chemical reaction among gases and vapors; and removal of particles from the air by agglomeration, sedimentation, or impaction.

Before the advent of humans and their works, there must have been particles in the air from natural sources. These certainly included all the particulate forms of condensed water vapor; the condensed and reacted forms of natural organic vapors; salt particles resulting from the evaporation of water from sea spray; wind-borne pollen, fungi, molds, algae, yeasts, rusts, bacteria, and debris from live and decaying plant and animal life; particles eroded by the wind from beaches, desert, soil, and rock; particles from volcanic and other geothermal eruption and from forest fires started by lightning; and particles entering the troposphere from outer space. As mentioned earlier, the true natural background concentration will never be known because when it existed humans were not there to measure it, and by the time humans started measuring PM levels in the air, they had already been polluting the atmosphere with particles resulting from their presence on earth for several million years. The best that can be done now is to assume that the particulate levels at remote places—the middle of the sea, the poles, and the mountaintops—approach the true background concentration. The very act of going to a remote location to make a measurement implies some change in the atmosphere of that remote location attributable to the means people used to travel and to maintain themselves while obtaining the measurements. PM is measured on a dry basis, thereby eliminating from the measurement not only water droplets and snowflakes but also all vapors, both aqueous and organic, that evaporate or are desiccated from the PM during the drying process. Since different investigators and investigative processes employ different drying procedures and definitions of dryness, it is important to know the procedures and definition employed when comparing data.

Although many of the air pollutants discussed in this book are best classified by their chemical composition, particles are first classified according to their physical properties. PM is a common physical classification of particles found in the air, such as dust, dirt, soot, smoke, and liquid droplets.[3] Unlike other US criteria pollutants [O_3, CO, SO_2, NO_2 and lead (Pb)], PM is not a specific chemical entity but is a mixture of particles from different sources and of different sizes, compositions, and properties. However, the chemical composition of PM is very important and highly variable. In fact, knowing what a particle is made of tells us much about its source, e.g. receptor models use chemical composition and morphology of particles as a means to trace pollutants back to the source.

The chemical composition of tropospheric particles includes inorganic ions, metallic compounds, elemental carbon, organic compounds, and crustal (e.g. carbonates and compounds of alkali and rare earth elementals) substances. For example, the mean 24 h $PM_{2.5}$ concentration measured near Baltimore, Maryland in 1999 was composed of 38% sulfate, 13% ammonium, 2% nitrate, 36% organic carbon, 7% elemental carbon, and 4% crustal matter.[4] In addition, some atmospheric particles can be hygroscopic, i.e., they contain particle-bound water. The organic fraction can be particularly difficult to characterize, since it often contains thousands of organic compounds.

The size of a particle is determined by how the particle is formed. For example, combustion can generate very small particles, while coarse particles are often formed by mechanical processes (see Fig. 2.3). If particles are sufficiently small and of low mass, they can be suspended in the air for long periods of time. Larger particles (e.g. $>10\,\mu m$ aerodynamic diameter) are found in smoke or soot (see Fig. 2.4), while very small particles ($<2.5\,\mu m$) may be apparent only indirectly, such as the manner in which they diffuse, diffract, absorb, and reflect light (see Fig. 2.5).

Sources of particles are highly variable. They may be emitted directly to the air from stationary sources, such as factories, power plants, and open burning, and from moving vehicles (known as "mobile sources"), first by direct emissions from internal combustion engines, but also when these and other

[3] UK Department of Environment, Food, and Rural Affairs, Expert Panel on Air Quality Standards, 2004, *Airborne Particles: What Is the Appropriate Measurement on Which to Base a Standard? A Discussion Document.*

[4] Bonne, G., Mueller, P., Chen, L. W., Doddridge, B. G., Butler, W. A., Zawadzki, P. A., Chow, J. C., Tropp, R. J., and Kohl, S. *Proceeding of the PM2000: Particulate Matter and Health Conference, Composition of PM2.5 in the Baltimore-Washington Corridor*, pp. W17–W18. Air & Waste Management Association, Washington, DC, January 2000.

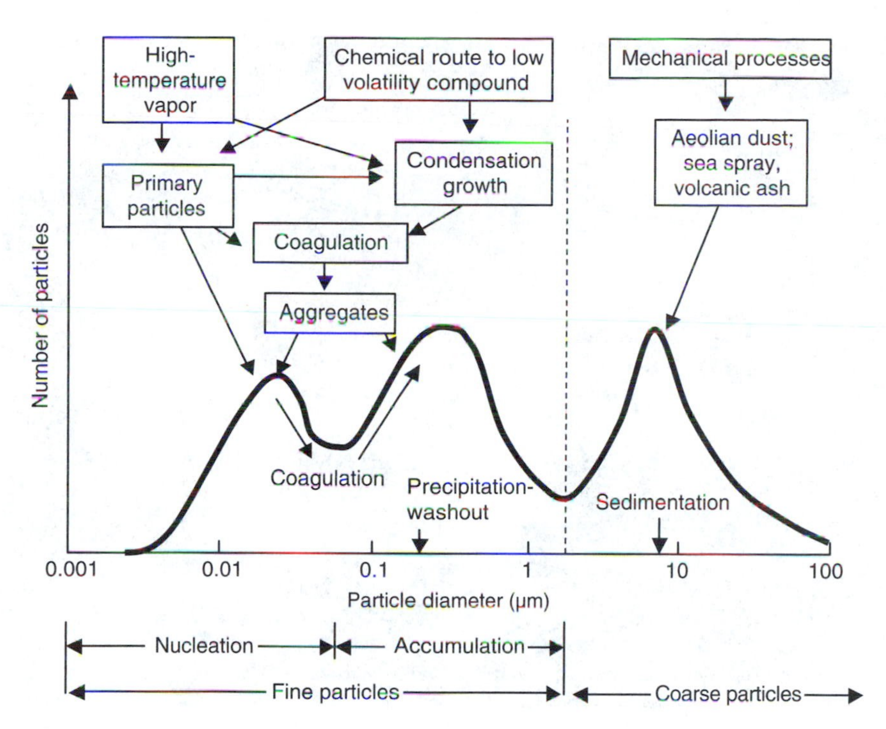

Fig. 2.3 Prototypical size distribution of tropospheric particles with selected sources and pathways of how the particles are formed. Dashed line is approximately 2.5 μm diameter. Adapted from: United Kingdom Department of Environment, Food, and Rural Affairs, Expert Panel on Air Quality Standards, 2004, *Airborne Particles: What Is the Appropriate Measurement on which to Base a Standard? A Discussion Document.*

particles are re-entrained due to the movement of vehicles (e.g. in a "near-road" situation). Area or non-point sources of particles include construction, agricultural activities such as plowing and tilling, mining, and forest fires.

Particles may also form from gases that have been previously emitted, such as when gases released from burning fuels react with sunlight and water vapor. A common production of such "secondary particles" occurs when gases undergo chemical reactions in the atmosphere involving O_2 and water vapor (H_2O). Photochemistry can be an important step in secondary particle formation, resulting when chemical species like ozone (O_3) are involved in step reactions with radicals, e.g. the hydroxyl (OH) and nitrate (NO_3) radicals. Photochemistry also incurs in the presence of air pollutant gases like sulfur dioxide (SO_2), nitrogen oxides (NO_x), and organic gases emitted by anthropogenic and natural sources. In addition, nucleation of particles from low-vapor pressure gases emitted from sources or formed in the atmosphere, condensation of low-vapor pressure gases on aerosols already present in the atmosphere, and coagulation of aerosols can contribute to the formation of particles. The chemical composition, transport, and fate of particles are directly associated with the characteristics of the surrounding gas.

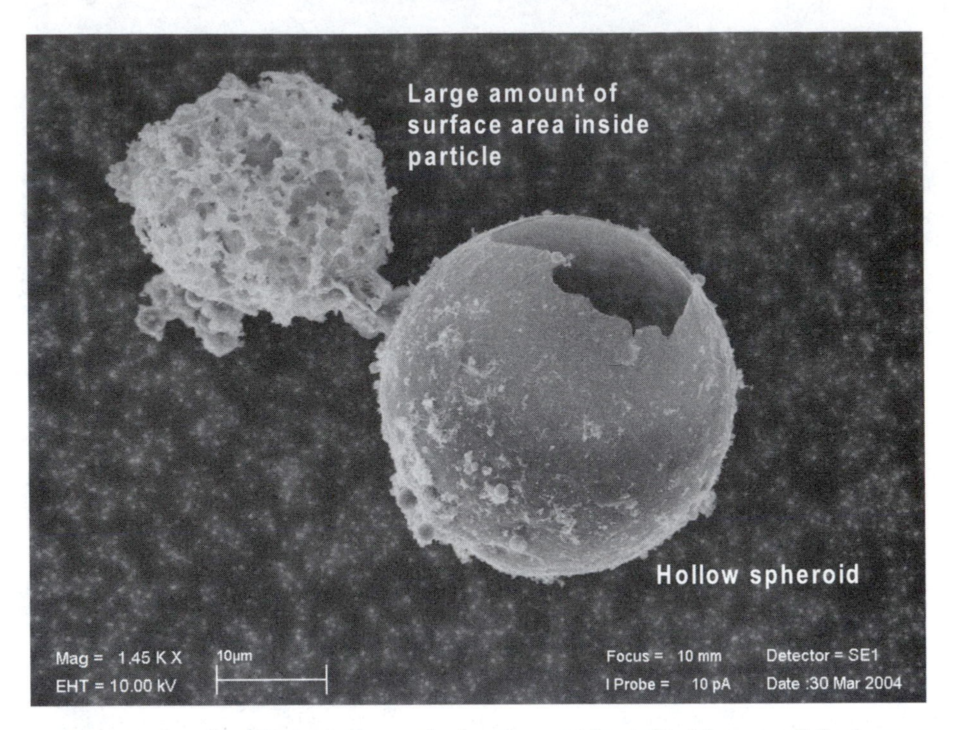

Fig. 2.4. Scanning electron micrograph of coarse particles emitted from an oil-fired power plant. Diameters of the particles are greater than 20 μm optical diameter. Both particles are hollow, so their aerodynamic diameter is significantly smaller than if they were solid. *Source*: Source characterization study by Stevens, R., Lynam, M., and Proffitt, D., 2004. Photo courtesy of Willis, R., ManTech Environmental Technology, Inc., 2004; used with permission.

The term "aerosol" is often used synonymously with PM. An aerosol can be a suspension of solid or liquid particles in air, and an aerosol includes both the particles and all vapor or gas phase components of air.

As mentioned, very small particles may remain suspended for some time, so they are can be particularly problematic from a pollutant transport perspective because their buoyancy allows them to travel longer distances. Smaller particles are also challenging because they are associated with numerous health effects (mainly because they can penetrate more deeply into the respiratory system than larger particles).

Generally, the mass of PM falling in two size categories is measured, i.e. $\leqslant 2.5\,\mu m$ diameter, and $\geqslant 2.5\,\mu m$ and $\leqslant 10\,\mu m$ diameter. These measurements are taken by instruments (see Fig. 2.6) with inlets using size exclusion mechanisms to segregate the mass of each size fraction (i.e. "dichotomous" samplers). Particles with diameters $\geqslant 10\,\mu m$ are generally of less concern, since these particles rarely travel long distances; however, they are occasionally measured if a large particulate emitting source (e.g. a coal mine) is nearby, since these particles rarely travel long distances.

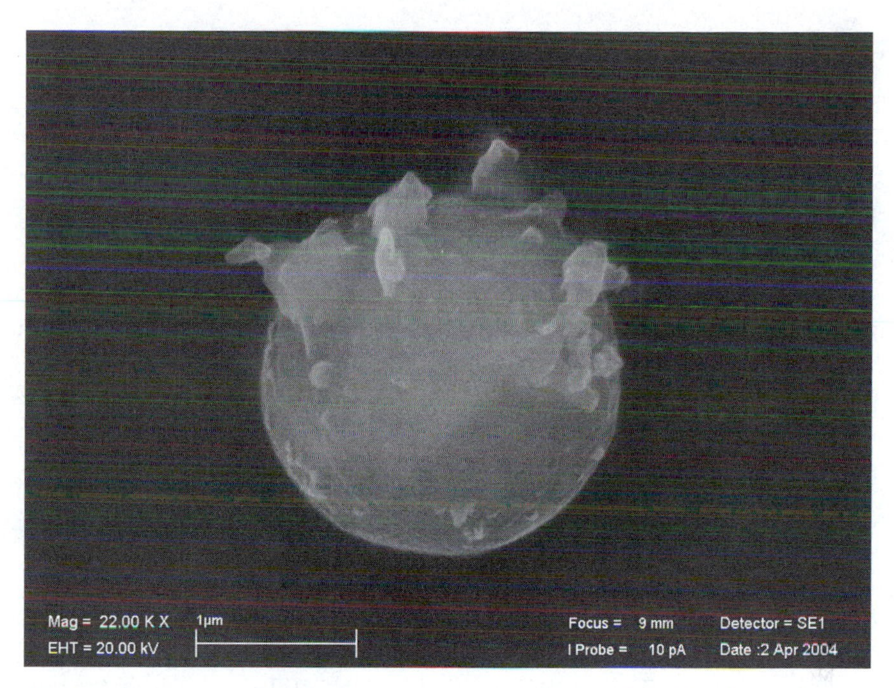

Fig. 2.5. Scanning electron micrograph of spherical aluminosilicate fly ash particle emitted from an oil-fired power plant. Diameter of the particle is approximately 2.5 μm. Photo courtesy of Willis, R., ManTech Environmental Technology, Inc., 2004; used with permission.

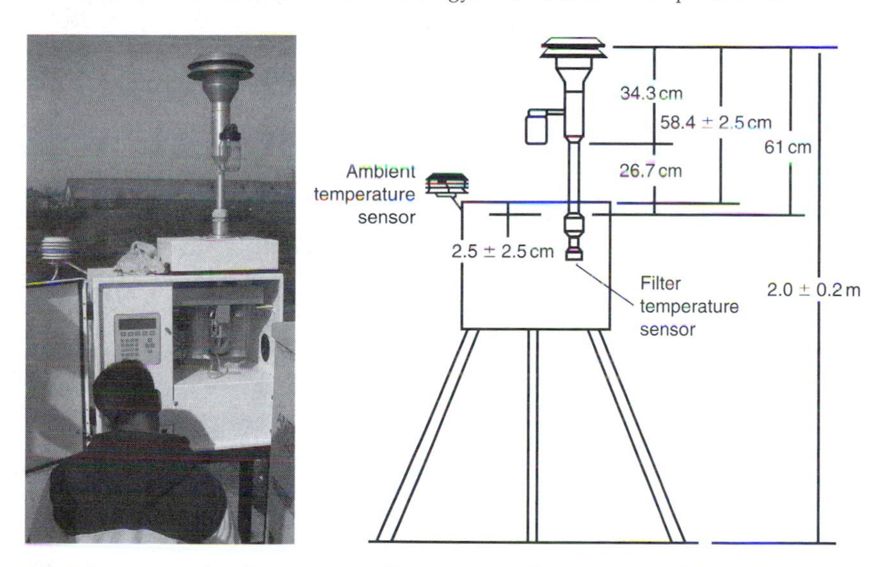

Fig. 2.6. Photo and schematic of sampling device used to measure particles with aerodynamic diameters ≤2.5 μm. Each sampler has an inlet (top) that takes in particles ≤10 μm. An impacter downstream in the instrument cuts the size fraction to ≤2.5 μm, which is collected on Teflon filter. The filter is weighed before and after collection. The Teflon allows construction for other analyses, e.g. X-ray fluorescence to determine inorganic composition of the particles. Quartz filters would be used if any subsequent carbon analyses are needed. Photo and schematic courtesy of US EPA.

Mass can be determined for a predominantly spherical particle by microscopy, either optical or electron, by light scattering and Mie theory, by the particle's electrical mobility, or by its aerodynamic behavior. However, since most particles are not spherical, PM diameters are often described using an equivalent diameter, i.e., the diameter of a sphere that would have the same fluid properties. Another term, optical diameter, is the diameter of a spherical particle that has an identical refractive index as the particle. Optical diameters are used to calibrate the optical particle sizing instruments, which scatter the same amount of light into the solid angle measured. Diffusion and gravitational settling are also fundamental fluid phenomena used to estimate the efficiencies of PM transport, collection, and removal processes, such as in designing PM monitoring equipment and ascertaining the rates and mechanisms of how particles infiltrate and deposit in the respiratory tract.

Only for very small diameter particles is diffusion sufficiently important that the Stokes diameter is often used. The Stokes diameter for a particle is the diameter of a sphere with the same density and settling velocity as the particle. The Stokes diameter is derived from the aerodynamic drag force caused by the difference in velocity of the particle and the surrounding fluid. Thus, for smooth, spherical particles, the Stokes diameter is identical to the physical or actual diameter. The aerodynamic diameter (D_{pa}) for all particles greater than 0.5 μm can be approximated[5] as the product of the Stokes particle diameter (D_{ps}) and the square root of the particle density (ρ_p):

$$D_{pa} = D_{ps}\sqrt{\rho_p} \tag{2.2}$$

The units of the diameters are in μm and the units of density are in $g\,cm^{-3}$.

Fine particles (<2.5 μm) generally come from industrial combustion processes (such as the particles in Fig. 2.4) and from vehicle exhaust. As mentioned, this smaller sized fraction has been closely associated with increased respiratory disease, decreased lung functioning, and even premature death, probably due to their ability to bypass the body's trapping mechanisms, such as cilia in the lungs, and nasal hair filtering. Some of the diseases linked to PM exposure include aggravation of asthma, chronic bronchitis, and decreased lung function.

In addition to health impacts, PM is also a major contributor to reduced visibility, including near national parks and monuments. Also, particles can be transported long distances and serve as vehicles on which contaminants are able to reach water bodies and soils. Acid deposition, for example, can be as dry or wet precipitation. Either way, particles play a part in acid rain. In the first, the dry particles enter ecosystems and potentially reduce the pH of receiving waters. In the latter, particles are washed out of the atmosphere and, in the

[5] Aerosol textbooks provide methods to determine the aerodynamic diameter of particles less than 0.5 μm. For larger particles gravitational settling is more important and the aerodynamic diameter is often used.

process; lower the pH of the rain. The same transport and deposition mechanisms can also lead to exposures to persistent organic contaminants like dioxins and organochlorine pesticides, and heavy metals like mercury that have sorbed in or on particles.

In addition to their inherent toxicity, particles can function as vehicles for transporting and transforming chemical contaminants. For example, compounds that are highly sorptive (e.g. those with large K_{oc} partitioning coefficients) can use particles as a means for long-range transport. Also, charge differences between the particle and ions (particularly metal cations) will also make particles a means by which contaminants are transported.

There are ways of measuring PM other than by weight per unit volume of air. They include a count of the total number of particles in a unit volume of air, a count of the number of particles of each size range, the weight of particles of each size range, and similar measures based on the surface area and volume of the particles rather than on their number or weight. Some particles in the air are so small that they cannot be seen by an optical microscope, individually weighing so little that their presence is masked in gravimetric analysis by the presence of a few large particles.

The mass of a spherical particle is

$$w = \frac{4}{3}\pi p r^3 \tag{2.3}$$

where w is the particle mass (g), r is the particle radius (cm), and p is the particle density (g cm^{-3}).

The size of small particles is measured in microns (μm). One micron is one-millionth of a meter or 10 000 Å (angstrom units)—the units used to measure the wavelength of light (visible light is between 3000 and 8000 Å) (Fig. 2.2) [2]. Compare the weight of a 10-μm particle near the upper limit of those found suspended in the air and a 0.1-μm particle which is near the lower limit. If both particles have the same density (p) the smaller particle will have one-millionth the weight of the larger one. This is because the radius term is cubed. The usual gravimetric procedures can scarcely distinguish a 0.1-μm particle in the presence of a 10-μm particle. To measure the entire size range of particles in the atmosphere, several measurement techniques must therefore be combined, each one most appropriate for its size range (Table 2.4). Thus, the smallest particles—those only slightly larger than a gas molecule—are measured by the electric charge they carry and by electron microscopy. The next larger size range is measured by electron microscopy or by the ability of these particles to act as nuclei upon which water vapor can be condensed in a cloud chamber. (The water droplets are measured rather than the particles themselves.) The still larger size range is measured by electron or optical microscopy; and the largest size range is measured gravimetrically, either as suspended particles separated from the air by a sampling device or as sedimented particles falling out of the air into a receptacle.

TABLE 2.4

Particle Size Ranges and Their Methods of Measurement

Particle size range (μm)	Ions	Nuclei	Visability	Suspended or settleable; nonairborne	Dispersion aerosol	Condensation aerosol	Pollen and spores	Sedimentation, diffusion, and settling
10^{-4}–10^{-3}	Small	—	—	Suspended	—	Gas molecules	—	Diffusion
10^{-3}–10^{-2}	Intermediate and large	Aitken nuclei	Electron microscope	Suspended	—	Vapor molecules	—	Diffusion
10^{-2}–10^{-1}	Large	Aitken and condensation nuclei	Electron microscope	Suspended	—	Fume-mist	—	Diffusion
Air pollution								
10^{-1}–10^{0}	—	Condensation nuclei	Microscope: electron and optical	Suspended	Dust-mist	Fume-mist	—	Diffusion and sedimentation
10^{0}–10^{1}	—	—	Microscope: optical	Suspended and settleable	Dust-mist	Fume-mist	—	Sedimentation
10^{1}–10^{2}	—	—	Eye, sieves	Settleable	Dust-mist	Mist-fog	Pollen and spores	—
10^{2}–10^{3}	—	—	Eye, sieves	Nonairborne	Dust-spray	Drizzle-rain	—	Sedimentation
10^{3}–10^{4}	—	—	Eye, sieves	Nonairborne	Sand-rocks	Rain	—	Sedimentation

1. *Fibers*

Generally, when environmental scientists discuss particles, they mean those that are somewhat spherical or angular like soil particles. Particles that are highly elongated are usually differentiated as "fibers." Such elongation is expressed as a particle's aspect ratio, i.e. the ratio of the length to width. Fibers generally have aspect ratios greater than 3:1. Environmentally important fibers include fiberglass, fabrics, and minerals (see Figs. 2.7 and 2.8). Exposure to fiberglass and textile fibers is most commonly found in industrial settings, such as it has been associated with the health problems of textile workers exposed to fibrous matter in high doses for many years. For example, chronic exposure to cotton fibers has led the ailment, byssinosis, also referred to as "brown lung disease," which is characterized by the narrowing of the lung's airways. However, when discussing fibers, it is highly likely that first contaminant to come to mind is asbestos, a group of highly fibrous minerals with separable, long, and thin fibers. Separated asbestos fibers are strong enough and flexible enough to be spun and woven. Asbestos fibers are heat resistant, making them useful for many industrial purposes. Because of their durability, asbestos fibers that get into lung tissue will remain for long periods of time.

2. *Asbestos: The Fiber of Concern*

There are two general types of asbestos, *amphibole* and *chrysotile*. Some studies show that amphibole fibers stay in the lungs longer than chrysotile, and this tendency may account for their increased toxicity.

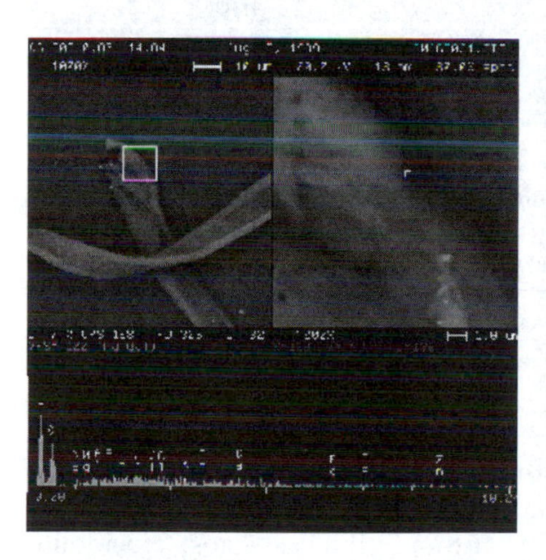

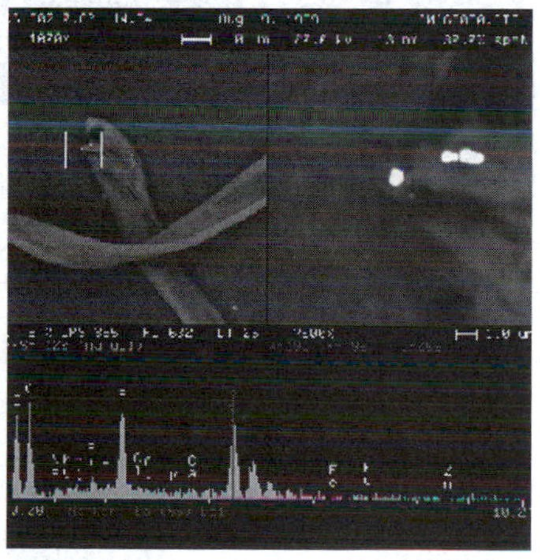

Fig. 2.7. Scanning electron micrograph of cotton fibers. Acquired using an Aspex Instruments, Ltd., Scanning electron microscope. *Source*: US Environmental Protection Agency, 2004. Note the different chemical composition at different locations of the same fiber as indicated by X-ray diffraction (XRD) spectrometry. Each peak at the bottom of the left and right micrographs indicates a different chemical element; the higher the peak, the greater the concentration of that element. Photo courtesy of Conner, T.; used with permission.

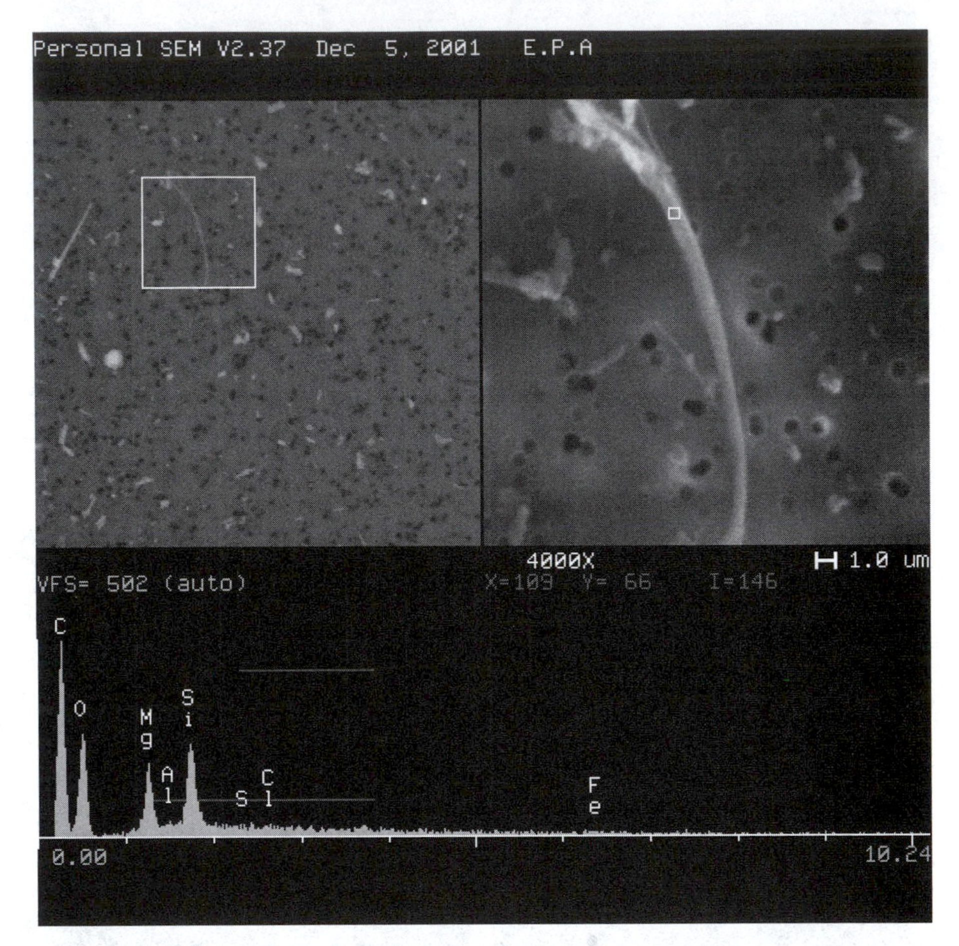

Fig. 2.8. Scanning electron micrograph of fibers in dust collected near the World Trade Center, Manhattan, NY, in September 2001. Acquired using an Aspex Instruments, Ltd., Scanning electron microscope. The bottom of the micrograph represents the elemental composition of the highlighted 15-μm long fiber by energy dispersive spectroscopy (EDS). This composition (i.e. O, Si, Al, and Mg) and the morphology of the fibers indicate they are probably asbestos. The EDS carbon peak results from the dust being scanned on a polycarbonate filter. *Source*: US Environmental Protection Agency, 2004. Photo courtesy of Conner, T.; used with permission.

Generally, health regulations classify asbestos into six mineral types: chrysotile, a serpentine mineral with long and flexible fibers; and five amphiboles, which have brittle crystalline fibers. The amphiboles include actinolite asbestos, tremolite asbestos, anthophyllite asbestos, crocidolite asbestos, and amosite asbestos (see Fig. 2.9).

3. Asbestos Routes of Exposure

Ambient air concentrations of asbestos fibers are about 10^{-5}–10^{-4} fibers per milliliter (fibers mL^{-1}), depending on location. Human exposure to concentrations much higher than 10^{-4} fibers mL^{-1} is suspected of causing health

Fig. 2.9. Scanning electron micrograph of asbestos fibers (amphibole) from a former vermiculite-mining site near Libby, Montana. *Source*: US Geological Survey and US Environmental Protection Agency, Region 8, Denver, CO.

effects.[6] Asbestos fibers are very persistent and resist chemical degradation (i.e. they are inert under most environmental conditions) so their vapor pressure is nearly zero meaning they do not evaporate, nor do they dissolve in water. However, segments of fibers do enter the air and water as asbestos-containing rocks and minerals that are weathered naturally or when extracted during mining operations. One of the most important exposures is when manufactured products (e.g. pipe wrapping and fire-resistant materials) begin to wear down. Small diameter asbestos fibers may remain suspended in the air for a long time and be transported advectively by wind or water

[6] For more information on asbestos exposure, see the Public Health Statement on Asbestos: ATSDR, 2001, Public Health Statement for Asbestos, http://www.atsdr.cdc.gov/toxprofiles/phs61.html.

before sedimentation. Like particles, heavier fibers settle more quickly. Asbestos seldom moves substantially via soil. The fibers are generally not broken down to other compounds in the environment and will remain virtually unchanged over long periods. Although most asbestos is highly persistent, chrysotile, the most commonly encountered form, may break down slowly in acidic environments. Asbestos fibers may break into shorter strands and, therefore, increased number of fibers, by mechanical processes (e.g. grinding and pulverization). Inhaled fibers may become trapped in the lungs and with chronic exposures build up over time. Some fibers, especially chrysotile, can be removed from or degraded in the lung with time.

Type of Control Dependent on Particle Characteristics

Recall from Figure 2.3 that numerous physical processes are at work in the formation of particles in the troposphere. These processes give a clue as to how to control particle emissions.

By measuring each portion of the particle size spectrum by the most appropriate method, a composite diagram of the size distribution of the atmospheric aerosol can be produced. Figure 2.10 shows that there are separate size distributions with respect to the number, surface area, and volume (or mass) of the particles. The volume (mass) distribution is called *bimodal* because of its separate maxima at about 0.2 and 10 μm, which result from

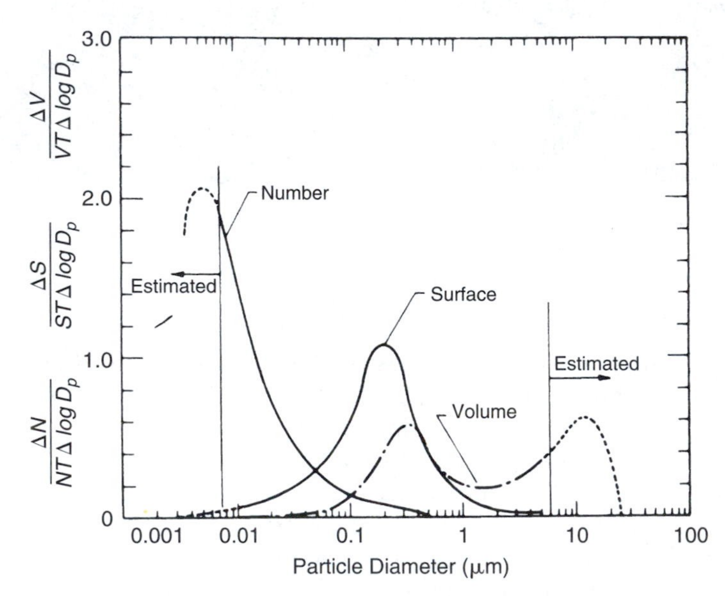

Fig. 2.10. Grand average number (N), surface area (S), and volume (V) distribution of Los Angeles smog. The linear ordinate normalized by total number (NT), area (ST), or volume (VT) is used so that the apparent area under the curves is proportional to the quantity in that size range. *Source*: Corn, M., Properties of non-viable particles in the air, in *Air Pollution*, 3rd ed., Vol. 1 (Stern, A. C., ed.), p. 123. Academic Press, New York, 1976.

different mechanisms of particle formation. The mode with the 0.2-μm maximum results from coagulation and condensation formation mechanisms. As mentioned, these particles are created in the atmosphere by chemical reaction among gases and vapors. They are called *fine* particles to differentiate them from the particles in the 10-μm maximum mode, which are called *coarse*. These fine particles are primarily sulfates, nitrates, organics, ammonium, and lead compounds. The mode with the 10-μm maximum are particles introduced to the atmosphere as solids from the surface of the earth and the seas, plus particles from the coagulation–condensation mode which have grown larger and moved across the saddle between the modes into the larger size mode. These are primarily silicon, iron, aluminum, sea salt, and plant particles. Thus, there is a dynamism that creates small particles, allows them to grow larger, and eventually allows the large particles to be scavenged from the atmosphere by sedimentation (in the absence of precipitation), plus washout and rainout when there is precipitation.

Understanding these mechanisms is the key to controlling air pollution. Designing and operating pollution control equipment effectively must account for the number, surface characteristics, volume, and shape of particles.

The majority of particles in the atmosphere are spherical in shape because they are formed by condensation or cooling processes or they contain core nuclei coated with liquid. Liquid surface tension draws the material in the particle into a spherical shape. Other important particle shapes exist in the atmosphere; e.g. asbestos is present as long fibers and fly ash can be irregular in shape.

The methods just noted tell something about the physical characteristics of atmospheric PM but nothing about its chemical composition. One can seek this kind of information for either individual particles or all particles *en masse*. Analysis of particles *en masse* involves analysis of a mixture of particles of many different compounds. How much of each element or radical, anion, or cation is present in the mixture can be determined. Specific organic compounds may be separated and identified. Individual particles may be analyzed by electron microscopy

Much of the concern about PM in the atmosphere arises because particles of certain size ranges can be inhaled and retained by the human respiratory system. There is also concern because PM in the atmosphere absorbs and scatters incoming solar radiation. For a detailed discussion of the human respiratory system and the defenses it provides against exposure of the lungs to PM, see Chapter 11.

V. CONCEPTS

A. Sources and Sinks

The places from which pollutants emanate are called *sources*. There are natural as well as anthropogenic sources of the permanent gases considered to be pollutants. These include plant and animal respiration and the decay of what was once living matter. Volcanoes and naturally caused forest fires are other natural sources. The places to which pollutants disappear from the air are called *sinks*. Sinks include the soil, vegetation, structures, and water bodies, particularly the oceans. The mechanisms whereby pollutants are removed from the atmosphere are called *scavenging mechanisms*, and the measure used for the aging of a pollutant is its *half-life*—the time it takes for half of the quantity of pollutant emanating from a source to disappear into its various sinks. Fortunately, most pollutants have a short enough half-life (i.e. days rather than decades) to prevent their accumulation in the air to the extent that they substantially alter the composition of unpolluted air shown in Table 2.1. Several gases do appear to be accumulating in the air to the extent that measurements have documented the increase in concentration from year to year. The best-known example is carbon dioxide (Fig. 2.11; see also Fig. 15.1). Other accumulating gases are nitrous oxide (N_2O), methane (CH_4), CFCs, and other halocarbons. All of these gases have complex roles in climate change processes, particularly global warming concerns. CFCs are chemically very stable compounds in the troposphere and have half-lives from tens of years to over 100 years. One of the sinks for CFCs is transport to the stratosphere, where shortwave UV radiation photodissociates the molecules, releasing chlorine (Cl) atoms. These Cl atoms are projected to reduce the steady-state stratospheric ozone concentration, in turn increasing the penetration of harmful UV radiation to the earth's surface.

Oxidation, either atmospheric or biological, is a prime removal mechanism for inorganic as well as organic gases. Inorganic gases, such as nitric oxide (NO), nitrogen dioxide (NO_2), hydrogen sulfide (H_2S), sulfur dioxide

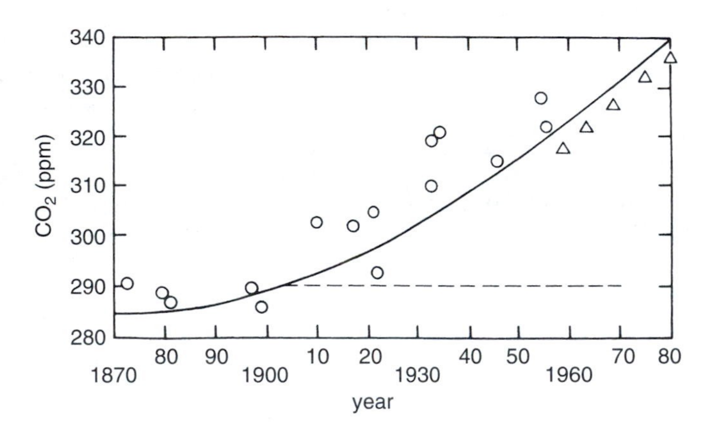

Fig. 2.11. Average CO_2 concentration: North Atlantic Region ○?, Pacific Region Δ. (The dashed line is the nineteenth-century base value: 290 ppm.) *Source*: Combination of data from Callender, G. C., *Tellus*, 10, 243 (1958), and Council on Environmental Quality, *Global Energy Futures and the Carbon Dioxide Problem*. Superintendent of Documents, US Government Printing Office, Washington, DC, 1981 (see also Fig. 15.1).

(SO_2), and sulfur trioxide (SO_3), may eventually form corresponding acids:

$$NO + \frac{1}{2}O_2 \rightarrow NO_2 \quad (2.4)$$

$$4NO_2 + 2H_2O + O_2 \rightarrow 4HNO_3 \quad (2.5)$$

$$H_2S + \frac{2}{3}O_2 \rightarrow SO_2 + H_2O \quad (2.6)$$

$$SO_2 + \frac{1}{2}O_2 \rightarrow SO_3 \quad (2.7)$$

$$SO_3 + H_2O \rightarrow H_2SO_4 \quad (2.8)$$

Oxidation of SO_2 is slow in a mixture of pure gases, but the rate is increased by light, NO_2, oxidants, and metallic oxides which act as catalysts for the reaction. The formed acids can react with PM or ammonia to form salts.

B. Receptors

A *receptor* is something which is adversely affected by polluted air. A receptor may be a person or animal that breathes the air and whose health may be adversely affected thereby, or whose eyes may be irritated or whose skin made dirty. It may be a tree or plant that dies, or the growth yield or appearance of which is adversely affected. It may be some material such as paper, leather, cloth, metal, stone, or paint that is affected. Some properties of the atmosphere itself, such as its ability to transmit radiant energy, may be affected. Aquatic life in lakes and some soils are adversely affected by acidification via acidic deposition.

C. Transport and Dispersion

Transport is the mechanism that moves the pollution from a source to a receptor. The simplest source–receptor combination is that of an isolated point source and an isolated receptor. A point source may best be visualized as a chimney or stack emitting a pollutant into the air; the isolated point source might be the stack of a smelter standing by itself in the middle of a flat desert next to the body of ore it is smelting. The isolated receptor might be a resort hotel 5 miles distant on the edge of the desert. The effluent from the stack will flow directly from it to the receptor when the wind is along the line connecting them (Fig. 2.12). The wind is the means by which the pollution is transported from the source to the receptor. However, during its transit over the 5 miles between the source and the receptor, the plume does not remain a

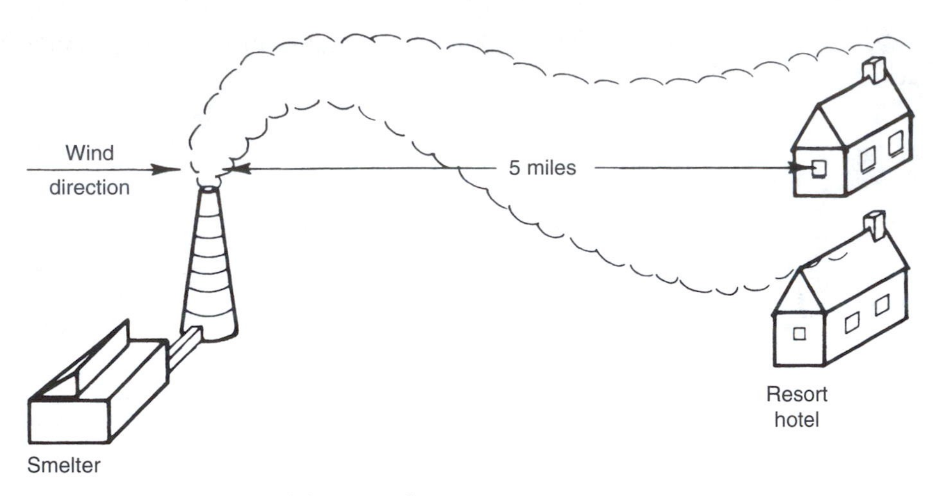

Fig. 2.12. Transport and dispersion from source to receptor.

cylindrical tube of pollution of the same diameter as the interior of the stack from which it was emitted. Instead, as it travels over the 5-mile distance, turbulent eddies in the air and in the plume move parcels from the edges of the plume into the surrounding air and move parcels of surrounding air into the plume. If the wind speed is greater than the speed of ejection from the stack, the wind will stretch out the plume until the plume speed equals wind speed. These two processes—mixing by turbulence and stretch-out of the plume, plus a third one—meandering (which means that the plume may not follow a true straight line between the source and the receptor, but may meander somewhat about that line as wind direction fluctuates from its mean value over the time of transit between the two points)—tend to make the concentration of the plume as it arrives at the receptor less than its concentration on release from the stack. The sum of all these processes is called *diffusion*. However, the term diffusion has a substantially different meaning in chemistry. Substances diffuse according to Fick's law of diffusion, wherein the concentration diminishes with distance from the source. This is known as a concentration gradient. Therefore, dispersion is the preferred term.

If the plume being transported is above the height where ground-based convective and turbulent processes will bring it down to the ground reasonably close to its origin, it may travel for hundreds of miles at that height before being brought to earth, by these processes, in a remote community. This is known as *long-range* or *long-distance transport*.

D. Significant Deterioration of Air Quality

It may be desirable to curtail transport of pollution to areas whose air is presently quite clean, even though, after such transport, the air quality of the area would be considerably cleaner than would be required by air quality standards. This concept is called prevention of significant deterioration (PSD) of the air quality in such areas. It requires definition of how much deterioration can be considered insignificant. PSD is needed:

1. to protect public health and welfare;
2. preserve, protect, and enhance the air quality in national parks, national wilderness areas, national monuments, national seashores, and other areas of special national or regional natural, recreational, scenic, or historic value;
3. to ensure that economic growth will occur in a manner consistent with the preservation of existing clean air resources; and
4. to view regulatory actions that allow certain levels of air pollutant releases from the perspective of the cumulative impact in any area.

Preserving air quality depends on knowing which sources are contributing to the deterioration. When the results of air pollution measurements are interpreted, one of the first questions asked by scientists, engineers, and policy makers is where did it come from? Sorting out the various sources of pollution is known as *source apportionment*. A number of tools are used to try to locate the sources of pollutants. A widely used approach is the "source–receptor model" or as it is more commonly known, the *receptor model.*

Using receptor models can be distinguished from the dispersion models in that dispersion models usually start from the source and estimate where the plume and its contaminants is heading (see Fig. 7.2). Conversely, receptor models are based upon measurements taken in the ambient environment and from these observations, make use of algorithms and functions to determine pollution sources. For a discussion of source–receptor relationships and modeling, see Discussion Box: Source Apportionment, Receptor Models, and Carbon Dating in Chapter 7 (p. 208).

E. Polluted Atmosphere

When is air polluted? This chapter presents principles by which materials are released into the atmosphere, move, transform, and are removed from the atmosphere. The definition of air pollutant or air pollution depends on the context of time, space, and impact for a particular set of circumstances. We have attempted to distinguish unpolluted from polluted air. However, the same chemical compounds or particles from a natural source (e.g. a volcano) elicit the same adverse effects as when they are emitted by anthropogenic sources.

Thus, unpolluted air is merely a benchmark to show the extent and trends of air pollution. Governments around the world have established and are continuously evaluating the impact of elevated levels of myriad gases and particulate material in the atmosphere. Clearly, these agencies are charged with addressing polluted air so that it is healthy to breathe, supports ecosystems, and supports other welfare uses, such as visibility and integrity of buildings and other structures. This leads to the need to consider the risks brought about by a polluted atmosphere.

REFERENCES

1. U.S. Environmental Protection Agency, SI:409 Basic Air Pollution Meteorology (2007): Accessed at: http://yosemite.epa.gov/oaqps/eogtrain.nsf/DisplayView/SI_409_0-5?OpenDocument
2. Lapple, C. E., *Stanford Res. Inst. J.* **5**, 95 (1961).

SUGGESTED READING

Bridgman, H. A., *Global Air Pollution: Problems for the 1990s*. Belhaven, London, 1990.

Graedel, T. E., and Crutzen, P. J., *Atmospheric Change: An Earth System Perspective*. W. H. Freeman, New York, 1993.

Seinfeld, J. H., *Atmospheric Chemistry and Physics of Air Pollution*. Wiley, New York, 1986.

Warneck, P., *Chemistry of the Natural Atmosphere*. Academic Press, San Diego, CA, 1988.

QUESTIONS

1. Prepare a graph showing the conversion factor from ppm (vol.) to $\mu g\, m^{-3}$ for compounds with molecular weights ranging from 10 to 200 at 25°C and 760 mmHg as well as at 0°C and 760 mmHg.
2. (a) Convert 0.2 ppm (vol.) NO and 0.15 ppm (vol.) NO_2 to $\mu g\, m^{-3}$ NO_x at 25°C and 760 mmHg. (b) Convert 0.35 ppm (vol.) NO_x to $\mu g\, m^{-3}$ at 25°C and 760 mmHg.
3. Prepare a table showing the weight in g and the surface area in m^3 of a 0.1-, 1.0-, 10.0-, and 100.0-μm-diameter spherical particle of unit density.
4. What is the settling velocity in $cm\, sec^{-1}$ in air at 25°C and 1 atm for a 100 mesh size spherical particle, i.e., one which just passes through the opening in the sieve (specific gravity = 2.0)?
5. How does the diameter of airborne pollen grains compare with the diameter of a human hair?
6. What are the principal chemical reactions that take place in the chemosphere to give it its name? How do they influence stratospheric and tropospheric chemical reactions?
7. What are the source and nature of the condensable organic vapors in unpolluted air?
8. Has the composition of the unpolluted air of the troposphere most probably always been the same as in Tables 2.1 and 2.2? Will Tables 2.1 and 2.2 most probably define unpolluted air in the year 2085? Discuss your answer.
9. Describe the apparatus and procedures used to measure atmospheric ions and nuclei.
10. What do the terms "dispersion" and "diffusion" have in common? How do they differ?
11. Why is it so difficult to establish a baseline of clean air?

3

Scales of the Air Pollution Problem

Ambient air pollution exists at all scales, from extremely local to global.[1] This chapter considers five different scales: local; urban; regional; continental; and global. The spheres of influence of the air pollutants themselves range from molecular (e.g. gases and nanoparticles) to planetary (e.g. diffusion of greenhouse gases throughout the troposphere). The local scale is up to about 5 km of the earth's surface. The urban scale extends to the order of 50 km. The regional scale is from 50 to 500 km. Continental scales are from 500 to several 1000 km. Of course, the global scale extends worldwide.

I. LOCAL

Local air pollution problems are usually characterized by one or several large emitters or a large number of relatively small emitters. The lower the release height of a source, the larger the potential impact for a given release. Carbon monoxide emitted from motor vehicles, which leads to high concentrations near roadways is an example. Any ground-level source, such as evaporation of volatile organic compounds from a waste treatment pond, will produce the highest concentrations near the source, with concentrations

[1] At an even smaller scale are *microenvironmental* (e.g. indoor air in a home, office, or garage) and *personal* (i.e. at the individual's breathing zone).

generally diminishing with distance. This phenomenon is known as a concentration gradient.

Large sources that emit high above the ground through stacks, such as power plants or industrial sources, can also cause local problems, especially under unstable meteorological conditions that cause portions of the plume to reach the ground in high concentrations.

There are many releases of pollutants from relatively short stacks or vents on the top of one- or two-story buildings. Under most conditions such releases are caught within the turbulent downwash downwind of the building. This allows high concentrations to be brought to the ground surface. Many different pollutants can be released in this manner, including compounds and mixtures that can cause odors. The modeling of the transport and dispersion of pollutants on this scale is discussed in Chapter 22.

Usually, the effects of accidental releases are confined to the local scale.

II. URBAN

Air pollution problems in urban areas generally are of two types. One is the release of primary pollutants (those released directly from sources). The other is the formation of secondary pollutants (those that are formed through chemical reactions of the primary pollutants).

Air pollution problems can be caused by individual sources on the urban scale as well as the local scale. For pollutants that are relatively nonreactive, such as carbon monoxide and particulate matter,[2] or relatively slowly reactive, such as sulfur dioxide, the contributions from individual sources combine to yield high concentrations. Since a major source of carbon monoxide is motor vehicles, "hot spots" of high concentration can occur especially near multilane intersections. The emissions are especially high from idling vehicles. The hot spots are exacerbated if high buildings surround the intersection, since the volume of air in which the pollution is contained is severely restricted. The combination of these factors results in high concentrations.

Tropospheric ozone is the dominant urban problem resulting from the formation of secondary pollutants. Many large metropolitan areas experience the formation of ozone from photochemical reactions of oxides of nitrogen and various species of hydrocarbons. These reactions are catalyzed by the ultraviolet light in sunlight and are therefore called photochemical reactions. Many metropolitan areas are in nonattainment for ozone; that is, they are not meeting the air quality standards. The Clean Air Act Amendments (CAAA) of 1990 recognize this as a major problem and have classified the various metropolitan areas to be in nonattainment according to the severity of the problem for that

[2] It should be noted that particulate matter is highly variable in its composition. Most of the particle may be nonreactive, but aerosols often contain chemically reactive substances, especially those sorbed onto the particle surface.

area. The CAAA sets timetables for the various classifications for reaching attainment with the National Ambient Air Quality Standards (NAAQS). Oxides of nitrogen, principally nitric oxide (NO), but also nitrogen dioxide (NO_2) are emitted from automobiles and from combustion processes. Hydrocarbons are emitted from many different sources. The various species have widely varying reactivities. Determining the emissions of the these chemical species from myriad sources as a basis for pollution control programs can be difficult, but methods continue to improve. These and other aspects of atmospheric chemistry are discussed in Chapter 7.

III. REGIONAL

At least three types of problems contribute to air pollution problems on the regional scale. The first is the blend of urban oxidant problems at the regional scale. Many major metropolitan are in close proximity to one another and continue to grow. Urban geographers refer to some of the larger urban aggregations as "megalopolises." As a result, the air from one metropolitan area, containing both secondary pollutants formed through reactions and primary pollutants, flows on to the adjacent metropolitan area. The pollutants from the second area are then added on top of the "background" from the first.

A second type of problem is the release of relatively slow-reacting primary air pollutants that undergo reactions and transformations during lengthy transport times. A common occurrence in environmental engineering is the direct relationship between time, spatial coverage, and chemical transformation. Thus, these protracted transport times result in transport distances over regional scales not only of the parent compounds but of numerous transformation byproducts. The gas, sulfur dioxide (SO_2), released primarily through combustion of fossil fuels (especially from coal and oil) is oxidized during long-distance transport to sulfur trioxide (SO_3):

$$2SO_2 + O_2 \rightarrow 2SO_3 \tag{3.1}$$

Although SO_2 is a gas, both gas phase and liquid phase oxidation of SO_2 occurs in the troposphere. The SO_3 in turn reacts with water vapor to form sulfuric acid:

$$SO_3 + H_2O \rightarrow H_2SO_4 \tag{3.2}$$

Sulfuric acid reacts with numerous compounds to form sulfates. These are fine (submicrometer) particulates.

Nitric oxide (NO) results from high-temperature combustion, both in stationary sources such as power plants or industrial plants in the production of process heat and in internal combustion engines in vehicles. The NO is oxidized in the atmosphere, usually rather slowly, or more rapidly if there is ozone present, to nitrogen dioxide (NO_2). NO_2 also reacts further with other constituents, forming nitrates, which is also in fine particulate form.

The sulfates and nitrates existing in the atmosphere as fine particulates, generally in the size range less than 1 μm, can be removed from the atmosphere by several processes. "Rain out" occurs when the particles serve as condensation nuclei that lead to the formation of clouds. The particles are then precipitated if the droplets grow to sufficient size to fall as raindrops. Another mechanism, known as "washout," also involves rain, but the particles in air are captured by raindrops falling through the air. Both mechanisms contribute to "acid rain," which results in the sulfate and nitrate particles reaching lakes and streams, and increasing their acidity. As such, acid rain is both a regional and continental problem (see discussion in the next section).

A third type of regional problem is visibility, which may be reduced by specific plumes or by the regional levels of particulate matter that produce various intensities of haze. The fine sulfate and nitrate particulates just discussed are largely responsible for reduction of visibility (see Chapter 14). This problem is of concern in locations of natural beauty, where it is desirable to keep scenic vistas as free of obstructions to the view as possible. Regional haze is a type of visibility impairment that is caused by the emissions of air pollutants from numerous sources across a broad region. The CAAA provides special protections for such areas; the most restrictive denoted as mandatory Federal Class I areas that cover 156 national parks and wilderness areas. On July 1, 1999, EPA published regulations to address regional haze in these areas. Decreased visibility can also impair safety, especially in aviation.

IV. CONTINENTAL

In a relatively small continental area such as Europe, there is little difference between what would be considered regional scale and continental scale. However, on larger continents there would be a substantial difference. Perhaps of greatest concern on the continental scale is that the air pollution policies of a nation are likely to create impacts on neighboring nations. Acid rain in Scandinavia has been considered to have had impacts from Great Britain and Western Europe. Japan has considered that part of their air pollution problem, especially in the western part of the country, has origins in China and Korea. For decades, Canada and the United States have cooperated in studying and addressing the North American acid rain problem.

A. The Science of Acid Rain

Acid rain is a regional air pollution concern, affecting surface waters in large parts of North America and Europe. We hear the term frequently, but what is it? Some correctly point out that in fact most of the rain falling in North America has been acidic since even before industrialization. The main source of natural acidity is carbon dioxide (CO_2) gas that is dissolved into water droplets in the atmosphere and the resulting carbonic acid (H_2CO_3) is ionized, lowering the water's pH. Given the mean partial pressure of CO_2 in the air is 3.0×10^{-4} atm, it is possible to calculate the pH of water in equilibrium with the air at 25°C, and the concentration of all species present in this solution. We can also assume that the mean concentration of CO_2 in the troposphere is 350 ppm, but this concentration is rising by some estimates at a rate of 1 ppm per year. The concentration of the water droplet's CO_2 in water in equilibrium with air is obtained from the partial pressure of Henry's law constant[3]:

$$p_{CO_2} = K_H[CO_2]_{aq} \tag{3.3}$$

The change from carbon dioxide in the atmosphere to carbonate ions in water droplets follows a sequence of equilibrium reactions:

$$CO_{2(g)} \xleftrightarrow{K_H} CO_{2(aq)} \xleftrightarrow{K_r} H_2CO_{3(aq)} \xleftrightarrow{K_{a1}} HCO^-_{3(aq)} \xleftrightarrow{K_{a2}} CO^{2-}_{3(aq)} \tag{3.4}$$

The Henry's law constant is a function of a substance's solubility and vapor pressure.

A more precise term for acid rain is acid deposition, which comes in two forms: wet and dry. Wet deposition refers to acidic rain, fog, and snow. The dry deposition fraction consists of acidic gases or particulates. The severity of ecological effects stemming from acid deposition depends on many factors, especially the strength of the acids and the buffering capacity of the soils. Note that this involves every species in the carbonate equilibrium reactions of Eq. 3.4 (see Fig. 3.1). The processes that release carbonates increase the buffering capacity of natural soils, which moderates the effects of acid rain. Thus, carbonate-rich soils like those in the Central North America are able to withstand even elevated acid deposition compared to the thin soil areas, such as those in the Canadian Shield, the New York Finger Lakes region, and much of Scandinavia.

The concentration of carbon dioxide (CO_2) is constant, since the CO_2 in solution is in equilibrium with the air that has a constant partial pressure of CO_2. And the two reactions and ionization constants for carbonic acid are:

$$H_2CO_3 + H_2O \leftrightarrow HCO_3^- + H_3O^+ \quad K_{a1} = 4.3 \times 10^{-7} \tag{3.5}$$

and;

$$HCO_3^- + H_2O \leftrightarrow CO_3^{-2} + H_3O^+ \quad K_{a2} = 4.7 \times 10^{-11} \tag{3.6}$$

[3] For a complete explanation of the Henry's law constant, including how it is calculated and example problems (see Chapter 6).

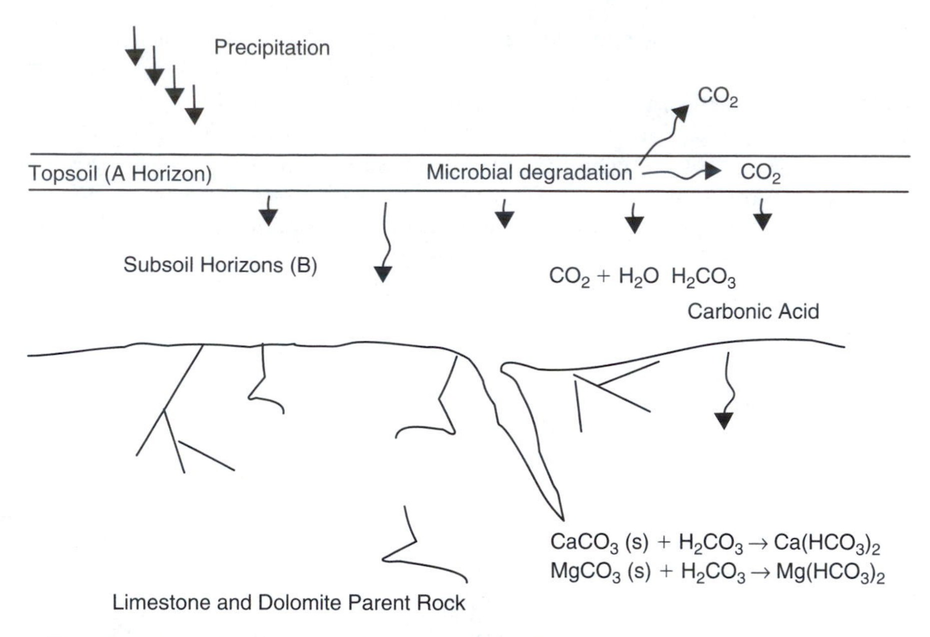

Fig. 3.1. Biogeochemistry of carbon equilibrium. The processes that release carbonates are responsible for much of the buffering capacity of natural soils against the effects of acid rain.

K_{a1} is four orders of magnitude greater than K_{a2}, so the second reaction can be ignored for environmental acid rain considerations. The solubility of gases in liquids can be described quantitatively by Henry's law, so for CO_2 in the atmosphere at 25°C, we can apply the Henry's law constant and the partial pressure to find the equilibrium. The K_H for $CO_2 = 3.4 \times 10^{-2}\,\text{mol}\,\text{L}^{-1}\,\text{atm}^{-1}$. We can find the partial pressure of CO_2 by calculating the fraction of CO_2 in the atmosphere. Since the mean concentration of CO_2 in the earth's troposphere is 350 ppm by volume in the atmosphere, the fraction of CO_2 must be 350 divided by 1 000 000 or 0.000350 atm.

Thus, the carbon dioxide and carbonic acid molar concentration can now be found:

$$[CO_2] = [H_2CO_3] = 3.4 \times 10^{-2}\,\text{mol}\,\text{L}^{-1}\,\text{atm}^{-1} \times 0.000350\,\text{atm} = 1.2 \times 10^{-5}\,\text{M}$$

At equilibrium, $[H_3O^+] = [HCO^-]$.

Taking this and our carbon dioxide molar concentration, gives us:

$$K_{a1} = 4.3 \times 10^{-7} = \frac{[HCO_3^-][H_3O^+]}{CO_2} = \frac{[H_3O^+]^2}{1.2 \times 10^{-5}}$$

$$[H_3O^+]^2 = 5.2 \times 10^{-12}$$

$$[H_3O^+] = 2.6 \times 10^{-6}\,\text{M}$$

Since pH, by definition is the negative log of the hydronium ion, $[H_3O^-]$, concentration (that is, $pH = -\log_{10}[H_3O^+] \approx -\log_{10}[H^+]$), then we calculate the pH. Thus, the droplet pH is about 5.6.

In the next section we are concerned with global air pollution. A big concern is possible warming of the troposphere. However, in addition to the radiant effect of carbon dioxide, what will happen to the acidity of precipitation with increasing concentrations of the gas? For example, if the concentration of CO_2 in the atmosphere increases to 400 ppm, what will happen to the pH of "natural rain"?

The new molar concentration would be 3.4×10^{-2} mol L^{-1} atm^{-1} $\times$ 0.000400 atm $= 1.4 \times 10^{-5}$ M, so $4.3 \times 10^{-7} = [H_3O^+]^2/1.4 \times 10^{-5}$ and $[H_3O^+]^2 = 6.0 \times 10^{-12}$ and $[H_3O^+] = 3.0 \times 10^{-6}$ M. So, the droplet pH would be about 5.5. This means that the incremental increase in atmospheric carbon dioxide can be expected contribute to greater acidity in natural rainfall. But, how is it then that some rain can have a pH of 2 or even lower?

Much of the pH reduction occurs as the result of interactions with a few specific air contaminants. For example, sulfur oxides produced in the burning of fossil fuels (especially coal) is a major contributor to low pH in rain. In fact, CO_2 is not the principal concern of acid rain. Other processes have led to dramatic increases in droplet acidity, especially sulfuric and nitric acid formation from the ionization of the oxides of sulfur and nitrogen, which further decreases this pH level to those harmful to aquatic organisms. Acid precipitation can result from the emissions of plumes of strong acids, such as sulfuric acid (H_2SO_4) or hydrochloric acid (HCl) in the forms of acid mists, but most commonly the pH drop is the result of secondary reactions of the acid gases SO_2 and NO_2:

$$SO_2 + \tfrac{1}{2}O_2 + H_2O \rightarrow \text{several intermediate reactions} \rightarrow \{2H^+ + SO_4^{2-}\}_{aq} \quad (3.7)$$

$$2NO_2 + \tfrac{1}{2}O_2 + H_2O \rightarrow \text{several intermediate reactions} \rightarrow \{2H^+ + NO_3^-\}_{aq} \quad (3.8)$$

In its simplest terms, SO_2 is emitted from the combustion of fuels containing sulfur, the reaction being

$$S + O_2 \xrightarrow{\text{heat}} SO_2 \quad (3.9)$$

The sulfur dioxide is then photochemically oxidized:

$$SO_2 + O \xrightarrow{\text{sunlight}} SO_3 \quad (3.10)$$

The sulfur trioxide is reduced to form sulfuric acid, which is ionized in water:

$$SO_3 + H_2 \rightarrow H_2SO_4 \rightarrow 2H^+ + SO_4^{2-} \quad (3.11)$$

Sulfur oxides do not literally produce sulfuric acid in the clouds, but the concept is the same.[4] The precipitation from air containing high concentrations of sulfur oxides is poorly buffered and its pH readily drops.

Nitrogen oxides, emitted mostly from automobile exhaust, but also from any other high-temperature combustion, contribute to the acid mix in the atmosphere. The chemical reactions that apparently occur with nitrogen are:

$$N_2 + O_2 \rightarrow 2NO \tag{3.12}$$

$$NO + O_3 \rightarrow NO_2 + O_2 \tag{3.13}$$

$$NO_2 + O_3 + H_2O \rightarrow 2HNO_3 + O_2 \rightarrow 2H^+ + 2NO_3^- + O_2 \tag{3.14}$$

Thus, the principal acid is nitric acid, HNO_3.

Hundreds of lakes in North America and Scandinavia have become so acidic that they no longer can support fish life. In a recent study of Norwegian lakes, more than 70% of the lakes having a pH of less than 4.5 contained no fish, and nearly all lakes with a pH of 5.5 and above contained fish. The low pH not only affects fish directly, but also contributes to the release of potentially toxic metals such as aluminum, thus magnifying the problem.

In North America, acid rain has already decimated the populations of numerous fish and many plants in 50% of the high mountain lakes in the Adirondacks. The pH in certain lakes has reached such levels of acidity as to replace the trout and native plants with acid-tolerant rough fish species and mats of algae.

The deposition of atmospheric acid on freshwater aquatic systems prompted the EPA to suggest a limit from 10 to 20 kg SO_4^{2-} per hectare per year. Putting this limit into the context of existing emissions, the state of Ohio alone has total annual emissions are 2.4×10^6 metric tons of gaseous sulfur dioxide (SO_2) per year. If all of this is converted to SO_4^{2-} and is deposited on the State of Ohio, the total would be 360 kg per hectare per year.[5] Similar calculations for the sulfur emissions for northeastern United States indicate that the rate of sulfur emission is 4 to 5 times greater than the rate of deposition. What happens to all of the excess sulfur? Certainly, as the reactions above indicate, much of it is oxidized into acidic chemical species. Canada and the United States have been discussing this for over a decade. Likewise, much of the problem in Scandinavia can be traced to the use of tall stacks in Great Britain and the lowland countries of continental Europe. British industries for some years simply built increasingly tall stacks as a method of air pollution control, reducing the immediate ground-level concentration, but emitting the

[4] Actually many acid–base systems in the environment are in equilibrium condition among acids, bases, ionic, and other chemical species. For example, in natural waters, carbonic acid which is responsible for most natural acidity in surface waters is indicated as $H_2CO_3^*$. The asterisk indicates that the actual acid is only part of the equilibrium, which also consists of CO_2 dissolved in the water, CO_2 dissolved in the air, and ionic forms, especially carbonates (CO_3^{2-}) and bicarbonates (HCO_3^-).

[5] US Environmental Protection Agency, Washington, DC.

same pollutants into the higher atmosphere. The air quality in the United Kingdom improved, but at the expense of acid rain at other parts of Europe. One of the difficulties in dealing with acid rain and global climate change is that they in essence are asking society to cut back on combustion. And, to many, combustion is tantamount to progress and development.

V. GLOBAL

The release of radioactivity from the accident at Chernobyl would be considered primarily a regional or continental problem. However, higher than usual levels of radioactivity were detected in the Pacific Northwest part of the United States soon after the accident. Likewise, persistent organic pollutants, such as polychlorinated biphenyls (PCBs) have been observed in Arctic mammals, thousands of miles from their sources. These observations demonstrate the effects of long-range transport.

A particularly noteworthy air pollution problem of a global nature is the release of chlorofluorocarbons used as propellants in spray cans and in air conditioners, and their effect on the ozone layer high in the atmosphere (see Chapter 15).

Some knowledge of the exchange processes between the stratosphere and the troposphere and between the Northern and Southern Hemispheres was learned in the late 1950s and early 1960s as a result of the injection of radioactive debris into the stratosphere from atomic bomb tests in the Pacific. The debris was injected primarily into the Northern Hemisphere in the stratosphere. The stratosphere is usually quite stable and resists vertical air exchange between layers. It was found that the exchange processes in the stratosphere between the Northern and Southern Hemispheres is quite slow. However, radioactivity did show up in the Southern Hemisphere within 3 years of the onset of the tests, although the levels remained much lower than those in the Northern Hemisphere. Similarly, the exchange processes between the troposphere into the stratosphere are quite slow. The main transfer from the troposphere into the stratosphere is injection through the tops of thunderstorms that occasionally penetrate the tropopause, the boundary between the troposphere and the stratosphere. Some transfer of stratospheric air downward also occurs through occasional gaps in the tropopause. Since the ozone layer is considerably above the troposphere, the transfer of chlorofluorocarbons upward to the ozone layer is expected to take place very slowly. Thus, there was a lag from the first release of these gases until an effect was seen. Similarly, with the cessation of use of these materials worldwide, there has been a commensurate lag between in restoration of the ozone layer (see Chapter 15).

Another global problem is climate change that is generated by excessive amounts of radiant gases (commonly known as greenhouse gases), especially methane (CH_4) and CO_2, which is not normally considered an air pollutant. But a portion of this radiation is intercepted by the carbon dioxide in the air and is reradiated both upward and downward. That which is radiated downward keeps the ground from cooling rapidly. As the carbon dioxide concentration continues to increase, the earth's temperature is expected to increase.

A natural air pollution problem that can cause global effects is injection into the atmosphere of fine particulate debris by volcanoes. The addition of this particulate matter has caused some spectacular sunsets throughout the earth. If sufficient material is released, it can change the radiation balance. Blocking of the incoming solar radiation will reduce the normal degree of daytime warming of the earth's surface. A "mini-ice age" was caused in the mid-1800s, when a volcanic mountain in the Pacific erupted. The summer of that year was much cooler than usual and snow occurred in July in New England.

Global phenomena like climate can be sensitive to the effects of air pollutants. Like most environmental systems climate varies in its response to different pollutants. In fact, some pollutants lead to cooling whereas others lead to heating of the troposphere. Currently, the major concern of much of the scientific community is with warming. For example, the International Panel of Climate Change (IPCC) reports that between 1970 and 2004, global greenhouse gas emissions increased by 70%. Models vary in predicting how this increase will affect mean global temperatures and other climatic variables, but many scientists and policy makers fear that delays in curbing the trend could lead to a substantial environmental and societal catastrophe.

SUGGESTED READING

40 *Code of Federal Regulations FR* 51.300–309.

Critchfield, H. J., *General Climatology*, 4th ed. Prentice-Hall, Englewood Cliffs, NJ, 1983.

Intergovernmental Panel on Climate Change. *Climate Change 2001: The Scientific Basis. Contribution of Working Group I to the Third Assessment Report of the Intergovernmental Panel on Climate Change* (Houghton, J. T., Ding, Y. Griggs, D. J., Noguer, M., van der Linden, P. J., Dai, X., Maskell, K., and Johnson, C. A., eds.). Cambridge University Press, Cambridge, United Kingdom and New York, NY, 2001.

Intergovernmental Panel on Climate Change. *IPCC Special Report on Carbon Dioxide Capture and Storage*. Prepared by Working Group III of the Intergovernmental Panel on Climate Change (Metz, B., Davidson, O., de Coninck, H. C., Loos, M., and Meyer, L. A., eds.). Cambridge University Press, Cambridge, United Kingdom and New York, NY, 2005.

Knap, A. H. (ed.), *The Long-Range Atmospheric Transport of Natural and Contaminant Substances*. Kluwer Academic Press, Hingham, MA, 1989.

Kramer, M. L., and Porch, W. M., *Meteorological Aspects of Emergency Response*. American Meteorological Society, Boston, MA, 1990.

Landsberg, H. E., *The Urban Climate*. Academic Press, New York, 1981.

US Environmental Protection Agency, 64 *Federal Register* 35714; July 1, 1999.

Vallero, D. A., *Paradigms Lost: Learning from Environmental Mistakes, Mishaps, and Misdeeds*. Butterworth-Heinemann/Elsevier, Burlington, MA, 2006.

QUESTIONS

1. What situation of emission has the potential to produce the greatest local problem?
2. What are the two types of air pollution problems found in urban areas?
3. What are the two primary gaseous pollutants that transform to fine-particle form during long-range transport?
4. In addition to the secondary pollutants discussed in this chapter, identify two more. How do these byproducts different from the parent chemical species?
5. If the concentration of CO_2 in the atmosphere increases to 500 ppm, what will happen to the pH of rain? Where might the biggest effects from such an increase be felt?
6. What are major concerns on the continental scale? What are some ways to ameliorate them?
7. What are two global issues and how might they best be addressed?

Part II

The Physics and Chemistry of Air Pollution

4

Air Pollution Physics

A thorough understanding of air pollution begins with physics. It is how we understand matter and energy. As such, it provides the basis for chemistry. So, how things move and how efficiently energy is transferred among compartments lie at the heart of understanding air pollution. This is not only true for air pollution control technologies and other engineered systems but also for how a pollutant will affect a receptor. Human health effects and exposure assessments require that the movement of and changes to contaminants be understood. This requires a solid grounding in the physical and chemical principles covered in this part. Likewise, ecological risk assessments require an understanding of physicochemical, thermodynamic, hydrological, and aerodynamic concepts to appreciate how chemical nutrients and contaminants cycle through the environment, how physical changes may impact receptors, and the many possible ways that the environment is put at risk. In a very basic way, we must consider the first principles to have rigorous environmental assessments and to give reliable information to the engineers and decision makers who will respond to health and ecological risks. Good policy stands on quality science.

Let us note a few areas of environmental risk that are heavily dependent upon physics. First, energy is often described as a system's capacity to do work, so getting things done in the environment is really an expression of how efficiently energy is transformed from one form to another. Energy and matter relationships determine how things move in the environment. That is why we begin our discussion of applied physics with environmental transport. The physical movement of contaminants among environmental *compartments* is central to risk assessment. After a contaminant is released, physical processes will go to work on transporting the contaminant and allow for *receptors* (like people and ecosystems) to be exposed. Transport is one of the two processes (the other is *transformation*) that determine a contaminant's *fate* in the environment. Figure 4.1 shows the major steps needed to study a contaminant as it moves through the environment.

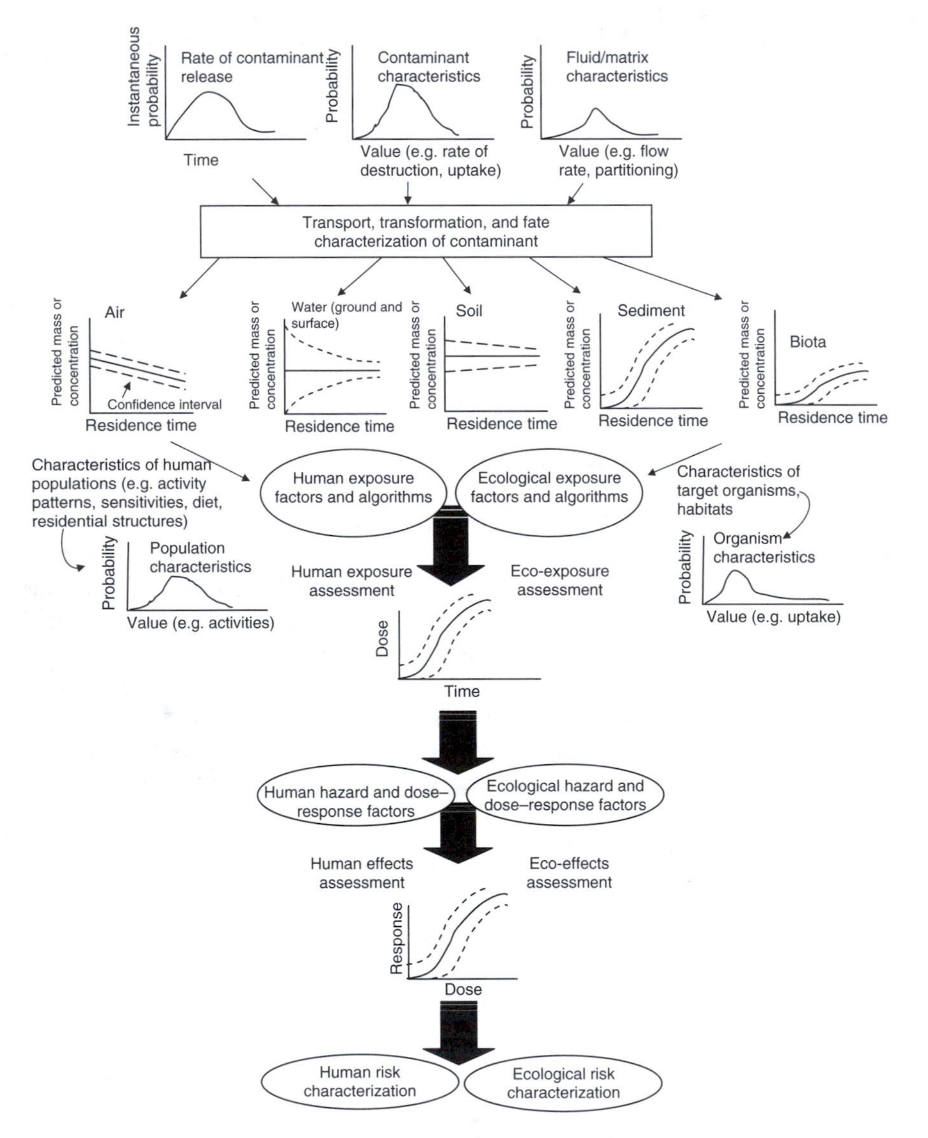

Fig. 4.1. Importance of transport, transformation, and fate processes to human and ecological exposure and risk assessments. Source for ecological exposure components: Suter, G., Predictive risk assessment of chemicals, in *Ecological Risk Assessment* (Suter, G., ed.), Lewis Publishers, Chelsea, MI, 1993.

Scientists and engineers are particularly interested in the role of fluids in air pollution. A fluid is a substance that cannot support a shearing force, so fluids take the form of their containing vessel. The obvious fluid that is important at all scales, from molecular to global, is the air itself. Next, arguably, is water. However, fluid properties, statics, and dynamics are involved in every aspect of air pollution, from characterizing emissions from a source to the biological response to the released pollutants. To identify a hazard and dose–response associated with the chemical, the fluid properties must be understood. For example, if a contaminant's fluid properties make it insoluble in water and blood, then the target tissues may be more likely to be the lipids. If a chemical is easily absorbed, the hazard may be higher. However, if it does not change phases under certain cellular conditions, it could be more or less toxic, depending on the organ.

In determining exposures, in addition to the transport phenomena mentioned earlier, the fluid properties of a pollutant is a crucial factor as to where the contaminant is likely to be found in the environment (e.g. in the air as a vapor, sorbed to a particle, dissolved in water, or taken up by biota).

Physics and chemistry are interrelated; in fact, for air pollution it is arguably best to refer to physicochemical processes. The literature is ripe with lists of chemical compounds that have been associated with diseases. Much of risk assessment follows the toxicological paradigm (which is really an enhancement of the pharmacological paradigm), wherein chemicals are considered from a dose–response perspective (see Fig. 4.1). Indeed, air pollution exposure assessments rely on analytical chemistry in determining the presence and quantity of a chemical contaminant in the environment. Also, toxicology begins with an understanding of the chemical characteristics of an agent, followed by investigations of how the agent changes after release and up to the adverse effect. So, whether the chemistry is inorganic or organic, whether it is induced, mediated, or complemented by biological processes or is simply abiotic chemistry, and whether it is applied at the molecular or global scale, every risk assessment, every intervention, and every engineering activity is a chemical expression.

Several of the topics considered in transport provide a transition from physics to atmospheric chemistry; especially phase partitioning and fluid properties. In fact, no air pollution chemistry discussion is complete without discussions of mass balance and partitioning.

I. MECHANICS OF AIR POLLUTION

Mechanics is the field of physics concerned with the motion and the equilibrium of bodies within particular frames of reference. Air pollution scientists make use of the mechanical principles in practically every aspect of pollution, from the movement of fluids that carry contaminants to the forces within substances that affect their properties to the relationships between matter and energy within organisms and ecosystems. Engineering mechanics is important because it includes statics and dynamics. Fluid mechanics is a particularly important branch of the mechanics of air pollution.

Statics is the branch of mechanics that is concerned with bodies at rest with relation to some frame of reference, with the forces between the bodies, and with the equilibrium of the system. It addresses rigid bodies that are at rest or moving with constant velocity. Hydrostatics is a branch of statics that is essential to environmental science and engineering in that it is concerned with the equilibrium of fluids (liquids and gases) and their stationary interactions with solid bodies, such as pressure. While many fluids are considered by environmental assessments, the principal fluids are water and air.

Dynamics is the branch of mechanics that deals with forces that change or move bodies. It is concerned with accelerated motion of bodies. It is an especially important science and engineering discipline because it is fundamental to understanding the movement of contaminants through the environment. Dynamics is sometimes used synonymously with kinetics. However, we will use the engineering approach and treat kinetics as one of the two branches of dynamics, with the other being kinematics. Dynamics combines the properties of the fluid and the means by which it moves. This means that the continuum fluid mechanics varies by whether the fluid is viscous or inviscid, compressible or incompressible, and whether flow is laminar or turbulent. For example, the properties of the two principal environmental fluids, i.e. water in an aquifer and an air mass in the troposphere, are shown in Table 4.1. Thus, in air pollution mechanics, turbulent systems are quite common. However, within smaller systems, such as control technologies, laminar conditions can be prominent.

Dynamics is divided into kinematics and kinetics. Kinematics is concerned with the study of a body in motion independent of forces acting on the body. That is, kinematics is the branch of mechanics concerned with motion of bodies with reference to force or mass. This is accomplished by studying the geometry of motion irrespective of what is causing the motion. Therefore, kinematics relates position, velocity, acceleration, and time.

Aerodynamics and hydrodynamics are the important branches of environmental mechanics. Both are concerned with deformable bodies and with the motion of fluids. Therefore, they provide an important underlying aspect of contaminant transport and movements of fluids, and consider fluid properties

TABLE 4.1

Contrasts Between Plume in Ground Water and Atmosphere

	Ground water plume	Air mass plume
General flow type	Laminar	Turbulent
Compressibility	Incompressible	Compressible
Viscosity	Low viscosity (1×10^{-3} kg m^{-1} s^{-1} at 288°K)	Very low viscosity (1.781×10^{-5} kg m^{-1} s^{-1} at 288°K)

such as compressibility and viscosity. These are key to understanding movement of contaminants within plumes, flows in vents and pipes, and design of air pollution control systems.

Kinetics is the study of motion and the forces that cause motion. This includes analyzing force and mass as they relate to translational motion. Kinetics also considers the relationship between torque and moment of inertia for rotational motion.

A key concept for environmental dynamics is that of linear momentum, the product of mass and velocity. A body's momentum is conserved unless an external force acts upon a body. Kinetics is based on Newton's *first law of motion*, which states that a body will remain in a state of rest or will continue to move with constant velocity unless an unbalanced external force acts on it. Stated as the *law of conservation of momentum*, linear momentum is unchanged if no unbalanced forces act on a body. Or, if the resultant external force acting on a body is zero, the linear momentum of the body is constant.

Kinetics is also based upon Newton's *second law of motion*, which states that the acceleration of a body is directly proportional to the force acting upon that body, and inversely proportional to the body's mass. The direction of acceleration is the same as the force of direction. The equation for the second law is:

$$F = \frac{\mathrm{d}p}{\mathrm{d}t} \tag{4.1}$$

where p is the momentum.

Newton's *third law of motion* states that for every acting force between two bodies, there is an equal but opposite reacting force on the same line of action, or:

$$F_{\mathrm{reacting}} = -F_{\mathrm{acting}} \tag{4.2}$$

Another force that is important to environmental systems is *friction*, which is a force that always resists motion or an impending motion. Friction acts parallel to the contacting surfaces. When bodies come into contact with one another, friction acts in the direction opposite to that that is bringing the objects into contact.

II. FLUID PROPERTIES

Air pollutants may move within one environmental compartment, such as a source within a home. Most often, however, pollutants move among numerous compartments, such as when a contaminant moves from the source to the atmosphere, until it is deposited to the soil and surface waters, where it is taken up by plants, and eaten by animals.[1]

[1] While this book strives to compartmentalize the science discussions among physics, chemistry, and biology, complex topics like transport require that all three of the sciences be considered together. So, while the focus and language of this chapter is predominantly on physical transport, it would be wise to interject chemical and biological topics to explain the concepts properly.

The general behavior of contaminants after they are released is shown in Fig. 4.2. The movement of pollutants is known as *transport*. This is half of the often cited duo of environmental "fate and transport." *Fate* is an expression of what contaminant becomes after all the physical, chemical, and biological processes of the environment have acted. It is the ultimate site a pollutant

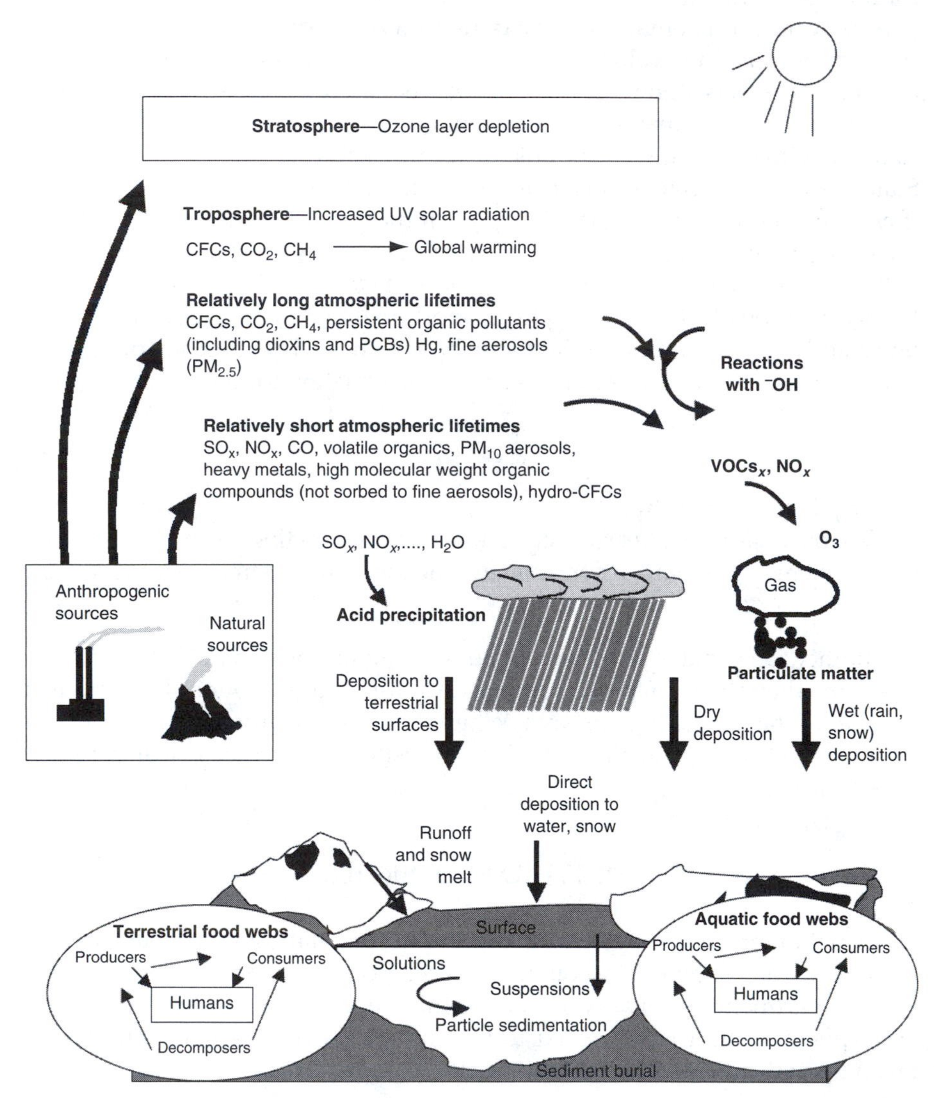

Fig. 4.2. The physical movement and accumulation of contaminants after release. *Sources*: Commission for Environmental Cooperation of North America, 2002, The Sound Management of Chemicals (SMOC) Initiative of the Commission for Environmental Cooperation of North America: Overview and Update, Montreal, Canada; Adapted in Vallero, D.A., *Environmental Contaminants: Assessment and Control*, Elsevier Academic Press, Burlington, MA, 2004.

after its release. The pollutant will undergo numerous changes in location and form before reaching its fate. Throughout the contaminant's journey it will be physically transported and undergo coincidental chemical processes, known as *transformations*, such as photochemical and biochemical reactions.[2]

Physical transport is influenced by the kinematics and mechanics of fluids. In addition, it is important to identify when these processes reach equilibrium, such as when a chemical is sequestered and stored. Fate is often described according to environmental media or compartments.

Understanding air pollutant transport begins with the characteristics of environmental fluids. A fluid is a collective term that includes all liquids and gases. A liquid is matter that is composed of molecules that move freely among themselves without separating from each other. A gas is matter composed of molecules that move freely and are infinitely able to occupy the space with which they are contained at a constant temperature. A fluid is a substance that will deform continuously upon the application of a shear stress; i.e., a stress in which the material on one side of a surface pushes on the material on the other side of the surface with a force parallel to the surface.

Fluids are generally divided into two types: *ideal* and *real*. The former has zero viscosity and, thus, no resistance to shear (explained below). An ideal fluid is incompressible and flows with uniform velocity distributions. It also has no friction between moving layers and no turbulence (i.e. eddy currents). On the contrary, a real fluid has finite viscosity, has non-uniform velocity distributions, is compressible, and experiences friction and turbulence. Real fluids are further subdivided according to their viscosities. A *Newtonian fluid* is one that has a constant viscosity at all shear rates at a constant temperature and pressure. Water and most solvents are Newtonian fluids. However, environmental engineers are confronted with non-Newtonian fluids, i.e. those with viscosities not constant at all shear rates. Sites contaminated with drilling fluids and oils have large quantities of non-Newtonian fluids onsite.

Physicists use the term "particle" to mean a theoretical point that has a rest-mass and location, but no geometric extension. We can observe this particle as it moves within the fluid as a representation of where that portion of the fluid is going and at what velocity. Another important concept is that of the *control volume*, which is an arbitrary region in space that is defined by boundaries. The boundaries may be either stationary or moving. The control volume is used to determine how much material and at what rate the material is moving through the air, water, or soil. The third concept, which is included in the definition of a fluid, is *stress*. The forces acting on a fluid may be *body forces* or *surface forces*. The former are forces that act on every particle within the fluid, occurring without actually making physical contact, such as

[2] Fate may also include some remediation reactions, such as thermal and mechanical separation processes, but in discussions of fate and transport, the reactions are usually those that occur in the ambient environment. The treatment and control processes usually fall under the category of environmental engineering.

gravitation force. The latter are forces that are applied directly to the fluid's surface by physical contact.

Stress represents the total force per unit area acting on a fluid at any point within the fluid volume. So, stress at any point P is

$$\sigma(P) = \lim_{\delta A \to 0} \frac{\delta F}{\delta A} \tag{4.3}$$

where $\sigma(P)$ is the vector stress at point P, δA is the infinitesimal area at point P, and δF is the force acting on δA.

Fluid properties are characteristics of the fluid that are used to predict how the fluid will react when subjected to applied forces. If a fluid is considered to be infinitely divisible, that is, it is made up of many molecules that are constantly in motion and colliding with one another, this fluid is in *continuum*. That is, a fluid acts as though it has no holes or voids, meaning its properties are continuous (i.e. temperature, volume, and pressure fields are continuous). If we make the assumption that a fluid is a continuum we can consider the fluid's properties to be functions of position and time. So, we can represent the fluid properties as two fields. The density field shows:

$$\rho = \rho(x, y, z, t) \tag{4.4}$$

where ρ is the density of the fluid, x, y, z are the coordinates in space, and t is the time.

The other fluid field is the velocity field:

$$\vec{v} = \vec{v}(x, y, z, t) \tag{4.5}$$

Thus, if the fluid properties and the flow characteristics at each position do not vary with time, the fluid is said to be at *steady flow*:

$$\rho = \rho(x, y, z) \text{ or } \frac{\partial \rho}{\partial t} = 0 \tag{4.6}$$

and

$$\vec{v} = \vec{v}(x, y, z) \text{ or } \frac{\partial \vec{v}}{\partial t} = 0 \tag{4.7}$$

Conversely, a *time-dependent flow* is considered to be an *unsteady flow*. Any flow with unchanging magnitude and direction of the velocity vector $\vec{v}$ is considered to be a *uniform flow*.

Fluids, then, can be classified according to observable physical characteristics of flow fields. A continuum fluid mechanics classification is shown in Fig. 4.3. Laminar flow is in layers, while turbulent flow has random movements of fluid particles in all directions. In incompressible flow, the variations in density are assumed to be constant, while the compressible flow has density variations, which must be included in flow calculations. Viscous flows must account for viscosity while inviscid flows assume viscosity is zero.

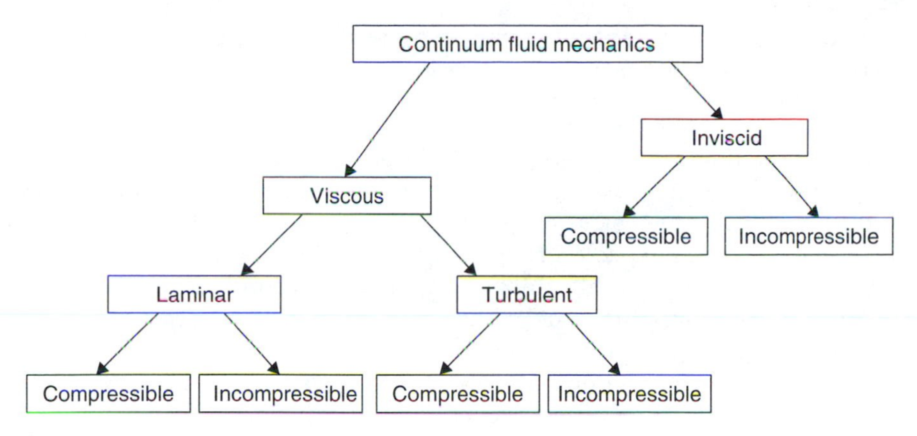

Fig. 4.3. Classification of Fluids Based on Continuum Fluid Mechanics. *Source*: Research and Education Association, *The Essentials of Fluid Mechanics and Dynamics I*. REA, Piscataway, NJ, 1987.

The velocity field is very important in environmental modeling, especially in modeling plumes in the atmosphere and in groundwater, since the velocity field is a way to characterize the motion of fluid particles and provides the means for computing these motions. The velocity field may be described mathematically using Eq. (4.5). This is known as the *Eularian* viewpoint.

Another way to characterize the fluid movement (i.e. flow) is to follow the particle (sometimes referred to as a "parcel") as it moves, using time functions that correspond to each particle as shown in Fig. 4.4. This *random walk* of the particle provides what is known as the *Lagrangian* viewpoint, which is expressed mathematically as:

$$\vec{v} = [x(t), y(t), z(t)] \tag{4.8}$$

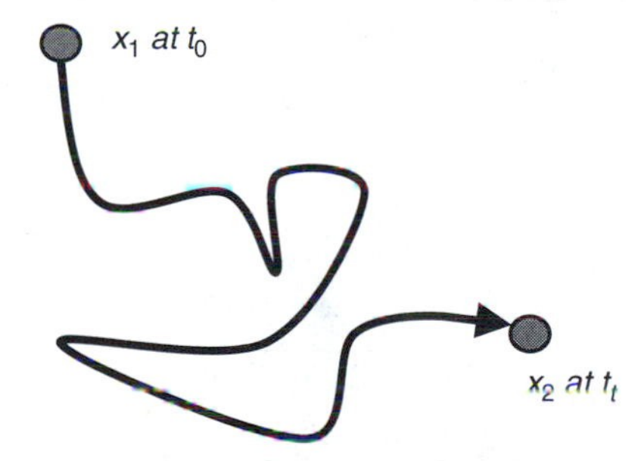

Fig. 4.4. Random walk representing the movement of a particle; i.e. a hypothetical point that is moving in a random path during time interval ($t_0 - t_t$). This is the theoretical basis for Lagrangian air plume models.

A Lagrangian plume model characterizes the plume by calculating the air dispersion from statistics of the trajectories of a large number of the particles (enough to represent the whole plume).

Velocity The time rate of change of a fluid particle's position in space is the fluid velocity (V). This is a vector field quantity. Speed (V) is the magnitude of the vector velocity V at some given point in the fluid, and average speed ($\bar{V}$) is the mean fluid speed through a control volume's surface. Therefore, velocity is a vector quantity (magnitude and direction), while speed is a scalar quantity (magnitude only). The standard units of velocity and speed are meter per second ($\mathrm{m\,s^{-1}}$).

Obviously, velocity is important to determine pollution, such as mixing rates after a pollutant is emitted into a plume. The distinction between velocity and speed is seldom made in air pollution.

Pressure A force per unit area is pressure (p):

$$p = \frac{F}{A} \tag{4.9}$$

So, p is a type of stress that is exerted uniformly in all directions. It is common to use pressure instead of force to describe the factors that influence the behavior of fluids. The standard unit of p is the Pascal (P), which is equal to $1\,\mathrm{N\,m^{-2}}$. Therefore, pressure will vary when the area varies, as shown in Fig. 4.5. In this example, the same weight (force) over different areas leads to different pressures, much higher pressure when the same force is distributed over a smaller area.

For a liquid at rest, the medium is considered to be a continuous distribution of matter. However, when considering p for a gas, the pressure is an average of the forces against as the vessel walls (i.e. gas pressure). Fluid pressure is a measure of energy per unit volume per the *Bernoulli equation*, which states that the static pressure in the flow plus one half of the density times

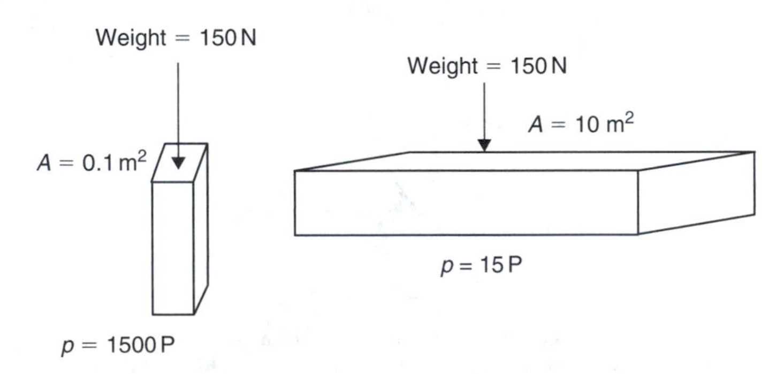

Fig. 4.5. Difference in pressure with same weight over different areas.

the velocity squared is equal to a constant throughout the flow, referred to as the total pressure of the flow:

$$p + \frac{1}{2}pV^2 + pgh = \text{constant} \quad (4.10)$$

where p is the pressure, V is the fluid velocity, h is the elevation, and g is the gravitational acceleration.

This also means that, in keeping with the conservation of energy principle, a flowing fluid will maintain the energy, but velocity and pressure can change. In fact, velocity and pressure will compensate for each other to adhere to the conservation principle, as stated in the Bernoulli equation:

$$p_1 + \frac{1}{2}pV_1^2 + pgh_1 = p_2 + \frac{1}{2}pV_2^2 + pgh_2 \quad (4.11)$$

This is shown graphically in Fig. 4.6. The so-called "Bernoulli effect" occurs when increased fluid speed leads to decreased internal pressure.

In environmental applications, fluid pressure is measured against two references: *zero pressure* and *atmospheric pressure*. *Absolute pressure* is compared to true zero pressure and *gage pressure* is reported in reference to atmospheric pressure. To be able to tell which type of pressure is reported, the letter "a" and the letter "g" are added to units to designate whether the pressure is absolute or gage, respectively. So, it is common to see pounds per square inch designated as "psia" or inches of water as "in wg". If no letter is designated, the pressure can be assumed to be absolute pressure.

When a gage measurement is taken, and the actual atmospheric pressure is known, absolute and gage pressure are related:

$$p_{\text{absolute}} = p_{\text{gage}} + p_{\text{atmospheric}} \quad (4.12)$$

Barometric and atmospheric pressure are synonymous. A negative gage pressure implies a *vacuum* measurement. A reported vacuum quantity is to be subtracted from the atmospheric pressure. So, when a piece of equipment

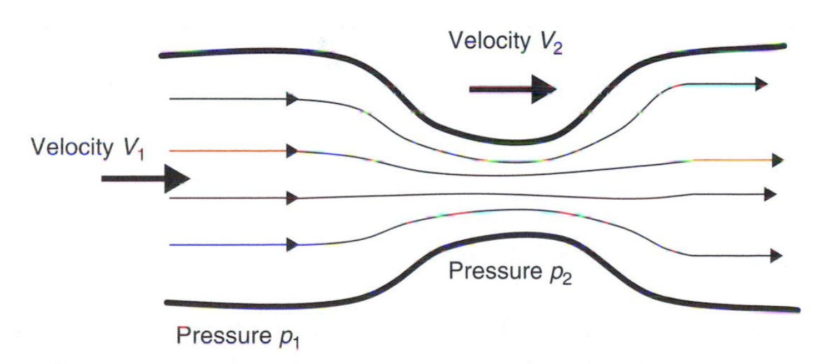

Fig. 4.6. Bernoulli principle and the effect of relationship between pressure, energy, area, and velocity. As the cross-sectional area of flow decreases, the velocity increases and the pressure decreases.

is operating with 20 kPa vacuum, the absolute pressure is 101.3 kPa − 20 kPa = 81.3 kPa. (*Note*: The standard atmospheric pressure = 101.3 kPa = 1.013 bars). Thus, the relationship between vacuums, which are always given as positive numbers, and absolute pressure is:

$$p_{\text{absolute}} = p_{\text{atmospheric}} - p_{\text{vacuum}} \tag{4.13}$$

Pressure is used throughout this text, as well as in any discussion of physics, chemistry, and biology. Numerous units are used. The preferred unit in this book is the kPa, since the standard metric unit of pressure is the Pascal, which is quite small. See Fig. 4.7 for a comparison of relative size of pressure units commonly used in environmental assessments, research studies, and textbooks.

Acceleration Any discussion of potential and kinetic energies includes acceleration due to gravity. In many ways, it seems that acceleration was a major reason for Isaac Newton's need to develop the calculus.[3] Renaissance

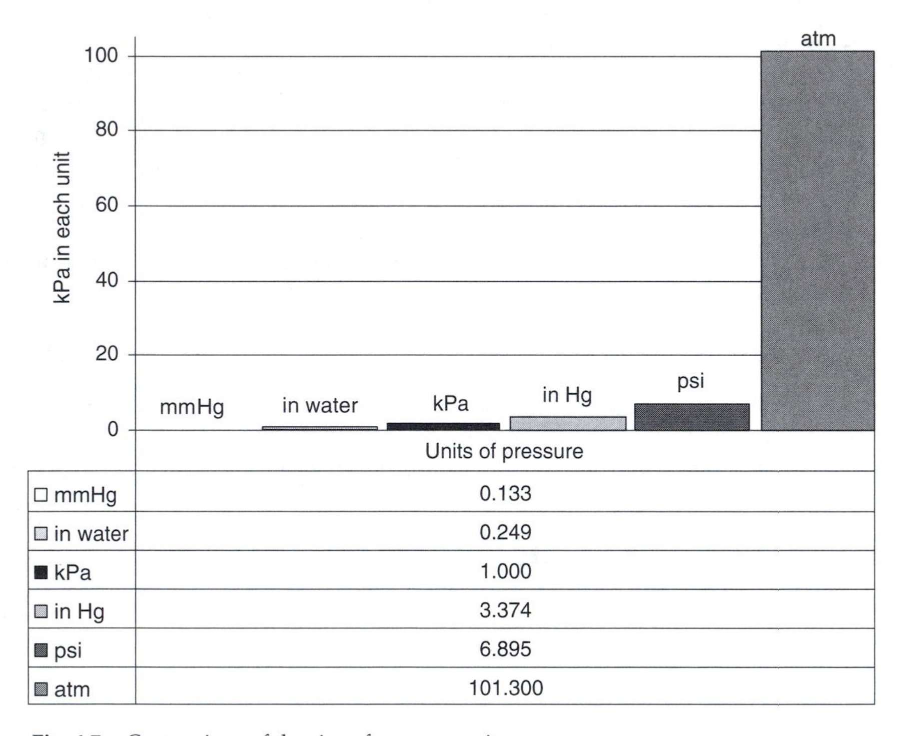

Unit	kPa
mmHg	0.133
in water	0.249
kPa	1.000
in Hg	3.374
psi	6.895
atm	101.300

Fig. 4.7. Comparison of the size of pressure units.

[3]Newton simultaneously invented the calculus with Gottfried Wilhelm Leibnitz (1646–1716) in the seventeenth century. Both are credited with devising the symbolism and the system of rules for computing derivatives and integrals, but their notation and emphases differed. A debate rages on who did what first, but both of these giants had good reason to revise the language of science; i.e., mathematics, to explain motion.

scientists (Galileo, Kepler, *et al.*) may well have understood the concept of acceleration, but needed the structure brought by the calculus. The calculus is the mathematics of change, which is what acceleration is all about.

Acceleration is the time rate of change in the velocity of a fluid particle. In terms of calculus, it is a second derivative. That is, it is the derivative of the velocity function. And a derivative of a function is itself a function, giving its rate of change. This explains why the second derivative must be a function showing the rate of change of the rate of change. This is obvious when one looks at the units of acceleration: length per time per time ($\mathrm{m\,s^{-2}}$).

FLUID ACCELERATION EXAMPLE

If a fluid is moving at the constant velocity of $4\,\mathrm{m\,s^{-1}}$, what is the rate of change of the velocity? What is the second derivative of the fluid's movement?

The function $s = f(t)$ shows the distance the fluid has moved (s) after t seconds. If the fluid is traveling at $4\,\mathrm{m\,s^{-1}}$, then it must travel 4 meters for each second, or $4t$ meters after t seconds. The rate of change of distance (how fast the distance is changing) is the speed. We know that this is $4\,\mathrm{m\,s^{-1}}$. So:

$$s = f(t) = 4t \tag{4.14}$$

and

$$\mathrm{d}s/\mathrm{d}t = f'(t) = 4$$

In acceleration, we are interested in the rate of change of the rate of change. This is the rate of change of the fluid velocity. Since the fluid is moving at constant velocity, it is not accelerating.

So acceleration = 0.

This is another way of saying that when we differentiate for a second time (called the *second derivative*), we find it is zero.

Displacement, Velocity and Acceleration The three concepts just discussed can be combined to describe fluid movement. If we are given the function $f(t)$ as the displacement of a particle in the fluid at time t, the derivative of this function $f'(t)$ represents the velocity. The second derivative $f''(t)$ represents the acceleration of the particle at time t:

$$s = f(t) \tag{4.15}$$

$$v = \mathrm{d}s/\mathrm{d}t = f'(t) \tag{4.16}$$

$$a = \mathrm{d}^2s/\mathrm{d}t^2 = f''(t) \tag{4.17}$$

AIR POLLUTION MATHEMATICS: STATIONARY POINTS IN A FLUID

The derivative of a function can be described graphically (see Fig. 4.8). If the derivative is zero, the function is flat and must therefore reside where the graph is turning. We are able to identify the turning points of a function by differentiating and setting the derivative equal to zero. Turning points may be of three types: minima (Fig. 4.8(a)), maxima (Fig. 4.8(b)) and points of inflexion (Fig. 4.8(c)). The graph shows how the derivatives are changing around each of these stationary points.

Near the point where the derivative is changing from negative to positive, it is increasing. In other words the rate of change in velocity is positive. So, the derivative of the derivative; i.e., second derivative, must be positive. When the second derivative is positive at a given turning point, this is the minimum point. Likewise, at the maximum negative to positive means that the derivative is decreasing; i.e., the rate of change is negative. This means when the second derivative is negative at a given turning point, this must be a maximum point.

At the inflection points, the rate of change is neither positive nor negative; i.e., the rate of change is zero. Keep in mind that zero is also a possible value for the second derivative at a maximum or minimum.

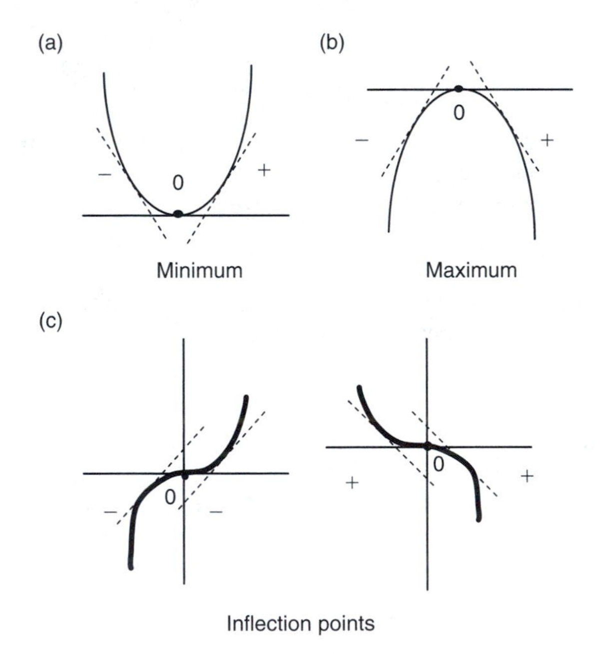

Fig. 4.8. Stationary points important to displacement, velocity, and acceleration of particles in a fluid.

Density The relationship between mass and volume is important in both environmental physics and chemistry, and is a fundamental property of fluids. The density (ρ) of a fluid is defined as its mass per unit volume. Its metric units are kg m^{-3}. The density of an ideal gas is found using the specific gas constant and applying the ideal gas law:

$$\rho = p(RT) - 1 \tag{4.18}$$

where p is the gas pressure, R is the specific gas constant, and T is the absolute temperature.

So, the specific gas constant must be known to calculate gas density. For example, the R for air is 287 J kg^{-1}K^{-1}. The specific gas constant for methane (R_{CH_4}) is 518 J kg^{-1}K^{-1}.

Density is a very important fluid property for environmental situations. For example, a first responder must know the density of substances in an emergency situation. If a substance is burning, whether it is of greater or lesser density than water will be one of the factors on how to extinguish the fire. If the substance is less dense than water, the water will be more likely to settle below the layer of water, making water a poor choice for fighting the fire. So, any flammable substance with a density less than water (see Table 4.2), such as benzene or acetone, will require fire-extinguishing substances other than water. For substances heavier than water, like carbon disulfide, water may be a good choice.

TABLE 4.2

Densities of Selective Fluids Important in Air Pollution Science and Engineering

Fluid	Density (kg m^{-3}) at 20°C unless otherwise noted
Air at standard temperature and pressure (STP) = 0°C and 101.3 N m^{-2}	1.29
Air at 21°C	1.20
Ammonia	602
Diethyl ether	740
Ethanol	790
Acetone	791
Gasoline	700
Kerosene	820
Turpentine	870
Benzene	879
Pure water	1000
Seawater	1025
Carbon disulfide	1274
Chloroform	1489
Tetrachloromethane (carbon tetrachloride)	1595
Lead (Pb)	11 340
Mercury (Hg)	13 600

UNITS IN HANDBOOKS AND REFERENCE MANUALS

Standardization is important in air pollution science and engineering, but in the "real world," information is reported in many different formats. In fact, engineers seem to resist the use of SI units more than their colleagues in the basic sciences. This may, at least in part, be due to the historic inertia of engineering, where many equations were derived from English units. When an equation is based on one set of units and is only reported in those units, it can take much effort to convert them to SI units. Exponents in many water quality, water supply, and sludge equations have been empirically derived from studies that applied English units.

Some equations may use either English or SI units, such as the commonly used Hazen–Williams formula for mean velocity flow (v) in pressure pipes is:

$$v = 1.318C \cdot r^{0.63} \cdot s^{0.54} \tag{4.19}$$

where r is the hydraulic radius in feet or meters, s is the slope of the hydraulic grade line (head divided by length), and C is the friction coefficient (a function of pipe roughness).

The exponents apply without regard to units. Other formulae, however, require that a specific set of units be used. An example is the fundamental equation for kinetic energy. Two different equations are needed when using either the SI system or the English system, which requires the gravitation conversion constant (g_c) in the denominator. These are, respectively:

$$E_{\text{kinetic}} = \frac{mv^2}{2} \tag{4.20}$$

$$E_{\text{kinetic}} = \frac{mv^2}{2g_c} \text{(in ft-lbf)} \tag{4.21}$$

Two other important physical equations, potential energy and pressure require the insertion of their denominators:

$$E_{\text{potential}} = \frac{mgz}{g_c} \text{(in ft-lbf)} \tag{4.22}$$

$$p = \frac{\rho g h}{g_c} (\text{in ft-lbf ft}^{-2}) \tag{4.23}$$

where g is the gravitational acceleration, ρ is density, and h is the height.

With this in mind, it sometimes better to simply apply the formulae using English units and convert to metric or SI units following the calculation. In other words, rather than try to change the exponent or coefficient to address the difference in feet and meters, just use the units called for in the empirically derived equation. After completing the calculation, convert the answer to the correct units. This may seem contrary to the need to standardize units, but it may save time and effort in the long run. Either way, it is mathematically acceptable dimension analysis.

Another variation in units is how coefficients and constants are reported. For example, the octanol-water coefficient seems to be reported more often as $\log K_{ow}$ than simply as K_{ow}. This is usually because the ranges of K_{ow} values can be so large. One compound may have a coefficient of 0.001, while another has one of 1000. Thus, it may be more manageable to report the $\log K_{ow}$ values as -3 and 3 respectively.

Further, chemists and engineers are comfortable with the "p" notation as representative of the negative log. This could be because pH and pOH are common parameters. So, one may see the negative logarithm used with units in handbooks. For example, vapor pressure is sometimes reported as a negative log.

Therefore, examples and problems in handbooks and reference manuals make use of several different units as they are encountered in the environmental literature.

Specific volume The reciprocal of a substance's density is known as its specific volume (υ). This is the volume occupied by a unit mass of a fluid. The units of υ are reciprocal density units ($m^3 kg^{-1}$). Stated mathematically, this is:

$$\upsilon = \rho^{-1} \tag{4.24}$$

Specific weight The weight of a fluid per its volume is known as specific weight (γ). Civil engineers sometimes use the term interchangeably with density. A substance's γ is not an absolute fluid property because it depends on the fluid itself and the local gravitational force:

$$\gamma = g\rho \tag{4.25}$$

The units are the same as those for density; e.g. $kg\,m^{-3}$.

Mole Fraction In a composition of a fluid made up of two or more substances ($A, B, C, \ldots$), the mole fraction ($x_A, x_B, x_C, \ldots$) is number of moles of each substance divided by the total number of moles for the whole fluid:

$$x_A = \frac{n_A}{n_A + n_B + n_C + \ldots} \tag{4.26}$$

The mole fraction value is always between 0 and 1. The mole fraction may be converted to mole percent as:

$$x_{A\%} = x_A \times 100 \tag{4.27}$$

For gases, the mole fraction is the same as the volumetric fraction of each gas in a mixture of more than one gas.

Mole Fraction Example

112 g of $MgCl_2$ are dissolved in 1 L of water. The density of this solution is 1.089 g cm^{-3}. What is the mole fraction of $MgCl_2$ in the solution at standard temperature and pressure.

Solution

The number of moles of $MgCl_2$ is determined from its molecular weight:

$$\frac{112\,g}{95.22\,g} = 1.18\,\text{mol}$$

Next, we calculate the number of moles of water:

Mass of water $= 1.00\,\text{L} \times (1000\,\text{cm}^{-3}\,\text{L}^{-1}) \times (1.00\,\text{g}\,\text{cm}^{-3}) = 1000\,\text{g}$ water

and

$$\text{Moles of water} = \frac{1000\,g}{18.02\,g\,\text{mol}^{-1}} = 55.49\,\text{mol}$$

$$\text{Thus, } x_{MgCl_2} = \frac{1.18\,\text{mol}}{55.49 + 1.18} = 0.021.$$

The mol% of $MgCl_2$ is 2.1%.

Compressibility The fractional change in a fluid's volume per unit change in pressure at constant temperature is the fluid's coefficient of compressibility. Gases, like air are quite compressible. That is, gases have large variations in density (ρ), much larger than in liquids. However any fluid can be compressed in response to the application of pressure (p). For example, water's compressibility at 1 atm is $4.9 \times 10^{-5}\,\text{atm}^{-1}$. This compares to the lesser compressibility of mercury ($3.9 \times 10^{-6}\,\text{atm}^{-1}$) and the much greater compressibility of hydrogen ($1.6 \times 10^{-3}\,\text{atm}^{-1}$).

A fluid's bulk modulus, E (analogous to the modulus of elasticity in solids) is a function of stress and strain on the fluid (see Fig. 4.9), and is a description of its compressibility. It is defined according to the fluid volume (V):

$$E = \frac{\text{stress}}{\text{strain}} = -\frac{dp}{dV/V_1} \tag{4.28}$$

E is expressed in units of pressure (e.g. kPa). Water's $E = 2.2 \times 10^6$ kPa at 20°C.

Surface Tension and Capillarity Surface tension effects occur at liquid surfaces (interfaces of liquid–liquid, liquid–gas, liquid–solid). Surface tension, σ, is the force in the liquid surface normal to a line of unit length drawn in the surface. Surface tension decreases with temperature and depends on the contact fluid. Surface tension is involved in capillary rise and drop. Water has a very high σ value (approximately 0.07 N m^{-2} at 200°C). Of the environmental fluids, only mercury has a higher σ (see Table 4.3).

The high surface tension creates a type of skin on a free surface, which is how an object more dense than water (e.g. a steel needle) can "float" on a still water surface. It is the reason insects can sit comfortably on water surfaces. Surface tension is somewhat dependent on the gas that is contacting the free surface. If not indicated, it is usually safe to assume that the gas is the air in the troposphere.

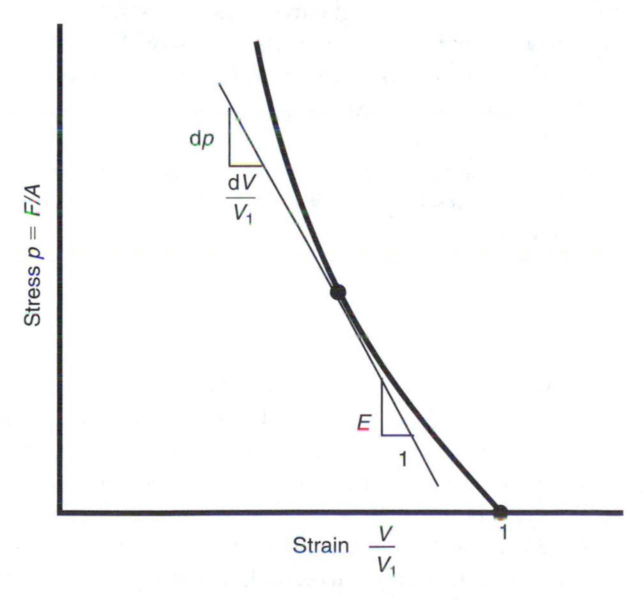

Fig. 4.9. Stress and strain on a fluid, and the bulk modulus of fluids.

TABLE 4.3

Surface Tension (Contact with Air) of Selected Fluids Important to Air Pollution

Fluid	Surface tension, σ ($N\,m^{-1}$ at 20°C)
Acetone	0.0236
Benzene	0.0289
Ethanol	0.0236
Glycerin	0.0631
Kerosene	0.0260
Mercury	0.519
n-Octane	0.0270
Tetrachloromethane	0.0236
Toluene	0.0285
Water	0.0728

Capillarity is a particularly important fluid property in the design of air pollutant sampling and analytical equipment, such as capillary action in filters and sorption traps used to collect gases and in capillary tubes used in chromatography. Capillarity is also an important factor in physiological response to air pollutants. For example, morphological changes have been observed in lung tissue exposed to ozone. These changes, in part, are due to ozone's effect on the alveolar capillaries and the degradation of capillary endothelia. This leads to less efficient oxygen exchange between the inhaled air and blood, which in turn leads to respiratory and cardiovascular stress.

Capillary rise occurs for two reasons, its adhesion to a surface, plus the cohesion of water molecules to one another. Higher relative surface tension causes a fluid rise to in a tube (or a pore) that is indirectly proportional to the diameter of the tube. In other words, capillarity is greater the smaller the inside diameter of the tube (see Fig. 4.10). The rise is limited by the weight of the fluid in the tube. The rise ($h_{\text{capillary}}$) of the fluid in a capillary is expressed as (Fig. 4.11 displays the variables):

$$h_{\text{capillary}} = \frac{2\sigma \cos\lambda}{\rho_{\text{w}} g R} \tag{4.29}$$

where σ is the fluid surface tension ($g\,s^{-2}$), λ is the angle of meniscus (concavity of fluid) in capillary (degrees), ρ_w is the fluid density ($g\,cm^{-3}$), g is the gravitational acceleration ($cm\,s^{-1}$), and R is the radius of capillary (cm).

The contact angle indicates whether cohesive or adhesive forces are dominant in the capillarity. When λ values are greater than 90°, cohesive forces are dominant; when $\lambda < 90°$, adhesive forces dominate. Thus, λ is dependent on both the type of fluid and the surface to which it comes into contact. For example, water–glass $\lambda = 0°$; ethanol–glass $\lambda = 0°$; glycerin–glass $\lambda = 19°$; kerosene–glass $\lambda = 26°$; water–paraffin $\lambda = 107°$; and mercury–glass $\lambda = 140°$.

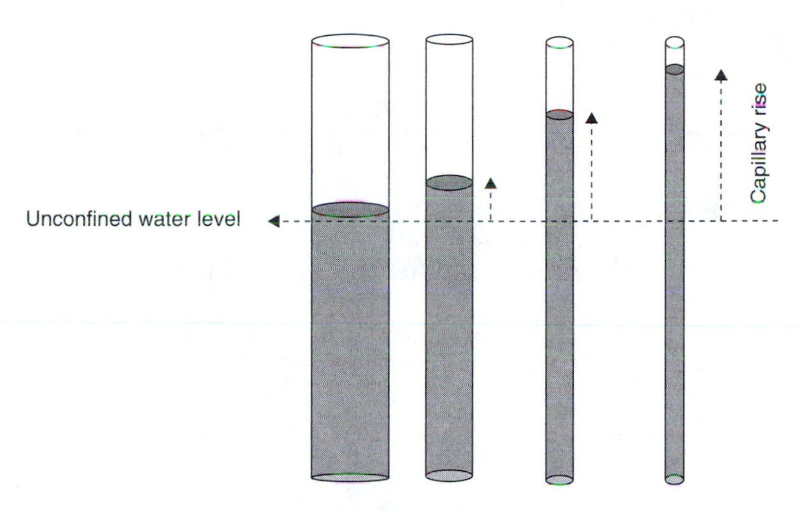

Fig. 4.10. Capillary rise of water with respect to diameter of conduit.

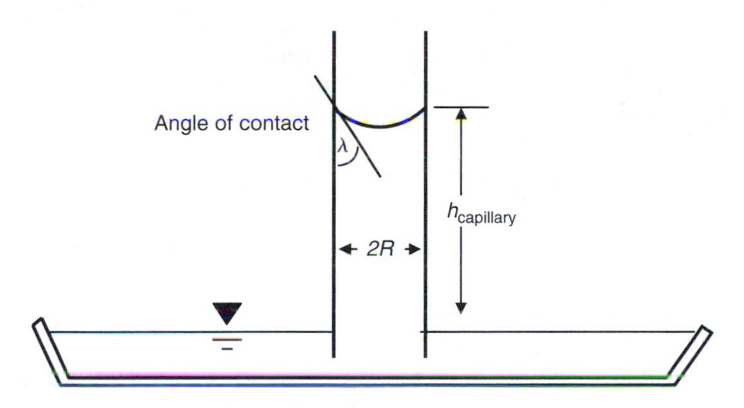

Fig. 4.11. Rise of a fluid in a capillary. In this example, adhesive forces within the fluid are dominant, so the meniscus is concave (i.e. a valley). This is the case for most fluids. However, if cohesive forces dominate, such as the extremely cohesive liquid mercury, the meniscus will be convex (i.e. a hill).

Capillarity Example

What is the rise of contaminated water (i.e. a solution of water and soluble and insoluble contaminants) in a sorption material used to collect gaseous air pollutants that has an average pore space diameter of 0.1 cm, at 18°C and a density of 0.999 g cm^{-3}, under surface tension of 50 g s^{-1} if the angle of contact of the meniscus is 30°?

What would happen if the average pore space were 0.01 cm, with all other variables remaining as stated?

Answer

$$h_{\text{capillary}} = \frac{2\sigma \cos \lambda}{\rho_w g R} = \frac{2 \times 80 \times \cos 30}{0.999 \times 980 \times 0.05} \text{cm} \cong 0.25 \text{cm}$$

If the pore space were 0.01 cm in diameter, the rise would be 2.5 cm. However, it is likely that the angle of contact would have also decreased since the angle is influenced by the diameter of the column (approaching zero with decreasing diameter).

Also note that since the solution is not 100% water, the curvature of the meniscus will be different, so the contact angle λ will likely be greater (i.e. less curvature) than the meniscus of water alone. The lower surface tension of the mixture also means that the capillary rise will be less.

Viscosity How much a fluid resists flow when it is acted on by an external force, especially a pressure differential or gravity, is the fluid's viscosity. This a crucial fluid property used in numerous environmental applications, including air pollution plume characterization.

Recall from Bernoulli's equation (4.10) and Fig. 4.6 that if a fluid is flowing in a long, horizontal conduit with constant cross-sectional area, the pressure along the pipe must be constant. But why if we measure the pressure as the fluid moves in the conduit, would there be a *pressure drop*? A pressure difference is needed to push the fluid through the conduit to overcome the drag force exerted by the conduit walls on the layer of fluid that is making contact with the walls. Since the drag force exerted by each successive layer of the fluid on each adjacent layer that is moving at its own velocity, then a pressure difference is needed (see Fig. 4.12). The drag forces are known as *viscous forces*. Thus, the fluid velocity is not constant across the conduit's diameter, owing to the viscous forces. The greatest velocity is at the center (furthest away from the walls), and the lowest velocity is found at the walls. In fact, at the point of contact with walls, the fluid velocity is zero.

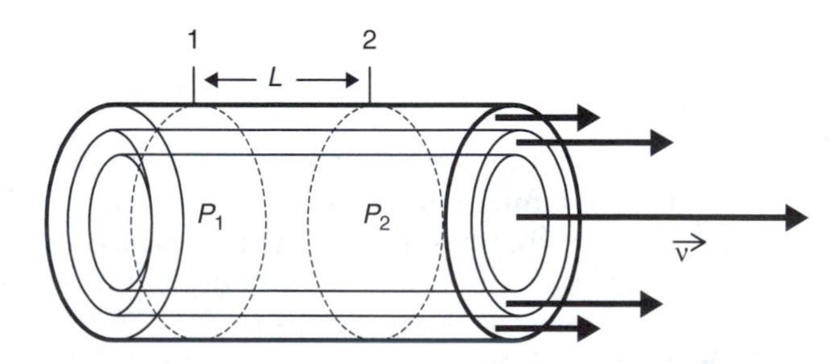

Fig. 4.12. Viscous flow through a horizontal conduit. The highest velocity is at the center of the conduit. As the fluid approaches the wall, the velocity declines and approaches zero.

So, if P_1 is the pressure at point 1, and P_2 is the pressure at point 2, with the two points separated by distance L, the pressure drop (ΔP) is proportional to the flow rate:

$$\Delta P = P_1 - P_2 \tag{4.30}$$

and

$$\Delta P = P_1 - P_2 = I_v R \tag{4.31}$$

where I_v is the volume flow rate, and R is the proportionality constant representing the resistance to the flow. This resistance R depends on the length (L) of pipe section, the pipe's radius, and the fluid's viscosity.

Viscosity Example 1

Workers are being exposed to an air pollutant known to decrease blood pressure in the capillaries, small arteries, and major arteries and veins after the blood is pumped from the aorta. If high-dose studies show an acute drop in the gage pressure of the circulatory system from 100 torr to 0 torr at a volume flow of $0.7\,\mathrm{L\,s^{-1}}$, give the total resistance of the circulatory system.

Answer

Solving for R from Eq. (4.31), and converting to SI units gives us:

$$\begin{aligned} R &= \Delta P\,(I_v)^{-1} \\ &= (100\,\mathrm{torr})(0.7\,\mathrm{L\,s^{-1}})^{-1}\,(133.3\,\mathrm{Pa})\,(1\,\mathrm{torr})^{-1}\,(1\,\mathrm{L})\,(10^3\mathrm{cm^{-3}})^{-1} \\ &\quad (1\,\mathrm{cm^{-3}})\,(10^{-6}\mathrm{m^{-3}}) \\ &= 1.45 \times 10^7\,\mathrm{Pa\,s\,m^{-3}} \\ &= 1.45 \times 10^7\,\mathrm{N\,s\,m^{-5}} \end{aligned}$$

Two types of viscosity are important in air pollution: absolute viscosity and kinematic viscosity.

Absolute viscosity Physicists define the fluid's coefficient of viscosity by assuming that the fluid is confined between two parallel, rigid plates with equal area. The absolute viscosity of a fluid can be measured by a number of ways, but engineers commonly use the *sliding plate viscometer test*. The test applies two plates separated by the fluid to be measured (see Fig. 4.13).

For Newtonian fluids, the force applied in the viscometer test has been found to be in direct proportion to the velocity of the moving plate and inversely proportional to the length of separation of the two plates:

$$\frac{F}{A} \propto \frac{dv}{dy} \tag{4.32}$$

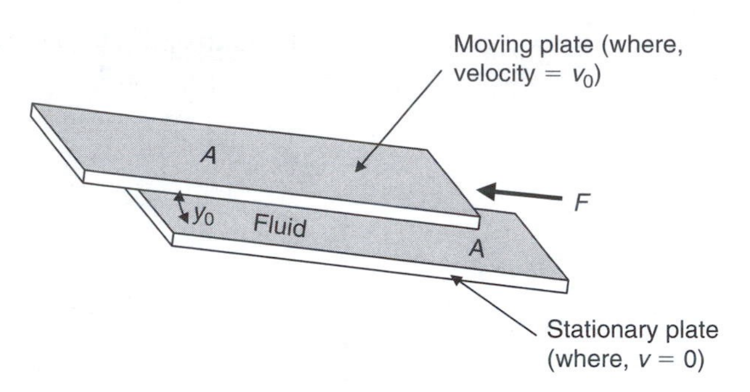

Fig. 4.13. The sliding plate viscometer. A fluid of thickness y_0 is placed between two plates of area A. The top plate moves at the constant velocity v_0 by the exertion of force F. *Source*: Lindeberg, M., *Civil Engineering Reference Manual for the PE Exam*, 8th ed. Professional Publications, Inc., Belmont, CA, 2001.

Using a constant this proportionality can become an equality:

$$\frac{F}{A} = \mu \frac{dv}{dy} \tag{4.33}$$

This equation is known at *Newton's law of viscosity*. Fluids that conform to this law are referred to as Newtonian fluids.[4] The constant, μ, is the fluid's *absolute viscosity*. The μ is also known as the *coefficient of viscosity*, but environmental texts often refer to μ as *dynamic viscosity*. The term *fluidity* is the reciprocal of dynamic velocity.

The inverse relationship between viscosity and fluidity should make sense after some thought. Since the definition of viscosity is the resistance to flow when an external force is applied, then it makes sense that if a substance does a poor job resisting the flow, the substance must have a lot of fluidity. An electrical analogy might be that of conductivity and resistance. If copper wire has much less resistance to electrical flow than does latex rubber, we say that copper must be a good conductor. Likewise, if water at 35°C is less effective at resisting flow downhill (i.e. gravity is applying our force) than is motor oil at the same temperature, we say that the water has less dynamic viscosity than the motor oil. And, we also say that the water has more fluidity than the oil. Before the modern blends of multi-viscosity motor oils, the temperature–viscosity relationship was part of the seasonal rituals of the oil change. Less viscous motor oil (e.g. 10 W) had to used in a car's engine to prepare for the lower temperatures in winter, so that the starter could "turn over" the engine (less viscous oil = less resistance the force of the starter moving the pistons). Conversely, in preparing for summer, a higher viscosity

[4] See discussion of Newtonian and non-Newtonian fluids earlier in this chapter.

motor oil (commonly 40 W)[5] would be used because the high temperatures in the engine would allow the oil to "blow out" through the piston rings or elsewhere (because the oil was not doing a good job of resisting the force applied by the pistons and shot out of the engine). The newer oil formulations (e.g. 10–40 W) maintain a smaller range of viscosities, so automobile owners worry less about the viscosity.

The $\frac{F}{A}$ term is known as the *shear stress*, τ, of the fluid. The $\frac{dv}{dy}$ term is known as the *velocity gradient* or the *rate of shear formation*.[6] So, the shear stress is linear; i.e., it can be expressed as a straight line (in the form $y = mx + b$):

$$\tau = \mu \frac{dv}{dy} \tag{4.34}$$

The relationship between the two sides of this equality determines the types of fluids, as shown in Fig. 4.14). Most fluids encountered in environmental studies are Newtonian, including water, all gases, alcohols, and most solvents. Most solutions also behave as Newtonian fluids. Slurries, muds, motor grease and oils, and many polymers behave has *pseudoplastic fluids*; i.e., viscosities decrease with increasing velocity gradient. They are easily pumped, since higher pumping rates lead to a less viscous fluid. Some slurries behave as *Bingham fluids* (e.g. behave like toothpaste or bread dough), where the shear formation is resisted up to a point. For example, depending on their chemical composition, slurries used in wet scrubbers to collect gases (e.g. sulfur dioxide), may behave as Bingham fluids. The rare *dilatant fluids* are sometimes encountered in environmental engineering applications, such as clay slurries used as landfill liners and when starches and certain paints and coatings are spilled. These can be difficult fluids to remove and clean up, since their viscosities increase with increasing velocity gradient, so pumping these fluids at higher rates can lead to their becoming almost solid with a sufficiently high shear rate. *Plastic fluids* (see Fig. 4.15) require the application of a finite force before any fluid movement.

Categorizing and characterizing fluids according to their behavior under shear stress and velocity gradient is not absolute. For example, a Bingham fluid can resist shear stresses indefinitely so long as they are small, but these fluids will become pseudoplastic at higher stresses. Even if all conditions remain constant, viscosity can also change with time. A *rheopectic fluid* is one where viscosity increases with time, and a *thixotropic fluid* is one that has decreasing viscosity with time. Those fluids that do not change with time are referred to as *time-independent fluids*. Colloidal materials, like certain components of sludges, sediments, and soils, act like thixotropic fluids. That is, they

[5] In the 1960s, many of us went further. We used 50 W or even higher viscosity racing formulas even if we never really allowed our cars to ever reach racing temperatures! We often applied the same logic for slicks, glass packs, four-barrel carburetors, and other racing equipment that was really never needed, but looked and sounded awesome!

[6] The $dv\,dy^{-1}$ term is also known as the *rate of strain* and the *shear rate*.

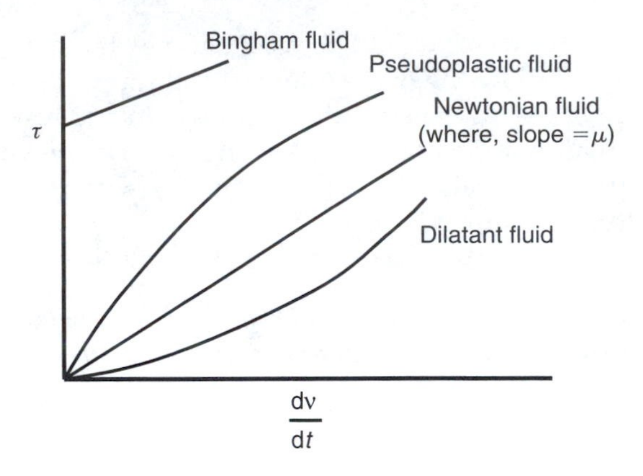

Fig. 4.14. Hypothetical fluid types according to shear stress (τ) behavior relative to velocity gradient. *Source*: Lindeberg, M., *Civil Engineering Reference Manual for the PE Exam*, 8th ed. Professional Publications, Inc., Belmont, CA, 2001.

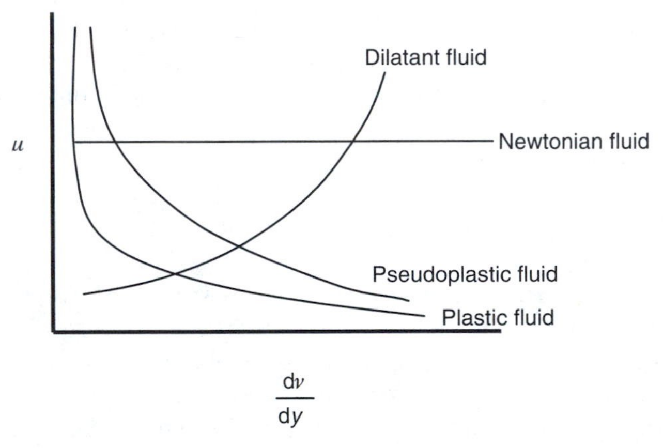

Fig. 4.15. Hypothetical fluid types according to viscosity (μ) and shear rate (velocity). *Source*: Lindeberg, M., *Civil Engineering Reference Manual for the PE Exam*, 8th ed. Professional Publications, Inc., Belmont, CA, 2001.

experience a decrease in viscosity when the shear is increased. However, there is no *hysteresis*, meaning that the viscosity does not return to the original state with the ceasing of the agitation.

There is a seeming paradox between viscosity and temperature. As a general rule, temperature is inversely proportional to viscosity of liquids, but temperature is directly proportional to the viscosity of gases. Viscosity of liquids is predominantly caused by molecular cohesion. These cohesive forces decrease with increasing temperature, which is why viscosity decreases with increasing temperature. Gas viscosity is mainly kinetic molecular in its origin, so increasing temperature means that more collisions will occur between

TABLE 4.4

Absolute Viscosity of Fluids Important to Health and Environmental Studies. ***Source*****: Tipler, P.,** ***Physics for Scientists and Engineers*****, Vol. 1. W.H. Freeman and Co., New York, 1999.**

Fluid	Temperature (°C)	Absolute viscosity, μ (Pa s)
Water	0	1.8×10^{-3}
	20	1×10^{-3}
	60	6.5×10^{-2}
Whole human blood	37	4×10^{-3}
SAE 10 motor oil	30	2×10^{-1}
Glycerin	0	10
	20	1.4
	60	8.1×10^{-2}
Air	20	1.8×10^{-5}

molecules. The more the gas is agitated, the greater the viscosity, so gas velocity increases with increasing temperatures.

The viscosity of liquids increases only slightly with increasing pressure. Under environmental conditions, absolute viscosity can be considered to be independent of pressure.

Absolute viscosity units are mass per length per time (e.g. $\text{g cm}^{-1}\text{s}^{-1}$). The coefficients for some common fluids are provided in Table 4.4. Note the importance of temperature in a substance's absolute viscosity, e.g. the several orders of magnitude decrease with only a 20°C increase in glycerin.

Viscosity Example 2

A liquid with the absolute viscosity of $3 \times 10^{-5}\,\text{g s cm}^{-1}$ flows through rectangular tube in an air sampling device. The velocity gradient is $0.5\,\text{m s}^{-1}\text{cm}^{-1}$. What is the shear stress in the fluid at this velocity gradient?

Answer

$$\tau = \mu \frac{dv}{dy}$$
$$= (3 \times 10^{-5}\,\text{g s cm}^{-1})(0.5\,\text{m s}^{-1}\,\text{cm}^{-1})(100\,\text{cm m}^{-1})$$
$$= 1.5 \times 10^{-3}\,\text{g cm}^{-2}$$

Kinematic viscosity The ratio of absolute viscosity to mass density is known as *kinematic viscosity* (ν):

$$\nu = \mu \rho^{-1} \tag{4.35}$$

The units of ν are area per second (e.g. $cm^2 s^{-1}$ = stoke). Because kinematic viscosity is inversely proportional to a fluid's density, ν is highly dependent on temperature and pressure. Recall that absolute viscosity is only slightly affected by pressure. Table 4.5 can be used to convert most of the units of μ and ν.

Laminar versus Turbulent Flow: The Reynolds Number At a sufficiently high velocity, a fluid's flow ceases to be laminar and becomes turbulent. A dimensionless Reynolds number (N_R) is used to differentiate types of flow. The N_R is expressed as the ratio of inertial to viscous forces in a fluid:

$$N_R = \frac{\text{Inertial forces}}{\text{Viscous forces}} \quad (4.36)$$

TABLE 4.5

Viscosity Units and Conversions

Multiply:	By:	To obtain:
Absolute viscosity (μ)		
centipoise (cP)	1.0197×10^{-4}	$kg f s m^{-2}$
cP	2.0885×10^{-5}	$lb f\text{-}s ft^{-2}$
cP	1×10^{-3}	Pa s
Pa s	2.0885×10^{-3}	$lb f\text{-}s ft^{-2}$
Pa s	1000	cP
$dyne s cm^{-2}$	0.10	Pa s
$lbf\text{-}s ft^{-2}$	478.8	poise (P)
$slug ft^{-1} s^{-1}$	47.88	Pa s
Kinematic viscosity (ν)		
$ft^2 s^{-1}$	9.2903×10^4	centistoke (cSt)
$ft^2 s^{-1}$	9.2903×10^{-2}	$m^2 s^{-1}$
$m^2 s^{-1}$	10.7639	$ft^2 s^{-1}$
$m^2 s^{-1}$	1×10^6	cSt
cSt	1×10^{-6}	$m^2 s^{-1}$
cSt	1.0764×10^{-5}	$ft^2 s^{-1}$
μ to ν		
cP	$1/\rho$ ($g cm^{-3}$)	cSt
cP	$6.7195 \times 10^{-4}/\rho$ in $lbm ft^{-3}$	cSt
$lbf\text{-}s ft^{-2}$	$32.174/\rho$ in $lbm ft^{-3}$	$ft^2 s^{-1}$
$kgf s m^{-2}$	$9.807/\rho$ in $kg m^{-3}$	$m^2 s^{-1}$
Pa s	$1000/\rho$ in $g cm^{-3}$	cSt
ν to μ		
cSt	ρ in $g cm^{-3}$	cP
cSt	1.6×10^{-5}	Pa s
$m^2 s^{-1}$	$0.10197 \times \rho$ in $kg m^{-3}$	$kg f s m^{-2}$
$m^2 s^{-1}$	$1000 \times \rho$ in $g cm^{-3}$	Pa s
$ft^2 s^{-1}$	$3.1081 \times 10^{-2} \times \rho$ in $lb m ft^{-3}$	$lb f\text{-}s ft^{-2}$
$ft^2 s^{-1}$	$1.4882 \times 10^3 \times \rho$ in $lb m ft^{-3}$	cP

The inertial forces are proportional to the velocity and density of the fluid, as well as to the diameter of the conduit in which the fluid is moving. An increase in any of these factors will lead to a proportional increase in the momentum of the flowing fluid. We know from our previous discussion that the coefficient of viscosity or absolute viscosity (μ) represents the total viscous force of the fluid, so, N_R can be calculated as:

$$N_R = \frac{D_e v \rho}{\mu} \tag{4.37}$$

where D_e is the conduit's equivalent diameter, which is a so-called "characteristic dimension"[7] which evaluates the fluid flow as a physical length.

It is actually the inside diameter (i.d.) of the conduit, vent, or pipe. Recall that $\mu \rho^{-1}$ is the kinematic viscosity υ, so the Reynolds number can be stated as the relationship between the size of the conduit, the average fluid velocity ν, and υ:

$$N_R = \frac{D_e \nu}{\upsilon} \tag{4.38}$$

When fluids move at very low velocities, the bulk material moves in discrete layers parallel to one another. The only movement across the fluid layers is molecular motion, which creates viscosity. Such a flow is *laminar* (see Fig. 4.16). Laminar flow is more common in water than in air, especially in low velocity systems like ground water.

With increasing fluid velocity, the bulk movement changes, forming eddy currents that create three-dimensional mixing across the flow stream. This is known as *turbulent* flow. Most pollution control equipment and atmospheric plumes are subjected to turbulent flow (see Fig. 4.17).

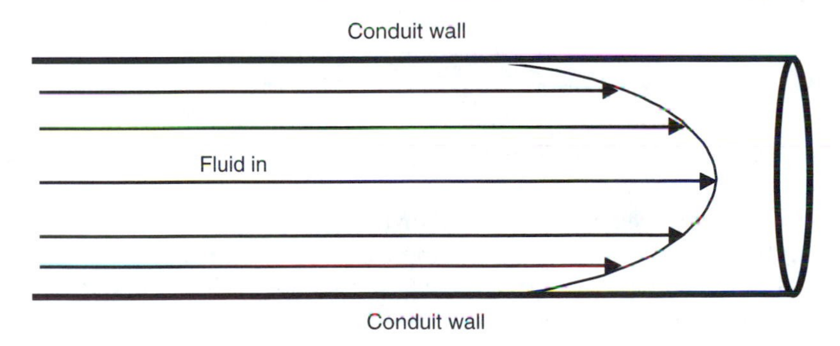

Fig. 4.16. Laminar flow in closed conduit.

[7]Other equivalent diameters for fully flowing conduits are the annulus, square, and rectangle. Equivalent diameters for partial flows in conduits are the half-filled circle, rectangle, wide and shallow stream, and trapezoid. For calculations of these diameters, see Lindeberg, M., *Civil Engineering Reference Manual for the PE Exam*, 8th ed. Professional Publications, Inc., Belmont, CA, 2001.

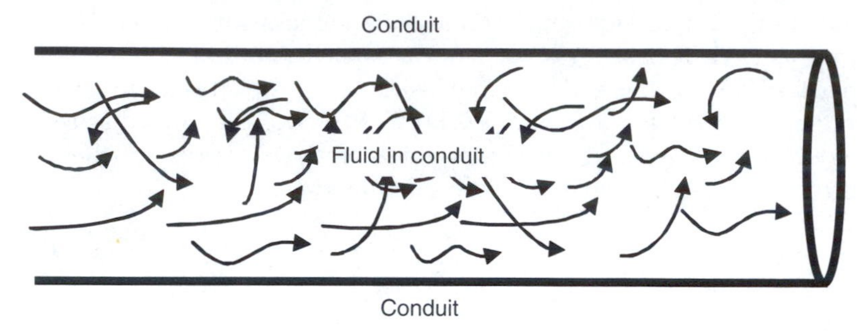

Fig. 4.17. Turbulent flow in a closed conduit.

Flows in closed conduits with Reynolds numbers under 2100 are usually laminar.[8] Due to the relatively low velocities associated with this type of flow, they are mainly encountered with liquids such as water moving through underground strata and blood flowing in arteries. In open atmospheric conditions, such as a plume of an air pollutant, laminar flow is quite rare. Flows with Reynolds numbers greater than 4000 are usually turbulent. The range of N_R values between these thresholds are considered "critical flows" or "transitional flows," that show properties of both laminar and turbulent flow in the flow streams. Usually, if the flow is in the transition region, engineers will design equipment as if the flow were turbulent, as this is the most conservative design assumption.

Under laminar conditions, the fluid particles adhere to the conduit wall. The closer to the wall that a particle gets, the more likely it will adhere to the wall. Laminar flow is, therefore, parabolic and its velocity at the conduit wall is zero (see Fig. 4.16). Laminar flow velocity is greatest at the pipe's center (v_{max} in the Fig. 4.18), and is twice the value of the average velocity, $v_{average}$:

$$v_{average} = \frac{\dot{V}}{A} = \frac{v_{max}}{2} \text{(laminar)} \tag{4.39}$$

where $\dot{V}$ is the volumetric fluid velocity and A is the cross-sectional area of the pipe.

Turbulent flow velocity, on the other hand, has no relationship with the proximity to the wall due to the mixing (see Fig. 4.17). So, all fluid particles in a turbulent system are assumed to share the same velocity (as depicted in Fig. 4.18), known as the average velocity or bulk velocity:

$$v_{average} = \frac{\dot{V}}{A} \tag{4.40}$$

[8] The literature is not consistent on the exact Reynolds numbers as thresholds for laminar versus turbulent flow. Another value used by engineers is 2300.

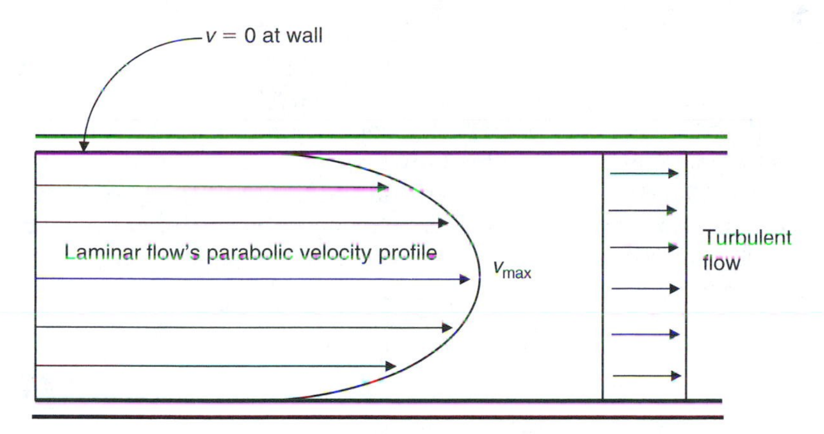

Fig. 4.18. Velocity distributions of laminar and turbulent flows.

There is a thin layer of turbulent flow near the wall of the conduit where the velocity increases from zero to $v_{average}$, known as the *boundary layer*. In fact, no flow is entirely turbulent and there is some difference between the centerline velocity and $v_{average}$. However, for many environmental applications the assumption of consistently mixed flow is acceptable.

Reynolds Number Example 1

Find the Reynold's number of water flowing in a 0.2 i.d. m pipe at $0.1\,\mathrm{m\,s^{-1}}$. Assume that the water's coefficient of viscosity is $8 \times 10^{-3}\,\mathrm{N\,s\,m^{-3}}$ and density is $1000\,\mathrm{kg\,m^{-3}}$.

Solution

Use Eq. (4.38).

$$N_R = \frac{D_c v \rho}{\mu} = \frac{(1000\,\mathrm{kg\,m^3})(0.1\,\mathrm{m\,s^{-1}})(0.2\,\mathrm{m})}{8 \times 10^{-3}\,\mathrm{N\,s\,m^{-2}}} = 2500$$

Reynolds Number Example 2

How is this flow characterized? Assuming this flow is representative of a cooling tower in a power plant, what kind of flow should be assumed in selecting pumps and other equipment?

Solution

Since the N_R is greater than 2100, but less than 4000, the flow is considered transitional or critical. Therefore, the conservative design calls for an assumption that the flow is turbulent.

The fundamental fluid properties and physical principles discussed in this chapter are important in all environmental sciences. In Chapter 5, we will extend these and other physical concepts specifically to the atmosphere.

QUESTIONS

1. Give an example in air pollution of a clearly physical phenomenon, an example of a clearly chemical phenomenon, and an example of a phenomenon that has qualities of both, that is a physicochemical phenomenon.
2. How does the Bernoulli effect play a role in air pollution engineering?
3. Why is density important in environmental measurements?
4. Why is an understanding of capillarity important to air pollution? Give three possible reasons.
5. Give a scenario where viscosity of a fluid needs to be understood to protect air quality. What is the importance of the Reynolds number in your scenario?

5

The Physics of the Atmosphere

The atmosphere serves as the medium through which air pollutants are transported and dispersed. While being transported, the pollutants may undergo chemical reactions and, in addition to removal by chemical transformations, may be removed by physical processes such as gravitational settling, impaction, and wet removal.

This chapter provides an introduction to basic concepts of meteorology necessary to an understanding of air pollution meteorology without specific regard to air pollution problems.

I. ENERGY

All of the energy that drives the atmosphere is derived from a minor star in the universe—our sun. The planet that we inhabit, earth, is 150 million km from the sun. The energy received from the sun is radiant energy—electromagnetic radiation (discussed in Chapter 7). The electromagnetic spectrum is shown in Fig. 5.1. Although this energy is, in part, furnished to the atmosphere, it is primarily received at the earth's surface and redistributed by several processes.

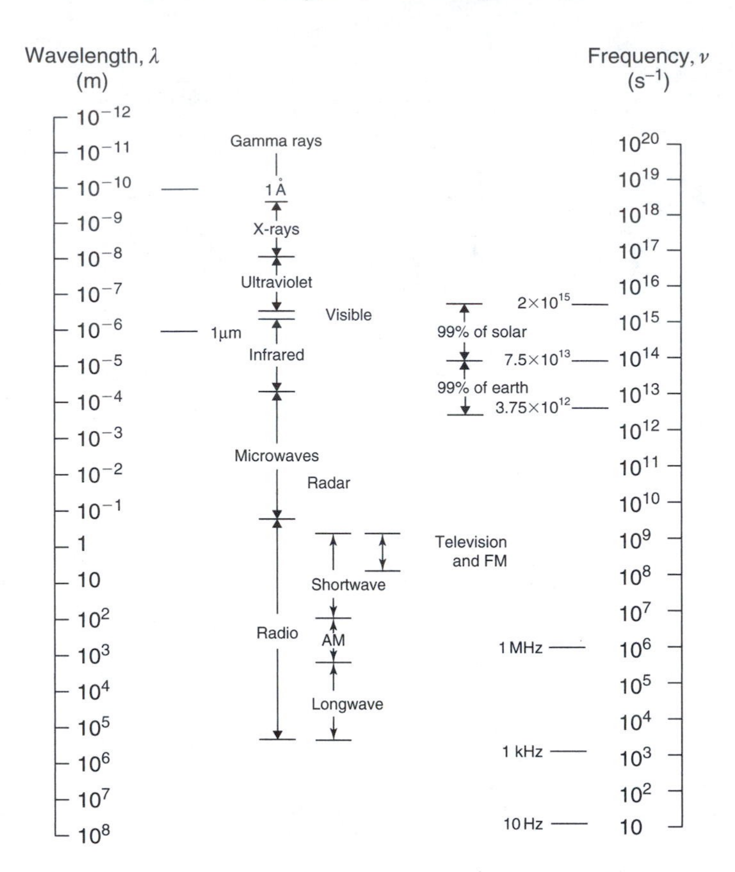

Fig. 5.1. Electromagnetic spectrum. Note the regions of solar and earth radiation.

The earth's gravity keeps the thin layer of gases that constitute the atmosphere from escaping. The combination of solar heating and the spin of the earth causes internal pressure forces in the atmosphere, resulting in numerous atmospheric motions. The strength of the sun's radiation, the distance of the earth from the sun, the mass and diameter of the earth, and the existence and composition of the atmosphere combine to make the earth habitable. This particular combination of conditions would not be expected to occur frequently throughout the universe.

As noted in Chapter 2, the atmosphere is approximately 76% nitrogen, 20% oxygen, 3% water, 0.9% argon, and 0.03% carbon dioxide; the rest consists of relatively inert gases such as neon, helium, methane, krypton, nitrous oxide, hydrogen, and xenon. Compared with the average radius of the earth, 6370 km, the atmosphere is an incredibly thin veil; 90% is below 12 km and 99% below 30 km. In spite of its thinness, however, the total mass of the atmosphere is about 5×10^{18} kg. Therefore, its heat content and energy potential are very large.

A. Radiation from a Blackbody

Blackbody is the term used in physics for an object that is a perfect emitter and absorber of radiation at all wavelengths. Although no such object exists in nature, the properties describable by theory are useful for comparison with materials found in the real world. The amount of radiation, or radiant flux over all wavelengths (F), from a unit area of a blackbody is dependent on the temperature of that body and is given by the Stefan–Boltzmann law:

$$F = \sigma T^4 \tag{5.1}$$

where σ is the Stefan–Boltzmann constant and equals 8.17×10^{-11} cal cm^{-2} min^{-1} deg^{-4} and T is the temperature in degrees K. Radiation from a blackbody ceases at a temperature of absolute zero, 0 K.

In comparing the radiative properties of materials to those of a blackbody, the terms *absorptivity* and *emissivity* are used. Absorptivity is the amount of radiant energy absorbed as a fraction of the total amount that falls on the object. Absorptivity depends on both frequency and temperature; for a blackbody it is 1. Emissivity is the ratio of the energy emitted by an object to that of a blackbody at the same temperature. It depends on both the properties of the substance and the frequency. Kirchhoff's law states that for any substance, its emissivity at a given wavelength and temperature equals its absorptivity. Note that the absorptivity and emissivity of a given substance may be quite variable for different frequencies.

As seen in Eq. (5.1), the total radiation from a blackbody is dependent on the fourth power of its absolute temperature. The frequency of the maximum intensity of this radiation is also related to temperature through Wien's displacement law (derived from Planck's law):

$$v_{\max} = 1.04 \times 10^{11}\, T \tag{5.2}$$

where frequency v is in s^{-1} and the constant is in s^{-1} K^{-1}.

The radiant flux can be determined as a function of frequency from Planck's distribution law for emission:

$$E_v \mathrm{d}v = c_1 v^3 \left[\exp(c_2 v/T) - 1\right]^{-1} \mathrm{d}v \tag{5.3}$$

where

$$c_1 = 2\pi h/c^2$$
$$h = 6.55 \times 10^{-27} \text{ erg s (Planck's constant)}$$
$$c = 3 \times 10^8 \text{ m s}^{-1} \text{ (speed of light)}$$
$$c_2 = h/k$$

and

$$k = 1.37 \times 10^{-16} \text{ erg K}^{-1} \text{ (Boltzmann's constant)}$$

The radiation from a blackbody is continuous over the electromagnetic spectrum. The use of the term black in blackbody, which implies a particular color, is quite misleading, as a number of nonblack materials approach blackbodies in behavior. The sun behaves almost like a blackbody; snow radiates in the infrared nearly as a blackbody. At some wavelengths, water vapor radiates very efficiently. Unlike solids and liquids, many gases absorb (and reradiate) selectively in discrete wavelength bands, rather than smoothly over a continuous spectrum.

B. Incoming Solar Radiation

The sun radiates approximately as a blackbody, with an effective temperature of about 6000 K. The total solar flux is 3.9×10^{26} W. Using Wien's law, it has been found that the frequency of maximum solar radiation intensity is $6.3 \times 10^{14}\ s^{-1}$ ($\lambda = 0.48\,\mu m$), which is in the visible part of the spectrum; 99% of solar radiation occurs between the frequencies of $7.5 \times 10^{13}\ s^{-1}$ ($\lambda = 4\,\mu m$) and $2 \times 10^{15}\ s^{-1}$ ($\lambda = 0.15\,\mu m$) and about 50% in the visible region between $4.3 \times 10^{14}\ s^{-1}$ ($\lambda = 0.7\,\mu m$) and $7.5 \times 10^{14}\ s^{-1}$ ($\lambda = 0.4\,\mu m$). The intensity of this energy flux at the distance of the earth is about 1400 W m^{-2} on an area normal to a beam of solar radiation. This value is called the *solar constant.* Due to the eccentricity of the earth's orbit as it revolves around the sun once a year, the earth is closer to the sun in January (perihelion) than in July (aphelion). This results in about a 7% difference in radiant flux at the outer limits of the atmosphere between these two times.

Since the area of the solar beam intercepted by the earth is πE^2, where E is the radius of the earth, and the energy falling within this circle is spread over the area of the earth's sphere, $4\pi E^2$, in 24 h, the average energy reaching the top of the atmosphere is 338 W m^{-2}. This average radiant energy reaching the outer limits of the atmosphere is depleted as it attempts to reach the earth's surface. Ultraviolet radiation with a wavelength less than 0.18 μm is strongly absorbed by molecular oxygen in the ionosphere 100 km above the earth; shorter X-rays are absorbed at even higher altitudes above the earth's surface. At 60–80 km above the earth, the absorption of 0.2–0.24 μm wavelength radiation leads to the formation of ozone; below 60 km there is so much ozone that much of the 0.2–0.3 μm wavelength radiation is absorbed. This ozone layer in the lower mesosphere and the top of the stratosphere shields life from much of the harmful ultraviolet radiation. The various layers warmed by the absorbed radiation reradiate in wavelengths dependent on their temperature and spectral emissivity. Approximately 5% of the total incoming solar radiation is absorbed above 40 km. Under clear sky conditions, another 10–15% is absorbed by the lower atmosphere or scattered back to space by the atmospheric aerosols and molecules; as a result, only 80–85% of the incoming radiation reaches the earth's surface. With average cloudiness, only about 50% of the incoming radiation reaches the earth's surface, because of the additional interference of the clouds.

C. Albedo and Angle of Incidence

The portion of the incoming radiation reflected and scattered back to space is the *albedo.* The albedo of clouds, snow, and ice-covered surfaces is around

TABLE 5.1

Percent of Incident Radiation Reflected by a Water Surface (Albedo of Water)[a]

Angle of incidence	Percent reflected	Percent absorbed
90	2.0	98.0
70	2.1	97.9
50	2.5	97.5
40	3.4	96.6
30	6.0	94.0
20	13.0	87.0
10	35.0	65.0
5	58.0	42.0

[a] Adapted from Fig. 3.13 of Battan [1].

0.5–0.8, that of fields and forests is 0.03–0.3, and that of water is 0.02–0.05 except when the angle of incidence becomes nearly parallel to the water surface. Table 5.1 shows the albedo of a water surface as a function of the angle of incidence. The albedo averaged over the earth's surface is about 0.35.

Although events taking place on the sun, such as sunspots and solar flares, alter the amount of radiation, the alteration is almost entirely in the X-ray and ultraviolet regions and does not affect the amount in the wavelengths reaching the earth's surface. Therefore, the amount of radiation from the sun that can penetrate to the earth's surface is remarkably constant.

In addition to the effect of albedo on the amount of radiation that reaches the earth's surface, the angle of incidence of the radiation compared to the perpendicular to the surface affects the amount of radiation flux on an area. The flux on a horizontal surface S_h is as follows:

$$S_h = S \cos Z \tag{5.4}$$

where S is the flux through an area normal to the solar beam and Z is the zenith angle (between the local vertical, the zenith, and the solar beam).

Because of the tilt of the earth's axis by 23.5° with respect to the plane of the earth's revolution around the sun, the north pole is tilted toward the sun on June 22 and away from the sun on December 21 (Fig. 5.2). This tilt causes the solar beam to have perpendicular incidence at different latitudes depending on the date. The zenith angle Z is determined from:

$$\cos Z = \sin \phi \sin \delta + \cos \phi \cos \delta \cos \eta \tag{5.5}$$

where ϕ is latitude (positive for Northern Hemisphere, negative for Southern Hemisphere), δ is solar declination (see Table 5.2), and η is hour angle, 15° × the number of hours before or after local noon.

The solar azimuth ω is the angle between south and the direction toward the sun in a horizontal plane:

$$\sin \omega = (\cos \delta \sin \eta)/\sin Z \tag{5.6}$$

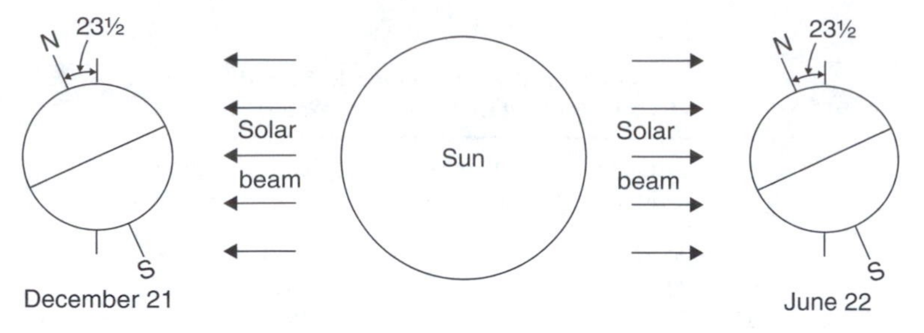

Fig. 5.2. Orientation of the earth to the solar beam at the extremes of its revolution around the sun.

TABLE 5.2

Solar Declination[a]

Date	Declination degree	Date	Declination degree
January 21	−20.90	July 21	20.50
February 21	−10.83	August 21	12.38
March 21	0.0	September 21	1.02
April 21	11.58	October 21	−10.42
May 21	20.03	November 21	−19.75
June 21	23.45	December 21	−23.43

[a] Adapted from Table 2.1 of Byers [2].

Since many surfaces receiving sunlight are not horizontal, a slope at an angle i from the horizontal facing an azimuth ω' degrees from south experiences an intensity of sunlight (neglecting the effects of the atmosphere) of

$$S_s = S[\cos Z \cos i + \sin Z \sin i \cos(\omega - \omega')] \tag{5.7}$$

Here ω and ω' are negative to the east of south and positive to the west.

At angles away from the zenith, solar radiation must penetrate a greater thickness of the atmosphere. Consequently, it can encounter more scattering due to the presence of particles and greater absorption due to this greater thickness.

D. Outgoing Longwave Radiation

Because most ultraviolet radiation is absorbed from the solar spectrum and does not reach the earth's surface, the peak of the solar radiation which reaches the earth's surface is in the visible part of the spectrum. The earth reradiates nearly as a blackbody at a mean temperature of 290 K. The resulting infrared radiation extends over wavelengths of 3–80 μm, with a peak at around 11 μm. The atmosphere absorbs and reemits this longwave radiation primarily because of water vapor but also because of carbon dioxide in the atmosphere. Because of the absorption spectrum of these gases, the atmosphere is mostly opaque to wavelengths less than 7 μm and greater than 14 μm and partly opaque between 7 and 8.5 μm and between 11 and 14 μm. The atmosphere loses heat to space directly through the nearly transparent window between 8.5 and 11 μm and also through the absorption and successive reradiation by layers of the atmosphere containing these absorbing gases.

Different areas of the earth's surface react quite differently to heating by the sun. For example, although a sandy surface reaches fairly high temperatures on a sunny day, the heat capacity and conductivity of sand are relatively low; the heat does not penetrate more than about 0.2–0.3 m and little heat is stored. In contrast, in a body of water, the sun's rays penetrate several meters and slowly heat a fairly deep layer. In addition, the water can move readily and convection can spread the heat through a deeper layer. The heat capacity of water is considerably greater than that of sand. All these factors combine to allow considerable storage of heat in water bodies.

E. Heat Balance

Because of the solar beam's more direct angle of incidence in equatorial regions, considerably more radiation penetrates and is stored by water near the equator than water nearer the poles. This excess is not compensated for by the outgoing longwave radiation, yet there is no continual buildup of heat in equatorial regions. The first law of thermodynamics requires that the energy entering the system (earth's atmosphere) be balanced with that exiting the system. Figure 5.3 shows the annual mean incoming and outgoing radiation averaged over latitude bands. There is a transfer of heat poleward from the equatorial regions to make up for a net outward transfer of heat

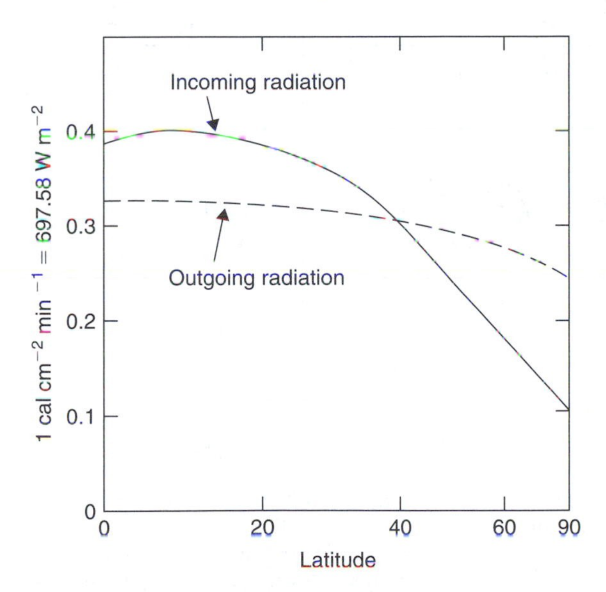

Fig. 5.3. Annual mean radiation by latitude. Note that the latitude scale simulates the amount of the earth's surface area between the latitude bands. Incoming radiation is that absorbed by earth and atmosphere. Outgoing radiation is that leaving the atmosphere. *Source*: After Byers [2] (Note: 1 cal cm^{-2} min^{-1} = 697.58 W m^{-2}).

near the poles. This heat is transferred by air and ocean currents as warm currents move poleward and cool currents move equatorward. Considerable heat transfer occurs by the evaporation of water in the tropics and its condensation into droplets farther poleward, with the release of the heat of condensation. Enough heat is transferred to result in no net heating of the equatorial regions or cooling of the poles. The poleward flux of heat across various latitudes is shown in Table 5.3.

Taking the earth as a whole over a year or longer, because there is no appreciable heating or cooling, there is a heat balance between the incoming solar radiation and the radiation escaping to space. This balance is depicted as bands of frequency of electromagnetic radiation in Fig. 5.4.

TABLE 5.3

Poleward Flux of Heat Across Latitudes (10^{19} kcal year^{-1})[a]

Latitude	Flux	Latitude	Flux
10	1.21	50	3.40
20	2.54	60	2.40
30	3.56	70	1.25
40	3.91	80	0.35

[a] Adapted from Table 12 of Sellers [3].

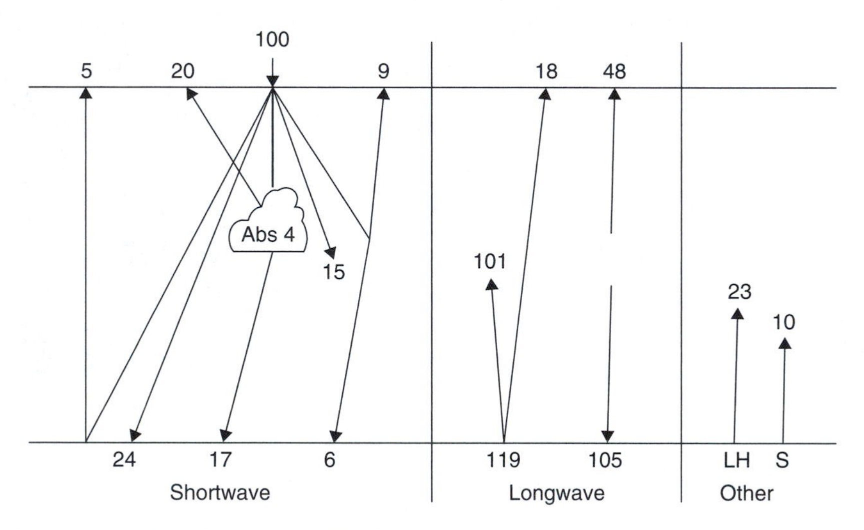

Fig. 5.4. Radiation heat balance. The 100 units of incoming shortwave radiation are distributed: reflected from earth's surface to space, 5; reflected from cloud surfaces to space, 20; direct reaching earth, 24; absorbed in clouds, 4; diffuse reaching earth through clouds, 17; absorbed in atmosphere, 15; scattered to space, 9; scattered to earth, 6. The longwave radiation comes from (1) the earth radiating 119 units: 101 to the atmosphere and 18 directly to space, and (2) the atmosphere radiating 105 units back to earth and 48 to space. Additional transfers from the earth's surface to the atmosphere consist of latent heat (LH), 23; and sensible (S) heat, 10. *Source*: After Lowry [4].

II. MOTION

Vertical air motions affect both weather and the mixing processes of importance to air pollution. Upward vertical motions can be caused by lifting over terrain, lifting over weather fronts, and convergence toward low-pressure centers. Downward vertical motions can be caused by sinking to make up for divergence near high-pressure centers. One must know whether the atmosphere enhances or suppresses these vertical motions to assess their effects. When the atmosphere resists vertical motions, it is called *stable*; when the atmosphere enhances vertical motions, it is called *unstable* or in a state of *instability*.

In incompressible fluids, such as water, the vertical structure of temperature very simply reveals the stability of the fluid. When the lower layer is warmer and thus less dense than the upper layer, the fluid is unstable and convective currents will cause it to overturn. When the lower layer is cooler than the upper layer, the fluid is stable and vertical exchange is minimal. However, because air is compressible, the determination of stability is somewhat more complicated. The temperature and density of the atmosphere normally decrease with elevation; density is also affected by moisture in the air.

The relationship between pressure p, volume V, mass m, and temperature T is given by the equation of state:

$$pV = RmT \tag{5.8}$$

where R is a specific gas constant equal to the universal gas constant divided by the gram molecular weight of the gas. Since the density ρ is m/V, the equation can be rewritten as

$$p = R\rho T \tag{5.9}$$

or considering specific volume $\alpha = 1/\rho$ as

$$\alpha p = RT \tag{5.10}$$

These equations combine Boyle's law, which states that when temperature is held constant the volume varies inversely with the pressure, and the law of Guy-Lussac, which states that when pressure is held constant the volume varies in proportion to the absolute temperature.

A. First Law of Thermodynamics

If a volume of air is held constant and a small amount of heat Δh is added, the temperature of the air will increase by a small amount ΔT. This can be expressed as

$$\Delta h = c_{\mathrm{v}}\, \Delta T \tag{5.11}$$

where c_{v} is the specific heat at constant volume. In this case, all the heat added is used to increase the internal energy of the volume affected by the temperature. From the equation of state (Eq. 5.8), it can be seen that the pressure will increase.

If, instead of being restricted, the volume of air considered is allowed to remain at an equilibrium constant pressure and expand in volume, as well as change temperature in response to the addition of heat, this can be expressed as

$$\Delta h = c_v \, \Delta T + p \, \Delta v \tag{5.12}$$

By using the equation of state, the volume change can be replaced by a corresponding pressure change:

$$\Delta h = c_p \Delta T + v \, \Delta p \tag{5.13}$$

where c_p is the specific heat at constant pressure and equals $c_v + R_d$, where R_d is the gas constant for dry air.

B. Adiabatic Processes

An adiabatic process is one with no loss or gain of heat to a volume of air. If heat is supplied or withdrawn, the process is *diabatic* or *nonadiabatic*. Near the earth's surface, where heat is exchanged between the earth and the air, the processes are diabatic.

However, away from the surface, processes frequently are adiabatic. For example, if a volume (parcel) of air is forced upward over a ridge, the upward-moving air will encounter decreased atmospheric pressure and will expand and cool. If the air is not saturated with water vapor, the process is called *dry adiabatic*. Since no heat is added or subtracted, Δh in Eq. (5.13) can be set equal to zero, and introducing the hydrostatic equation

$$-\Delta p = \rho g \, \Delta z \tag{5.14}$$

and combining equations results in

$$-\Delta T / \Delta z = g / c_p \tag{5.15}$$

Thus air cools as it rises and warms as it descends. Since we have assumed an adiabatic process, $-\Delta T/\Delta z$ defines γ_d, the dry adiabatic process lapse rate, a constant equal to 0.0098 K/m, is nearly 1 K/100 m or 5.4°F/1000 ft.

If an ascending air parcel reaches saturation, the addition of latent heat from condensing moisture will partially overcome the cooling due to expansion. Therefore, the saturated adiabatic lapse rate (of cooling) γ_w is smaller than γ_d.

C. Determining Stability

By comparing the density changes undergone by a rising or descending parcel of air with the density of the surrounding environment, the enhancement or suppression of the vertical motion can be determined. Since pressure decreases with height, there is an upward-directed pressure gradient force. The force of gravity is downward. The difference between these two forces is the buoyancy force. Using Newton's second law of motion, which indicates that a net force equals an acceleration, the acceleration a of an air parcel at a particular position is given by

$$a = g(T_p - T_e)/T_p \tag{5.16}$$

where g is the acceleration due to gravity (9.8 m s^{-2}), T_p is the temperature of an air parcel that has undergone a temperature change according to the process lapse rate, and T_e is the temperature of the surrounding environment at the same height. (Temperatures are expressed in degrees Kelvin.)

Figure 5.5 shows the temperature change undergone by a parcel of air forced to rise 200 m in ascending a ridge. Assuming that the air is dry, and therefore that no condensation occurred, this figure also represents the warming of the air parcel if the flow is reversed so that the parcel moves downslope from B to A.

Comparing the temperature of this parcel to that of the surrounding environment (Fig. 5.6), it is seen that in rising from 100 to 300 m, the parcel

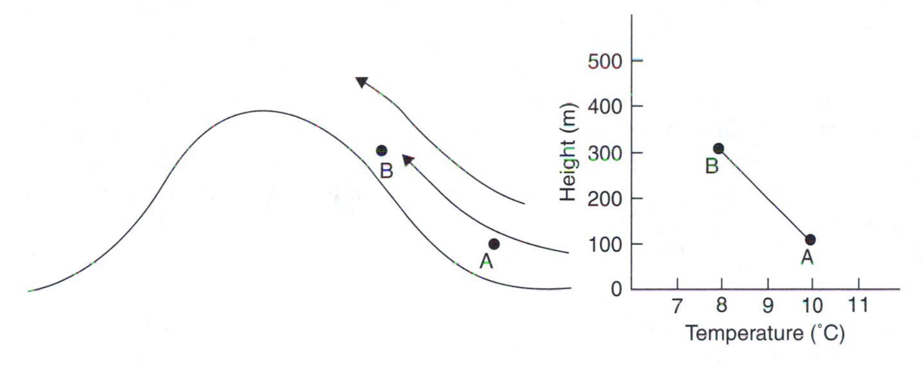

Fig. 5.5. Cooling of ascending air. Dry air forced to rise 200 m over a ridge cools adiabatically by 2°C.

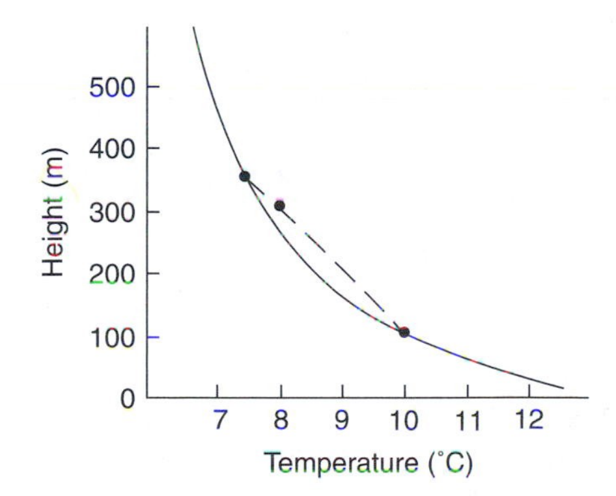

Fig. 5.6. Temperature of a parcel of air forced to rise 200 m compared to the superadiabatic environmental lapse rate. Since the parcel is still warmer than the environment, it will continue to rise.

undergoes the temperature change of the dry adiabatic process lapse rate. The dashed line is a dry adiabatic line or dry adiabat. Suppose that the environmental temperature structure is shown by the solid curve. Since the lapse rate of the surrounding environment in the lowest 150–200 m is steeper than the adiabatic lapse rate (superadiabatic)—that is, since the temperature drops more rapidly with height—this part of the environment is thermally unstable. At 300 m the parcel is 0.2°C warmer than the environment, the resulting acceleration is upward, and the atmosphere is enhancing the vertical motion and is unstable. The parcel of air continues to rise until it reaches 350 m, where its temperature is the same as that of the environment and its acceleration drops to zero. However, above 350 m the lapse rate of the surrounding environment is not as steep as the adiabatic lapse rate (subadiabatic), and this part of the environment is thermally stable (it resists upward or downward motion).

If the temperature structure, instead of being that of Fig. 5.6, differs primarily in the lower layers, it resembles Fig. 5.7, where a temperature inversion (an increase rather than a decrease of temperature with height) exists. In the forced ascent of the air parcel up the slope, dry adiabatic cooling produces parcel temperatures that are everywhere cooler than the environment; acceleration is downward, resisting displacement; and the atmosphere is stable.

Thermodynamic diagrams which show the relationships between atmospheric pressure (rather than altitude), temperature, dry adiabatic lapse rates, and moist adiabatic lapse rates are useful for numerous atmospheric thermodynamic estimations. The student is referred to a standard text on meteorology (see Suggested Reading) for details. In air pollution meteorology, the thermodynamic diagram may be used to determine the current mixing height (the top of the neutral or unstable layer). The mixing height at a given time may be estimated by use of the morning radiosonde ascent plotted on a thermodynamic chart. The surface temperature at the given time is plotted on the diagram. If a dry adiabat is drawn through this temperature, the height

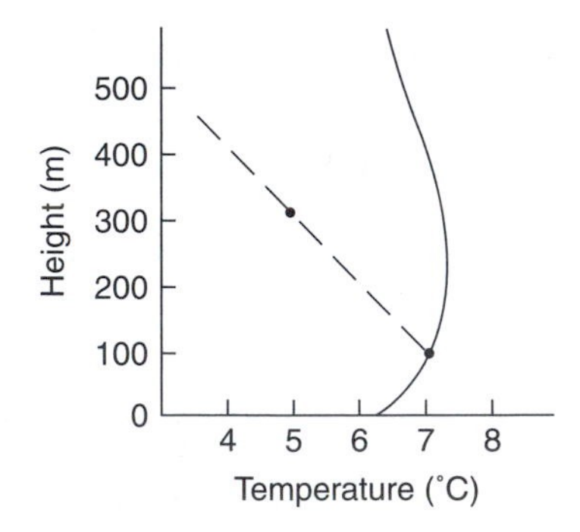

Fig. 5.7. Temperature of a parcel of air forced to rise 200 m compared to an inversion environmental lapse rate. Since the parcel is cooler than the environment, it will sink back to its original level.

aboveground at the point where this dry adiabat intersects the morning sounding is the mixing height for that time. The mixing height for the time of maximum temperature is the maximum mixing height. Use of this sounding procedure provides an approximation because it assumes that there has been no significant advection since the time of the sounding.

D. Potential Temperature

A useful concept in determining stability in the atmosphere is *potential temperature*. This is a means of identifying the dry adiabat to which a particular atmospheric combination of temperature and pressure is related. The potential temperature θ is found from:

$$\theta = T(1000/p)^{0.288} \tag{5.17}$$

where T is temperature and p is pressure (in millibars, mb). This value is the same as the temperature that a parcel of dry air would have if brought dry adiabatically to a pressure of 1000 mb.

If the potential temperature decreases with height, the atmosphere is unstable. If the potential temperature increases with height, the atmosphere is stable. The average lapse rate of the atmosphere is about 6.5°C km^{-1}; that is, the potential temperature increases with height and the average state of the atmosphere is stable.

E. Effect of Mixing

The mixing of air in a vertical layer produces constant potential temperature throughout the layer. Such mixing is usually mechanical, such as air movement over a rough surface. In Fig. 5.8 the initial temperature structure is subadiabatic (solid line). The effect of mixing is to achieve a mean potential temperature throughout the layer (dashed line), which in the lower part

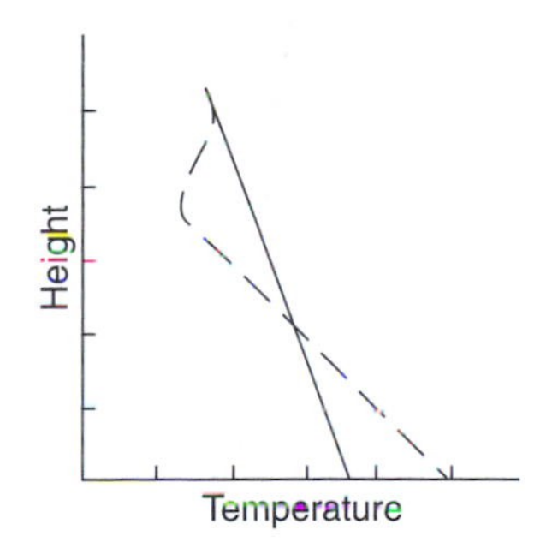

Fig. 5.8. Effect of forced mixing (dashed line) on the environmental subadiabatic lapse rate (solid line). Note the formation of an inversion at the top of the mixed layer.

is dry adiabatic. The bottom part of the layer is warmed; the top is cooled. Note that above the vertical extent of the mixing, an inversion is formed connecting the new cooled portion with the old temperature structure above the zone of mixing. If the initial layer has considerable moisture, although not saturated, cooling in the top portion of the layer may decrease the temperature to the point where some of the moisture condenses, forming clouds at the top. An example of this is the formation of an inversion and a layer of stratus clouds along the California coast.

F. Radiation or Nocturnal Inversions

An inversion caused by mixing in a surface layer was just discussed above. Inversions at the surface are caused frequently at night by radiational cooling of the ground, which in turn cools the air near it.

G. Subsidence Inversions

There is usually some descent (subsidence) of air above surface high-pressure systems. This air warms dry adiabatically as it descends, decreasing the relative humidity and dissipating any clouds in the layer. A subsidence inversion forms as a result of this sinking. Since the descending air compresses as it encounters the increased pressures lower in the atmosphere, the top portion of the descending layer will be further warmed due to its greater descent than will the bottom portion of the layer (Fig. 5.9). Occasionally a subsidence inversion descends all the way to the surface, but usually its base is well above the ground.

Inversions are of considerable interest in relation to air pollution because of their stabilizing influence on the atmosphere, which suppresses the vertical motion that causes the vertical spreading of pollutants.

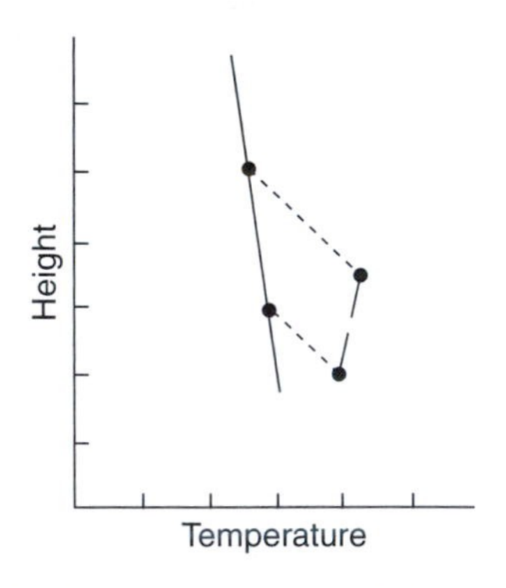

Fig. 5.9. Formation of a subsidence inversion in subsiding (sinking) air. Note the vertical compression of the sinking layer which is usually accompanied by horizontal divergence.

III. ENERGY-MOTION RELATIONSHIPS

The atmosphere is nearly always in motion. The scales and magnitude of these motions extend over a wide range. Although vertical motions certainly occur in the atmosphere and are important to both weather processes and the movement of pollutants, it is convenient to consider wind as only the horizontal component of velocity.

On the regional scale (hundreds to thousands of kilometers), the winds are most easily understood by considering the balance of various forces in the atmosphere. The applicable physical law is Newton's second law of motion, $F = ma$; if a force F is exerted on a mass m, the resulting acceleration a equals the force divided by the mass. This can also be stated as the rate of change of momentum of a body, which is equal to the sum of the forces that act on the body. It should be noted that all the forces to be discussed are vectors; that is, they have both magnitude and direction. Although Newton's second law applies to absolute motion, it is most convenient to consider wind relative to the earth's surface. These create some slight difficulties, but they can be rather easily managed.

A. Pressure Gradient Force

Three forces of importance to horizontal motion are the pressure gradient force, gravity, and friction. Atmospheric pressure equals mass times the acceleration of gravity. Considering a unit volume, $p = \rho g$; the gravitational force on the unit volume is directed downward. Primarily because of horizontal temperature gradients, there are horizontal density gradients and consequently horizontal pressure gradients. The horizontal pressure gradient force $p_{\mathrm{h}} = \Delta p / \rho \Delta x$, where Δp is the horizontal pressure difference over the distance Δx. The direction of this force and of the pressure difference measurement is locally perpendicular to the lines of equal pressure (isobars) and is directed from high to low pressure.

B. Coriolis Force

If the earth were not rotating, the wind would blow exclusively from high to low pressure. Close to the earth, it would be slowed by friction between the atmosphere and the earth's surface but would maintain the same direction with height. However, since the earth undergoes rotation, there is an apparent force acting on horizontal atmospheric motions when examined from a point of reference on the earth's surface. For example, consider a wind of velocity $10\,\mathrm{m\,s^{-1}}$ blowing at time 1 in the direction of the 0° longitude meridian across the north pole (Fig. 5.10). The wind in an absolute sense continues to blow in this direction for 1 h, and a parcel of air starting at the pole at time 1 travels 36 km in this period. However, since the earth turns 360° every 24 h, or $15°\,\mathrm{h^{-1}}$, it has rotated 15° in the hour and we find that at time 2 (60 min

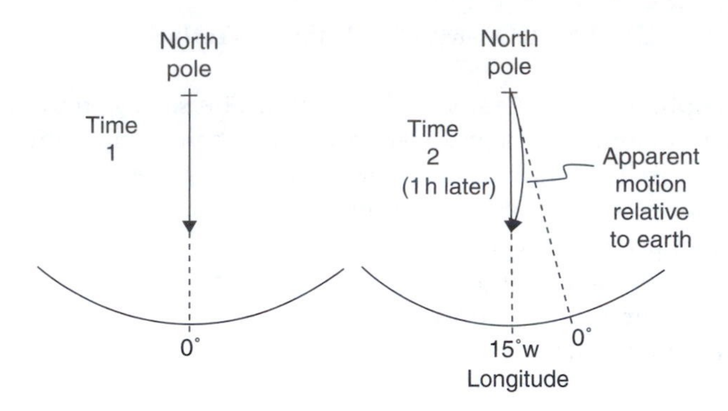

Fig. 5.10. Effect of the coriolis force. The path of air moving from the north pole to the south as viewed from space is straight; as viewed from the earth's surface it is curved.

after time 1) the 15° meridian is now beneath the wind vector. As viewed from space (the absolute frame of reference), the flow has continued in a straight line. However, as viewed from the earth, the flow has undergone an apparent deflection to the right. The force required to produce this apparent deflection is the coriolis force and is equal to $D = vf$ where f, the coriolis parameter, equals $2\Omega \sin \phi$. Here Ω is the angular speed of the earth's rotation, $2\pi/(24 \times 60 \times 60) = 7.27 \times 10^{-5}\,\mathrm{s}^{-1}$, and ϕ is the latitude. It is seen that f is maximal at the poles and zero at the equator. The deflecting force is to the right of the wind vector in the Northern Hemisphere and to the left in the Southern Hemisphere. For the present example, the deflecting force is $1.45 \times 10^{-3}\,\mathrm{ms}^{-2}$, and the amount of deflection after the 36-km movement in 1 h is 9.43 km.

C. Geostrophic Wind

Friction between the atmosphere and the earth's surface may generally be neglected at altitudes of about 700 m and higher. Therefore, large-scale air currents closely represent a balance between the pressure gradient force and the coriolis force. Since the coriolis force is at a right angle to the wind vector, when the coriolis force is equal in magnitude and opposite in direction to the pressure gradient force, a wind vector perpendicular to both of these forces occurs, with its direction along the lines of constant pressure (Fig. 5.11). In the Northern Hemisphere, the low pressure is to the left of the wind vector (Buys Ballot's law); in the Southern Hemisphere, low pressure is to the right. The geostrophic velocity is

$$v_g = -\Delta p/\rho f \,\Delta d \tag{5.18}$$

When the isobars are essentially straight, the balance between the pressure gradient force and the coriolis force results in a geostrophic wind parallel to the isobars.

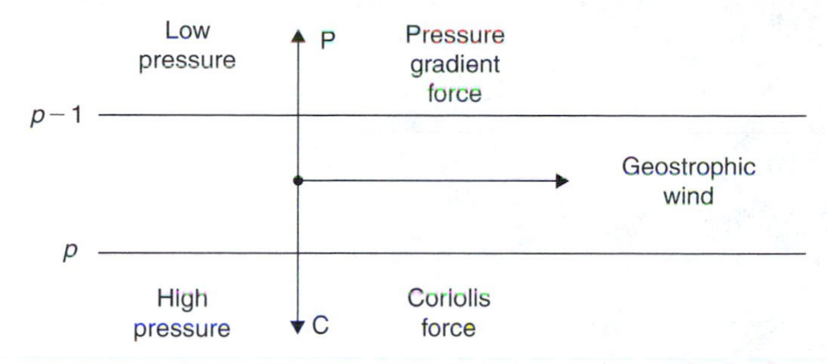

Fig. 5.11. Balance of forces resulting in geostrophic wind.

D. Gradient Wind

When the isobars are curved, an additional force, a centrifugal force outward from the center of curvature, enters into the balance of forces. In the case of curvature around low pressure, a balance of forces occurs when the pressure gradient force equals the sum of the coriolis and centrifugal forces (Fig. 5.12) and the wind continues parallel to the isobars. In the case of curvature around high pressure, a balance of forces occurs when the sum of the pressure gradient and centrifugal forces equals the coriolis force (Fig. 5.13). To maintain a given gradient wind speed, a greater pressure gradient force (tighter spacing of the isobars) is required in the flow around low-pressure systems than in the flow around high-pressure systems.

E. The Effect of Friction

The frictional effect of the earth's surface on the atmosphere increases as the earth's surface is approached from aloft. Assuming that we start with geostrophic balance aloft, consider what happens to the wind as we move downward toward the earth. The effect of friction is to slow the wind velocity, which in turn decreases the coriolis force. The wind then turns toward low pressure until the resultant vector of the frictional force and the coriolis

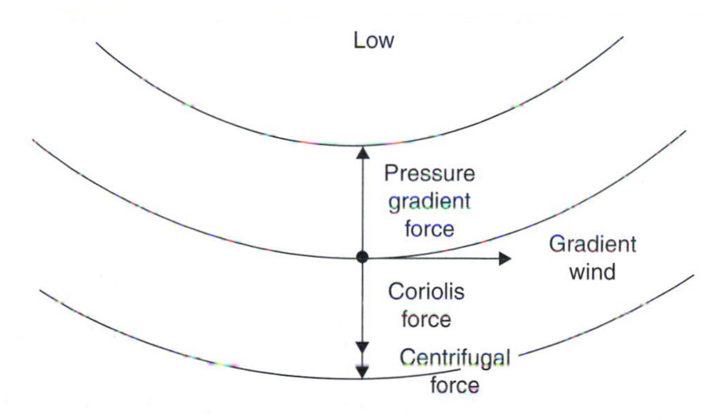

Fig. 5.12. Balance of forces resulting in gradient wind around low pressure.

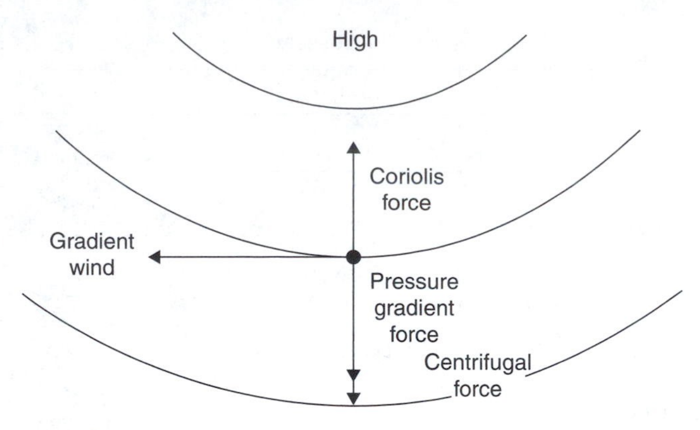

Fig. 5.13. Balance of forces resulting in gradient wind around high pressure. Note that the wind speed is greater for a given pressure gradient force than that around low pressure.

force balances the pressure gradient force (Fig. 5.14). The greater the friction, the slower the wind and the greater the amount of turning toward low pressure. The turning of the wind from the surface through the friction layer is called the *Ekman spiral*. A radial plot, or hodograph, of the winds through the friction layer is shown diagrammatically in Fig. 5.15.

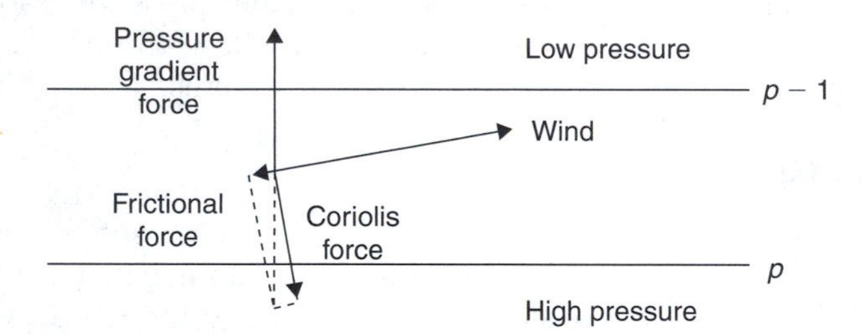

Fig. 5.14. Effect of friction on the balance of forces, causing wind to blow toward low pressure.

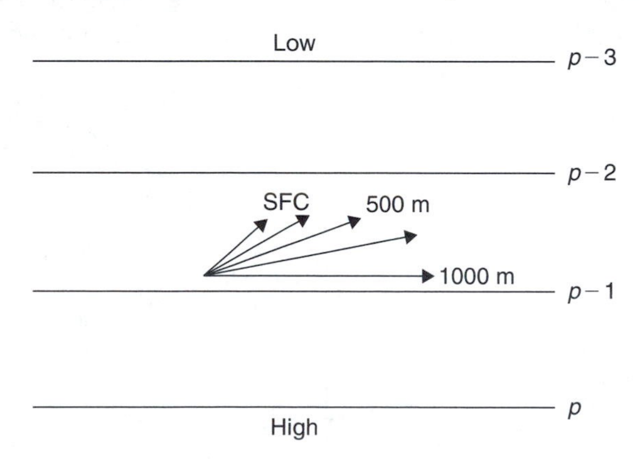

Fig. 5.15. Hodograph showing variation of wind speed and direction with height above ground. SFC: surface wind.

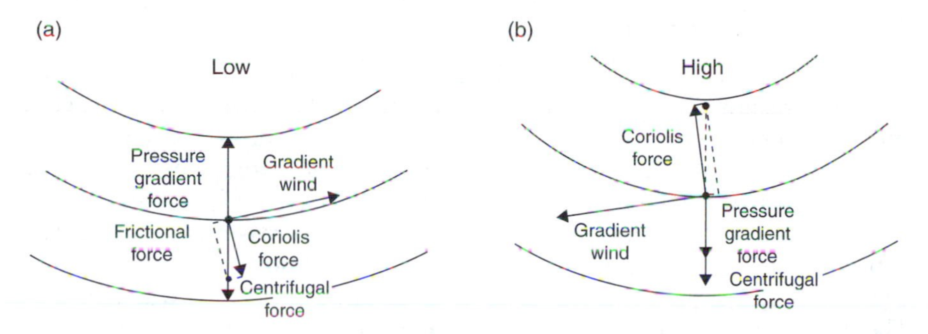

Fig. 5.16. Effect of friction upon gradient wind around (a) low and (b) high pressures.

Note that this frictional effect will cause pollutants released at two different heights to tend to move in different directions.

In the friction layer where the isobars are curved, the effect of frictional drag is added to the forces discussed under gradient wind. The balance of the pressure gradient force, the coriolis deviating force, the centrifugal force, and the frictional drag in the vicinity of the curved isobars results in wind flow around low pressure and high pressure in the Northern Hemisphere, as shown in Fig. 5.16.

F. Vertical Motion: Divergence

So far in discussing motion in the atmosphere, we have been emphasizing only horizontal motions. Although of much smaller magnitude than horizontal motions, vertical motions are important both to daily weather formation and to the transport and dispersion of pollutants.

Persistent vertical motions are linked to the horizontal motions. If there is divergence (spreading) of the horizontal flow, there is sinking (downward vertical motion) of air from above to compensate. Similarly, converging (negative divergence) horizontal air streams cause upward vertical motions, producing condensation and perhaps precipitation in most air masses, as well as transport of air and its pollutants from near the surface to higher altitudes.

IV. LOCAL WIND SYSTEMS

Frequently, local wind systems are superimposed on the larger-scale wind systems just discussed. These local flows are especially important to air pollution since they determine the amount of a pollutant that will come in contact with the receptor. In fact, local conditions may dominate when the larger-scale flow becomes light and indefinite. Local wind systems are usually quite significant in terms of the transport and dispersion of air pollutants.

A. Sea and Land Breezes

The sea breeze is a result of the differential heating of land and water surfaces by incoming solar radiation. Since solar radiation penetrates several meters of a body of water, it warms very slowly. In contrast, only the upper few centimeters of land are heated, and warming occurs rapidly in response to solar heating. Therefore, especially on clear summer days, the land surface heats rapidly, warming the air near the surface and decreasing its density. This causes the air to rise over the land, decreasing the atmospheric pressure near the surface relative to the pressure at the same altitude over the water surface. The rising air increases the pressure over the land relative to that above the water at altitudes of approximately 100–200 m. The air that rises over the land surface is replaced by cooler air from over the water surface. This air, in turn, is replaced by subsiding air from somewhat higher layers of the atmosphere over the water. Air from the higher-pressure zone several hundred meters above the surface then flows from over the land surface out over the water, completing a circular or cellular flow (Fig. 5.17). Any general flow due to large-scale pressure systems will be superimposed on the sea breeze and may either reinforce or inhibit it. Ignoring the larger-scale influences, the strength of the sea breeze will generally be a function of the temperature excess of the air above the land surface over that above the water surface.

Just as heating in the daytime occurs more quickly over land than over water, at night radiational cooling occurs more quickly over land. The pressure pattern tends to be the reverse of that in the daytime. The warmer air tends to rise over the water, which is replaced by the land breeze from land to water, with the reverse flow (water to land) completing the circular flow at altitudes somewhat aloft. Frequently at night, the temperature differences between land and water are smaller than those during the daytime, and therefore the land breeze has a lower speed.

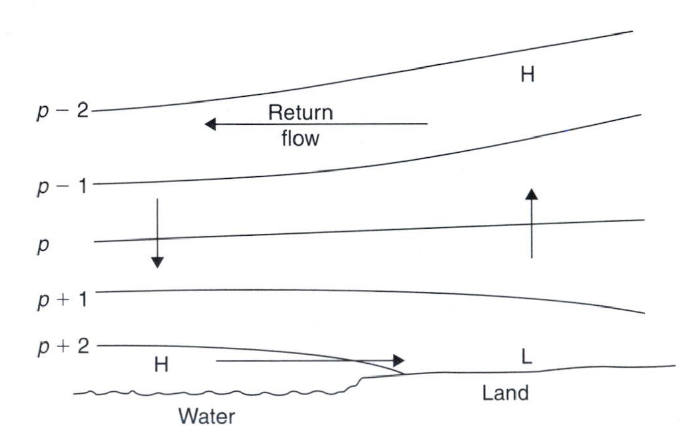

Fig. 5.17. Sea breeze due to surface heating over land, resulting in thermals, and subsidence over water.

B. Mountain and Valley Winds

Solar heating and radiational cooling influence local flows in terrain situations. Consider midday heating of a south-facing mountainside. As the slope heats, the air adjacent to the slope warms, its density is decreased, and the air attempts to ascend (Fig. 5.18). Near the top of the slope, the air tends to rise vertically. Along each portion of the slope farther down the mountain, it is easier for each rising parcel of air to move upslope, replacing the parcel ahead of it rather than rising vertically. This upslope flow is the valley wind.

At night when radiational cooling occurs on slopes, the cool dense air near the surface descends along the slope (Fig. 5.19). This is the downslope wind. To compensate for this descending air, air farther from the slope that is cooled very little is warmer relative to the descending air and rises, frequently resulting in a closed circular path. Where the downslope winds occur on opposite slopes of a valley, the cold air can accumulate on the valley floor. If there is any slope to the valley floor, this pool of cold air can move down the valley, resulting in a drainage or canyon wind.

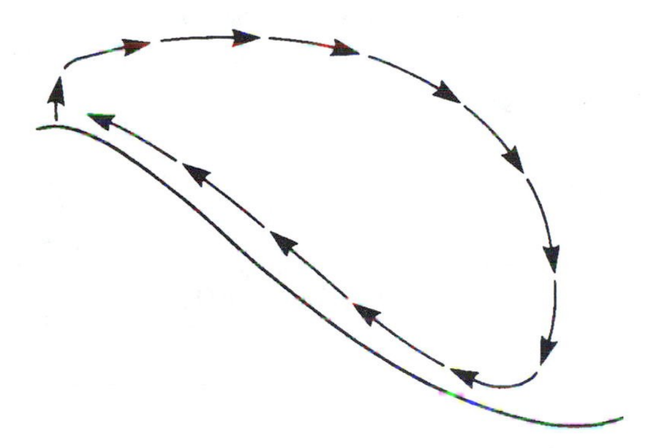

Fig. 5.18. Upslope wind (daytime) due to greater solar heating on the valley's side than in its center.

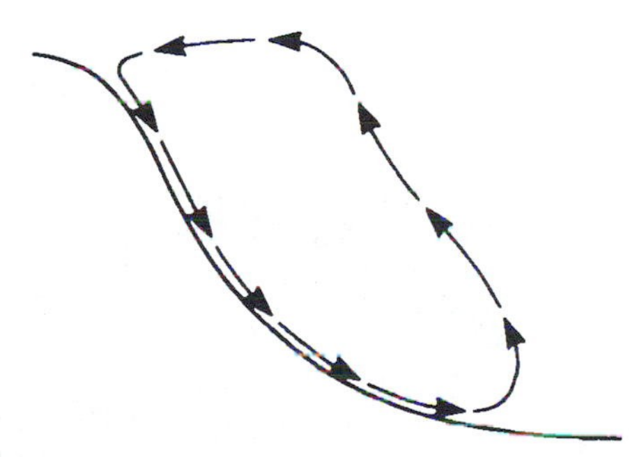

Fig. 5.19. Downslope wind (night) due to more rapid radiational cooling on the valley's slope than in its center.

Different combinations of valley and mountain slope, especially with some slopes nearly perpendicular to the incoming radiation and others in deep shadow, lead to many combinations of wind patterns, many nearly unique. Also, each local flow can be modified by the regional wind at the time which results from the current pressure patterns. Table 5.4 gives characteristics of eight different situations depending on the orientation of the ridgeline and valley with respect to the sun, wind direction perpendicular or parallel to the ridgeline, and time of day. Figure 5.20 shows examples of some of the mountain and valley winds listed in Table 5.4. These are rather idealized circulations compared to observed flows at any one time.

The effect of solar radiation is different with valley orientation. An east–west valley has only one slope that is significantly heated—the south-facing slope may be near normal with midday sunshine. A north–south valley will have both slopes heated at midday. The effect of flow in relation to valley orientation is such that flows perpendicular to valleys tend to form circular eddies and encourage local flows; flows parallel to valleys tend to discourage local flows and to sweep clean the valley, especially with stronger wind speeds.

Keep in mind that the flows occurring result from the combination of the general and local flows; the lighter the general flow, the greater the opportunity for generation of local flows.

TABLE 5.4

Generalized Mesoscale Windflow Patterns Associated with Different Combinations of Wind Direction and Ridgeline Orientation

Wind direction relative to ridgeline	Time of day	Ridgeline orientation: East–west	Ridgeline orientation: North–south
Parallel	Day	1[a] South-facing slope is heated—single helix	2 Upslope flow on both heated slopes—double helix
	Night	3 Downslope flow on both slopes—double helix	4 Downslope flow on both slopes—double helix
Perpendicular	Day	South-facing slope is heated	6 Upslope flow on both heated slopes—stationary eddy on one side of valley
		5a North wind—stationary eddy fills valley	
		5b South wind—eddy suppressed, flow without separation	
	Night	7 Indefinite flow—extreme stagnation in valley bottom	8 Indefinite flow—extreme stagnation in valley bottom

[a] Numbers refer to Fig. 5.20.

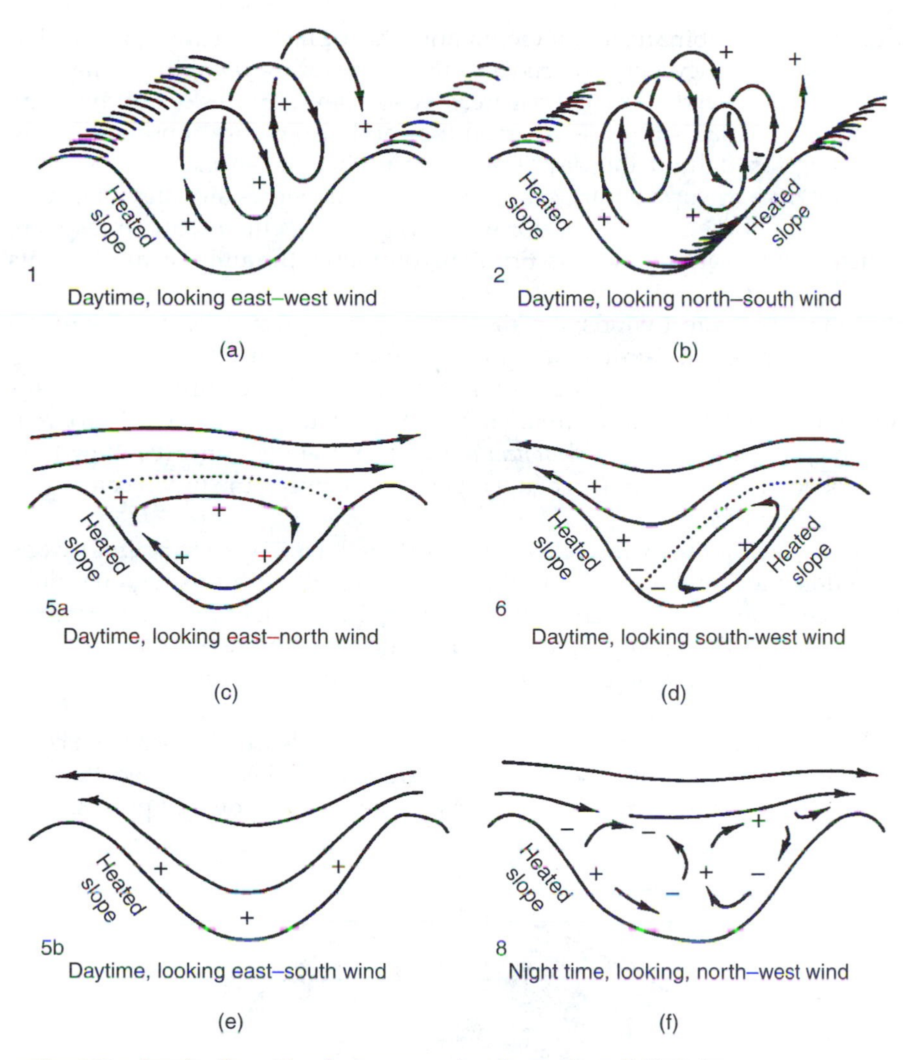

Fig. 5.20. Local valley–ridge flow patterns (numbers refer to Table 5.4).

Complicated terrain such as a major canyon with numerous side canyons will produce complicated and unique flows, especially when side canyon drainage flows reinforce the drainage flow in the main valley.

C. Urban–Rural Circulations

Urban areas have roughness and thermal characteristics different from those of their rural surroundings. Although the increased roughness affects both the vertical wind profile and the vertical temperature profile, the effects due to the thermal features are dominant. The asphalt, concrete, and steel of urban areas heat quickly and have a high heat-storing capability compared to the soil and vegetation of rural areas. Also, some surfaces of buildings are normal to the sun's rays just after sunrise and also before sunset, allowing warming throughout the day. The result is that the urban area becomes warmer than its surroundings during the day and stores sufficient heat that reradiation of the stored heat during the night keeps the urban atmosphere considerably warmer than its rural surroundings throughout most nights with light winds.

Under the lightest winds, the air rises over the warmest part of the urban core, drawing cooler air from all directions from the surroundings (Fig. 5.21). Subsidence replaces this air in rural areas, and a closed torus (doughnut)-shaped circulation occurs with an outflow above the urban area. This circulation is referred to as the *urban heat island*. The strength of the resulting flow is dependent on the difference in temperature between the urban center and its surroundings.

When the regional wind allows the outflow to take place in primarily one direction and the rising warm urban air moves off with this regional flow, the circulation is termed the *urban plume* (Fig. 5.22). Under this circumstance, the inflow to the urban center near the surface may also be asymmetric, although it is more likely to be symmetric than the outflow at higher altitudes.

The urban area also gives off heat through the release of gases from combustion and industrial processes. Compared to the heat received through solar radiation and subsequently released, the combustion and process heat

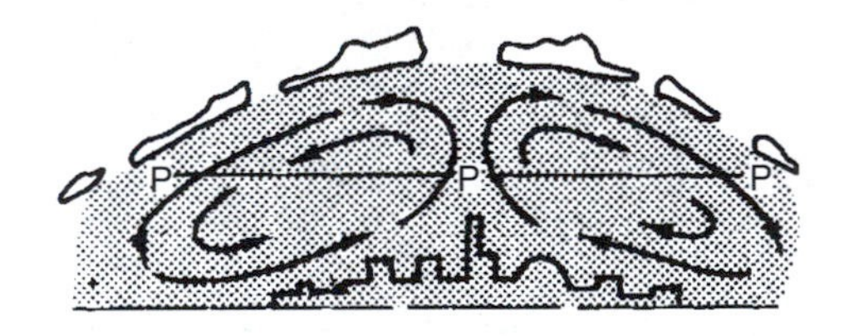

Fig. 5.21. Urban heat island (light regional wind).

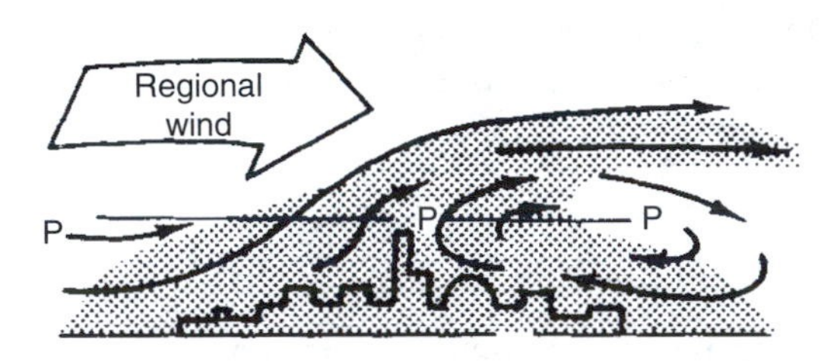

Fig. 5.22. Urban plume (moderate regional wind).

is usually quite small, although it may be 10% or more in major urban areas. It can be of significance in the vicinity of a specific local source, such as a steam power plant (where the release of heat is large over a small area) and during light-wind winter conditions.

D. Flow Around Structures

When the wind encounters objects in its path such as an isolated structure, the flow usually is strongly perturbed and a turbulent wake is formed in the vicinity of the structure, especially downwind of it. If the structure is semi-streamlined in shape, the flow may move around it with little disturbance. Since most structures have edges and corners, generation of a turbulent wake is quite common. Figure 5.23 shows schematically the flow in the vicinity of a cubic structure. The disturbed flow consists of a cavity with strong turbulence and mixing, a wake extending downwind from the cavity a distance equivalent to a number of structure side lengths, a displacement zone where flow is initially displaced before entering the wake, and a region of flow that is displaced away from the structure but does not get caught in the wake. Wind tunnels, water channels, and/or towing tanks are extremely useful in studying building wake effects.

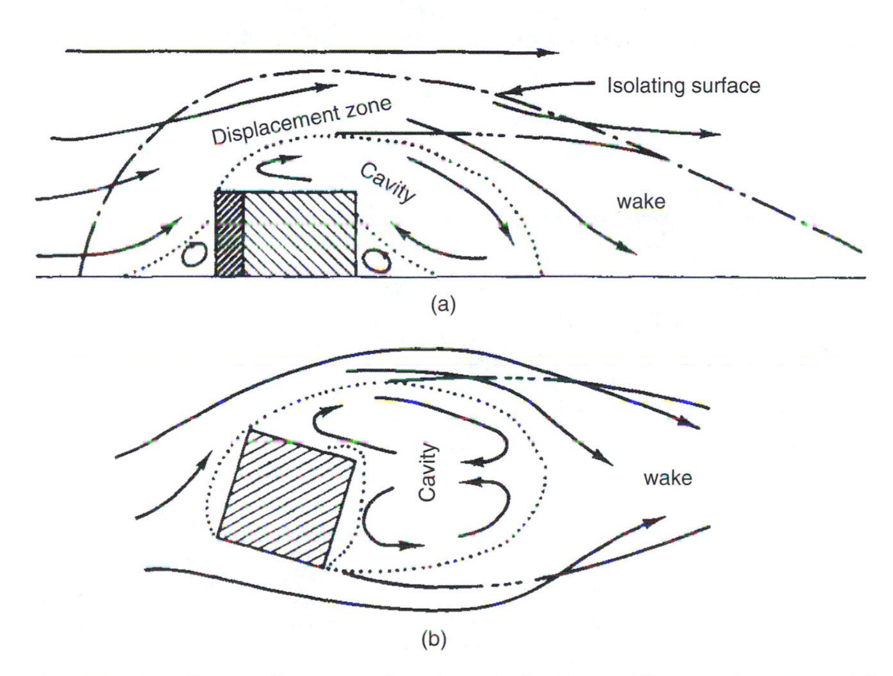

Fig. 5.23. Aerodynamic flow around a cube: (a) side view and (b) plan view. *Source*: After Halitsky [5].

V. GENERAL CIRCULATION

Atmospheric motions are driven by the heat from incoming solar radiation and the redistribution and dissipation of this heat to maintain constant temperatures on the average. The atmosphere is inefficient, because only about 2% of the received incoming solar radiation is converted to kinetic energy, that is, air motion; even this amount of energy is tremendous compared to that which humans are able to produce. As was shown in Section I, a surplus of radiant energy is received in the equatorial regions and a net outflux of energy occurs in the polar regions. Many large-scale motions serve to transport heat poleward or cooler air toward the equator.

If the earth did not rotate or if it rotated much more slowly than it does, a meridional (along meridians) circulation would take place in the troposphere (Fig. 5.24). Air would rise over the tropics, move poleward, sink over the poles forming a subsidence inversion, and then stream equatorward near the earth's surface. However, since the earth's rotation causes the apparent deflection due to the coriolis force, meridional motions are deflected to become zonal (along latitude bands) before moving more than 30°. Therefore, instead of the single cell consisting of dominantly meridional motion (Fig. 5.24), meridional transport is accomplished by three cells between the equator and the pole (Fig. 5.25). This circulation results in subsidence inversions and high pressure where there is sinking toward the earth's surface and low pressure where there is upward motion.

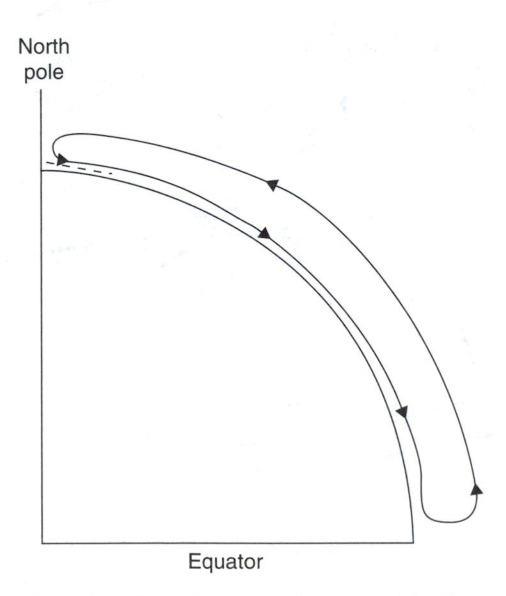

Fig. 5.24. Meridional single-cell circulation (on the sunny side of a nonrotating earth).

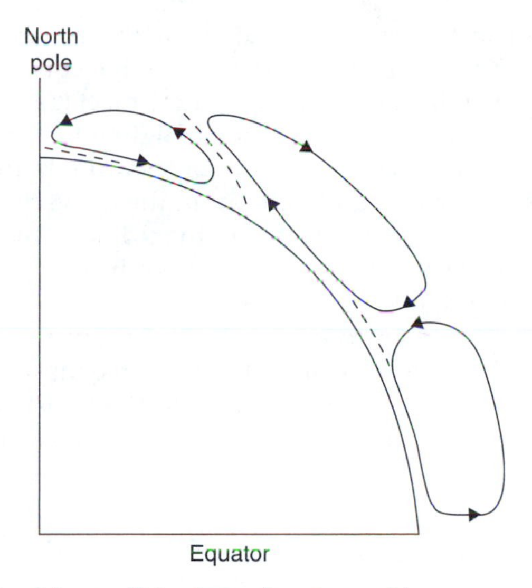

Fig. 5.25. Meridional three-cell circulation (rotating earth).

A. Tropics

Associated with the cell nearest the equator are surface winds moving toward the equator which are deflected toward the west. In the standard terminology of winds, which uses the direction from which they come, these near-surface winds are referred to as *easterlies* (Fig. 5.26), also called *trade*

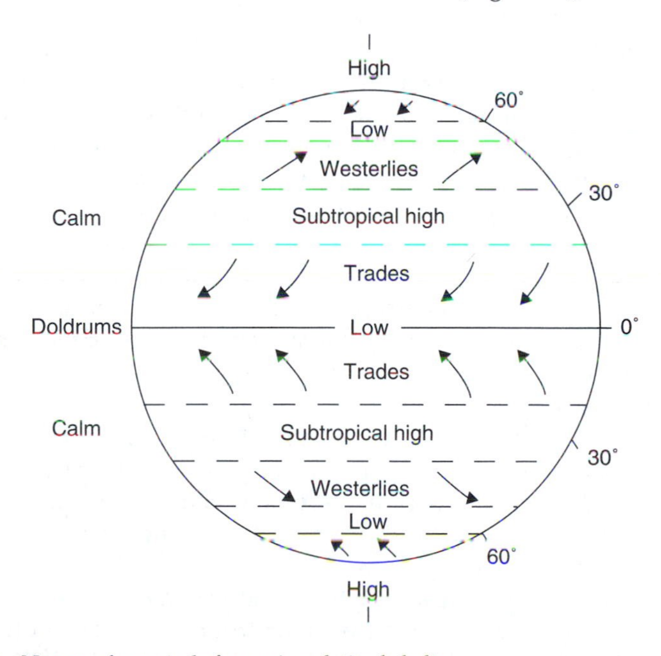

Fig. 5.26. Near-surface winds for various latitude belts.

winds. Since most of the earth's surface near the equator is ocean, these winds absorb heat and moisture on their way to the equator.

Where the trade winds from each hemisphere meet is a low-pressure zone, the *intertropical convergence zone*. This zone of light winds or doldrums shifts position with the season, moving slightly poleward into the summer hemisphere. The rising air with high humidity in the convective motions of the convergence zone causes heavy rainfall in the tropics. This giant convective cell, or Hadley cell, absorbs heat and the latent heat of evaporation at low levels, releasing the latent heat as the moisture condenses in the ascending air. Some of this heat is lost through infrared radiation from cloud tops. The subsiding air, which warms adiabatically as it descends in the vicinity of 30° latitude (horse latitudes), feeds warm air into the mid-latitudes. Although the position of the convergence zone shifts somewhat seasonally, the Hadley cell circulation is quite persistent, resulting in a fairly steady circulation.

B. Mid-latitudes

Because at higher latitudes the coriolis force deflects wind to a greater extent than in the tropics, winds become much more zonal (flow parallel to lines of latitude). Also in contrast to the persistent circulation of the tropics, the mid-latitude circulations are quite transient. There are large temperature contrasts, and temperature may vary abruptly over relatively short distances (frontal zones). In these regions of large temperature contrast, potential energy is frequently released and converted into kinetic energy as wind. Near the surface there are many closed pressure systems—cyclones and anticyclones, which are quite mobile, causing frequent changes in weather at any given location. In contrast to the systems near the earth's surface, the motions aloft (above about 3 km) have few closed centers and are mostly waves moving from west to east. The core where speeds are highest in this zonal flow is the jet stream at about 11–14 km aboveground (Fig. 5.27). Where the jet stream undergoes acceleration, divergence occurs at the altitude of the jet stream. This, in turn, promotes convergence near the surface and encourages cyclogenesis (formation of cyclonic motion). Deceleration of the jet stream conversely causes convergence aloft and subsidence near the surface, intensifying high-pressure systems. The strength of the zonal flow is determined by the zonal index, which is the difference in average pressure of two latitude circles such as 35° and 55°. A high index thus represents strong zonal flow; a low index indicates weak zonal flow. A low index is frequently accompanied by closed circulations which provide a greater degree of meridional flow. In keeping with the transient behavior of the mid-latitude circulation, the zonal index varies irregularly, cycling from low to high in periods ranging from 20 to 60 days.

The jet stream is caused by strong temperature gradients, so it is not surprising that it is frequently above the polar front, which lies in the convergence zone between the mid-latitude loop of the general circulation and the

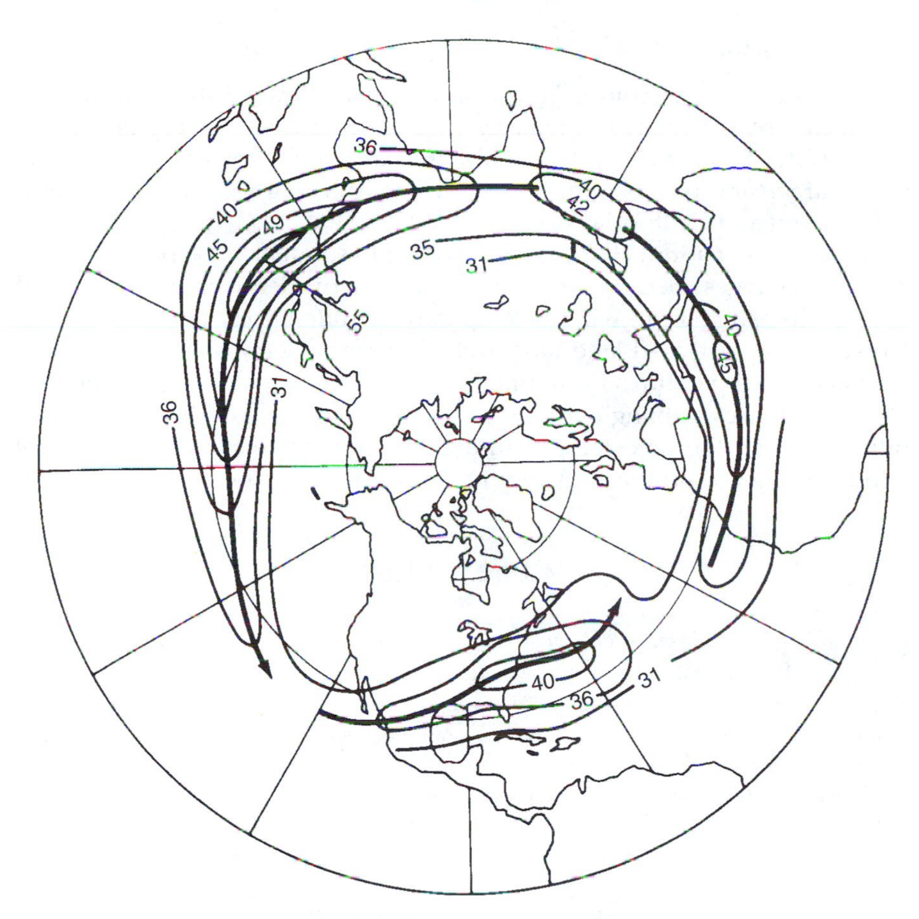

Fig. 5.27. Average position and strength of the jet stream in January between 11 and 14 km above the earth's surface (speeds are in m s^{-1}). *Source:* After Battan [1].

loop nearest the poles (Fig. 5.25). The positions of both the polar front and the jet stream are quite variable, shifting poleward with surface surges of warm air and moving toward the equator with outbreaks of cold air.

C. Polar Region

The circulation cells nearest the poles include rising air along the polar front, movement toward the poles aloft, sinking in the polar regions causing subsidence inversions, and flow toward the equator near the earth's surface. These motions contribute to the heat balance as the moisture in the air rising over the polar front condenses, releasing the heat that was used to evaporate the water nearer the equator. Also, the equatorward-moving air is cold and will be warmed as it is moved toward the tropics.

D. Other Factors

Of considerable usefulness in transporting heat toward the poles are the ocean currents. They are particularly effective because of the high heat content of water. Significant poleward-moving currents are the Brazil, Kuroshio, and Gulf Stream currents. Currents returning cold water toward the equator are the Peru and California currents.

The pressure pattern changes from winter to summer in response to temperature changes. Because most of the Southern Hemisphere consists of ocean, the summer-to-winter temperature differences are moderated. However, the increased landmass in the Northern Hemisphere allows high continental temperatures in summer, causing small equator-to-pole temperature differences; cooling over the continents in winter produces more significant equator-to-pole temperature differences, increasing the westerly winds in general and the jet stream in particular.

REFERENCES

1. Battan, L. J., "Fundamentals of Meteorology." Prentice-Hall, Englewood Cliffs, NJ, 1979.
2. Byers, H. R., "General Meteorology," 4th ed. McGraw-Hill, New York, 1974.
3. Sellers, W. D., "Physical Climatology." University of Chicago Press, Chicago, 1965.
4. Lowry, W., "Weather and Life: An Introduction to Biometeorology." Academic Press, New York, 1970.
5. Halitsky, J., Gas diffusion near buildings, *in* "Meteorology and Atomic Energy—1968" (Slade, D., ed.), TID-24190, pp. 221–255. US Atomic Energy Commission, Oak Ridge, TN, 1968.

SUGGESTED READING

Ahrens, C. D., *Meteorology Today: An Introduction to Weather, Climate, and the Environment*, Eighth Edition, Brooks/Cole, Thomson Learning Inc., 2007.

Arya, S. P., "Introduction to Micrometeorology," International Geophysics Series, Vol. 42. Academic Press, Troy, MO, 1988.

Critchfield, H. J., "General Climatology," 4th ed. Prentice-Hall, Englewood Cliffs, NJ, 1983.

Landsberg, H. E., "The Urban Climate." Academic Press, New York, 1981.

Neiburger, M., Edinger, J. G., and Bonner, W. D., "Understanding Our Atmospheric Environment." Freeman, San Francisco, CA, 1973.

Petterssen, S., "Introduction to Meteorology," 3rd ed. McGraw-Hill, New York, 1969.

Stull, R. B., "An Introduction to Boundary Layer Meteorology." Kluwer Academic Press, Hingham, MA, 1989.

Wallace, J. M., and Hobbs, P. V., "Atmospheric Science—An Introductory Survey." Academic Press, Orlando, FL, 1977.

Wanta, R. C., and Lowry, W. P., The meteorological setting for dispersal of air pollutants, *in* "Air Pollution," 3rd ed., Vol. I, "Air Pollutants, Their Transformation and Transport" (Stern, A. C., ed.). Academic Press, New York, 1976.

Yau, M. K. and Rogers, R. R. *Short Course in Cloud Physics*, Third Edition, Butterworth-Heinemann, Burlington, MA, 1989.

QUESTIONS

1. Verify the intensity of the energy flux from the sun in cal cm^{-2} min^{-1} reaching the outer atmosphere of the earth from the total solar flux of 5.6×10^{27} cal min^{-1} and the fact that the earth is 1.5×10^8 km from the sun. (The surface area of a sphere of radius r is $4\pi r^2$.)
2. Compare the difference in incoming radiation on a horizontal surface at noon on June 22 with that at noon on December 21 at a point at 23.5°N latitude.
3. What is the zenith angle at 1000 local time on May 21 at a latitude of 36°N?
4. At what local time is sunset on August 21 at 40°S latitude?
5. Show the net heating of the atmosphere, on an annual basis, by determining the difference between heat entering the atmosphere and heat radiating to the earth's surface and to space (see Fig. 5.4).
6. If the universal gas constant is 8.31×10^{-2} mb m^{-3} (g mole)$^{-1}$ K^{-1} and the gram molecular weight of dry air is 28.9, what is the mass of a cubic meter of air at a temperature of 293 K and an atmospheric pressure of 996 mb?
7. On a particular day, temperature can be considered to vary linearly with height between 28°C at 100 m aboveground and 26°C at 500 m aboveground. Do you consider the layer between 100 and 500 m aboveground to be stable or unstable?
8. What is the potential temperature of air having a temperature of 288 K at a pressure of 890 mb?
9. Using Wien's displacement law, determine the mean effective temperature of the earth–atmosphere system if the resulting longwave radiation peaks at 11 μm. Contrast the magnitude of the radiant flux at 11 μm with that at 50 μm.
10. What accompanies horizontal divergence near the earth's surface? What effect is this likely to have on the thermal stability of this layer?
11. At what time of day and under what meteorological conditions is maximum ground-level pollution likely to occur at locations several kilometers inland from a shoreline industrial complex whose pollutants are released primarily from stacks of moderate height (about 40–130 m)?
12. When the regional winds are light, at what time of day and what location might high ground-level concentrations of pollutants occur from the low-level sources (less than 20 m) of a town in a north–south-oriented valley whose floor slopes down to the north? Can you answer this question for sources releasing in the range 50–70 m aboveground?
13. Consider the air movement in the Northern Hemisphere (See Fig. 5-27). Explain how persistent organic compounds, such as polychlorinated biphenyls (PCBs) might be found in the tissues of polar bears and in human mother's milk in the arctic regions.
14. New York City is located in a complex geographic and meteorological situation. Explain how local wind systems and heat balances may account for air pollution in and around the city.

6

Air Pollution Systems and Processes

I. CHEMICAL PROCESSES IN AIR POLLUTION

A. Chemical Characteristics

Air pollution processes are a function of both the chemical characteristics of the compartment (e.g. air) and those of the contaminant. The inherent properties of the air pollutant are influenced and changed by the extrinsic properties of the air and other media in which the pollutant resides in the environment. Thus, Table 6.1 describes both sets of properties.

B. Chemical Reactions in the Environment

Five categories of chemical reactions take place in the environment or in systems that ultimately lead to contamination (such as closed systems where toxic chemicals, like pesticides, are synthesized before being used and released into the ambient, environment, or thermal systems where precursor compounds form new contaminants, like dioxins and furans). The categories of chemical reactions are as follows:

1. Synthesis or combination:

$$A + B \rightarrow AB \quad (6.1)$$

In combination reactions, two or more substances react to form a single substance. Two types of combination reactions are important in environmental systems, i.e. formation and hydration.

TABLE 6.1

Physical, Chemical, and Biological Processes Important to the Fate and Transport of Contaminants in the Environment

Process	Description	Physical phases involved	Major mechanisms at work	Outcome of process	Factors included in process
Advection	Transport by turbulent flow; mass transfer	Aqueous, gas	Mechanical	Transport due to mass transfer	Concentration gradients, porosity, permeability, hydraulic conductivity, circuitousness or tortuosity of flow paths
Dispersion	Transport from source	Aqueous, gas	Mechanical	Concentration gradient and dilution driven	Concentration gradients, porosity, permeability, hydraulic conductivity, circuitousness or tortuosity of flow paths
Diffusion	Fick's law (concentration gradient)	Aqueous, gas, solid	Mechanical	Concentration gradient-driven transport	Concentration gradients
Liquid separation	Various fluids of different densities and viscosities are separated within a system	Aqueous	Mechanical	Recalcitrance due to formation of separate gas and liquid phases (e.g. gasoline in water separates among benzene, toluene, and xylene)	Polarity, solubility, K_d, K_{ow}, K_{oc}, coefficient of viscosity, density
Density stratification	Distinct layers of differing densities and viscosities	Aqueous	Physical/Chemical	Recalcitrance or increased mobility in transport of lighter fluids (e.g. light non-aqueous phase liquids, LNAPLs) that float at water table in groundwater, or at atmospheric pressure in surface water	Density (specific gravity)

(continued)

TABLE 6.1 *(Continued)*

Process	Description	Physical phases involved	Major mechanisms at work	Outcome of process	Factors included in process
Migration along flow paths	Faster through large holes and conduits, e.g. path between interstices of sorbant packing in air stripping towers	Aqueous, gas	Mechanical	Increased mobility through fractures	Porosity, flow path diameters
Sedimentation	Heavier compounds settle first	Solid	Chemical, physical, mechanical, varying amount of biological	Recalcitrance due to deposition of denser compounds	Mass, density, viscosity, fluid velocity, turbulence (R_N)
Filtration	Retention in mesh	Solid	Chemical, physical, mechanical, varying amount of biological	Recalcitrance due to sequestration, destruction, and mechanical trapping of particles	Surface charge, soil, particle size, sorption, polarity
Volatilization	Phase partitioning to vapor	Aqueous, gas	Physical	Increased mobility as vapor phase of contaminant migrates to soil gas phase and atmosphere	Vapor pressure (P^0), concentration of contaminant, solubility, temperature
Dissolution	Co-solvation, attraction of water molecule shell	Aqueous	Chemical	Various outcomes due to formation of hydrated compounds (with varying solubilities, depending on the species)	Solubility, pH, temperature, ionic strength, activity
Absorption	Retention on solid surface	Solid	Chemical, physical, varying amount of biological	Partitioning of lipophilic compounds into soil organic matter, and penetration into an aerosol	Polarity, surface charge, Van der Waals attraction, electrostatics, ion exchange, solubility, K_d, K_{ow}, K_{oc}, coefficient of viscosity, density

Adsorption	Retention on solid surface	Solid	Chemical, physical, varying amount of biological	Recalcitrance due to ion exchanges and charge separations on a particle's surface	Polarity, surface charge, Van der Waals attraction, electrostatics, ion exchange, solubility, K_d, K_{ow}, K_{oc}, coefficient of viscosity, density
Complexation	Reactions with matrix (e.g. soil compounds like humic acid) that form covalent bonds	Solid	Chemical, varying amount of biological	Recalcitrance and transformation due to reactions with soil organic compounds to form residues (bound complexes)	Available oxidants/reductants, soil organic matter content, pH, chemical interfaces, available O_2, electrical interfaces, temperature
Oxidation/Reduction	Electron loss and gain	All	Chemical, physical, varying amount of biological	Destruction or transformation due to mineralization of simple carbohydrates to CO_2 and water from respiration of organisms	Available oxidants/reductants, soil organic matter content, pH, chemical interfaces, available O_2, electrical interfaces, temperature
Ionization	Complete co-solvation leading to separation of compound into cations and anions	Aqueous	Chemical	Dissolution of salts into ions	Solubility, pH, temperature, ionic strength, activity
Hydrolysis	Reaction of water molecules with contaminants	Aqueous	Chemical	Various outcomes due to formation of hydroxides (e.g. aluminum hydroxide) with varying solubilities, depending on the species)	Solubility, pH, temperature, ionic strength, activity
Photolysis	Reaction catalyzed by electromagnetic (EM) energy (sunlight)	Gas (major phase)	Chemical, physical	Photooxidation of compounds with hydroxyl radical upon release to the atmosphere	Free radical concentration, wavelength, and intensity of EM radiation

(*continued*)

TABLE 6.1 *(Continued)*

Process	Description	Physical phases involved	Major mechanisms at work	Outcome of process	Factors included in process
Biodegradation	Microbially mediated, enzymatically catalyzed reactions	Aqueous, solid	Chemical, biological	Various outcomes, including destruction and formation of daughter compounds (degradation products) intracellularly and extracellularly	Microbial population (count and diversity), pH, temperature, moisture, biofilm, acclimation potential of available microbes, nutrients, appropriate enzymes in microbes, available and correct electron acceptors (i.e. oxygen for aerobes, others for anaerobes)
Activation	Metabolic, detoxification process that renders a compound more toxic	Aqueous, gas, solid, tissue	Biochemical	Phase 1 or 2 metabolism, e.g. oxidation for epoxides on aromatics	Available detoxification and enzymatic processes in cells
Metal catalysis	Reactions sped up in the presence of certain metallic compounds (e.g. noble metal oxides in the degradation of nitric acid).	Aqueous, gas, solid, and biotic	Chemical (especially reduction and oxidation)	Same reaction, but faster	Species and oxidation state of metal

Formation reactions are those where elements combine to form a compound. Examples include the formation of ferric oxide and the formation of octane:

$$4\,Fe\,(s) + 3\,O_2\,(g) \rightarrow 2\,Fe_2O_3\,(s) \tag{6.2}$$

$$8\,C\,(s) + 9\,H_2\,(g) \rightarrow C_8H_{18}\,(l) \tag{6.3}$$

Hydration reactions involve the addition of water to synthesize a new compound, for example, when calcium oxide is hydrated to form calcium hydroxide, and when phosphate is hydrated to form phosphoric acid:

$$CaO\,(s) + H_2O\,(l) \rightarrow Ca\,(OH)_2\,(s) \tag{6.4}$$

$$P_2O_5\,(s) + 3\,H_2O\,(l) \rightarrow 2\,H_3PO_4\,(aq) \tag{6.5}$$

2. Decomposition (often referred to as "degradation" when discussing organic compounds in toxicology, environmental sciences, and engineering):

$$AB \rightarrow A + B \tag{6.6}$$

In decomposition, one substance breaks down into two or more new substances, such as in the decomposition of carbonates. For example, calcium carbonate breaks down into calcium oxide and carbon dioxide:

$$CaCO_3\,(s) \rightarrow CaO\,(s) + CO_2\,(g) \tag{6.7}$$

3. Single replacement (or single displacement):

$$A + BC \rightarrow AC + B \tag{6.8}$$

This commonly occurs when one metal ion in a compound is replaced with another metal ion, such as when trivalent chromium replaces monovalent silver:

$$3\,AgNO_3\,(aq) + Cr\,(s) \rightarrow Cr(NO_3)_3\,(aq) + 3\,Ag\,(s) \tag{6.9}$$

4. Double replacement (also metathesis or double displacement):

$$AB + CD \rightarrow AD + CB \tag{6.10}$$

In metathetic reactions, metals are exchanged between the salts. These newly formed salts have different chemical and physical characteristics from those of the reagents and are commonly encountered in metal precipitation reactions, such as when lead is precipitated (indicated by the "(s)" following $PbCl_2$), as in the reaction of a lead salt with an acid like potassium chloride:

$$Pb(ClO_3)_2\,(aq) + 2\,KCl\,(aq) \rightarrow PbCl_2\,(s) + 2\,KClO_3\,(aq) \tag{6.11}$$

5. Complete or efficient combustion (thermal oxidation) occurs when an organic compound is oxidized in the presence of heat (indicated by Δ):

$$a(CH)_x + bO_2 \xrightarrow{\Delta} cCO_2 + dH_2O \tag{6.12}$$

Combustion is the combination of O_2 in the presence of heat (as in burning fuel), producing CO_2 and H_2O during complete combustion of organic compounds, such as the combustion of octane:

$$C_8H_{18}\,(l) + 17\,O_2\,(g) \xrightarrow{\Delta} 8\,CO_2\,(g) + 9\,H_2O\,(g) \tag{6.13}$$

Complete combustion may also result in the production of molecular nitrogen (N_2) when nitrogen-containing organics are burned, such as in the combustion of methylamine:

$$4CH_3NH_2\,(l) + 9\,O_2\,(g) \xrightarrow{\Delta} 4\,CO_2\,(g) + 10\,H_2O\,(g) + 2\,N_2\,(g) \quad (6.14)$$

Incomplete combustion can produce a variety of compounds. Some are more toxic than the original compounds being oxidized, such as polycyclic aromatic hydrocarbons (PAHs), dioxins, furans, and CO. The alert reader will note at least two observations about these categories. First, all are kinetic, as denoted by the one-directional arrow ($\rightarrow$). Second, in the environment, many processes are incomplete, such as the common problem of incomplete combustion and the generation of new compounds in addition to carbon dioxide and water.

With respect to the first observation, indeed, many equilibrium reactions take place. However, as mentioned in previous discussions, getting to equilibrium requires a kinetic phase. So, upon reaching equilibrium, the kinetic reactions (one-way arrows) would be replaced by two-way arrows ($\leftrightarrow$). Changes in the environment or in the quantities of reactants and products can invoke a change back to kinetics.

Incomplete reactions are very important sources of environmental contaminants. For example, these reactions generate *products of incomplete combustion* (PICs), such as CO, PAHs, dioxins, furans, and hexachlorobenzene (HCB). However, even the products of complete combustion are not completely environmentally acceptable. Both carbon dioxide and water are a greenhouse gases. They are both essential to life on earth, but excessive amounts of CO_2 are strongly suspected of altering climate, especially increasing mean surface temperatures on earth. Thus, CO_2 is considered by many to be an "air pollutant."

II. AIR POLLUTION CHEMODYNAMICS

Environmental chemodynamics is concerned with how chemicals move and change in the environment. Up to this point we have discussed some of the fluid properties that have a great bearing on the movement and distribution of contaminants to, from, and within the atmosphere. However, we are now ready to consider three specific partitioning relationships that control the "leaving" and "gaining" of pollutants among compartments, especially air, particles, surfaces, and organic tissues. These concepts may be applied to estimating and modeling where a contaminant will go after it is released. These relationships are sorption, solubility, volatilization, and organic carbon–water partitioning, which are respectively expressed by coefficients of sorption (distribution coefficient, K_D, or solid–water partition coefficient, K_p), dissolution or solubility coefficients, air–water partitioning (and the Henry's Law, K_H, constant), and organic carbon–water (K_{oc}) partitioning.

In chemodynamics, the environment is subdivided into finite compartments. Thermodynamically, the mass of the contaminant entering and the mass leaving a control volume must be balanced by what remains within the control volume (ala the conservation laws). Likewise, within that control volume, each compartment may be a gainer or loser of the contaminant mass, but the overall mass must balance. The generally inclusive term for these compartmental changes is known as fugacity or the "fleeing potential" of a substance. It is the propensity of a chemical to escape from one type of environmental compartment to another. Combining the relationships between and among all of the partitioning terms is one means of modeling chemical transport in the environment.[1] This is accomplished by using thermodynamic principles and, hence, *fugacity* is a thermodynamic term.

The simplest chemodynamic approach addresses each compartment where a contaminant is found in discrete phases of air, water, soil, sediment, and biota. This can be seen graphically in Fig. 6.1. However, a complicating factor in environmental chemodynamics is that even within a single compartment, a contaminant may exist in various phases (e.g. dissolved in water and sorbed to a particle in the solid phase). The physical interactions of the contaminant at the interface between each compartment is a determining factor in the fate of the pollutant. Within a compartment, a contaminant may remain unchanged (at least during the designated study period), or it may move physically, or it may be transformed chemically into another substance. Actually, in many cases all three mechanisms will take place. A mass fraction will remain unmoved and unchanged. Another fraction remains unchanged but is transported to a different compartment. Another fraction becomes chemically transformed with all remaining products staying in the compartment where they were generated. And, a fraction of the original contaminant is transformed and then moved to another compartment. So, upon release from a source, the contaminant moves as a result of thermodynamics.

Fugacity requires that at least two phases must be in contact with the contaminant. For example, the K_{ow} value is an indication of a compound's likelihood to exist in the organic versus aqueous phase. This means that if a substance is dissolved in water and the water comes into contact with another substance, e.g. octanol, the substance will have a tendency to move from the water to the octanol. Its octanol–water partitioning coefficient reflects just how much of the substance will move until the aqueous and organic solvents (phases) will reach equilibrium. So, for example, in a spill of equal amounts of the polychlorinated biphenyl (PCB), decachlorobiphenyl

[1] Fugacity models are valuable in predicting the movement and fate of environmental contaminants within and among compartments. This discussion is based on work by one of the pioneers in this area, Don MacKay and his colleagues at the University of Toronto. See, for example, MacKay, D., and Paterson, S., Evaluating the fate of organic chemicals: a level III fugacity model. *Environ. Sci. Technol.* **25**, 427–436 (1991).

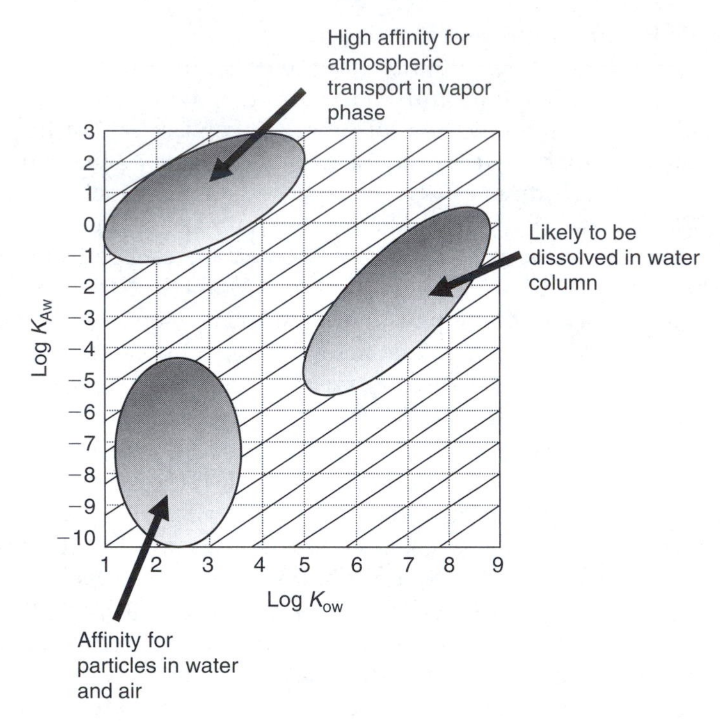

Fig. 6.1. Relationship between air–water partitioning and octanol–water partitioning and affinity of classes of contaminants for certain environmental compartments. A source of this information and format is: van de Meent, D., McKone, T., Parkerton, T., Matthies, M., Scheringer, M., Wania, F., Purdy, R., and Bennett, D., Persistence and transport potential of chemicals in a multimedia environment, in *Proceedings of the SETAC Pellston Workshop on Criteria for Persistence and Long-Range Transport of Chemicals in the Environment*, Fairmont Hot Springs, British Columbia, Canada, July 14–19, 1998. Society of Environmental Toxicology and Chemistry, Pensacola, FL, 1999.

(log K_{ow} of 8.23), and the pesticide chlordane (log K_{ow} of 2.78), the PCB has much greater affinity for the organic phases than does the chlordane (more than five orders of magnitude). This does not mean that a great amount of either of the compounds is likely to stay in the water column, since they are both hydrophobic, but it does mean that they will vary in the time and mass of each contaminant moving between phases. The rate (kinetics) is different, so the time it takes for the PCB and chlordane to reach equilibrium will be different. This can be visualized by plotting the concentration of each compound with time (see Fig. 6.2). When the concentrations plateau, the compounds are at equilibrium with their phase.

When phases contact one another, a contaminant will escape from one to another until the contaminant reaches equilibrium among the phases that are in contact with one another. Kinetics takes place until equilibrium is achieved.

We can now consider the key partitioning factors needed for a simple chemodynamic model.

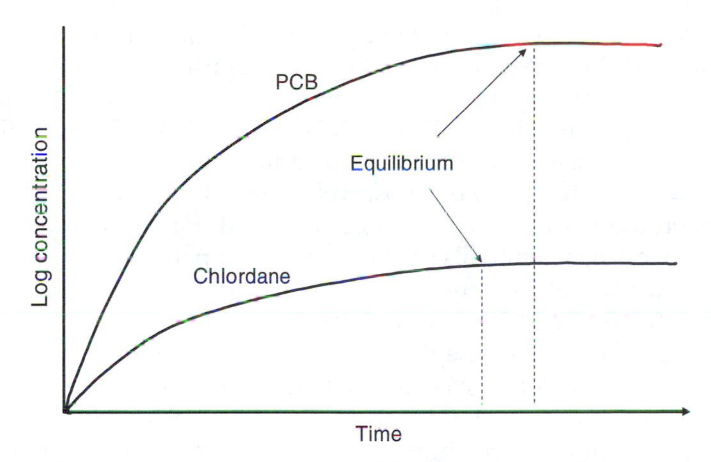

Fig. 6.2. Relative concentrations of a PCB and chlordane in octanol with time.

A. Partitioning to Solids: Sorption

Some experts spend a lot of time differentiating the various ways that a contaminant will attach to or permeate into surfaces of solid phase particles. Others, including many environmental engineers, lump these processes together into a general phenomenon called *sorption*, which is the process in which a contaminant or other solute becomes associated, physically or chemically, with a solid sorbent.

The physicochemical transfer[2] of a chemical, A, from liquid to solid phase is expressed as

$$A_{(solution)} + \text{solid} = \text{A-solid} \tag{6.15}$$

The interaction of the solute (i.e. the chemical being sorbed) with the surface of a solid surface can be complex and dependent on the properties of the chemical and the water. Other fluids are often of such small concentrations that they do not determine the ultimate solid–liquid partitioning. While, it is often acceptable to consider "net" sorption, let us consider briefly the four basic types or mechanisms of sorption:

1. *Adsorption* is the process wherein the chemical in solution attaches to a solid surface, which is a common sorption process in clay and organic constituents in soils. This simple adsorption mechanism can occur on clay particles where little carbon is available, such as in groundwater.
2. *Absorption* is the process that often occurs in porous materials so that the solute can diffuse into the particle and be sorbed onto the inside surfaces of the particle. This commonly results from short-range electrostatic interactions between the surface and the contaminant.

[2] Lyman, W., Transport and transformation processes, in *Fundamentals of Aquatic Toxicology: Effects, Environmental Fate, and Risk Assessment* (Rand, G., ed.), 2nd ed., Chapter 15. Taylor & Francis, Washington, DC, 1995.

3. *Chemisorption* is the process of integrating a chemical into porous materials surface via chemical reaction. In an airborne particle, this can be the result of a covalent reaction between a mineral surface and the contaminant.
4. *Ion exchange* is the process by which positively charged ions (cations) are attracted to negatively charged particle surfaces and negatively charged ions (anions) are attracted to positively charged particle surfaces, causing ions on the particle surfaces to be displaced. Particles undergoing ion exchange can include soils, sediment, airborne particulate matter, or even biota, such as pollen particles. Cation exchange has been characterized as being the second most important chemical process on earth, after photosynthesis. This is because the cation exchange capacity (CEC), and to a lesser degree anion exchange capacity (AEC) in tropical soils, is the means by which nutrients are made available to plant roots. Without this process, the atmospheric nutrients and the minerals in the soil would not come together to provide for the abundant plant life on planet earth.[3]

These four types of sorption are a mix of physics and chemistry. The first two are predominantly controlled by physical factors, and the second two are combinations of chemical reactions and physical processes. We will spend a bit more time covering these specific types of sorption when we consider the surface effects of soils. Generally, sorption reactions affect the following three processes[4] in aquatic systems:

1. The chemical contaminant's transport in water due to distributions between the aqueous phase and particles.
2. The aggregation and transport of the contaminant as a result of electrostatic properties of suspended solids.
3. Surface reactions such as dissociation, surface catalysis, and precipitation of the chemical contaminant.

Therefore, the difference between types of sorption is often important in air pollution control technologies. For example, collecting a contaminant on the surface of a sorbant depends solely on adsorption, while applying a control technology that traps the pollutant within the sorbant makes uses of absorption. Many technologies make use of both, as well as cation and anion exchange and chemisorption (see discussion below). Researchers attempt to parcel out which of the mechanisms are prominent in a given situation, since they vary according to the contaminant and the substrate.

When a contaminant enters the soil, some of the chemical remains in soil solution and some of the chemical is adsorbed onto the surfaces of the soil particles. Sometimes this sorption is strong due to cations adsorbing to the negatively charged soil particles. In other cases the attraction is weak. Sorption of

[3] Professor Daniel Richter of Duke University's Nicholas School of the Environment has waxed eloquently on this subject.

[4] See Westfall, J., Adsorption mechanisms in aquatic surface chemistry, in *Aquatic Surface Chemistry*. Wiley-Interscience, New York, NY, 1987.

chemicals on solid surfaces needs to be understood because they hold onto contaminants, not allowing them to move freely with the pore water or the soil solution. Therefore sorption slows that rate at which contaminants move downwardly through the soil profile.

Contaminants will eventually establish a balance between the mass on the solid surfaces and the mass that is in solution. Recall that molecules will migrate from one phase to another to maintain this balance. The properties of both the contaminant and the soil (or other matrix) will determine how and at what rates the molecules partition into the solid and liquid phases. These physicochemical relationships, known as *sorption isotherms*, are found experimentally. Figure 6.3 shows three isotherms for pyrene from experiments using different soils and sediments.

The x-axis shows the concentration of pyrene dissolved in water, and the y-axis shows the concentration in the solid phase. Each line represents the relationship between these concentrations for a single soil or sediment. A straight-line segment through the origin represents the data well for the range of concentrations shown. Not all portions of an isotherm are linear, particularly at high concentrations of the contaminant. Linear chemical partitioning can be expressed as

$$S = K_D C_W \tag{6.16}$$

where S is the concentration of contaminant in the solid phase (mass of solute per mass of soil or sediment); C_W is the concentration of contaminant in the liquid phase (mass of solute per volume of pore water); and K_D is the

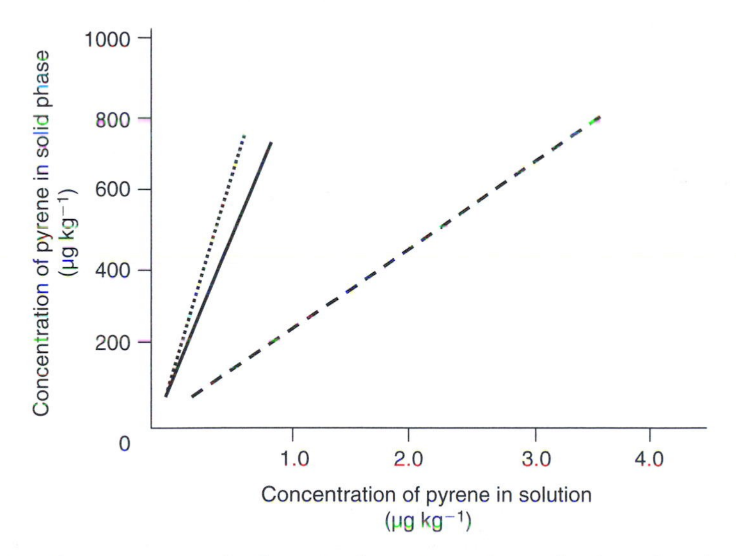

Fig. 6.3. Three experimentally determined sorption isotherms for the, polycyclic aromatic hydrocarbon, pyrene. *Source*: Hassett, J., and Banwart, W., 1989, The sorption of nonpolar organics by soils and sediments, in *Reactions and Movement of Organic Chemicals in Soils* (Sawhney, B., and Brown, K., eds.), p. 35. Soil Science Society of America. Special Publication 22.

partition coefficient (volume of pore water per mass of soil or sediment) for this contaminant in this soil or sediment.

For many soils and chemicals, the partition coefficient can be estimated using

$$K_D = K_{OC}\,OC \tag{6.17}$$

where K_{OC} is the organic carbon partition coefficient (volume of pore water per mass of organic carbon) and OC is the soil organic matter (mass of organic carbon per mass of soil).

This relationship is a very useful tool for estimating K_D from the known K_{OC} of the contaminant and the organic carbon content of the soil horizon of interest. The actual derivation of K_D is

$$K_D = C_S\,(C_W)^{-1} \tag{6.18}$$

where C_S is the equilibrium concentration of the solute in the solid phase and C_W is the equilibrium concentration of the solute in the water.

Therefore, K_D is a direct expression of the partitioning between the aqueous and solid (e.g. particle) phases. A strongly sorbed chemical like a dioxin or the banned pesticide dichlorodiphenyltrichloroethane (DDT) can have a K_D value exceeding 10^6. Conversely, a highly hydrophilic, miscible substance like ethanol, acetone, or vinyl chloride, will have K_D values less than 1. This relationship between the two phases demonstrated by Eq. (6.18) and Fig. 6.4 is roughly what environmental scientists call the Freundlich sorption isotherm:

$$C_{sorb} = K_F C^n \tag{6.19}$$

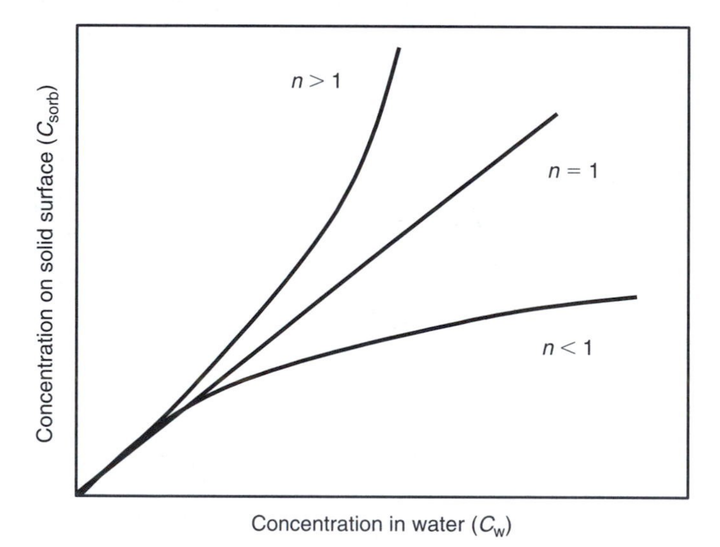

Fig. 6.4. Hypothetical Freundlich isotherms with exponents (n) less than, equal to, and greater than 1, as applied to the equation $C_{sorb} = K_F C^n$. *Sources*: Schwarzenbach, R., Gschwend, P., and Imboden, D., *Environmental Organic Chemistry*. John Wiley & Sons Inc., New York, NY, 1993; and Hemond, H. F., and Fechner-Levy, E. J., *Chemical Fate and Transport in the Environment*. Academic Press, San Diego, CA, 2000.

where C_{sorb} is the concentration of the sorbed contaminant, i.e. the mass sorbed at equilibrium per mass of sorbent, and K_F is the Freundlich isotherm constant. The exponent determines the linearity or order of the reaction. Thus, if $n = 1$, then the isotherm is linear; meaning the more of the contaminant in solution, the more would be expected to be sorbed to surfaces. For values of $n < 1$, the amount of sorption is in smaller proportion to the amount of solution and, conversely, for values of $n > 1$, a greater proportion of sorption occurs with less contaminant in solution. These three isotherms are shown in Fig. 6.4. Also note that if $n = 1$, then Eq. (6.20) and the Freundlich sorption isotherm are identical.

Research has shown that when organic matter content is elevated in the particle, the amount of a contaminant that is sorbed is directly proportional to the particle's organic matter content. This allows us to convert the K_D values from those that depend on specific soil or sediment conditions to those that are substrate-independent sorption constants, K_{OC}:

$$K_{OC} = K_D\,(f_{OC})^{-1} \tag{6.20}$$

where f_{OC} is the dimensionless weight fraction of organic carbon in the particle. K_{OC} and K_D have units of mass per volume. Table 6.2 provides the $\log K_{OC}$ values that are calculated from chemical structure and those measured empirically for several organic compounds, and compares them to the respective K_{ow} values.

B. Partitioning to the Liquid Phase: Dissolution

Unless otherwise stated, one can usually assume that when a compound is described as insoluble, such a statement means that the compound is *hydrophobic*. However, in environmental matters, it can be dangerous to make such assumptions, since such as wide range of scientific and engineering disciplines are involved. A good resource for contaminant solubilities for water, dimethylsulfoxide (DMSO), ethanol, acetone, methanol, and toluene is the National Toxicology Program's *Chemical Solubility Compendium*[5] and the program's Health and Safety reports.[6] The latter are updated frequently and provide useful data on properties of toxic chemicals.

Most characterizations of contaminants will describe solubility in water and provide values for aqueous solubility, as well as a substance's solubility in other organic solvents, such as methanol or acetone.

[5] Keith, L., and Walters, D., *National Toxicology Program's Chemical Solubility Compendium*. Lewis Publishers Inc., Chelsea, MI, 1992.

[6] http://ntp-db.niehs.nih.gov/htdocs/Chem_Hs_Index.html

TABLE 6.2

Calculated and Experimental Organic Carbon Coefficients (K_{oc}) for Selected Contaminants Found at Hazardous Waste Sites Compared to Octanol-Water Coefficients (K_{ow})

		Calculated		Measured	
Chemical	log K_{ow}	log K_{oc}	K_{oc}	log K_{oc}	K_{oc} (geomean)
Benzene	2.13	1.77	59	1.79	61.7
Bromoform	2.35	1.94	87	2.10	126
Carbon tetrachloride	2.73	2.24	174	2.18	152
Chlorobenzene	2.86	2.34	219	2.35	224
Chloroform	1.92	1.60	40	1.72	52.5
Dichlorobenzene, 1,2- (*o*)	3.43	2.79	617	2.58	379
Dichlorobenzene, 1,4- (*p*)	3.42	2.79	617	2.79	616
Dichloroethane, 1,1-	1.79	1.50	32	1.73	53.4
Dichloroethane, 1,2-	1.47	1.24	17	1.58	38.0
Dichloroethylene, 1,1-	2.13	1.77	59	1.81	65
Dichloroethylene, *trans*-1,2-	2.07	1.72	52	1.58	38
Dichloropropane, 1,2-	1.97	1.64	44	1.67	47.0
Dieldrin	5.37	4.33	21 380	4.41	25 546
Endosulfan	4.10	3.33	2138	3.31	2040
Endrin	5.06	4.09	12 303	4.03	10 811
Ethylbenzene	3.14	2.56	363	2.31	204
Hexachlorobenzene	5.89	4.74	54 954	4.90	80 000
Methyl bromide	1.19	1.02	10	0.95	9.0
Methyl chloride	0.91	0.80	6	0.78	6.0
Methylene chloride	1.25	1.07	12	1.00	10
Pentachlorobenzene	5.26	4.24	17 378	4.51	32 148
Tetrachloroethane, 1,1,2,2-	2.39	1.97	93	1.90	79.0
Tetrachloroethylene	2.67	2.19	155	2.42	265
Toluene	2.75	2.26	182	2.15	140
Trichlorobenzene, 1,2,4-	4.01	3.25	1778	3.22	1659
Trichloroethane, 1,1,1-	2.48	2.04	110	2.13	135
Trichloroethane, 1,1,2-	2.05	1.70	50	1.88	75.0
Trichloroethylene	2.71	2.22	166	1.97	94.3
Xylene, *o*-	3.13	2.56	363	2.38	241
Xylene, *m*-	3.20	2.61	407	2.29	196
Xylene, *p*-	3.17	2.59	389	2.49	311

Source: US Environmental Protection Agency, 1996, Soil Screening Program.

The process of *co-solvation* is a very important mechanism by which a highly lipophilic and hydrophobic compound enters water. If a compound is hydrophobic and nonpolar, but is easily dissolved in acetone or methanol, it may well end up in the water because these organic solvents are highly miscible in water. The organic solvent and water mix easily, and a hydrophobic compound will remain in the water column because it is dissolved in the organic solvent, which in turn has mixed with the water. Compounds like PCBs and dioxins may be transported as co-solutes in water by this means. So, the combination of hydrophobic compounds being sorbed to suspended materials and co-solvated in organic co-solvents that are miscible in water can mean that they are able to move in water and receptors can be exposed through the water pathways.

Solubility is determined from saturation studies. In other words, in the laboratory at a certain temperature, as much of the solute is added to a solvent until the solvent can no longer dissolve the substance being added. So, if compound A has a published solubility of $10\,\mathrm{mg\,L^{-1}}$ in water at 20°C, this means that the 1 L of water could only dissolve 10 mg of that substance. If, under identical conditions, compound B has a published aqueous solubility of $20\,\mathrm{mg\,L^{-1}}$, this means that 1 L of water could dissolve 20 mg of compound B, and that compound B has twice the aqueous solubility of compound A.

Actually, solutions are in "dynamic equilibrium" because the solute is leaving and entering the solution at all times, but the average amount of solute in solution is the same. The functional groups on a molecule determine whether it will be more or less polar. So, compounds with hydroxyl groups are more likely to form H-bonds with water. Thus, methane is less soluble in water than methanol. Also, since water interacts strongly with ions, salts are usually quite hydrophilic. The less the charge of the ion, the greater the solubility in water.

C. Partitioning to the Gas Phase: Volatilization

In its simplest connotation, volatilization is a function of the concentration of a contaminant in solution and the contaminant's partial pressure.

Henry's law states that the concentration of a dissolved gas is directly proportional to the partial pressure of that gas above the solution:

$$p_\mathrm{a} = K_\mathrm{H}[c] \tag{6.21}$$

where, K_H is the Henry's law constant; p_a is the partial pressure of the gas; and $[c]$ is the molar concentration of the gas

or,

$$p_\mathrm{a} = K_\mathrm{H}C_\mathrm{W} \tag{6.22}$$

where C_W is the concentration of gas in water.

So, for any chemical contaminant we can establish a proportionality between the solubility and vapor pressure. Henry's law is an expression of this proportionality between the concentration of a dissolved contaminant and its partial pressure in the headspace (including the open atmosphere) at equilibrium. A dimensionless version of the partitioning is similar to that of sorption, except that instead of the partitioning between solid and water phases, it is between the air and water phases (K_{AW}):

$$K_{AW} = \frac{C_A}{C_W} \tag{6.23}$$

where C_A is the concentration of gas A in the air.

The relationship between the air/water partition coefficient and Henry's law constant for a substance is:

$$K_{AW} = \frac{K_H}{RT} \tag{6.24}$$

where R is the gas constant ($8.21 \times 10^{-2}\,\text{L atm mol}^{-1}\text{K}^{-1}$) and T is the temperature (°K).

Henry's law relationships work well for most environmental conditions. It represents a limiting factor for systems where a substance's partial pressure is approaching zero. At very high partial pressures (e.g. 30 Pa) or at very high contaminant concentrations (e.g. >1000 ppm), Henry's law assumptions cannot be met. Such vapor pressures and concentrations are seldom seen in ambient environmental situations, but may be seen in industrial and other source situations. Thus, in modeling and estimating the tendency for a substance's release in vapor form, Henry's law is a good metric and is often used in compartmental transport models to indicate the fugacity from the water to the atmosphere.

HENRY'S LAW EXAMPLE

At 25°C, the log Henry's Law constant ($\log K_H$) for 1,2-dimethylbenzene (C_8H_{10}) is $0.71\,\text{L atm mol}^{-1}$ and the log octanol–water coefficient ($\log K_{ow}$) is 3.12. The $\log K_H$ for the pesticide parathion ($C_{10}H_{14}NO_5PS$) is $-3.42\,\text{L atm mol}^{-1}$, but its $\log K_{ow}$ is 3.81. Explain how these substances can have similar values for octanol–water partitioning yet so different Henry's law constants. What principle physicochemical properties account for much of this difference?

Answer

Both dimethylbenzene and parathion have an affinity for the organic phase compared to the aqueous phase. Since Henry's law constants are a function of both vapor pressure (P^0) and water solubility, and both compounds have similar octanol–water coefficients, the difference in the Henry's law characteristics must be mainly attributable to the compounds' respective water solubilities, their vapor pressures, or both.

Parathion is considered "semivolatile" because its vapor pressure at 20°C is only 1.3×10^{-3} kPa. Parathion's solubility[7] in water is 12.4 mg L^{-1} at 25°C.

1,2-Dimethylbenzene is also known as *ortho*-xylene (*o*-xylene). The xylenes are simply benzenes with more two methyl groups. The xylenes have very high vapor pressures of 4.5×10^2 kPa, and water solubilities[8] of about 200 mg L^{-1} at 25°C.

Thus, since both the solubilities are relatively low, it appears that the difference in vapor pressures is responsible for the large difference in the Henry' law constants, i.e. the much larger tendency of the xylene to leave the water and enter the atmosphere. Some of this tendency may result from the higher molecular weight of the parathion, but is also attributable to the additional functional groups on the parathion benzene than the two methyl groups on the xylene (see Fig. 6.5).

Another way to look at the chemical structures is to see them as the result of adding increasing complex functional groups, i.e., moving from the unsubstituted benzene to the single methylated benzene (toluene) to *o*-xylene to parathion. The substitutions result in respective progressively decreasing vapor pressures:

$$\text{Benzene's } P^0 \text{ at } 20°\text{C} = 12.7 \text{ kPa}$$

$$\text{Toluene's } P^0 \text{ at } 20°\text{C} = 3.7 \text{ kPa}$$

$$o\text{-Xylene's } P^0 \text{ at } 20°\text{C} = 0.9 \text{ kPa}$$

$$\text{Parathion's } P^0 \text{ at } 20°\text{C} = 1.3 \times 10^{-3} \text{ kPa}$$

The effect of these functional group additions on vapor pressure is even more obvious when seen graphically (Fig. 6.6)

[7] Meister, R. (ed.), *Farm Chemicals Handbook '92*. Meister Publishing Company, Willoughby, OH, 1992.

[8] National Park Service, US Department of the Interior, Environmental Contaminants Encyclopedia, 1997. O-Xylene Entry: http://www.nature.nps.gov/toxic/xylene_o.pdf

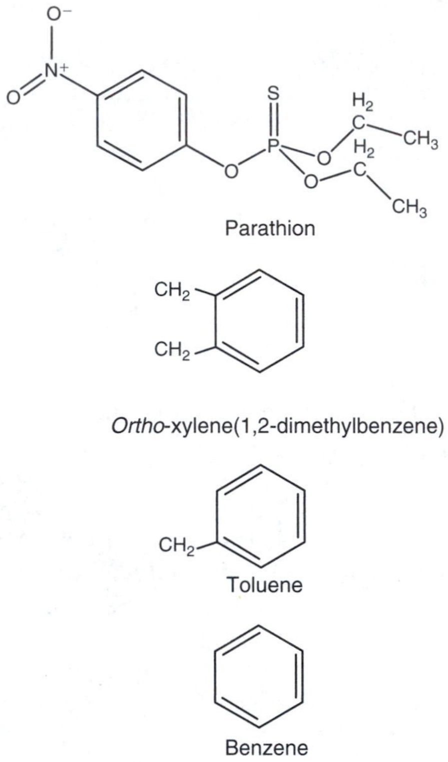

Parathion

Ortho-xylene(1,2-dimethylbenzene)

Toluene

Benzene

Fig. 6.5. Molecular structure of the pesticide parathion and the solvents *ortho*-xylene, toluene, and benzene.

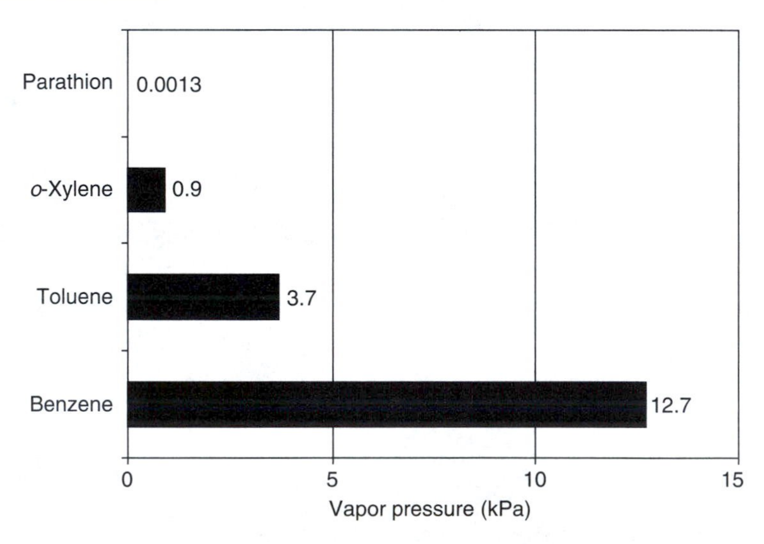

Fig. 6.6. Effect of functional group substitutions on vapor pressure of four organic aromatic compounds.

It is important to keep in mind that Henry's law constants are highly dependent on temperature, since both vapor pressure and solubility are also temperature dependent. So, when using published K_H values, one must compare them isothermically. Also, when combining different partitioning coefficients in a model or study, it is important to either use only values derived at the same temperature (e.g. sorption, solubility, and volatilization all at 20°C), or to adjust them accordingly. A general adjustment is an increase of a factor of 2 in K_H for each 8°C temperature increase.

Also, any sorbed or otherwise bound fraction of the contaminant will not exert a partial pressure, so this fraction should not be included in calculations of partitioning from water to air. For example, it is important to differentiate between the mass of the contaminant in solution (available for the K_{AW} calculation) and that in the suspended solids (unavailable for K_{AW} calculation). This is crucial for many hydrophobic organic contaminants, where they are most likely not to be dissolved in the water column (except as co-solutes), with the largest mass fraction in the water column being sorbed to particles.

The relationship between K_H and K_{ow} is also important. It is often used to estimate the *environmental persistence*, as reflected in the chemical *half-life* ($T_{1/2}$) of a contaminant. However, many other variables determine the actual persistence of a compound after its release. Note in the table, for example, that benzene and chloroform have nearly identical values of K_H and K_{ow}, yet benzene is far less persistent in the environment. We will consider these other factors in the next chapters, when we discuss abiotic chemical destruction and biodegradation.

With these caveats in mind, however, relative affinity for a substance to reside in air and water can be used to estimate the potential for the substance to partition not only between water and air, but more generally between the atmosphere and biosphere, especially when considering the long-range transport of contaminants (e.g. across continents and oceans).[9] Such long-range transport estimates make use of both atmospheric $T_{1/2}$ and K_H. The relationship between octanol–water and air–water coefficients can also be an important part of predicting a contaminant's transport. For example, Fig. 6.1 provides some general classifications according to various substances' K_H and K_{ow} relationships. In general, chemicals in the upper left-hand group have a great affinity for the atmosphere, so unless there are contravening factors, this is where to look for them. Conversely, substances with relatively low K_H and K_{ow} values are less likely to be transported long distance in the air.

Partitioning to Organic Tissue: Relatively hydrophobic substances frequently have a strong affinity for fatty tissues (i.e. those containing high K_{ow} compounds). Therefore, such contaminants can be sequestered and

[9] See Mackay, D., and Wania, F., Transport of contaminants to the arctic: partitioning, processes and models. *Sci. Total Environ.* **160/161**, 26–38, 1995.

can accumulate in organisms. In other words, certain chemicals are very *bioavailable* to organisms that may readily take them up from the other compartments. Bioavailability is an expression of the fraction of the total mass of a compound present in a compartment that has the potential of being absorbed by the organism. *Bioaccumulation* is the process of uptake into an organism from the abiotic compartments. *Bioconcentration* is the concentration of the pollutant within an organism above levels found in the compartment in which the organism lives. So, for a fish to bioaccumulate DDT, the levels found in the total fish or in certain organs (e.g. the liver) will be elevated above the levels measured in the ambient environment. In fact, DDT is known to bioconcentrate many orders of magnitude in fish. A surface water DDT concentration of 100 ppt in water has been associated with 10 ppm in certain fish species (a concentration of 10 000 times!). Thus the straightforward equation for the *bioconcentration factor* (BCF) is the quotient of the concentration of the contaminant in the organism and the concentration of the contaminant in the host compartment. So, for a fish living in water, the BCF is

$$\mathrm{BCF} = \frac{C_{\mathrm{organism}}}{C_{\mathrm{w}}} \tag{6.25}$$

The BCF is applied to an individual organism that represents a genus or some other taxonomical group. However, considering the whole food chain and trophic transfer processes, in which a compound builds up as a result of predator/prey relationships, the term *biomagnification* is used. Some compounds that may not appreciably bioconcentrate within lower trophic state organisms may still become highly concentrated. For example, even though plankton have a small BCF (e.g. 10), if subsequently higher-order organisms sequester the contaminant at a higher rate, by the time top predators (e.g. alligators, sharks, panthers, and humans) may suffer from the continuum of biomagnification, with levels many orders of magnitude higher than what is found in the abiotic compartments.

For a substance to bioaccumulate, bioconcentrate, and biomagnify, it must be at least somewhat persistent. If an organism's metabolic and detoxification processes are able to degrade the compound readily, it will not be present (at least in high concentrations) in the organism's tissues. However, if an organism's endogenous processes degrade a compound into a chemical species that is itself persistent, the metabolite or degradation product will bioaccumulate, and may bioconcentrate, and biomagnify. Finally, cleansing or *depuration* will occur if the organism that has accumulated a contaminant enters an abiotic environment that no longer contains the contaminant. However, some tissues have such strong affinities for certain contaminants that the persistence within the organism will remain long after the source of the contaminant is removed. For example, the piscivorous birds, such as the Common Loon (*Gavia immer*), decrease the concentrations of the metal mercury (Hg) in their bodies by translocating the metal to feathers and eggs.

So, every time the birds molt or lay eggs they undergo Hg depuration. Unfortunately, when the birds continue to ingest mercury that has bioaccumulated in their prey (fish), they often have a net increase in tissue Hg concentrations because the bioaccumulation rate exceeds the depuration rate.[10]

Bioconcentration can vary considerably in the environment. The degree to which a contaminant builds up in an ecosystem, especially in biota and sediments, is related to the compound's persistence. For example, a highly persistent compound often possesses chemical structures that are also conducive to sequestration by fauna. Such compounds are generally quite often lipophilic, have high K_{ow} values, and usually low vapor pressures. This means that they may bind to the organic molecules in living tissues and may resist elimination and metabolic process, so that they build up over time. However, the bioaccumulation and bioconcentration can vary considerably, both among biota and within the same species of biota. For example, the pesticide mirex has been shown to exhibit BCFs of 2600 and 51 400 in pink shrimp and fathead minnows, respectively. The pesticide endrin has shown an even larger interspecies variability in BCF values, with factors ranging from 14 to 18 000 recorded in fish after continuous exposure. Intraspecies BCF ranges may also be high, e.g., oysters exposed to very low concentrations of the organometallic compound, tributyl tin, exhibit BCF values ranging from 1000 to 6000.[11]

Even the same compound in a single medium, e.g. a lake's water column or sediment, will show large BCF variability among species of fauna in that compartment. An example is the so-called "dirty dozen" compounds. This is a group of *persistent organic pollutants* (POPs) that have been largely banned, some for decades, but that are still found in environmental samples throughout the world. As might be expected from their partitioning coefficients, they have concentrated in sediment and biota.

The worst combination of factors is when a compound is persistent in the environment, builds up in organic tissues, and is toxic. Such compounds are referred to as *persistent bioaccumulating toxic* substances (PBTs). Recently, the United Nations Environmental Program (UNEP) reported on the concentrations of the persistent and toxic compounds. Each region of the world was evaluated for the presence of these compounds. For example, the North American report[12] includes scientific assessments of the nature and scale of environmental threats posed by persistent toxic compounds. The results of these assessments are summarized in Tables 6.3 (organic compounds) and 6.4 (organometallic compounds). In the US, mining and mineral extraction

[10] Schoch, N., and Evers, D., Monitoring Mercury in Common Loons: New York Field Report, 1998–2000. Report BRI 2001-01 submitted to US Fish Wildlife Service and New York State Department of Environmental Conservation, BioDiversity Research Institute, Falmouth, ME, 2002.

[11] United Nations Environmental Program, Chemicals: North American Regional Report, Regionally Based Assessment of Persistent Toxic Substances, Global Environment Facility, 2002.

[12] United Nations Environmental Program, 2002.

TABLE 6.3

Summary of Persistent and Toxic Organic Compounds in North America, Identified by the United Nations as Highest Priorities for Regional Actions

Compound	Properties	Persistence/Fate	Toxicity*
Aldrin 1,2,3,4,10,10-Hexachloro-1,4,4a,5,8,8a-hexahydro-1,4-endo,exo-5,8-dimethanonaphthalene ($C_{12}H_8Cl_6$).	Solubility in water: 27 μg L^{-1} at 25°C; Vapor pressure: 2.31 × 10^{-5} mmHg at 20°C; log K_{ow}: 5.17–7.4.	Readily metabolized to dieldrin by both plants and animals. Biodegradation is expected to be slow and it binds strongly to soil particles, and is resistant to leaching into groundwater. Classified as moderately persistent with $T_{1/2}$ in soil ranging from 20–100 days.	Toxic to humans. Lethal dose for an adult estimated to be about 80 mg kg^{-1} body weight. Acute oral LD_{50} in laboratory animals is in the range of 33 mg kg^{-1} body weight for guinea pigs to 320 mg kg^{-1} body weight for hamsters. The toxicity of aldrin to aquatic organisms is quite variable, with aquatic insects being the most sensitive group of invertebrates. The 96-h LC_{50} values range from 1–200 μg L^{-1} for insects, and from 2.2–53 μg L^{-1} for fish. The maximum residue limits in food recommended by the World Health Organization (WHO) varies from 0.006 mg kg^{-1} milk fat to 0.2 mg kg^{-1} meat fat. Water quality criteria between 0.1 and 180 μg L^{-1} have been published.
Dieldrin 1,2,3,4,10,10-Hexachloro-6,7-epoxy-1,4,4a,5,6,7,8,8a-octahydroexo-1,4-endo-5,8-dimethanonaphthalene ($C_{12}H_8Cl_6O$).	Solubility in water: 140 μg L^{-1} at 20°C; vapor pressure: 1.78 × 10^{-7} mmHg at 20°C; log K_{ow}: 3.69–6.2.	Highly persistent in soils, with a $T_{1/2}$ of 3–4 years in temperate climates, and bioconcentrates in organisms.	Acute toxicity for fish is high (LC_{50} between 1.1 and 41 mg L^{-1}) and moderate for mammals (LD_{50} in mouse and rat ranging from 40 to 70 mg kg^{-1} body weight). Aldrin and dieldrin mainly affect the central nervous system but there is no direct evidence that they cause cancer in humans. The maximum residue limits in food recommended by WHO varies from 0.006 mg kg^{-1} milk fat and 0.2 mg kg^{-1}

			poultry fat. Water quality criteria between 0.1 and 18 μg L^{-1} have been published.
Endrin 3,4,5,6,9,9-Hexachloro-1a,2,2a,3,6,6a,7,7a-octahydro-2,7:3,6-dimethanonaphth[2,3-b] oxirene ($C_{12}H_8Cl_6O$).	Solubility in water: 220–260 μg L^{-1} at 25°C; vapor pressure: 7×10^{-7} mmHg at 25°C; log K_{ow}: 3.21–5.34.	Highly persistent in soils ($T_{1/2}$ of up to 12 years have been reported in some cases). BCFs of 14 to 18 000 have been recorded in fish, after continuous exposure.	Very toxic to fish, aquatic invertebrates, and phytoplankton; the LC_{50} values are mostly less than 1 μg L^{-1}. The acute toxicity is high in laboratory animals, with LD_{50} values of 3–43 mg kg^{-1}, and a dermal LD_{50} of 6–20 mg kg^{-1} in rats. Long-term toxicity in the rat has been studied over 2 years and a NOEL of 0.05 mg kg^{-1} bw day^{-1} was found.
Chlordane 1,2,4,5,6,7,8,8-Octachloro-2,3,3a,4,7,7a-hexahydro-4,7-methanoindene ($C_{10}H_6Cl_8$).	Solubility in water: 180 μg L^{-1} at 25°C; vapor pressure: 0.3×10^{-5} mmHg at 20°C; log K_{ow}: 4.4–5.5.	Metabolized in soils, plants, and animals to heptachlor epoxide, which is more stable in biological systems and is carcinogenic. The $T_{1/2}$ of heptachlor in soil is in temperateregions 0.75–2 years. Its high partition coefficient provides the necessary conditions for bioconcentrating in organisms.	Acute toxicity to mammals is moderate (LD_{50} values between 40 and 119 mg kg^{-1} have been published). The toxicity to aquatic organisms is higher and LC_{50} values down to 0.11 μg L^{-1} have been found for pink shrimp. Limited information is available on the effects in humans and studies are inconclusive regarding heptachlor and cancer. The maximum residue levels recommended by FAO/WHO are between 0.006 mg kg^{-1} milk fat and 0.2 mg kg^{-1} meat or poultry fat.
Dichlorodiphenyltrichloroethane (DDT) 1,1,1-Trichloro-2,2-bis-(4-chlorophenyl)-ethane ($C_{14}H_9Cl_5$).	Solubility in water: 1.2–5.5 μg L^{-1} at 25°C; vapor pressure: 0.02×10^{-5} mmHg at 20°C; log K_{ow}: 6.19 for pp-DDT, 5.5 for pp-DDD and 5.7 for pp-DDE.	Highly persistent in soils with a $T_{1/2}$ of about 1.1–3.4 years. It also exhibits high BCFs (in the order of 50 000 for fish and 500 000 for bivalves). In the environment, the parent DDT is metabolized mainly to DDD and DDE.	Lowest dietary concentration of DDT reported to cause egg shell thinning was 0.6 mg kg^{-1} for the black duck. LC_{50} of 1.5 mg L^{-1} for largemouth bass and 56 mg L^{-1} for guppy have been reported. The acute toxicity of DDT for mammals is moderate with an LD_{50} in rat of 113–118 mg kg^{-1} body weight. DDT has been shown to have an estrogen-like

(*continued*)

TABLE 6.3 (*Continued*)

Compound	Properties	Persistence/Fate	Toxicity*
			activity and possible carcinogenic activity in humans. The maximum residue level in food recommended by WHO/FAO, ranges from 0.02 mg kg^{-1} milk fat to 5 mg kg^{-1} meat fat. Maximum permissible DDT residue levels in drinking water (WHO) is 1.0 μg L^{-1}.
Toxaphene Polychlorinated bornanes and camphenes ($C_{10}H_{10}Cl_8$).	Solubility in water: 550 μg L^{-1} at 20°C; vapor pressure: 0.2–0.4 mmHg at 25°C; log K_{ow}: 3.23–5.50.	Half-life in soil from 100 days up to 12 years. It has been shown to bioconcentrate in aquatic organisms (BCF of 4247 in mosquito fish and 76 000 in brook trout).	Highly toxic in fish, with 96-h LC_{50} values in the range of 1.8 μg L^{-1} in rainbow trout to 22 μg L^{-1} in bluegill. Long-term exposure to 0.5 μg L^{-1} reduced egg viability to zero. The acute oral toxicity is in the range of 49 mg kg^{-1} body weight in dogs to 365 mg kg^{-1} in guinea pigs. In long-term studies NOEL in rats was 0.35 mg kg^{-1}bw day^{-1}, LD_{50} ranging from 60 to 293 mg kg^{-1}bw. For toxaphene, there exists strong evidence of the potential for endocrine disruption. Toxaphene is carcinogenic in mice and rats and is of carcinogenic risk to humans, with a cancer potency factor of 1.1 mg kg^{-1} day^{-1} for oral exposure.
Mirex 1,1a,2,2,3,3a,4,5,5a,5b,6-Dodecachloroacta-hydro-1,3,4-metheno-1H-cyclobuta[cd] pentalene ($C_{10}Cl_{12}$).	Solubility in water: 0.07 μg L^{-1} at 25°C; vapor pressure: 3×10^{-7} mmHg at 25°C; log K_{ow}: 5.28.	Among the most stable and persistent pesticides, with a $T_{1/2}$ in soils of up to 10 years. BCFs of 2600 and 51 400 have been observed in pink shrimp and fathead minnows, respectively.	Acute toxicity for mammals is moderate with an LD_{50} in rat of 235 mg kg^{-1} and dermal toxicity in rabbits of 80 mg kg^{-1}. Mirex is also toxic to fish and can affect their behavior (LC_{50} (96 h) from 0.2 to 30 mg L^{-1} for rainbow trout and bluegill,

		Capable of undergoing long-range transport due to its relativevolatility (VPL = 4.76 Pa; $H = 52\,Pa\,m^3\,mol^{-1}$).	respectively). Delayed mortality of crustaceans occurred at $1\,\mu g\,L^{-1}$ exposure levels. There is evidence of its potential for endocrine disruption and possibly carcinogenic risk to humans.
Hexachlorobenzene (HCB) (C_6H_6).	Solubility in water: $50\,\mu g\,L^{-1}$ at 20°C; vapor pressure: 1.09×10^{-5} mmHg at 20°C; log K_{ow}: 3.93–6.42.	Estimated "field half-life" of 2.7–5.7 years. HCB has a relatively high bioaccumulation potential and long $T_{1/2}$ in biota.	LC_{50} for fish varies between 50 and $200\,\mu g\,L^{-1}$. The acute toxicity of HCB is low with LD_{50} values of $3.5\,mg\,g^{-1}$ for rats. Mild effects of the [rat] liver have been observed at a daily dose of 0.25 mg HCB kg^{-1} bw. HCB is known to cause liver disease in humans (porphyria cutanea tarda) and has been classified as a possible carcinogen to humans by IARC.
Polychlorinated biphenyls (PCBs) ($C_{12}H_{(10-n)}Cl_n$, where n is within the range of 1–10).	Water solubility decreases with increasing chlorination: 0.01–0.0001 $\mu g\,L^{-1}$ at 25°C; vapor pressure: $1.6–0.003 \times 10^{-6}$ mmHg at 20°C; log K_{ow} : 4.3–8.26.	Most PCB congeners, particularly those lacking adjacent unsubstituted positions on the biphenyl rings (e.g. 2,4,6-, 2,3,6- or 2,3,6-substituted on both rings) are extremely persistent in the environment. They are estimated to have $T_{1/2}$ ranging from 3 weeks to 2 years in air and, with the exception of mono- and dichlorodiphenyl, more than6 years in aerobic soils and sediments. PCBs also have extremely long $T_{1/2}$ in adult fish,for example, an 8-year study of eels found that the $T_{1/2}$ of CB153 was more than 10 years.	LC_{50} for the larval stages of rainbow trout is $0.32\,\mu g\,L^{-1}$ with a NOEL of $0.01\,\mu g\,L^{-1}$. The acute toxicity of PCB in mammals is generally low and LD_{50} values in rat of $1\,g\,kg^{-1}$ bw. IARC has concluded that PCBs are carcinogenic to laboratory animals and probably also for humans. They have also been classified as substances for which there is evidence of endocrine disruption in an intact organism.

(*continued*)

TABLE 6.3 (*Continued*)

Compound	Properties	Persistence/Fate	Toxicity*
Polychlorinated dibenzo-p-dioxins (PCDDs) and *Polychlorinated dibenzofurans (PCDFs)* ($C_{12}H_{(8-n)}Cl_nO_2$) and PCDFs ($C_{12}H_{(8-n)}Cl_nO$) may contain between 1 and 8 chlorine atoms. Dioxins and furans have 75 and 135 possible positional isomers, respectively.	Solubility in water: in the range 550–0.07 ng L^{-1} at 25°C; vapor pressure: 2–0.007 × 10^{-6} mmHg at 20°C; log K_{ow}: in the range 6.60–8.20 for tetra- to octa-substituted congeners.	PCDD/Fs are characterized by their lipophilicity, semivolatility, and resistance to degradation ($T_{1/2}$ of TCDD in soil of 10–12 years) and to long-range transport. They are also known for their ability to bioconcentrate and biomagnify under typical environmental conditions.	Toxicological effects reported refers to the 2,3,7,8-substituted compounds (17 congeners) that are agonist for the AhR. All the 2,3,7,8-substituted PCDDs and PCDFs plus dioxin-like PCBs (DLPCBs) (with no chlorine substitution at the ortho positions) show the same type of biological and toxic response. Possible effects include dermal toxicity, immunotoxicity, reproductive effects and teratogenicity, endocrine disruption, and carcinogenicity. At the present time, the only persistent effect associated with dioxin exposure in humans is chloracne. The most sensitive groups are fetus and neonatal infants. Effects on the immune systems in the mouse have been found at doses of 10 ng kg^{-1} bw day^{-1}, while reproductive effects were seen in rhesus monkeys at 1–2 ng kg^{-1} bw day^{-1}. Biochemical effects have been seen in rats down to 0.1 ng kg^{-1} bw day^{-1}. In a re-evaluation of the TDI for dioxins, furans (and planar PCB), the WHO decided to recommend a range of 1–4 TEQ pg kg^{-1} bw, although more recently the acceptable intake value has been set monthly at 1–70 TEQ pg kg^{-1} bw.

Atrazine 2-Chloro-4-(ethylamino)-6-(isopropylamino)-s-triazine ($C_{10}H_6Cl_8$).	Solubility in water: 28 mg L^{-1} at 20°C; vapor pressure: 3.0 × 10^{-7} mmHg at 20°C; log K_{ow}: 2.34.	Does not adsorb strongly to soil particles and has a lengthy $T_{1/2}$ (60 to > 100 days). Atrazine has a high potential for groundwatercontamination despite its moderate solubility in water.	Oral LD_{50} is 3090 mg kg^{-1} in rats, 1750 mg kg^{-1} in mice, 750 mg kg^{-1} in rabbits, and 1000 mg kg^{-1} in hamsters. The dermal LD_{50} in rabbits is 7500 mg kg^{-1} and greater than 3000 mg kg^{-1} in rats. Atrazine is practically nontoxic to birds. The LD_{50} is greater than 2000 mg kg^{-1} in mallard ducks. Atrazine is slightly toxic to fish and other aquatic life. Atrazine has a low level of bioaccumulation in fish. Available data regarding atrazine's carcinogenic potential are inconclusive.
Hexachlorocyclohexane (HCH) 1,2,3,4,5,6-hexachlorocyclohexane (mixed isomers) ($C_6H_6Cl_6$).	γ-HCH (lindane): solubility in water: 7 mg L^{-1} at 20°C; vapor pressure: 3.3 × 10^{-5} mmHg at 20°C; log K_{ow}: 3.8.	Lindane and other HCH isomers are relatively persistent in soils and water, with half-lives generally greater than 1 and 2 years, respectively. HCHs are much less bioaccumulative than other organochlorines of concern because of their relatively low lipophilicity. On the contrary, their relatively high vapor pressures, particularly of the α-HCH isomer, determine their long-range transport in the atmosphere.	Lindane is moderately toxic for invertebrates and fish, with LC_{50} values of 20–90 μg L^{-1}. The acute toxicity for mice and rats is moderate with LD_{50} values in the range of 60–250 mg kg^{-1}. Lindane resulted to have no mutagenic potential in a number of studies but an endocrine disrupting activity.
Chlorinated paraffins (CPs) Polychlorinated alkanes ($C_xH_{(2x-y+2)}Cl_y$). Manufactured by chlorination of liquid *n*-alkanes or paraffin wax and contain from 30% to 70%	Properties largely dependent on the chlorine content. Solubility in water: 1.7–236 μg L^{-1} at 25°C; vapor pressure: very low, highest for short chains – those with	May be released into the environment from improperly disposed metal-working fluids or polymers containing CPs. Loss of CPs by leaching from paints and coatings may also	Acute toxicity of CPs in mammals is low with reported oral LD_{50} values ranging from 4 to 50 g kg^{-1} bw, although in repeated dose experiments, effects on the liver have been seen at doses of 10–100 mg kg^{-1} bw day^{-1}. Short-chain

(*continued*)

TABLE 6.3 (*Continued*)

Compound	Properties	Persistence/Fate	Toxicity*
chlorine. The products are often divided into three groups depending on chain length: short chain (C10–C13), medium (C14–C17), and long (C18–C30) chain lengths.	50% chlorine: 1.6×10^2 mmHg at 40°C; log K_{ow} in the range from 5.06 to 8.12.	paints and coatings may also contribute to environmental contamination. Short-chain CPs with less than 50% chlorine content seem to be degraded under aerobic conditions. The medium-and long-chain products are degraded more slowly. CPs are bioaccumulated and both uptake and elimination are faster for thesubstances with low chlorine content.	and mid-chain grades have been shown, in laboratory tests, to show toxic effects on fish and other forms of aquatic life after long-term exposure. The NOEL appears to be in the range of 2–5 μg L^{-1} for the most sensitive aquatic species tested.
Chlordecone or Kepone Chemical name: 1,2,3,4,5,5,6,7,9,10,10-dodecachlorooctahydro-1,3,4-metheno-2H-cyclobuta(cd) pentalen-2-one ($C_{10}Cl_{10}O$).	Solubility in water: 7.6 mg L^{-1} at 25°C; vapor pressure: less than 3×10^{-5} mmHg at 25°C; log K_{ow}: 4.50	Estimated $T_{1/2}$ in soils is between 1 and 2 years, whereas in air is much higher, up to 50 years. Not expected to hydrolyze, biodegrade in the environment. Also direct photodegradation and vaporization from what not significant. General population exposure to chlordecone mainly through the consumption of contaminated fish and seafood.	Workers exposed to high levels of chlordecone over a long period (more than 1 year) have displayed harmful effects on the nervous system, skin, liver, and male reproductive system (likely through dermal exposure to chlordecone, although they may have inhaled or ingested some as well). Animal studies with chlordecone have shown effects similar to those seen in people, as well as harmful kidney effects, developmental effects, and effects on the ability of females to reproduce. There are no studies available on whether chlordecone is carcinogenic in people. However, studies in mice and rats have shown that ingesting chlordecone can cause liver, adrenal gland, and kidney tumors. Very highly toxic for

			some species such as Atlantic menhaden, sheepshead minnow, or Donaldson trout with LC_{50} between 21.4 and 56.9 mg L^{-1}.
Endosulfan 6,7,8,9,10,10-Hexachloro-1,5,5a,6,9,9a-hexahydro-6,9-methano-2,4,3-benzodioxathiepin-3-oxide ($C_9H_6Cl_6O_3S$).	Solubility in water: 320 μg L^{-1} at 25°C; vapor pressure: 0.17×10^{-4} mmHg at 25°C; log K_{ow}: 2.23–3.62.	Moderately persistent in soil, with a reported average field $T_{1/2}$ of 50 days. The two isomers have different degradation times in soil ($T_{1/2}$ of 35 and 150 days for α- and β-isomers, respectively, in neutral conditions). It has a moderate capacity to adsorb to soils and it is not likely to leach to groundwater. In plants, endosulfan is rapidly broken down to the corresponding sulfate, on most fruits and vegetables, 50% of the parent residue is lost within 3–7 days.	Highly to moderately toxic to bird species (Mallards: oral LD_{50} 31–243 mg kg^{-1}) and it is very toxic to aquatic organisms (96-h LC_{50} rainbow trout 1.5 μg L^{-1}). It has also shown high toxicity in rats (oral LD_{50}: 18–160 mg kg^{-1}, and dermal: 78–359 mg kg^{-1}). Female rats appear to be 4–5 times more sensitive to the lethal effects of technical-grade endosulfan than male rats. The α-isomer is considered to be more toxic than the β-isomer. There is a strong evidence of its potential for endocrine disruption.
Pentachlorophenol (PCP) (C_6Cl_5OH).	Solubility in water: 14 mg L^{-1} at 20°C; vapor pressure: 16×10^{-5} mmHg at 20°C; log K_{ow}: 3.32–5.86.	Photodecomposition rate increases with pH ($T_{1/2}$ 100 h at pH 3.3 and 3.5 h at pH 7.3). Complete decomposition in soil suspensions takes > 72 days, other authors report $T_{1/2}$ in soils of about 45 days. Although enriched through the food chain, it is rapidly eliminated after discontinuing the exposure ($T_{1/2}$ 10–24 h for fish).	Acutely toxic to aquatic organisms. Certain effects on human health. 24 h LC_{50} values for trout were reported as 0.2 mg L^{-1}, and chronic toxicity effects were observed at concentrations down to 3.2 μg L^{-1}. Mammalian acute toxicity of PCP is moderate–high. LD_{50} oral in rat ranging from 50 to 210 mg kg^{-1}bw have been reported. LC_{50} ranged from 0.093 mg L^{-1} in rainbow trout (48 h) to 0.77–0.97 mg L^{-1} for guppy (96 h) and 0.47 mg L^{-1} for fathead minnow (48 h).

(continued)

TABLE 6.3 *(Continued)*

Compound	Properties	Persistence/Fate	Toxicity*
Hexabromobiphenyl (HxBB) ($C_{12}H_4Br_6$). A congener of the class polybrominated biphenyls (PBBs)	Solubility in water: 11 $\mu g\,L^{-1}$ at 25°C; vapor pressure: mmHg at 20°C; log K_{ow}: 6.39.	Strongly adsorbed to soil and sediments and usually persist in the environment. Resists chemical and biological degradation. Found in sediment samples from the estuaries of large rivers and has been identified in edible fish.	Few toxicity data are available from short-term tests on aquatic organisms. The LD_{50} values of commercial mixtures show a relatively low order of acute toxicity (LD_{50} range from > 1 to 21.5 $g\,kg^{-1}$ body weight in laboratory rodents). Oral exposure of laboratory animals to PBBs produced body weight loss, skin disorders, and nervous system effects, and birth defects. Humans exposed through contaminated food developed skin disorders, such as acne and hair loss. PBBs exhibit endocrine disrupting activity and possible carcinogenicity to humans.
Polybrominated diphenyl ethers (PBDEs) ($C_{12}H_{(10-n)}Br_nO$, where $n = 1-10$). As in the case of PCBs the total number of congeners is 209, with a predominance in commercial mixtures of the tetra-, penta- and octa-substituted isomers.	Solubility in water: mg L^{-1} at 25°C; vapor pressure: 3.85 up to 13.3×10^{-3} mmHg at 20–25°C; log K_{ow}: 4.28–9.9.	Biodegradation does not seem to be an important degradation pathway, but that photodegradation may play a significant role. Have been found in high concentrations in marine birds and mammals from remote areas. The half-lives of PBDE components in rat adipose tissue vary between 19 and 119 days, the higher values being for the more highly brominated congeners.	Lower (tetra- to hexa-) PBDE congeners likely to be carcinogens, endocrine disruptors, and/or neurodevelopmental toxicants. Studies in rats with commercial penta BDE indicate a low acute toxicity via oral and dermal routes of exposure, with LD_{50} values > 2000 $mg\,kg^{-1}$bw. In a 30-day study with rats, effects on the liver could be seen at a dose of 2 $mg\,kg^{-1}$bw day^{-1}, with a NOEL at 1 $mg\,kg^{-1}$bw day^{-1}. The toxicity to Daphnia magna has also been investigated and LC_{50} was found to be 14 $\mu g\,L^{-1}$ with a NOEC of 4.9 $\mu g\,L^{-1}$. Although data on toxicology is limited, they have potential endocrine disrupting properties, and there are concerns over the health effects of exposure.

Polycyclic aromatic hydrocarbons (PAHs) A group of compounds consisting of two or more fused aromatic rings.	Solubility in water: 0.000 14–2.1 mg L^{-1} at 25°C; vapor pressure: ranges from relatively volatile (e.g. naphthalene, 9×10^{-2} mmHg at 20°C) to semivolatile (e.g. benzo(a)pyrene, 5×10^{-9} mmHg at 25°C) to nearly nonvolatile (e.g. Indeno(1,2,3-cd)pyrene, 1×10^{-10} mmHg at 25°; log K_{ow}: 4.79–8.20	Persistence of the PAHs varies with their molecular weight. The low molecular weight PAHs are most easily degraded. The reported $T_{1/2}$ of naphthalene, anthracene, and benzo(e)pyrene in sediment are 9, 43, and 83 h, respectively, whereas for higher molecular weight PAHs' $T_{1/2}$ are up to several years in soils and sediments. The BCFs in aquatic organisms frequently range between 100 and 2000, and it increases with increasing molecular size. Due to their wide distribution, the environmental pollution by PAHs has aroused global concern.	Acute toxicity of low PAHs is moderate with an LD_{50} of naphthalene and anthracene in rat of 490 and 18 000 mg kg^{-1} body weight respectively, whereas the higher PAHs exhibit higher toxicity and LD_{50} of benzo(a)anthracene in mice is 10 mg kg^{-1} body weight. In Daphnia pulex, LC_{50} for naphthalene is 1.0 mg L^{-1}, for phenanthrene 0.1 mg L^{-1} and for benzo(a)pyrene is 0.005 mg L^{-1}. The critical effect of many PAHs in mammals is their carcinogenic potential. The metabolic actions of these substances produce intermediates that bind covalently with cellular DNA. IARC has classified benz[a]anthracene, benzo[a]pyrene, and dibenzo[a, h]anthracene as probable carcinogenic to humans. Benzo[b]fluoranthene and indeno[1,2,3-c,d]pyrene were classified as possible carcinogens to humans.
Phthalates Includes a wide family of compounds. Among the most common contaminants are: dimethylphthalate (DMP), diethylphthalate (DEP), dibutylphthalate (DBP), benzylbutylphthalate (BBP), di(2-ethylhexyl)phthalate (DEHP)($C_{24}H_{38}O_4$) and dioctylphthalate (DOP).	Properties of phthalic acid esters vary greatly depending on the alcohol moieties. log K_{ow} 1.5–7.1.	Ubiquitous pollutants, in marine, estuarine and freshwater sediments, sewage sludges, soils,and food. Degradation ($T_{1/2}$) values generally range from 1 to 30 days in freshwaters.	Acute toxicity of phthalates is usually low: the oral LD_{50} for DEHP is about 26–34 g kg^{-1}, depending on the species; for DBP reported LD_{50} values following oral administration to rats range from 8 to 20 g kg^{-1} body weight; in mice, values are approximately 5–16 g kg^{-1} body weight. In general, DEHP is not toxic for aquatic communities at the low levels usually present. In animals, high levels of DEHP damaged the liver and

(*continued*)

TABLE 6.3 (*Continued*)

Compound	Properties	Persistence/Fate	Toxicity*
			kidney and affected the ability to reproduce. There is no evidence that DEHP causes cancer in humans but they have been reported as endocrine disrupting chemicals. The Environmental Protection Agency (EPA) proposed a maximum admissible concentration (MAC) of 6 μg L^{-1} of DEHP in drinking water.
Nonyl- and octyl-phenols NP: $C_{15}H_{24}O$; OP: $C_{14}H_{22}O$.	log K_{ow}: 4.5 (NP) and 5.92 (OP).	NP and OP are the end degradation products of APEs under both aerobic and anaerobic conditions. Therefore, the major part is released to water and concentrated in sewage sludges. NPs and *t*-OP are persistent in the environment with $T_{1/2}$ of 30–60 years in marine sediments, 1–3 weeks in estuarine waters and 10–48 h in the atmosphere. Due to their persistence they can bioaccumulate to a significant extent in aquatic species. However, excretion and metabolism are rapid.	Acute toxicity values for fish, invertebrates, and algae range from 17–3000 μg L^{-1}. In chronic toxicity tests the lowest NOEC are 6 μg L^{-1} in fish and 3.7 μg L^{-1} in invertebrates. The threshold for vitellogenin induction in fish is 10 μg L^{-1} for NP and 3 μg L^{-1} for OP (similar to the lowest NOEC). Alkylphenols are endocrine disrupting chemicals also in mammals.

Perfluorooctane sulfonate (PFOS) ($C_8F_{17}SO_3$)	Solubility in water: 550 mg L^{-1} in pure water at 24–25°C; the potassium salt of PFOS has a low vapor pressure, 3.31×10^{-4} Pa at 20°C. Due to the surface-active properties of PFOS, the log K_{ow} cannot be measured.	Does not hydrolyze, photolyze, or biodegrade under environmental conditions. It is persistent in the environment and has been shown to bioconcentrate in fish. It has been detected in a number of species of wildlife, including marine mammals. Animal studies show that PFOS is well absorbed orally and distributes mainly in the serum and the liver. The half-life in serum is 7.5 days in adult rats and 200 days in cynomolgus monkeys. The half-life in humans is, on average, 8.67 years (range 2.29–21.3 years, SD = 6.12).	Moderate acute toxicity to aquatic organisms, the lowest LC_{50} for fish is a 96-h LC_{50} of 4.7 mg L^{-1} to the fathead minnow (*Pimephales promelas*) for the lithium salt. For aquatic invertebrates, the lowest EC_{50} for freshwater species is a 48-h EC_{50} of 27 mg L^{-1} for *Daphnia magna* and for saltwater species, a 96-h LC_{50} value of 3.6 mg L^{-1} for the Mysid shrimp (*Mysidopsis bahia*). Both tests were conducted on the potassium salt. The toxicity profile of PFOS is similar among rats and monkeys. Repeated exposure results in hepatotoxicity and mortality; the dose–response curve is very steep for mortality. PFOS has shown moderate acute toxicity by the oral route with a rat LD_{50} of 251 mg kg^{-1}. Developmental effects were also reported in prenatal developmental toxicity studies in the rat and rabbit, although at slightly higher dose levels. Signs of developmental toxicity in the offspring were evident at doses of 5 mg kg^{-1} day^{-1} and above in rats administered PFOS during gestation. Significant decreases in fetal body weight and significant increases in external and visceral anomalies, delayed ossification, and skeletal variations were observed.

(*continued*)

TABLE 6.3 (*Continued*)

Compound	Properties	Persistence/Fate	Toxicity*
			An NOAEL of $1\,mg\,kg^{-1}\,day^{-1}$ and an LOAEL of $5\,mg\,kg^{-1}\,day^{-1}$ for developmental toxicity were indicated. Studies on employees conducted at PFOS manufacturing plants in the US and Belgium showed an increase in mortality resulting from bladder cancer and an increased risk of neoplasms of the male reproductive system, the overall category of cancers and benign growths, and neoplasms of the gastrointestinal tract.

$T_{1/2}$: chemical half-life; LD_{50}: lethal dose to 50% of tested organism; LC_{50}: lethal concentration to 50% of tested organism; BCF: bioconcentration factor; NOEL: no observable effect level; NOEC: no observable effect concentration; DDD: 1,1-dichloro-2,2-bis(*p*-chlorophenyl)ethane; DDE: 1,1-dichloro-2,2-bis(*p*-chlorophenyl)ethylene; VPL: vapor pressure lowering; IARC: International Agency for Research on Cancer; TCDD: tetrachlorodibenzo-*p*-dioxin; AhR: aryl hydrocarbon receptor; TDI: tolerable daily intake; TEQ: toxic equivalent; APEs: alkylphenol ethoxylates; NOAEL: no observed adverse effect level; LOAEL: lowest observed adverse effect level.

Source: United Nations Environmental Program, 2002, Chemicals: North American Regional Report, Regionally Based Assessment of Persistent Toxic Substances, Global Environment Facility.

TABLE 6.4

Summary of Persistent and Metallic Compounds in North America, Identified by the United Nations as Highest Priorities for Regional Actions

Compound	Properties	Persistence/Fate	Toxicity*
Compounds of tin (Sn) Organ tin compounds comprise mono-,di-, tri-, and tetra-butyl and triphenylene tin compounds. They conform to the following general formula $(n\text{-}C_4H_9)_nSn-X$and $(C_6H_5)3Sn-X$, where X is an anion or a group linked covalently through a heteroatom.	log K_{ow}: 3.19–3.84. In sea water and under normal conditions, tributyl tin exists as three species (hydroxide, chloride, and carbonate).	Under aerobic conditions, tributyl tin takes 30–90 days to degrade, but in anaerobic soils may persist for more than 2 years Due to low water solubility it binds strongly to suspended material and sediments. Tributyl tin is lipophilic and accumulates in aquatic organisms. Oysters exposed to very low concentrations exhibit BCF values ranging from 1000 to 6000.	Tributyl tin is moderately toxic and all breakdown products are even less toxic. Its impact on the environment was discovered in the early 1980s in France with harmful effects in aquatic organisms, such as shell malformations of oysters, imposex in marine snails, and reduced resistance to infection (e.g. in flounder). Gastropods react adversely to very low levels of tributyl tin (0.06–2.3 $\mu g\,L^{-1}$). Lobster larvae show a nearly complete cessation of growth at just 1.0 $\mu g\,L^{-1}$ tributyl tin. In laboratory tests, reproduction was inhibited when female snails exposed to 0.06–0.003 $\mu g\,L^{-1}$ tributyl tin developed male characteristics. Large doses of tributyl tin have been shown to damage the reproductive and central nervous systems, bone structure, and the liver bile duct of mammals.
Compounds of mercury (Hg) The main compound of concern is methyl mercury ($HgCH_3$).		Mercury released into the environment can either stay close to its source for long periods, or be widely dispersed on a regionalor even worldwide basis. Not only are methylated mercury compounds toxic, but highly bioaccumulative as well.	Long-term exposure to either inorganic or organic mercury can permanently damage the brain, kidneys, and developing fetus. The most sensitive target of low-level exposure to metallic and organic mercury from short- or long-term exposures is likely the nervous system.

(*continued*)

TABLE 6.4 (*Continued*)

Compound	Properties	Persistence/Fate	Toxicity*
		The increase in mercury as it rises in the aquatic food chain results in relatively high levels of mercury in fish consumed by humans. Ingested elemental mercury is only 0.01% absorbed, but methyl mercury is nearly 100% absorbed from the gastrointestinal tract. The biological $T_{1/2}$ of Hg is 60 days.	
Compounds of lead (Pb) Alkyl lead compounds may be confined to tetramethyl lead (TML, $Pb(CH_3)_4$) and tetraethyl lead (TEL, $Pb(C_2H_5)_4$).	Solubility in water: 17.9 mg L^{-1} (TML) and 0.29 mg L^{-1} (TEL) at 25°C; vapor pressure: 22.5 and 0.15 mmHg at 20°C for TML and TEL, respectively.	Under environmental conditions, dealkylation produces less alkylated forms and finally inorganic Pb. However, there is limited evidence that under some circumstances, natural methylation of Pb salts may occur. Minimal bioaccumulations have been observed for TEL in shrimps (650), mussels (120), and plaice (130) and for TML in shrimps (20), mussels (170), and plaice (60).	Exposure to Pb and its compounds have been associated with cancer in the respiratory and digestive systems of workers in lead battery and smelter plants. However, tetra-alkyl lead compounds have not been sufficiently tested for the evidence of carcinogenicity. Acute toxicity of TEL and TML are moderate in mammals and high for aquatic biota. LD_{50} (rat, oral) for TEL is 35 mg Pb kg^{-1} and 108 mg Pb kg^{-1} for TML. LC_{50} (fish, 96 h) for TEL is 0.02 mg kg^{-1} and for TML is 0.11 mg kg^{-1}.

$T_{1/2}$: Chemical half-life; LD_{50}: Lethal dose to 50% of tested organism; LC_{50}: Lethal concentration to 50% of tested organism; BCF: Bioconcentration factor; NOEL: No observable effect level; NOEC: No observable effect concentration.

Source: United Nations Environmental Program, 2002, Chemicals: North American Regional Report, Regionally Based Assessment of Persistent Toxic Substances, Global Environment Facility.

activities contribute the large quantity of PBTs, with energy production the second largest source category (see Fig. 6.7). Organometallic compounds, especially lead and its compounds comprise the lion's share of PBTs in the US. And, the second largest quantity is represented by another metal, mercury and its compounds (see Fig. 6.8).

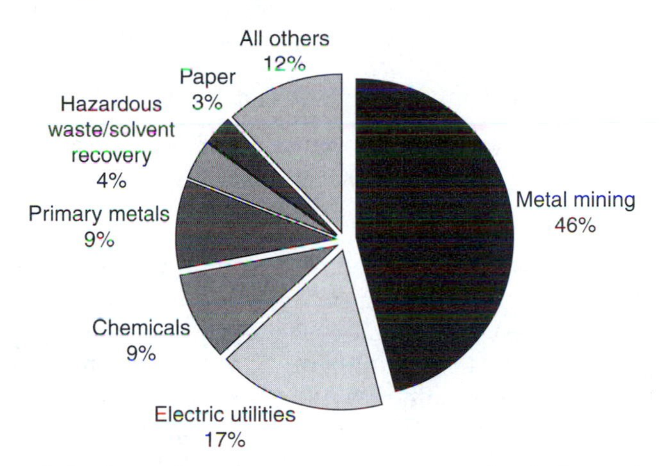

Fig. 6.7. Total US releases of contaminants in 2001, as reported to the Toxic Release Inventory. Total releases = 2.8 billion kg. *Note*: Off-site releases include metals and metal compounds transferred off-site for solidification/stabilization and for wastewater treatment, including publicly owned treatment works. Off-site releases do not include transfers to disposal sent to other TRI facilities that reported the amount as an on-site release. *Source*: US Environmental Protection Agency.

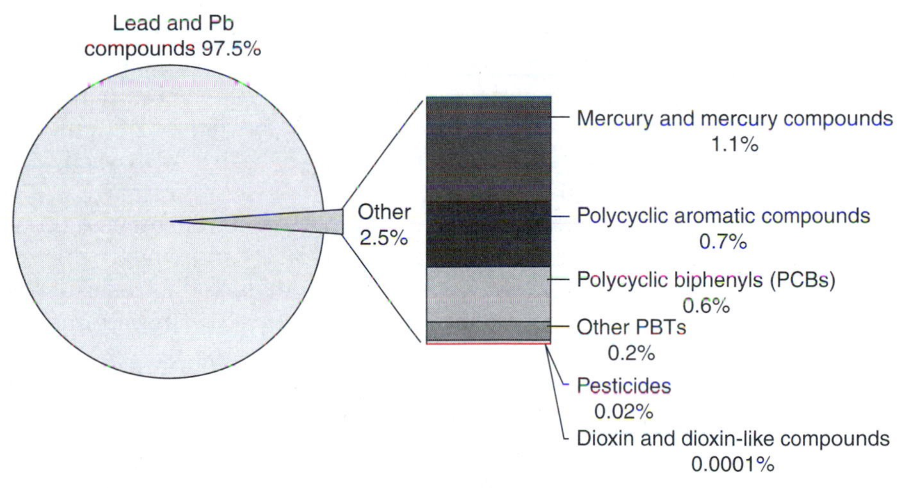

Fig. 6.8. Total US releases of PBTs in 2001, as reported in the Toxic Release Inventory (TRI). Total releases = 206 million kg. *Note*: Off-site releases include metals and metal compounds transferred off-site for solidification/stabilization and for wastewater treatment, including publicly owned treatment works. Off-site releases do not include transfers to disposal sent to other TRI facilities that reported the amount as an on-site release. *Source*: US Environmental Protection Agency.

The sources of PBTs are widely varied. Many are intentionally manufactured to serve some public need, such as the control of pests that destroy food and spread disease. Other PBTs are generated as unintended byproducts, such as the PICs. In either case, there are often measures and engineering controls available that can prevent PBT releases, rather than having to deal with them after they have found their way into the various environmental compartments.

D. Concentration-Based and Fugacity-Based Transport Models

Let us now combine these phase and compartmental distributions into a simple fugacity-based, chemodynamic transport model. Such models are classified into three types:

- *Level 1 model*: This model is based on an equilibrium distribution of fixed quantities of contaminants in a closed environment (i.e. conservation of contaminant mass). No chemical or biological degradation, advection (discussed in detail later), and no transport among compartments (such as sediment loading or atmospheric deposition to surface waters).
  A Level 1 calculation describes how a given quantity of a contaminant will partition among the water, air, soil, sediment, suspended particles, and fauna, but does not take into account chemical reactions. Early Level 1 models considered an area of 1 km^2 with 70% of the area covered in surface water. Larger areas are now being modeled (e.g. about the size of the state of Ohio).
- *Level 2 model*: This model relaxes the conservation restrictions of Level 1 by introducing direct inputs (e.g. emissions) and advective sources from air and water. It assumes that a contaminant is being continuously loaded at a constant rate into the control volume, allowing the contaminant loading to reach steady state and equilibrium between contaminant input and output rates. Degradation and bulk movement of contaminants (advection) is treated as a loss term. Exchanges between and among media are not quantified.

  Since the Level 2 approach is a simulation of a contaminant being continuously discharged into numerous compartments and which achieves a steady-state equilibrium, the challenge is to deduce the losses of the contaminant due to chemical reactions and advective (non-diffusive) mechanisms.

  Reaction rates are unique to each compound and are published according to reactivity class (e.g. fast, moderate, or slow reactions), which allows modelers to select a class of reactivity for the respective contaminant to insert into transport models. The reactions are often assumed to be first order, so the model will employ a first-order rate constant for each compartment in the environmental system (e.g. x mol h^{-1} in water, y mol h^{-1} in air, z mol h^{-1} in soil). Much uncertainty is associated with the reactivity class and rate constants, so it is best to use rates published in the literature based on experimental and empirical studies, wherever possible.

Advection flow rates in Level 2 models are usually reflected by residence times in the compartments. These residence times are commonly set to 1 h in each medium, so the advection rate (G_i) is volume of the compartment divided by the residence time (t):

$$G_i = V t^{-1} \tag{6.26}$$

- *Level 3 model*: Same as Level 2, but does not assume equilibrium between compartments, so each compartment has its own fugacity. Mass balance applies to whole system and each compartment within the system. It includes mass transfer coefficients, rates of deposition and re-suspension of contaminant, rates of diffusion (discussed later), soil runoff, and area covered. All of these factors are aggregated into an intermedia transport term (D) for each compartment.

  The assumption of equilibrium in Level 1 and 2 models is a simplification, and often a gross oversimplification of what actually occurs in environmental systems. When, the simplification is not acceptable, kinetics must be included in the model. Numerous diffusive and non-diffusive transport mechanisms are included in Level 3 modeling. For example, values for the various compartments' unique intermedia transport velocity parameters (in length per time dimensions) are applied to all contaminants being modeled (these are used to calculate the *D values* mentioned above).

E. Kinetics Versus Equilibrium

Since, Level 3 models do not assume equilibrium conditions, a word about chemical kinetics is in order at this point. Chemical kinetics is the description of the rate of a chemical reaction.[13] This is the rate at which the reactants are transformed into products. This may take place by abiotic or by biological systems, such as microbial metabolism. Since a rate is a change in quantity that occurs with time, the change we are most concerned with is the change in the concentration of our contaminants into new chemical compounds:

$$\text{Reaction rate} = \frac{\text{change in product concentration}}{\text{corresponding change in time}} \tag{6.27}$$

[13] Although "kinetics" in the physical sense and the chemical sense arguably can be shown to share many common attributes, for the purposes of this discussion, it is probably best to treat them as two separate entities. Physical kinetics, as discussed in Chapter 4, is concerned with the dynamics of material bodies and the energy in a body owing to its motions. Chemical kinetics address rates of chemical reactions. The former is more concerned with mechanical dynamics, the latter with thermodynamics.

and

$$\text{Reaction rate} = \frac{\text{change in reactant concentration}}{\text{corresponding change in time}} \tag{6.28}$$

In environmental degradation, the change in product concentration will be decreasing proportionately with the reactant concentration, so, for contaminant X, the kinetics looks like

$$\text{Rate} = -\frac{\Delta(X)}{\Delta t} \tag{6.29}$$

The negative sign denotes that the reactant concentration (the parent contaminant) is decreasing. It stands to reason then that the degradation product Y resulting from the concentration will be increasing in proportion to the decreasing concentration of the contaminant X, and the reaction rate for Y is

$$\text{Rate} = \frac{\Delta(Y)}{\Delta t} \tag{6.30}$$

By convention, the concentration of the chemical is shown in parentheses to indicate that the system is not at equilibrium. $\Delta(X)$ is calculated as the difference between an initial concentration and a final concentration:

$$\Delta(X) = \Delta(X)_{\text{final}} - \Delta(X)_{\text{initial}} \tag{6.31}$$

So, if we were to observe the chemical transformation[14] of one isomer of the compound butane to different isomer over time, this would indicate the kinetics of the system, in this case the homogeneous gas phase reaction of *cis*-2-butene to *trans*-2-butene (see Fig. 6.9 for the isomeric structures). The transformation is shown in Fig. 6.10. The rate of reaction at any time is the negative of the slope of the tangent to the concentration curve at that specific time (see Fig. 6.11).

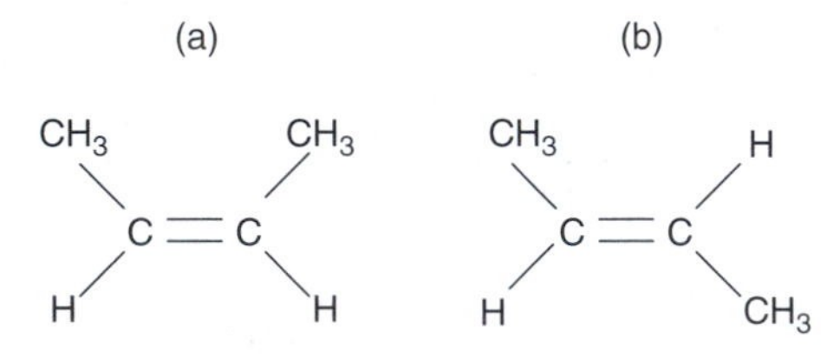

Fig. 6.9. Two isomers of butane: (a) *cis*-2-butene and (b) *trans*-2-butene.

[14] This example was taken from Spencer, J., Bodner, G., and Rickard, L., *Chemistry: Structure and Dynamics*, 2nd ed. John Wiley & Sons, New York, NY, 2003.

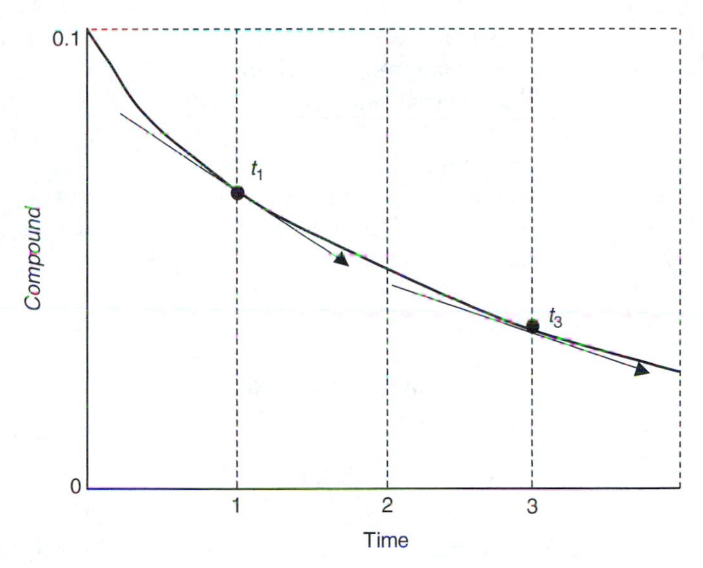

Fig. 6.10. The kinetics of the transformation of a compound. The rate of reaction at any time is the negative of the slope of the tangent to the concentration curve at that time. The rate is higher at t_1 than at t_3. This rate is concentration dependent (first order).

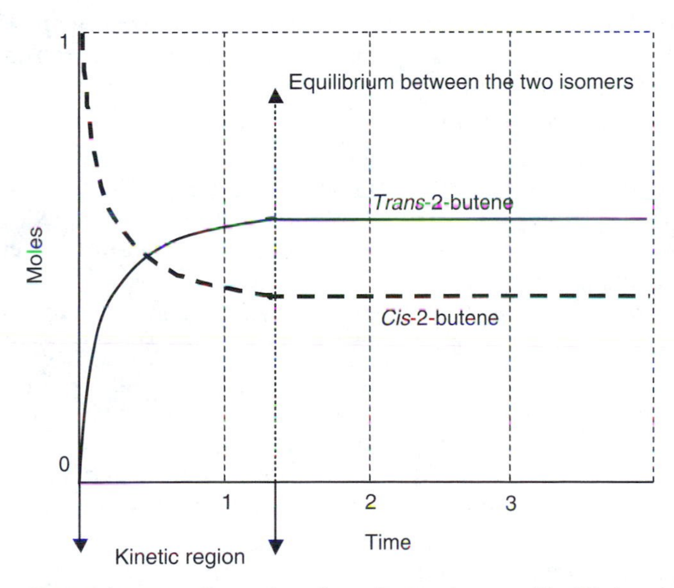

Fig. 6.11. Change in respective moles of two butene isomers. Equilibrium is reached at about 1.3 time units. The concentrations of the isomers depend on the initial concentration of the reactant (*cis*-2-butene). The actual time that equilibrium is reached depends on environmental conditions, such as temperature and other compounds present; however, at a given temperature and conditions, the ratio of the equilibrium concentrations will be the same, no matter the amount of the reactant at the start.

For a reaction to occur, the molecules of the reactants must meet (collide). So, high concentrations of a contaminant are more likely to collide than low concentrations. Thus, the reaction rate must be a function of the concentrations of the reacting substances. The mathematical expression of this function is known as the "rate law." The rate law can be determined experimentally for any contaminant. Varying the concentration of each reactant independently and then measuring the result will give a concentration curve. Each reactant has a unique rate law (this is one of a contaminant's physicochemical properties). So, let us consider the reaction of reactants A and B which yield product C (i.e., $A + B \rightarrow C$), where the reaction rate increases in accord with the increasing concentration of either A or B. This means that if we triple the amount of A, the rate of this whole reaction triples. Thus, the rate law for such a reaction is

$$\text{Rate} = k[A][B] \tag{6.32}$$

However, let us consider another reaction $X + Y \rightarrow Z$, in which the rate is only increased if the concentration of X is increased (changing the Y concentration has no effect on the rate law). In this reaction, the rate law must be

$$\text{Rate} = k[X] \tag{6.33}$$

Thus, the concentrations in the rate law are the concentrations of reacting chemical species at any specific point in time during the reaction. The rate is the velocity of the reaction at that time. The constant k in the equations above is the *rate constant*, which is unique for every chemical reaction and is a fundamental physical constant for a reaction, as defined by environmental conditions (e.g. pH, temperature, pressure, type of solvent). The rate constant is defined as the rate of the reaction when all reactants are present in a 1 molar (M) concentration, so the rate constant k is the rate of reaction under conditions standardized by a unit concentration.

We can demonstrate the rate law by drawing a concentration curve for a contaminant that consists of an infinite number of points at each instant of time, so an instantaneous rate can be calculated along the concentration curve. At each point on the curve the rate of reaction is directly proportional to the concentration of the compound at that moment in time. This is a physical demonstration of *kinetic order*. The overall kinetic order is the sum of the exponents (powers) of all the concentrations in the rate law. So for the rate $k[A][B]$, the overall kinetic order is 2. Such a rate describes a second-order reaction because the rate depends on the concentration of the reactant raised to the second power. Other decomposition rates are like $k[X]$, and are first-order reactions because the rate depends on the concentration of the reactant raised to the first power.

The kinetic order of each reactant is the power that its concentration is raised in the rate law. So, $k[A][B]$ is first order for each reactant and $k[X]$ is first order for X and zero order for Y. In a zero-order reaction, compounds degrade at a constant rate and are independent of reactant concentration.

Further, if we plot the change in the number of moles with respect to time, we would see the point at which kinetics ends and equilibrium begins. This simple example applies to any chemical kinetics process, but the kinetics is complicated in the "real world" by the ever changing conditions of the atmosphere, industrial processes, ecosystems, tissues, and human beings.

REFERENCES

1. MacKay, D., and Paterson, S., Evaluating the fate of organic chemicals: a level III fugacity model. *Environ. Sci. Technol.* **25**, 427–436 (1991).
2. Lyman, W., Transport and transformation processes, in *Fundamentals of Aquatic Toxicology: Effects, Environmental Fate, and Risk Assessment* (Rand, G., ed.), 2nd ed., Chapter 15. Taylor & Francis, Washington, DC, 1995.
3. Westfall, J., Adsorption mechanisms in aquatic surface chemistry, in *Aquatic Surface Chemistry*. Wiley-Interscience, New York, 1987.
4. Hassett, J., and Banwart, W., The sorption of nonpolar organics by soils and sediments, in *Reactions and Movement of Organic Chemicals in Soils* (Sawhney, B., and Brown, K., eds.), p. 35. Soil Science Society of America, Special Publication No. 22 (1989).
5. Schwarzenbach, R., Gschwend, P., and Imboden, D., *Environmental Organic Chemistry*. John Wiley & Sons Inc., New York, 1993.
6. Hemond, H. F., and Fechner-Levy, E. J., *Chemical Fate and Transport in the Environment*. Academic Press, San Diego, CA, 2000.
7. Keith, L., and Walters, D., *National Toxicology Program's Chemical Solubility Compendium*. Lewis Publishers Inc., Chelsea, MI, 1992.
8. MacKay, D., and Wania, F., Transport of contaminants to the arctic: partitioning, processes and models. *Sci. Total Environ.* **160/161**, 26–38 (1995).
9. Schoch, N., and Evers, D., *Monitoring Mercury in Common Loons: New York Field Report, 1998–2000*. Report BRI 2001-01 submitted to US Fish Wildlife Service and New York State Department of Environmental Conservation, Biodiversity Research Institute, Falmouth, ME, 2002.
10. United Nations Environmental Program, *Chemicals: North American Regional Report, Regionally Based Assessment of Persistent Toxic Substances, Global Environment Facility*, 2002.
11. Spencer, J., Bodner, F., and Rickard, L., *Chemistry: Structure and Dynamics*, 2nd ed. John Wiley & Sons, New York, 2003.

References 1–11 are generally resources used by the author and, as such, are not cross-referenced in the text.

SUGGESTED READING

Agency for Toxic Substances and Disease Registry, US Public Health Service, *Toxicological Profiles for Numerous Chemicals*. Accessible at http://www.atsdr.cdc.gov/toxpro2.html#bookmark01.

Air and Waste Management Association, in *Air Pollution Engineering Manual* (Davis, W. T., ed.), 2nd ed., John Wiley & Sons, Inc., New York, NY, 2000.

Spencer, J., Bodner, G., and Rickard, L., *Chemistry: Structure and Dynamics*, 2nd ed. John Wiley & Sons, New York, 2003.

Vallero, D. A. *Environmental Contaminants: Assessment and Control*. Elsevier Academic Press, Amsterdam, The Netherlands, 2004.

QUESTIONS

1. What are the advantages of using Henry's law constants versus vapor pressure to estimate the likelihood of a contaminant moving to the atmosphere?
2. Look up the physicochemical characteristics of benzo(a)pyrene. Where do you believe it will fall on Fig. 6.1? What will be some of the likely routes it will take to reach the atmosphere?
3. What types of chemical reactions would you expect to occur in a copper smelter? What types of reactions would you expect in the atmosphere immediately downwind from the stack of the smelter? What reactions would you expect on the particles that are deposited 10 m from the stack? ... 1000 m from the stack?
4. Consider three products of incomplete combustion. What practices can a facility take to improve (i.e. lower) emissions of these compounds?
5. Consider the products of complete combustion. What practices can a facility take to improve (i.e. lower) emissions of these compounds?
6. Explain what might account for some of the differences between calculated and measured coefficients in Table 6.2. What, if any, relationship can you draw between K_{ow} and K_{oc}?
7. What might be some differences between a soil particle and an airborne aerosol? What are some similarities? How might these differences affect the chemodynamics of the two particle types?

7

Characterizing Air Pollution

Discussions of environmental fate and transport must always consider both physical processes associated with chemical reactions and other chemical processes. Thus, air pollution physics must be considered mutually with air pollution chemistry. Any complete discussion of the physical process of solubility, for example, must include a discussion of chemical phenomenon *polarity*. Further, any discussion of polarity must include a discussion of electronegativity. Likewise, discussions of sorption and air–water exchanges must consider both chemical and physical processes. Such is the nature of environmental science; all concepts are interrelated.

I. RELATIONSHIP BETWEEN PHYSICS AND CHEMISTRY

The interconnectedness between physical and chemical processes is evident in Table 6.1, which lists some of the most important processes involved in the fate of environmental contaminants. This chapter highlights basic physicochemical processes that affect air pollution.

It is important to bear in mind that the air pollution processes are a function of both the chemical characteristics of the compartment (e.g. air) and those of the contaminant. The inherent properties of the air pollutant are influenced and changed by the extrinsic properties of the air and other media in which the pollutant resides in the environment. Thus, to characterize air pollution, the physical and chemical properties of the pollutant and the air must be considered together.

II. BASIC CHEMICAL CONCEPTS

Environmental chemistry is the discipline that concerns itself with how chemicals are formed, how they are introduced into the environment, how they change after being introduced, where they end up in organisms and other receptors, and the effects they have (usually the damage they do) once they get there. To cover these concepts, environmental chemistry must address the processes in effect in every environmental compartment. This is evident by the diverse subdisciplines within environmental chemistry, including atmospheric chemistry. There are even fields such as environmental physical chemistry (such as environmental photochemistry), environmental analytical chemistry (including environmental separation sciences and chromatography), and environmental chemical engineering (including fields addressing environmental thermodynamics).

The element is a material substance that has decomposed chemically to its simplest form. These are what appear on the periodic table of elements (Fig. 7.1). Elements may be further broken down only by nuclear reactions, where they are released as subatomic particles. Such particles are important sources of pollution and often are environmental contaminants. An atom is

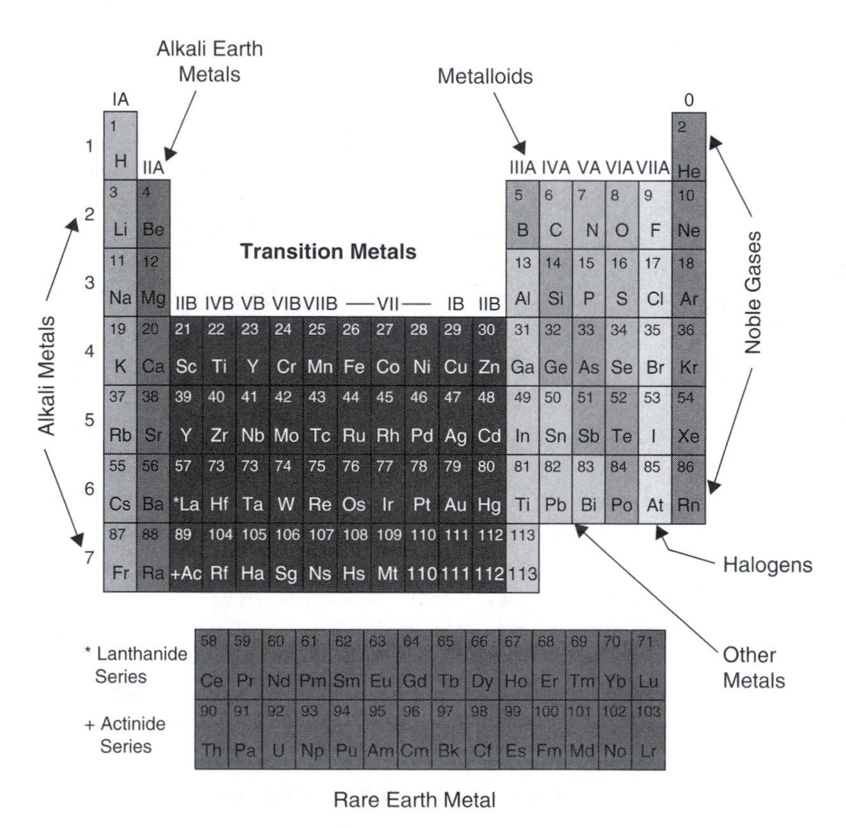

Fig. 7.1. Periodic table of elements.

the smallest part of an element that can enter into a chemical reaction. The molecule, which may also be an atom, is the smallest subdivision of an element that is able to exist as a natural state of matter. The nucleus of an atom, consisting of protons and neutrons (hydrogen has only one proton in its nucleus), account for virtually all of the atomic mass, or the atomic mass unit (amu). The term nucleon is inclusive of protons and neutrons (i.e. the particles comprising the atom's nucleus). An amu is defined as one-twelfth of the mass of carbon (C^{12}), or 1.66×10^{-27} kg. The atomic weight of an element listed in most texts and handbooks is the relative atomic weight, which is the total number of nucleons in the atom. So, for example, oxygen (O) has an atomic mass of 16. The atomic number (Z) is the number of protons in the nucleus. The chemical nomenclature for atomic weight A and number of element E is in the form:

$$_{z}\mathrm{E}^{A} \tag{7.1}$$

However, since an element has only one atomic number, Z is usually not shown. For example, the most stable form of carbon is seldom shown as $_{12}C^{12}$, and is usually indicated as C^{12}.

Elements may have different atomic weights if they have different numbers of neutrons (the number of electrons and protons of stable atoms must be the same). The elements' forms with differing atomic weights are known as *isotopes*. All atoms of a given element have the same atomic number, but atoms of a given element may contain different numbers of neutrons in the nucleus. An element may have numerous isotopes. Stable isotopes do not undergo natural radioactive decay, whereas radioactive isotopes involve spontaneous radioactive decay, as their nuclei disintegrate. This decay leads to the formation of new isotopes or new elements. The stable product of an element's radioactive decay is known as a radiogenic isotope. For example, lead (Pb; $Z = 82$) has four naturally occurring isotopes of different masses (^{204}Pb, ^{206}Pb, ^{207}Pb, ^{208}Pb). Only the isotope ^{204}Pb is stable. The isotopes ^{206}Pb and ^{207}Pb are daughter (or progeny) products from the radioactive decay of uranium (U), while ^{208}Pb is a product from thorium (Th) decay. Owing to the radioactive decay, the heavier isotopes of lead will increase in abundance compared to ^{204}Pb.

The kinds of chemical reactions for all isotopes of the same element are the same. However, the rates of reactions may vary. This can be an important factor, for example, in dating material. Such processes have been used to ascertain the sources of pollution (see Discussion Box: Carbon Dating).

Radiogenic isotopes are useful in determining the relative age of materials. The length of time necessary for the original number of atoms of a radioactive element in a rock to be reduced by half (*radioactive half-life*) can range from a few seconds to billions of years. Scientists use these "radioactive clocks" by the following procedure[1] by:

1. Extracting and purifying the radioactive parent and daughter from the relevant rock or mineral.

[1] US Geological Survey, Radiogenic Isotopes and the Eastern Mineral Resources Program of the US Geological Survey, 2003.

2. Measuring variations in the masses of the parent and daughter isotopes.
3. Combining the abundances with the known rates of decay to calculate an age.

Radiogenic isotopes are being increasingly used as *tracers*, the movement of substances through the environment. Radiogenic isotope tracer applications using Pb, strontium (Sr), and neodymium (Nd), among others make use of the fact that these are heavy isotopes, in contrast to lighter isotopes such as hydrogen (H), oxygen (O), and sulfur (S). Heavy isotopes are relatively unaffected by changes in temperature and pressure during transport and accumulation, variations in the rates of chemical reactions, and the coexistence of different chemical species available in the environment. Chemical reactions and processes involving Pb, for example, will not discriminate among the naturally occurring isotopes of this element on the basis of atomic mass differences (^{204}Pb, ^{206}Pb, ^{207}Pb, ^{208}Pb).

Long-term monitoring data are frequently not available for environmental systems, so indirect methods, like radiogenic isotope calculations must be used. For example, in sediments, chronological scales can be determined by the distribution of radioactive isotopes in the sediment, based on the isotopes' half-lives.[2] The age of the sediment containing a radioactive isotope with a known half-life can be calculated by knowing the original concentration of the isotope and measuring the percentage of the remaining radioactive substance. For this process to work the chemistry of the isotope must be understood, the half-life known, and the initial amount of the isotope per unit substrate accurately estimated. The only change in concentration of the isotope must be entirely attributable to radioactive decay, with a reliable means for measuring the concentrations. The effective range covers approximately eight half-lives. The four isotopes meeting these criteria (^{137}Cs, ^{7}Be, ^{14}C, and ^{210}Pb) are being used to measure the movement (e.g. deposition and lateral transport) for the past 150 years. The following summarizes the uses and potential uses of these four radioisotopes in dating recent sediment. This process also lends itself to differentiating sources of air pollutants (e.g. that caused by recent human activities from that of geological or biological origins, as discussed in the box below).

The process is analogous to an hourglass (see Fig. 7.2), where the number of grains of sand in the top reservoir represents the parent isotope and the sand in the bottom reservoir represents the daughter isotopes. A measurement of the ratio of the number of sand grains in the two reservoirs will give the length of time that the sand has been flowing, which represents the process of radioactive decay. For deposited material like an aerosol deposited from the troposphere onto the soil or taken up by flora, the counting begins when the aerosol is deposited so that the carbon or other element is taken up into the plant (t_0), and the exchange between the water and particle ceases. As the particles and plant life are subsequently buried, the parent isotope decays to the daughter products.

[2] US Geological Survey, Short-Lived Isotopic Chronometers: A Means of Measuring Decadal Sedimentary Dynamics, FS-073-98, 2003.

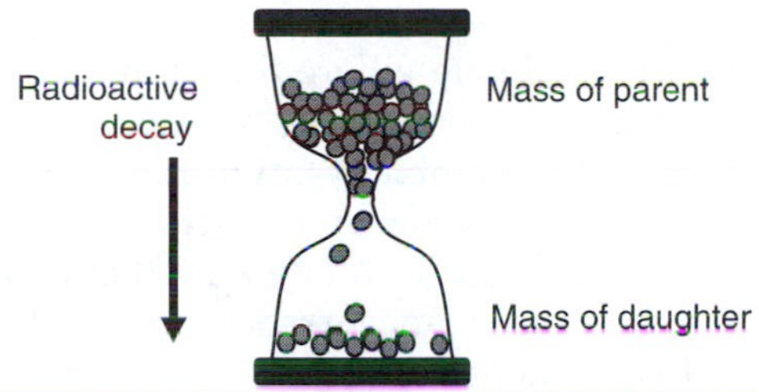

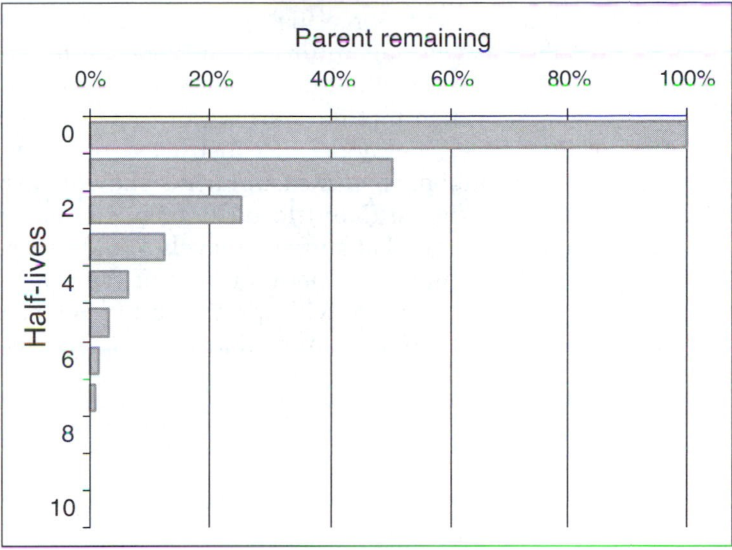

Fig. 7.2. Radio-dating is a function of the radioactive decay of specific isotopes in a substance. The hourglass analogy holds, where the number of grains of sand in the top reservoir represents the parent isotope and the sand in the bottom reservoir represents the daughter isotopes. A measurement of the ratio of the number of sand grains in the two reservoirs will give the length of time that the sand has been flowing (radioactive decay). *Source*: US Geological Survey, *Short-Lived Isotopic Chronometers: A Means of Measuring Decadal Sedimentary Dynamics*, FS-073-98, 2003.

SOURCE APPORTIONMENT, RECEPTOR MODELS, AND CARBON DATING

When the results of air pollution measurements are interpreted, one of the first questions asked by scientists, engineers, and policy makers is where did it come from? Sorting out the various sources of pollution is known as *source apportionment*. A number of tools are used to try to locate the sources of pollutants. A widely used approach is the "source-receptor model" or as it is more commonly known, the *receptor model*.

Receptor models are often distinguished from the atmospheric and hydrologic dispersion models. For example, dispersion models usually start from the source and estimate where the plume and its contaminants is heading (see Fig. 7.3). Conversely, receptor models are based on measurements taken in the ambient environment and from these observations, make use of algorithms and functions to determine pollution sources. One common approach is the mathematical "back trajectory" model. Often, chemical co-occurrences are applied. So, it may be that a certain fuel is frequently contaminated with a conservative and, hopefully, unique element. Some fuel oils, for example, contain trace amounts of the element vanadium. Since there are few other sources of vanadium in most ambient atmospheric environments, its presence is a strong indication that the burning of fuel oil is a most likely source of the plume. The model, if constructed properly, can even quantify the contribution. So, if measurements show that sulfur dioxide (SO_2) concentrations are found to be $10\,\mu g\,m^{-3}$ in an urban area, and vanadium is also found at sufficient levels to indicate home heating systems are contributing a certain amount of the SO_2 to the atmosphere, the model will correlate with the amount of SO_2 coming from home heating systems. If other combustion sources, e.g. cars and

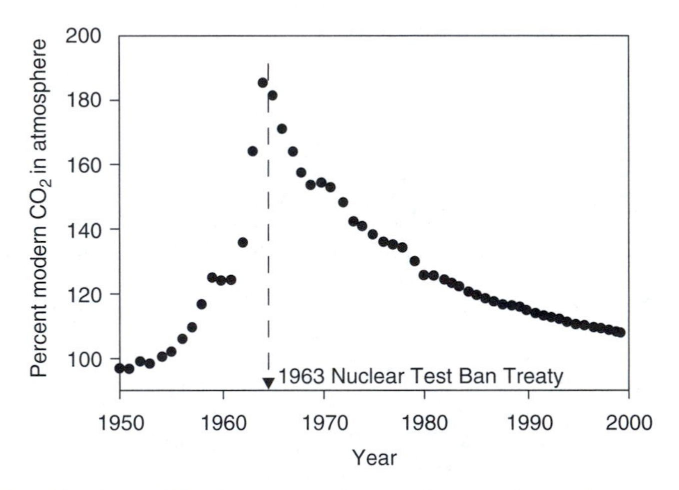

Fig. 7.3. Biospheric 14C enhancement of atmospheric modern carbon as a result of radiocarbon additions from nuclear testing and nuclear power generation. The plot indicates the time record of ^{14}C in the biosphere. The ^{14}C content of northern hemisphere biomass carbon was doubled in 1963, but since the cessation of atmospheric nuclear testing, the excess ^{14}C is now nearing natural, cosmic ray background levels. Fraction of modern carbon relative standard uncertainties are typically $<0.5\%$. *Source*: National Institute of Standards and Technology, A critical evaluation of interlaboratory data on total, elemental, and isotopic carbon in the carbonaceous particle reference material, NIST SRM 1649a. **107** (3) (2002); and Lewis, C., Klouda, G., and Ellenson, W., *Cars or Trees: Which Contribute More to Particulate Matter Air Pollution?* US Environmental Protection Agency, Science Forum, Washington, DC, 2003.

power plants, also have unique trace elements associated with their SO_2 emissions, further SO_2 source apportionment can occur, so that the total may look something like Table 7.1.

Receptor models need tracers that are sufficiently sensitive and specific to identify sources. We also mentioned that they be "conservative." This, perhaps, needs some explanation. A conservative tracer is a chemical that strongly resists chemical reactions but has transport properties similar to those of other, more reactive chemicals (i.e. the contaminant of concern). A good tracer is one that moves exactly with the fluid (i.e. the air), so if a chemical is reactive, its mass diminishes and does not allow direct interpretation of the advective movement of the plume in an air mass. A conservative tracer also makes for a sufficiently stable compound to sample form the atmosphere and to analyze in the laboratory.

One very promising development for such tracers is the comparison of carbon isotopes. Since combustion involves the oxidation of organic matter, which always contains carbon, it stands to reason that if there were a way to distinguish "old carbon" from "new carbon" we may have a reliable means of differentiating fossil fuels from *biogenic* hydrocarbon sources (e.g. volatile organic carbons released from coniferous trees, including pinene). As the name implies, fossil fuels are made up of carbon deposited long ago and until now, the carbon has been sequestered. During that time the ratio of the isotopes of carbon has changed. So, the ratios can tell us whether the carbon we are measuring had been first sequestered a few years ago or many thousands of years ago.

TABLE 7.1

Hypothetical Source Apportionment of Measured Sulfur Dioxide Concentrations

Source	Distance from measurement (km)	SO_2 concentration contributed to ambient measurement ($\mu g\ m^{-3}$)	Percent contribution to measured SO_2
Coal-fired electric generating station	25	3.0	30
Coal-fired electric generating station	5	2.0	20
Mobile sources (cars, trucks, trains, and planes)	0–10	1.5	15
Oil refinery	30	1.5	15
Home heating (fuel oil)	0–1	1.0	10
Unknown	Not applicable	1.0	10
Total		10.0	100

Naturally occurring radioactive carbon (^{14}C) is present at very low concentrations in all biotic (living) matter. The ^{14}C concentrations result from plants' photosynthesis of atmospheric carbon dioxide (CO_2), which contains all of the natural isotopes of carbon. However, no ^{14}C is found in fossil fuels since all of the carbon has had sufficient time to undergo radioactive decay. Studies have begun to take advantage of this dichotomy in ratios. For example, they have begun to address an elusive contributor to particulate matter (PM), i.e. *biogenic* hydrocarbons. In the summer months, biogenic aerosols are formed from gas-to-particle atmospheric conversions of volatile organic compounds (VOCs) that are emitted by vegetation.[3] New methods for estimating the contribution of biogenic sources of VOCs and PM are needed because current estimates of the importance of biogenic aerosols as contributors to total summertime PM have very large ranges (from negligible to dominant). There are large uncertainties in both the conversion mechanisms, and the amount and characteristics of biogenic VOC emissions.

The good news seems to be that direct experimental estimates can be gained by measuring the quantity of ^{14}C in a PM sample. The method depends on the nearly constant fraction of ^{14}C relative to ordinary carbon (^{12}C) in all living and recently living material, and its absence in fossil fuels. The fine fraction of PM ($PM_{2.5}$) summertime samples are available from numerous locations in the United States, from which ^{14}C measurements, can be conducted. Some recent studies have shown that the carbonaceous biogenic fraction may be contributing as much as one-half of the particles formed from VOCs.

The method for measuring and calculating the isotope ratios is straightforward. The percent of modern carbon (pMC) equals the percentage of ^{14}C in a sample of unknown origin relative to that in a sample of living material, and this pMC is about equal to the percentage of carbon in a sample that originated from non-fossil (i.e. biogenic) sources. So, for sample[4] X:

$$pMC_x = \frac{\left({}^{14}C/{}^{13}C\right)X}{0.95 \cdot \left({}^{14}C/{}^{13}C\right)_{SRM4990B}} \times 100 \tag{7.2}$$

Where, the numerator is the ratio measured in the $PM_{2.5}$ sample, and the denominator is the ratio measured using the method specified by the

[3] Lewis, C., Klouda, G., and Ellenson, W., *Cars or Trees: Which Contribute More to Particulate Matter Air Pollution?* US Environmental Protection Agency, Science Forum, Washington, DC, 2003.

[4] This is the carbon component of a fine particulate sample ($PM_{2.5}$), such as those measured at ambient air monitoring stations. The ratios are calculated according to the National Bureau of Standards, Oxalic Acid Standard Reference Method SRM 4990B.

National Institute of Standards and Testing (NIST) for modern carbon.[5] Further:

$$pMC_{Fossilfuel} = 0 \quad (7.3)$$

Thus, for a sample X, the biogenic fraction is:

$$\%BiogenicC_x = \frac{pMC_x}{pMC_{Biogenic}} \times 100 \quad (7.4)$$

The 0.95 correction is needed to address the increasing in radiocarbon due to nuclear weapons testing in the 1950s and 1960s (see Fig. 7.2) and to calibrate the measurements with the standard used for radiocarbon dating (i.e. wood from 1890).[6] Although the levels have dropped since the 1963 test ban treaty, they are still elevated above the pre-1950s background level.

III. EXPRESSIONS OF CHEMICAL CHARACTERISTICS

The gravimetric fraction of an element in a compound is the fraction by mass of the element in that compound. This is found by a gravimetric (or ultimate) analysis of the compound. The empirical formula of a compound provides the relative number of atoms in the compound. The empirical formula is found by dividing the gravimetric fractions (percent elemental composition) by atomic weights of each element in the compound, and dividing all of the gravimetric fraction-to-atomic weight ratios by the smallest ratio.

EMPIRICAL FORMULA DEVELOPMENT EXAMPLE

An air sampling stainless steel canister was evacuated by the local fire department and brought to the environmental laboratory for analysis. The person who brought in the sample said that the sample was taken near a site where a rusty 55-gallon drum was found by some children in creek near their school. The children and neighbors reported an unpleasant smell near the site where the drum was found.

[5] National Bureau of Standards, Oxalic Acid Standard Reference Method SRM 4990B.

[6] The defined reference standard for ^{14}C is 0.95 times the ^{14}C specific activity of the original NBS Oxalic Acid Standard Reference Material (SRM 4990B), adjusted to a ^{13}C delta value of −19.09‰. This is "modern" carbon. It approximates wood grown in 1890 that was relatively free of CO_2 from fossil sources. Due to the anthropogenic release of radiocarbon from nuclear weapons testing and nuclear power generation, oxalic acid from plant material grown after World War II is used currently to standardize ^{14}C measurements contains more ^{14}C than 1890 wood.

The gravimetric analysis of the gas in the canister indicated the following elemental compositions:

Carbon: 40.0%
Hydrogen: 6.7%
Carbon: 53.3%

Solution

First, divide the elemental percentage compositions by the respective atomic weights:

$$C: \frac{40.0}{12} = 3.3$$

$$H: \frac{6.7}{1} = 6.7$$

$$O: \frac{53.3}{16} = 3.3$$

Next, divide every ratio by the smallest ratio (3.3):

$$C: \frac{3.3}{3.3} = 1$$

$$H: \frac{6.7}{3.3} = 2$$

$$O: \frac{3.3}{3.3} = 1$$

So, the empirical formula is CH_2O or HCHO. This is formaldehyde, a toxic substance.

Preliminary Interpretation

The challenge of formaldehyde, however, is that it comes from many sources, including emissions from factories and automobiles, and even natural sources. However, since the drum seems to be a likely source, the liquid contents should be analyzed (and a search for additional drums should begin immediately—the illegal dumping often is not limited to a single unit).

The first likelihood is that the liquid is formalin, a mixture that contains formaldehyde. The high vapor pressure of the formaldehyde may be causing it to leave the solution and move into the air.

Since children are in the area and there may be a relatively large amount of the substance, steps must be taken to prevent exposures and to remove the formaldehyde.

A. The Periodic Table

The periodic table (Fig. 7.1) follows the periodic law, which states that the properties of elements depend on the atomic structure and vary systematically according to atomic number. The elements in the table are arranged according to increasing atomic numbers from left to right.

An element shares many physicochemical properties with its vertical neighbors, but differs markedly from its horizontal neighbors. For example, oxygen (O) will chemically bind and react similarly to sulfur (S) and selenium (Se), but behaves very differently from nitrogen (N) and fluorine (F). Elements in the horizontal rows, known as periods, grow increasingly different with the distance moved to the left or right. So, O differs physically and chemically more from boron (B) than O does from F, and O is a very different from lithium (e.g. O is a nonmetal and Li is a light metal).

The groups (vertical columns) are designated by numerals (often Roman numerals). For example, O is a group VIA element and gold (Au) is in group IB. The A and B designations are elemental families. Elements within families share many common characteristics. Within families, elements with increasing atomic weights become more metallic in their properties.

Metals (elements to the left of the periodic table) form positive ions (*cations*), are reducing agents, have low electron affinities, and have positive valences (oxidation numbers). Nonmetals (on the right side of the table) form negative ions (*anions*), are oxidizing agents, have high electron affinities, and have negative valences. Metalloids have properties of both metals and nonmetals. However, two environmentally important metalloids, arsenic (As) and antimony (Sb) are often treated as heavy metals in terms of fate, transport, and toxicity.

Some common period table chemical categories are:

- *Metals*: Every element except the nonmetals
- *Heavy metals*: Metals near the center of the table
- *Light metals*: Groups I and II
- *Alkaline earth metals*: Group IIA
- *Alkali metals*: Group IA
- *Transition metals*: All Group VIII and B families
- *Actinons*: Elements 90–102
- *Rare earths*: Lanthanons (Lanthanides), Elements 58–71
- *Metalloids*: Elements separating metals and nonmetals, Elements 5, 14, 32, 33, 51, 52, and 84
- *Nonmetals*: Elements 2, 5–10, 14–18, 33–36, 52–54, 85, and 86
- *Halogens*: Group VIIA
- *Noble gases*: Inert elements, Group 0

It is important to keep in mind that every element in the table has environmental relevance. In fact, at some concentration, every element except those generated artificially by fission in nuclear reactors are found in soils. Thus, it would be absurd to think of how to "eliminate" them. This is a common misconception, especially with regard to heavy metal and metalloid contamination.

For example, mercury (Hg) and lead (Pb) are known to be important contaminants that cause neurotoxic and other human health effects and environmental pollution. However, the global mass balance of these metals does not change, only their locations and forms (i.e. *speciation*). So, protecting health and ecological resources is a matter of reducing and eliminating exposures and changing the form of the compounds of these elements so that they are less mobile and less toxic. The first place to start such a strategy is to consider the oxidation states, or valence, of elements.

IV. ELECTROMAGNETIC RADIATION, ELECTRON DENSITY, ORBITALS, AND VALENCE

Quantum mechanics tells us that the energy of a photon of light can cause an electron to change its energy state, so that the electron is disturbed from its original state. *Electromagnetic radiation* (EMR) is related to atomic structure. Much that is known about atomic structure, especially an atom's arrangement of electrons around its nucleus, is from what scientists have learned about the relationship between matter and different types of EMR. One principle is that EMR has properties of both a *particle* and a *wave*. Particles have a definite mass and occupy space (i.e. they conform to the classic description of matter). Waves have no mass but hold energy with them as they travel through space. Waves have four principle characteristics, i.e. speed (v), frequency (ν), wavelength (λ), and amplitude. These are demonstrated in Fig. 7.4.

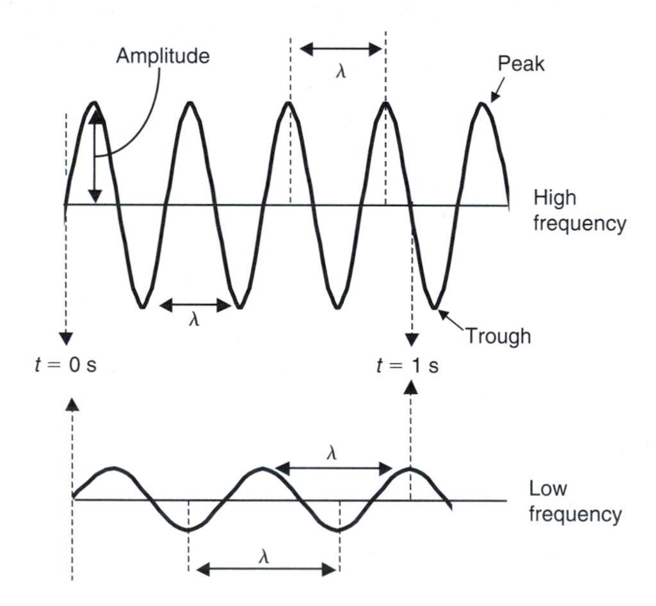

Fig. 7.4. EMR. The amplitude of the wave in the top chart is higher than that in lower chart. The bottom wave is 2.5 cycles per seconds (2.5 Hz). The top wave is 3.5 Hz, so the bottom wave has a 1 Hz lower frequency than the top wave.

Measuring ν in cycles per second (hertz, Hz) and λ in meters, the product gives the velocity of the wave moving through space:

$$v = \nu \lambda \tag{7.5}$$

For example, if a certain light's λ is 10^{-7} m and its ν is 10^{15} Hz, then the velocity of that light is $10^{8}\,\mathrm{m\,s^{-1}}$.

ELECTROMAGNETIC RADIATION

Most of the time, when someone mentions the term "environmental contaminant," it calls to mind some chemical compound. However, contaminants may also come in the form of biological or physical agents. Biological contaminants may be pathogenic bacteria or viruses that adversely affect health, or introduced species (e.g. the zebra mussel or kudzu) that harm ecosystems. Physical agents are often the least likely to come to mind. A common physical contaminant is energy. Life depends on energy, but like most resources, when it comes in the wrong form and quantity, it may be harmful. EMR is comprised of wave functions that are propagated by simultaneous periodic variations in electrical and magnetic field intensities (see Fig. 7.3). Natural and many anthropogenic sources produce EMR energy in the form of waves, which are oscillating energy fields that can interact with an organism's cells. The waves are described according to their wavelength and frequency, and the energy that they produce.

Wave frequency is the number of oscillations that passes a fixed point per unit of time, measured in cycles per second (cps). 1 cps = 1 hertz (Hz). Thus, the shorter the wavelength, the higher the frequency. For example, the middle of the amplitude modulated (AM) radio broadcast band has a frequency of one million hertz (i.e. 1 megahertz = 1 MHz) and a wavelength of about 300 m. Microwave ovens use a frequency of about 2.5 billion hertz (i.e. 2.5 gigahertz = 2.5 GHz) and a wavelength of 12 cm. So, the microwave, with its shorter wavelength has a much higher frequency.

An EMR wave is made of tiny packets of energy called photons. The energy in each photon is directly proportional to the frequency of the wave. So the higher the frequency, the more energy there will be in each photon. Biological tissue and cellular material is affected in part by the intensity of the field and partly by quantity of energy in each photon.

At low frequencies EMR waves are known as electromagnetic fields and at high frequencies EMR waves are referred to as electromagnetic radiations. Also, the frequency and energy determines whether an EMR will be ionizing or non-ionizing radiation. Ionizing radiation consists of high frequency electromagnetic waves (e.g. X-rays and gamma rays), having sufficient photon energy to produce ionization (producing positive and negative electrically charged atoms or parts of molecules) by breaking bonds of molecules. The general term non-ionizing radiation is the portion of the electromagnetic spectrum where photon energies are not strong enough to break atomic bonds. This segment of the spectrum includes ultraviolet (UV) radiation, visible light, infrared radiation, radio waves, and microwaves, along with static electrical and magnetic fields. Even at high intensities, non-ionizing radiation cannot ionize atoms in biological systems, but such radiation has been associated with other effects, such as cellular heating, changes in chemical reactions and rates, and the induction of electrical currents within and between cells.

Of course, not every EMR effect causes harm to an organism; such as when a mammal may respond to EMR by increasing blood flow in the skin in response to slightly greater heating from the sun. Life as we know it depends on various EMR wavelengths and frequencies, including the conversion of visible and UV wavelengths to infrared by the earth's surface which warms the planet. Photosynthesis depends on incoming light. EMR also induces positive health effects, such as the sun's role in helping the body produce vitamin D. Unfortunately, certain direct or indirect responses to EMR may lead to adverse effects, including skin cancer.

The data supporting UV as a contaminant are stronger than those associated with more subtle fears that sources, like high-energy power transmission lines and cell phones may be producing health effects. The World Health Organization (WHO) is addressing the health concerns raised about exposure to radio frequency (RF) and microwave fields, intermediate frequencies (IF), extremely low frequency (ELF) fields, and static electric and magnetic fields. IF and RF fields produce heating and the induction of electrical currents, so it is highly plausible that this is occurring to some extent in cells exposed to IF and RF fields. Fields at frequencies above about 1 MHz primarily cause heating by transporting ions and water molecules through a medium. Even very low energy levels generate a small amount of heat, but this heat is carried away by the body's normal thermoregulatory processes. However, some studies indicate that exposure to fields too weak to cause heating may still produce adverse health consequences, including cancer and neurological disorders (i.e. memory loss).

Since, electrical currents already exist in the body as a normal part of the biochemical reactions and metabolic process, the fear is that should electromagnetic fields induce sufficiently high currents, the additive effects may overload the system and engender adverse biological effects.

ELF electric fields exist when a charge is generated, but hardly any of the electric field penetrates into the human body. At very high field strengths they can feel like one's skin is "crawling" or their hair is raised. Some studies, however, have associated low level ELF electric fields with elevated incidence of childhood cancer or other diseases, while other studies have not been able to establish a relationship. The WHO is recommending that more focused research be conducted to improve health risk assessments. ELF magnetic fields also exist whenever an electric current is flowing. However, unlike the ELF electric fields, magnetic fields readily penetrate an organism's tissue with virtually no attenuation. Again the epidemiology is mixed, with some studies associating ELF fields with cancer, especially in children, and others finding no such association. WHO continues to conduct studies in this area.

The primary action in biological systems by these static electrical and magnetic fields is by inducing electrical and magnetic reliable data to come to any conclusions about chronic effects associated with long-term exposure to static magnetic fields at levels found in the working environment.

The challenge of EMR is similar to that of chemical contamination. The exposure and risks associated with this hazard is highly uncertain. The key decision is whether sufficient scientific evidence exists to encourage limits on certain types of activities (either as an individual or as a public agency) where adverse effects may be occurring. Just walk in a mall, on a campus, or into a restaurant. Chances are you will see a number of people using cell phones. Drive through a neighborhood. Chances are you will see some overhead power lines. The challenge is deciding what level of evidence linking EMR to adverse effects would be sufficient to require actions to protect public health.

Quantum mechanics is the basis for numerous chemical analyses (e.g. mass spectrometry) of air pollutants. An atom's electrons occupy orbitals where the electrons contain various amounts of energy. Electrons vary in the spatial orientations and average distances from the nucleus, so that electrons occupying inner orbitals are closer to the nucleus (see Fig. 7.5). The electron's velocity or position changes. If we can measure the electron's position we will not know its velocity, and if we measure the electron's velocity, we will not be able to know its position. We are uncertain about the electron's simultaneous position and velocity. This is the basis for the Heisenberg Uncertainty Principle, which tells us that the more we know about an electron's position, the less we can know about its velocity. Further, to keep an electron from escaping from an atom, the electron must maintain a minimum velocity, which corresponds back to the uncertainty of the electron's position in the

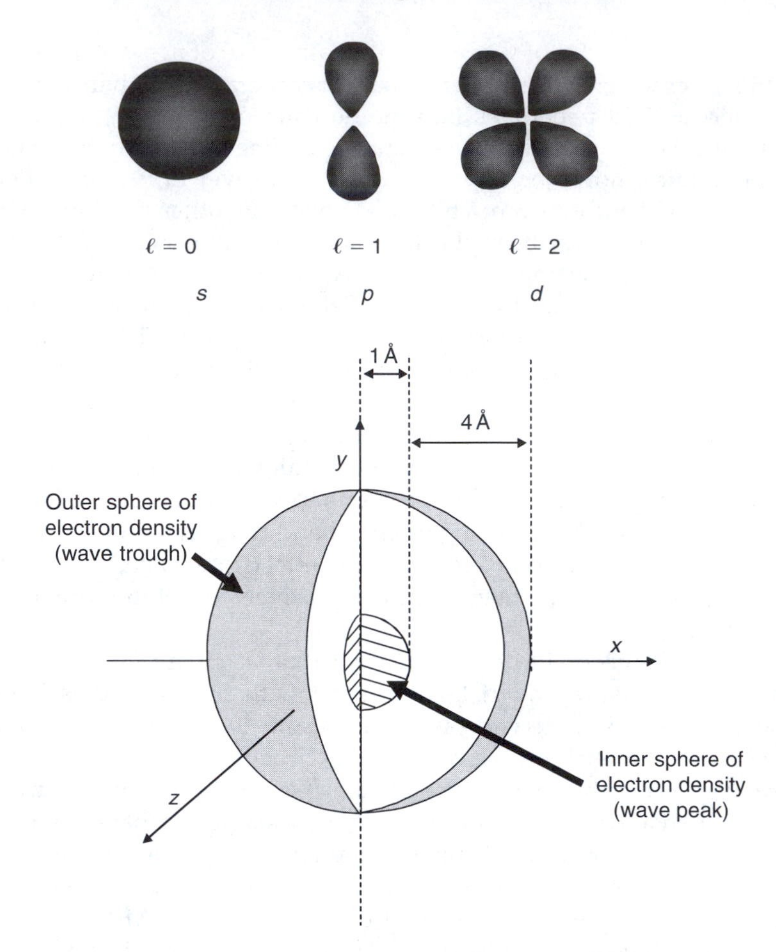

Fig. 7.5. Two-dimensional orbital shapes, showing three angular quantum numbers (ℓ) and subshells *s*, *p*, and *d* (top). The three-dimensional 2*s* orbital (bottom) is shown as a cutaway view into the atom, the peak (inner sphere) and trough (outer concentric sphere) of the electron wave. The orbital is two concentric spheres of electron densities. *Source*: Loudon, G., *Organic Chemistry*, 3rd ed. Benjamin/Cummings Publishing Company, Inc., Redwood City, CA, 1995.

atom. Since, the uncertainty of the position is actually the whole atom, i.e. the electron can be anywhere in the atom, chemists refer to an electron as a "cloud of electron density" within the atom rather than describing an electron as a finite particle. So, the electron orbitals are envisioned as regions in space where the electrons are statistically most likely to be located. We do not know where an electron is, but we know where it might be; that is, somewhere in the electron cloud.

The Schrödinger model applies three coordinates to locate electrons, known as quantum numbers. The coordinates are principal (n), angular (ℓ), and magnetic (m_l) quantum numbers. These characterize the shape, size, and orientation of the electron cloud orbitals of the atom. The principal quantum number n gives the size of the orbital. A relative size of $n = 2$ is larger than a cloud with the size $n = 1$. Energy is needed to excite an electron to make it move from a position closer to the nucleus (e.g. $n = 1$) to a position further from the nucleus ($n = 2$, 3, or higher). So, n is an indirect expression of an orbital's energy level.

The angular quantum number ℓ maps the shape of the cloud. A spherical orbital has an $\ell = 0$. Polar shaped orbitals have $\ell = 1$. Cloverleaf orbitals have $\ell = 2$. See Fig. 7.4 for renderings of these shapes. The magnetic quantum number $m_<$ describes the orientation of the orbital in space. Orbitals with the same value of n form a shell. Within a shell, orbitals are divided into subshells labeled by their ℓ value. The commonly used two-character description of shells (e.g. 2*p* or 3*d*) exemplifies the shell and subshell. For example, 2*p* indicates the shell ($n = 2$) and the subshell (p). Subshells are indicted by:

s: $\ell = 0$
p: $\ell = 1$
d: $\ell = 2$
f: $\ell = 3$
g: $\ell = 4$
h: $\ell = 5$

The number of subshells in any shell will equal the n for the shell. So, for example, the $n = 2$ shell contains two subshells (i.e. 2*s* and 2*p*) and $n = 5$ shell contains five subshells (i.e. 5*s*, 5*p*, 5*d*, 5*f*, and 5*g* orbitals).

The electrons occupying the outermost shell are known as valence electrons. Valence is the number of bonds that an element can form, which is related to number of electrons in the outermost shell. The arrangement of the electrons in the outermost (i.e. valence) determines the ultimate chemical behavior of the atom. The outer electrons become involved in transfer to and sharing with shells in other atoms, i.e. forming new compounds and ions. Note that the number of valence electrons in an "A" group in the periodic table (except helium, whose shell is filled with 2) is equal to the group number. So, sodium (Na), a Group 1A element, has one valence electron. Carbon (C) is a Group 4A compound, so it has four valence electrons. Chlorine (Cl), fluorine (F), and the other halogens of Group 7A have seven valence electrons. The noble gases, except He, in Group 8A, have eight valence electrons.

The noble gases are actually the only elements that exist as individual atoms, because the noble gases have no valence electrons. Conversely, carbon has four electrons in its outermost shell, so it has just as many electrons to gain or to lose (i.e. 4 is just as close to 8, for a newly filled shell, as it to 0, for the loss of a shell), so there are many ways for it to reach chemical stability. This is one of the reasons that so many subtly, but profoundly different compounds, i.e. organic compounds, are in existence. Most atoms combine by chemical bonding to other atoms, creating molecules.

Thus, the outermost electrons tell the story of how readily an element will engage in a chemical reaction and the type of reaction that will occur. The oxidation number is the electrical charge assigned to an atom. The sum of the oxidation numbers is equal to the net charge. Table 7.2 shows the oxidation numbers of certain atoms that form contaminants and nutrients in the environment. Table 7.3 gives the oxidation numbers for environmentally important radicals, i.e. groups of atoms that combine and behave as a single chemical unit. An atom will gain or lose valence electrons to form a stable ion that have the same number of electrons as the noble gas nearest the atom's atomic number. For example, Na with a single valence electron and a total of 11 electrons, will tend to lose an electron to form Na^{+}, the sodium

TABLE 7.2

Oxidation Numbers for Atoms Important to Air Pollution

Atom	Chemical Symbol	Oxidation Number(s)
Aluminum	Al	+3
Antimony	Sb	−3, +3, +5
Arsenic	As	−3, 0, +3, +5
Barium	Ba	+2
Boron	B	+3
Calcium	Ca	+2
Carbon	C	+2, +3, +4, −4
Chlorine	Cl	−1
Chromium	Cr	+2, +3, +6
Cobalt	Co	+2, +3
Copper	Cu	+1, +2
Fluorine	F	−1
Gold	Au	+1, +3
Hydrogen	H	+1
Iron	Fe	+2, +3
Lead	Pb	0, +2, +4
Lithium	Li	+1
Magnesium	Mg	+2
Manganese	Mn	+2, +3, +4, +6, +7
Mercury	Hg	0, +1, +2
Nickel	Ni	+2, +3
Nitrogen	N	−3, +2, +3, +4, +5
Oxygen	O	−2
Phosphorus	P	−3, +3, +5
Plutonium	Pu	+3, +4, +5, +6
Potassium	K	+1
Radium	Ra	+2
Radon	Rn	0 (noble gas)
Selenium	Se	−2, +4, +6
Silver	Ag	+1
Sodium	Na	+1
Sulfur	S	−2, +4, +6
Tin	Sn	+2, +4
Uranium	U	+3, +4, +5, +6
Zinc	Zn	+2

TABLE 7.3

Oxidation Numbers for Radicals Important to Air Pollution

Radical	Chemical symbol	Oxidation number(s)
Acetate	$C_2H_3O_2$	−1
Acrylate	$CHCO_2$	−1
Ammonium	NH_4	+1
Bicarbonate	HCO_3	−1
Borate	BO_3	−3
Carbonate	CO_3	−2
Chlorate	ClO_3	−1
Chlorite	ClO_2	−1
Chromate	CrO_4	−2
Cyanide	CN	−1
Dichromate	Cr_2O_7	−2
Hydroperoxide	HO_2	−1
Hydroxide	OH	−1
Hypochlorite	ClO	−1
Nitrate	NO_3	−1
Nitrate	NO_2	−1
Perchlorate	ClO_4	−1
Permanganate	MnO_4	−1
Phosphate	PO_4	−3
Sulfate	SO_4	−2
Sulfite	SO_3	−2
Thiocyanate	SCN	−1
Thiosulfate	S_2O_3	−2

cation. This ion has the same number of electrons (eight) as the nearest noble gas, neon (Ne). Fluorine, with seven valence electrons and nine total electrons, tends to gain (accept) an electron to form a 10-electron fluorine anion, F^- that, like the sodium ion, has the same number of electrons as Ne (eight).

Noble gases have an octet (i.e. group of eight) of electrons in their valence shells, meaning that the tendency of an atom to gain or to lose its valence electrons to form ions in the noble gas arrangement is called the "octet rule."[7] Chemical species of atoms are particularly stable when their outermost shells contain eight electrons.

Elements combine to form compounds. Two-element compounds are known as binary compounds. Three-element compounds are ternary or tertiary compounds. The representation of the relative numbers of each element is a chemical formula. Compounds are formed according to the law of definite proportions. That is, a pure compound must always be composed of the same elements that are always combined in a definite proportion by mass. Also, compounds form consistently with the law of multiple proportions, which states that when two elements combine to make more than one compound, the

[7] Obviously, for the atoms near He, it is the "duet rule."

combining mass of each element must always exist as small integer ratios to one another. The sum of all oxidation numbers must equal zero in a stable, neutral compound. The simplest example is water. Oxygen's −2 valence is balanced by the two hydrogen's +1 valences.

COMPOUND FORMATION EXAMPLE

Is $PbC_2H_3O_2$ a valid compound?

Answer

Consulting Tables 7.3 and 7.4, we find that lead (Pb) has two common oxidation numbers (+2 and +4), and that acetate ($C_2H_3O_2$) has an oxidation number of −1, so the molecular formula given is *not* valid.

Lead acetate is $Pb(C_2H_3O_2)_2$. Two atoms of acetate are needed to balance the +2 valence of Pb. The molecule is often called lead (II) acetate to show that in this instance it is the "divalent" form of lead that has reacted with the acetate radical. Incidentally, lead (II) compounds are suspected human carcinogens, based on experiments conducted on laboratory animals. Lead acetate is also very acutely toxic and may be fatal if swallowed, and is harmful if inhaled or absorbed through the skin. Like other lead compounds, long-term exposure may harm the central nervous system, blood, and gastrointestinal tract.

A. Physicochemical Processes in the Formation of Air Toxics

Atoms and molecules combine in many ways, according to the reactions described in Chapter 6. Let us consider some of the important processes under which toxic compounds are formed.

1. Combustions Reactions

Dioxins and furans are important air pollutants. Dioxin formation is illustrative of the complex set of combustion reactions that can generate toxic air pollutants. Chlorinated dioxins have 75 different forms and there are 135 different chlorinated furans, simply by the number and arrangement of chlorine atoms on the molecules. The compounds can be separated into groups that have the same number of chlorine atoms attached to the furan or dioxin ring. Each form varies in its chemical, physical, and toxicological characteristics (see Fig. 7.6).

Dioxins are highly toxic compounds that are created unintentionally during combustion processes. The most toxic form is the 2,3,7,8-tetrachlorodibenzo-*p*-dioxin (TCDD) isomer. Other isomers with the 2,3,7,8 configuration are also

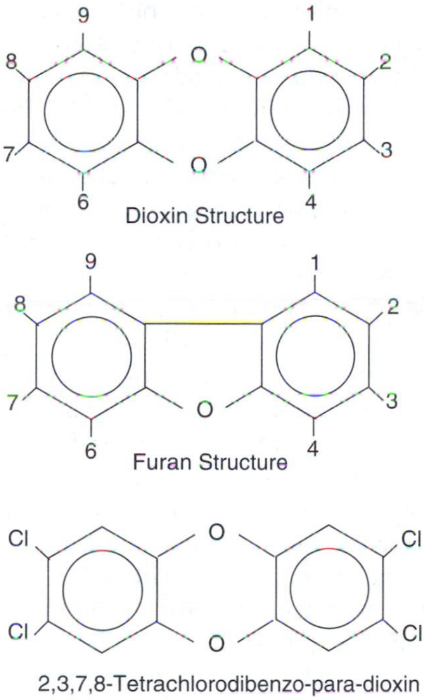

Dioxin Structure

Furan Structure

2,3,7,8-Tetrachlorodibenzo-para-dioxin

Fig. 7.6. Molecular structures of dioxins and furans. Bottom structure is of the most toxic dioxin congener, tetrachlorodibenzo-*p*-dioxin (TCDD), formed by the substitution of chlorine for hydrogen atoms at positions 2, 3, 7, and 8 on the molecule.

considered to have higher toxicity than the dioxins and furans with different chlorine atom arrangements.

What is currently known about the conditions needed to form these compounds has been derived from studying full-scale municipal solid waste incinerators, and the experimental combustion of fuels and feeds in the laboratory. Most of the chemical and physical mechanisms identified by these studies can relate to combustion systems in which organic substances combusted in the presence of chlorine (Cl). Incinerators of chlorinated wastes are the most common environmental sources of dioxins, accounting for about 95% of the volume.

The emission of dioxins and furans from combustion processes may follow three general physicochemical pathways. The first pathway occurs when the feed material going to the incinerator contains dioxins and/or furans and a fraction of these compounds survives thermal breakdown mechanisms, and pass through to be emitted from vents or stacks. This is not considered to account for a large volume of dioxin released to the environment, but it may account for the production of dioxin-like, coplanar polychlorinated biphenyls (PCBs).

The second process is the formation of dioxins and furans from the thermal breakdown and molecular rearrangement of precursor compounds, such as the chlorinated benzenes, chlorinated phenols (such as pentachlorophenol, PCP), and PCBs, which are chlorinated aromatic compounds with structural resemblances to the chlorinated dioxin and furan molecules. Dioxins appear to form after the precursor has condensed and adsorbed onto the surface of particles, such as fly ash. This is a heterogeneous process, where the active sorption sites on the particles allow for the chemical reactions, which are catalyzed by the presence of inorganic chloride compounds and ions sorbed to the particle surface. The process occurs within the temperature range, 250–450°C, so most of the dioxin formation under the precursor mechanism occurs away from the high-temperature zone in the incinerator, where the gases and smoke derived from combustion of the organic materials have cooled during conduction through flue ducts, heat exchanger and boiler tubes, air pollution control equipment or the vents and the stack.

The third means of synthesizing dioxins is *de novo* within the so-called "cool zone" of the incinerator, wherein dioxins are formed from moieties different from those of the molecular structure of dioxins, furans, or precursor compounds. Generally, these can include a wide range of both halogenated compounds like polyvinyl chloride (PVC), and non-halogenated organic compounds like petroleum products, non-chlorinated plastics (polystyrene), cellulose, lignin, coke, coal, and inorganic compounds like particulate carbon, and hydrogen chloride gas. No matter which *de novo* compounds are involved, however, the process needs a chlorine donor (a molecule that "donates" a chlorine atom to the precursor molecule). This leads to the formation and chlorination of a chemical intermediate that is a precursor. The reaction steps after this precursor is formed can be identical to the precursor mechanism discussed in the previous paragraph.

De novo formation of dioxins and furans may involve even more fundamental substances than those moieties mentioned above. For example, dioxins may be generated[8] by heating of carbon particles absorbed with mixtures of magnesium–aluminum silicate complexes when the catalyst copper chloride ($CuCl_2$) is present (see Table 7.4 and Fig. 7.7). The *de novo* formation of chlorinated dioxins and furans from the oxidation of carbonaceous particles seems to occur at around 300°C. Other chlorinated benzenes, chlorinated biphenyls, and chlorinated naphthalene compounds are also generated by this type of mechanism.

Other processes generate dioxin pollution. A source that has been greatly reduced in the last decade is the paper production process, which formerly used chlorine bleaching. This process has been dramatically changed, so that most paper mills no longer use the chlorine bleaching process. Dioxin is also produced in the making of PVC plastics, which may follow chemical and physical mechanisms similar to the second and third processes discussed above.

Since dioxin and dioxin-like compounds are lipophilic and persistent, they accumulate in soils, sediments, and organic matter and can persist in solid

[8] Stieglitz, L., Zwick, G., Beck, J., Bautz, H., and Roth, W., *Chemosphere* **19**, 283 (1989).

TABLE 7.4

De Novo **Formation of Chlorinated Dioxins and Furans after Heating Mg–Al Silicate, 4% Charcoal, 7% Cl, 1% $CuCl_2 \cdot H_2O$ at 300°C**

	Concentrations (ng g^{-1})				
	Reaction time (h)				
Compound	0.25	0.5	1	2	4
Tetrachlorodioxin	2	4	14	30	100
Pentachlorodioxin	110	120	250	490	820
Hexachlorodioxin	730	780	1600	2200	3800
Heptachlorodioxin	1700	1840	3500	4100	6300
Octachlorodioxin	800	1000	2000	2250	6000
Total chlorinated dioxins	*3342*	*3744*	*7364*	*9070*	*17 020*
Tetrachlorofuran	240	280	670	1170	1960
Pentachlorofuran	1360	1670	3720	5550	8300
Hexachlorofuran	2500	3350	6240	8900	14 000
Heptachlorofuran	3000	3600	5500	6700	9800
Octachlorofuran	1260	1450	1840	1840	4330
Total chlorinated furans	*8360*	*10 350*	*17 970*	*24 160*	*38 390*

Source: Stieglitz, L., Zwick, G., Beck, J., Bautz, H., and Roth, W. *Chemosphere* **19**, 283 (1989).

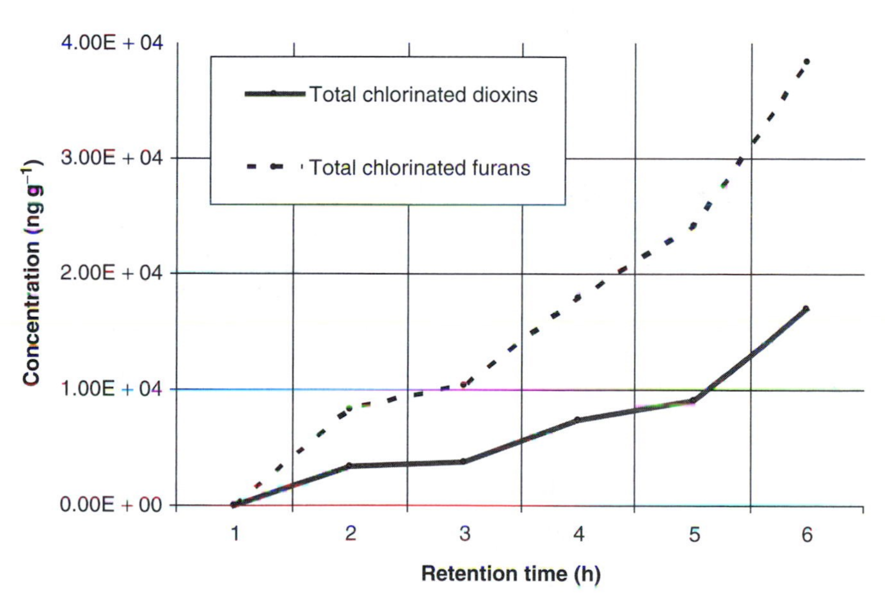

Fig. 7.7. *De novo* formation of chlorinated dioxins and furans after heating Mg–Al silicate, 4% charcoal, 7% Cl, 1% $CuCl_2$ H_2O at 300°C. *Source*: Stieglitz, L., Zwick, G., Beck, J., Bautz, H., and Roth,W. *Chemosphere* **19**, 283 (1989).

and hazardous waste disposal sites.[9] These compounds are semivolatile, so they may migrate away from these sites and transported in the atmosphere either as aerosols (solid and liquid phase) or as gases (the portion of the compound that volatilizes). Therefore, the engineer must take great care in removal, and remediation efforts not to unwittingly cause releases from soils and sediments via volatilization or via perturbations, such as landfill and dredging operations.

Dioxin demonstrates the complexity of air pollution's physicochemical processes. However, at the most basic level only a few types of chemical reactions dominate in the environment. These include ionization, acid–base, precipitation, and oxidation–reduction (redox).

2. Ionization

When a salt is dissolved in water, it dissociates into ionic forms. Notwithstanding their variability in doing so, under the right conditions all complexes can become dissolved in water. Ions that are dissolved in a solution can react with one another, and can form solid complexes and compounds. Actually, a salt compound does not exist in water. For example, when the salts calcium sulfate and sodium chloride are added to water, it is commonly held that $CaSO_4$ and NaCl are in the water. However, what is really happening is that the metals and nonmetals are associated with one another as:

$$CaSO_4(s) \Leftrightarrow Ca^{2+}(aq) + SO_4^{2-}(aq) \tag{7.6}$$

and

$$NaCl(s) \Leftrightarrow Na^{+}(aq) + Cl^{-}(aq) \tag{7.7}$$

Thus, all four of the dissociated ions are free and no longer associated with each other. That is, the Na, Ca, Cl, and SO_4 ions are "unassociated" in the water. The Na and the Cl are no longer linked to each other as they were before the compound was added to the water.

Even though atoms are neutral, in the process of losing or gaining electrons, they become electrically charged, i.e. they become *ions*. An atom that loses one or more electrons is positively charged, known as a *cation*. For example, the sodium atom loses one electron and becomes the monovalent potassium cation:

$$K - e^{-} \rightarrow K^{+} \tag{7.8}$$

When the mercury atom loses two electrons, it becomes the divalent mercury cation:

$$Hg - 2e^{-} \rightarrow Hg^{2+} \tag{7.9}$$

[9] For discussion of the transport of dioxins, see Koester, C. J., and Hites, R. A., Wet and dry deposition of chlorinated dioxins and furans, *Environ. Sci. Technol.* **26**, 1375–1382 (1992); and Hites, R. A., Atmospheric transport and deposition of polychlorinated dibenzo-*p*-dioxins and dibenzofurans, EPA/600/3-91/002. Research Triangle Park, NC, 1991.

When the chromium atom loses three electrons it becomes the trivalent chromium cation:

$$Cr - 3e^- \rightarrow Cr^{3+}$$

Conversely, an atom that gains electrons becomes a negatively charged ion, known as an *anion*. For example, when chlorine gains an electron it becomes the chlorine anion:

$$Cl + e^- \rightarrow Cl^- \quad (7.10)$$

When sulfur gains two electrons, it becomes the divalent sulfide anion:

$$S + 2e^- \rightarrow S^{2-} \quad (7.11)$$

Note that the Greek prefix (mono-, di-, tri-, ...) denoting the valence is the number of electrons that the ion differs from neutrality.

Reactions between ions, known as ionic reactions, frequently occur as ions of water-soluble salts can react in aqueous solution to form salts that are nearly insoluble in water. This causes them to separate into insoluble precipitates:

$$\text{(Ions in Solution 1)} + \text{(Ions in Solution 2)} \rightarrow \text{Precipitate} + \text{Unreacted ions} \quad (7.12)$$

3. Solubility and Electrolytes

The aqueous solubility of contaminants ranges from completely soluble in water to virtually insoluble. Solubility equilibrium is the phenomenon that keeps molecules dissolved in a solvent. Solubility and precipitation are in a way, two sides of the same coin. There is truth in the old chemists' pun, "If you're not part of the solution, you are part of the precipitate!" Solubilities typically are quantitatively expressed in mass of solute per volume of solvent (e.g. $mg\,L^{-1}$), and sometimes expressed by the adjectives "soluble," "slightly soluble," or "insoluble."

In the solid phase, a salt is actually a collection of ions in a lattice, where the ions are surrounded by one another. However, when the salt is dissolved in water, the ions become surrounded by the water, rather than by the other ions. Each ion now has its own coordinating water envelope or "hydration sphere," i.e. a collection of water molecules surrounding it. An ion-association reaction is an ion–ion interaction between ions in an electrolyte (i.e. ion-containing) solution.[10] So, when the salt lattice enters the

[10] The Swedish chemist, Svante Arrhenius, is credited with establishing the relationship between electrical and chemical properties of molecules. He observed that particular chemical compounds (later to be known as electrolytes) conduct electricity when they are dissolved in water, while other chemicals do not. He also saw that certain chemicals are involved in seemingly instantaneous reactions, while others took much longer to react. Finally, he observed that particular chemical compounds showed extremely strange colligative properties while others were consistent with Raoult's Law, which states that the solvent's vapor pressure in an ideal solution is equal to the product of the mole fraction of the solvent and the vapor pressure of the pure solvent. The four colligative properties of solutions are the elevation of boiling point, the depression of freezing point, the decreasing of vapor pressure, and osmotic pressure.

water, the ions assemble into couplets of separate oppositely charged ions, i.e. cations and anions, the so-called "ion pairs." The pairs are held together by electrostatic attraction. Ion association is the reverse of dissociation where the ions separate from a compound into free ions. Ions (or molecules) surrounded by water exist in the aqueous phase. So, ionic compounds that are soluble in water break apart (i.e. dissociate) into their ionic components, i.e. anions and cations (see Fig. 7.8).

Solutions may contain nonelectrolytes, strong electrolytes, and/or weak electrolytes. Nonelectrolytes do not ionize. They are nonionic molecular compounds that are neither acids nor bases. Sugars, alcohols, and most other organic compounds are nonelectrolytic. Some inorganic compounds are also nonelectrolytes.

Weak electrolytes only partially dissociate in water. Most weak electrolytes dissociate less than 10%, i.e. greater than 90% of these substances remain undissociated. Organic acids, such as acetic acid, are generally weak electrolytes.

An example of dissociation is strontium carbonate dissolved in water:

$$SrCO_3(s) \rightarrow Sr^{2+}\,(aq) + CO_3^{2-}\,(aq) \tag{7.13}$$

Conversely, the reverse reaction forms a solid; that is, it returns from the solution to again form the lattice of ions surrounding ions. This is a precipitation reaction. In our Sr example, the carbonate species is precipitated:

$$Sr^{2+}\,(aq) + CO_3^{2-}\,(aq) \rightarrow SrCO_3\,(s) \tag{7.14}$$

The *ionic product* (Q) is a measure of the ions present in the solvent. The *solubility product constant* (K_{sp}) is the ionic product when the system is in equilibrium. So, solubility is often expressed as the specific K_{sp}, the equilibrium constant for dissolution of the substance in water. Since K_{sp} is another type of chemical equilibrium, it is a state of balance of opposing reversible chemical reactions which proceed at constant and equal rates, resulting in no net change in the system (hence the symbol, $\leftrightarrow$). Like sorption, Henry's law and other equilibrium constants, solubility follows Le Chatelier's Principle, which states that in a balanced equilibrium, if one or more factors change,

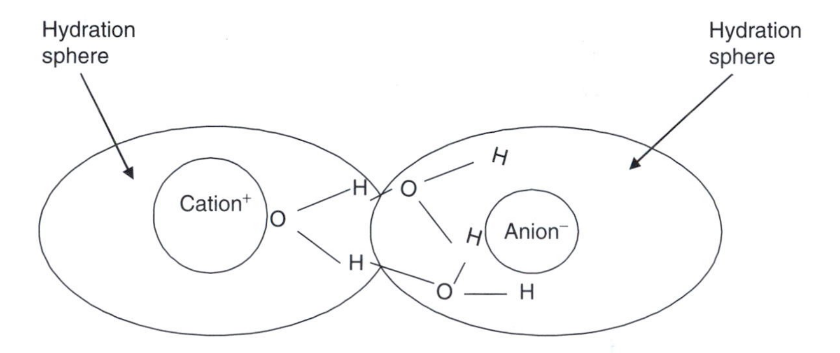

Fig. 7.8. Ion pairs. *Source*: Evangelou, V., *Environmental Soil and Water Chemistry: Principles and Applications*. John Wiley & Sons, New York, NY, 1998.

TABLE 7.5

Solubility Product Constant versus Solubility for Four Types of Salts

Salt	Example	Solubility product, K_{sp}	Solubility, S
AB	$CaCO_3$	$[Ca_{2+}][CO_3^2] = 4.7 \times 10^{-9}$	$(K_{sp})^{1/2} = 6.85 \times 10^{-5}$M
AB_2	$Zn(OH)_2$	$[Zn^{2+}][OH^-]^2 = 4.5 \times 10^{-17}$	$(K_{sp}/4)^{1/3} = 2.24 \times 10^{-6}$M
AB_3	$Cr(OH)_3$	$[Zn^{3+}][OH^-]^3 = 6.7 \times 10^{-31}$	$(K_{sp}/27)^{1/4} = 1.25 \times 10^{-8}$M
A_3B_2	$Ca_3(PO_4)_2$	$[Ca^{2+}]^3[PO_4^3]^2 = 1.3 \times 10^{-32}$	$(K_{sp}/108)^{1/5} = 1.64 \times 10^{-7}$M

Source: US Army Corps of Engineers, *Engineering and Design: Precipitation/Coagulation/Flocculation*, Chapter 2, EM 1110-1-4012, 2001.

the system will readjust to reach equilibrium. K_{sp} values and the resulting solubility calculations for some important reactions are shown in Table 7.5. K_{sp} constants for many reactions can be found in engineering handbooks.

Since $SrCO_3$ is a highly insoluble salt (aqueous solubility = 6×10^{-3} mg L^{-1}), its equilibrium constant for the reaction is quite small:

$$K_{sp} = [Sr^{+2}][CO_3^{2-}] = 1.6 \times 10^{-9} \quad (7.15)$$

The small K_{sp} value of the constant reflects the low concentration of dissolved ions. So, as the number of dissolved ions approaches zero, the compound is being increasingly insoluble in water. This does not mean that that an insoluble product cannot be dissolved, but that chemical treatment is needed. In this case, strontium carbonate requires the addition of an acid to *solubilize* the Sr^{+2} ion.

To precipitate a compound, the product of the concentration of the dissolved ions in the equilibrium expression must exceed the value of the K_{sp}. The concentration of each of these ions does not need to be the same. For example, if $[Sr^{+2}]$ is 1×10^{-5} molar, the carbonate ion concentration must exceed 0.0016 molar for precipitation to occur because $(1 \times 10^{-5}) \times (1.6 \times 10^{-4}) = 1.6 \times 10^{-9}$ (i.e. the K_{sp} for strontium carbonate).

DISSOCIATION AND PRECIPITATION REACTION EXAMPLE

An environmental analytical chemist adds 100 mL of 0.050 M NaCl to 200 mL of 0.020 M $Pb(NO_3)_2$. Will the lead chloride that is formed precipitate from the 300 mL sample?

Solution

Calculate the *ion product* (*Q*) and compare it to the K_{sp} for the reaction:

$$PbCl_2(s) \rightarrow Pb^{2+}(aq) + 2Cl^-(aq)$$

When the two solutions are mixed, the unassociated ions are formed as:

$$[Pb^{2+}] = 0.2\,L \times 2.0 \times 10^{-2}M/0.3\,L = 1.3 \times 10^{-2}M$$

and

$$[Cl^-] = 0.1\,L \times 5.0 \times 10^{-2}M/0.3\,L = 1.7 \times 10^{-2}M$$

The value for the ion product is calculated as:

$$Q = [Pb^{2+}][Cl^-]^2 = [1.3 \times 10^{-2}][1.7 \times 10^{-2}]^2 = 3.8 \times 10^{-7}$$

The K_{sp} for this reaction is 1.6×10^{-5}.

$Q < K_{sp}$, so no precipitate will be formed. If the ion product were greater than the K_{sp} a precipitate would have been formed.

Environmental conditions can affect solubility. One such instance is the *common ion effect*. Compared to a solution in pure water, an ion's solubility is decreased in an aqueous solution that contains a common ion (i.e. one of the ions that make up the compound). This allows a precipitate to form if the K_{sp} is exceeded. For example, soluble sodium carbonate (Na_2CO_3) in solution with strontium ions can cause the precipitation of strontium carbonate, since the carbonate ions from the sodium salt are contributing to their overall concentration in solution and reversing the solubility equilibrium of the "insoluble" compound, strontium carbonate:

$$Na_2CO_3\ (s) \rightarrow 2\ Na^+(aq) + CO_3^{2-}\ (aq) \tag{7.16}$$

$$SrCO_3\ (s) \rightarrow Sr^+(aq)^{2+}\ CO_3^{2-}\ (aq) \tag{7.17}$$

Also, a complexing agent or *ligand* may react with the cation of a precipitate, enhancing the solubility of the compound. In addition, several metal ions are weakly acidic and readily hydrolyze in solution. For example, hydrolyzing ferric ion (Fe^{3+}) forms a hydroxide and hydrogen ion:

$$Fe^{3+} + H_2O \rightarrow Fe(OH)^{2+} + H^+ \tag{7.18}$$

When such metal ions hydrolyze, they produce a less soluble complex. The solubility of the salt is inversely related to the pH of the solution, with solubility increasing as the pH decreases. The minimum solubility is found under acidic conditions when the concentrations of the hydrolyzed species approach zero.

IONIZATION EXAMPLE

What exactly is the pH scale? Why do pH values range from 0 to 14?

Answer

Water ionizes. It does not exist only as molecular water (H_2O), but also includes hydrogen (H^+) and hydroxide (OH^-) ions:

$$H_2O \leftrightarrow H^+ + OH^- \tag{7.19}$$

The negative logarithm of the molar concentration of hydrogen ions, i.e. $[H^+]$ in a solution (usually water in the environmental sciences), is referred to as pH. This convention is used because the actual number of ions is extremely small. Thus pH is defined as:

$$pH = -\log_{10}[H^+] = \log_{10}([H^+]^{-1}) \quad (7.20)$$

The brackets refer to the molar concentrations of chemicals, and in this case it is the ionic concentration in moles of hydrogen ions per liter. The reciprocal relationship of molar concentrations and pH means that the more hydrogen ions you have in solution the lower your pH value will be.

Likewise, the negative logarithm of the molar concentration of hydroxide ions, i.e. $[OH^-]$ in a solution is pOH:

$$pOH = -\log_{10}[OH^-] = \log_{10}([OH^-]^{-1}) \quad (7.21)$$

The relationship between pH and pOH is constant. At 25°C, this constant is:

$$K = [H^+]\,[OH^-] = 10^{-14} \quad (7.22)$$

When expressed as a negative log, one can see that the pH and pOH scales are reciprocal to one another and that they both range from 0 to 14. Thus, a pH 7 must be neutral (just as many hydrogen ions as hydroxide ions).

Upon further investigation, the log relationship means that for each factor pH unit change there is a factor of 10 change in the molar concentration of hydrogen ions. Thus, a pH 2 solution has 100 000 times more hydrogen ions than neutral water (pH 7), or $[H^+] = 10^{12}$ versus $[H^+] = 10^7$, respectively.

B. Environmental Acid and Base Chemistry

For many air pollution situations, an acid is considered to be any substance that causes hydrogen ions to be produced when dissolved in water. Conversely, a base is any substance that produces hydroxide ions when dissolved in water. Acids are proton donors and bases are proton acceptors (this is known as the Brönsted–Lowry model). Acids are electron-pair acceptors and bases are electron-pair donors (following the Lewis model). Actually, the H^+ is a bare proton, having lost its electron, so it is highly reactive. In reality, the acid produces a hydronium ion:

$$[H^+\,(H_2O)] = H_3O^+ \quad (7.23)$$

When acids react with bases, a double-replacement reaction takes place, resulting in neutralization. The products are water and a salt. One mole of acid neutralizes precisely one mole of base. Being electrolytes, a strong acid dissociates and ionizes 100% into H_3O^+ and anions. These anions are the acid's specific conjugate base. A strong base also dissociates and ionizes completely. The ionization results in hydroxide ions and cations, known as the base's conjugate acid. Weak acids and weak bases dissociate less than 100% into the respective ions.

There are four strong acids, that are important in air pollution. These are hydrochloric acid (HCl), nitric acid (HNO_3), sulfuric acid (H_2SO_4), and perchloric acid ($HClO_4$). Many weak acids are also important, such as carbonic acid, acetic acid, and phosphoric acid.

Strong bases include sodium hydroxide (NaOH) and potassium hydroxide (KOH). Weak bases include ammonia (NH_3), which dissolves in water to become ammonium hydroxide (NH_4OH) and organic amines (i.e. compounds with the radical: $-NH$).

Nonmetal oxides, such as carbonate (CO_2) and sulfate (SO_2), are generally acidic, e.g. forming carbonic acid and sulfuric acid, respectively in water, while metal oxides like those of calcium (e.g. CaO) and magnesium (e.g. MgO), are generally basic. These two metal oxides, for example, form calcium hydroxide (CaOH) and magnesium hydroxide (MgOH) in water, respectively.

The principal environmental metric for acidity and basicity in assessing and controlling air pollution is pH. As mentioned, the "p" in pH represents the negative log and the "H" represents the hydrogen ion or hydronium ion molar concentration:

$$pH = -\log[H^+] \text{ and } [H^+] = 10^{-pH} \tag{7.24}$$

The autoionization of water into its hydrogen ion and hydroxide ions is an equilibrium constant, i.e. the water dissociation equilibrium constant (K_w). At 25 °C, the molar concentration of the product of these ions is 1.0×10^{-14}. That is:

$$K_w = [H^+][OH^-] = 1.0 \times 10^{-14} \tag{7.25}$$

Thus, pH ranges from 0 to 14, with 7 being neutral. Values below 7 are acidic and values above 7 are basic.

STRONG ACID/BASE EXAMPLE 1

An air pollution analytical laboratory is using an aqueous solution of sulfuric acid, which is 0.05 M. What is the hydrogen ion and chlorine ion molar concentration of the solution?

Answer

Since this is a strong acid, it should ionize and dissociate completely. Thus:

$[H^+] = 0.05$ and $[Cl^-] = 0.05$, so none of the associated acid remains.

Strong Acid/Base Example 2

What is the pH of the solution above? What is the $[OH^-]$ of the solution above?

Answer

Since $[H^+] = 0.05$, and $pH = -\log[H^+] = -\log 0.05\,M = 1.3$

Also, recall that:

$$K_w = [H^+][OH^-] = 1.0 \times 10^{-14}$$

$$\text{So, } [OH^-] = \frac{K_w}{[H^+]} = \frac{10^{-4}}{0.05} = 2.0 \times 10^{-13}$$

Thus, even with a very high relative concentration of hydrogen ions in the acidic solution, there is still a small amount of hydroxide ion concentration.

Characterizing an acid or base as strong or weak has nothing to do with the concentration (i.e. molarity) of the solution, and everything to do with the extent to which the acid or base dissociates when it enters water. In other words, whether at a concentration of 6.0 M or at 0.00001 M, sulfuric acid will completely ionize in the water, but acetic acid will not completely ionize at any concentration. To demonstrate this, Table 7.6 shows the pH for a number of acids and bases, all with the same molar concentration. The strongest acids are at the top and the strongest bases are at the bottom. The weak acids and bases are in the middle of the table.

Most acid–base reactions in the atmosphere involve weak substances. In fact the amount of ionization in most environmental reactions, especially those in the ambient environment (as opposed to those in chemical engineering and laboratory reactors), are quite weak, usually well below 10% dissociation.[11]

[11] Note that by this definition, water itself is a weak acid in that it ionizes (autoionizes into 10^{-14} molar concentration of ions) into hydroxide and hydronium ions. Hydronium is the hydrogen ion bound to a water molecule. The importance of water's ionization in virtually all biological processes should not be underestimated. At 25°C, there are 55.35 mol water per liter. So, since half of the ions are hydronium ions, this means:

$$\frac{1.0 \times 10^{-7}\ M\,H_3O^+}{55.35\ M\,H_2O} = 1.8 \times 10^{-9} \text{ hydronium ions per water molecule!}$$

Even this small ratio provides enough H^+ given the amount of water available in the hydrological cycle and the highly reactive nature of each hydrogen ion.

TABLE 7.6

The Experimentally Derived pH Values for 0.1 M Solutions of Acids and Bases at 25°C

Compound	pH
HCl	1.1
H_2SO_4	1.2
H_3PO_4	1.5
CH_3COOH	2.9
H_2CO_3 (in saturated solution)	3.8
HCN	5.1
NaCl	6.4
H_2O (distilled)	7.0
$NaCH_3CO_2$	8.4
$NaSO_3$	9.8
NaCN	11.0
NH_3 (aqueous)	11.1
$NaPO_4$	12.0
NaOH	13.0

Source: Spencer, J., Bodner, G., and Rickard, L., *Chemistry: Structure and Dynamics*, 2nd ed. John Wiley & Sons, New York, NY, 2003.

So, for every 1000 molecules of a weak acid, only a few, say 50, molecules of the acid will dissociate into hydronium ions in the water. So, taking acetic acid as an example, the acid–base equilibrium reaction is:

$$CH_3COOH\ (aq) + H_2O\ (l) \Leftrightarrow H_3O^+\ (aq) + CH_3COO^-\ (aq) \qquad (7.26)$$

The acetate ion (CH_3COO^-) is the reaction's conjugate base and the hydronium ion is the active acid chemical species. Because the reaction is in equilibrium all for species exist together, so we can establish another equilibrium constant for acid reactions, i.e. the acid constant (K_a):

$$K_a = \frac{[H_3O^+][CH_3COO^-]}{[CH_3COOH]} \qquad (7.27)$$

At 25°C, the K_a for acetic acid is 1.8×10^{-5} (see Table 7.7). If the percent dissociation in an acid reaction is known, the product of this percentage and the initial acid concentration will give the molar concentration of hydrogen ions, $[H^+]$. For example, if a 0.1 M solution of cyanic acid (HOCN) is 2.8% ionized, the $[H^+]$ can be found. We know that HOCN is a weak acid because the percent ionization is less than 100. In fact, it is well below 10%. This means that the hydrogen ion molar concentration is

$$[H^+] = 2.8\% \times 0.1\,M = 0.0028 \text{ or } 2.8 \times 10^{-3}$$

Published K_a constants show that the HOCN constant at 25°C is 3.5×10^{-3}, so environmental conditions, likely temperature, are slightly affecting the pH of the solution. Remember that all equilibrium constants are temperature dependent.

Weak bases follow the exact same protocol as weak acids, with a base equilibrium constant, K_b. Some important acid and base equilibrium constants are provided in Tables 7.7 and 7.8.

TABLE 7.7

Equilibrium Constants for Selected Environmentally Important Weak Monoprotic (Single H Atom) Acids and Bases at 25°C

Monoprotic acid	Dissociation reaction	K_a
Hydrofluoric acid	$HF + H_2O \Leftrightarrow H_3O^+ + F^-$	7.2×10^{-4}
Nitrous acid	$HNO_2 + H_2O \Leftrightarrow NO_2^- + H_3O^+$	4.0×10^{-4}
Lactic acid	$CH_3CH\ (OH)\ CO_2H + H_2O \Leftrightarrow CH_3CH\ (OH)\ CO_2^- + H_3O^+$	1.38×10^{-4}
Benzoic acid	$C_6H_5CO_2H + H_2O \Leftrightarrow C_6H_5CO_2^- + H_3O^+$	6.4×10^{-5}
Acetic acid	$HC_2H_3O_2 + H_2O \Leftrightarrow C_2H_3O_2^- + H_3O^+$	1.8×10^{-5}
Propionic acid	$CH_3CH_2CO_2H + H_2O \Leftrightarrow CH_3CH_2CO_2^- + H_3O^+$	1.3×10^{-5}
Hypochlorous acid	$HOCl + H_2O \Leftrightarrow OCl^- + H_3O^+$	3.5×10^{-8}
Hypobromous acid	$HOBr + H_2O \Leftrightarrow OBr^- + H_3O^+$	2×10^{-9}
Hydrocyanic acid	$HCN + H_2O \Leftrightarrow CN^- + H_3O^+$	6.2×10^{-10}
Phenol	$HOC_6H_5 + H_2O \Leftrightarrow OC_6H_5^- + H_3O^+$	1.6×10^{-10}
Base	Dissociation reaction	K_b
Dimethylamine	$(CH_3)_2NH + H_2O \Leftrightarrow (CH_3)_2NH_2^+ + OH^-$	5.9×10^{-5}
Methylamine	$CH_3NH_2 + H_2O \Leftrightarrow CH_3\ NH_3^+ + OH^-$	7.2×10^{-4}
Ammonia	$NH_3 + H_2O \Leftrightarrow NH_3^+ + OH^-$	1.8×10^{-5}
Hydrazine	$H_2NNH_2 + H_2O \Leftrightarrow H_2NNH_3^+ + OH^-$	1.2×10^{-6}
Analine	$C_6H_5NH_2 + H_2O \Leftrightarrow C_6H_5NH_3^+ + OH^-$	4.0×10^{-10}
Urea	$H_2NCONH_2 + H_2O \Leftrightarrow H_2NCONH_3^+ + OH^-$	1.5×10^{-14}

Source: Casparian, A., Chemistry, in *How to Prepare for the Fundamentals of Engineering (FE/EIT) Exam* (Olia, M., ed.). Barron's Educational Series, Inc., Hauppauge, NY, 2000.

TABLE 7.8

Equilibrium Constants for Selected Environmentally Important Polyprotic (Two or More H Atom) Acids at 25°C

Acid	Dissociation reactions	K_{a1}	K_{a2}	K_{a3}
Sulfuric acid	$H_2SO_4 + H_2O \Leftrightarrow HSO_4^- + H_3O^+$	1.0×10^3		
	$HSO_4^- + H_2O \Leftrightarrow SO_4^{2-} + H_3O^+$		1.2×10^{-2}	
Hydrogen sulfide	$H_2S + H_2O \Leftrightarrow HS^- + H_3O^+$	1.0×10^{-7}		
	$HS^- + H_2O \Leftrightarrow S^{2-} + H_3O^+$		1.3×10^{-13}	
Phosphoric acid	$H_3PO_4 + H_2O \Leftrightarrow HPO_4^- + H_3O^+$	7.1×10^{-3}		
	$HPO_4^- + H_2O \Leftrightarrow HPO_4^{2-} + H_3O^+$		6.3×10^{-8}	
	$HPO_4^{2-} + H_2O \Leftrightarrow PO_4^{3-} + H_3O^+$			4.2×10^{-13}
Carbonic acid	$H_2CO_3 + H_2O \Leftrightarrow HCO_3^- + H_3O^+$	4.5×10^{-7}		
	$HCO_3^- + H_2O \Leftrightarrow CO_3^{2-} + H_3O^+$		4.7×10^{-11}	

Source: Casparian, A., Chemistry, in *How to Prepare for the Fundamentals of Engineering (FE/EIT) Exam* (Olia, M. ed.). Barron's Educational Series, Inc., Hauppauge, NY, 2000.

Because the atmosphere contains relatively large amounts of carbon dioxide (on average about 350 ppm), the CO_2 becomes dissolved in surface water and in soil water (because CO_2 is a common soil gas). Thus, one of the most important environmental acid–base reactions[12] is the dissociation of CO_2:

$$CO_2 + H_2O \Leftrightarrow H_2CO_3^* \tag{7.28}$$

The asterisk (*) denotes that this compound is actually the sum of two compounds, i.e. the dissolved CO_2 and the reaction product, carbonic acid H_2CO_3.

Since the carbonic acid that is formed is a diprotic acid (i.e. it has two hydrogen atoms), an additional equilibrium step reaction occurs in water. The first reactions forms bicarbonate and hydrogen ions:

$$H_2CO_3^* \Leftrightarrow HCO_3^- + H^+$$

Followed by a reaction that forms carbonate and hydrogen:

$$HCO_3^- \Leftrightarrow CO_3^{2-} + H^+ \tag{7.29}$$

Each of the two-step reactions has its own acid equilibrium constant (K_{a1} and K_{a2}, respectively), as shown in Table 7.9. For a triprotic acid, there would be three unique constants. Note that the constants decrease substantially with each step. In other words, most of the hydrogen ion production occurs in the first step.

Numerous reactions can be predicted from the relative strength of acids and bases, since their strength results from how well the protons via the hydronium ion is transferred from the acid and the electron is transferred via the hydroxide ion from the base. If an acid is weak, its conjugate base must be strong, and if an acid is strong, its conjugate base must be weak. Likewise, if a base is weak, its conjugate acid must be strong, and if the base is strong, its conjugate acid must be weak. Our tables show actual K_a and K_b constants, however, many sources value for pK_a and pK_b values. Since the "p" denotes negative logarithm, the larger the pK_a, the weaker the acid, and the larger the pK_b, the weaker the base.

An example of how the K_a is an indicator of relative strength of reactants and products is that of hydrocyanic acid:

Stronger base (OH⁻) — Weaker base (CN⁻)

$$HCN + OH^- \Leftrightarrow CN^- + H_2O \tag{7.30}$$

Stronger acid (HCN) ($K_a = 6.2 \times 10^{-10}$) — Weaker acid ($H_2O$) ($K_a = 1.8 \times 10^{-16}$)

[12] For an excellent discussion of carbon dioxide equilibrium in water see Hemond, H. F., and Fechner-Levy, E. J., *Chemical Fate and Transport in the Environment*. Academic Press, San Diego, CA, 2000.

The ratio of the K_a constant values of the two acids is a direct way to quantify the equilibrium. In the hydrocyanic acid instance above, the ratio is $\frac{6.2 \times 10^{-10}}{1.8 \times 10^{-16}} \cong 4 \times 10^6$.

This large quotient indicates that the equilibrium is quite far to the right. So, if HCN is dissolved in a hydroxide solution (e.g. NaOH), the resulting reaction will produce much greater amounts of the cyanide ion (CN^-) than the amount of both the hydroxide ion (OH^-) and molecular HCN. Conversely, for an aqueous solution of sodium cyanide (NaCN), the water will only react with a tiny amount of CN^-.

Among the properties of water, one of the most important environmentally is that it can behave as either an acid or a base, i.e. water is an amphoteric compound. That is one of the reasons that water is sometimes shown as HOH. When water acts as a base its $pK_b = -1.7$. When water acts as an acid its $pK_a = 15.7$.

Many air pollution reactions occur in water (e.g. in droplets, on aerosols, and even in the water within organisms), so the relationship between conjugate acid–base equilibrium and pH is important. The Henderson–Hasselbach equation states this relationship:

$$pK_a = pH + \log \frac{[HA]}{[A^-]} \tag{7.31}$$

Thus, Henderson–Hasselbach tells us that when the pH of an aqueous solution equals the pK_a of an acidic component, the concentrations of the conjugate acids and bases must be equal (since the log of 1 = 0). If pH is 2 or more units lower than pK_a, the acid concentration will be greater than 99%. Conversely, when pH is greater than pK_a by 2 or more units, the conjugate base concentration will account for more than 99% of the solution.[13]

This means that mixtures of acidic and non-acidic compounds can be separated with a pH adjustment, which has strong implications for the design and operation of air pollution control equipment. The application of this principle is also important to the transformation of environmental contaminants in the form of weak organic acids or bases, because these compounds in their non-ionized form are much more lipophilic, meaning they will be absorbed more easily through the skin than when they exist in ionized forms. As a general rule, the smaller the pK_a for an acid and the larger the pK_b for a base, the more extensive will be the dissociation in aqueous environments at normal pH values, and the greater compound's electrolytic nature.

[13] US Environmental Protection Agency, *Dermal Exposure Assessment: Principles and Applications*, Interim Report, EPA/600/8-91/011B, Washington, DC, 1992.

ACID RAIN EXAMPLE

Acid-base reactions are demonstrated by the acid rain problem. Dissolved CO_2 causes most rainfall to be slightly acidic:

$$CO_2(aq) + H_2O(l) \Leftrightarrow H^+(aq) + HCO_3^-(aq)\ K_a = 4.3 \times 10^{-7} \quad (7.32)$$

Acidic precipitation, popularly known as "acid rain" contains the strong acids H_2SO_4 and HNO_3, mainly from the combustion of fossil fuels that contain sulfur (the air contains molecular nitrogen that is oxidized during internal combustion in engines, i.e. "mobile sources" and in any high temperature furnaces, like those in power plants). The sulfuric acid is stepped process. Recall that the contaminant released from stacks is predominantly sulfur dioxide when the elemental sulfur oxidized:

$$S + O_2 \rightarrow SO_2\,(g) \quad (7.33)$$

The sulfur dioxide in turn is oxidized to sulfur trioxide:

$$SO_2(g) + \tfrac{1}{2}O_2(g) \rightarrow SO_3(g) \quad (7.34)$$

The sulfur trioxide then reacts with atmospheric water (vapor, clouds, on particles) to form sulfuric acid:

$$SO_3\,(g) + H_2O \rightarrow H_2SO_4\,(aq) \quad (7.35)$$

The net result is to increase the acidity of the rain, which is a threat to aquatic life, and metallic and carbonate materials (including artwork, statues, and buildings). Near pollution sources, rainwater pH can be found to be less than 3 (i.e. 10 000 times more acidic than neutral).

What are the molar concentrations of $[H^+]$ and $[OH^-]$ of rainwater at pH 3.7 at 25°C? When SO_2 dissolves in water, sulfurous acid (H_2SO_3, $K_{a1} = 1.7 \times 10^{-2}$, $K_{a2} = 6.4 \times 10^{-8}$) is formed. What is the reaction when sulfurous acid donates a proton to a water molecule? What is the Brönsted–Lowry acid and base in this reaction?

Surface waters have a natural buffering capacity, especially in regions where there is limestone which gives rise to dissolved calcium (e.g. central Kansas is less at risk of acid rain's effects than are the Finger Lakes of New York). What is the reaction of a minute amount of acid rain containing sulfuric acid reaching a lake containing carbonate (CO_3^{2-}) ions?

Solution

Since, $pH = -\log[H^+]$, then $[H^+] = 10^{-pH} = 10^{-3.7}\ 2.0 \times 10^{-4}M$ in aqueous solution at 25°C.

The sulfurous acid proton donation reaction is:

$$H_2SO_3\,(aq) + H_2O\,(l) \Leftrightarrow HSO_3^-\,(aq) + H_3O^+\,(aq) \quad (7.36)$$

H_3O^+ is the stronger acid (i.e. $K_{a1} < 1$) and HSO_3^- is the stronger base.

Regarding the buffered water system, CO_3^{2-} is the conjugate base of the weak acid HCO_3^-, so the former can react with the strong acid H_3O^+ in the sulfuric acid solution:

$$CO_3^{2-}\,(aq) + H_3O^+\,(aq) \Leftrightarrow HCO_3^-\,(aq) + H_2O\,(l) \tag{7.37}$$

The $K = 1/K_{a2}$ of H_2CO_3 (making for a large K)

Similarly HSO_4^- also reacts with CO_3^{2-}:

$$CO_3^{2-}\,(aq) + HSO_4^-\,(aq) \Leftrightarrow SO_4^{2-}\,(aq) + HCO_3^-\,(aq) \tag{7.38}$$

The $K = (K_{a2}$ of $H_2SO_4)/(K_{a2}$ of $H_2CO_3)$ (*thus, large K*)

So, a HCO_3^-/CO_3^{2-} buffer is produced.

An excess of acid rain will consume all the CO_3^{2-} and HCO_3^-, converting all to H_2CO_3 (and completely eliminating the buffer).

These are important and representative reactions of the challenging global problem of acid rain.

1. Hydrolysis

All of the acid–base reactions involve chemical species reacting with water. It is worth noting that even some so-called neutral compounds have acidic or basic properties. For example, metal[14] acetates $(MeC_2H_3O_2)_x$) can dissolve in water and actually react with the water to form weak acids and hydrogen ions. The resulting solutions are generally slightly basic (pH > 7), since the acetate ion is conjugate base of the weak acid. This process is known as hydrolysis (i.e. lysis or breaking apart water molecules).

Likewise, some compounds produce weak bases when dissolved in water. For example, when ammonium chloride (NH_4Cl) is dissolved in water, the solution becomes more acidic. The NH_4^+ cation serves as the conjugate base of ammonium hydroxide (NH_4OH), or molecular ammonia (NH_3) is the conjugate base of the NH_4^+ cation. Oxide gases (e.g. carbon dioxide and sulfur dioxide) can hydrolyzed to form hydronium ions and anions:

$$CO_2\,(g) + H_2O\,(l) \Leftrightarrow H_3O^+\,(aq) + HCO_3^-(aq) \tag{7.39}$$

And,

$$SO_2\,(g) + H_2O\,(l) \Leftrightarrow H_3O^+\,(aq) + HSO_3^-(aq) \tag{7.40}$$

[14] The abbreviation "Me" is commonly used for "metals" when a number of metals exhibit identical or similar properties and behavior (e.g. Hg, Cd, Pb, and Mn may behave similarly on an aerosol's surface, catalyzing certain reactions, or they may also exhibit similar physiological processes, such as interferences with neurons and calcium gates in the brain and central nervous system). Rather than listing every metal, the chemical formulae will display Me.

These reactions also take place in organic compounds, where the organic molecule, RX, reacts with water to form a covalent bond with OH and cleaves the covalent bond of the leaving group (X) in RX, which displaces X with the hydroxide ion and an ion formed from the leaving group:

$$RX + H_2O \rightarrow ROH + H^+ + X^- \quad (7.41)$$

Amides, epoxides, carbonates, esters, organic halides, nitriles, urea compounds, and esters of organophosphate compounds are functional groups that are susceptible to hydrolysis. The process involves an electron-rich nucleus seeker (i.e. a nucleophile) attacking an electron-poor electron seeker (i.e. an electrophile) to displace the leaving group (such as a halogen). This is why hydrolysis is one of the methods of dechlorination, a detoxification process for hazardous chlorinated hydrocarbons.

Environmental engineers use hydrolysis to eliminate or to reduce the toxicity of hazardous contaminants by abiotic transformation and biotransformation, especially bacterial. For example, the highly toxic methyl isocynate, infamous as the contaminant that led to loss of life and health effects in the Bhopal, India incident, can be transformed hydrolytically.

Two factors are particularly enhancing to hydrolysis. Microbial mediation, including enzymatic activity can catalyze hydrolytic reactions. This occurs both in the ambient environment, such as the hydrolysis of inorganic and organic compounds by soil bacteria, as well as in engineered systems, such as acclimating those same bacteria to the treatment of chlorinated organic compounds in solid and liquid wastes.

The second factor is pH. Hydrolysis can be affected by specific acid and base catalysis. In acid catalysis, this H^+ ion catalyzes the reaction; and in base catalysis the OH^- ion serves as the catalyst. The effect of temperature on hydrolysis can be profound. Each 10°C incremental temperature increase results in a hydrolysis rate constant change by a factor of 2.5.[15]

Metal (Me) hydrolysis is a special case. Cations in water act like Lewis acids in that they are prone to accept electrons, while water behaves like a Lewis base because it makes its oxygen's two unshared electrons available to the cations. When strong water–metal (acid–base) interactions take place, H^+ dissociates and hydronium ions form in a prototypical reaction:

$$Me^{n+} + H_2O \Leftrightarrow MeOH^{(n+1)} + H^+ \quad (7.42)$$

Although reactions such as these have obvious applications to contaminant transformations in surface water, soil water, and groundwater, they occur in any medium where water is present. Water is present in the atmosphere, so hydrolysis occurs in clouds and fog, as well as in the water fraction hygroscopic nuclei. Water is also present in all living things, so hydrolysis is a common process in metabolism (particularly in the first phase, as discussed later)

[15] Knox, R., Sabatini, D., and Canter, L., *Subsurface Transport and Fate Processes*. Lewis Publishers, Boca Raton, FL, 1993.

and other organic processes. Thus, hydrolysis is important in numerous environmental and toxicological processes.

2. Photolysis

The sun's EMR at UV and visible wavelengths can induce chemical reactions directly and indirectly. Direct photolysis is the process where sunlight adds activation energy needed to transform a compound. Indirect photolysis is the process by which an intermediate compound is energized, which in turn transfers energy to another compound.

Contaminants are photochemically degraded in both atmospheric and aquatic environments. Photolysis can combine or interchange with other processes, such as in the degradation pathways for chlorinated organic contaminants. For example, the degradation pathway for 1,4-dichlorobenzene in air is reaction with photochemically generated OH^- radicals and oxides of nitrogen. However, in soil and water, the degradation is mainly microbial biodegradation, leading to very different end products (see Fig. 7.9).

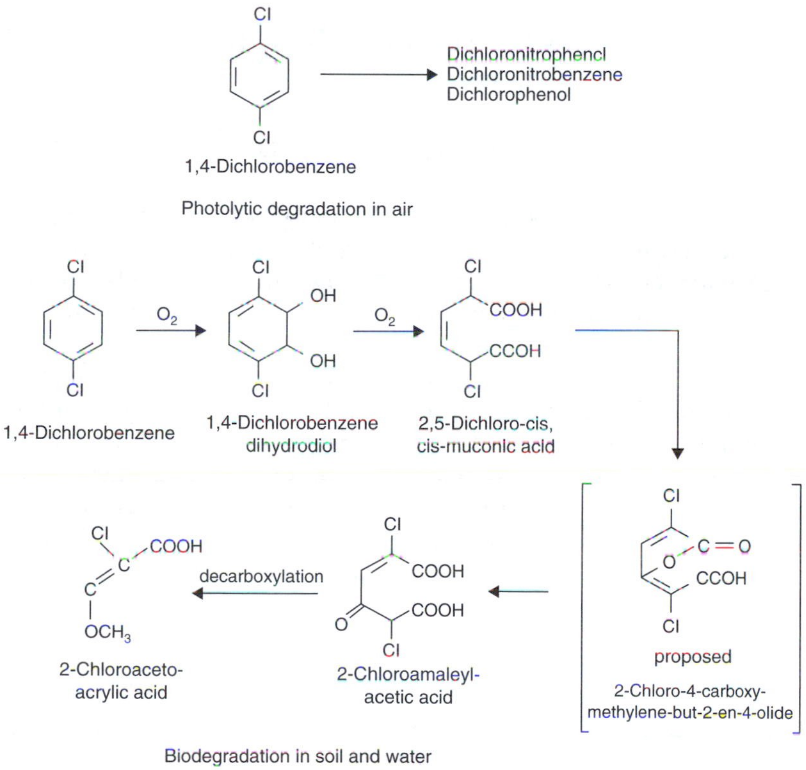

Fig. 7.9. Different 1,4-dichlorobenzene reactions according to environmental media. *Source*: Agency for Toxic Substances and Disease Registry, 1998, Toxicological Profile for 1,4-Dichlorobenzene, http://www.atsdr.cdc.gov/toxprofiles/tp10.html.

3. Precipitation Reactions

Dissolved ions may react with one another to form a solid phase compound under environmental conditions of temperature and pressure. Salts are compounds that form when metals react with nonmetals, such as sodium chloride (NaCl). They may also form from cation and anion combinations, such as ammonium nitrate (NH_4NO_3). Precipitation is both a physical and chemical process, wherein soluble metals and inorganic compounds change into insoluble metallic and inorganic salts. In other words, the dissolved forms become solids. Such reactions in which soluble chemical species become insoluble products are known as "precipitation reactions."

Chemical precipitation occurs within a defined pH and temperature range unique for each metallic salt. Usually in such reactions, an alkaline reagent is added to the solution, thereby raising the solution pH. The higher pH often decreases the solubility of the metallic constituent, bringing about the precipitation (see Fig. 7.10). For example, adding caustic soda (NaOH) decreases the amount of soluble nickel, producing the much less water-soluble species nickel hydroxide precipitate (recall that "s" denotes a solid precipitate):

$$Ni^{2+} + NaOH \Leftrightarrow Na^{+} + Ni\,(OH)_2\,(s) \tag{7.43}$$

V. ORGANIC CHEMISTRY

Carbon is an amazing element. It can bond to itself and to other elements in a myriad of ways. In fact, it can form single, double, and triple bonds with itself. This makes for millions of possible organic compounds. An organic compound is a compound that includes at least one carbon-to-carbon or carbon-to-hydrogen covalent bond.

The majority of air pollutants are organic.[16] Organic compounds can be further classified into two basic groups: aliphatics and aromatics. Hydrocarbons are the most fundamental type of organic compound. Unsubstituted hydrocarbons contain only the elements carbon and hydrogen. Aliphatic compounds are classified into a few chemical families. Each carbon normally forms four covalent bonds. Alkanes are hydrocarbons that form chains with each link comprised of the carbon. A single link is CH_4, methane. The carbon

[16] This is true in terms of the actual number of chemical compounds. By far, most contaminants are organic. However, in terms of total mass of reactants and products in the biosphere, inorganic compounds represent a greater mass. For example, most hazardous waste sites are contaminated with organic contaminants, but large-scale waste represented by mining, extraction, transportation, and agricultural activities have larger volumes and masses of metals and inorganic substances.

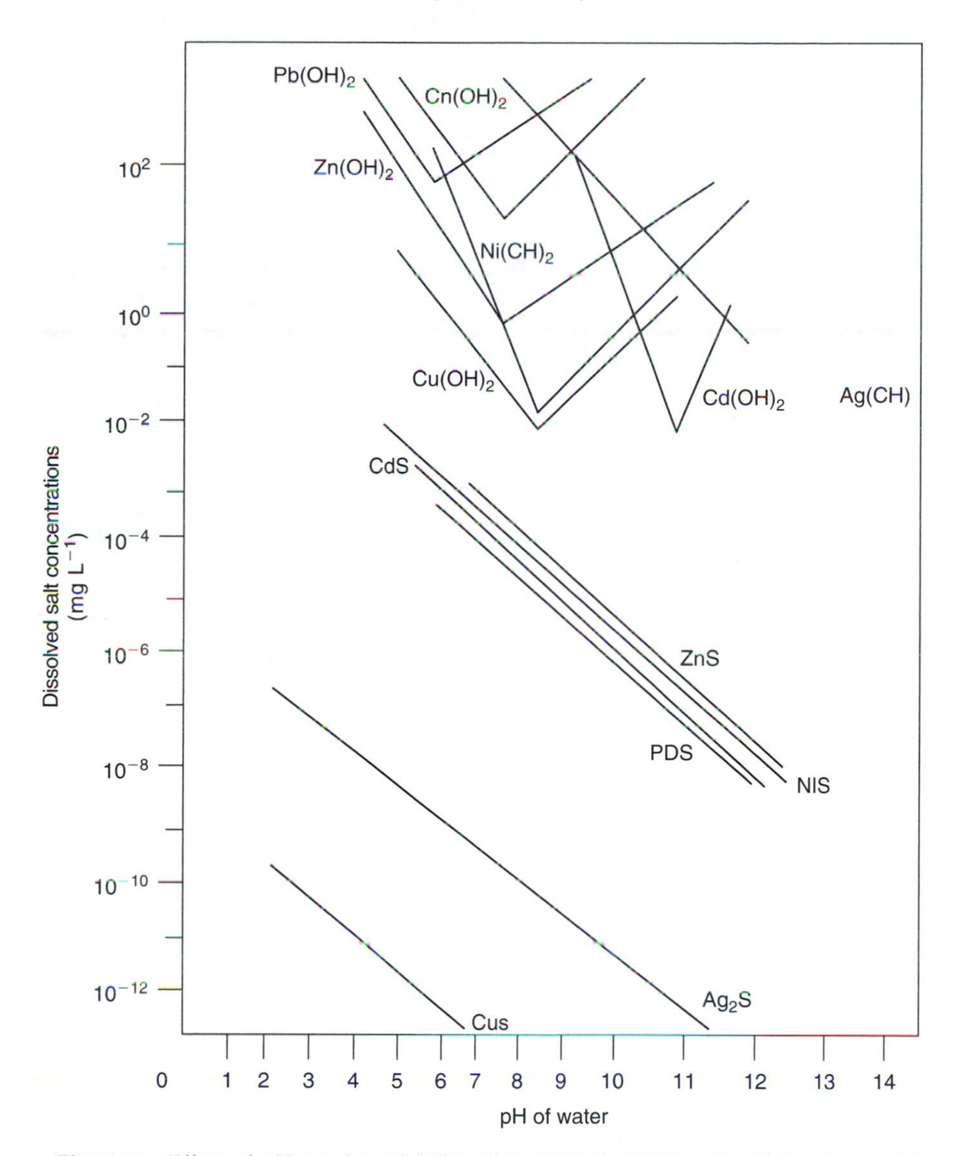

Fig. 7.10. Effect of pH on the solubility of metal hydroxides and sulfides. *Source*: US Environmental Protection Agency, *Summary Report: Control and Treatment Technology for the Metal Finishing Industry; Sulfide Precipitation*, Report No. EPA 625/8-80-003, Washington, DC, 1980.

chain length increases with the addition of carbon atoms. For example, ethane's structure is:

```
    H   H
    |   |
H — C — C — H
    |   |
    H   H
```

And the protypical alkane structure is:

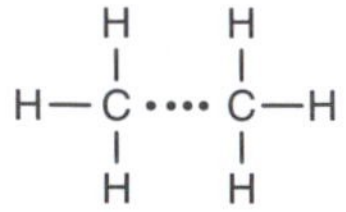

```
    H     H
    |     |
H—C····C—H
    |     |
    H     H
```

The alkanes contain a single bond between each carbon atoms, and include the simplest organic compound, methane (CH_4), and its derivative "chains" such as ethane (C_2H_6) and butane (C_4H_{10}). Alkenes contain at least one double bond between carbon atoms. For example, 1,3-butadiene's structure is CH_2—CH—CH—CH_2. The numbers "1" and "3" indicate the position of the double bonds. The alkynes contain triple bonds between carbon atoms, the simplest being ethyne, CH≡CH, which is commonly known as acetylene (the gas used by welders).

The aromatics are all based on the six-carbon configuration of benzene (C_6H_6). The carbon–carbon bond in this configuration shares more than one electron, so that benzene's structure allows for resonance among the double and single bonds, i.e. the actual benzene bonds flip locations. Benzene is the average of two equally contributing resonance structures.

The term "aromatic" comes from the observation that many compounds derived from benzene are highly fragrant, such as vanilla, wintergreen oil, and sassafras. Aromatic compounds, thus, contain one or more benzene rings. The rings are planar, that is, they remain in the same geometric plane as a unit. However, in compounds with more than one ring, such as the highly toxic PCBs, each ring is planar, but the rings that are bound together may or may not be planar. This is actually a very important property for toxic compounds. It has been shown that some planar aromatic compounds are more toxic than their non-planar counterparts, possibly because living cells may be more likely to allow planar compounds to bind to them and to produce nucleopeptides that lead to biochemical reactions associated with cellular dysfunctions, such as cancer or endocrine disruption.

Both the aliphatic and aromatic compounds can undergo substitutions of the hydrogen atoms. These substitutions render new properties to the compounds, including changes in solubility, vapor pressure, and toxicity. For example, halogenation (substitution of a hydrogen atom with a halogen) often makes an organic compound much more toxic. Thus, trichloroethane is a highly carcinogenic liquid that has been found in drinking water supplies, whereas non-substituted ethane is a gas with relatively low toxicity. This is also why one of the means for treating chlorinated hydrocarbons and aromatic compounds involves dehalogenation techniques.

The important functional groups that are part of many organic compounds are shown in Table 7.9.

Structures of organic compounds can induce very different physical and chemical characteristics, as well as change the bioaccumulation and toxicity of these compounds. For example the differences between the estradiol

TABLE 7.9

Structures of Organic Compounds

Chemical class	Functional group
Alkanes	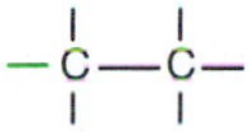 $-\overset{\vert}{\underset{\vert}{C}}-\overset{\vert}{\underset{\vert}{C}}-$
Alkenes	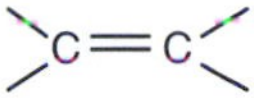 $>C=C<$
Alkynes	$-C\equiv C-$
Aromatics	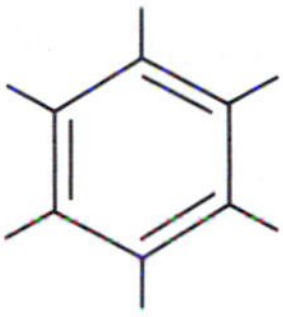 (benzene ring with six substituent bonds)
Alcohols	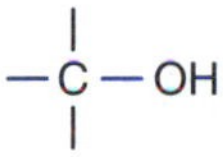 $-\overset{\vert}{\underset{\vert}{C}}-OH$
Amines	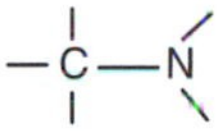 $-\overset{\vert}{\underset{\vert}{C}}-N<$
Aldehydes	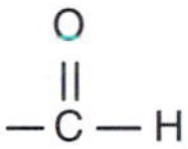 $-\overset{O}{\overset{\Vert}{C}}-H$
Ether	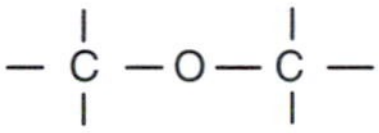 $-\overset{\vert}{\underset{\vert}{C}}-O-\overset{\vert}{\underset{\vert}{C}}-$
Ketones	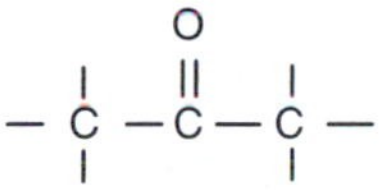 $-\overset{\vert}{\underset{\vert}{C}}-\overset{O}{\overset{\Vert}{C}}-\overset{\vert}{\underset{\vert}{C}}-$
Carboxylic acids	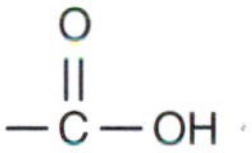 $-\overset{O}{\overset{\Vert}{C}}-OH$
Alkyl halides[a]	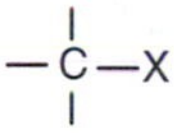 $-\overset{\vert}{\underset{\vert}{C}}-X$
Phenols (aromatic alcohols)	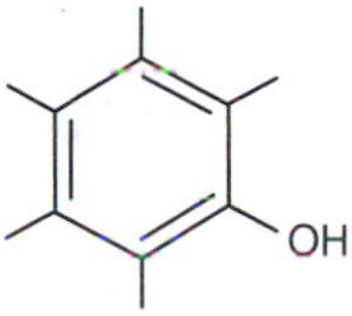 (benzene ring with OH substituent) OH
Substituted aromatics (substituted benzene derivatives):	
Nitrobenzene	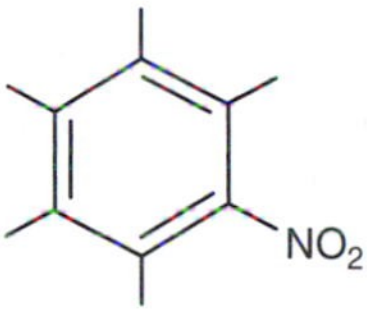 (benzene ring with NO_2 substituent) NO_2

(continued)

TABLE 7.9 (*Continued*)

Chemical class	Functional group
Monosubstituted alkylbenzenes	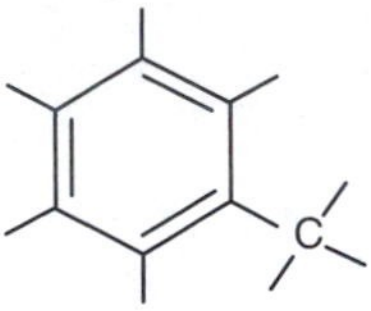 C
Toluene (simplest monosubstituted alky benzene)	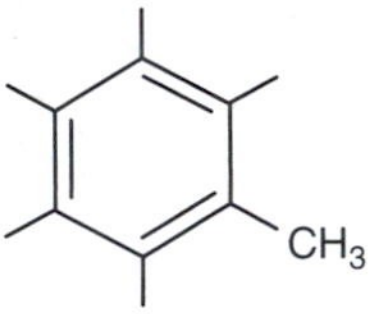 CH_3
Polysubstituted alkylbenzenes:	
1,2-alkyl benzene (also known as ortho or *o*-...)	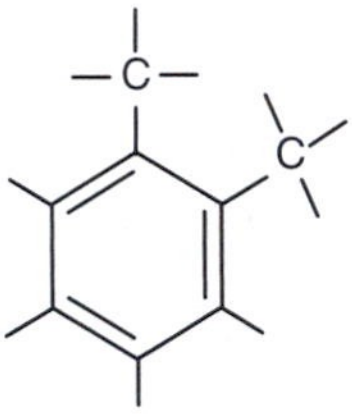 C, C
1,2-xylene or *ortho*-xylene (*o*-xylene)	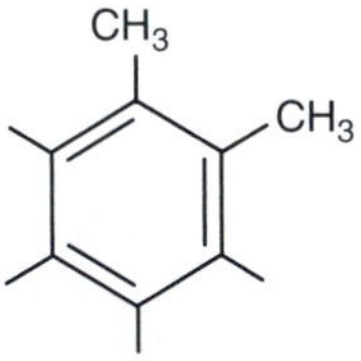 CH_3, CH_3
1,3-xylene or *meta*-xylene (*m*-xylene)	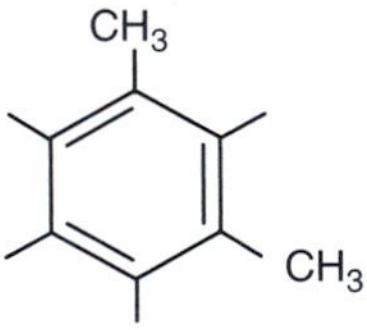 CH_3, CH_3
1,4-xylene or *para*-xylene (*p*-xylene)	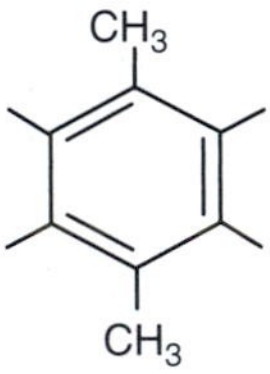 CH_3, CH_3
Hydroxyphenols do not follow general nomenclature rules for substituted benzenes:	
Catechol (1,2-hydroxiphenol)	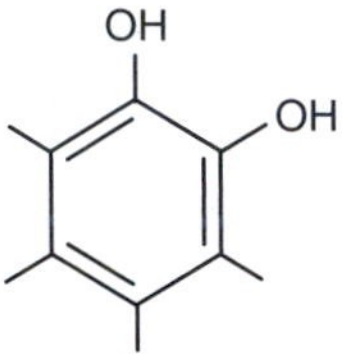 OH, OH

(*continued*)

TABLE 7.9 (*Continued*)

Chemical class	Functional group
Resorcinol (1,3-hydroxiphenol)	OH OH
Hydroquinone (1,4-hydroxiphenol)	OH OH

[a]The letter "X" commonly denotes a halogen, e.g. fluorine, chlorine, or bromine, in organic chemistry. However, in this text, since it is an amalgam of many scientific and engineering disciplines, where "x" often means an unknown variable and horizontal distance on coordinate grids, this rule is sometimes violated. Note that when consulting manuals on the physicochemical properties of organic compounds, such as those for pesticides and synthetic chemistry, the "X" usually denotes a halogen.

and a testosterone molecule may seem small but they cause significant differences in the growth and reproduction of animals. The very subtle differences between an estrogen and an androgen, female and male hormones respectively, can be seen in these structures. But look at the dramatic differences in sexual and developmental changes that these compounds induce in organisms!

Incremental to a simple compound, such as substituting chlorine atoms for three hydrogen atoms on the ethane molecule, can make for large differences (see Table 7.10). Replacing two or three hydrogen atoms with chlorine atoms makes for differences in toxicities between the non-halogenated form and the chlorinated form. The same is true for the simplest aromatic, benzene. Substituting a methyl group for one of the hydrogen atoms forms toluene.

The lessons here are many. There are uncertainties in using surrogate compounds to represent whole groups of chemicals (since a slight change can change the molecule significantly). However, this points to the importance of "green chemistry" and computational chemistry, as tools to prevent dangerous chemical reaching the marketplace and the environment before they are manufactured. Subtle differences in molecular structure can render molecules safer, while maintaining the characteristics that make them useful in the first place, including their market value.

TABLE 7.10

Incremental Differences in Molecular Structure Leading to Changes in Physicochemical Properties and Hazards

Compound	Physical state at 25°C	$-\log P^0$ solubility in H_2O at 25°C ($\text{mol}L^{-1}$)	$-\log$ vapor pressure at 25°C (atm)	Worker exposure limits (parts per million)	Regulating agency
Methane, CH_4	Gas	2.8	−2.4	25	Canadian Safety Association
Tetrachloromethane (carbon tetrachloride), CCl_4	Liquid	2.2	0.8	2 short-term exposure limit (STEL) = 60 min	National Institute of Occupation Health Sciences (NIOSH)
Ethane, C_2H_6	Gas	2.7	−1.6	None (simple asphyxiant)	Occupational Safety and Health Administration (OSHA)
Trichloroethane, $C_2H\,Cl_3$	Liquid	2.0	1.0	450 STEL (15 min)	OSHA
Benzene, C_6H_6	Liquid	1.6	0.9	5 STEL	OSHA
Phenol, C_6H_6O	Liquid	0.2	3.6	10 ppm	OSHA
Toluene C_7H_8	Liquid	2.3	1.4	150 STEL	UK Occupational & Environmental Safety Services

ORGANIC STRUCTURE EXAMPLE 1

If the aqueous solubility as expressed as $-\log S$ (in mol L^{-1}) of 1-butanol, 1-hexanol, 1-octanol, and 1-nonanol are 0.1, 0.9, 2.4, and 3.1, respectively, what does this tells you about the length of carbon chains and the solubility of alcohols? (Remember that this expression of solubility is a negative log!)

Answer and Discussion

Recall that solubility is expressed as a negative logarithm, so lengthening the carbon chain *decreases polarity and, therefore, aqueous solubility* since "like dissolves like" and water is very polar. Thus, as shown in the Fig. 7.11, butanol is orders of magnitude more hydrophilic than is nonanol.

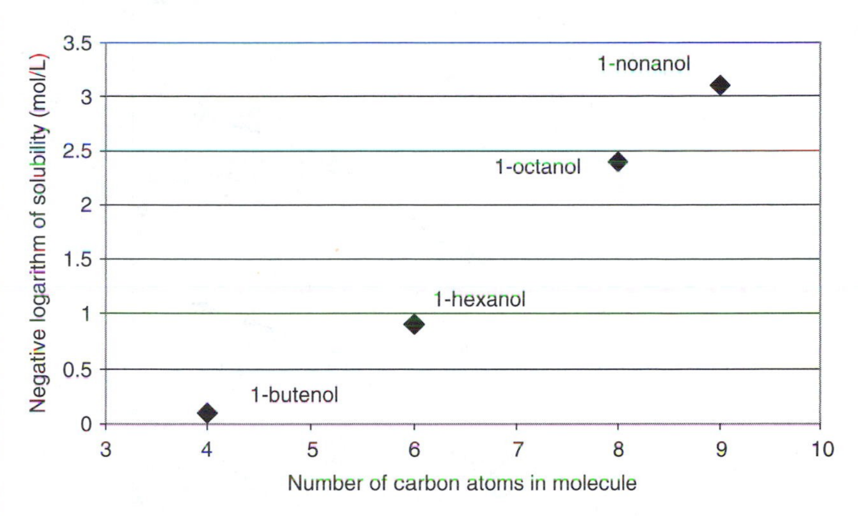

Fig. 7.11. Aqueous solubility of four alcohols.

ORGANIC STRUCTURE EXAMPLE 2

Consider the polarity of the four alcohols in #2 above. If *n*-butane, *n*-hexane, *n*-octane, and *n*-nonane's aqueous solubilities as expressed as $-\log P^0$ are, respectively: 3.0, 3.8, 5.2, and 5.9 mol L^{-1}, what effect does the substitution of a hydroxyl functional group for the hydrogen atom have on an alkane's polarity? Can this effect also be observed in aromatics? (*Hint*: Compare *S* for benzene, toluene to phenol in the table in #1 above; and recall that phenol is a hydroxylated benzene.)

Answer

There is a direct relationship (see Fig. 7.12) between the increase in polarity and hydrophilicity when alkanes are hydroxylated into alcohols. This is why alcohols are miscible in water. This can be an important fact, especially in anaerobic treatment processes where microbes reduce organic compounds and in the process generate alcohols (and ultimately methane and water). Hydroxylation of an aromatic, as indicated by comparing the solubilities of benzene and phenol, also increases polarity and hydrophilicity.

It should be noted that solubility may be modified in environmental systems by using compounds known as "surfactants" (see Fig. 7.13). Surfactants can be very effective in solubilizing adsorbed hydrophobic compounds in soils. The increased solubility and dispersion of hydrocarbons and aromatic compounds with very low aqueous solubility

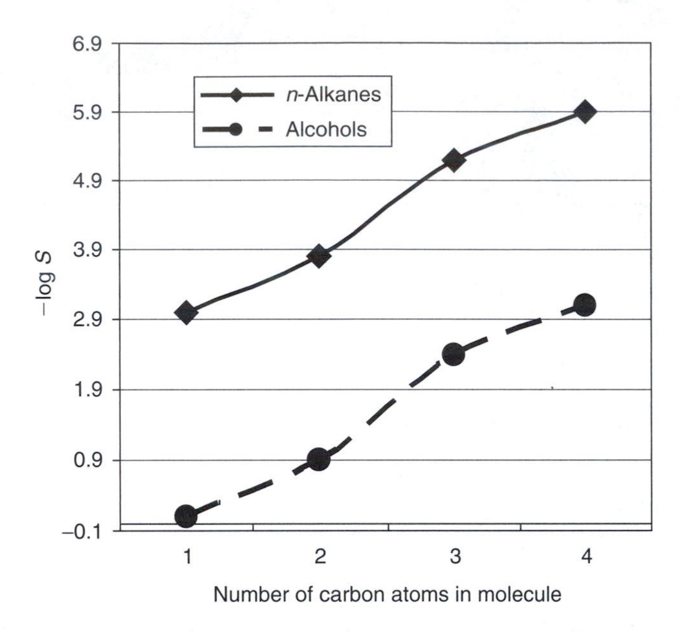

Fig. 7.12. Aqueous solubility of selected aliphatic compounds.

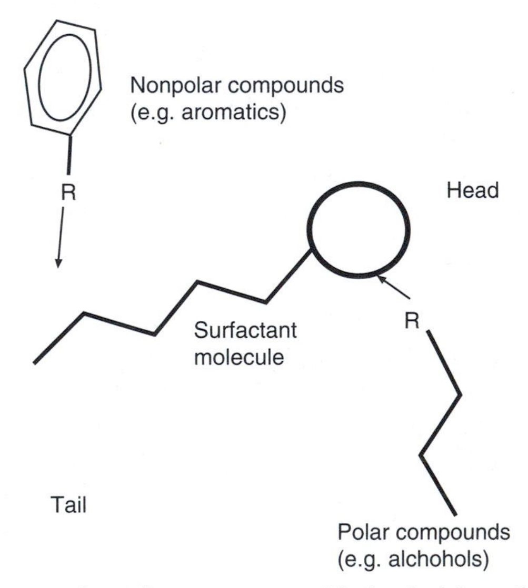

Fig. 7.13. Structure of a surfactant prototype. The head of the molecule is relatively polar and the tail is relatively nonpolar, so in the aqueous phase the more hydrophilic compounds will react with the surfactant molecule at the head position, while the more lipophilic compounds will react at the tail position. The result is that a greater amount of the organic compounds will be in solution when surfactants are present. *Source*: Vallero, D., *Engineering the Risks of Hazardous Wastes*. Butterworth-Heinemann, Boston, MA, 2003.

enhances desorption and bioavailability. Unfortunately, a widely used group of surfactants, i.e. alkylphenolethoxylates (APEs) have been banned in Europe because their breakdown products can be very toxic to aquatic organisms. The APEs have been particularly of concern because of their hormonal effects, particularly their estrogenicity. So, one must keep in mind not only the physical and chemical advantages of such compounds, but also any ancillary risks that they may introduce. This is a classic case of competition among values (i.e. easier and more effective cleanup versus additive health risks from the cleanup chemical).

1. *Isomers*

Isomers are compounds with identical chemical formulae, but different structures. They are very important in air pollution chemistry, because even slightly different structures can evoke dramatic differences in chemical and physical properties. So, isomers may exhibit different chemodynamic behavior and different toxicities. For example, the three isomers of pentane (C_5H_{12}) are shown in Fig. 7.14. The difference in structure accounts for significant physical differences. For example, the boiling points for *n*-pentane, isopentane, and neopentane at 1 atm are 36.1°C, 27.8°C, and 9.5°C, respectively. Thus, neopentane's lower boiling point means that this isomer has a lower vapor pressure, which makes it is more likely to enter the atmosphere than the other two isomers under the same environmental conditions.

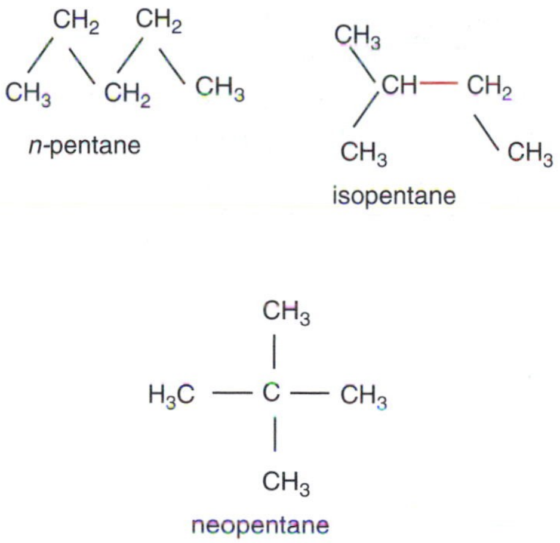

$CH_3-CH_2-CH_2-CH_2-CH_3$

n-pentane

$(CH_3)_2CH-CH_2-CH_3$

isopentane

$C(CH_3)_4$

neopentane

Fig. 7.14. Isomers of pentane. Although the chemical composition is identical, the different molecular arrangements result in molecules that exhibit very dissimilar physical, chemical, and biological properties. *Source*: Vallero, D., *Environmental Contaminants: Assessment and Control*. Academic Press, Burlington, MA, 2004.

Optical isomers or chiral forms of the same compound are those that are mirror images to each other. These difference may make one, e.g. the left-handed form, virtually nontoxic and easily biodegradable, yet the right-handed form may be toxic and persistent.

VI. INTRODUCTION TO ATMOSPHERIC CHEMISTRY

Atmospheric chemistry encompasses all of the chemical transformations occurring in the various atmospheric layers from the troposphere to beyond the stratosphere. Air pollution chemistry represents the subset of these atmospheric chemical processes which have a direct impact on human beings, vegetation, and surface water bodies. Classification of atmospheric chemical processes as either human made (anthropogenic) or natural is useful but not precise. For example, the trace gases nitric oxide (NO) and sulfur dioxide (SO_2) have both anthropogenic and natural sources, and their atmospheric behavior is independent of their source. A vivid example was the 1980 Mt. St. Helen's volcanic eruption in Washington, a gigantic point source for SO_2 and PM in the atmosphere. This natural source was of such magnitude as to become first a regional air pollution problem and subsequently a global atmospheric chemical problem.

A. Types of Atmospheric Chemical Transformations

The chemical transformations occurring in the atmosphere are best characterized as oxidation processes. Reactions involving compounds of carbon (C), nitrogen (N), and sulfur (S) are of most interest. The chemical processes in the troposphere involve oxidation of hydrocarbons, NO, and SO_2 to form oxygenated products such as aldehydes, nitrogen dioxide (NO_2), and sulfuric acid (H_2SO_4). These oxygenated species become the secondary products formed in the atmosphere from the primary emissions of anthropogenic or natural sources (Fig. 7.15).

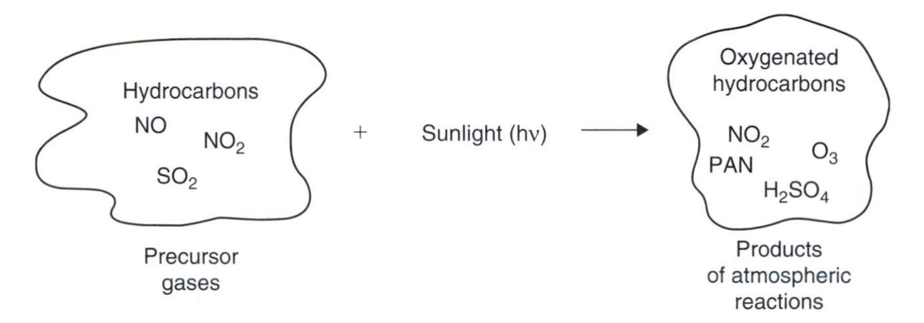

Fig. 7.15. Precursor-product relationship of atmospheric chemical reactions.

Solar radiation influences the chemical processes in the atmosphere by interacting with molecules that act as photoacceptors. Free radicals are formed by the photodissociation of certain types of molecules. Free radicals are neutral fragments of stable molecules and are very reactive. Examples are O, atomic oxygen; H, atomic hydrogen; OH, the hydroxyl radical; and HO_2, the hydroperoxy radical. In areas with photochemical smog, the principal photoacceptors are aldehydes, NO_2, nitrous acid (HNO_2), and ozone. The photodissociation process is energy dependent, and only photons with sufficient energy are capable of causing photodissociation. The wavelength dependence of solar radiation is discussed in Chapter 5.

The reactivity of chemical compounds will differ because of their structure and molecular weight. Hydrocarbon compounds have been ranked according to their rate of reaction with various types of oxidizing species such as OH, NO_3, and O_3.[17] The role of hydrocarbons, along with oxides of nitrogen, in the formation of ozone is very complex. Ozone formation is a function of the mixture of hydrocarbons present and the concentration of NO_x, $[NO_x]$ (= [NO] + $[NO_2]$). The concept of an incremental reactivity scale permits accessing the increment of ozone formation per incremental change in a single hydrocarbon component.[18] Incremental reactivity is determined by calculating the ozone formation potential in a baseline scenario using a simple mixture of hydrocarbons representing an urban atmosphere. Then for each hydrocarbon species of interest, the ozone formation is recalculated with incremental hydrocarbons added to the mixture. From this approach, the $\Delta[O_3]/\Delta[HC]$ values represent the impact of a specific hydrocarbon on urban photochemical smog formation.

The vapor pressure of a compound is important in determining the upper limit of its concentration in the atmosphere. High vapor pressures will permit higher concentrations than low vapor pressures. Examples of organic compounds are methane and benzo[*a*]pyrene. Methane, with a relatively high vapor pressure, is always present as a gas in the atmosphere; in contrast, benzo[*a*]pyrene, with a relatively low vapor pressure, is adsorbed on PM and is therefore not present as a gas. Vapor pressure also affects the rate of evaporation of organic compounds into the atmosphere and the conversion of atmospheric gases to PM, e.g. SO_2 to the aerosol H_2SO_4.[19]

Atmospheric chemical reactions are classified as either photochemical or thermal. Photochemical reactions are the interactions of photons with species which result in the formation of products. These products may undergo

[17] Pitts Jr., J. N., Winer, A. M., Darnall, K. R., Lloyd, A. C., and Doyle, G. J., in *Proceedings of the International Conference on Photochemical Oxidant Pollution and Its Control*, Vol. II (Dimitriades, B., ed.), EPA-600/3-77-OOlb, pp. 687–707. US Environmental Protection Agency, Research Triangle Park, NC, 1977.

[18] Carter, W. P. L., *Development of Ozone Reactivity Scales for Volatile Organic Compounds*, EPA 600/3-91-050. US Environmental Protection Agency, August 1991.

[19] National Research Council, *Ozone and Other Photochemical Oxidants*. National Academy of Sciences, Washington, DC, 1977.

further chemical reaction. These subsequent chemical reactions are called *thermal* or *dark* reactions.

Finally, atmospheric chemical transformations are classified in terms of whether they occur as a gas (homogeneous), on a surface, or in a liquid droplet (heterogeneous). An example of the last is the oxidation of dissolved sulfur dioxide in a liquid droplet. Thus, chemical transformations can occur in the gas phase, forming secondary products such as NO_2 and O_3; in the liquid phase, such as SO_2 oxidation in liquid droplets or water films; and as gas-to-particle conversion, in which the oxidized product condenses to form an aerosol.

B. Role of Solar Radiation in Atmospheric Chemistry

The time required for atmospheric chemical processes to occur is dependent on chemical kinetics. Many of the air quality problems of major metropolitan areas can develop in just a few days. Most gas-phase chemical reactions in the atmosphere involve the collision of two or three molecules, with subsequent rearrangement of their chemical bonds to form molecules by combination of their atoms. Consider the simple case of a bimolecular reaction of the following type:

$$B + C \rightarrow \text{products} \tag{7.44}$$

$$\text{Rate of reaction} = k[B][C] \tag{7.45}$$

where $k = A \exp[-E_a/RT]$ and k, the rate constant, is dependent on the frequency factor (A), the temperature (T), the activation energy of the reaction (E_a) and the ideal gas constant (R). The frequency factor, A, is of the same order of magnitude for most gas reactions. For $T = 298$ K the rate of reaction is strongly dependent on the activation energy E_a as shown in Table 7.11.[20] When E_a is >30 kJ/mol the rates become very small, limiting the overall rate of reaction. Table 7.12 contains the activation energies for bimolecular collisions of different

TABLE 7.11

Values of exp ($-E_a/RT$) as a Function of the Activation Energy for $T = 298$ K[a]

E_a	exp ($-E_a/RT$)
1.0	0.67
3.0	0.30
10.0	0.0177
30.0	5.51×10^{-6}
100.0	2.95×10^{-18}
300.0	2.56×10^{-53}

[a] In SI units, $R = 8.31434\ \text{kJ}^{-1}\text{mol}^{-1}$.

[20] Campbell, I. M., *Energy and the Atmosphere*. Wiley, New York, 1977.

TABLE 7.12

Activation Energies for Atmospheric Reactions

Reaction	E_a (kJ mol^{-1})
$N_2 + O_2 \rightarrow N_2O + O$	538
$CO + O_2 \rightarrow CO_2 + O$	251
$SO_2 + NO_2 \rightarrow SO_3 + NO$	106
$O + H_2S \rightarrow OH + HS$	6.3
$O + NO_2 \rightarrow NO + O_2$	<1
$HO_2 + NO \rightarrow NO_2 + OH$	<1

Source: Campbell, I. M., *Energy and the Atmosphere*, pp. 212–213. Wiley, New York, 1977.

TABLE 7.13

Hydrocarbon Compounds Identified in Ambient Air Samples from St. Petersburg, Florida

Acetaldehyde	*m*-Ethyltoluene	Methylcyclohexane	Propene
Acetylene	*o*-Ethyltoluene	3-Methylhexane	*n*-Propylbenzene
1,3-Butadiene	*p*-Ethyltoluene	2-Methylpentane	Toluene
n-Butane	*n*-Heptane	Nonane	2,2,4 Trimethylpentane
trans-2-Butene	Isobutane	*n*-Pentane	*m*-Xylene
Cyclopentane	Isobutylene	1-Pentene	*o*-Xylene
n-Decane	Isopentane	*cis*-2-Pentene	*p*-Xylene
2,3-Dimethylpentane	Isopropyl benzene	*trans*-2-Pentene	1,2,4-rimethylbenzene
Ethane	Limonene	*alpha*-Pinene	1,3,5-rimethylbenzene
Ethylbenzene	Methane	*beta*-Pinene	
Ethylene	2-Methyl-1-butene	Propane	

Source: Lonneman, W. A., Seila, R. L., and Bufalini, J. J. *Environ. Sci. Technol.* **12**, 459–463 (1978).

molecular species. For the first three reactions between molecular species, E_a is >100 kJ, but for the last three reactions $E_a < 10$ kJ. The last three reactions involve the participation of free radical or atomic species. The activation energies of reactions involving atomic or free radicals are very small, permitting chemical transformations on a short timescale.

C. Gas-Phase Chemical Reaction Pathways

The complexity of the atmospheric chemical reactions occurring in major metropolitan areas can be staggering. Urban atmospheres are characterized as complex mixtures of hydrocarbons and oxides of sulfur and nitrogen. Table 7.13 show the hydrocarbons identified in the urban air of St. Petersburg, Florida.[21] The interactions among this large number of compounds can be understood by studying simpler systems. Figure 7.16 shows

[21] Lonneman, W. A., Seila, R. L., and Bufalini, J. J. *Environ. Sci. Technol.* **12**, 459–463 (1978).

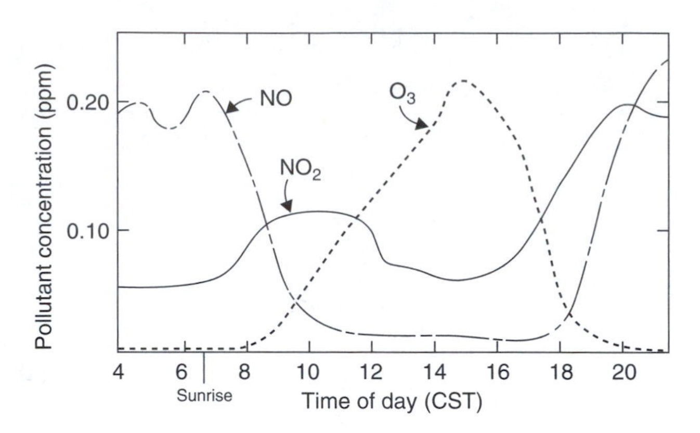

Fig. 7.16. NO–NO_2–O_3 ambient concentration profiles from average of four Regional Air Monitoring Stations (RAPS) in downtown St. Louis, Missouri (USA) on October 1, 1976. *Source*: RAPS, Data obtained from the 1976 data file for the Regional Air Pollution Study Program. US Environmental Protection Agency, Research Triangle Park, NC, 1976.

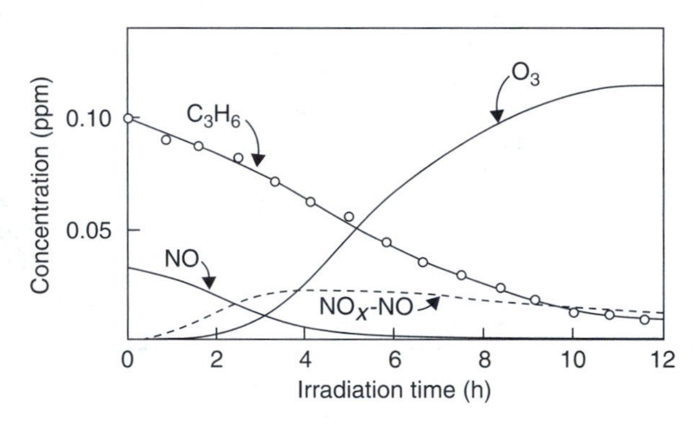

Fig. 7.17. Concentration versus time profiles of propene, NO, NO_x-NO, and O_3 from smog chamber irradiation; $k_1 = 0.16\ \text{min}^{-1}$. *Source*: Akimoto, H., Sakamaki, F., Hoshino, M., Inoue, G., and Oduda, M., *Environ. Sci. Technol.* **13**, 53–58 (1979).

the diurnal patterns of NO, NO_2, and O_3 for St. Louis, Missouri.[22] These diurnal patterns are interrelated. The concentration profiles of Fig. 7.16 are the result of a combination of atmospheric chemical and meteorological processes. To uncouple this combination of factors, laboratory (smog chamber) studies such as those of the propene-NO_x system (Fig. 7.17) have been undertaken.[23] These profiles show chemical transformations separated from meteorological processes.

[22] Data obtained from the 1976 data file of the Regional Air Pollution Study Program. US Environmental Protection Agency, Research Triangle Park, NC, 1976.

[23] Akimoto, H., Sakamaki, F., Hoshino, M., Inoue, G., and Oduda, M. *Environ. Sci. Technol.* **13**, 53–58 (1979).

Similar chemical steps occur in the ambient air and in laboratory smog chamber simulations. Initially, hydrocarbons and nitric oxide are oxidized to form nitrogen dioxide, ozone, and other oxidation products such as peroxyacyl nitrate (PAN) and aldehydes. The complete process is very complicated, with many reaction steps.

The principal components of atmospheric chemical processes are hydrocarbons, oxides of nitrogen, oxides of sulfur, oxygenated hydrocarbons, ozone, and free radical intermediates. Solar radiation plays a crucial role in the generation of free radicals, whereas water vapor and temperature can influence particular chemical pathways. Table 7.14 lists a few of the components of each of these classes. Although more extensive tabulations may be found in *Atmospheric Chemical Compounds*,[24] those listed in Table 7.14 are sufficient for an understanding of smog chemistry. The major undesirable

TABLE 7.14

Classes and Examples of Atmospheric Compounds

Hydrocarbons	Oxygenated hydrocarbons
Alkenes	Aldehydes
Ethene C_2H_4	Formaldehyde HCHO
Propene C_3H_6	Acetylaldehyde CH_3CHO
trans-2-Butene	Other aldehydes RCHO
Alkanes	Acids
Methane CH_4	Formic acid HCOOH
Ethane C_2H_6	Acetic acid CH_3COOH
Alkynes	Alcohols
Acetylene C_2H_2	Methanol CH_3OH
Aromatics	
Toluene C_6H_6	
m-Xylene C_6H_{10}	
Oxides of nitrogen	Oxides of sulfur
Nitric oxide NO	Sulfur dioxide SO_2
Nitrogen dioxide NO_2	Sulfur trioxide SO_3
Nitrous acid HNO_2	Sulfuric acid H_2SO_4
Nitric acid HNO_3	Ammonium bisulfate $(NH_4)HSO_4$
Nitrogen trioxide NO_3	Ammonium sulfate $(NH_4)_2SO_4$
Dinitrogen pentoxide N_2O_5	
Ammonium nitrate NH_4NO_3	
Free radicals	Oxidants
Atomic oxygen O	PAN $CH_3COO_2NO_2$
Atomic hydrogen H	Ozone O_3
Hydroxyl OH	
Hydroperoxyl HO_2	
Acyl RCO	
Peroxyacyl $RCOO_2$	

[24] Graedel, T. E., Hawkins, D. T., and Claxton, L. D., *Atmospheric Chemical Compounds: Sources, Occurrence, and Bioassay*. Academic Press, Orlando, FL, 1986.

components of photochemical smog are NO_2, O_3, SO_2, H_2SO_4, PAN, and aldehydes. Air quality standards have been established in several countries for SO_2, NO_2, and O_3; H_2SO_4 contributes to acidic deposition and reduction in visibility; and PAN and aldehydes can cause eye irritation and plant damage if their concentrations are sufficiently high.

1. Photoabsorption of Solar Radiation

Solar radiation initiates the formation of free radicals. According to the elementary quantum theory of atoms and molecules, the internal energy of molecules is composed of electronic energy states. Molecules interact with solar radiation by absorbing photons. This absorption process causes the molecule to undergo a transition from the ground electronic state to an excited state. The change in energy between the two states corresponds to a quantum or photon of solar radiation. The frequencies v of absorption are expressed by Planck's law:

$$E = hv = hc/\lambda \tag{7.46}$$

where h is Planck's constant, c is the speed of light, and v and λ are the frequency and wavelength of the light of the photon, respectively. The spectrum of solar radiation in the lower troposphere starts at ~295 nm and increases. Photons of shorter wavelength and higher energy are absorbed in the upper atmosphere and therefore do not reach the lower troposphere.

Molecules and atoms interact with photons of solar radiation under certain conditions to absorb photons of light of various wavelengths. Figure 14.4 shows the absorption spectrum of NO_2 as a function of the wavelength of light from 240 to 500 nm. This molecule absorbs solar radiation from 295 nm through the visible region. The absorption of photons at these different wavelengths causes the NO_2 molecule to enter an excited state. For longer wavelengths, transitions only in the rotational–vibrational states occur, whereas for shorter wavelengths changes in electronic states may occur. The process of photoabsorption for NO_2 is expressed as

$$NO_2 + hv \rightarrow NO_2^* \tag{7.47}$$

where hv represents the photon of solar radiation of energy and NO_2^* is the NO_2 molecule in the excited state. The excited NO_2 molecule can follow several pathways:

Fluorescence $$NO_2^* \rightarrow NO_2 + hv \tag{7.48}$$

Collisional deactivation where X_2 is

N_2 or O_2 $$NO_2^* + X_2 \rightarrow NO_2 + X_2 \tag{7.49}$$

Direct reaction $$NO_2^* + O_2 \rightarrow NO_3 + O \tag{7.50}$$

Photodissociation $$NO_2^* \rightarrow NO + O \tag{7.51}$$

In the case of NO_2, for each photon absorbed below 400 nm, photodissociation occurs. For other photoabsorbers, HNO_2 and aldehydes, the photodissociation process leads to the formation of free radicals.

2. Nitric Oxide, Nitrogen Dioxide, and Ozone Cycles

Three relatively simple reactions can describe the interrelationships among these components.

$$NO_2 + h\upsilon \rightarrow NO + O \tag{7.52}$$

$$O + O_2 + M \rightarrow O_3 + M \tag{7.53}$$

$$NO + O_3 \rightarrow NO_2 + O_2 \tag{7.54}$$

Equation (7.52) shows the photochemical dissociation of NO_2. Equation (7.53) shows the formation of ozone from the combination of O and molecular O_2 where M is any third-body molecule (principally N_2 and O_2 in the atmosphere). Equation (7.54) shows the oxidation of NO by O_3 to form NO_2 and molecular oxygen. These three reactions represent a cyclic pathway (Fig. 7.18) driven by photons represented by $h\upsilon$. Throughout the daytime period, the flux of solar radiation changes with the movement of the sun. However, over short-time periods (~10 min) the flux may be considered constant, in which case the rate of Eq. (7.52) may be expressed as

$$\text{Rate} = k_1[NO_2] \tag{7.55}$$

where k_1 is a function of time of day. Expressions for the time rate of change for each of the components may be written. If this cycle reaches a steady state, the change in concentration with time no longer occurs, so that d[conc]/dt is equal to zero.

$$d[NO]/dt = -k_1[NO_2] + k_3[NO][O_3] \tag{7.56}$$

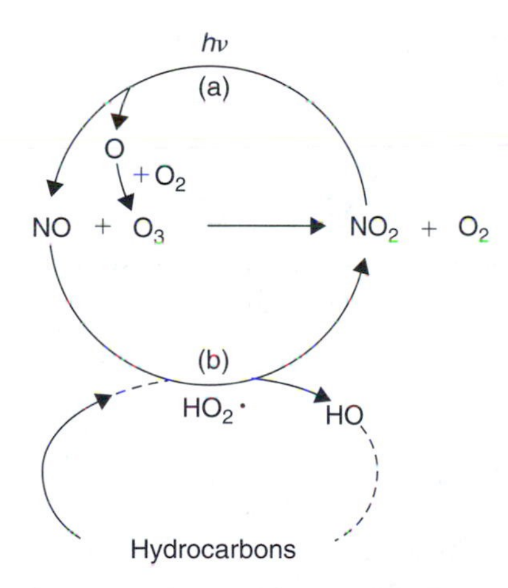

Fig. 7.18. Photochemical cycle of NO, NO_2, O_3, and free radicals.

$$d[O]/dt = k_1[NO_2] - k_2[O][O_2][M] \quad (7.57)$$

$$d[O_3]/dt = k_2[O][O_2][M] - k_3[NO][O_3] \quad (7.58)$$

From Eq. (7.56), it is possible to obtain an expression for the relationship of NO, NO_2, and O_3:

$$d[NO]/dt = O; \quad k_1[NO_2] = k_3[NO][O_3] \quad (7.59)$$

$$[O_3] = k_1[NO_2]/k_3[NO] \quad (7.60)$$

Equation (7.60) is called the *photostationary state expression* for ozone. Upon examination, one sees that the concentration of ozone is dependent on the ratio NO_2/NO for any value of k_1. The maximum value of k_1 is dependent on the latitude, time of year, and time of day. In the United States, the range of k_1 is from 0 to 0.55 min^{-1}. Table 7.15 illustrates the importance of the NO_2/NO ratio with respect to how much ozone is required for the photostationary state to exist. The conclusion to be drawn from this table is that most of the NO must be converted to NO_2 before O_3 will build up in the atmosphere. This is also seen in the diurnal ambient air patterns shown in Fig. 7.15 and the smog chamber simulations shown in Fig. 7.16. It is apparent that without hydrocarbons, the NO is not converted to NO_2 efficiently enough to permit the buildup of O_3 to levels observed in urban areas.

The cycle represented by Eqs. (7.52–7.54) is illustrated by the upper loop (a) in Fig. 7.17. In this figure, the photolysis of NO_2 by a photon forms an NO and an O_3 molecule. If no other chemical reaction is occurring, these two species react to form NO_2, which can start the cycle over again. In order for the O_3 concentration to build up, oxidizers other than O_3 must participate in the oxidation of NO to form NO_2. This will permit the NO_2/NO ratio to build up and steady-state O_3 concentrations as represented by Eq. (7.60) to achieve typical ambient values. The other oxidizers in the atmosphere are free radicals. In the lower loop (b) of Fig. 7.17, a second pathway for NO oxidation is shown, with free radicals participating. These free radicals are derived from the participation of hydrocarbons in atmospheric chemical reactions.

TABLE 7.15

$[O_3]$ Predicted from Photostationary State Approximation as a Function of Initial $[NO_2]$[a]

$[NO_2]$ (ppm)	$[NO_2]_{final}$ (ppm)	$[O_3]_{final}$ (ppm)	$[NO_2]/[NO]$
0.1	0.064	0.036	1.78
0.2	0.145	0.055	2.64
0.3	0.231	0.069	3.35
0.4	0.319	0.081	3.94
0.5	0.408	0.092	4.43

[a] $k_1 = 0.5\ min^{-1}$; $k_3 = 24.2\ ppm^{-1}min^{-1}$.

3. *Role of Hydrocarbons*

The important hydrocarbon classes are alkanes, alkenes, aromatics, and oxygenates. The first three classes are generally released to the atmosphere, whereas the fourth class, the oxygenates, is generally formed in the atmosphere. Propene will be used to illustrate the types of reactions that take place with alkenes. Propene reactions are initiated by a chemical reaction of OH or O_3 with the carbon–carbon double bond. The chemical steps that follow result in the formation of free radicals of several different types which can undergo reaction with O_2, NO, SO_2, and NO_2 to promote the formation of photochemical smog products.

Ozone Reaction with Propene A schematic diagram of the O_3 reaction with propene (Fig. 7.19) is based on the work of Atkinson and Lloyd.[25] The molozonide formed by addition of ozone to the double bond decomposes to form an aldehyde and an energy-rich (‡) biradical. In the case of propene, two sets of products are formed. Along the pathway on the right, approximately 40% of the biradicals (HĊHOO·)‡ form a thermalized biradical (ĊH$_2$OO·).* (*The dots represent unpaired electrons.) The remainder undergo rearrangement to form

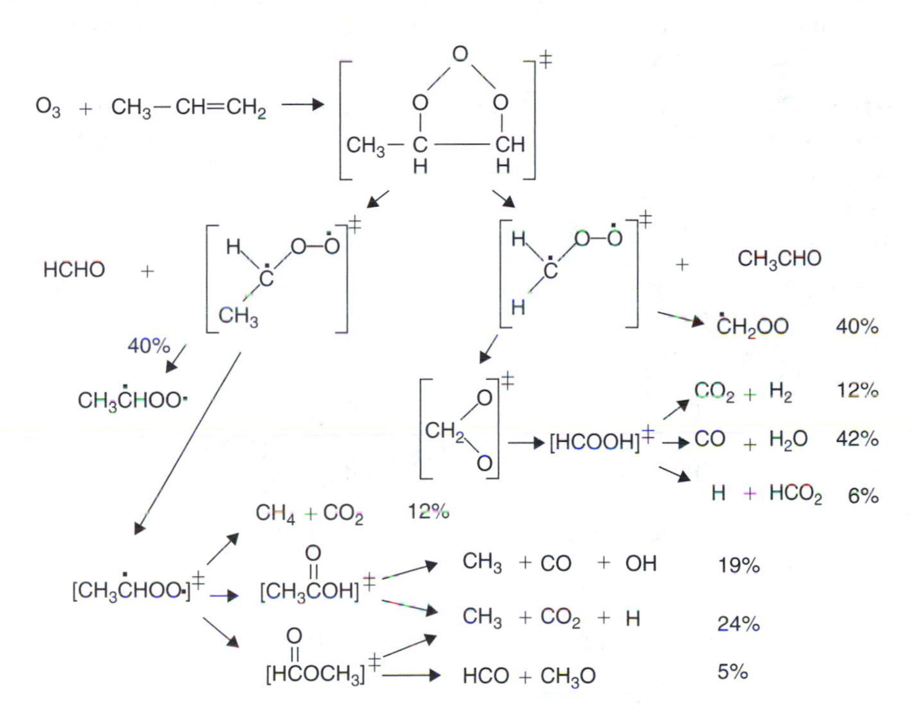

Fig. 7.19. Ozone–propene reaction pathways showing oxidation products.

[25] Atkinson, R., and Lloyd, A. C. *J. Phys. Chem. Ref. Data* **13**, 315–444 (1984).

energy-rich acetic acid (HCOOH)‡, which subsequently decomposes to form H_2O, CO, CO_2, H_2, H, and HCO_2 radicals with percentages assigned to each pathway. The larger biradical ($CH_3\dot{C}HOO\cdot$)‡ follows a slightly different pathway. Approximately 40% forms a thermalized biradical ($CH_3\dot{C}HOO\cdot$). Of the remaining 60%, a portion decomposes to CH_4 and CO_2 and two additional energy-rich species (CH_3COOH)‡ and ($CHOOCH_3$)‡. These two unstable species decompose as shown to form CH_3, OH, H, HCO, CH_3O, CO, and CO_2.

Alkyl radicals, R, react very rapidly with O_2 to form alkylperoxy radicals. H reacts to form the hydroperoxy radical HO_2. Alkoxy radicals, RO, react with O_2 to form HO_2 and R′CHO, where R′ contains one less carbon. This formation of an aldehyde from an alkoxy radical ultimately leads to the process of hydrocarbon chain shortening or clipping upon subsequent reaction of the aldehyde. This aldehyde can undergo photodecomposition forming R, H, and CO; or, after OH attack, forming CH(O)OO, the peroxyacyl radical.

Hydroxyl Radical Addition to Propene As shown in Fig. 7.20, hydroxyl radicals primarily add to either of the carbon atoms which form the double bond. The remaining carbon atom has an unpaired electron which combines with molecular oxygen, forming an RO_2 radical. There are two types of RO_2 radicals labeled C_3OHO_2 in Fig. 7.20. Each of these RO_2 radicals reacts with NO to form NO_2, and an alkoxy radical reacts with O_2 to form formaldehyde, acetaldehyde, and HO_2.

Aldehyde Photolysis and Reactions Aldehydes undergo two primary reactions: photolysis and reaction with OH radicals. These reactions lead to formation of CO, H, and R radicals.

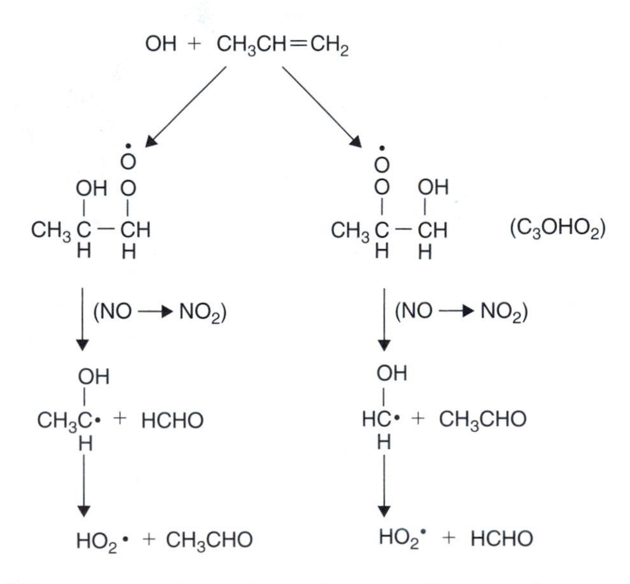

Fig. 7.20. OH–propene reaction pathways showing oxidation products.

Radical Reactions with Nitric Oxide and Nitrogen Dioxide Alkylperoxy (RO_2) and peroxyacyl (RC(O)OO) radicals react with NO to form NO_2. The alkylperoxy radicals (RO_2) react with NO_2 to form pernitric acid-type compounds, which decompose thermally as the temperature increases. The peroxyacyl radical reacts with NO_2 to form PAN-type compounds, which also decompose thermally.

Radical Oxidation of Sulfur Dioxide Sulfur dioxide is oxidized in the atmosphere eventually to form sulfate compounds. The oxidation process includes both homogeneous and heterogeneous pathways. The free radicals produced from the degradation of hydrocarbons can and do react with SO_2 in the gas phase. Both OH and HO_2 oxidize SO_2 to reactive intermediates such as HSO_3 and SO_3.[26] These intermediates combine rapidly with water vapor in the atmosphere to form sulfuric acid aerosol. This type of process is dependent on atmospheric conditions. In urban areas with existing photochemical smog problems, the homogeneous oxidation of SO_2 by free radicals is probably dominant during the daytime.

VII. HETEROGENEOUS REACTIONS

Heterogeneous reactions are defined as those involving the gas–liquid or gas–solid phases. The chemistry of NO_2 and SO_2 has a heterogeneous component in the atmosphere. Heterogeneous reactions involve the dissolving of NO_2 and SO_2 in water droplets, with subsequent chemical reactions occurring to form HNO_3 and H_2SO_4 in the liquid phase. The heterogeneous oxidation of SO_2 in liquid droplets and water films is also a major pathway for conversion to sulfate in wet plumes and during humid or foggy conditions.

VIII. SCAVENGING AND REMOVAL FROM THE ATMOSPHERE

The atmosphere is a dynamic system, with gases and PM entering, undergoing transformation, and leaving. Atmospheric chemical processes of oxidation transform gases into more highly oxidized products, e.g., NO to NO_2 to HNO_3, hydrocarbons to aldehydes, and SO_2 to sulfate particles. The removal of material from the atmosphere involves two processes: wet and dry deposition. The water solubility of gases influences the extent of removal by wet versus dry deposition. Gases such as SO_2 and NO_2 are sufficiently soluble to dissolve in water associated with in-cloud formation of rain droplets. These soluble gases may be removed by wet deposition of liquid droplets in the form of rain or fog. Less soluble gases such as O_3 and hydrocarbon vapors are removed by transport to the surface of the earth, where they diffuse to vegetation, materials, or water bodies (see Chapter 14).

[26] Calvert, J. G., Su, F., Bottenheim, J. W., and Strausz, O. P. *Atmos. Environ.* **12**, 197–226 (1978).

REFERENCES

1. US Geological Survey, *Radiogenic Isotopes and the Eastern Mineral Resources Program of the US Geological Survey*, 2003.
2. US Geological Survey, *Short-Lived Isotopic Chronometers: A Means of Measuring Decadal Sedimentary Dynamics*, FS-073-98, 2003.
3. Lewis, C., Klouda, G., and Ellenson, W., *Cars or Trees: Which Contribute More to Particulate Matter Air Pollution?* US Environmental Protection Agency, Science Forum, Washington, DC, 2003.
4. National Bureau of Standards, *Oxalic Acid Standard Reference Method*, SRM 4990B.
5. Stieglitz, L., Zwick, G., Beck, J., Bautz, H., and Roth, W., *Chemosphere* **19**, 283 (1989).
6. Koester, C. J., and Hites, R. A., Wet and dry deposition of chlorinated dioxins and furans, *Environ. Sci. Technol.* **26**, 1375–1382 (1992); and Hites, R. A., Atmospheric transport and deposition of polychlorinated dibenzo-*p*-dioxins and dibenzofurans, EPA/600/3-91/002. Research Triangle Park, NC, 1991.
7. Evangelou, V., *Environmental Soil and Water Chemistry: Principles and Applications*. John Wiley & Sons, New York, 1998.
8. US Army Corps of Engineers, *Engineering and Design: Precipitation/Coagulation/Flocculation*, Chapter 2, EM 1110-1-4012, 2001.
9. Spencer, J., Bodner, G., and Rickard, L., *Chemistry: Structure and Dynamics*, 2nd ed. John Wiley & Sons, New York, 2003.
10. Casparian, A., Chemistry, in *How to Prepare for the Fundamentals of Engineering (FE/EIT) Exam* (Olia, M., ed.). Barron's Educational Series, Inc., Hauppauge, NY, 2000.
11. Hemond, H. F., and Fechner-Levy, E. J., *Chemical Fate and Transport in the Environment*. Academic Press, San Diego, CA, 2000.
12. US Environmental Protection Agency, *Dermal Exposure Assessment: Principles and Applications*. Interim Report, EPA/600/8-91/011B, Washington, DC, 1992.
13. Knox, R., Sabatini, D., and Canter, L., *Subsurface Transport and Fate Processes*. Lewis Publishers, Boca Raton, FL, 1993.
14. US Environmental Protection Agency, *Summary Report: Control and Treatment Technology for the Metal Finishing Industry; Sulfide Precipitation*. Report No. EPA 625/8-80-003, Washington, DC, 1980.
15. Vallero, D., *Engineering the Risks of Hazardous Wastes*. Butterworth-Heinemann, Boston, MA, 2003.
16. Vallero, D., *Environmental Contaminants: Assessment and Control*. Academic Press, Burlington, MA, 2004.
17. Pitts Jr., J. N., Winer, A. M., Darnall, K. R., Lloyd, A. C., and Doyle, G. J., in *Proceedings of the International Conference on Photochemical Oxidant Pollution and Its Control*, Vol. II (Dimitriades, B., ed.), EPA-600/3-77-OOlb, pp. 687–707. US Environmental Protection Agency, Research Triangle Park, NC, 1977.
18. Carter, W. P. L., *Development of Ozone Reactivity Scales for Volatile Organic Compounds*, EPA 600/3-91-050. US Environmental Protection Agency, August 1991.
19. National Research Council, *Ozone and Other Photochemical Oxidants*. National Academy of Sciences, Washington, DC, 1977.
20. Campbell, I. M., *Energy and the Atmosphere*. Wiley, New York, 1977.
21. Lonneman, W. A., Seila, R. L., and Bufalini, J. J. *Environ. Sci. Technol.* **12**, 459–463 (1978).
22. Data obtained from the 1976 data file of the Regional Air Pollution Study Program. US Environmental Protection Agency, Research Triangle Park, NC, 1976.
23. Akimoto, H., Sakamaki, F., Hoshino, M., Inoue, G., and Oduda, M. *Environ. Sci. Technol.* **13**, 53–58 (1979).

24. Graedel, T. E., Hawkins, D. T., and Claxton, L. D., *Atmospheric Chemical Compounds: Sources, Occurrence, and Bioassay*. Academic Press, Orlando, FL, 1986.
25. Atkinson, R., and Lloyd, A. C. *J. Phys. Chem. Ref. Data* **13**, 315–444 (1984).
26. Calvert, J. G., Su, F., Bottenheim, J. W., and Strausz, O. P. *Atmos. Environ.* **12**, 197–226 (1978).

SUGGESTED READING

National Research Council, *Rethinking the Ozone Problem in Urban and Regional Air Pollution*. National Academy Press, Washington, DC, 1991.

Sloane, C. S., and Tesche, T. W., *Atmospheric Chemistry: Models and Predictions for Climate and Air Quality*. Lewis Publishers, Chelsea, MI, 1991.

Warneck, P., *Chemistry of the Natural Atmosphere*. Academic Press, San Diego, CA, 1988.

QUESTIONS

1. What adjustments would have to be made if a non-conservative tracer were used to characterize a plume? Why might such a tracer need to be used instead of a conservative tracer?
2. Review the following table with data from a southeaster US city. What is the contribution of non-fossil-fuel sources to particulate-laden carbon during the summer? Why are concurrent measurements of OC/EC ratios important to characterizing biogenic secondary organic aerosols (SOA), especially as an indicator of biogenic sources (i.e. from other than human activities) being a significant non-fossil-fuel contributor? Identify and describe confounders (e.g. dates) that should be explained (Table 7.16).

TABLE 7.16

Air Mass Origin, $PM_{2.5}$ Total Carbon Concentration, Organic to Elemental Carbon Ratio (OC/EC), and Percent Modern Carbon (pMC), and Biogenic Percentage for Total Carbon (Bio TC)

Start date 1999	Start time (h)	Duration (h)	Air mass origin	TC[a] ($\mu g\,m^{-3}$)	OC/EC[b]	pMC_{TC}[b,c] (%)	Bio TC[b,d] (%)
June 21	1900	11.5	ENE	8.2	12.6 ± 1.9	75 ± 1	69 ± 1
June 22	0700	11.5	S	8.0	15.5 ± 2.7	66 ± 1	61 ± 1
June 28	0700	10.5	SW	2.9	10.3 ± 3.0	66 ± 2	61 ± 2
June 29	0700	11.5	NW	3.2	15.8 ± 5.6	80 ± 2	73 ± 2
June 29	1900	11.5	NW	4.3	5.4 ± 0.8	56 ± 1	51 ± 1
July 1	0700	5.5	S	3.9	10.2 ± 4.0	68 ± 3	62 ± 3
July 1	1250	5.7	S	3.9	29.2 ± 27.3[e]	67 ± 3	61 ± 3
July 1	1900	11.5	SSW	2.8	9.9 ± 2.7	75 ± 2	69 ± 2
July 4	1900	11.5	S	7.6	3.6 ± 0.4	64 ± 1	59 ± 1
						66 ± 1[f]	61 ± 1[f]
July 5	0700	5.5	S	7.7	11.8 ± 2.1	63 ± 1	58 ± 1
July 5	1300	5.5	S	7.9	13.1 ± 3.5	64 ± 1	59 ± 1
July 6	1900	11.5	WNW	4.6	5.7 ± 0.8	65 ± 1	60 ± 1
July 7	0700	11.5	N	5.6	14.0 ± 2.9	66 ± 1	61 ± 1
July 8	1900	11.5	S	5.0	8.4 ± 1.4	70 ± 1	64 ± 1

(continued)

TABLE 7.16 (*Continued*)

Start date 1999	Start time (h)	Duration (h)	Air mass origin	TC[a] ($\mu g\,m^{-3}$)	OC/EC[b]	pMC_{TC}[b,c] (%)	Bio TC[b,d] (%)
July 9	0700	11.5	SW	4.8	17.5 ± 4.8	78 ± 1	72 ± 1
July 10	1900	11.5	N	3.8	11.2 ± 2.7	75 ± 1	69 ± 1
July 11	0700	5.5	NE	3.5	7.1 ± 2.3	70 ± 3	64 ± 3
July 11	1300	5.5	NE	4.1	7.1 ± 2.3	78 ± 3	72 ± 3
July 12	1900	11.5	NNE	5.7	15.0 ± 2.2	79 ± 1	72 ± 1
July 13	0700	5.5	N	3.6	7.9 ± 2.8	65 ± 3	60 ± 3
July 13	1300	5.5	NNW	4.1	15.1 ± 7.7	67 ± 2	61 ± 2

Source: Lewis, C. W., Klouda, G. A., and Ellenson, W. D., Radiocarbon measurement of the biogenic contribution to summertime $PM_{2.5}$ ambient aerosol in Nashville, TN. *Atmos. Environ.* **39**(35), 6053–6061.

[a]Estimated uncertainty, ±15%.

[b]Uncertainty calculated by usual error propagation using the standard deviations of the measured value and the blank used to correct the measured value.

[c]Corrected for TC composite ($n = 3$) filter blank: pMC = 43 ± 1, concentration = 1.7 ± 0.2 μg TC cm^{-2}.

3. Consider a combustion reaction other than the ones described in this chapter and explain how products of incomplete combustion may form.
4. Consider an acid–base reaction other than those described in this chapter and explain how an air toxic may form in a water droplet.
5. What wavelength band of solar radiation leads to photodissociation for nitrogen dioxide? What determines the lower limit?
6. Equation (7.60) describes the steady-state $[O_3]$ in the presence of solar radiation. If $k_1 = 0.3\ min^{-1}$ and $k_3 = 30\,ppm^{-1}min^{-1}$, which sets of conditions given in the table below are consistent with $[O_3] = 0.1$ ppm?

[NO] (ppm)	$[NO_2]$ (ppm)
0.005	0.050
0.125	0.125
0.047	0.47
0.47	0.047
0.001	0.010

7. What does the answer to question 6 indicate about the functional dependence of ozone concentrations on the absolute magnitude of $[NO_x] = [NO] + [NO_2]$?
8. How are free radicals formed, and why are they so reactive?
9. From Fig. 7.19, how many molecules of NO can be oxidized to NO_2 by the reaction of one OH free radical with one propene molecule?
10. List various classification approaches used to account for hydrocarbon reactivity.
11. Classify the compounds in Table 7.15 into precursors (reactants) and products of atmospheric reaction processes.
12. List two photoacceptors in addition to nitrogen dioxide that provide an initial source of free radicals.
13. Describe the role of hydrocarbons in photochemical oxidant formation.
14. What are the molar concentrations of $[H^+]$ and $[OH^-]$ of rainwater at pH 3.7 at 25°C?

8

Air Quality

Air is a vital resource, so its quality must fall within a tightly bound range. This quality is the level needed to protect public health. In addition, the quality must be able to support other life, notably diverse and sustainable ecosystems. Finally, air must be of a quality to protect welfare, including the prevention of the corrosion of materials, preservation of vistas, and protection of agricultural crops.

The terms *ambient air, ambient air pollution, ambient levels, ambient concentrations, ambient air monitoring, ambient air quality,* etc. occur frequently in air pollution parlance. The adjective "ambient" distinguishes pollution of the air outdoors (i.e. ambient air pollution) from contamination of the air indoors by the same substances.

The air inside a factory building can be polluted by release of contaminants from industrial processes to the air of the workroom. This is a major cause of occupational disease. Prevention and control of such contamination are part of the practice of industrial hygiene. To prevent exposure of workers to such contamination, industrial hygienists use industrial ventilation systems that remove the contaminated air from the workroom and discharge it, either with or without treatment to remove the contaminants, to the ambient air outside the factory building.

The air inside a home, office, or public building is the subject of much interest and is referred to as indoor air pollution or indoor air quality (see Chapter 3). These interior spaces may be contaminated by such sources as fuel-fired cooking or space heating ranges, ovens, or stoves that discharge their gaseous and aerosol combustion products to the room; by solvents evaporated from coatings, paints, adhesives, cleaners, or other products; by formaldehyde, radon, and other products emanating from building materials; and by other pollutant sources indoors [1]. If some of these sources exist inside a building, the pollution level of the indoor air might be higher than that of the outside air. However, if none of these sources are inside the building, the pollution level inside would be expected to be lower than the ambient concentration outside because of the ability of the surfaces inside the building—walls, floors, ceilings, furniture, and fixtures—to adsorb or react with gaseous pollutants and to attract and retain particulate pollutants, thereby partially removing them from the air breathed by occupants of the building. This adsorption and retention would occur even if doors and windows were open, but the difference between outdoor and indoor concentrations would be even greater if they were closed, in which case air could enter the building only by infiltration through cracks and walls.

For example, volatile organic compounds, such as the carbonyls (aldehydes and ketones), are frequently found in residences. Various sources of carbonyls are often present inside homes, but little is known about their indoor source strengths; that is, the amount generated indoors versus outdoors. A recent study[1] using a database established in the relationships of indoor, outdoor, and personal air (RIOPA) estimated indoor source strengths of 10 carbonyls and outdoor contributions to measured indoor concentrations of these carbonyls. A mass balance model was applied to analyze paired indoor and outdoor carbonyl concentrations simultaneously measured in 234 RIOPA homes. Table 8.1 shows indoor and outdoor concentrations of carbonyls. Scatter plots of paired indoor and outdoor carbonyl concentrations are shown in Fig. 8.1.

Table 8.2 indicates that the study found variations in the estimated decay rate (rate of chemical breakdown) constants across the RIOPA homes.

Across-home estimates of indoor source strengths are presented in Fig. 8.2.

Myriad factors influence indoor air pollutant concentrations, but this study found that house volume and indoor source strength were the most influential parameters in determining indoor concentrations for compounds with strong indoor sources, whereas the outdoor concentration had the most significant impact for compounds mainly generated from outdoor sources. Interestingly the air exchange rates and the decay rate constant were the least sensitive parameters in the model used to determine the indoor concentrations for the RIOPA homes.

Many materials used and dusts generated in buildings and other enclosed spaces are allergenic to their occupants. Occupants who do not smoke are exposed to tobacco and its associated gaseous and particulate emissions

[1] Liu, W., Zhang, J., Zhang, L., Turpin, B. J., Weisel, C. P., Morandi, M. T., Stock, T. H., Colome, S., and Korn, L. R., 2006. Estimating contributions of indoor and outdoor sources to indoor carbonyl concentrations in three urban areas of the United States. *Atmos. Environ.* **40**(12), 2202–2214 (2006).

TABLE 8.1

Residential Indoor and Outdoor Carbonyl Concentrations ($\mu g\,m^{-3}$) Measured in 234 Homes

	Indoor samples ($N = 353$)				Outdoor samples ($N = 353$)				
Compounds	5 percentile	Median	95 percentile	% above MDL	5 percentile	Median	95% percentile	% above MDL	*P*-value[a]
Formaldehyde	12.5	20.1	32.5	100	2.21	6.42	9.95	100	<0.001
Acetaldehyde	7.53	18.6	50.2	100	1.47	5.44	14.9	98	<0.001
Acetone	0.98	8.08	45.8	97	MDL	4.19	19.5	91	<0.001
Acrolein	MDL	0.59	5.54	71	MDL	0.46	4.58	68	0.125
Propionaldehyde	0.23	1.74	3.65	97	0.05	1.37	3.68	95	<0.001
Crotonaldehyde	MDL	0.44	2.51	72	MDL	0.26	2.03	63	0.002
Benzaldehyde	0.98	2.92	5.25	98	MDL	1.88	4.21	94	<0.001
Glyoxal	1.12	2.53	4.37	100	0.44	1.81	3.48	99	<0.001
Methylglyoxal	1.13	2.75	4.77	99	0.26	2.05	3.99	96	<0.001
Hexaldehyde	1.63	3.81	9.94	100	0.23	2.01	4.69	99	<0.001

[a] *P* values based on Wilcoxon signed-ranked test for residential indoor and outdoor carbonyl concentration differences in medians.

$P < 0.05$ indicates the difference is significant at $\alpha = 0.05$.

MDL: measurement detection limit.

Source: Liu, W., Zhang, J., Zhang, L., Turpin, B. J., Weisel, C. P., Morandi, M. T., Stock, T. H., Colome, S., and Korn, L. R., Estimating contributions of indoor and outdoor sources to indoor carbonyl concentrations in three urban areas of the United States. *Atmos. Environ.* **40**(12), 2202–2214 (2006).

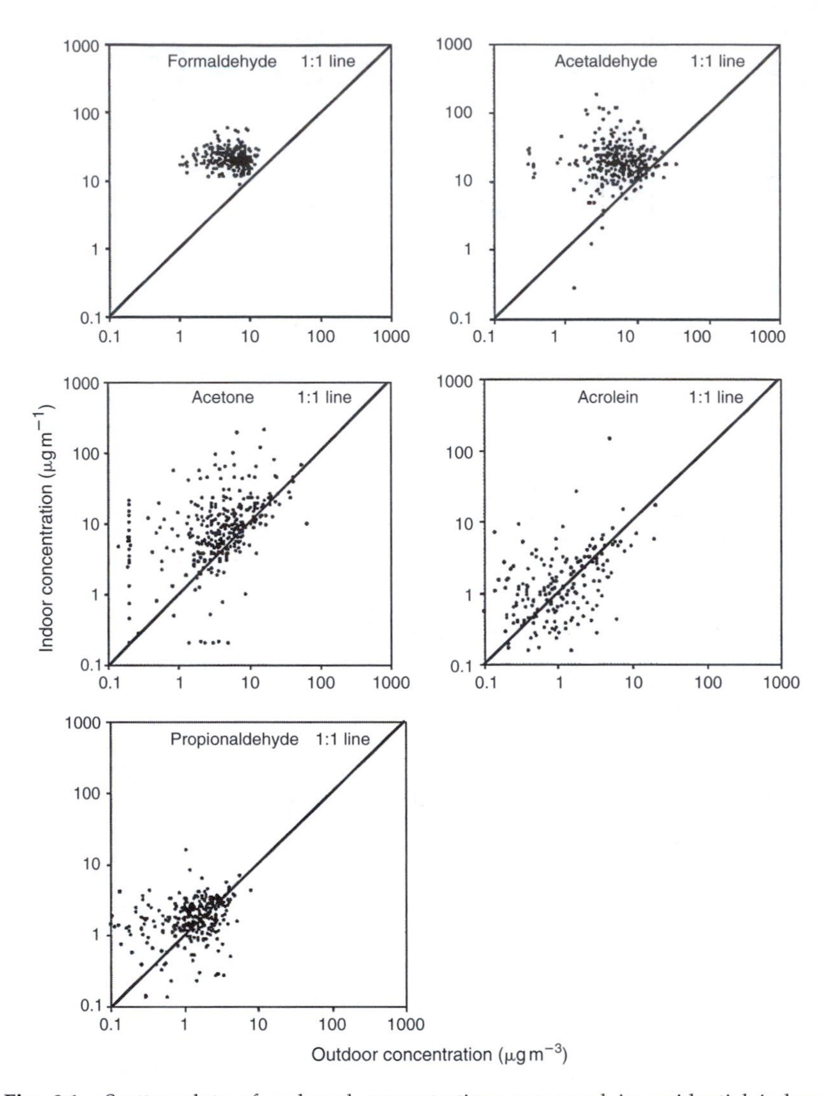

Fig. 8.1. Scatter plots of carbonyl concentrations measured in residential indoor and outdoor air of homes (indoor concentration versus outdoor concentration, $N = 353$). *Source*: Liu. W., Zhang, J., Zhang, L., Turpin, B. J., Weisel, C. P., Morandi, M. T., Stock, T. H., Colome, S., and Korn, L. R., Estimating contributions of indoor and outdoor sources to indoor carbonyl concentrations in three urban areas of the United States. *Atmos. Environ.* **40**(12), 2202–2214 (2006).

from those who do. This occurs to a much greater extent indoors than in the outdoor air. Many ordinances have been established to limit or prohibit smoking in public and work places. Attempts have been made to protect occupants of schoolrooms from infections and communicable diseases by using ultraviolet light or chemicals to disinfect the air. These attempts have been unsuccessful because disease transmission occurs instead outdoors

TABLE 8.2

Estimated Carbonyl Decay Rate Constants Using the Data Collected in 234 Homes

Compounds (k, $1\,h^{-1}$)	N	5th Percentile	25th Percentile	Median	75th Percentile	95th Percentile	Ratio 95%/5%
Acetaldehyde	19	0.023	0.057	0.23	0.66	1.4	61.3
Acetone	75	0.011	0.078	0.34	0.97	4.0	360
Acrolein	153	0	0.072	0.54	3.2	14	NA
Propionaldehyde	121	0.011	0.085	0.39	1.4	2.0	183
Crotonaldehyde	152	0	0.098	0.62	3.9	12	NA
Benzaldehyde	79	0.019	0.082	0.24	0.61	1.5	78.9
Glyoxal	71	0.032	0.053	0.15	0.48	8.1	253
Methylglyoxal	76	0.020	0.062	0.17	0.45	1.2	60
Hexaldehyde	29	0.032	0.073	0.32	1.2	3.6	113

Source: Liu, W., Zhang, J., Zhang, L., Turpin, B. J., Weisel, C. P., Morandi, M. T., Stock, T. H., Colome, S., and Korn, L. R., Estimating contributions of indoor and outdoor sources to indoor carbonyl concentrations in three urban areas of the United States. *Atmos. Environ.* **40**(12), 2202–2214 (2006).

and in unprotected rooms. There is, of course, a well-established technology for maintaining sterility in hospital operating rooms and for manufacturing operations in pharmaceutical, semiconductor and similar plants.

I. AVERAGING TIME

The variability inherent in the transport and dispersion process, the time variability of source strengths, and the scavenging and conversion mechanisms in the atmosphere cause pollutants to have an effective half-life. In turn, these factors result in variability in the concentration of a pollutant arriving at a receptor. Thus, a continuous record of the concentration of a pollutant at a receptor, as measured by an instrument with rapid response, might look like Fig. 8.3(a). If, however, instead of measuring with a rapid-response instrument, the measurement at the receptor site was made with sampling and analytical procedures that integrated the concentration arriving at the receptor over various time periods, e.g., 15 min, 1 h, or 6 h, the resulting information would look variously like Figs. 8.3(b)–(d), respectively. It should be noted that from the information in Fig. 8.3(a), it is possible to derive mathematically the information in Figs. 8.3(b)–(d), and it is possible to derive the information in Figs. 8.3(c) and (d) from that in Fig. 8.3(b). The converse is not true. With only the information from Fig. 8.3(d) available, Figs. 8.3(a)–(c) could never be constructed, nor could Figs. 8.3(a) and (b) be constructed from Fig. 8.3(c), nor Fig. 8.3(a) from Fig. 8.3(b). In these examples, the time intervals involved in Figs. 8.3(b)–(d)—15 min, 1 h, and 6 h, respectively—are the *averaging times* of the measurement of pollutant exposure at the receptor.

The averaging time of the rapid-response record [Fig. 8.3(a)] is an inherent characteristic of the instrument and the data acquisition system. It can become

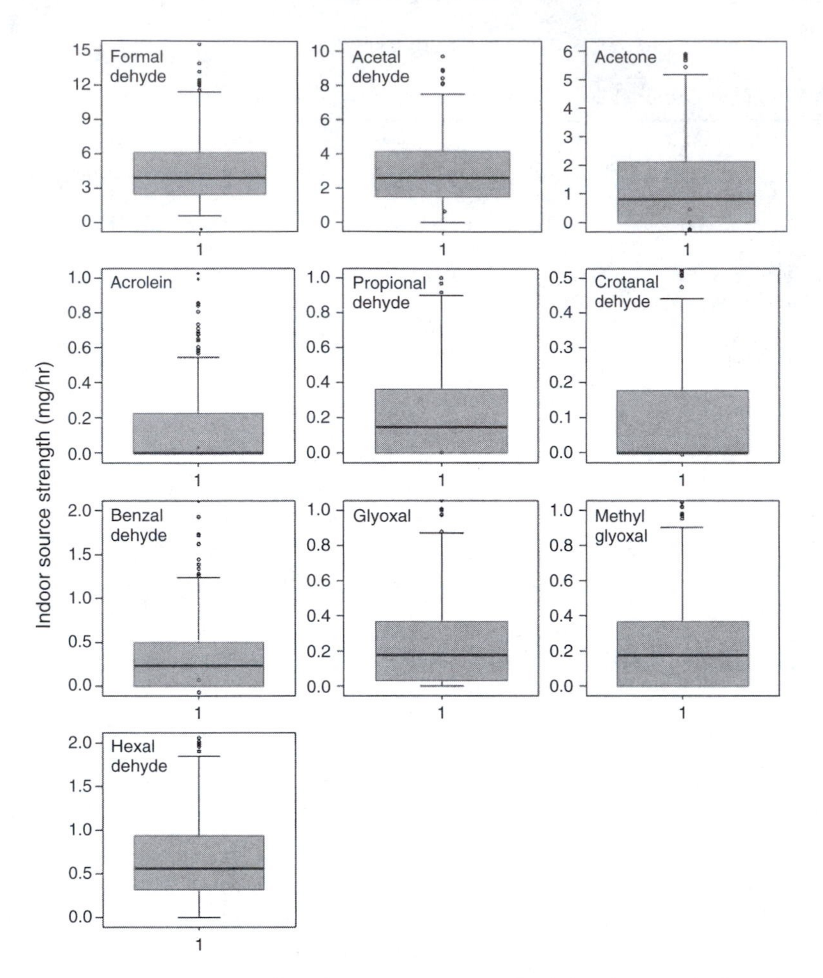

Fig. 8.2. Estimated indoor carbonyl source strengths in the RIOPA homes. The box plots summarize the median, lower quartile, upper quartile, lower range, and upper range of the distribution of indoor source strengths. "°" represents outliers with more than 1.5 box lengths from the upper or lower edge of the box. Some outliers were not presented in the plots because of the scale limit. *Source*: Liu, W., Zhang, J., Zhangm L., Turpin, B.J., Weisel, C.P., Morandi, M.T., Stock, T.H., Colome, S., and Korn, L.R., Estimating contributions of indoor and outdoor sources to indoor carbonyl concentrations in three urban areas of the United States. *Atmos. Environ.* **40**(12), 2202–2214 (2006).

almost an instantaneous record of concentration at the receptor. However, in most cases this is not desirable, because such an instantaneous record cannot be put to any practical air pollution control use. What such a record reveals is something of the turbulent structure of the atmosphere, and thus it has some utility in meteorological research. In communications science parlance, an instantaneous recording has too much *noise* (see Fig. 8.4), preventing a proper interpretation of the measurements. It is therefore necessary to filter or damp out the noise in order to extract the useful information about pollution

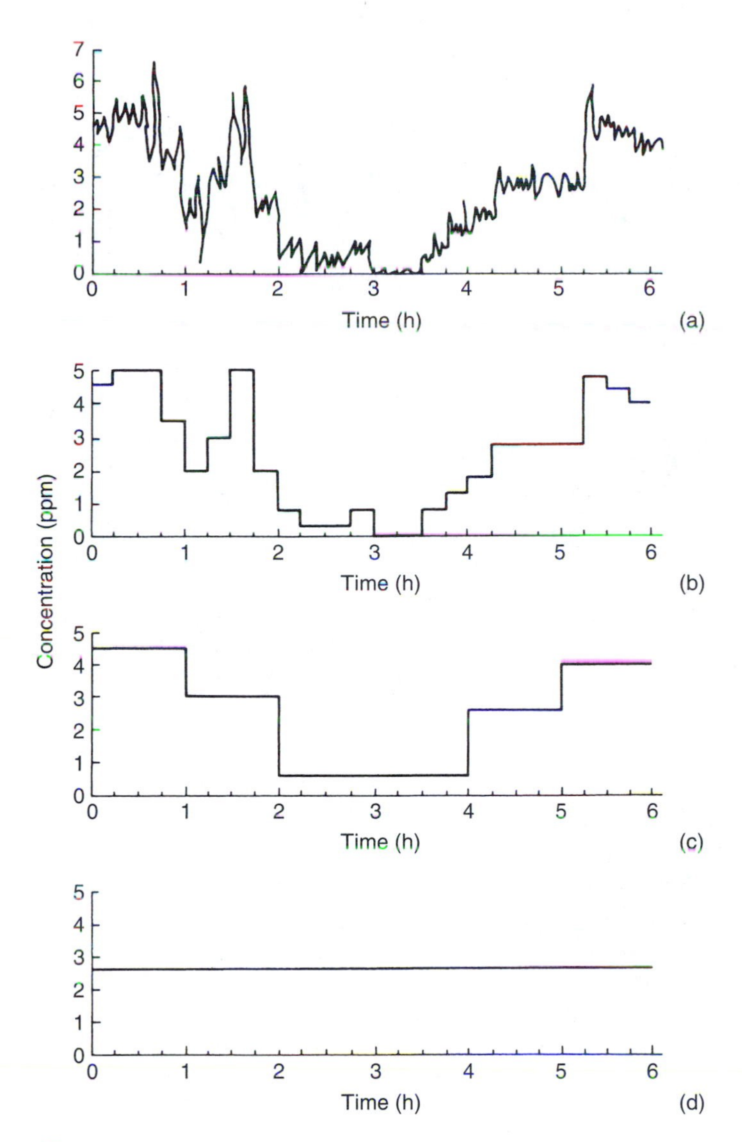

Fig. 8.3. The same atmosphere measured by (a) a rapid-response instrument and by sampling and analytical procedures that integrate the concentration arriving at the receptor over a time period of (b) 15 min, (c) 1 h, and (d) 6 h.

concentration at the receptor that the signal is trying to reveal. This damping is achieved by building time lags into the response of the sampling, analysis, and recording systems (or into all three); by interrogating the instantaneous output of the analyzer at discrete time intervals, e.g., once every minute or once every 5 min, and recording only this extracted information; or by a combination of damping and periodic interrogation.

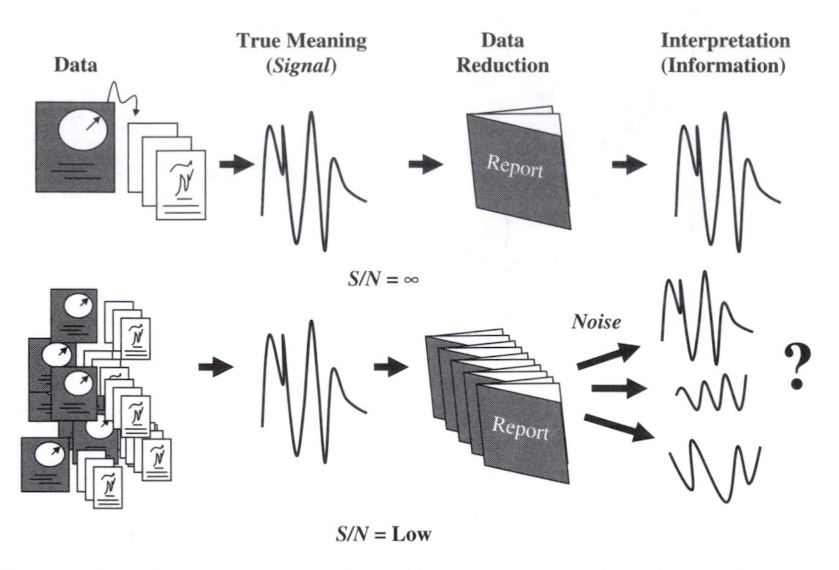

Fig. 8.4. Scientific communication of air quality measurements depends on high quality data that is properly reduced and interpreted in order to become useful information for users. However, even data that are properly collected may become noisy if their volume is so large that they are not able to be interpreted in a timely manner. This leads to a low signal to noise (S/N) ratio.

II. CYCLES

The most significant of the principal cyclic influences on variability of pollution concentration at a receptor is the diurnal cycle (Fig. 8.5). First, there is a diurnal pattern to source strength. In general, emissions from almost all categories of sources are less at night than during the day. Factories and businesses shut down or reduce activity at night. There is less automotive, aircraft, and railroad traffic, use of electricity, cooking, home heating, and refuse burning at night. Second, there is a diurnal pattern to transport and diffusion that will be discussed in detail later in this book.

The next significant cycle is the weekend–weekday cycle. This is entirely a source strength cycle associated with the change in the pattern of living on weekends as compared with weekdays.

Finally, there is the seasonal cycle associated with the difference in climate and weather over the four seasons: winter, spring, summer, and fall (Fig. 8.6). The climatic changes affect source strength, and the weather changes affect transport and dispersion.

On an annual basis, some year-to-year changes in source strength may be expected as a community, a region, a nation, or the world increases in population or changes its patterns of living. Source strength will be reduced if control efforts or changes in technology succeed in preventing more pollution emission than would have resulted from increases in population (Fig. 8.7). These changes are called *trends.* Although an annual trend in

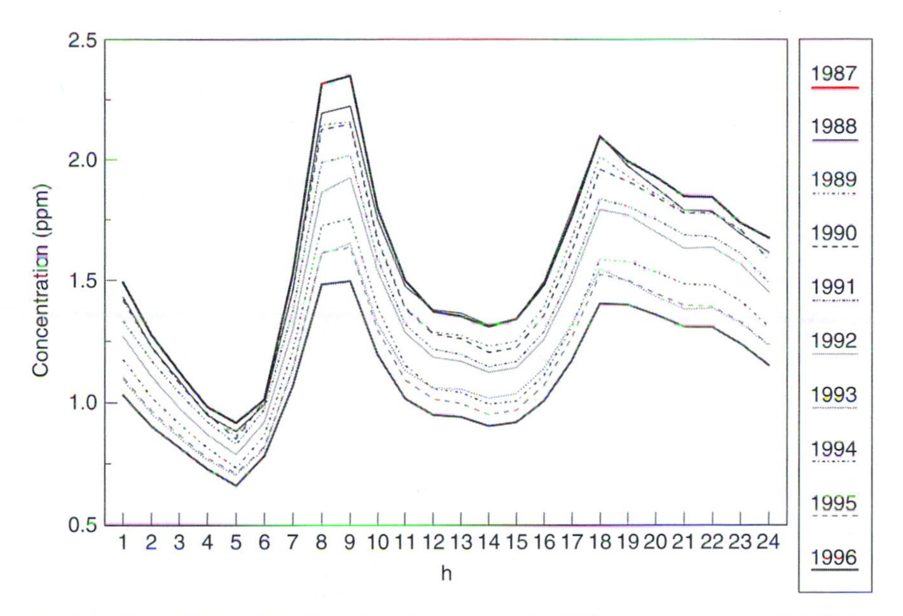

Fig. 8.5. Diurnal patterns in hourly carbon monoxide (CO) concentrations averaged across 2778 monitoring stations located throughout the United States for the 10-year period, 1987–1996. The trends are derived from the composite average of these direct measurements. *Source*: US Environmental Protection Agency, *National Air Quality and Emissions Trends Report*, 1997, http://www.epa.gov/oar/aqtrnd97/chapter2.pdf; accessed February 18, 2007.

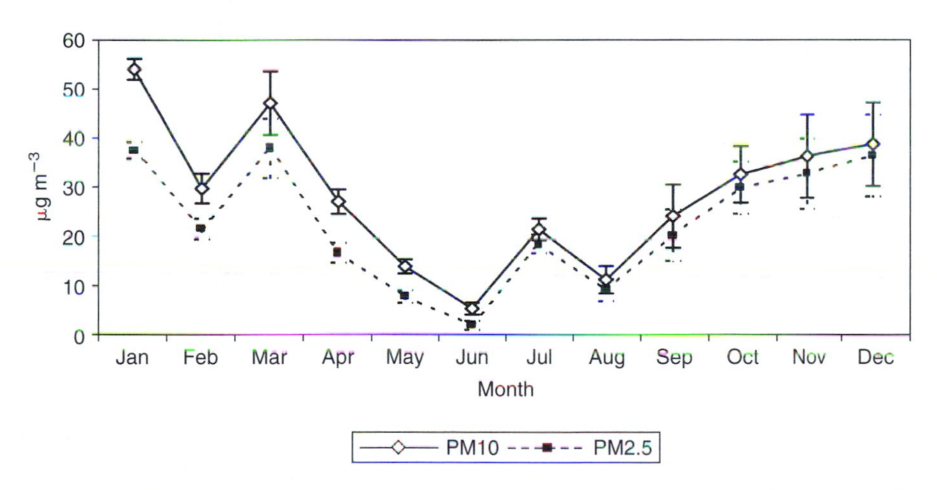

Fig. 8.6. Seasonal variations of particle concentrations in the Hyderabad, India (178 10′–178 50′N latitude and 788 10′–788 50′E longitude). The climate of the study area is of semi-arid type with total rainfall of 700 mm occurring mostly during monsoon season corresponding to July–October. Measurements were gathered approximately at hourly intervals for 4–5 days, with the sampling duration of 6 min during January–December, 2003. *Source*: Latha, K. M., and Badarinath, K. V. S., Seasonal variations of PM_{10} and $PM_{2.5}$ particles loading over tropical urban environment. *Int. J. Environ. Health Res.* **15** (1), 63–68 (2005).

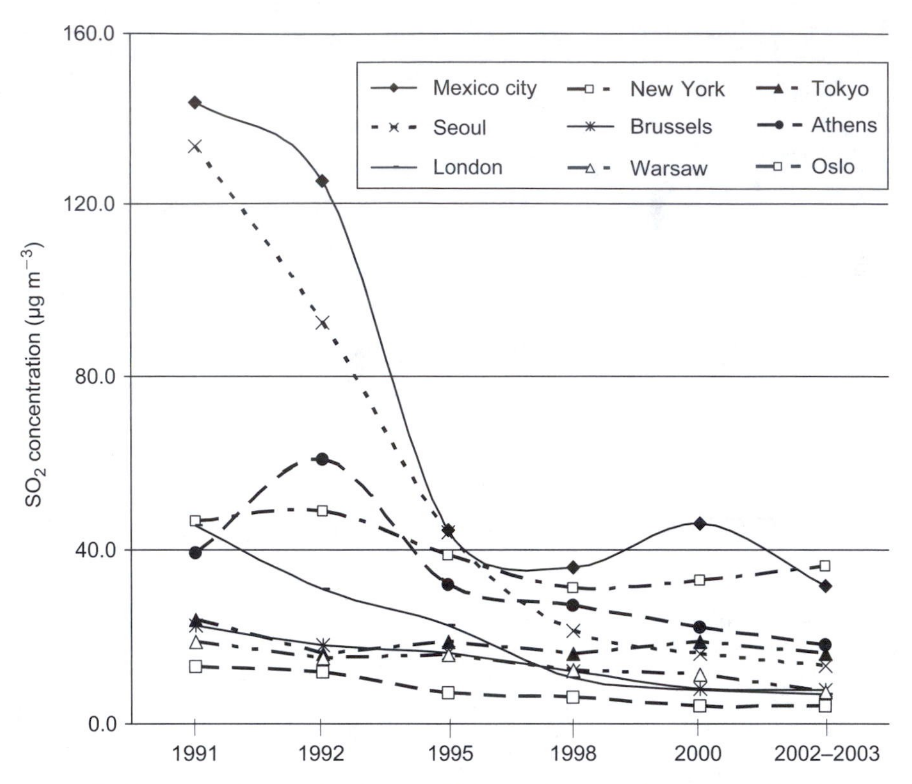

Fig. 8.7. Urban trends in sulfur dioxide (SO_2) concentrations. Data source: *United Nations Environmental Programme, 2006, Geodata Portal*, http://geodata.grid.unep.ch/download/archive/2006_yearbook/airpollution_indicators_cities.xls; accessed on October 11, 2006.

source strength is expected, no such trend is expected in climate or weather, even though each year will have its own individuality with respect to its weather. However, there is at present an ongoing debate as to whether global mean temperatures will continue to increase due to increasing concentration of global greenhouse gases, especially carbon dioxide and methane. If so, this would be a climatic trend.

Other examples of trends come from Great Britain, where the emission of industrial smoke was reduced from 1.4 million tons per year in 1953 to 0.1 million tons per year in 1972; domestic smoke emission was reduced from 1.35 million tons per year in 1953 to 0.58 million tons per year in 1972; and the number of London fogs (smogs) capable of reducing visibility at 9 a.m. to less than 1 km was reduced from 59 per year in 1946 to 5 per year in 1976.

Annual trends in urban ozone are much more subtle because of the complex interaction among precursors (hydrocarbons and oxides of nitrogen) and meteorology (including solar radiation) (Fig. 8.8).

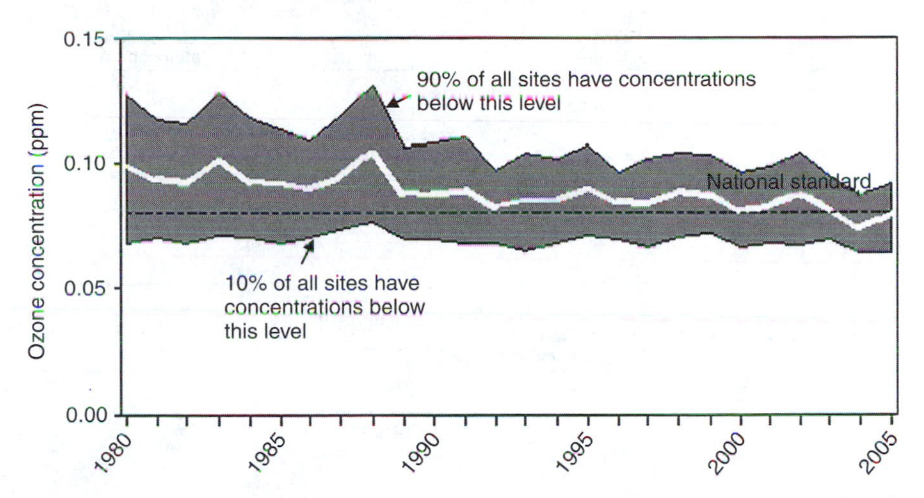

Fig. 8.8. In the United States, average ozone concentrations have been decreasing, but the rate of decrease for 8-h levels slowed during the 1990s. Concentrations are based on the annual fourth-highest daily maximum 8-h average ozone concentration. The gray area around the trend line (white) represents the distribution of air pollution levels among the 286 trend sites, displaying the middle 80%. *Source*: US Environmental Protection Agency, 2006, *National Trends in Ozone Levels*, http://www.epa.gov/airtrends/ozone.html#oznat; accessed on October 11, 2006.

III. PRIMARY AND SECONDARY POLLUTANTS

A substantial portion of the gas and vapors[2] emitted to the atmosphere in appreciable quantity from anthropogenic sources tends to be relatively simple in chemical structure: carbon dioxide, carbon monoxide, sulfur dioxide, and nitric oxide from combustion processes; hydrogen sulfide, ammonia, hydrogen chloride, and hydrogen fluoride from industrial processes. The solvents and gasoline fractions that evaporate are alkanes, alkenes, and aromatics with relatively simple structures.

In addition, more complex molecules such as polycyclic aromatic hydrocarbons (PAHs) and dioxins are released from industrial processes and combustion sources and are referred to as toxic pollutants. Substances such as these, emitted directly from sources, are called *primary pollutants*. They are certainly not innocuous, as will be seen when their adverse effects are discussed in later chapters. However, the primary pollutants do not, of themselves, produce all

[2] The terms gas and vapor are often used synonymously; however, a vapor is a gaseous substance that is a solid or liquid at standard temperature and pressure (STP), i.e. at 0°C and 101.325 kPa. Thus, CO_2 is a gas, but ethanol is predominantly a liquid under these conditions. Note that not all of ethanol is liquid, since it is vaporizing. However, with increasing temperature and constant pressure, a greater amount of ethanol will move to the gas phase (according to Henry's law). That is, at lower temperatures only a small number of ethanol molecules are sufficiently excited to move to the gas phase, but with increasing temperature, the number of molecules escaping increases. Thus, when we measure these ethanol molecules in the gas phase, we consider this to be a vapor measurement.

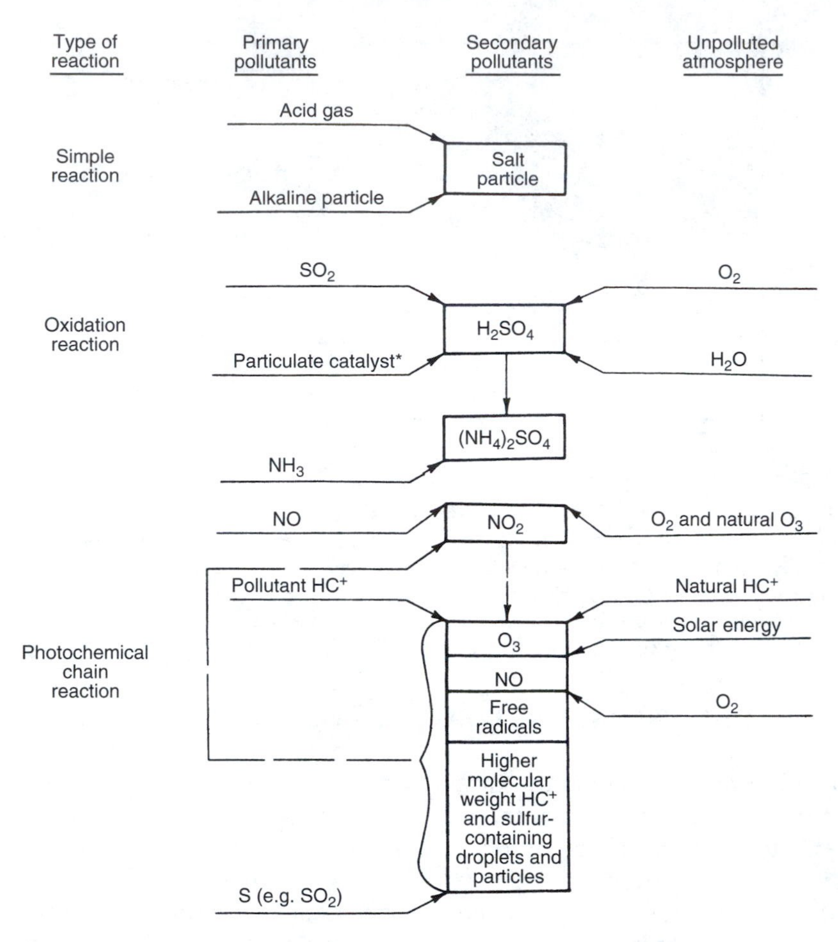

Fig. 8.9. Primary and secondary pollutants. *Reaction can occur without catalysis (HC^+, hydrocarbons).

of the adverse effects of air pollution. Chemical reactions may occur among the primary pollutants and the constituents of the unpolluted atmosphere (Fig. 8.9). The atmosphere may be viewed as a reaction vessel into which are poured reactable constituents and in which are produced a tremendous array of new chemical compounds, generated by gases and vapors reacting with each other and with the particles in the air. The pollutants manufactured in the air are called *secondary pollutants*; they are responsible for most of the smog, haze, and eye irritation and for many of the forms of plant and material damage attributed to air pollution. In air pollution parlance, the primary pollutants that react are termed the *precursors* of the secondary pollutants. With the knowledge that each secondary pollutant arises from specific chemical reactions involving specific primary reactants, we must control secondary pollutants by controlling how much of each primary pollutant is allowed to be emitted. Note that the use of primary and secondary pollutants used here is different from the regulatory definitions of primary and secondary pollutants. The Clean Air Act sets primary standards to protect public health and secondary standards to protect welfare.

IV. MEASUREMENT SYSTEMS

Many methods of air quality measurement have inherent averaging times. In selecting methods for measuring air quality or assessing air pollution effects, this fact must be borne in mind (Table 8.3). Thus, an appropriate way to assess the influence of air pollution on metals is to expose identical specimens at different locations and compare their annual rates of corrosion among the several locations. Since soiling is mainly due to the sedimentation of particulate matter from the air, experience has shown that this can be conveniently measured by exposing open-topped receptacles to the atmosphere for a month and weighing the settled solids. Human health seems to be related to day-to-day variation in pollutant level.

Because a filter sample includes particles both larger and smaller than those retained in the human respiratory system (see Chapter 11, Section III), various types of samplers are used which allow measurement of the size ranges of particles retained in the respiratory system. Some of these are called *dichotomous samplers* because they allow separate measurement of the respirable and nonrespirable fractions of the total. *Size-selective samplers* rely on impactors, miniature cyclones, and other means. The United States has selected the size fraction below an aerodynamic diameter of 2.5 μm ($PM_{2.5}$) for compliance with the air quality standard for airborne particulate matter. However, there continues to be concern coarse-sized particles (>2.5 to <10 μm).

Because an agricultural crop can be irreparably damaged by an excursion of the level of several gaseous pollutants lasting just a few hours, recording such an excursion requires a measuring procedure that will give hourly data. The least expensive device capable of doing this is the *sequential sampler*, which will allows a sequence of 1- or 2-h samples day after day for as long as the bubblers in the sampler are routinely serviced and analyzed.

TABLE 8.3

Air Quality Measurement

Measure of averaging time	Cyclic factor measured	Measurement method with same averaging time	Effect with same averaging time
Year	Annual trend	Metal specimen	Corrosion
Month	Seasonal cycle	Dustfall	Soiling
Day	Weekly cycle	Dichotomous	Human health
Hour	Diurnal cycle	Sequential sampler	Vegetation damage
Minute	Turbulence	Continuous instrument	Irritation (odor)

As discussed, the hourly data from sequential samplers can be combined to yield daily, monthly, and annual data.

None of the foregoing methods will tell the frequency or duration of exposure of any receptor to irritant or odorous gases when each such exposure may exceed the irritation or odor response threshold for only minutes or seconds. The only way that such an exposure can be measured instrumentally is by an essentially continuous monitoring instrument, the record from which will yield not only this kind of information but also all the information required to assess hourly, daily, monthly, and annual phenomena. Continuous monitoring techniques may be used at a particular location or involve remote sensing techniques, such as the so-called open path systems (e.g. open path infrared or differential optical absorption spectrometers, DOAS) (see Fig. 8.10 and Table 8.4). All of these techniques are considered to be "active" monitoring devices, since they involve mechanical movement of air through the filters and sensors. Passive systems are also gaining use in detecting air pollution, especially for detecting those gases where diffusion

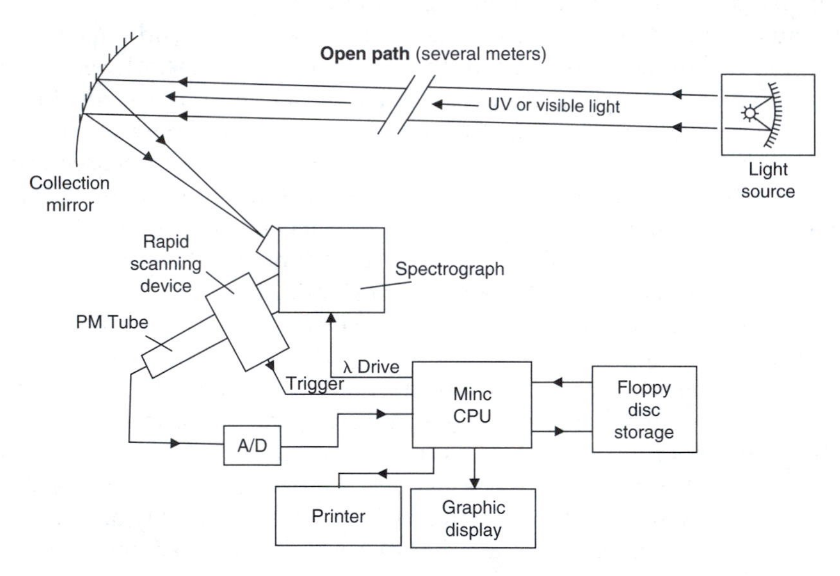

Fig. 8.10. Schematic of an open path air measurement device: differential optical adsorption spectrometer (DOAS), which measures the absorption through an atmospheric path (typically 0.5–1.5 km) of two closely spaced wavelengths of light from an artificial source. One wavelength is chosen to match an absorption line of the specific compound of interest, and the other is near that line to account for atmospheric effects (EPA/600/P-93/004aF, pp. 3–98). The term absorption line is used to mean a wavelength that a given atom or molecule absorbs more than it does other wavelengths. For example, if ozone is the compound of interest, a wavelength would be identified which ozone absorbs more than other wavelengths. As in other open path systems (e.g. FTIR) a beam of electromagnetic radiation (in this case, visible or ultraviolet light) containing this particular wavelength is directed at a segment (open path) of the atmosphere, and the amount of the wavelength absorbed would be measured and calibrated against a known concentration of the compound of interest. *Source*: US Environmental Protection Agency, *Air Quality Criteria for Ozone and Photochemical Oxidants* (EPA/600/P-93/004aF), July 1996.

TABLE 8.4

Minimum Detection Level Defined as the Lowest Concentration of a Species that DOAS Devices Can Distinguish from Background Noise

Compound	Minimum detection limit in parts per billion volume (ppbv)	Reference
p-Xylene	0.3	Axelsson, H. *et al.*, Measurement of aromatic hydrocarbons with the DOAS technique. *Appl. Spect.* **49** (9), 1254–1260 (1995).
Ethylbenzene	2	Axelsson, H. *et al.*, 1258 (1995)
1,2,3-Trimethylbenzene	6	Axelsson, H. *et al.*, 1258 (1995)
NO_2	4	Biermann, H. W., *et al.*, Simultaneous absolute measurements of gaseous nitrogen species in urban ambient air by long pathlength infrared and ultraviolet-visible spectroscopy, *Atmos. Environ.* **22** (8), 1545–1554 (1988).
NO_3	0.02	Biermann, H. W., *et al.*, 1551 (1988)
Ozone (O_3)	3	Stevens, R. K., *et al.*, A long path DOAS and EPA-approved fixed-point methods intercomparison. *Atmos. Environ.* **27B** (2), 231–236 (1993).
SO_2	10	Stevens, R. K. *et al.*, 234 (1993)

Source: US Environmental Protection Agency.

(following Fick's laws) is well understood. In passive systems, passive monitors are exposed to the air over a defined time period and the amount of the gas that has diffused is measured and integrated for that time period to give the concentration, often reported as a mean for that reporting period (e.g. a daily or weekly mean).

V. AIR QUALITY LEVELS

A. Levels

Air quality levels vary between concentrations so low that they are less than the minimum detectable values of the instruments we use to measure them and maximum levels that are the highest concentrations ever measured. Figure 8.11 gives national data (1980–2005) for CO in the United States. The mean chemical composition and atmospheric concentration of suspended particulate matter (total, coarse, and fine) measured in the United States in 1980 are shown in Table 8.5. The percentages do not add up to 100% because they exclude the oxygen (except for the nitrate and sulfate components),

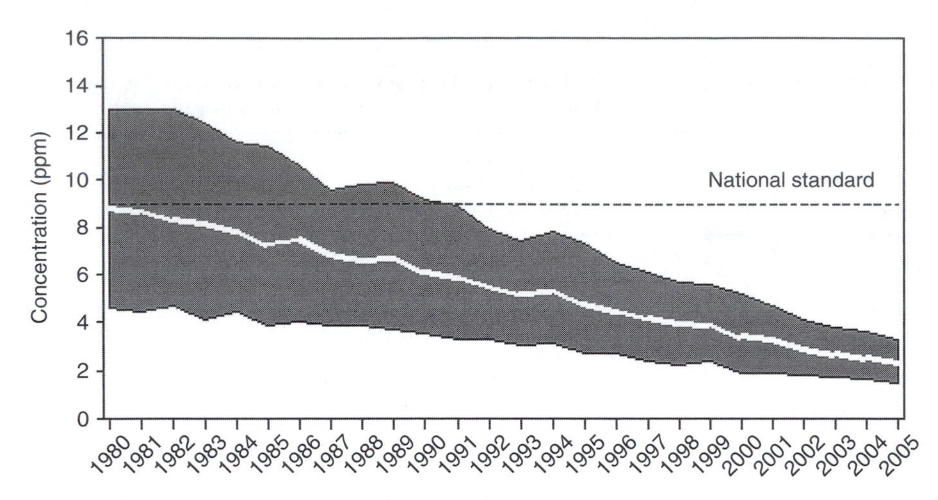

Fig. 8.11. Trend in ambient concentrations of carbon monoxide, based on the annual second maximum 8-h average concentrations from 152 sites across the United States. The gray band shows the distribution of air pollution levels among the trend sites, displaying the middle 80%. The white line represents the average among all the trend sites. Ninety percent of sites have concentrations below the top line, while 10% of sites have concentrations below the bottom line. The 16-year trend shows a 74% improvement overall. *Source*: US Environmental Protection Agency, 2007, *Air Trends*, http://www.epa.gov/airtrends/carbon.html; accessed on February 18, 2007.

nitrogen (except for the nitrate component), hydrogen, and other components of the compounds of the listed elements in the form in which they actually exist in the atmosphere; for example, the most common form of particulate sulfur and sulfate in the atmosphere is $(NH_4)_2(SO_4)$. The table indicates that about 30% of the mass of particulate matter is in the fine fraction ($>2.5\,\mu m$), 30% is in the coarse fraction $>15\,\mu m$), and 40% is coarser still—between 15 and *ca.* $50\,\mu m$.

It is a general rule that the larger the population base, the dirtier the air will be. Urban air in the United States is cleaner than it was in the 1950s and 1960s because of the tremendous efforts made to clean up the air in the intervening decades (Table 8.6).

B. Display of Air Quality Data

Information about air quality usually starts with observed measurements, which are either presented "as is," but more often as a *sample* representing a larger population. The data are also frequently enhanced using various air quality models. Inferential statistics are applied to the observed sample so that the condition of the population (e.g. an airshed) can be inferred from the limited number of observations (e.g. from State and Local Ambient Air Monitoring Sites (SLAMS); see Fig. 8.12).

TABLE 8.5

Mean Chemical Composition and Atmospheric Concentrations of Suspended Particulate Matter Sampled by the United States Environmental Protection Agency's Inhalable Particle and National Air Surveillance Networks—$\mu g\, m^{-3}$ and Percentage of Total Mass Sampled, 1980

Type of sample	Urban				Rural			
Number of samples			745		2255[a]			133[b]
Particle size	Coarse, 15–2.5 μm		Fine, less than 2.5 μm		All less than *ca.* 50 μm			
	Mean value of 745 values	%	Mean value of 745 values	%	Mean value of 2255 values	%	Mean value of 133 values	%
All (total mass)	21.655	100.00	22.680	100.00	74.990	100.00	36.504[a]	100.00
Aluminum	1.797	8.30	0.353	1.56	—	—	—	—
Antimony	0.051	0.24	0.050	0.22	—	—	—	—
Arsenic	0.003	0.01	0.004	0.02	0.005[c]	0.01	0.003[d]	0.01
Barium	0.060	0.28	0.060	0.26	0.273	0.36	0.281	0.77
Beryllium[e]	—	—	—	—	(0.095)	—	(0.084)	—
Bromine	0.019	0.09	0.077	0.34	—	—	—	—
Cadmium	0.006	0.03	0.007	0.03	0.002	0.01	0.001	0.01
Calcium	1.503	6.94	0.340	1.50	—	—	—	—
Chlorine	0.440	2.03	0.155	0.68	—	—	—	—
Chromium	0.008	0.04	0.006	0.03	0.013[c]	0.02	0.015[d]	0.05
Cobalt	—	—	—	—	0.001[c]	0.01	0.001[d]	0.01
Copper	0.019	0.09	0.026	0.12	0.143	0.19	0.136	0.37
Iron	0.743	3.43	0.205	0.90	0.923	1.23	0.254	0.70
Lead	0.083	0.38	0.314	1.38	0.353	0.47	0.066	0.18
Manganese	0.021	0.10	0.013	0.06	0.031	0.04	0.008	0.02
Mercury	0.003	0.01	0.003	0.01	—	—	—	—
Molybdenum	—	—	—	—	0.002	0.01	0.001	0.01
Nickel	0.004	0.02	0.007	0.03	0.007	0.01	0.002	0.01
Phosphorus	0.056	0.26	0.021	0.09	—	—	—	—
Potassium	0.222	1.03	0.156	0.69	—	—	—	—
Selenium	0.001	0.01	0.002	0.01	—	—	—	—
Silicon	2.561	11.83	0.360	1.59	—	—	—	—
Strontium	0.246	0.21	0.051	0.22	—	—	—	—
Sulfur	0.339	1.56	2.056	9.07	—	—	—	—
Tin	0.006	0.03	0.006	0.03	—	—	—	—
Titanium	0.042	0.19	0.015	0.07	—	—	—	—
Vanadium	0.008	0.04	0.010	0.04	0.015	0.02	0.004	0.01
Zinc	0.038	0.18	0.067	0.30	0.147	0.20	0.114	0.31
Nitrate	0.699	3.23	1.071	4.72	4.647	6.20	2.341	6.41
Sulfate	0.706	3.26	5.30	23.37	10.811	14.42	8.675	23.77
Sum of percentages	—	43.82[f]	—	47.34[f]	—	23.20	—	32.64

[a] Except for arsenic, chromium, and cobalt where the number of samples was 1245.
[b] Except for arsenic, chromium, and cobalt where the number of samples was 30.
[c] Except for arsenic, chromium, and cobalt where the mean total mass was 76.647 $\mu g\, m^{-3}$.
[d] Except for arsenic, chromium, and cobalt where the mean total mass was 30.367 $\mu g\, m^{-3}$.
[e] $ng\, m^{-3}$.
[f] Sulfur is counted twice: as sulfur and as sulfate. Some of this sulfur exists as sulfides, sulfites, and forms other than sulfate.

TABLE 8.6

Distribution of Cities by Population Class and Particulate Matter Concentration, 1957–1967

	Average particulate matter concentration ($\mu m\, m^{-3}$)										
Population class	Less than 40	40–59	60–79	80–99	100–119	120–139	140–159	160–179	180–199	More than 200	Total
Over 3 million	—	—	—	—	—	—	1	—	1	—	2
1–3 million	—	—	—	—	—	—	2	1	—	—	3
0.7–1 million	—	—	1	—	2	—	4	—	—	—	7
400–700 000	—	—	—	4	5	6	1	1	1	—	18
100–400 000	—	3	7	30	24	17	12	3	2	1	99
50–100 000	—	2	20	28	16	12	6	5	1	3	93
25–50 000	—	5	24	12	12	10	2	1	2	3	71
10–25 000	—	7	18	19	9	5	2	3	1	—	64
Under 10 000	1	5	7	15	11	2	1	2	—	—	44
Total	1	22	77	108	79	52	31	16	8	7	401

Source: US Environmental Protection Agency, *Air Quality Data from 1967* (Rev. 1971). Office of Air Programs, Publication No. APTD 0471, Research Triangle Park, NC, 1971.

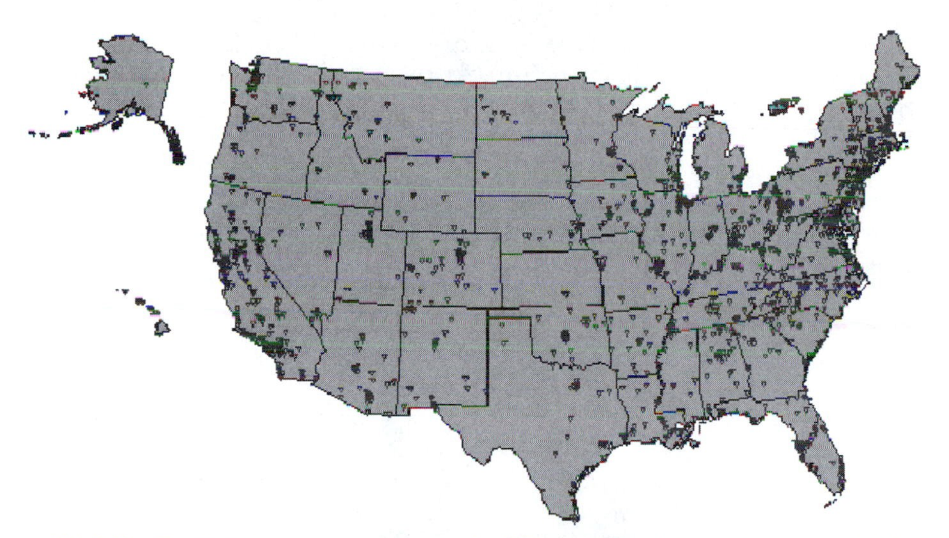

Fig. 8.12. The SLAMS consist of a network of ~4000 monitoring stations whose size and distribution is largely determined by the needs of State and local air pollution control agencies to meet their respective State Implementation Plan (SIP) requirements. *Source*: US Environmental Protection Agency 2006, *The Ambient Air Monitoring Program*, http://epa.gov/oar/oaqps/qa/monprog.html#SLAMS; accessed on October 30, 2006.

Often, a relatively small number of samples represent a large population. Air quality data are highly variable. For example, they vary diurnally (within the same 24-h interval), seasonally, and year-over-year.

A sample of n individual values is:

$$\{x\} = x_1, x_2, x_3, \ldots, x_n \tag{8.1}$$

where { } indicates a vector of individual values, and the n individual observations are a sample taken from the set of possible outcomes of the population. Another important concept used to characterize air quality data is *probability*, which is the likelihood that some outcome will occur. Thus, the likelihood ranges from not all (0%) to 100%. Accordingly, the range of probability is between 0 and 1.

Once data are collected, they are displayed graphically. Drawing a frequency histogram from the observed data is often a first step. This gives an estimate of the data's *probability distribution*. The data are grouped into discrete intervals, with the number of samples falling in each interval shown by a bar (intervals on the x-axis and frequency of occurrence on the y-axis). For example, a stack test measuring the concentration of mercury (Hg) from a fossil fuel power plant is shown in Fig. 8.13.

The intervals must be mutually exclusive of the next interval. In other words, a value must fall in one and only one interval. Thus, in the stack test, the intervals are $0 < 1$; $1 < 2$; $2 < 3$; …; $14 < 15\,\mu g\,m^{-3}$ Hg in the stack gas. This means that the value of $2.5\,\mu g\,m^{-3}$ Hg falls in the $2 < 3$ interval (shown as "3" in the figure) and that the value of $11.99\,\mu g\,m^{-3}$ Hg falls in the $11 < 12$ interval (shown as "12" in the figure). So, the relative number of occurrences of an event is shown as $P(E_1)$, meaning the probability of event E_1, which is equal to n_1/n if the event happened to occur n_1 times in n samples. For example, samples of mercury $8 < 9\,\mu g\,m^{-3}$ occurred 28 times out of the total 130 samples taken in the stack test. This means that 21.5% of the samples fell within the $8 < 9\,\mu g\,m^{-3}$ interval. Finding this relative frequency of each interval yields a relative frequency of occurrence histogram (see Fig. 8.14).

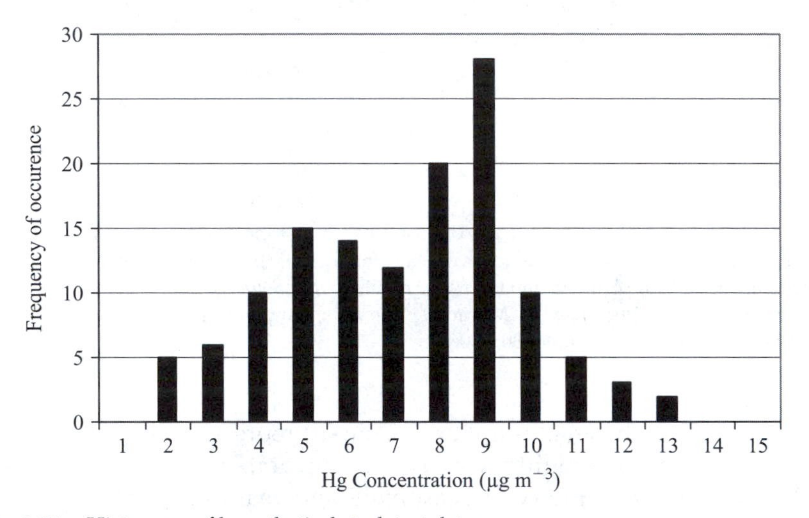

Fig. 8.13. Histogram of hypothetical stack test data.

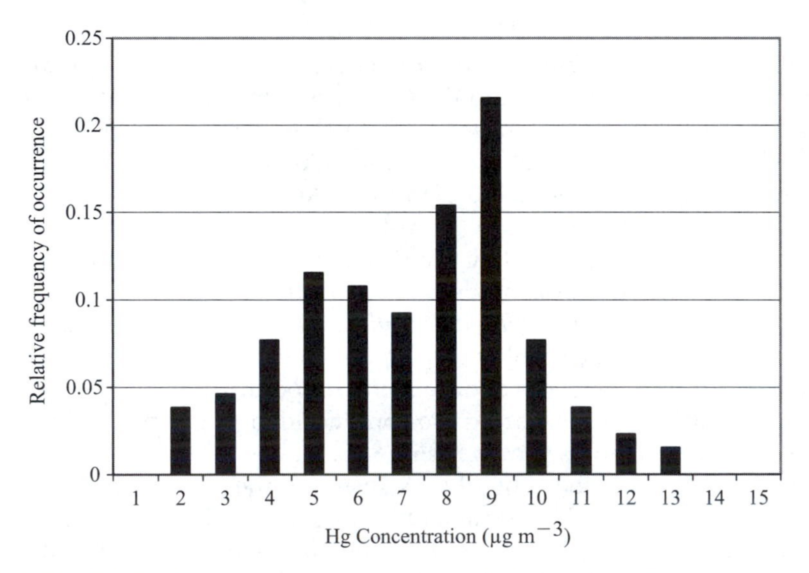

Fig. 8.14. Relative frequency of occurrence of hypothetical stack test data.

A very useful format in which to display air quality data for analysis is that of Fig. 8.15, which has as its abscissa averaging time expressed in two different time units and, as its ordinate, concentration of the pollutant at the receptor. This type of chart is called an *arrowhead chart* and includes enough information to characterize fully the variability of concentration at the receptor.

To understand the meaning of the information given, let us concentrate on the data for 1-h averaging time. In the course of a year there will be 8760 such values, one for each hour. If all 8760 are arrayed in decreasing value, there will be one maximum value and one minimum value. (For some pollutants the minimum value is indefinite if it is below the minimum detectable value of the analytical method or instrument employed.) In this array, the value 2628 from the maximum will be the value for which 30% of all values are greater and 70% are lower. Similarly, the value 876 from the maximum will be the one for which 10% of all values are greater and 90% are lower. The 1% value will be between the 87th and 88th values from the maximum, and the 0.1% value will lie between the 8th and 9th values in the array.

The 50% value, which is called the *median value,* is not the same as the average value, which is also called the *arithmetic mean value*. Both are types of measures of central tendency. The median value is the middle value of the data set; half values exceed the median, while the other half are less than the median. The arithmetic mean value is obtained by adding all 8760 values and then dividing the total by 8760. The arithmetic average value obtained for other averaging times, e.g., by adding all 365 24-h values and dividing the

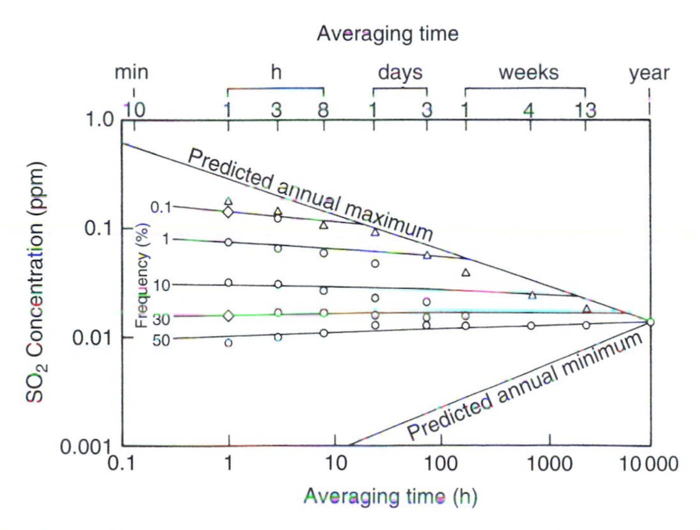

Fig. 8.15. Sulfur dioxide concentration versus averaging time and frequency for 1980 at US National Aerometric Data Bank (NADB) Site 264280007 HO1, 8227 S. Broadway, St. Louis, Missouri. *Source*: Chart courtesy of Dr. Larsen, R. I., US Environmental Protection Agency, Research Triangle Park, NC; see also Fig. 21.13.

total by 365, will be the same and will equal the annual arithmetic average value. The median value will equal the arithmetic average value only if the distribution of all values allows this to occur. For example:

1 2 3 4 5 6 7 8 9 10 Median = 5.5 Mean = 5.5

1 3 3 3 5 6 7 7 8 12 Median = 5.5 Mean = 5.5

In most instances, however, the mean is different from the median. Consider the following example. A monitor collects the following naphthalene daily average values ($\mu g\,m^{-3}$):

100 m to the north: 0.1, 0.2, 0.6, 0.7, 0.5, 0.1, 0.7

100 m to the south: 5.2, 3.3, 11.1, 12.7, 5.8, 7.7, 11.8

To find the median, the values are shown in ascending order (Table 8.7).

Since the number of values is odd (i.e. seven), we simply select the middle value (the fourth). Had the number of values been even, the two middle values would be averaged to give the mean (e.g. if the two middle values are 1.1 and 1.3, the median would be 1.2). Thus, the median of the two data sets are 0.5 and 7.7 $\mu g\,m^{-3}$, respectively. However, the mean ($\bar{x}$) is found as the quotient of the sum of the values divided by the number of values:

$$\bar{x} = \frac{\sum_{i=1}^{n} x_i}{n} \tag{8.2}$$

The north data set's mean, then, is

$$\bar{x} = \frac{2.9}{7} = 0.47$$

TABLE 8.7

Daily Naphthalene Measurements ($\mu g\,m^{-3}$)

North	South
0.1	3.3
0.1	5.2
0.2	5.8
0.5	*7.7*
0.6	11.1
0.7	11.8
0.7	12.7

The south data set's mean is:

$$\bar{x} = \frac{57.6}{7} = 8.2$$

So, this is the common instance of a skewed data set. In the north data, the median is larger than the mean, but in the south data set, the mean is larger than the median.

If air quality data at a receptor for any one averaging time are lognormally distributed, these data will plot as a straight line on log probability graph paper (Fig. 8.16) which bears a note $S_g = 2.35$. S_g is the standard geometric deviation about the geometric mean (the geometric mean is the Nth root of the product of the n values of the individual measurements):

$$S_g = \exp\sum_{i=1}^{n}\left[\frac{(\ln X_i - \overline{\ln X_i})^2}{n-1}\right]^{1/2} \tag{8.3}$$

where $X_i = X_1, X_2, \ldots, X_n$ are the individual measurements and n is the number of measurements.

Further discussion of the significance of mean values and standard deviations can be found in Chapters 22 and 36 and in any textbook on statistics.

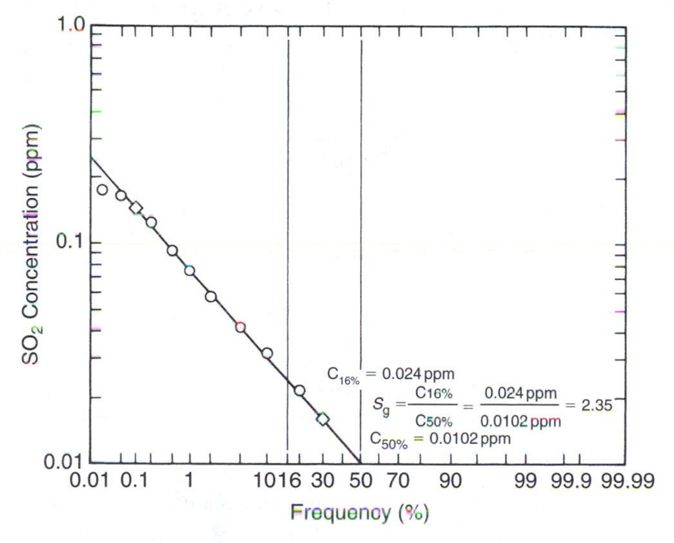

Fig. 8.16. Frequency of 1-h average sulfur dioxide concentrations equal to or greater than stated values during 1980 at US National Aerometric Data Bank (NADB) Site 264280007 HO1, 8227 S. Broadway, St. Louis. Missouri. *Source*: Chart courtesy of Larsen, R. I., US Environmental Protection Agency, Research Triangle Park, NC.

C. Adverse Responses to Air Quality Levels

The objective of air pollution control is to prevent adverse responses by all receptor categories exposed to the atmosphere: human, animal, vegetable, and material. These adverse responses have characteristic response times: short-term (i.e. seconds or minutes), intermediate-term (i.e. hours or days), and long-term (i.e. months or years) (Table 8.8). For there to be no adverse responses, the pollutant concentration in the air must be lower than the concentration level at which these responses occur. Figure 8.16 illustrates the relationship between these concentration levels. This figure displays response curves, which remain on the concentration duration axes because they are characteristic of the receptors, not of the actual air quality to which the receptors are exposed. The odor response curve, e.g. to hydrogen sulfide, shows that a single inhalation requiring approximately 1 s can establish the presence of the odor but that, due to odor fatigue, the ability to continue to recognize that odor can be lost in a matter of minutes. Nasopharyngeal and eye irritation, e.g. by ozone, is similarly subject to acclimatization due to tear and mucus production. The three visibility lines correlate with the concentration of suspended particulate matter in the air. Attack of metal, painted surfaces, or nylon hose is shown by a line starting at 1 s and terminating in a matter of minutes (when the acidity of the droplet is neutralized by the material attacked).

Vegetation damage can be measured biologically or socioeconomically. Using the latter measure, there is a 0% loss when there is no loss of the sale value of the crops or ornamental plants but a 100% loss if the crop is damaged to the extent that it cannot be sold. These responses are related to dose, i.e. concentration times duration of exposure, as shown by the percent loss

TABLE 8.8

Examples of Receptor Category Characteristic Response Times

	Characteristic response times		
Receptor category	Short-term (seconds–minutes)	Intermediate-term (hours–days)	Long-term (months–years)
Human	Odor, visibility, nasopharyngeal and eye irritation	Acute respiratory disease	Chronic respiratory disease and lung cancer
Animal, vegetation	Field crop loss and ornamental plant damage	Field crop loss and ornamental plant damage	Fluorosis of livestock, decreased fruit and forest yield
Material	Acid droplet pitting and nylon hose destruction	Rubber cracking, silver tarnishing, and paint blackening	Corrosion, soiling, and materials deterioration

curves on the chart. A number of manifestations of material damage, e.g. rubber cracking by ozone, require an exposure duration long enough for the adverse effects to be significant economically. That is, attack for just a few seconds or minutes will not affect the utility of the material for its intended use, but attack for a number of days will.

The biological response line for acute respiratory disease is a dose–response curve which for a constant concentration becomes a duration–response curve. The shape of such a curve reflects the ability of the human body to cope with short-term, ambient concentration respiratory exposures and the overwhelming of the body's defenses by continued exposure.

Fluorosis of livestock is not induced until there has been a long enough period of deposition of a high enough ambient concentration of fluoride to increase the level of fluoride in the forage. Since the forage is either eaten by livestock or cut for hay at least once during the growing season, the duration of deposition ends after the growing season. The greater the duration of the season, the greater the time for deposition, hence the shape of the line labeled "fluorosis." Long-term vegetation responses—decreased yield of fruit and forest—and long-term material responses—corrosion, soiling, and material deterioration—are shown on the chart as having essentially the same response characteristics as human chronic respiratory disease and lung cancer.

The relationship of these response curves to ambient air quality is shown by lines A, B, and C, which represent the maximum or any other chosen percentile line from a display such as Fig. 8.17, which shows actual air quality.

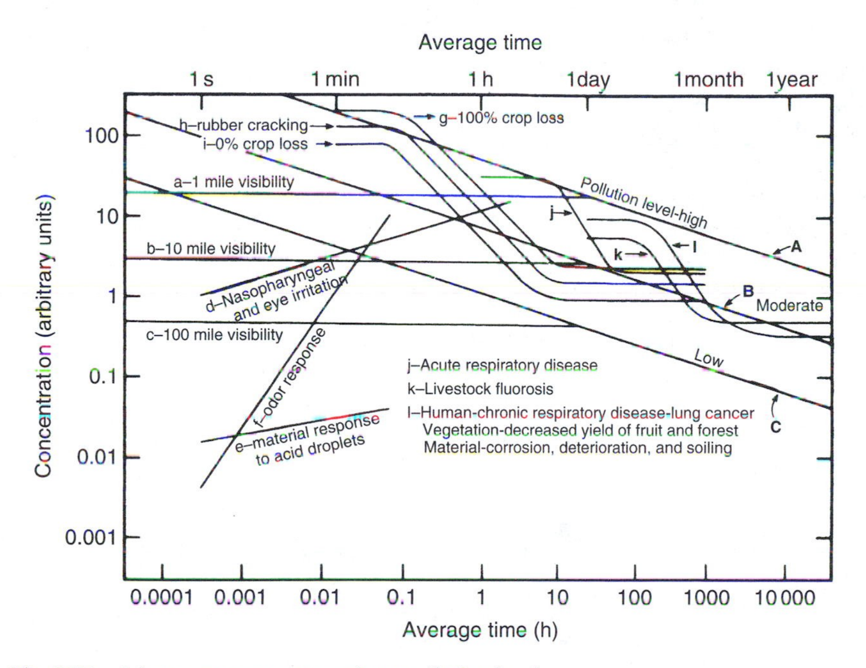

Fig. 8.17. Adverse responses to various pollution levels.

Where the air quality is poor (line A), essentially all the adverse effects displayed will occur. Where the air quality is good (line C), most of the intermediate and long-term adverse effects displayed will not occur. Where the air quality is between good and poor, some of the intermediate and long-term adverse effects will occur, but in an attenuated form compared with those of poor air quality. These concepts will be referred to later in this text when air quality standards are discussed.

D. Air Quality Indices

Air quality indexes (AQIs) have been devised for categorizing the air quality measurements of several individual pollutants by one composite number. The index used by the US Environmental Protection Agency (EPA) as authorized by Section 319 of the Clean Air Act is called the *Pollutant Standards Index* (PSI). In July 1999, the EPH replaced the PSI with the AQI (Table 8.9). Comparable values for international standards are shown in Table 8.10.

The changes from the PSI to the AQI include a new category "unhealthy for sensitive groups;" new breakpoints for the ozone (O_3) sub-index in terms of 8-h average O_3 concentrations, a new sub-index for fine particulate matter ($PM_{2.5}$), and conforming changes to the sub-indices for coarse particulate matter (PM_{10}), carbon monoxide (CO), and sulfur dioxide (SO_2).

The AQI is designed to convey information to the public regarding daily air quality and its associated health risks. The scientific understanding of pollutants, especially particulate matter and O_3, has increased steadily. This is reflected in the National Ambient Air Quality Standards (NAAQS) promulgated in 1997. In particular, much more is now known about nature of the relationships between exposure to ambient concentrations of these pollutants and the health effects likely to be experienced, especially near the level of the NAAQS. These criteria pollutants have no threshold below which health effects are not likely to occur. Rather, there appears to be a continuum of effects potentially extending down to ambient background levels.

Sensitivity varies by pollutant. Thus, when the AQI exceeds 100, the sensitive groups of concern to be at greatest risk include:

- *Ozone*: Children and people with asthma.
- *$PM_{2.5}$*: People with respiratory or heart disease, the elderly and children.
- *PM_{10}*: People with respiratory disease.
- *Carbon monoxide*: People with heart disease.
- *Sulfur dioxide*: People with asthma.
- *Nitrogen dioxide*: Children and people with respiratory disease.

The precautionary statements for the NAAQS pollutants are shown in Table 8.11.

TABLE 8.9

AQI for Criteria Air Pollutants

	Air Quality Index (AQI)						
Category	Good	Moderate	Unhealthy for sensitive groups[a]	Unhealthy	Very unhealthy	Hazardous	
Index value	0–50	51–100	101–150	151 – 200	201–300	301–400	401–500
Pollutant			Concentration ranges				
Carbon monoxide (ppm)	0–4.4	4.5–9.4	9.5–12.4	12.5–15.4	15.5–30.4	30.5–40.4	40.5–50.4
Nitrogen dioxide (ppm)	—	—	—	—	0.65–1.24	1.25–1.64	1.65–2.04
Ozone (average concentration for 1 h in ppm)	—	—	0.125–0.164	0.165 – 0.204	0.205–0.404	0.405–0.504	0.505–0.604
Ozone (average concentration for 8 h in ppm)	0–0.064	0.065–0.084	0.085–0.104	0.105–0.124	0.125–0.374	—	—
$PM_{2.5}$ ($\mu g\,m^{-3}$)	0–15.4	15.5–40.4	40.5–65.4	65.5–150.4	150.5–250.4	250.5–350.4	350.5–500.4
PM_{10} ($\mu g\,m^{-3}$)	0–54	55–154	155–254	255–354	355–424	425–504	505–604
Sulfur dioxide (ppm)	0–0.034	0.035–0.144	0.145–0.224	0.225–0.304	0.305–0.604	0.605–0.804	0.805–1.004

[a] Each category corresponds to a different level of health concern. The six levels of health concern and what they mean are:

- "Good": The AQI value for your community is between 0 and 50. Air quality is considered satisfactory, and air pollution poses little or no risk.
- "Moderate": The AQI for your community is between 51 and 100. Air quality is acceptable; however, for some pollutants there may be a moderate health concern for a very small number of people. For example, people who are unusually sensitive to ozone may experience respiratory symptoms.
- "Unhealthy for Sensitive Groups": When AQI values are between 101 and 150, members of sensitive groups may experience health effects. This means they are likely to be affected at lower levels than the general public. For example, people with lung disease are at greater risk from exposure to ozone, while people with either lung disease or heart disease are at greater risk from exposure to particle pollution. The general public is not likely to be affected when the AQI is in this range.
- "Unhealthy": Everyone may begin to experience health effects when AQI values are between 151 and 200. Members of sensitive groups may experience more serious health effects.
- "Very unhealthy: AQI values between 201 and 300 trigger a health alert, meaning everyone may experience more serious health effects.
- "Hazardous": AQI values over 300 trigger health warnings of emergency conditions. The entire population is more likely to be affected.

Source: US Environmental Protection Agency.

TABLE 8.10

WHO Air Quality Guidelines (AQGs) for Particulate Matter, Ozone, Nitrogen Dioxide and Sulfur Dioxide

	PM_{10} ($\mu g\,m^{-3}$)	$PM_{2.5}$ ($\mu g\,m^{-3}$)	Basis for the selected level
Iinterim target-1 (IT-1)	70	35	These levels are associated with about a 15% higher long-term mortality risk relative to the AQG level
Interim target-2 (IT-2)	50	25	In addition to other health benefits, these levels lower the risk of premature mortality by approximately 6% (2–11%) relative to the IT-1 level
Interim target-3 (IT-3)	30	15	In addition to other health benefits, these levels reduce the mortality risk by approximately 6% (2–11%) relative to the IT-2 level
AQG	20	10	These are the lowest levels at which total, cardiopulmonary and lung cancer mortality have been shown to increase with more than 95% confidence in response to long-term exposure to $PM_{2.5}$

WHO AQGs and interim targets for particulate matter[a]: 24-h concentrations[b]

	PM_{10} ($\mu g\,m^{-3}$)	$PM_{2.5}$ ($\mu g\,m^{-3}$)	Basis for the selected level
Interim target-1 (IT-1)	150	75	Based on published risk coefficients from multi-centre studies and meta-analyzes (about 5% increase of short-term mortality over the AQG value)
Interim target-2 (IT-2)	100	50	Based on published risk coefficients from multi-centre studies and meta-analyzes (about 2.5% increase of short-term mortality over the AQG value)
Interim target-3 (IT-3)[c]	75	37.5	Based on published risk coefficients from multi-centre studies and meta-analyzes (about 1.2% increase in short-term mortality over the AQG value)
AQG	50	25	Based on relationship between 24-h and annual PM levels

WHO AQG and interim target for ozone: 8-h concentrations

	Daily maximum 8-h mean ($\mu g\,m^{-3}$)	Basis for the selected level
High levels	240	Significant health effects; substantial proportion of vulnerable populations affected
Interim target-1 (IT-1)	160	Important health effects; does not provide adequate protection of public health. Exposure to this level of ozone is associated with: • Physiological and inflammatory lung effects in healthy exercising young adults exposed for periods of 6.6 h

(*continued*)

TABLE 8.10 (*Continued*)

	Daily maximum 8-h mean ($\mu g\,m^{-3}$)	Basis for the selected level
		• Health effects in children (based on various summer camp studies in which children were exposed to ambient ozone levels) • An estimated 3–5% increase in daily mortality[d] (based on findings of daily time-series studies)
AQG	100	Provide adequate protection of public health, though some health effects may occur below this level. Exposure to this level of ozone is associated with: • An estimated 1–2% increase in daily mortality[d] (based on findings of daily time-series studies) • Extrapolation from chamber and field studies based on the likelihood that real-life exposure tends to be repetitive and chamber studies exclude highly sensitive or clinically compromised subjects, or children • Likelihood that ambient ozone is a marker for related oxidants
WHO AQGs for nitrogen dioxide: annual mean		
	Annual mean ($\mu g\,m^{-3}$)	
AQG	40	Recent indoor studies have provided evidence of effects on respiratory symptoms among infants at NO_2 concentrations below 40 $\mu g\,m^{-3}$. These associations cannot be completely explained by co-exposure to PM, but it has been suggested that other components in the mixture (such as organic carbon and nitrous acid vapor) might explain part of the observed association
WHO AQGs for nitrogen dioxide: 1-h mean		
	1-h mean ($\mu g\,m^{-3}$)	
AQG	200	Epidemiological studies have shown that bronchitic symptoms of asthmatic children increase in association with annual NO_2 concentration, and that reduced lung function growth in children is linked to elevated NO_2 concentrations within communities already at current North American and European urban ambient air levels. A number of recently published studies have demonstrated that NO_2 can have a higher spatial variation than other traffic-related air pollutants, for example, particle mass. These studies also found adverse effects on the health of children living in metropolitan areas characterized by higher levels of NO_2 even in cases where the overall city-wide NO_2 level was fairly low. Since the existing WHO AQG short-term NO_2 guideline value of 200 m^{-3}(1-h) has not been challenged by more recent studies, it is retained

(continued)

TABLE 8.10 (*Continued*)

WHO AQGs and interim targets for SO_2: 24-h and 10-min concentrations			
	24-h average ($\mu g\, m^{-3}$)	10-min average ($\mu g\, m^{-3}$)	Basis for the selected level
Interim target-1 (IT-1)[e]	125	—	
Interim target-2 (IT-2)	50	—	Intermediate goal based on controlling either motor vehicle emissions, industrial emissions and/or emissions from power production. This would be a reasonable and feasible goal for some developing countries (it could be achieved within a few years) which would lead to significant health improvements that, in turn, would justify further improvements (such as aiming for the AQG value)
AQG	20	500	

[a]The use of $PM_{2.5}$ guideline value is preferred.
[b]99th percentile (3 days per year).
[c]For management purposes. Based on annual average guideline values precise number to be determined on basis of local frequency distribution of daily means. The frequency distribution of daily $PM_{2.5}$ or PM_{10} values usually approximates to a log-normal distribution.
[d]Deaths attributable to ozone. Time-series studies indicate an increase in daily mortality in the range of 0.3–0.5% for every $10\, \mu g\, m^{-3}$ increment in 8-h ozone concentrations above an estimated baseline level of $70\, \mu g\, m^{-3}$.
[e]Formerly the WHO Air Quality Guideline (WHO, 2000).
Source: World Health Organization. 2006. Report No. WHO/SDE/PHE/OEH/06.02.

TABLE 8.11

Cautionary Statements for Criteria Air Pollutants in the United States

Index values	Levels of health concern	Cautionary statements[a]			
		Ozone	Particulate matter	Carbon monoxide	Sulfur dioxide
0–50	Good	None	None	None	None
51–100[a]	Moderate	Unusually sensitive people should consider reducing prolonged or heavy exertion outdoors	Unusually sensitive people should consider reducing prolonged or heavy exertion	None	People with asthma should consider reducing exertion outdoors
101–150	Unhealthy for sensitive groups	Active children and adults, and people with lung disease, such as asthma, should reduce prolonged or heavy exertion outdoors	People with heart or lung disease, older adults, and children should reduce prolonged or heavy exertion	People with heart disease, such as angina, should reduce heavy exertion and avoid sources of CO, such as heavy traffic	Children, asthmatics, and people with heart or lung disease should reduce exertion outdoors
151–200	Unhealthy	Active children and adults, and people with lung disease, such as asthma, should avoid prolonged or heavy exertion outdoors. Everyone else, especially children, should reduce prolonged or heavy exertion outdoors	People with heart or lung disease, older adults, and children should avoid prolonged or heavy exertion. Everyone else should reduce prolonged or heavy exertion	People with heart disease, such as angina, should reduce moderate exertion and avoid sources of CO, such as heavy traffic	Children, asthmatics, and people with heart or lung disease should avoid outdoor exertion. Everyone else should reduce exertion outdoors

(continued)

TABLE 8.11 (*Continued*)

Index values	Levels of health concern	Cautionary statements[a]			
		Ozone	Particulate matter	Carbon monoxide	Sulfur dioxide
201–300	Very unhealthy	Active children and adults, and people with lung disease, such as asthma, should avoid all outdoor exertion. Everyone else, especially children, should avoid prolonged or heavy exertion outdoors	People with heart or lung disease, older adults, and children should avoid all physical activity outdoors. Everyone else should avoid prolonged or heavy exertion	People with heart disease, such as angina, should avoid exertion and sources of CO, such as heavy traffic	Children, asthmatics, and people with heart or lung disease should remain indoors. Everyone else should avoid exertion outdoors
301–500	Hazardous	Everyone should avoid all physical activity outdoors	People with heart or lung disease, older adults, and children should remain indoors and keep activity levels low. Everyone else should avoid all physical activity outdoors	People with heart disease, such as angina, should avoid exertion and sources of CO, such as heavy traffic. Everyone else should reduce heavy exertion	

[a]Nitrogen dioxide can cause respiratory problems in children and adults who have respiratory diseases, such as asthma. The AQI for nitrogen dioxide is not included here because ambient nitrogen dioxide concentrations in the US have been below the national air quality standard for the past several years. These concentrations are sufficiently low so as to pose little direct threat to human health. Nitrogen dioxide, however, is a concern because it plays a significant role in the formation of tropospheric ozone, particulate matter, haze, and acid rain.

As concentrations increase, the proportion of people prone to experience health effects and the seriousness of these effects are expected to increase. Thus, the 1997 standards were intended to include an ample margin of safety (as required by Section 109(b) of the Clean Air Act Amendments. The margin includes special concern about protection the health of sensitive individuals. However, they were not considered risk free and exposures to ambient concentrations just below the numerical level of the standards may be problematic for the most sensitive individuals. On the other hand, exposures to levels just above the NAAQS are not expected to be associated with health concerns for most healthy individuals. Such is the complicated nature of individual response to air pollutants, and one of the objectives of the revised index is to provide sufficient information to allow sensitive people to avoid unhealthy exposures.

REFERENCES

1. Samet, J. M., and Spengler, J. D. (eds.), *Indoor Air Pollution: A Health Perspective*. Johns Hopkins University Press, Baltimore, MD, 1991.
2. *Environmental Protection Agency Air Quality Index Reporting*, 40 CFR Port 58, Final Rule, August 4, 1999.

SUGGESTED READING

Benedick, R. E., *Ozone Diplomacy: New Directions in Safeguarding the Planet*. Harvard University Press, Cambridge, MA, 1991.

Brooks, B. O., and Davis, W. F., *Understanding Indoor Air Quality*. CRC Press, Boca Raton, FL, 1992.

Council on Environmental Quality, Annual Reports, Washington, DC, 1984–1992.

QUESTIONS

1. How does the range of concentrations of air pollutants of concern to the industrial hygienist differ from that of concern to the air pollution specialist? To what extent are air sampling and analytical methods in factories and in the ambient air the same or different?
2. Using the data in Fig. 8.3, draw the variation in concentration over the 6-h period as it would appear using sampling and analytical procedures which integrate the concentrations arriving at the receptor over 30 min and 2 h, respectively.
3. Sketch the appearance of a stripchart record measuring one pollutant for a week to show the weekday–weekend cycle.
4. Draw a chart showing the most probable trend of the concentration of airborne particles of horse manure in the air of a large midwestern US city from 1850 to 1950.
5. Describe an air quality measurement system used to assess the levels and types of aeroallergens.
6. Using the data of Fig. 8.15, determine the frequency with which a 30-min average SO_2 concentration of 0.2 ppm would probably have been exceeded at site 26428007 HO1 in St. Louis, Missouri in 1980.
7. Prepare a table describing air quality levels in your community, or in the nearest community to you that has such data available.
8. If you were to prepare a figure representing the relationship between air quality levels and the effects caused by these levels, what changes would you make in Fig. 8.17?
9. Discuss the extent and usefulness of dissemination by the media of the AQI values in the communities in which you have lived.
10. What do these findings tell you about the source strength of carbonyls shown in Tables 8.1 and 8.2 and Figures 8.1 and 8.2? How and why do you think these source strengths differ for various carbonyls? What actions and controls would you recommend to reduce exposures to each of these carbonyls?

9

The Philosophy of Air Pollution Control

I. STRATEGY AND TACTICS: THE AIR POLLUTION SYSTEM

Since primary pollutants may have the dual role of causing adverse effects in their original unreacted form and of reacting chemically to form secondary pollutants, air pollution control consists mainly of reducing the emission of primary pollutants to the atmosphere. Air pollution control has two major aspects: strategic and tactical. The former is the long-term reduction of pollution levels at all scales of the problem from local to global. This aspect is called *strategic* in that long-term strategies must be developed. Goals can be set for air quality improvement 5, 10, or 15 years ahead and plans made to achieve these improvements. One notable example is the requirement of the Clean Air Act Amendments (CAAA) of 1990 to reduce the emissions of hazardous air pollutants (so-called "HAPs"). The law requires the US Environmental Protection Agency (EPA) to establish an inventory to track progress in reducing HAPs in ambient air. The first step of this strategy was given in Section 112(d) of CAAA, which requires EPA to promulgate technology-based emission standards (maximum achievable control technology, i.e. MACT). For major sources of HAPs, Section 112(f) requires standards to address risks remaining after implementation of MACT standards, known as residual risks standards. Section 112(c)(3) and Section 112(k) of the CAAA requires EPA to address the emissions and risks of HAPs from area sources and to show a 75% reduction in cancer incidence of emissions from stationary sources of HAPs since 1990. To monitor the success in meeting these strategies in reducing emissions and human health, EPA compiles the National Emissions Inventory (NEI) for HAPs. The EPA previously compiled a baseline 1990 and 1996 National Toxics Inventory (NTI) and 1999 NEI for HAPs and has recently completed version 3 of the 2002 NEI.

The NEI includes major point and area sources, non-point area and other sources, and mobile source estimates of emissions. Stationary major sources of HAPs are defined as sources that have the potential to emit 10 tons per year or more of any single HAP or 25 tons per year or more of any combination of HAPs [2]. Stationary area sources of HAPs are defined as sources that have the potential to emit less than 10 tons per year or more of any single HAP or less than 25 tons per year or more of any combination of HAPs. Mobile sources include onroad vehicles, non-road equipment, and aircraft/locomotive/commercial marine vessels (ALM). EPA has developed the National Air Toxics Assessment (NATA) to estimate the magnitude of HAP emissions reductions and demonstrate reduced public risk from HAP emissions attributable to CAA toxics programs [2].

Emissions of HAPs emissions fell through the 1990s in response to the MACT standards and mobile source regulations. However, area and other source emissions have increased because EPA has not yet fully implemented its area source program as required by Section 112c(3) and 112(k) of the CAAA. Figure 9.1 and Table 9.1 present emissions trends for the sum of 188 HAPs by source sectors. Toxicity-weighted emissions have also declined between 1990 and 2002 for cancer and non-cancer respiratory and neurological effects. Figures 9.2–9.4 and Tables 9.2–9.4 provide the toxicity-weighted

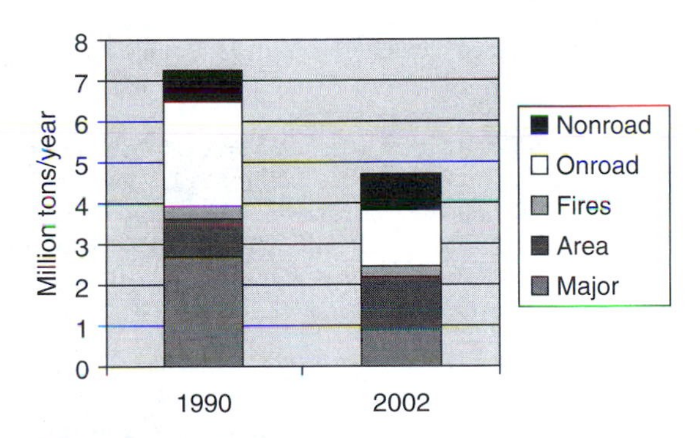

Fig. 9.1. Trends in US emissions of HAPs. *Note*: Aircraft, Locomotive and Commercial Marine Vessels is not reported. *Source*: A. Pope and M. Strum, *1990–2002* NEI HAP Trends: Success of CAA Air Toxic Programs in Reducing HAP Emissions and Risk, 16th Annual International Emission Inventory Conference, *Emission Inventories: "Integration, Analysis, and Communications"*, Raleigh, NC, May 14–17, 2007.

TABLE 9.1.

Sum of Emissions of the 188 HAPs listed in the CAAA of 1990

Sector	1990 Emissions	2002 Emissions	% Reduction
TOTAL	**7.24 milion tons**	**4.7 million tons**	**35**
Major	2.69 million tons	0.89 million tons	67
Area	0.91 million tons	1.29 million tons	−42
Fires (Wildfires & Prescribed Burns)	0.34 million tons	0.28 million tons	18
Onroad Mobile	2.55 million tons	1.36 million tons	47
Non-road Mobile	0.75 million tons[b]	0.86 million tons	−15
ALM Mobile[a]		0.02 million tons	

[a] ALM – Aircraft, Locomotive and Commercial Marine Vessels
[b] 1990 Non-road Mobile includes ALM
Source: A. Pope and M. Strum, 1990–2002 NEI HAP Trends: Success of CAA Air Toxic Programs in Reducing HAP Emissions and Risk, 16th Annual International Emission Inventory Conference, *Emission Inventories: "Integration, Analysis, and Communications"*, Raleigh, NC, May 14–17, 2007.

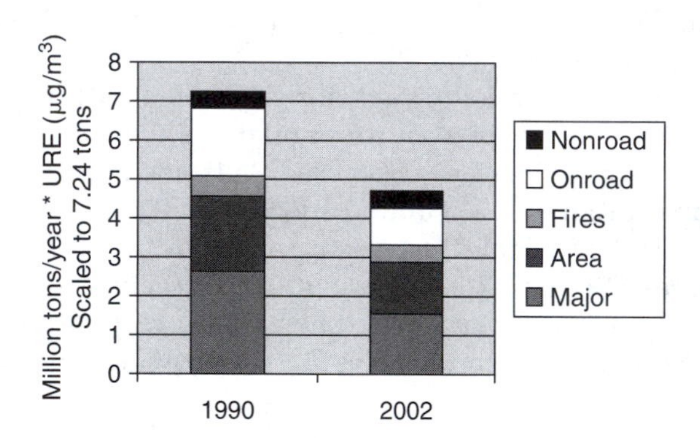

Fig. 9.2. Trends in scaled cancer toxicity-weighted US emissions of HAPs. *Note*: Unit risk estimate (URE) is upper bound risk estimate of an individual's probability of contracting cancer over a lifetime of exposure to a concentration of 1 μg of the pollutant per cubic meter of air. For example, if an URE is 1.5×10^{-6} per $\mu g\,m^{-3}$, 1.5 excess tumors are expected to develop per 1 000 000 people if they are exposed daily for a lifetime to 1 μg of chemical in 1 cubic meter of air. Aircraft, Locomotive and Commercial Marine Vessels is not reported. *Source*: A. Pope and M. Strum, 1990–2002 NEI HAP Trends: Success of CAA Air Toxic Programs in Reducing HAP Emissions and Risk. 16th Annual International Emission Inventory Conference, *Emission Inventories: "Integration, Analysis, and Communications"*, Raleigh, NC, May 14–17, 2007.

emissions scaled to the sum of 7.24 million tons for 1990 total emissions. Table 9.5 identifies the sectors contributing the larger share of the pollutants shown in Figure 9.5. In 2002, benzene accounts for 28% of cancer risks in the toxicity-weighted NEI, whereas manganese accounts for 77% of non-cancer

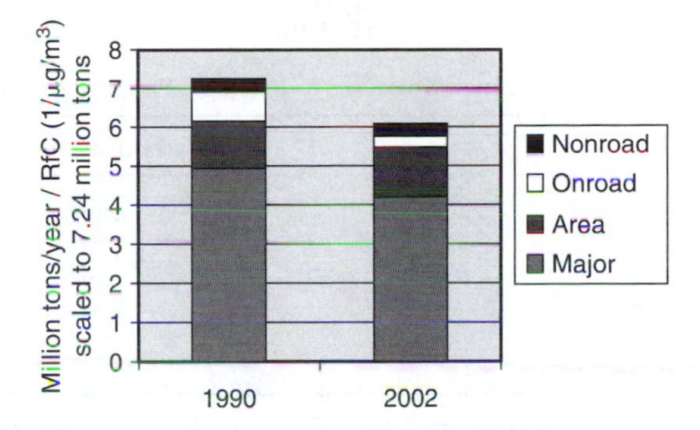

Fig. 9.3. Trends in scaled non-cancer neurological toxicity-weighted US emissions of HAPs. *Note*: RfC is the reference concentration, i.e. the level below which no adverse effect is expected. Aircraft, Locomotive and Commercial Marine Vessels is not reported. Fires is not reported. *Source*: A. Pope and M. Strum, 1990–2002 NEI HAP Trends: Success of CAA Air Toxic Programs in Reducing HAP Emissions and Risk. 16th Annual International Emission Inventory Conference, *Emission Inventories: "Integration, Analysis, and Communications"*, Raleigh, NC, May 14–17, 2007.

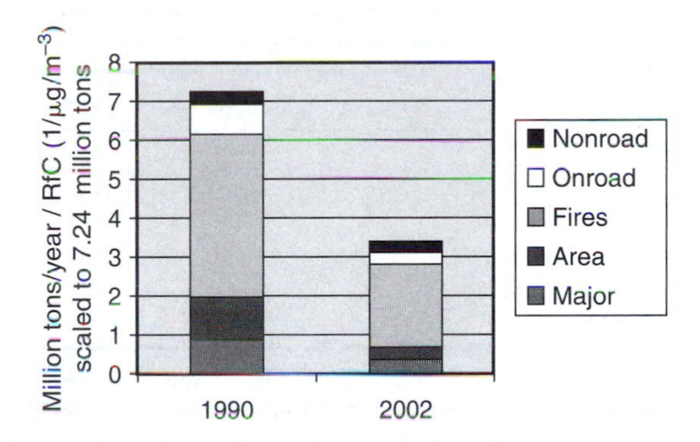

Fig. 9.4. Trends in scaled non-cancer respiratory toxicity-weighted US emissions of HAPs. *Note*: RfC is the reference concentration, i.e. the level below which no adverse effect is expected. Aircraft, Locomotive and Commercial Marine Vessels is not reported. *Source*: A. Pope and M. Strum, 1990–2002 NEI HAP Trends: Success of CAA Air Toxic Programs in Reducing HAP Emissions and Risk. 16th Annual International Emission Inventory Conference, *Emission Inventories: "Integration, Analysis, and Communications"*, Raleigh, NC, May 14–17, 2007.

neurological effects in the toxicity-weighted NEI, and acrolein accounts for 90% of non-cancer respiratory effects in the toxicity-weighted NEI [2].

There can be a regional strategy to affect planned reductions at the urban and local scales; a state or provincial strategy to achieve reductions at the state, provincial, urban, and local scales; and a national strategy to achieve them at national and lesser scales. The continental and global scales require an international strategy for which an effective instrumentality is being developed.

TABLE 9.2

Trends in Scaled Cancer Toxicity-Weighted US Emissions of HAPs

Sector	% Reduction from 1990 to 2002
TOTAL	**36**
Major	44
Area	31
Fires (Wildfires & Prescribed Burns)	−4
Onroad Mobile	49
Non-road Mobile	7.5

Source: A. Pope and M. Strum, 1990–2002 NEI HAP Trends: Success of CAA Air Toxic Programs in Reducing HAP Emissions and Risk. 16th Annual International Emission Inventory Conference, *Emission Inventories: "Integration, Analysis, and Communications"*, Raleigh, NC, May 14–17, 2007.

TABLE 9.3

Trends in Scaled Non-cancer Neurological Toxicity-Weighted US Emissions of HAPs

Sector	% Reduction from 1990 to 2002
TOTAL	**16**
Major	15
Area	−7
Fires (Wildfires & Prescribed Burns)	0
Onroad Mobile	62
Non-road Mobile	12

Source: A. Pope and M. Strum, 1990–2002 NEI HAP Trends: Success of CAA Air Toxic Programs in Reducing HAP Emissions and Risk. 16th Annual International Emission Inventory Conference, *Emission Inventories: "Integration, Analysis, and Communications"*, Raleigh, NC, May 14–17, 2007.

The other major aspect of air pollution reduction is the control of short-term episodes on the urban scale. This aspect is called *tactical* because, prior to an episode, a scenario of tactical maneuvers must be developed for application on very short notice to prevent an impending episode from becoming a disaster. Since an episode usually varies from a minimum of about 36 h to a maximum of 3 or 4 days, temporary controls on emissions much more severe than are called for by the long-term strategic control scenario must be implemented rapidly and maintained for the duration of the episode. After the weather conditions that gave rise to the episode have passed, these temporary episode controls can be relaxed and controls can revert to those required for long-term strategic control.

The mechanisms by which a jurisdiction develops its air pollution control strategies and episode control tactics are outlined in Fig. 9.6. Most of the boxes

TABLE 9.4

Trends in Scaled Non-cancer Respiratory Toxicity-weighted US Emissions of HAPs

Sector	% Reduction from 1990 to 2002
TOTAL	**54**
Major	69
Area	64
Fires (Wildfires & Prescribed Burns)	50
Onroad Mobile	58
Non-road Mobile	35

Source: A. Pope and M. Strum, *1990–2002 NEI HAP* Trends: Success of CAA Air Toxic Programs in Reducing HAP Emissions and Risk. 16th Annual International Emission Inventory Conference, *Emission Inventories: "Integration, Analysis, and Communications"*, Raleigh, NC, May 14–17, 2007.

TABLE 9.5

Source Sector Contribution to Pollutants in Fig. 9.5

HAP	Largest sector
Acrolein	Fires
Arsenic	Major
Benzene	Mobile
1,3-Butadiene	Mobile
Chlorine	Major
Hexavalent chromium	Major
Cyanide	Area
Hydrochloric acid	Major
Manganese	Major
Toluene	Mobile
Xylenes	Mobile

in the figure have already been discussed—sources, pollutant emitted, transport and diffusion, atmospheric chemistry, pollutant half-life, air quality, and air pollution effects. To complete an analysis of the elements of the air pollution system, it is necessary to explain the several boxes not yet discussed.

II. EPISODE CONTROL

The distinguishing feature of an air pollution episode is its persistence for several days, allowing continued buildup of pollution levels. Consider the situation of the air pollution control officer who is expected to decide when to use the stringent control restrictions required by the episode control tactics

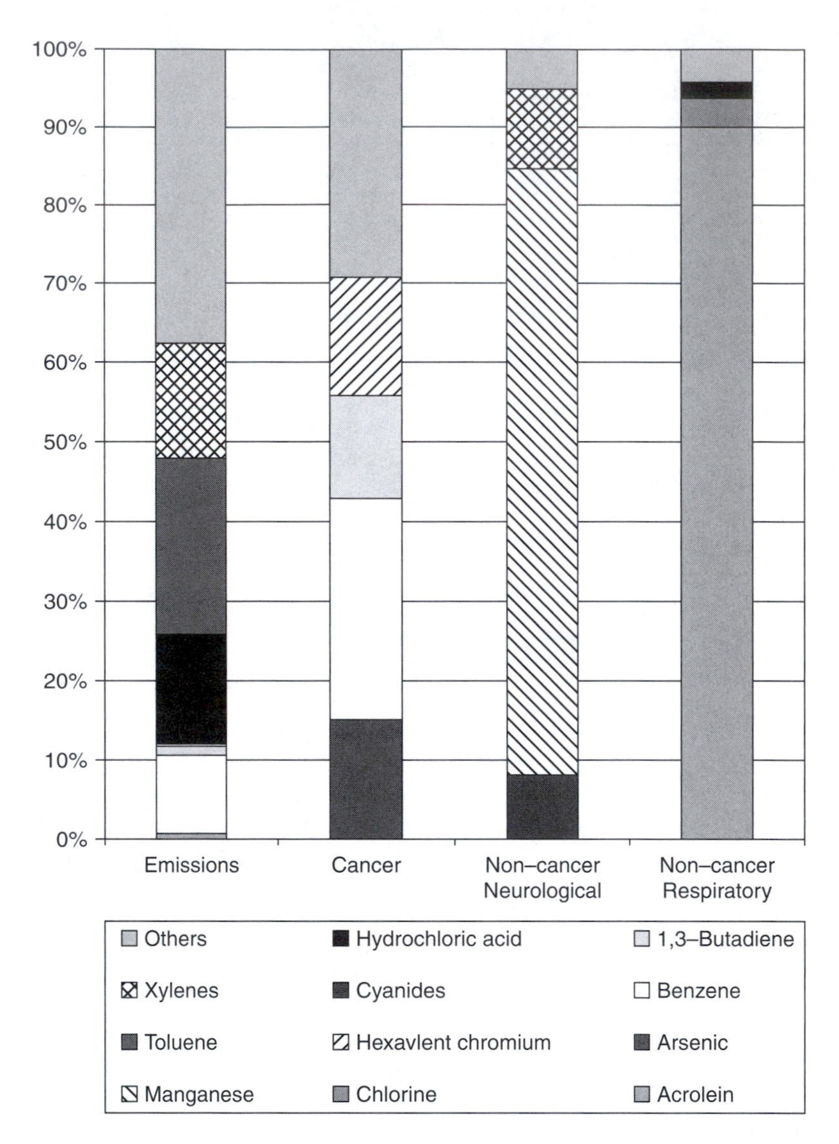

Fig. 9.5. Percent Contribution of US HAPs to 2002 emissions and toxicity-weighted emissions. *Source*: A. Pope and M. Strum, 1990–2002 NEI HAP Trends: Success of CAA Air Toxic Programs in Reducing HAP Emissions and Risk. 16th Annual International Emission Inventory Conference, *Emission Inventories: "Integration, Analysis, and Communications"*, Raleigh, NC, May 14–17, 2007.

scenario (Fig. 9.7 and Table 9.6). If these restrictions are imposed and the episode does not mature, i.e. the weather improves and blows away the pollution without allowing it to accumulate for another 24 h or more, the officer will have required for naught a very large expenditure by the community and a serious disruption of the community's normal activities. Also, part of the

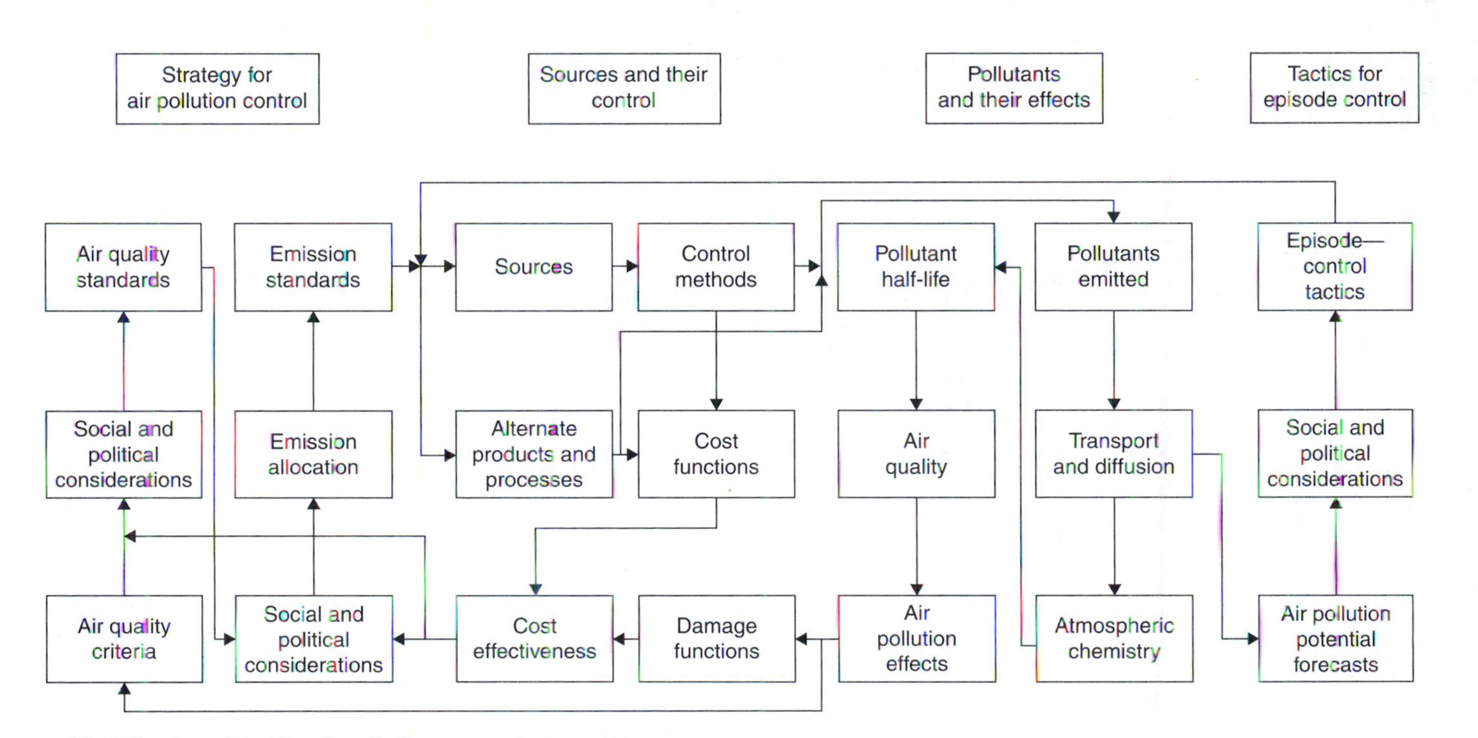

Fig. 9.6. A model of the air pollution management system.

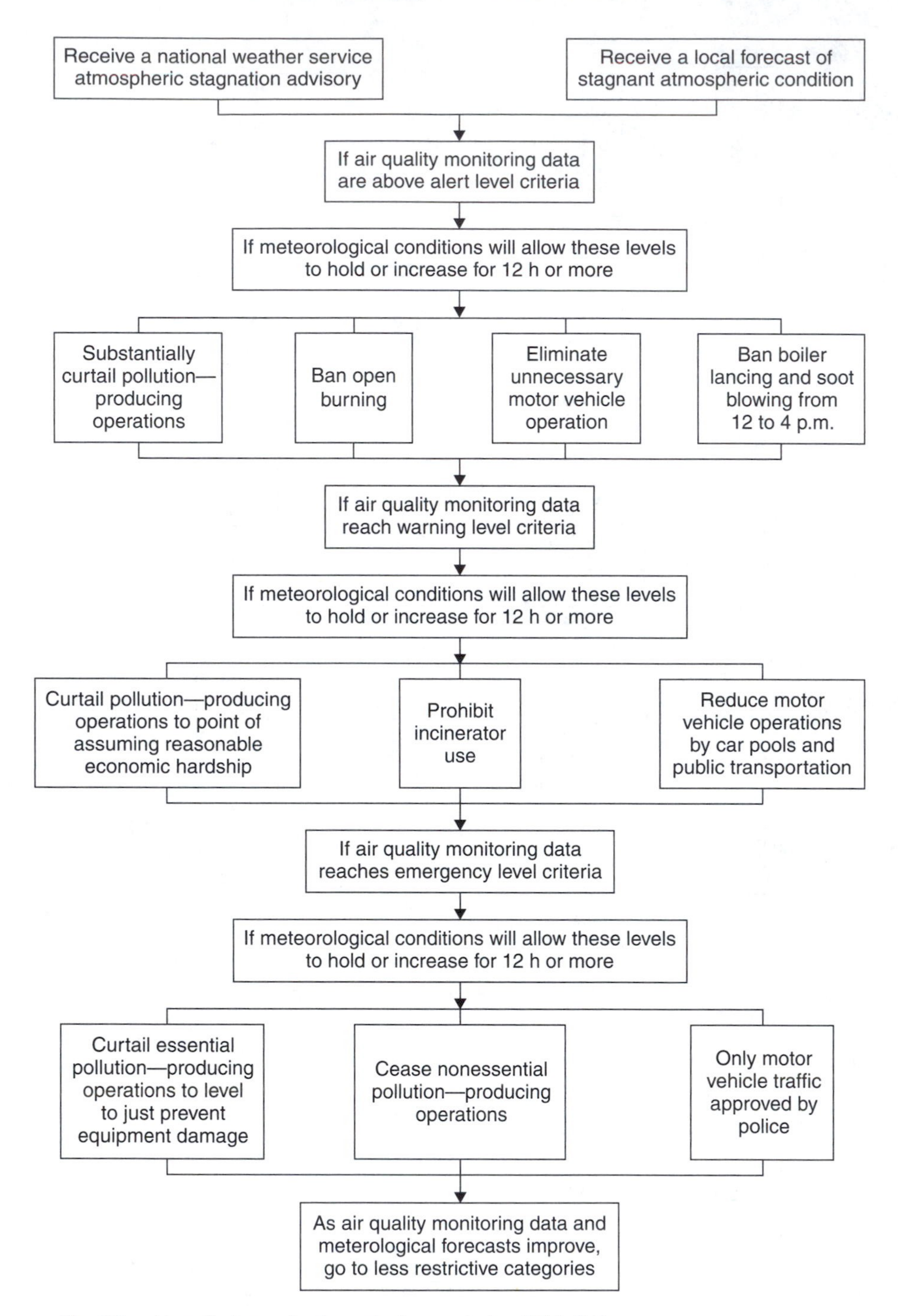

Fig. 9.7. Air pollution episode control scenario (see Table 9.6).

TABLE 9.6

United States Alert, Warning, and Emergency Level Criteria[a,b,c]

Alert level criteria
SO_2
800 μg m^{-3} (0.3 ppm), 24-h average
PM_{10}
350 μg m^{-3}, 24-h average
CO
17 mg m^{-3} (15 ppm), 8-h average
Ozone
400 μg m^{-3} (0.2 ppm), 1-h average
NO_2
1130 μg m^{-3} (0.6 ppm), 1-h average; 282 μg m^{-3} (0.15 ppm), 24-h average
Warning level criteria
SO_2
1600 μg m^{-3} (0.6 ppm), 24-h average
PM_{10}
420 μg m^{-3}, 24-h average
CO
34 mg m^{-3} (30 ppm), 8-h average
Ozone
800 μg m^{-3} (0.4 ppm), 1-h average
NO_2
2260 μg m^{-3} (1.2 ppm), 1-h average; 565 μg m^{-3} (0.15 ppm), 24-h average
Emergency level criteria
SO_2
2100 μg m^{-3} (0.8 ppm), 24-h average
PM_{10}
500 μg m^{-3}, 24-h average
CO
46 mg m^{-3} (15 ppm), 8-h average
Ozone
1000 μg m^{-3} (0.1 ppm), 1-h average
NO_2
3000 μgm^{-3} (0.6 ppm), 1-h average; 750 μg m^{-3} (0.15 ppm), 24-h average

[a] 2003 Code of Federal Regulations, Title 40-Protection of Environment, Chapter 1—Environmental Protection Agency, Appendix L, Example Regulations for Prevention of Air Pollution Emergency Episodes, 1.1 Episode Criteria, pp. 351–352.
[b] There is no criterion for lead, due to the chronic nature of the health effects of concern.
[c] *Note*: Append to each entry: Meterological conditions are such that pollutant concentrations can be expected to remain at the above levels for 12 or more hours or increase, or in the case of ozone the situation is likely to reoccur within the next 24 h unless control actions are taken.

officer's credibility in the community will be destroyed. If this happens more than once, the officer will be accused of crying wolf, and when a real episode occurs the warnings will be unheeded. If, however, the reverse situation occurs—i.e. the restrictions are not invoked and an episode does occur—there can be illness or possibly deaths in the community that could have been averted.

In deciding whether or not to initiate episode emergency plans, the control officer cannot rely solely on measurements from air quality monitoring stations, because even if pollutant concentrations rise toward acute levels over the preceding hours, these readings give no information on whether they will rise or fall during the succeeding hours.

The only way to avert this dilemma is for the community to develop and utilize its capability of forecasting the advent and persistence of the stagnation conditions during which an episode occurs and its capability of computing pollution concentration buildup under stagnation conditions. The details of how these forecasts and computations are made are discussed in Chapter 21, but at this point the foregoing discussion should explain the reason for the box in Fig. 9.6 labeled air pollution potential forecasts. The connecting box marked social and political considerations provides a place in the system for the public debates, hearings, and action processes necessary to decide, well in advance of an episode, what control tactics to use and when to call an end to the emergency. The public needs to be involved because alternatives have to be written into the scenario concerning where, when, and in what order to impose restrictions on sources. This should be done in advance and should be well publicized, because during the episode there is no time for public debate. In any systems analysis, the system must form a closed loop with feedback to keep the system under control. It will be noted that the system for tactical episode control is closed by the line connecting episode control tactics to sources, which means that the episode control tactics are to limit sources severely during the episode. Since it takes hours before emergency plans can be put into effect and their impact on pollution levels felt, it is possible that by the time the community responds, the situation has disappeared. Experience has shown that the time for community response is slowed by the need to write orders to close down sources and to respond in court to requests for judicial relief from such orders. To circumvent the former type of delay, orders can be written in blank in advance. To circumvent the latter type of delay, the agency's legal counsel must move as rapidly as the counsel seeking such relief.

III. AIR QUALITY MANAGEMENT CONTROL STRATEGY

Now let us consider the system for long-range strategy for air pollution control. The elements in this system that have not yet been discussed include several listed in Fig. 9.6 under sources and their control and all those listed under "Strategy for Air Pollution Control." Control of sources is affected in several ways. We can (1) use devices to remove all or part of the pollutant from the gases discharged to the atmosphere, (2) change the raw materials used in the pollution-producing process, or (3) change the operation of the

TABLE 9.7

Control Methods

I. Applicable to all emissions
   A. Decrease or eliminate production of emission
      1. Change specification of product
      2. Change design of product
      3. Change process temperature, pressure, or cycle
      4. Change specification of materials
      5. Change the product
   B. Confine the emissions
      1. Enclose the source of emissions
      2. Capture the emissions in an industrial exhaust system
      3. Prevent drafts
   C. Separate the contaminant from effluent gas stream
      1. Scrub with liquid

II. Applicable specifically to particulate matter emissions
   A. Decrease or eliminate particulate matter production
      1. Change to process that does not require blasting, blending, buffing, calcining[a], chipping, crushing, drilling, drying, grinding, milling, polishing, pulverizing, sanding, sawing, spraying, and tumbling, etc.
      2. Change from solid to liquid or gaseous material
      3. Change from dry to wet solid material
      4. Change particle size of solid material
      5. Change to process that does not require particulate material
   B. Separate the contaminant from effluent gas stream
      1. Gravity separator
      2. Centrifugal separator
      3. Filter
      4. Electrostatic precipitator

III. Applicable specifically to gaseous emissions
   A. Decrease or eliminate gas or vapor production
      1. Change to process that does not require annealing, baking, boiling, burning, casting, coating, cooking, dehydrating, dipping, distilling, expelling, galvanizing, melting, pickling, plating, quenching, reducing, rendering, roasting, and smelting, etc.
      2. Change from liquid or gaseous to solid material
      3. Change to process that does not require gaseous or liquid material
   B. Burn the contaminant to CO_2 and H_2O
      1. Incinerator
      2. Catalytic burner
   C. Adsorb the contaminant
      1. Activated carbon

[a] Calcining is the process of heating a substance to a temperature less than melting or fusing point, but sufficiently high to cause moisture loss, redox, and decomposition of carbonates and other compounds.

process so as to decrease pollutants emitted. These are control methods (Table 9.7). Such control methods have a cost associated with them and are the cost functions that appear in the system. There is always the option of seeking alternate products or processes which will provide the same utility to the public but with less pollutants emitted. Such products and processes have their own cost functions.

Just as it costs money to control pollution, it also costs the public money not to control pollution. All the adverse air pollution effects represent economic burdens on the public for which an attempt can be made to assign dollar values, i.e. the cost to the public of damage to vegetation, materials, structures, animals, the atmosphere, and human health. These costs are called damage functions. To the extent that there is knowledge of cost functions and damage functions, the cost effectiveness of control methods and strategies can be determined by their interrelationship. Cost effectiveness is an estimate of how many dollars worth of damage can be averted per dollar expended for control. It gives information on how to economically optimize an attack on pollution, but it gives no information on the reduction in pollution required to achieve acceptable public health and well-being. However, when these goals can be achieved by different control alternatives, it behooves us to utilize the alternatives that show the greatest cost effectiveness.

To determine what pollution concentrations in air are compatible with acceptable public health and well-being, use is made of air quality criteria, which are statements of the air pollution effects associated with various air quality levels. It is inconceivable that any jurisdiction would accept levels of pollution it recognizes as damaging to health. However, the question of what constitutes damage to health is judgmental and therefore debatable. The question of what damage to well-being is acceptable is even more judgmental and debatable. Because they are debatable, the same social and political considerations come into the decision-making process as in the previously discussed case of arriving at episode control tactics. Cost effectiveness is not a factor in the acceptability of damage to health, but it is a factor in determining acceptable damage to public well-being. Some jurisdictions may opt for a pollution level that allows some damage to vegetation, animals, materials, structures, and the atmosphere as long as they are assured that there will be no damage to their constituents' health. The concentration level the jurisdiction selects by this process is called an air quality standard. This is the level the jurisdiction says it wishes to maintain.

Adoption of air quality standards by a jurisdiction produces no air pollution control. Control is produced by the limitation of emission from sources, which, in turn, is achieved by the adoption and enforcement of emission standards. However, before emission standards are adopted, the jurisdiction must make some social and political decisions on which of several philosophies of emission standard development are to be utilized and which of the several responsible groups in the jurisdiction should bear the brunt of the control effort—its homeowners, landlords, industries, or institutions. This latter type of decision-making is called emission allocation. It will be seen in Fig. 9.6 that the system for strategic control is closed by the line connecting emission standards and sources, which means that long-range pollution control strategy consists of applying emission limitation to sources.

IV. ALTERNATIVE CONTROL STRATEGIES

There are several different strategies for air pollution control. The strategy just discussed and shown in Fig. 9.6 is called the *air quality management* strategy. It is distinguished from other strategies by its primary reliance on the development and promulgation of ambient air quality standards. This is the strategy in use in the United States. The second principal strategy is the *emission standard* strategy, also known as the *best practicable means of control* approach. In this strategy, neither air quality criteria nor ambient air quality standards are developed and promulgated. Either an emission standard is developed and promulgated or an emission limit on sources is determined on a case-by-case basis, representing the best practicable means for controlling emissions from those sources. This is the strategy in use in Great Britain. A third strategy controls pollution by adopting financial incentives (Table 9.8).

TABLE 9.8

Financial Incentives to Supplement or Replace Regulation

Taxes	
Sales taxes	On fuel, fuel additives, ingredients in fuel, pollution-producing equipment
Ultimate disposal taxes	On automobiles or other objects requiring ultimate disposal
Land-use taxes	For pollution-producing activities
Tax remission	
Corporate income tax	For investment in or operation of pollution control equipment; accelerated write-off of pollution control equipment
Property taxes	For pollution control equipment
Fines, effluent charges, and fees	
Fines	For violation of regulations
Effluent charges	For permission to emit excessive quantities of pollution; paid after emission
Fees	For permission to emit excessive quantities of pollution, paid before emission
Subsidies	
Direct	Governmental production (e.g. nuclear fuel)
Grants-in-aid	For pollution control installations
Indirect (low-interest bonds and loans)	For pollution control process or equipment development
Import restraints	
Duties and quotas	On materials, fuels, and pollution control-producing apparatus
Domestic production restraints	
Quotas, land, and offshore use restraints	On material and fuels

Source: From Stern, A. C., Heath, M. S., and Hufschmidt, M. M., A critical review of the role of fiscal policies and taxation in air pollution control, *Proceedings of the Third International Clean Air Congress*, Verein Deutscher Ingeniuere, Dusseldorf, 1974, pp. D-10–12.

This is usually but not necessarily in addition to the promulgation of air quality standards. Among the countries that have adopted tax, fee, or fine schedules on a national basis are Czechoslovakia, Hungary, Japan, The Netherlands, and Norway. There is additional discussion of this strategy in Chapter 29, Section III. A fourth strategy seeks to maximize cost effectiveness and is called the *cost–benefit* strategy. These strategies may result in lower emissions from existing processes or promote process modifications which reduce pollution generation.

V. ECONOMIC CONSIDERATIONS

The situation with regard to economic considerations has been so well stated in the First Report of the British Royal Commission on Environmental Pollution [1] that this section contains an extensive quotation from that report:

> Our survey of the activities of the Government, industry and voluntary bodies in the control of pollution discloses several issues which need further enquiry. The first and most difficult of these is how to balance the considerations which determine the levels of public and private expenditure on pollution control. Some forms of pollution bear more heavily on society than others; some forms are cheaper than others to control; and the public are more willing to pay for some forms of pollution control than for others. There are also short and long-term considerations: in the short-term the incidence of pollution control on individual industries or categories of labor may be heavy; but ... what may appear to be the cheapest policy in the short-term may prove in the long-term to have been a false economy.
>
> While the broad outlines of a general policy for protecting the environment are not difficult to discern, the economic information needed to make a proper assessment of the considerations referred to in the preceding paragraph ... seems to us to be seriously deficient. This is in striking contrast with the position regarding the scientific and technical data where, as our survey has shown, a considerable amount of information is already available and various bodies are trying to fill in the main gaps. The scientific and technical information is invaluable, and in many cases may be adequate for reaching satisfactory decisions, but much of it could be wasted if it were not supported by some economic indication of priorities and of the best means of dealing with specific kinds of pollution.
>
> So, where possible, we need an economic framework to aid decision making about pollution, which would match the scientific and technical framework we already have. This economic framework should include estimates of the way in which the costs of pollution, including disamenity costs, vary with levels of pollution; the extent to which different elements contribute to the costs; how variations in production and consumption affect the costs; and what it would cost to abate pollution in different ways and by different amounts. There may well be cases where most of the costs and benefits of abatement can be assessed in terms of money. Many of the estimates are likely to be speculative, but this is no reason for not making a start. There are other cases where most of the costs and benefits cannot be given a monetary value. In these cases decisions about pollution abatement must not await the results of a full economic calculation: they will have to be based largely on subjective judgments anyway. Even so, these subjective judgments should be supported by as much quantitative information as possible, just as decisions about health and education are supported by extensive statistical data. Further, even if decisions to abate pollution

are not based on rigorous economic criteria, it is still desirable to find the most economic way of achieving the abatement.

As air pollution management moves forward, economics has a major role in reducing pollution. Multimedia considerations are forcing a blend of traditional emission reduction approaches and innovative methods for waste minimization. These efforts are directed toward full cost accounting of the life cycle of products and residuals from the manufacturing, use, and ultimate disposal of materials.

REFERENCES

1. *Royal Commission on Environmental Pollution*. First Report, Command 4585, H.M. Stationery Office, London, 1971.
2. A. Pope and M. Strum, 1990–2002 NEI HAP Trends: Success of CAA Air Toxic Programs in Reducing HAP Emissions and Risk. 16th Annual International Emission Inventory Conference, *Emission Inventories: "Integration, Analysis, and Communications"*, Raleigh, NC, May 14–17, 2007.

SUGGESTED READING

Ashby, E., and Anderson, M., *The Politics of Clean Air*. Oxford University Press (Clarendon), New York, 1981.

Cohen, R. E., *Washington at Work: Back Rooms and Clean Air*. Macmillan, New York, 1992.

Crandall, R. W., *Controlling Industrial Pollution: The Economics and Politics of Clean Air*. Brookings Institution, Washington, DC, 1983.

Hoberg, G., *Pluralism by Design: Environmental Policy and the American Regulatory State*. Praeger, New York, 1992.

The Air Quality Strategy for England, Scotland, Wales and Northern Ireland: Working Together for Clean Air. Presented to Parliament by the Secretary of State for the Environment, Transport and the Regions by Command of Her Majesty, January 2000, http://www.defra.gov.uk/environment/airquality/strategy/pdf/foreword.pdf.

United Nations Economic Commission for Europe, Convention on long-range transboundary air pollution, *Strategies and Policies for Air Pollution Abatement*. United Nations, Geneva Switzerland, 1999.

QUESTIONS

1. Explain why certain important long-range air pollution control strategies will not suffice for short-term episode control, and vice versa.
2. Develop an episode control scenario for a single large coal-fired steam electric generating station.
3. In Fig. 9.6, the words "Social and political considerations" appear several times. Discuss these considerations for the various contexts involved.
4. Discuss the relative importance of air quality criteria and cost effectiveness in the setting of air quality standards.

5. The quotation in Section V contains the words "what may appear to be the cheapest policy in the short-term may prove in the long-term to have been a false economy." Give some examples of this.
6. Draw a simplified version of Fig. 9.6 with fewer than 10 boxes.
7. In the early 1970s, some experts expected the regulatory measures against air pollutants to be temporary until market forces supplanted them. Give examples of situations in which these experts were right and when they were wrong. What role do you believe that the marketplace will have in reducing air pollution in the future? Explain the obstacles and promise for at least one conventional pollutant (i.e. criteria pollutant) and one toxic pollutant (i.e. HAP).
8. How would one go about developing an air pollution damage function for human health?
9. The Organization for Economic Cooperation and Development (OECD) has been a proponent of the "polluter pays" principle. What is the principle and how can it be implemented?
10. Study Table 9.7 and determine whether there are any control methods that you believe should be added or deleted.

10

Sources of Air Pollution

I. GENERAL

The sources of air pollution are nearly as numerous as the grains of sand. In fact, the grains of sand themselves are air pollutants when the wind entrains them and they become airborne. We would class them as a natural air pollutant, which implies that such pollution has always been with us. Natural sources of air pollution are defined as sources not caused by people in their activities.

Consider the case in which someone has removed the ground cover and left a layer of exposed soil. Later the wind picks up some of this soil and transports it a considerable distance to deposit it at another point, where it affects other people. Would this be classed as a natural pollutant or an anthropogenic pollutant? We might call it natural pollution if the time span between when the ground cover was removed and when the material became airborne was long enough. How long would be long enough? The answers to such questions are not as simple as they first appear. This is one of the reasons why pollution problems require careful study and analysis before a decision to control them at a certain level can be made.

Fig. 10.1. Mt. St. Helens during the eruption on May 1980. *Source*: Photo by C. Rosenfeld, Oregon Air National Guard.

A. Natural Sources

An erupting volcano emits particulate matter. Pollutant gases such as SO_2, H_2S, and methane are also emitted. The emission from an eruption may be of such magnitude as to harm the environment for a considerable distance from the volcanic source. Clouds of volcanic particulate matter and gases have remained airborne for very long periods of time. The eruption of Mt. St. Helens in the state of Washington is a classic example of volcanic activity. Figure 10.1 is a photograph of Mt. St. Helens during the destructive eruption on May 18, 1980.

Accidental fires in forests and on the prairies are usually classified as natural sources even though they may have been originally ignited by human activities. In many cases foresters intentionally set fires in forestlands to burn off the residue, but lightning setting off a fire in a large section of forestland could only be classed as natural. A large uncontrolled forest fire, as shown in Fig. 10.2, is a frightening thing to behold. Such a fire emits large quantities of pollutants in the form of smoke, unburned hydrocarbons, carbon monoxide, carbon dioxide, oxides of nitrogen, and ash. Forest fires in the Pacific Northwest of the United States have been observed to emit a plume which caused reduction in visibility and sunlight as far away as 350 km from the actual fire.

Dust storms that entrain large amounts of particulate matter are a common natural source of air pollution in many parts of the world. Even a relatively small dust storm can result in suspended particulate matter readings one or two orders of magnitude above ambient air quality standards. Visibility reduction during major dust storms is frequently the cause of severe highway accidents and can even affect air travel. The particulate matter transferred by dust storms from the desert to urban areas causes problems to householders, industry, and automobiles. The materials removed by the air cleaner of an

Fig. 10.2. Uncontrolled forest fire. *Source*: Information and Education Section, Oregon Department of Forestry.

automobile are primarily natural pollutants such as road dust and similar entrained material.

The oceans of the world are an important natural source of pollutant material. The ocean is continually emitting aerosols to the atmosphere, in the form of salt particles, which are corrosive to metals and paints. The action of waves on rocks reduces them to sand, which may eventually become airborne. Even the shells washed up on the beach are eroded by wave and tidal action until they are reduced to such a small size that they too may become airborne.

An extensive source of natural pollutants is the trees and other plant life of the earth. Even though these green plants play a large part in the conversion of carbon dioxide to oxygen through photosynthesis, they are still the major source of hydrocarbons on the planet. The familiar blue haze over forested areas is nearly all from the atmospheric reactions of the volatile organics given off by the trees of the forest [1]. Another air pollutant problem, which can be attributed to plant life, is the pollens which cause respiratory distress and allergic reactions in humans.

Other natural sources, such as alkaline and saltwater lakes, are usually quite local in their effect on the environment. Sulfurous gases from hot springs also fall into this category in that the odor is extremely strong when close to the source but disappears a few kilometers away.

B. Anthropogenic Sources

1. Industrial Sources

The reliance of modern people on industry to produce their needs has resulted in transfer of the pollution sources from the individual to industry.

A soap factory will probably not emit as much pollution as did the sum total of all the home soap-cooking kettles it replaces, but the factory is a source that all soap consumers can point to and demand that it be cleaned up.

A great deal of industrial pollution comes from manufacturing products from raw materials—(1) iron and steel from ore, (2) lumber from trees, (3) gasoline and other fuels from crude oil, and (4) stone from quarries. Each of these manufacturing processes produces a product, along with several waste products which we term pollutants. Often, part or all of the polluting material can be recovered and converted into a usable product.

Industrial pollution is also emitted by industries that convert products to other products—(1) automobile bodies from steel, (2) furniture from lumber, (3) paint from solids and solvents, and (4) asphaltic paving from rock and oil.

Industrial sources are stationary, and each emits relatively consistent qualities and quantities of pollutants. A paper mill, for example, will be in the same place tomorrow that it is today, emitting the same quantity of the same kinds of pollutants unless a major process change is made. Control of industrial sources can usually be accomplished by applying known technology. The most effective regulatory control is that which is applied uniformly within all segments of industries in a given region, e.g., "Emission from all asphalt plant dryers in this region shall not exceed 230 mg of particulate matter per standard dry cubic meter of air."

2. *Utilities*

The utilities in our modern society are so much a part of our lives that it is hard to imagine how we survived without them. An electric power plant generates electricity to heat and light our homes in addition to providing power for the personal computer, television, refrigerator, and the recharging of the batteries for the laptop, cell phone, PDA, MP3 player, and electric toothbrush. When our homes were heated with wood fires, home-made candles were used for light, there were no television or entertainment appliances, and food was stored in a cellar, the total of the air pollution generated by all the individual sources probably exceeded that of the modern generating stations supplying today's energy. It is easy for citizens to point out the utility as an air pollution source without connecting their own use of the power to the pollution from the utility. Figure 10.3 illustrates an electric power plant.

Utilities are in the business of converting energy from one form to another and transporting that energy. If a large steam generating plant, producing 2000 MW, burns a million kilograms per hour of 4% ash coal, it must somehow dispose of 40 000 kg of ash per hour. Some will be removed from the furnaces by the ash-handling systems, but some will go up the stack with the flue gases. If 50% of the ash enters the stack and the fly ash collection system is 99% efficient, 200 kg of ash per hour will be emitted to the atmosphere. For a typical generating plant, the gaseous emissions would include 341 000 kg of oxides of

Fig. 10.3. Coal-fired electric generating plant.

sulfur per day and 185 000 kg of oxides of nitrogen per day. If this is judged as excessive pollution, the management decision can be to (1) purchase lower-ash or lower-sulfur coal, (2) change the furnace so that more ash goes to the ash pit and less goes up the stack, or (3) install more efficient air pollution control equipment. In any case, the cost of operation will be increased and this increase will be passed on to the consumer.

Another type of utility that is a serious air pollution source is the one that handles the wastes of modern society. An overloaded, poorly designed, or poorly operated sewage treatment plant can cause an air pollution problem which will arouse citizens to demand immediate action. In many countries around the world, open dumps still exist. These may catch fire and release harmful plumes of smoke and fumes. Even in more economically developed nations, landfills remain sources of dust and smoke, as are fires in abandoned mines shafts and in industrial waste disposal and reclamation sites. These are certainly sources of public complaint, even though it may be explained to the same public that it is the "cheapest" way to dispose of their solid waste. The public has shown its willingness to ban burning dumps and pay the additional cost of adequate waste disposal facilities to have a pollution-free environment.

3. *Personal Sources*

Even though society has moved toward centralized industries and utilities, we still have many personal sources of air pollution for which we alone can answer—(1) automobiles, (2) home furnaces, (3) home fireplaces and stoves, (4) backyard barbecue grills, and (5) open burning of refuse and leaves. Figure 10.4 illustrates the personal emissions of a typical US family.

The energy release and air pollution emissions from personal sources in the United States are greater than those from industry and utilities combined.

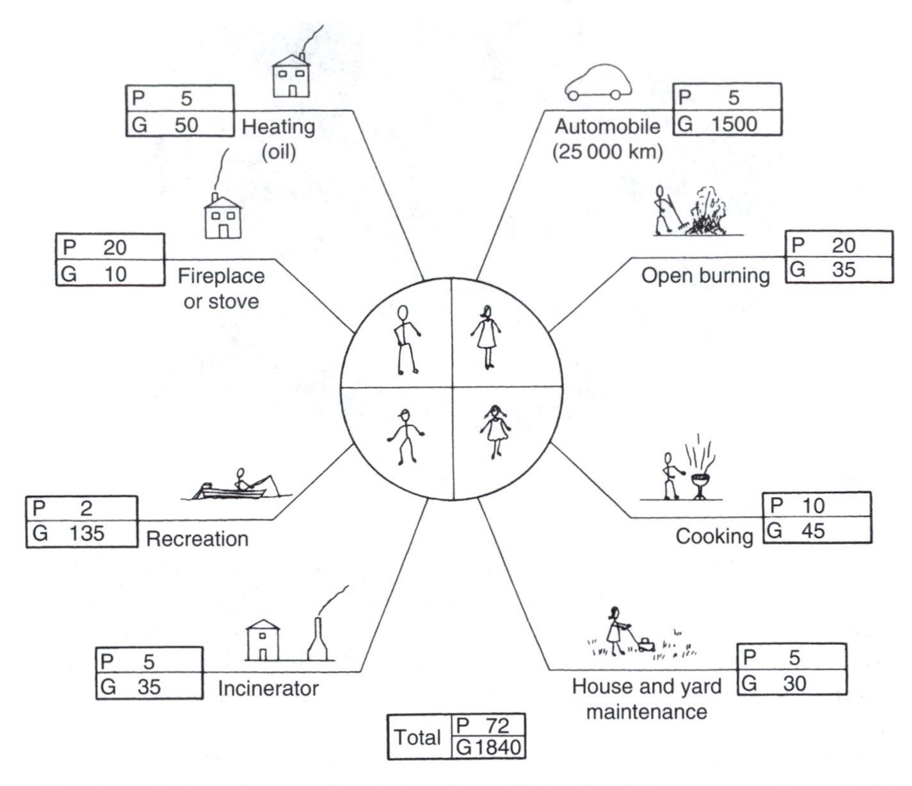

Fig. 10.4. Estimated personal emissions from US family of four persons. P, particulate matter in kilograms per year; G, gases in kilograms per year.

In any major city in the United States, the mass of pollutants emitted by the vast numbers of private automobiles exceeds that from any other source.

Control of these personal sources of pollution takes the form of (1) regulation (fireplaces and stoves may be used only when atmospheric mixing is favorable), (2) change of lifestyle (sell the automobile and ride public transportation), (3) change from a more polluting to a less polluting source (convert the furnace to natural gas), or (4) change the form of pollution (instead of burning leaves, haul them to the city landfill for composting). Whatever method is used for control of pollution from personal sources, it is usually difficult and unpopular to enforce. It is difficult to get citizens to believe that their new, highly advertised, shiny, unpaid-for automobiles are as serious a pollution problem as the smoking factory stack on the horizon. It is also a very ineffective argument to point out that the workers at that factory put more pollution into the air each day by driving their automobiles to and from work, or by mowing their lawns on Saturday, than does the factory with its visible plume of smoke.

II. COMBUSTION

Combustion is the most widely used, and yet one of the least understood, chemical reactions at our disposal. *Combustion* is defined as the rapid union of a substance with oxygen accompanied by the evolution of light and heat [2].

The economies of highly industrialized nations are heavily dependent on combustion. Much of the transportation by automobile, rail, and airlines is based on internal combustion engines that burn gasoline or diesel fuels. Small internal combustion engines are deceivingly important sources of air pollutants. In fact, with the advent of low emitting vehicles (LEVs), lawn mower usage for 1 h emits about as much as driving a car 100 miles (650 miles for a vehicle made before 1990). A push mower emits as much hourly pollution as 11 cars and a riding mower emits as much as 34 cars. Since small engine use exceeds three billion hours per year in the United States, the US EPA proposed a rule in 2007 to reduce small engine exhaust of hydrocarbons by 35% and oxides of nitrogen emissions by 60%. In addition, fuel evaporative emissions of hydrocarbons are expected to fall by 45% if the rule is implemented. To meet exhaust emission standards, manufacturers are expected to use catalytic converters for the first time in numerous small watercraft, lawn, and garden equipment. The rule also lays out fuel evaporative standards, national standards for vessels powered by sterndrive or inboard engines, and CO standards for gasoline-powered engines used in recreational watercraft [3]. Combustion is the relatively simple phenomenon of oxidizing a substance in the presence of heat. Chemically, efficient combustion is

$$(CH)_x + O_2 \rightarrow CO_2 + H_2O \quad (10.1)$$

Most thermal processes, however, do not reach complete combustion. More complex combustion reactions are shown in Table 10.1. They are usually oxygen limited, leading to the generation of a wide variety of compounds, many that are toxic. Other reactions besides combustion also produce air pollutants. Decomposition of a substance in the absence of oxygen is known as pyrolysis. In fact, a single fire can have pockets of both combustion and pyrolytic processes. This lack of homogeneity results in temperatures varying in both space and time. Plastic fires, for example, can release over 450 different organic compounds.[1] The relative amount of combustion and pyrolysis in a fire affects the actual amounts and types of compounds released.

Temperature is also important, but there is no direct relationship between temperature and pollutants released. For example, Fig. 10.5 shows that in a

[1] Levin, B.C., A summary of the NBS literature reviews on the chemical nature and toxicity of pyrolysis and combustion products from seven plastics: acrylonitrile–butadiene–styrenes; nylons; polyesters; polyethylenes; polystyrenes; poly(vinyl chlorides) and rigid polyurethane foams. *Fire Mater.* **11**, 143–157 (1987).

TABLE 10.1

Balanced Combustion Reactions for Selected Organic Compounds

Chlorobenzene	$C_6H_5Cl + 7O_2 \rightarrow 6CO_2 + HCl + 2H_2O$
Tetrachloroethene (TCE)	$C_2Cl_4 + O_2 + 2H_2O \rightarrow 2CO_2 + HCl$
Hexachloroethane (HCE)	$C_2Cl_6 + ½O_2 + 3H_2O \rightarrow 2CO_2 + 6HCl$
Post-chlorinated polyvinyl chloride (CPVC)	$C_4H_5Cl_3 + 4½O_2 \rightarrow 4CO_2 = 3HCl + H_2O$
Natural gas fuel (methane)	$CH_4 + 2O_2 \rightarrow CO_2 + 2H_2O$
PTEE Teflon	$C_2F_4 + O_2 \rightarrow CO_2 + 4HF$
Butyl rubber	$C_9H_{16} + 13O_2 \rightarrow 9CO_2 + 8H_2O$
Polyethylene	$C_2H_4 + 3O_2 \rightarrow 2CO_2 + 2H_2O$

Wood is considered to have the composition of $C_{6.9}H_{10.6}O_{3.5}$. Therefore, the combustion reactions are simple carbon and hydrogen combustion:

$$C + O_2 \rightarrow CO_2$$
$$H + 0.25O_2 \rightarrow 0.5H_2O$$

Source: US Environmental Protection Agency

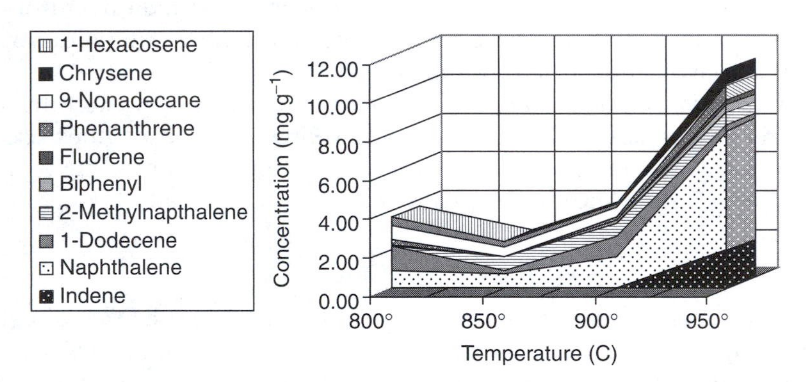

Fig. 10.5. Selected hydrocarbon compounds generated in a low-density polyethylene fire (pyrolysis) in four temperature regions. Data from Hawley-Fedder, R. A., Parsons, M. L., and Karasek, F. W., Products obtained during combustion of polymers under simulated incinerator conditions. *J. Chromatogr.* **314**, 263–272 (1984).

plastics fire (i.e. low-density polyethylene pyrolysis) compounds are generated at lower temperatures, but for others the optimal range is at higher temperatures. The aliphatic compounds in this fire (i.e. 1-dodecene, 9-nonadecane, and 1-hexacosene) are generated in higher concentrations at lower temperatures (about 800°C), and the aromatics need higher temperatures (see Figs. 10.5 and 10.6).

We use combustion primarily for heat by changing the potential chemical energy of the fuel to thermal energy. We do this in a fossil fuel-fired power plant, a home furnace, or an automobile engine. We also use combustion as a means of destruction for our unwanted materials. We reduce the volume of

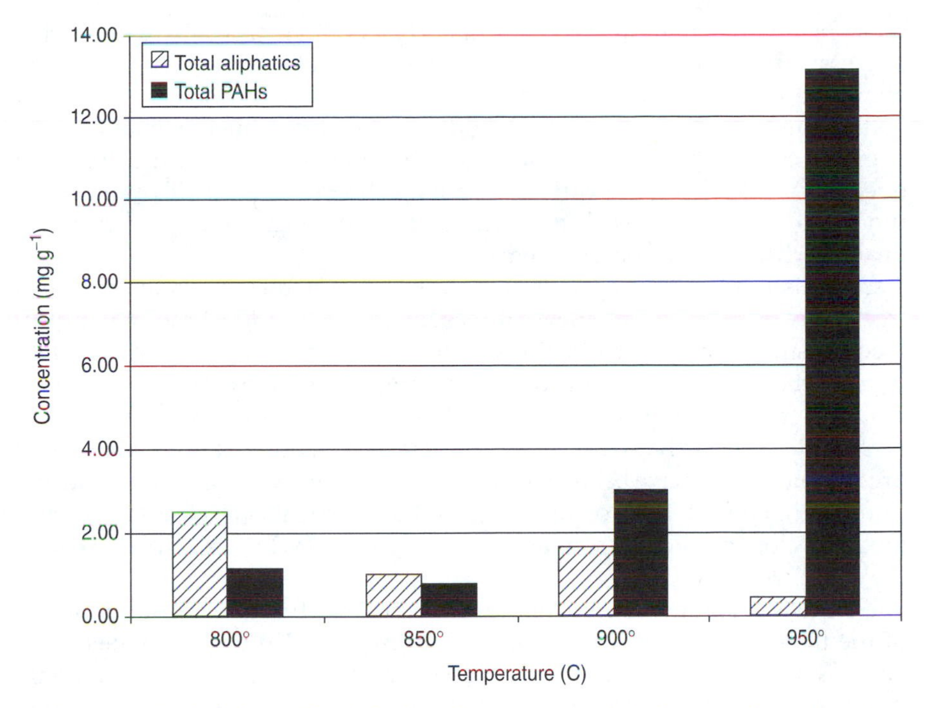

Fig. 10.6. Total aliphatic (chain) hydrocarbons versus polycyclic aromatic hydrocarbons (PAHs) generated in a low-density polyethylene fire (pyrolysis) in four temperature regions. Data from: Hawley-Fedder, R. A., Parsons, M. L., and Karasek, F. W., Products obtained during combustion of polymers under simulated incinerator conditions. *J. Chromatogr*. **314**, 263–272, 1984.

a solid waste by burning the combustibles in an incinerator. We subject combustible gases, with undesirable properties such as odors, to a high temperature in an afterburner system to convert them to less objectionable gases.

The efficient reaction can be seen as two combustion equations that are simple:

$$C + O_2 \rightarrow CO_2 \tag{10.2}$$

$$2H_2 + O_2 \rightarrow 2H_2O \tag{10.3}$$

They produce the products carbon dioxide and water, which are odorless and invisible.

The problems with the combustion reaction occur because the process also produces many other products, most of which are termed as *air pollutants*. These can be carbon monoxide, carbon dioxide, oxides of sulfur, oxides of nitrogen, smoke, fly ash, metals, metal oxides, metal salts, aldehydes, ketones, acids, polycyclic aromatic hydrocarbons (PAHs), hexachlorobenzene, dioxins, furans, volatile organic compounds (VOCs), and many others. Only in the past few decades have combustion engineers become concerned about these relatively small quantities of materials emitted from the combustion process. An automotive engineer, for example, was not overly concerned about the 1% of carbon monoxide in the exhaust of the gasoline engine. By getting this 1% to burn to carbon dioxide inside the combustion chamber, the engineer could expect an increase in gasoline mileage of something less than one-half of 1%. This 1% of carbon monoxide, however, is 10 000 ppm by volume, and a number of such magnitudes cannot be ignored by an engineer dealing with air pollution problems.

Combustion is extremely complicated but is generally considered to be a free radical chain reaction. Several reasons exist to support the free radical mechanism: (1) Simple calculations of the heats of disassociation and formation for the molecules involved do not agree with the experimental values obtained for heats of combustion. (2) A great variety of end products may be found in the exhaust from a combustion reaction. Many complicated organic molecules have been identified in the effluent from a system burning pure methane with pure oxygen. (3) Inhibitors, such as tetraethyl lead and methylcyclopentadienyl manganese tricarbonyl (MMT)[2], can greatly change the rate of reaction [3].

When visualizing a combustion process, it is useful to think of it in terms of the three Ts: time, temperature, and turbulence. Time for combustion to occur is necessary. A combustion process that is just initiated, and suddenly has its reactants discharged to a chilled environment, will not go to completion and will emit excessive pollutants. A high enough temperature must exist for the combustion reaction to be initiated. Combustion is an exothermic reaction (it gives off heat), but it also requires energy to be initiated. This is illustrated in Fig. 10.7.

[2] MMT $(CH_3C_5H_4)Mn(CO)_3$ is an octane enhancer, which was approved as a fuel additive in the US. Some decades ago, the US EPA determined that MMT, added at 1/32 grams per gallon of manganese (Mn), will not cause or contribute to regulated emissions failures of vehicles. Some have expressed concerns that the use of MMT may harm on-board diagnostic equipment (OBD) which monitors the performance of emissions control devices in the vehicle. Currently, the Agency believes the data collected is inconclusive with regard to OBD. After completing a 1994 risk evaluation on the use of MMT in gasoline, the US EPA was unable to determine if there is a risk to the public health from exposure to emissions of MMT gasoline. Like lead, manganese is also neurotoxic, so the US EPA has required the manufacturer perform testing to support a more definitive risk evaluation, including health pharmacokinetic (PK) studies and one emission characterization. Completed final reports for all of these studies have been submitted to EPA. These final reports can be found in the Federal Docket Management System (FDMS) at www.regulations.gov identified by docket number EPA-HQ-OAR-2004-0074. In addition to the already completed tests, the manufacturer is now in the process of developing physiologically based pharmacokinetic (PBPK) models for manganese which are being derived from data generated from the completed testing. The manufacturer anticipates that these PBPK models will be completed in 2008. The US EPA is presently reviewing exposure and risk information and may refine its risk evaluation or may ask for further testing based on the results of the submitted testing and resulting model now being developed, as well as any other available data. See http://www.epa.gov/otaq/regs/fuels/additive/mmt_cmts.htm.

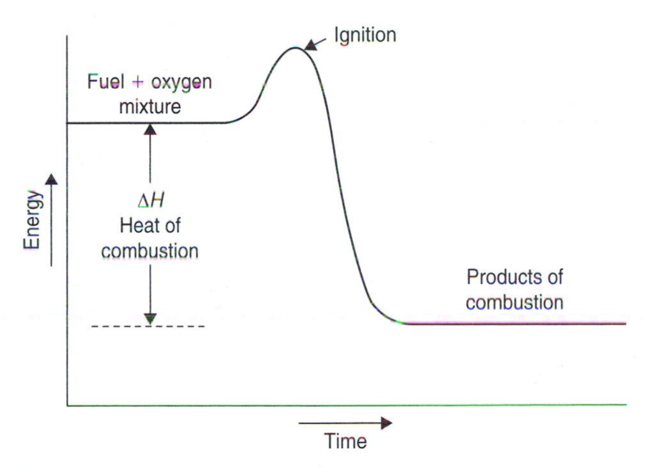

Fig. 10.7. Energies involved in combustion processes.

Turbulence is necessary to ensure that the reacting fuel and oxygen molecules in the combustion process are in intimate contact at the proper instant, when the temperature is high enough to cause the reaction to begin.

The physical state of the fuel for a combustion process dictates the type of system to be used for burning. A fuel may be composed of volatile material, fixed carbon, or both. The volatile material burns as a gas and exhibits a visible flame, whereas the fixed carbon burns without a visible flame in a solid form. If a fuel is in the gaseous state, such as natural gas, it is very reactive and can be fired with a simple burner.

If a fuel is in the liquid state, such as fuel oil, most of it must be vaporized to the gaseous state before combustion occurs. This vaporization can be accomplished by supplying heat from an outside source, but usually the liquid fuel is first atomized and then the finely divided fuel particles are sprayed into a hot combustion chamber to accomplish the gasification.

With a solid fuel, such as coal or wood, a series of steps are involved in combustion. These steps occur in a definite order, and the combustion device must be designed with these steps in mind. Figure 10.8 shows what happens to a typical solid fuel during the combustion process.

The cycle of operation of the combustion source is very important as far as emissions are concerned. A steady process, such as a large steam boiler, operates with a fairly uniform load and a continuous fuel flow. The effluent gases, along with any air pollutants, are discharged steadily and continually from the stack. An automobile engine, on the other hand, is a series of intermittent sources. The emissions from the automotive engine will be vastly different from those from the boiler in terms of both quantity and quality. A four-cylinder automotive engine operating at 2500 rpm has 5000 separate combustion processes started and completed each minute of its operation. Each of these lasts about 1/100 of a second from beginning to end.

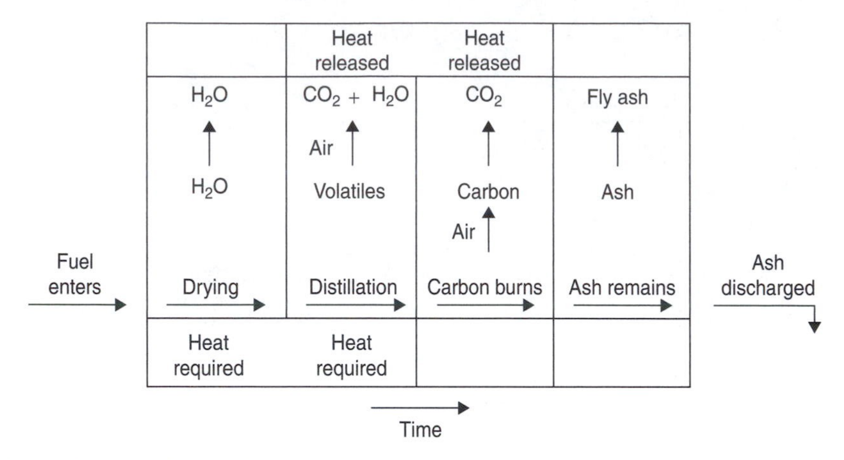

Fig. 10.8. Solid fuel combustion schematic.

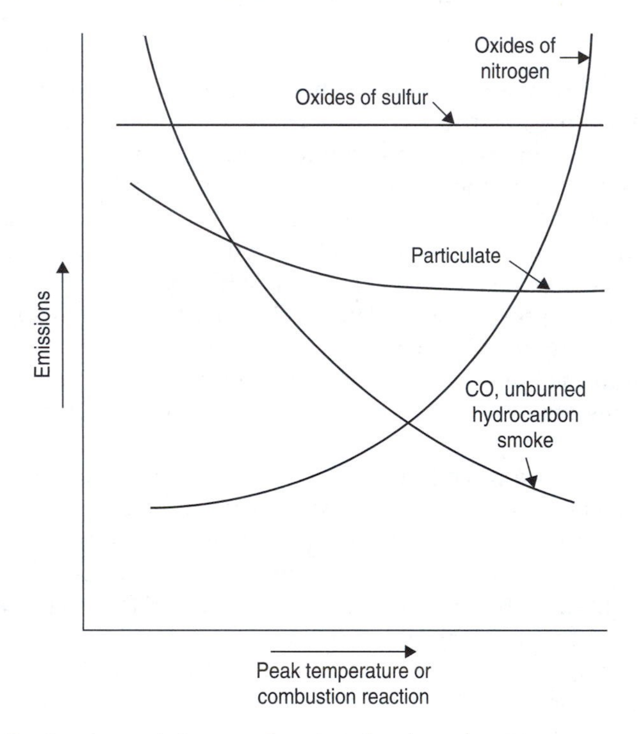

Fig. 10.9. Combustion emissions as a function of peak combustion temperatures.

The emissions from combustion processes may be predicted to some extent if the variables of the processes are completely defined. Figure 10.9 indicates how the emissions from a combustion source would be expected to vary with the temperature of the reaction. No absolute values are shown, as these will vary greatly with fuel type, independent variables of the combustion process, etc.

TABLE 10.2

Comparison of Combustion Pollutants[a]

Contaminant	Power plant emission (gm kg^{-1} fuel)			Refuse burning emission (gm kg^{-1} refuse)		Uncontrolled automotive emission (gm kg^{-1} fuel)	
	Coal	Oil	Gas	Open burning	Multiple chamber	Gasoline	Diesel
Carbon monoxide	Nil	Nil	Nil	50.0	Nil	165.0	Nil
Oxides of sulfur (SO_2)	(20)x	(20)x	(16)x	1.5	1.0	0.8	7.5
Oxides of nitrogen (NO_2)	0.43	0.68	0.16	2.0	1.0	16.5	16.5
Aldehydes and ketones	Nil	0.003	0.001	3.0	0.5	0.8	1.6
Total hydrocarbons	0.43	0.05	0.005	7.5	0.5	33.0	30.0
Total particulate	(75)y	(2.8)y	Nil	11	11	0.05	18.0

[a] x = percentage of sulfur in fuel; y = percentage of ash in fuel.

A comparison of typical emissions from various common combustion sources may be seen in Table 10.2.

III. STATIONARY SOURCES

Emissions from industrial processes are varied and often complex [4]. These emissions can be controlled by applying the best available technology. The emissions may vary slightly from one facility to another, using apparently similar equipment and processes, but in spite of this slight variation, similar control technology is usually applied [5]. For example, a method used to control the emissions from steel mill X may be applied to control similar emissions at plant Y. It should not be necessary for plant Y to spend excessive amounts for research and development if plant X has a system that is operating satisfactorily. The solution to the problem is often to look for a similar industrial process, with similar emissions, and find the type of control system used. That said, good engineering must always consider the details in any design. Thus, even seemingly small differences between two similar processes (e.g. quantity, flow rates, holding times, materials and reactor size and shape) can lead to dramatically different waste products, not to mention unforeseen safety and liability issues. Consequently, the following discussions are merely guidelines that must be adapted for each facility.

A. Chemical and Allied Products

The emissions from a chemical process can be related to the specific process. A plant manufacturing a resin might be expected to emit not only the resin being manufactured but also some of the raw material and some other products which may or may not resemble the resin. A plant manufacturing sulfuric acid can be expected to emit sulfuric acid fumes and SO_2. A plant manufacturing soap products could be expected to emit a variety of odors. Depending on the process, the emissions could be any one or a combination of dust, aerosols, fumes, or gases. The emissions may or may not be odorous or toxic. Some of the primary emissions might be innocuous but later react in the atmosphere to form an undesirable secondary pollutant. A flowchart and material balance sheet for the particular process are very helpful in understanding and analyzing any process and its emissions [6].

In any discussion of the importance of emissions from a particular process for an area, several factors must be considered—(1) the percentage of the total emissions of the area that the particular process emits, (2) the degree of toxicity of the emissions, and (3) the obvious characteristics of the source (which can be related to either sight or smell).

B. Resins and Plastics

Resins are solid or semisolid, water-insoluble,[3] organic substances with little or no tendency to crystallize. They are the basic components of plastics and are also used for coatings on paper, particleboard, and other surfaces that require a decorative, protective, or special-purpose finish. The common characteristic of resins is that heat is used in their manufacture and application, and gases are exhausted from these processes. Some of the gases that are economically recoverable may be condensed, but a large portion is lost to the atmosphere. One operation, coating a porous paper with a resin to form battery separators, was emitting to the atmosphere about 85% of the resin purchased. This resin left the stacks of the plant as a blue haze, and the odor was routinely detected more than 2 km away. Since most resins and their by-products have low-odor thresholds, disagreeable odor is the most common complaint against any operation using them.

C. Varnish and Paints

In the manufacture of varnish, heat is necessary for formulation and purification. The same may be true of operations preparing paints, shellac, inks, and other protective or decorative coatings. The compounds emitted to the atmosphere are gases, some with extremely low-odor thresholds. Acrolein, with an odor threshold of about $4000\,\mu g\,m^{-3}$, and reduced sulfur compounds, with odor thresholds of $2\,\mu g\,m^{-3}$, are both possible emissions from varnish cooking operations. The atmospheric emissions from varnish cooking appear

[3] However, the reactants and other materials used to produce or modify resins and plastics (e.g. polymers) may include chemicals with high aqueous solubility, such as the neurotoxic compound, acrylamide ($650\,000\,mg\,L^{-1}$). This means that such ancillary chemicals can find their way to the air via water vapor or water films on particles. They may also move to other environmental compartments (e.g. contaminating surface and ground water).

to have little or no recovery value, whereas some of the solvents used in paint preparation are routinely condensed for recovery and returned to the process. If a paint finish is baked to harden the surface by removal of organic solvents, the solvents must either be recovered, destroyed, or emitted to the atmosphere. The last course, emission to the atmosphere, is undesirable and may be prohibited by the air pollution control agency.

D. Acid Manufacture

Acids are used as basic raw materials for many chemical processes and manufacturing operations. Figure 10.10 illustrates an acid plant with its flow diagram. Sulfuric acid is one of the major inorganic chemicals in modern industry. The atmospheric discharges from a sulfuric acid plant can be expected to contain gases including SO_2 and aerosol mists, containing SO_3 and H_2SO_4, in the submicron to 10-μm size range. The aerosol mists are particularly damaging to paint, vegetation, metal, and synthetic fibers.

Other processes producing acids, such as nitric, acetic, and phosphoric acids, can be expected to produce acid mists from the processes themselves as well as various toxic and nontoxic gases. The particular process must be thoroughly studied to obtain a complete listing of all the specific emissions.

E. Soaps and Detergents

Soaps are made by reacting fats or oils with a base. Soaps are produced in a number of grades and types. They may be liquid, solid, granules, or powder. The air pollution problems of soap manufacture are primarily odors from the chemicals, greases, fats, and oils, although particulate emissions may

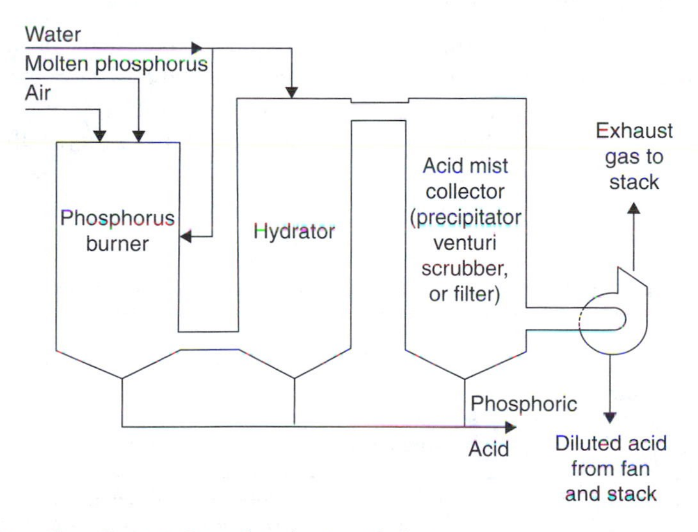

Fig. 10.10. Flow diagram for a phosphoric acid plant.

occur during drying and handling operations. Detergents are manufactured from base stocks similar to those used in petroleum refineries, so the air pollution problems are similar to those of refineries.

F. Phosphate Fertilizers

Phosphate fertilizers are prepared by beneficiation of phosphate rock to remove its impurities, followed by drying and grinding. The PO_4 in the rock may then be reacted with sulfuric acid to produce normal superphosphate fertilizer. Over 100 plants operating in the United States produce approximately a billion kilograms of phosphate fertilizer per year. Figure 10.11 is a flow diagram for a normal superphosphate plant which notes the pollutants emitted. The particulate and gaseous fluoride emissions cause greatest concern near phosphate fertilizer plants.

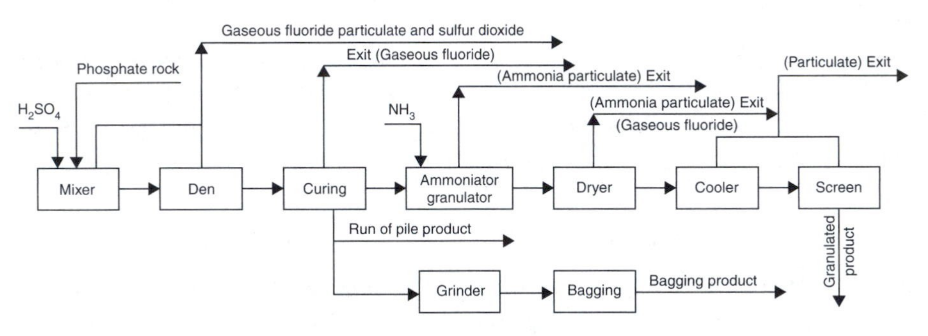

Fig. 10.11. Flow diagram for a normal superphosphate plant.

G. Other Inorganic Chemicals

Production of the large quantities of inorganic chemicals necessary for modern industrial processes can result in air pollutant emissions as undesirable by-products. Table 10.3 lists some of the more common inorganic chemicals produced, along with the associated air pollutants potentially emitted from the specific process [6].

TABLE 10.3

Miscellaneous Inorganic Chemicals and Associated Air Pollution Emissions

Inorganic chemical produced	Associated air pollution emissions
Calcium oxide (lime)	Lime dust
Sodium carbonate (soda ash)	Ammonia—soda ash dust
Sodium hydroxide (caustic soda)	Ammonia—caustic dust and mist
Ammonium nitrate	Ammonia—nitric oxides
Chlorine	Chlorine gas, hydrochloric acid (HCI)
Bromine	Bromine gas and compounds, chlorine gas, radionuclides

H. Petroleum and Coal

Petroleum and coal supply the majority of the energy in all industrial countries. This fact gives an indication of the vast quantities of materials handled and also hints at the magnitude of the air pollution problems associated with obtaining the resource, transporting it, refining it, and transporting it again. The emission problems from burning fossil fuel have been previously discussed.

1. Petroleum

Petroleum products are obtained from crude oil. In the process of getting the crude oil from the ground to the refinery, many possibilities for emission of hydrocarbon and reduced sulfur gaseous emissions occur. In many cases, these operations take place in relatively remote regions and affect only those employed by the industry, so that little or no control is attempted.

As shown in Fig. 10.12, an uncontrolled petroleum refinery is a potential source of large tonnages of atmospheric emissions. All refineries are odorous, the degree being a matter of the housekeeping practices around the refinery. Since refineries are essentially closed processes, emissions are not normally considered a part of the operation. Refineries do need pressure relief systems and vents, and emissions from them are possible. Most refineries use very strict control measures for economic as well as regulatory reasons. The recovery of 1% or 2% of a refinery throughput which was previously lost to the atmosphere can easily pay for the cost of the control equipment. The expense of the catalyst charge in some crackers and regenerators requires that the best possible control equipment be used to prevent catalyst emissions to the atmosphere.

Fig. 10.12. Uncontrolled petroleum refinery.

Potential air pollutants from a petroleum refinery could include: (1) hydrocarbons from all systems, leaks, loading, and sampling; (2) sulfur oxides from boilers, treaters, and regenerators; (3) carbon monoxide from regenerators and incinerators; (4) nitrogen oxides from combustion sources and regenerators; (5) odors from air and steam blowing, condensers, drains, and vessels; and (6) particulate matter from boilers, crackers, regenerators, coking, and incinerators.

Loading facilities must be designed to recover all vapors generated during filling of tank trucks or tanker ships. Otherwise these vapors will be lost to the atmosphere. Since they may be both odorous and photochemically reactive, serious air pollution problems could result. The collected vapors must be returned to the process or disposed of by some means.

An increasingly important concern is the likelihood of an uncontrolled fire caused by an accident or by a terrorist act at the refinery. This can be the source of large and highly toxic plumes. Thus, refineries need special measures for protection and security, coupled with vigorously diligent monitoring, operation, and maintenance. Every refinery also needs an up-to-date contingency plan.

2. Coal

The air pollution problems associated with combustion of coal are of major concern. These problems generally occur away from the coal mine. The problems of atmospheric emissions due to mining, cleaning, handling, and transportation of coal from the mine to the user are of lesser significance as far as the overall air pollution problems are concerned. Whenever coal is handled, particulate emission becomes a problem. The emissions can be either coal dust or inorganic inclusions. Control of these emissions can be relatively expensive if the coal storage and transfer facilities are located near residential areas. This is particularly problematic in developing nations. For example, China's economic and industrial expansion over the past decade has been largely supported by coal combustion (about two-thirds of energy use). This leads to three types of problems:

1. Coal is among the worst fuels in terms of production of greenhouse gases, mainly CO_2.
2. Coal often has high sulfur and ash content, so it is a major source of oxides of sulfur and particulate matter, respectively.
3. Coal contains heavy metals, particularly mercury, so it is also a major contributor to air toxics.

I. Primary Metals Industry

Metallurgical equipment has long been an obvious source of air pollution. The effluents from metallurgical furnaces are submicron-size dusts and fumes and hence are highly visible. The emissions from associated coke ovens are not only visible but odorous as well.

1. Ferrous Metals

Iron and steel industries have been concerned with emissions from their furnaces and cupolas since the industry started. Pressures for control have forced the companies to such a low level of permissible emissions that some of the older operations have been closed rather than spend the money to comply. Most of the companies controlling these operations have not gone out of business but rather have opened a new, controlled plant to replace each old plant. Table 10.4 illustrates the changes in the steelmaking processes that have occurred in the United States.

Air-polluting emissions from steelmaking furnaces include metal oxides, smoke, fumes, and dusts to make up the visible aerosol plume. They may also include gases, both organic and inorganic. If steel scrap is melted, the charge may contain appreciable amounts of oil, grease, and other combustibles that further add to the organic gas and smoke loadings. If the ore used has appreciable fluoride concentrations, the emission of both gaseous and particulate fluorides can be a serious problem.

Emissions from foundry cupolas are relatively small but still significant, in some areas. An uncontrolled 2-m cupola can be expected to emit up to 50 kg of dust, fumes, smoke, and oil vapor per hour. Carbon monoxide, oxides of nitrogen, and organic gases may also be expected. Control is possible, but the cost of the control may be prohibitive for the small foundry which only has one or two heats per week.

Steel-making is commonly associated with coke ovens. Coke is coal that has undergone pyrolysis, i.e. heated up to 1000–1400°C in the absence of oxygen, so it is not burned. This process intentionally releases gaseous components of the coal to produce nearly pure carbon [7].

Coke oven emissions are complex mixtures of gas, liquid, and solid phases, usually including a range of about 40 PAHs, as well as other products of incomplete combustion; notably formaldehyde, acrolein, aliphatic aldehydes, ammonia, carbon monoxide, nitrogen oxides, phenol, cadmium, arsenic, and

TABLE 10.4

Changes in Steel-making Processes in the United States

	Production by specific process (%)				
Year	Bessemer	Open hearth	Electric	Basic oxygen furnace	Total
1920	21	78	1	0	100
1940	6	92	2	0	100
1960	2	89	7	2	100
1970	1	36	14	48	100
1980	1	22	31	46	100
1990	0	4	37	59	100

mercury. More than 60 organic compounds have been collected near coke plants. A metric ton of coal yields up to 635 kg of coke, up to 90 kg of coke breeze (large coke particulates), 7–9 kg of ammonium sulfate, 27.5–34 L of coke oven gas tar, 55–135 L of ammonia liquor, and 8–12.5 L of light oil. Up to 35% of the initial coal charge is emitted as gases and vapors. Most of these gases and vapors are collected during by-product coke production. Coke oven gas is comprised of hydrogen, methane, ethane, carbon monoxide, carbon dioxide, ethylene, propylene, butylene, acetylene, hydrogen sulfide, ammonia, oxygen, and nitrogen. Coke oven gas tar includes pyridine, tar acids, naphthalene, creosote oil, and coal-tar pitch. Benzene, xylene, toluene, and solvent naphthas may be extracted from the light oil fraction. Coke production in the US increased steadily between 1880 and the early 1950s, peaking at 65 million metric tons in 1951. In 1976, the United States was second in the world with 48 million metric tons of coke, i.e. 14.4% of the world production. By 1990, the US produced 24 million metric tons, falling to fourth in the world. A gradual decline in production has continued; production has decreased from 20 million metric tons in 1997 to 15.2 million metric tons in 2002. Demand for blast furnace coke also has declined in recent years because technological improvements have reduced the amount of coke consumed per amount of steel produced by as much as 25%.

Obviously, the volatilized gases are air pollutants. In fact, coke ovens are the source of thousands of compounds, many that are toxic. Coke facilities have recognized that many of these gases are also economically valuable, so technologies exist and are being applied to recover, separate, and sell them for profit. Some of the processes shown in Table 10.4 no longer need coke to produce steel. This is an example of green engineering.

2. *Nonferrous Metals*

Around the turn of the century, one of the most obvious effects of industry on the environment was the complete destruction of vegetation downwind from copper, lead, and zinc smelters. This problem was caused by the smelting of the metallic sulfide ores. As the metal was released in the smelting process, huge quantities of sulfur were oxidized to SO_2, which was toxic to much of the vegetation fumigated by the plume. Present smelting systems go to great expense to prevent the uncontrolled release of SO_2, but in many areas the recovery of the ecosystem will take years and possibly centuries.

Early aluminum reduction plants were responsible for air pollution because of the fluoride emissions from their operations. Fluoride emissions can cause severe damage to vegetation and to animals feeding on such vegetation. The end result was an area surrounding the plant devoid of vegetation. Such scenes are reminiscent of those downwind from some of the uncontrolled copper smelters. New aluminum reduction plants are going to considerable expense to control fluoride emissions. Some of the older plants are finding that the cost of control will exceed the original capital investment in the entire facility. Where the problem is serious, control agencies have developed extensive sampling networks to monitor emissions from the plant of concern.

Emissions from other nonferrous metal facilities are primarily metal fumes or metal oxides of extremely small diameter. Zinc oxide fumes vary from 0.03 to 0.3 μm and are toxic. Lead and lead oxide fumes are extremely toxic and have been extensively studied. Arsenic, cadmium, bismuth, and other trace metals can be emitted from many metallurgical processes.

J. Stone and Clay Products

The industries which produce and handle various stone products emit considerable amounts of particulate matter at every stage of the operation. These particulates may include fine mineral dusts of a size to cause damage to the lungs. Depending on the type of rock, mineral fibers can also be released, notably asbestos. The threshold values for such dusts have been set quite low to prevent disabling diseases for the worker, including lung cancer, mesothelioma, pleural diseases, asbestosis, and silicosis.

In the production of clay, talc, cement, chalk, etc., an emission of particulate matter will usually accompany each process. These processes may involve grinding, drying, and sieving, which can be enclosed and controlled to prevent the emission of particles. In many cases, the recovered particles can be returned to the process for a net economic gain.

During the manufacture of glass, considerable dust, with particles averaging about 300 μm in size, will be emitted. Some dusts may also be emitted from the handling of the raw materials involved. Control of this dust to prevent a nuisance problem outside the plant is a necessity. When glass is blown or formed into the finished product, smoke and gases can be released from the contact of the molten glass with lubricated molds. These emissions are quite dense but of a relatively short duration.

K. Forest Products Industry

1. Wood Processing

Trees are classified as a renewable resource which is being utilized in most portions of the world on a sustained yield basis. A properly managed forest will produce wood for lumber, fiber, and chemicals indefinitely. Harvesting this resource can generate considerable dust and other particulates. Transportation over unpaved roads causes excessive dust generation. The cultural practice of burning the residue left after a timber harvest, called *slash burning*, is still practiced in some areas and is a major source of smoke, gaseous, and particulate air pollution in the localities downwind from the fire (see Fig. 10.13). Visibility reduction from such burning can be a serious problem.

Processing the harvested timber into the finished product may involve sawing, peeling, planing, sanding, and drying operations, which can release considerable amounts of wood fiber and lesser amounts of gaseous material to the

Fig. 10.13. Plume from slash burning in the Amazon region of South America. From the rainforests around the Amazon River (top) in Brazil, through the central highlands and into Bolivia to the southwest, numerous fires were burning throughout the region on September 8, 2002. The fires were detected by the Moderate Resolution Imaging Spectroradiometer (MODIS) on NASA's Terra satellite, and their locations are marked with red dots in this true-color image. Thick smoke and clouds are shrouding the highlands in the southern portion of the image. *Source*: Photo by J. Allen, based on data from the National Aeronautics and Space Administration, 2005, Goddard Space Flight, Space Visualization Studio, Maryland.

atmosphere. Control of wood fiber emissions from the pneumatic transport and storage systems can be a major problem of considerable expense for a plywood mill or a particleboard plant.

2. Pulp and Paper

Pulp and paper manufacture is increasing in the world at an exponential rate. The demand for paper will continue as new uses are found for this product. Since most paper is manufactured from wood or wood residue, it provides an excellent use for this renewable resource.

The most widely used pulping process is the kraft process, as shown in Fig. 10.14, which results in recovery and regeneration of the chemicals. This occurs in the recovery furnace, which operates with both oxidizing and reducing zones. Emissions from such recovery furnaces include particulate matter, very odorous reduced sulfur compounds, and oxides of sulfur. Bleaching has

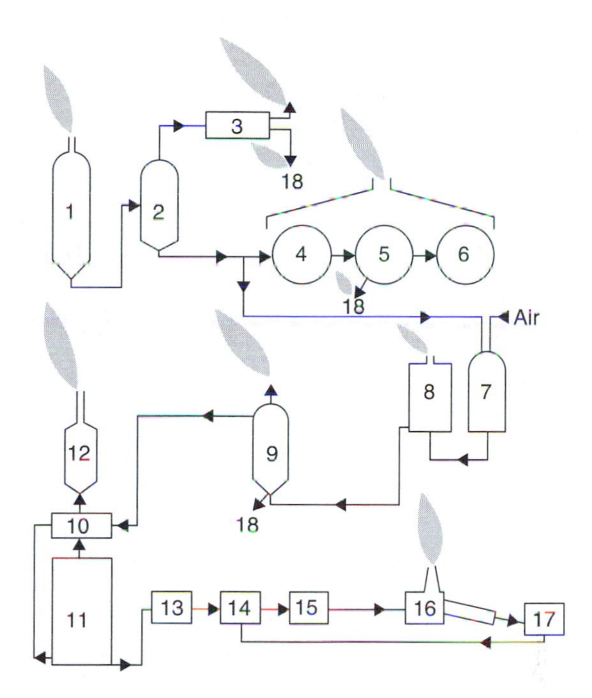

Fig. 10.14. Schematic diagram of the kraft pulping process [6]. 1, digester; 2, blow tank; 3, blow heat recovery; 4, washers; 5, screens; 6, dryers; 7, oxidation tower; 8, foam tank; 9, multiple effect evaporator; 10, direct evaporator; 11, recovery furnace; 12, electrostatic precipitator; 13, dissolver; 14, causticizer; 15, mud filter; 16, lime kiln; 17, slaker; 18, sewer.

been a substantial source of dioxins, but many processors have now modified their operations to use substantially less chlorine, greatly reducing the dioxin formation. If extensive and expensive control is not exercised over the kraft pulp process, the odors and aerosol emissions will affect a wide area. Odor complaints have been reported over 100 km away from these plants. A properly controlled and operated kraft plant will handle huge amounts of material and produce millions of kilograms of finished products per day, with little or no complaint regarding odor or particulate emissions.

L. Noxious Trades

As the name implies, these operations, if uncontrolled, can cause a serious air pollution problem. The main problem is the odors associated with the process. Examples of such industries are tanning works, confined animal feeding operations (CAFOs), rendering plants, and many of the food processing plants such as fish meal plants. In most cases, the emissions of particulates and gases from such plants are not of concern, only the odors. However, other air pollutants are also common (e.g. chromium in tanning and nitric oxide (NO) near CAFOs). Requiring these industries to locate away from the business or residential areas is no longer acceptable as a means of control.

TABLE 10.5

Emissions from Mobile Sources

Power plant type	Fuel	Major emissions	Vehicle type
Otto cycle	Gasoline	HC, CO, CO_2, NO_x	Auto, truck, bus, aircraft, marine, motorcycle, tractor
Two-stroke cycle	Gasoline	HC, CO, CO_2, NO_x, particulate	Motorcycle, outboard motor
Diesel	Diesel oil	NO_x, particulate, SO_x, CO_2	Auto, truck, bus, railroad, marine, tractor
Gas turbine (jet)	Turbine	NO_x, particulate, CO_2	Aircraft, marine, railroad
Steam	Oil, coal	NO_x, SO_x, particulate, CO_2	Marine

IV. MOBILE SOURCES

A mobile source of air pollution can be defined as one capable of moving from one place to another under its own power. According to this definition, an automobile is a mobile source and a portable asphalt batching plant is not. The regulatory definition seems to center around the internal combustion engine. Generally, mobile sources imply transportation, but sources such as construction equipment, gasoline-powered lawn mowers, and gasoline-powered tools are included in this category.

Mobile sources therefore consist of many different types of vehicles, powered by engines using different cycles, fueled by a variety of products, and emitting varying amounts of both simple and complex pollutants. Table 10.5 includes the more common mobile sources.

The predominant mobile air pollution source in all industrialized countries of the world is the automobile, powered by a four-stroke cycle (Otto cycle) engine and using gasoline as the fuel. In the United States, over 85 million automobiles were in use in 1990. If the 15 million gasoline-powered trucks and buses and the 4 million motorcycles are included, the United States total exceeds 100 million vehicles. The engine used to power these millions of vehicles has been said to be the most highly engineered machine of the century. When one considers the present reliability, cost, and life expectancy of the internal combustion engine, it is not difficult to see why it has remained so popular. A modern automotive engine traveling 100 000 km will have about 2.5×10^8 power cycles.

The emissions from a gasoline-powered vehicle come from many sources. Figure 10.15 illustrates what might be expected from an uncontrolled (1960 model) automobile and a controlled (1983 or later model) automobile if it complies with the 1983 federal standards [7]. With most of today's automobiles using unleaded gasoline, lead emissions are no longer a major concern in the West, but are still a problem in numerous other parts of the world.

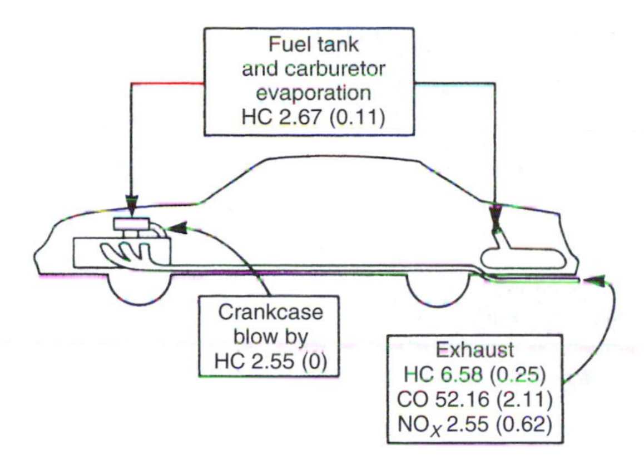

Fig. 10.15. Emissions from uncontrolled automobiles (and those meeting US EPA standards) in $g\,km^{-1}$.

V. AIR TOXICS SOURCES

Hazardous air pollutants are known in most air pollution control communities as toxic air pollutants or air toxics. These are the pollutants that cause or are strongly suspected to cause cancer or other serious health effects, including reproductive and developmental effects or birth defects. They are also often associated with adverse environmental and ecological effects. The Clean Air Act Amendments of 1990 require the US Environmental Protection Agency (EPA) to control 188 air toxics (see Table 10.6). Examples of toxic air pollutants include benzene, which is found in gasoline; perchloroethlyene, which is emitted from some dry cleaning facilities; and methylene chloride, which is used as a solvent and paint stripper by a number of industries.

The Clean Air Act allows for petitions to modify the list (e.g. in 2006, caprolactam, hydrogen sulfide, and methyl ethyl ketone were undergoing modification).

Most air toxics originate from human-made sources, including mobile and stationary sources, similar to the criteria pollutants discussed in this chapter. However, many are predominantly or nearly exclusively indoor air pollutants.

Two types of stationary sources routinely emit air toxics:

1. *Major sources* emit 10 tons per year of any of the listed toxic air pollutants, or 25 tons per year of a mixture of air toxics. Major sources may release air toxics from equipment leaks, when materials are transferred from one location to another, or during discharge through emission stacks or vents.
2. *Area sources* are smaller-size facilities that release lesser quantities of air toxics. These sources emit less than 10 tons per year of a single air toxic, or less than 25 tons per year of a combination of air toxics. Though emissions from individual area sources are often relatively small, collectively their emissions can be of concern—especially in heavily populated communities with many sources.

TABLE 10.6

Air Toxics Regulated by the US EPA

CAS number	Toxic air pollutant	Notes
75070	Acetaldehyde	
60355	Acetamide	
75058	Acetonitrile	
98862	Acetophenone	
53963	2-Acetylaminofluorene	
107028	Acrolein	
79061	Acrylamide	
79107	Acrylic acid	
107131	Acrylonitrile	
107051	Allyl chloride	
92671	4-Aminobiphenyl	
62533	Aniline	
90040	*o*-Anisidine	
1332214	Asbestos	
71432	Benzene (including benzene from gasoline)	
92875	Benzidine	
98077	Benzotrichloride	
100447	Benzyl chloride	
92524	Biphenyl	
117817	Bis(2-ethylhexyl)phthalate (DEHP)	
542881	Bis(chloromethyl)ether	
75252	Bromoform	
106990	1,3-Butadiene	
156627	Calcium cyanamide	
105602	Caprolactam	Visit EPA website for current status of modification: http://www.epa.gov/ttn/atw/pollutants/atwsmod.html
133062	Captan	
63252	Carbaryl	
75150	Carbon disulfide	
56235	Carbon tetrachloride	
463581	Carbonyl sulfide	
120809	Catechol	
133904	Chloramben	
57749	Chlordane	
7782505	Chlorine	
79118	Chloroacetic acid	
532274	2-Chloroacetophenone	
108907	Chlorobenzene	
510156	Chlorobenzilate	
67663	Chloroform	
107302	Chloromethyl methyl ether	
126998	Chloroprene	
1319773	Cresols/Cresylic acid (isomers and mixture)	

(*continued*)

TABLE 10.6 (*Continued*)

CAS number	Toxic air pollutant	Notes
95487	*o*-Cresol	
108394	*m*-Cresol	
106445	*p*-Cresol	
98828	Cumene	
94757	2,4-D, salts and esters	
3547044	DDE	
334883	Diazomethane	
132649	Dibenzofurans	
96128	1,2-Dibromo-3-chloropropane	
84742	Dibutylphthalate	
106467	1,4-Dichlorobenzene(*p*)	
91941	3,3-Dichlorobenzidene	
111444	Dichloroethyl ether (bis(2-chloroethyl)ether)	
542756	1,3-Dichloropropene	
62737	Dichlorvos	
111422	Diethanolamine	
121697	*N,N*-Diethyl aniline (*N,N*-dimethylaniline)	
64675	Diethyl sulfate	
119904	3,3-Dimethoxybenzidine	
60117	Dimethyl aminoazobenzene	
119937	3,3′-Dimethyl benzidine	
79447	Dimethyl carbamoyl chloride	
68122	Dimethyl formamide	
57147	1,1-Dimethyl hydrazine	
131113	Dimethyl phthalate	
77781	Dimethyl sulfate	
534521	4,6-Dinitro-*o*-cresol, and salts	
51285	2,4-Dinitrophenol	
121142	2,4-Dinitrotoluene	
123911	1,4-Dioxane (1,4-diethyleneoxide)	
122667	1,2-Diphenylhydrazine	
106898	Epichlorohydrin (l-chloro-2,3-epoxypropane)	
106887	1,2-Epoxybutane	
140885	Ethyl acrylate	
100414	Ethyl benzene	
51796	Ethyl carbamate (urethane)	
75003	Ethyl chloride (chloroethane)	
106934	Ethylene dibromide (dibromoethane)	
107062	Ethylene dichloride (1,2-dichloroethane)	
107211	Ethylene glycol	
151564	Ethylene imine (aziridine)	
75218	Ethylene oxide	
96457	Ethylene thiourea	
75343	Ethylidene dichloride (1,1-dichloroethane)	

(*continued*)

TABLE 10.6 (*Continued*)

CAS number	Toxic air pollutant	Notes
50000	Formaldehyde	
76448	Heptachlor	
118741	Hexachlorobenzene	
87683	Hexachlorobutadiene	
77474	Hexachlorocyclopentadiene	
67721	Hexachloroethane	
822060	Hexamethylene-1,6-diisocyanate	
680319	Hexamethylphosphoramide	
110543	Hexane	
302012	Hydrazine	
7647010	Hydrochloric acid	
7664393	Hydrogen fluoride (hydrofluoric acid)	
7783064	Hydrogen sulfide	Visit EPA website for current status of modification: http://www.epa.gov/ttn/atw/pollutants/atwsmod.html
123319	Hydroquinone	
78591	Isophorone	
58899	Lindane (all isomers)	
108316	Maleic anhydride	
67561	Methanol	
72435	Methoxychlor	
74839	Methyl bromide (bromomethane)	
74873	Methyl chloride (chloromethane)	
71556	Methyl chloroform (1,1,1-trichloroethane)	
78933	Methyl ethyl ketone (2-butanone)	Visit EPA website for current status of modification: http://www.epa.gov/ttn/atw/pollutants/atwsmod.html
60344	Methyl hydrazine	
74884	Methyl iodide (iodomethane)	
108101	Methyl isobutyl ketone (hexone)	
624839	Methyl isocyanate	
80626	Methyl methacrylate	
1634044	Methyl tert butyl ether	
101144	4,4-Methylene bis(2-chloroaniline)	
75092	Methylene chloride (dichloromethane)	
101688	Methylene diphenyl diisocyanate (MDI)	
101779	4,4-Methylenedianiline	
91203	Naphthalene	
98953	Nitrobenzene	
92933	4-Nitrobiphenyl	
100027	4-Nitrophenol	
79469	2-Nitropropane	
684935	*N*-Nitroso-*N*-methylurea	

(*continued*)

TABLE 10.6 (*Continued*)

CAS number	Toxic air pollutant	Notes
62759	*N*-Nitrosodimethylamine	
59892	*N*-Nitrosomorpholine	
56382	Parathion	
82688	Pentachloronitrobenzene (quintobenzene)	
87865	Pentachlorophenol	
108952	Phenol	
106503	*p*-Phenylenediamine	
75445	Phosgene	
7803512	Phosphine	
7723140	Phosphorus	
85449	Phthalic anhydride	
1336363	Polychlorinated biphenyls (aroclors)	
1120714	1,3-Propane sultone	
57578	beta-Propiolactone	
123386	Propionaldehyde	
114261	Propoxur (baygon)	
78875	Propylene dichloride (1,2-dichloropropane)	
75569	Propylene oxide	
75558	1,2-Propylenimine (2-methyl aziridine)	
91225	Quinoline	
106514	Quinone	
100425	Styrene	
96093	Styrene oxide	
1746016	2,3,7,8-Tetrachlorodibenzo-*p*-dioxin	
79345	1,1,2,2-Tetrachloroethane	
127184	Tetrachloroethylene (perchloroethylene)	
7550450	Titanium tetrachloride	
108883	Toluene	
95807	2,4-Toluene diamine	
584849	2,4-Toluene diisocyanate	
95534	*o*-Toluidine	
8001352	Toxaphene (chlorinated camphene)	
120821	1,2,4-Trichlorobenzene	
79005	1,1,2-Trichloroethane	
79016	Trichloroethylene	
95954	2,4,5-Trichlorophenol	
88062	2,4,6-Trichlorophenol	
121448	Triethylamine	
1582098	Trifluralin	
540841	2,2,4-Trimethylpentane	
108054	Vinyl acetate	
593602	Vinyl bromide	
75014	Vinyl chloride	

(*continued*)

TABLE 10.6 (*Continued*)

CAS number	Toxic air pollutant	Notes
75354	Vinylidene chloride (1,1-dichloroethylene)	
1330207	Xylenes (isomers and mixture)	
95476	*o*-Xylenes	
108383	*m*-Xylenes	
106423	*p*-Xylenes	
0	Antimony compounds	
0	Arsenic compounds (inorganic including arsine)	
0	Beryllium compounds	
0	Cadmium compounds	
0	Chromium compounds	
0	Cobalt compounds	
0	Coke oven emissions	
0	Cyanide compounds	X′CN where X = H′ or any other group where a formal dissociation may occur. For example KCN or $Ca(CN)_2$
0	Glycol ethers	Includes mono- and di-ethers of ethylene glycol, diethylene glycol, and triethylene glycol R-$(OCH_2CH_2)n$-OR′ where
0	Lead compounds	
0	Manganese compounds	
0	Mercury compounds	
0	Fine mineral fibers	Includes mineral fiber emissions from facilities manufacturing or processing glass, rock, or slag fibers (or other mineral derived fibers) of average diameter 1 μm or less
0	Nickel compounds	
0	Polycyclic organic matter	Includes organic compounds with more than one benzene ring, and which have a boiling point greater than or equal to 100°C
0	Radionuclides (including radon)	A type of atom which spontaneously undergoes radioactive decay
0	Selenium compounds	

Notes: For all listings above which contain the word "compounds" and for glycol ethers, the following applies: unless otherwise specified, these listings are defined as including any unique chemical substance that contains the named chemical (i.e., antimony, arsenic, etc.) as part of that chemical's infrastructure. n = 1, 2, or 3; R = alkyl or aryl groups; and R′ = R, H, or groups which, when removed, yield glycol ethers with the structure: R-$(OCH_2CH)n$-OH. Polymers are excluded from the glycol category.

The 1990 Clean Air Act Amendments direct the EPA to set standards for all major sources of air toxics (and some area sources that are of particular concern). The EPA published the initial list of source categories in 1992 in the *Federal Register* (57FR31576, July 16, 1992), which has since been revised and updated frequently. The list designates whether the sources are considered to be major or area sources.

VI. EMISSION INVENTORY

An *emission inventory* is a list of the amount of pollutants from all sources entering the air in a given time period. The boundaries of the area are fixed [8].

The tables of emission inventory are very useful to control agencies as well as planning and zoning agencies. They can point out the major sources whose control can lead to a considerable reduction of pollution in the area. They can be used with appropriate mathematical models to determine the degree of overall control necessary to meet ambient air quality standards. They can be used to indicate the type of sampling network and the locations of individual sampling stations if the areas chosen are small enough. For example, if an area uses very small amounts of sulfur-bearing fuels, establishing an extensive SO_2 monitoring network in the area would not be an optimum use of public funds. Emission inventories can be used for publicity and political purposes: "If natural gas cannot meet the demands of our area, we will have to burn more high-sulfur fuel, and the SO_2 emissions will increase by 8 tons per year."

The method used to develop the emission inventory does have some elements of error, but the other two alternatives are expensive and subject to their own errors. The first alternative would be to monitor continually every major source in the area. The second method would be to monitor continually the pollutants in the ambient air at many points and apply appropriate diffusion equations to calculate the emissions. In practice, the most informative system would be a combination of all three, knowledgeably applied.

The US Clean Air Act Amendments of 1990 [9] strengthened the emission inventory requirements for plans and permits in nonattainment areas. The amendments state:

> INVENTORY—Such plan provisions shall include a comprehensive, accurate, current inventory of actual emissions from all sources of the relevant pollutant or pollutants in such area, including such periodic revisions as the Administrator may determine necessary to assure that the requirements of this part are met.
>
> IDENTIFICATION AND QUANTIFICATION—Such plan provisions shall expressly identify and quantify the emissions, if any, of any such pollutant or pollutants which will be allowed, from the construction and operation of major new or modified stationary sources in each such area. The plan shall demonstrate to the satisfaction of the Administrator that the emissions quantified for this purpose will be consistent with the achievement of reasonable further progress and will not interfere with the attainment of the applicable national ambient air quality standard by the applicable attainment date.

A. Inventory Techniques

To develop an emission inventory for an area, one must (1) list the types of sources for the area, such as cupolas, automobiles, and home fireplaces;

(2) determine the type of air pollutant emission from each of the listed sources, such as particulates and SO_2; (3) examine the literature [10] to find valid emission factors for each of the pollutants of concern (e.g. "particulate emissions for open burning of tree limbs and brush are 10 kg per ton of residue consumed"); (4) through an actual count, or by means of some estimating technique, determine the number and size of specific sources in the area (the number of steelmaking furnaces can be counted, but the number of home fireplaces will probably have to be estimated); and (5) multiply the appropriate numbers from (3) and (4) to obtain the total emissions and then sum the similar emissions to obtain the total for the area.

A typical example will illustrate the procedure. Suppose we wish to determine the amount of carbon monoxide from oil furnaces emitted per day, during the heating season, in a small city of 50 000 population:

1. The source is oil furnaces within the boundary area of the city.
2. The pollutant of concern is carbon monoxide.
3. Emission factors for carbon monoxide are listed in various ways [10] (240 g per 1000 L of fuel oil, 50 g per day per burner, 1.5% by volume of exhaust gas, etc.). For this example, use 240 g per 1000 L of fuel oil.
4. Fuel oil sales figures, obtained from the local dealers association, average 40 000 L per day.
5. $\dfrac{240\,\text{g CO}}{1000\,\text{L}} \times \dfrac{40\,000\,\text{L}}{\text{day}} = 9.6\,\text{kg CO/day}$

B. Emission Factors

Valid emission factors for each source of pollution are the key to the emission inventory. It is not uncommon to find emission factors differing by 50%, depending on the researcher, variables at the time of emission measurement, etc. Since it is possible to reduce the estimating errors in the inventory to $\pm10\%$ by proper statistical sampling techniques, an emission factor error of 50% can be overwhelming. It must also be realized that an uncontrolled source will emit at least 10 times the amount of pollutants released from one operating properly with air pollution control equipment installed.

Actual emission data are available from many handbooks, government publications, and literature searches of appropriate research papers and journals. In addition, online support is available, especially the Technology Transfer Network/Clearinghouse for Inventories and Emissions Factors, accessible at http://www.epa.gov/ttn/chief/index.html. This site provides information on emission factors, inventory, modeling, along with a knowledge base for emissions monitoring.

It is always wise to verify the data, if possible, as to the validity of the source and the reasonableness of the final number. Some emission factors, which have been in use for years, were only rough estimates proposed by someone years ago to establish the order of magnitude of the particular source.

Emission factors must be also critically examined to determine the tests from which they were obtained. For example, carbon monoxide from an automobile will vary with the load, engine speed, displacement, ambient temperature, coolant temperature, ignition timing, carburetor adjustment, engine condition, etc. However, in order to evaluate the overall emission of carbon monoxide to an area, we must settle on an average value that we can multiply by the number of cars, or kilometers driven per year, to determine the total carbon monoxide released to the area.

C. Data Gathering

To compile the emission inventory requires a determination of the number and types of units of interest in the study area. It would be of interest, for example, to know the number of automobiles in the area and the number of kilometers each was driven per year. This figure would require considerable time and expense to obtain. Instead, it can be closely approximated by determining the liters of gasoline sold in the area during the year. Since a tax is collected on all gasoline sold for highway use, these figures can be obtained from the tax collection office.

Data regarding emissions are available from many sources. Sometimes the same item may be checked by asking two or more agencies for the same information. An example of this would be to check the liters of gasoline sold in a county by asking both the tax office and the gasoline dealers association. Sources of information for an emission inventory include: (1) city, county, and state planning commissions; (2) city, county, and state chambers of commerce; (3) city, county, and state industrial development commissions; (4) census bureaus; (5) national associations such as coal associations; (6) local associations such as the County Coal Dealers Association; (7) individual dealers or distributors of oil, gasoline, coal, etc.; (8) local utility companies; (9) local fire and building departments; (10) data gathered by air pollution control agencies through surveys, sampling, etc.; (11) traffic maps; and (12) insurance maps.

D. Data Reduction and Compilation

The final emission inventory can be prepared on a computer. This will enable the information to be stored on magnetic tape or disk so that it can be updated rapidly and economically as new data or new sources appear. The computer program can be written so that changes can easily be made. There will be times when major changes occur and the inventory must be completely changed. Imagine the change that would take place when natural gas first becomes available in a commercial–residential area which previously used oil and coal for heating.

To determine emission data, as well as the effect that fuel changes would produce, it is necessary to use the appropriate thermal conversion factor from one fuel to another. Table 10.7 lists these factors for fuels in common use.

TABLE 10.7

Thermal Conversion Factors for Fuels

Fuel	Joule $\times 10^6$
Bituminous coal	30.48 per kg
Anthracite coal	29.55 per kg
Wood	20.62 per kg
Distillate fuel oil	38.46 per kg
Residual fuel oil	41.78 per L
Natural gas	39.08 per m^3
Manufactured gas	20.47 per m^3

A major change in the emissions for an area will occur if control equipment is installed. This can be shown in the emission inventory to illustrate the effect on the community.

By keeping the emission inventory current and updating it at least yearly as fuel uses change, industrial and population changes occur, and control equipment is added, a realistic record for the area is obtained.

VII. AN INTERNATIONAL PERSPECTIVE: DIFFERENCES IN TIME AND SPACE

Western Civilization is often criticized for its disproportionate demand for fuel, especially nonrenewable fossil fuels. The way that a nation addresses energy use is an important measure of how advanced it is, not only in dealing with pollution, but the level of sophistication of its economic systems. As evidence, many poorer nations are confronted with the choice of saving sensitive habitat or allowing large-scale biomass burns. And, the combustion processes in developing countries are usually much less restrictive than those in more developed nations.

This chapter addresses some of the major sources of air pollution from the perspective of the highly industrialized nations. However, industrial processes can vary significantly between developed and underdeveloped nations. For example, in Canada and the United States, the major sources of dioxins, a chemical group comprising some of the world's most toxic and carcinogenic compounds, range from dispersed activities, such as trash burning, to heavy industry (see Fig. 10.16). The major sources of dioxin emissions in Latin America are distributed quite differently than those in Canada and US. Actual emission inventories are being developed, but preliminary information indicates that much of the dioxin produced in Mexico, for example, is from backyard burning, such as neighborhood scale brick making. Refractory in developing nations are often small-scale, neighborhood operations. Often, the heat sources used to reach refractory temperatures are furnaces with

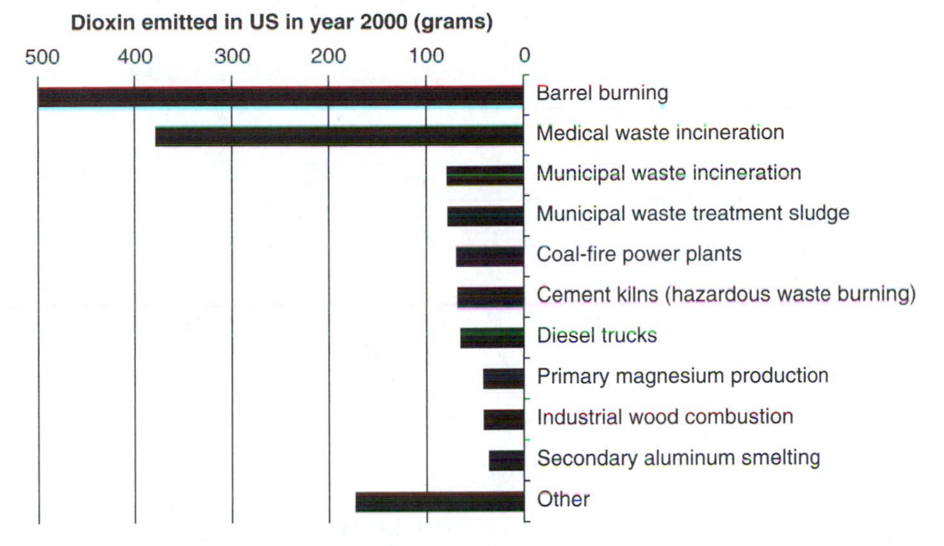

Fig. 10.16. Industrial categories of dioxin emitters in the United States in 2000. *Source*: US Environmental Protection Agency, 2005, The Inventory of Sources of Dioxin in the United States (External Review Draft); accessed on June 12, 2006 at: http://cfpub.epa.gov/ncea/cfm/recordisplay.cfm?deid=132080.

scrapped materials as fuel, especially petroleum-derived substances like automobile tires. This not only is important to the country in which the combustion occurs, but can be an international concern when the burning is near borders. This is the case for the metropolitan area of El Paso, Texas, and Ciudad Juarez, Mexico, with a combined population of two million. The cities are located in a deep canyon between two mountain ranges, which can contribute to thermal inversions in the atmosphere (see Fig. 10.17). The air quality has been characterized by the US EPA as seriously polluted, with brick making on the Mexican side identified as a major source.[4]

In Mexico, workers who make bricks are called *ladrilleros*. Many *ladrilleros* live in unregulated settlements known as *colonias* on the outskirts of Ciudad Juarez. The kilns, using the same design as those in Egypt thousands of years ago, are located within these neighborhoods, next to the small houses. The *ladrilleros* are not particular about the fuel, burning anything with caloric value, including scrap wood and old tires, as well as more conventional fuels like methane and butane. The dirtier fuels, like the tires, release large black plumes of smoke that contains a myriad of contaminants.

Children are at elevated risk of health problems when exposed to these plumes, since their lungs and other organs are undergoing prolific tissue growth. Thus, the *ladrilleros'* families have particularly elevated risks due to

[4] The major source of information about Rio Grande brick making is *Environmental Health Perspectives*, Vol. 104, Number 5, May 1996.

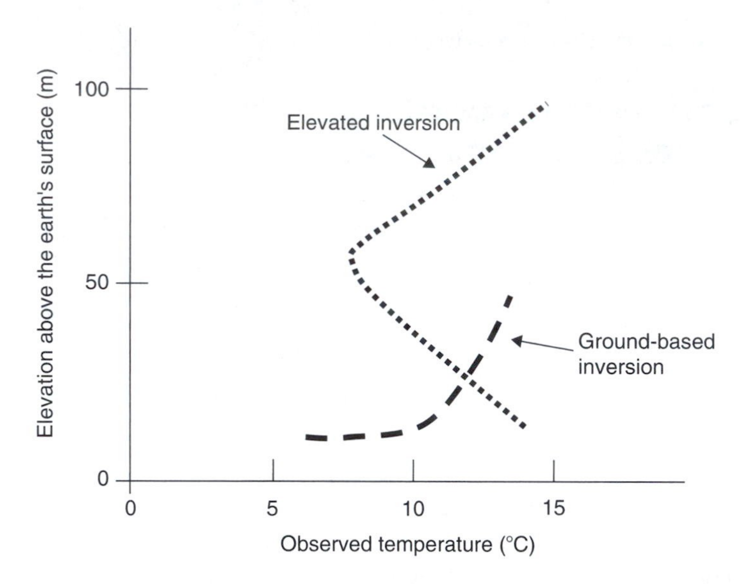

Fig. 10.17. Two types of thermal inversions that contribute to air pollution.

their frequent and high dose exposures. "The health impact is not only of concern to the worker but also the entire family, especially pregnant women and children who, because of their socioeconomic status, tend to be undernourished," according to Beatriz Vera, project coordinator for the US–Mexico Border Environment and Health Projects. She adds that "many times the entire family participates in the process. Sometimes children are put directly into the area where the kiln is fired."

The two nations' governments appear to recognize the problem, as do numerous nongovernmental organizations (NGOs). These have included Environmental Defense (ED), Physicians for Social Responsibility, the Federacion Mexicana de Asociaciones Privadas de Salud y Desarrollo Comunitario (FEMAP), and El Paso Natural Gas (EPNG). For example, FEMAP and EPNG offer courses to the *ladrilleros* from throughout the region on ways to use higher quality fuel, including improved safety and business practices as well. Often, however, even if the brick makers know about cleaner fuels, they cannot afford them. For example, they have used butane, but in 1994 the Mexican government started to phase out its subsidy and about the same time the peso was devalued, leading to a sharp increase in butane costs. The *ladrilleros* were forced to return to using the cheaper fuels. In the meantime the Mexican government banned the burning of tires, so much of the more recent tire burning has been done surreptitiously at night.

A number of solutions to the problem have been proposed, including more efficient kilns. However, arguably the best approach is to prevent the combustion in the first place. In fact, many of the traditional villages where bricks are now used were previously constructed with adobe. A return to such a noncombustion approach could hold the key. The lesson here is that often in developing countries the simpler, "low-tech" solutions are the most sustainable.

In the mid-1990s, the US EPA and the Texas Natural Resource Conservation Commission (TNRCC) conducted a study in the Rio Grande valley region to address concerns about the potential health impact of local air pollutants, and especially since little air quality information was available at the time. There are numerous "cottage industries," known as *maquiladoras*,[5] along both sides of the Texas–Mexico border as ascribed by the Rio Grande. In particular, the study addressed the potential for air pollution to move across the US–Mexican border into the southern part of Texas. Air pollution and weather data were collected for a year at three fixed sites near the border in and near Brownsville, Texas. The study found overall levels of air pollution to be similar to or even lower than other urban and rural areas in Texas and elsewhere and that transport of air pollution across the border did not appear to adversely impact air quality across the US border.

Many developing countries are evolving into industrialized nations. Not long ago in the US, the standard means of getting rid of household trash was the daily burn. Each evening, people in rural areas, small towns, and even larger cities, made a trip to the backyard, dumped the trash they had accumulated into a barrel,[6] and burned the contents. Also, burning was a standard practice elsewhere, such as intentional fires to remove brush, and the previously mentioned "cottage industries" like backyard smelters and metal recycling operations. Beginning in the 1960s and 1970s, the public acceptance

[5] The Coalition for Justice in the Maquiladoras, a cross-border group that organizes maquiladora workers, traces the term maquiladora to *maquilar*, a popular form of the verb maquinar that roughly means "to submit something to the action of a machine," as when rural Mexicans speak of maquilar with regard to the grain that is transported to a mill for processing. The farmer owns the grain; yet someone else owns the mill who keeps a portion of the value of the grain for milling. So, the origin of maquiladora can be found in this division of labor. The term has more recently been applied to the small factories opened by US companies to conduct labor-intensive jobs on the Mexican side of the border. Thus, maquilar has changed to include this process of labor, especially assembling parts from various sources, and the maquiladoras are those small assembling operations along the border.

While the maquiladoras have provided opportunities to entrepreneurs along the Mexico–US border, they have also given opportunity for the workers and their families to be exploited in the interests of profit and economic gain.

[6] Often, these barrels were the 55-gallon drum variety, so the first burning likely volatilized some very toxic compounds, depending on the residues remaining in the drum. These contents could have been solvents (including halogenated compounds like chlorinated aliphatics and aromatics), plastic residues (like phthalates), and petroleum distillates. They may even have contained substances with elevated concentrations of heavy metals, like mercury, lead, cadmium, and chromium. The barrels (drums) themselves were often perforated to allow for higher rates of oxidation (combustion) and to take advantage of the smokestack effect (i.e. driving the flame upward and pushing the products of incomplete combustion out of the barrel and into the plume). I recall that in the 1960s my neighbors not being happy about people in the neighborhood burning trash while their wash was drying on the clothesline. They would complain of ash (aerosols) blackening their clothes and the odor from the incomplete combustion products on their newly washed laundry. Both of these complaints are evidence that the plume leaving the barrel contained harmful contaminants.

and tolerance for open burning was waning. Local governments began to restrict and eventually to ban many fires. Often, these restrictions had multiple rationales, especially public safety (fires becoming out of control, especially during dry seasons) and public health (increasing awareness of the association between particulate matter in the air and diseases like asthma and even lung cancer).

Emissions depend on the source material being burned. For example, when halogenated organic compounds are burned, some highly toxic compounds are released. To illustrate, let us consider the processing of polyvinyl chloride (PVC). PVC is a polymer, like polyethylene. However, rather than a series of ethylenes in the backbone chain, a chlorine atom replaces the hydrogen on each of the ethylene groups by free radical polymerization of vinyl chloride. The first thing that can happen when PVC is heated is that the polymers become unhinged and chlorine is released (see Fig. 10.18). Also, the highly toxic and carcinogenic dioxins and furans can be generated from the thermal breakdown and molecular rearrangement of PVC in a heterogeneous process, i.e. the reaction occurs in more than one phase (in this case, in the solid and gas phases). The active sorption sites on the particles allow for the chemical reactions, which are catalyzed by the presence of inorganic chloride compounds and ions sorbed to the particle surface. The process occurs within the temperature range, 250–450°C, so most of the dioxin formation under the precursor mechanism occurs away from the high temperature of the fire, where the gases and smoke derived from combustion of the organic materials have cooled. Dioxins and furans may also form *de novo*, wherein dioxins are formed from moieties different from those of the molecular structure of dioxins, furans, or precursor compounds. The process needs a chlorine donor (a molecule that "donates" a chlorine atom to the precursor molecule). This leads to the formation and chlorination of a chemical intermediate that is a precursor.

In addition, PVC is seldom in a pure form. In fact, most wires have to be pliable and flexible. On its own, PVC is rigid, so plasticizers must be added, especially phthalates. These compounds have been associated with chronic effects in humans including endocrine disruption. Also, since PVC catalyzes its own decomposition, metal stabilizers have been added to PVC products. These have included lead, cadmium, and tin (e.g. butylated forms). Another very common class of toxic compounds released when plastics are burned are the PAHs.

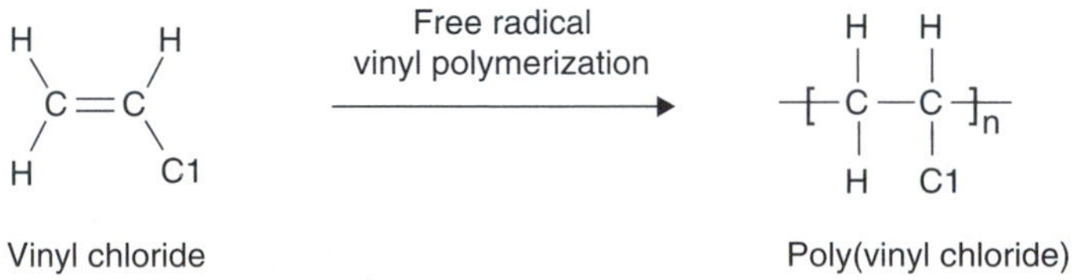

Free radical vinyl polymerization

Vinyl chloride

Poly(vinyl chloride)

Fig. 10.18. Free radical polymerization of vinyl chloride to form PVC.

Like open burning, emissions from industrial stacks can be pervasive and contained myriad toxic components. People living in the plume of industries like coke ovens experience the obnoxious smelling compounds, including metallic and sulfur compounds that volatilized during the conversion of coal to coke needed for steel manufacturing. While such areas continue to be industrialized, such ambient air quality as that in the 1960s is no longer tolerated in the West.

Other sources in developing countries are distributed, such as burning tire piles and junkyards. Again, this can be similar to conditions in the West a half century ago. And, they continue in some lower socioeconomic communities to this day. In fact, the combination of fires, wet muck (composed of soil, battery acid, radiator fluids, motor oil, corroded metal, and water), and oxidizing metals created a rather unique odor around the yards.

VIII. ODORS: MORE THAN JUST A NUISANCE

A number of the processes discussed in this chapter produce odors. Odors have often been associated with public health nuisances. In addition to the link between memory and olfactory centers, however, the nasal–neural connection is important to environmental exposure. This goes beyond nuisance and is an indication of potential adverse health effects. For example, nitric oxide (NO) is a neurotoxic gas released from many sources, such as confined animal feeding operations, breakdown of fertilizers after they are applied to the soil and crops, and emissions from vehicles. Besides being inhaled into the lungs, NO can reach the brain directly. The gas can pass through a thin membrane via the nose to the brain.

The nasal exposure is a different paradigm from that usually used to calculate exposure. In fact, most sources do not have a means for calculating exposures other than dermal, inhalation, and ingestion. People who live near swine and poultry facilities can be negatively affected when they smell odors from the facility. This is consistent with other research that has found that people experience adverse health symptoms more frequently when exposed to livestock odors. These symptoms include eye, nose, and throat irritation, headache, nausea, diarrhea, hoarseness, sore throat, cough, chest tightness, nasal congestion, palpitations, shortness of breath, stress, and drowsiness. There is quite a bit of diversity in response, with some people being highly sensitive to even low concentrations of odorant compounds while others are relatively unfazed even at much higher concentrations.

Actually, response to odors can be triggered by three different mechanisms. In the first mechanism, symptoms can be induced by exposure to odorant compounds at sufficiently high concentrations to cause irritation or other toxicological effects. The irritation, not the odor, evokes the health symptoms. The odor sensation is merely as an exposure indicator. In the second mechanism, symptoms of adverse effects result from odorants concentrations lower than

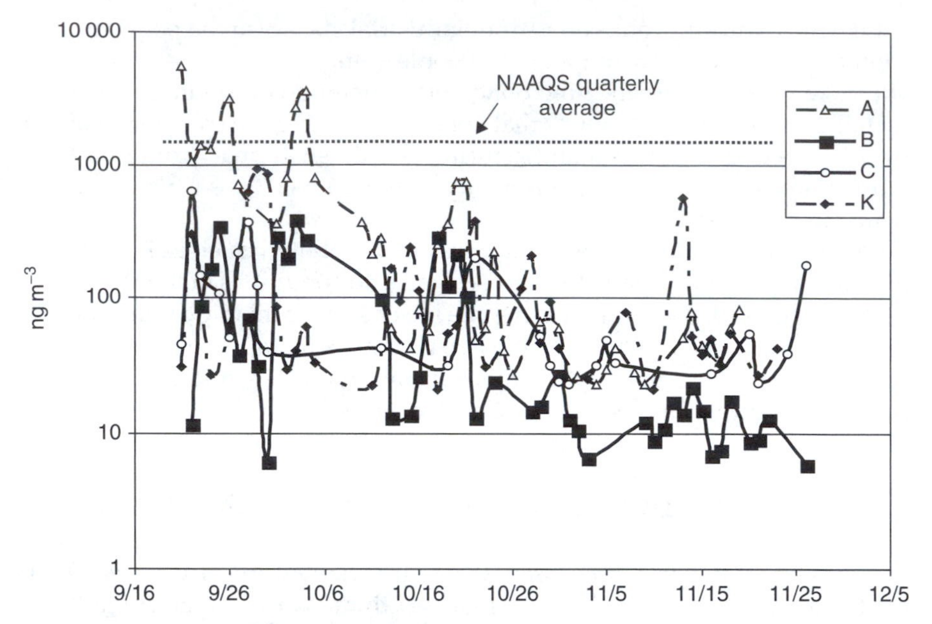

Fig. 10.19. Lead concentrations (composition of $PM_{2.5}$) at the World Trade Center site in 2001. NAAQS: The National Ambient Air Quality Standard for lead, i.e. $1.55\,\mu g\,m^{-3}$ averaged each quarter.

those eliciting irritation. This can be owing to genetic predisposition or conditioned aversion. In the third mechanism, symptoms can result from a coexisting pollutant, e.g. an endotoxin, which is a component of the odorant mixture.

Therefore, to address this new paradigm, better technologies will be needed. For example, innovative research is helping to make odor measurements more quantifiable, such as the development of "artificial noses."[7]

During the aftermath of the World Trade Center attacks, odors played a key role in response. People were reminded daily of the traumatic episode, even when the smoke from the fire was not visible. And, the odors were evidence of exposure to potentially toxic substances. In this instance, it was not NO, but volatile, semivolatile organic, and metallic compounds. The presence of the odors was one of many factors that kept New Yorkers on edge and until they significantly subsided, they continued to be a source of anxiety. Analysis of the samples confirmed my brain's olfactory–memory connection. The plume from the World Trade Center fires contained elevated concentrations of PAHs, dioxins, furans, VOCs (e.g. benzene), and particles containing metals (see Fig. 10.19). The elevated levels of these air pollutants persisted well after the attacks on September 11, 2001.

[7] For a survey of the state-of-the-science in electronic odor sensing technologies, see Nagle, H. T., Schiffman, S. S., and Gutierrez-Osuna, R., The how and why of electronic notes. *IEEE Spectr.* **35**(9), 22–34 (1998).

REFERENCES

1. Tombach, I., A critical review of methods for determining visual range in pristine and near-pristine areas, *Proceedings of the Annual Meeting of the Air Pollution Control Association*, Pittsburgh, PA, 1978.
2. Singer, J. G. (ed.), *Combustion, Fossil Power Systems*, 3rd ed. Combustion Engineering, Windsor, CT, 1981.
3. US Environmental Protection Agency, *Regulatory Announcement: "Proposed Emission Standards for New Nonroad Spark-Ignition Engines, Equipment, and Vessels."* EPA420-F-07-032, April 2007.
4. Popovich, M., and Hering, C., *Fuels and Lubricants*. Oregon State University Press, Corvallis, OR, 1978.
5. Straus, W., *Industrial Gas Cleaning*, 2nd ed. Pergamon, Oxford, 1975.
6. Buonicore, A. J., and Davis, W. T. (eds.), *Air Pollution Engineering Manual*. Van Nostrand Reinhold, New York, 1992.
7. Stern, A. C. (ed.), *Air Pollution*, 3rd ed., Vol. IV. Academic Press, New York, 1977.
8. National Toxicology Program, *Eleventh Report on Carcinogens, Coke Oven Emissions, Substance Profile*; http://ntp.niehs.nih.gov/ntp/roc/eleventh/profiles/s049coke.pdf (accessed on May 11, 2005).
9. Elston, J. C., *J. Air Pollut. Control Assoc.* **31**(5), 524–547 (1981).
10. Stern, A. C. (ed.), *Air Pollution*, 3rd ed., Vol. III. Academic Press, New York, 1977.
11. Public Law 101-549, 101st Congress—November 15, 1990, An act to amend the Clean Air Act to provide for attainment and maintenance of health protective national ambient air quality standards, and for other purposes.
12. US Environmental Protection Agency, *Compilation of Air Pollutant Emission Factors*, AP-42 and Supplements. Research Triangle Park, NC, 1973–1992.

SUGGESTED READING

Faith, W. L., and Atkission, A. A., *Air Pollution*, 2nd ed. Wiley, New York, 1972.

US Interagency Team proposes program to quantify effects of Kuwait oil fires, *J. Air Waste Manage. Assoc.* **41**(6), June 1991.

Wark, K., and Warner, C. F., *Air Pollution: Its Origin and Control*. IEPA, Dun-Donnelley, New York, 1976.

Oregon Department of Environmental Quality, *Wood Heating and Air Quality*, 1981 Annual Report, June 1982.

European Environment Agency, *EMEP/CORINAIR Emission Inventory Guidebook*. Technical Report No. 11/2006, 2006.

US Army Environmental Hygiene Agency, *Final Report—Kuwait Oil Fire Health Risk Assessment*. No. 39-26-L192-91, February 1994.

US Environmental Protection Agency, *Emission Factors and Emission Factors*, 2007, http://www.epa.gov/ttn/chief/ap42/index.html.

US Environmental Protection Agency, *TTNWeb—Technology Transfer Network*, http://www.epa.gov/ttn/.

US Navy Bureau of Medicine, *Silicosis and Operational Exposures to Dust and Sand*, 1990, http://www.gulflink.osd.mil/particulate_final/particulate_final_refs/n73en021/120396_sep96_decls12_0000001.htm.

QUESTIONS

1. Calculate the heat generated by dissociation and formation as one molecular weight of methane, CH_4, burns to carbon dioxide and water. How does this heating value compare to the tabular heating value for methane?
2. Many control districts have banned the use of private backyard incinerators. Would you expect a noticeable increase in air quality as a result of this action?
3. Show a free radical reaction which results in ethane in the effluent of a combustion process burning pure methane with pure oxygen.
4. A power plant burns oil that is 4% ash and 3% sulfur. At 50% excess air, what particulate ($mg\,m^{-3}$) would you expect?
5. Many control districts have very tight controls over petroleum refineries. Suppose these refineries produce 100 million liters of products per day and required air pollution control devices to recover all of the 2% which was previously lost. What are the savings in dollars per year at an average product cost of 10 cents per liter? How does this compare to the estimate that the refineries spent $400 million for control equipment over a 10-year period?
6. Suppose a 40 000-liter gasoline tank is filled with liquid gasoline with an average vapor pressure of 20 mm Hg. At 50% saturation, what weight of gasoline would escape to the atmosphere during filling?
7. If a major freeway with four lanes of traffic in one direction passes four cars per second at 100 km per hour during the rush period, and each car carries two people, how often would a commuter train of five cars carrying 100 passengers per car have to be operated to handle the same load? Assume the train would also operate at 100 km per hour.
8. An automobile traveling 50 km per hour emits 0.1% CO from the exhaust. If the exhaust rate is 80 m^3 per minute, what is the CO emission in grams per kilometer?
9. List the following in increasing amounts from the exhaust of an idling automobile: O_2, NO_x, SO_x, N_2, unburned hydrocarbons, CO_2, and CO.
10. What technologies can be adapted to moderate the increased air pollution associated with industrial development in developing nations? What are some of the important things to keep in mind when transferring such technologies to other cultures?
11. Identify two air pollutants wherein odor may be a promising indicator of air quality. What are the limitations of using odor beyond a screening level?
12. Consider the following information from an accident report filed with the US EPA and write a brief contingency plan that would prevent such an incident occurring at any refinery:

    At about 10:15 a.m., on October 16, 1995, an explosion and fire occurred at Plant No. 1 of the Pennzoil Products Company refinery in Rouseville, Pennsylvania. After the initial explosion, flames quickly engulfed a large area of the refinery, including areas under construction, storage trailers, a trailer where contractors took work breaks, and many storage tanks. The flames ignited several tanks containing naphtha and fuel oil. During the fire, several loud explosions could be heard as compressed gas cylinders and other sealed containers exploded.

    The explosions hurled some plant debris beyond the fenceline. Thick black smoke spread throughout the area. The fire forced Pennzoil employees and contractors at the plant, residents of the town of Rouseville and an elementary school, and the Pennzoil office across Route 8 from the facility, to evacuate. Firefighters extinguished the fire at about 12:30 p.m. that same day. Three workers were killed in the fire and three others were injured. Two of the injured died later as a result of their injuries. The fire resulted in extensive damage to the facility. Minor "sheening" was reported on the stream that runs past the refinery, but there were no reports of any materials spilled into the stream or environmental damage.

    A welding operation was in progress on a service stairway located between two waste liquid storage tanks (tanks 487 and 488) at the time of the incident. These tanks contained mixtures of waste hydrocarbons and water. A hot work (welding, cutting) permit had been prepared, as required by Occupational Safety and Health Administration (OSHA) standard, which included combustible gas detection prior to welding to ensure the safety of the work.

The EPA Chemical Accident Investigation Team (CAIT) identified the immediate cause of the fire and the conditions which triggered the serious consequences. The immediate cause of the fire was the ignition of flammable vapors in storage tank 487. Although the CAIT could not determine the exact mechanism, there are at least two likely scenarios: undetected flammable vapors emitted from tank 487 were ignited by an ignition source which then flashed back into the tank; or an electrical discharge in the tank 487, generated by the arc welding, ignited flammable vapors in the tank.

Part III

Risks from Air Pollution

11

Effects on Health and Human Welfare

I. AIR–WATER–SOIL INTERACTIONS

The harmful effects of air pollutants on human beings have been the major reason for efforts to understand and control their sources. During the past two decades, research on acidic deposition on water-based ecosystems and greenhouse gas emissions on climate has helped to reemphasize the importance of air pollutants in other receptors, such as soil-based ecosystems [1]. When discussing the impact of air pollutants on ecosystems, the matter of scale becomes important. We will discuss three examples of elements which interact with air, water, and soil media on different geographic scales. These are the carbon cycle on a global scale, the sulfur and nitrogen cycles on a regional scale, and the fluoride cycle on a local scale.

A. The Carbon Cycle: Global Scale

Human interaction with the global cycle is most evident in the movement of the element carbon. The burning of biomass, coal, oil, and natural gas to generate heat and electricity has released carbon to the atmosphere and oceans in the forms of CO_2 and carbonate. Because of the relatively slow reactions and removal rates of CO_2, its concentration has been increasing steadily since the beginning of the Industrial Revolution (Fig. 2.4).

In its natural cycle, CO_2 enters the global atmosphere from vegetative decay and atmospheric oxidation of methane and is removed from the atmosphere by photosynthesis and solution by water bodies. These natural sources and sinks of CO_2 have balanced over thousands of years to result in an atmospheric CO_2 concentration of about 200–250 ppm by volume. Over the past 200 years, however, the escalation in burning of fossil fuels has accompanied a steady increase in the atmospheric CO_2 concentration to its current value of ~360 ppm by volume, with projected concentrations over the next 50 years ranging up to 400–600 ppm by volume worldwide [2]. In raising the concentration of CO_2, humans are clearly interacting with nature on a global scale, producing a potential for atmospheric warming and subsequent changes in ocean depths and agricultural zones. This topic is currently subject to considerable research. Current research topics include further development of radiative–convective models, determination of global temperature trends, measurement of changes in polar ice packs, and refinement of the global carbon cycle.

B. The Sulfur and Nitrogen Cycles: Regional Scale

Human production of sulfur from fossil fuel and ore smelting has caused an observable impact on the regional scale (hundreds of kilometers). Considerable evidence suggests that long-range transport of SO_2 occurs in the troposphere. In transit, quantities of SO_2 are converted to sulfate, with eventual deposition by dry or wet processes on the surface far from the original source of SO_2. Sulfate deposition plays the principal role in acid deposition which results in lowering the pH of freshwater lakes and alters the composition of some soils. These changes affect the viability of some plant and aquatic species. The long-range transport of SO_2 and the presence of sulfates as fine particulate matter play a significant role in reduction of visibility in the atmosphere.

Sulfur is present in most fossil fuels, usually higher in coal than in crude oil. Prehistoric plant life is the source for most fossil fuels. Most plants contain S as a nutrient and as the plants become fossilized a fraction of the sulfur volatilizes (i.e. becomes a vapor) and is released. However, some sulfur remains in the fossil fuel and can be concentrated because much of the carbonaceous matter is driven off. Thus, the S-content of the coal is available to react with oxygen when the fossil fuel is combusted. In fact, the S-content of coal is an important characteristic in its economic worth; the higher the S-content the less it is worth. So, the lower the sulfur content and volatile constituents and the higher the carbon content makes for a more valuable coal. Since combustion is the combination of a substance (fuel) with molecular oxygen (O_2) in the presence of heat, the reaction for complete or efficient combustion of a hydrocarbon results in the formation of carbon dioxide and water:

$$(CH)_x + O_2 \xrightarrow{\Delta} CO_2 + H_2O \tag{11.1}$$

However, the fossil fuel contains other elements, including sulfur, so the side reaction forms oxides of sulfur:

$$S + O_2 \xrightarrow{\Delta} SO_2 \tag{11.2}$$

Actually, many other oxidized forms of sulfur can form during combustion, so air pollution experts refer to them collectively as SO_x, which is commonly seen in the literature.

Likewise, nitrogen compounds also form during combustion, but their sources are very different from those of sulfur compounds. In fact, the atmosphere itself is the source of much of the nitrogen leading to the formation of oxides of nitrogen (NO_x). Molecular nitrogen (N_2) makes up most of the gases in the earth's atmosphere (79% by volume). Because N_2 is relatively nonreactive under most atmospheric conditions, it seldom enters into chemical reactions, but under pressure and at very high temperatures, it will react with O_2:

$$N_2 + O_2 \xrightarrow{\Delta} 2NO \tag{11.3}$$

Approximately, 90–95% of the nitrogen oxides generated in combustion processes are in the form of nitric oxide (NO), but like the oxides of sulfur, other nitrogen oxides can form, especially nitrogen dioxide (NO_2), so air pollution experts refer to NO and NO_2 collectively as NO_x. In fact, in the atmosphere the emitted NO is quickly converted photochemically to nitrogen dioxide (NO_2). Such high temperature/high pressure conditions exist in internal combustion engines, like those in automobiles (known as "mobile sources"). Thus, NO_x is one of the major mobile source air pollutants. These conditions of high temperature and pressure can also exist in boilers such as those in power plants, so NO_x is also commonly found in high concentrations leaving fossil fuel power generating stations.

In addition to the atmospheric nitrogen, other sources exist, particularly the nitrogen in fossil fuels. The nitrogen oxides generated from atmospheric nitrogen are known as "thermal NO_x" since they form at high temperatures, such as near burner flames in combustion chambers. Nitrogen oxides that form from the fuel or feedstock are called "fuel NO_x." Unlike the sulfur compounds, a significant fraction of the fuel nitrogen remains in the bottom ash or in unburned aerosols in the gases leaving the combustion chamber, i.e. the fly ash. Nitrogen oxides can also be released from nitric acid processing plants and other types of industrial processes involving the generation and/or use of nitric acid (HNO_3).

Nitric oxide is a colorless, odorless gas and is essentially insoluble in water. Nitrogen dioxide has a pungent acid odor and is somewhat soluble in water. At low temperatures such as those often present in the ambient atmosphere, NO_2 can form the molecule NO_2–O_2N or simply N_2O_4 that consists of two identical simpler NO_2 molecules. This is known as a *dimer*. The dimer N_2O_4 is distinctly reddish-brown and contributes to the brown haze that is often associated with photochemical smog incidents.

Both NO and NO_2 are harmful and toxic to humans, although atmospheric concentrations of nitrogen oxides are usually well below the concentrations expected to lead to adverse health effects. The low concentrations owe to the moderately rapid reactions that occur when NO and NO_2 are emitted into the atmosphere. Much of the concern for regulating NO_x emissions is to suppress the reactions in the atmosphere that generate the highly reactive molecule ozone (O_3). Nitrogen oxides play key roles in important reactants in O_3 formation. Ozone forms photochemically (i.e. the reaction is caused or accelerated by light energy) in the lowest level of the atmosphere, known as the troposphere, where people live. Nitrogen dioxide is the principal gas responsible for absorbing sunlight needed for these photochemical reactions. So, in the presence of sunlight, the NO_2 that forms from the NO incrementally stimulates the photochemical smog-forming reactions because nitrogen dioxide is very efficient at absorbing sunlight in the ultraviolet portion of its spectrum. This is why ozone episodes are more common during the summer and in areas with ample sunlight. Other chemical ingredients, i.e. ozone precursors, in O_3 formation include volatile organic compounds (VOCs), and carbon monoxide (CO). Governments regulate the emissions of precursor compounds to diminish the rate at which O_3 forms.

The oxidized chemical species of sulfur and nitrogen [e.g. sulfur dioxide (SO_2) and nitrogen dioxide (NO_2)] form acids when they react with water. The lowered pH is responsible for numerous environmental problems (i.e. acid deposition). Many compounds contain both nitrogen and sulfur along with the typical organic elements (carbon, hydrogen and oxygen). The reaction for the combustion of such compounds, in general form, is

$$C_aH_bO_cN_dS_e + (4a + b - 2c)O_2 \xrightarrow{\Delta} aCO_2 + \left(\frac{b}{2}\right)H_2O + \left(\frac{d}{2}\right)N_2 + eS \quad (11.4)$$

Reaction (11.4) demonstrates the incremental complexity as additional elements enter the reaction. In the real world, pure reactions are rare. The environment is filled with mixtures. Reactions can occur in sequence, parallel or both. For example, a feedstock to a municipal incinerator contains myriad types of wastes, from garbage to household chemicals to commercial wastes, and even small (and sometimes) large industrial wastes that may be illegally dumped. For example, the nitrogen-content of typical cow manure is about 5 kg per metric ton (about 0.5%). If the fuel used to burn the waste also contains sulfur along with the organic matter, then the five elements will react according to the stoichiometry of Reaction (11.4).

Certainly, combustion specifically and oxidation generally are very important processes that lead to nitrogen and sulfur pollutants. But they are certainly not the only ones. In fact, we need to explain what oxidation really means. In the environment, oxidation *and* reduction occur. The formation of two sulfur dioxide and nitric oxide by acidifying molecular sulfur is a redox reaction:

$$S(s) + NO_3^-(aq) \rightarrow SO_2(g) + NO(g) \quad (11.5)$$

The designations in parentheses give the physical phase of each reactant and product: "s" for solid; "aq" for aqueous; and "g" for gas.

The oxidation half-reactions for this reaction are:

$$S \rightarrow SO_2 \tag{11.6}$$

$$S + 2H_2O \rightarrow SO_2 + 4H^+ + 4e^- \tag{11.7}$$

The reduction half-reactions for this reaction are:

$$NO_3^- \rightarrow NO \tag{11.8}$$

$$NO_3^- + 4H^+ + 3e^- \rightarrow NO + 2H_2O \tag{11.9}$$

Therefore, the balanced oxidation–reduction reactions are:

$$4NO_3^- + 3S + 16H^+ + 6H_2O \rightarrow 3SO_2 + 16H^+ + 4NO + 8H_2O \tag{11.10}$$

$$4NO_3^- + 3S + 4H^+ \rightarrow 3SO_2 + 4NO + 2H_2 \tag{11.11}$$

A reduced form of sulfur that is highly toxic and an important pollutant is hydrogen sulfide (H_2S). Certain microbes, especially bacteria, reduce nitrogen and sulfur using the N or S as energy sources through the acceptance of electrons. For example, sulfur-reducing bacteria can produce hydrogen sulfide (H_2S), by chemically changing oxidized forms of sulfur, especially sulfates (SO_4). To do so, the bacteria must have access to the sulfur, i.e. it must be in the water, which can be in surface or ground water, or the water in soil and sediment. These sulfur-reducers are often anaerobes, i.e. bacteria that live in water where concentrations of molecular oxygen (O_2) are deficient. The bacteria remove the O_2 molecule from the sulfate leaving only the S, which in turn combines with hydrogen (H) to form gaseous H_2S. In ground water, sediment and soil water, H_2S is formed from the anaerobic or nearly anaerobic decomposition of deposits of organic matter, e.g. plant residues. Thus, redox principles can be used to treat H_2S contamination, i.e. the compound can be oxidized using a number of different oxidants. Strong oxidizers, like molecular oxygen and hydrogen peroxide, most effectively oxidize the reduced forms of S, N or any reduced compound.

Ionization is also important in environmental reactions. This is due to the configuration of electrons in an atom. The arrangement of the electrons in the atom's outermost shell, i.e. valence, determines the ultimate chemical behavior of the atom. The outer electrons become involved in transfer to and sharing with shells in other atoms, i.e. forming new compounds and ions. An atom will gain or lose valence electrons to form a stable ion that have the same number of electrons as the noble gas nearest the atom's atomic number. For example, the nitrogen cycle (Fig. 11.1) includes three principal forms that are soluble in water under environmental conditions: the cation (positively charged ion) ammonium (NH_4^+) and the anions (negatively charged ions) nitrate (NO_3^-) and nitrite (NO_2^-). Nitrates and nitrites combine with various organic and

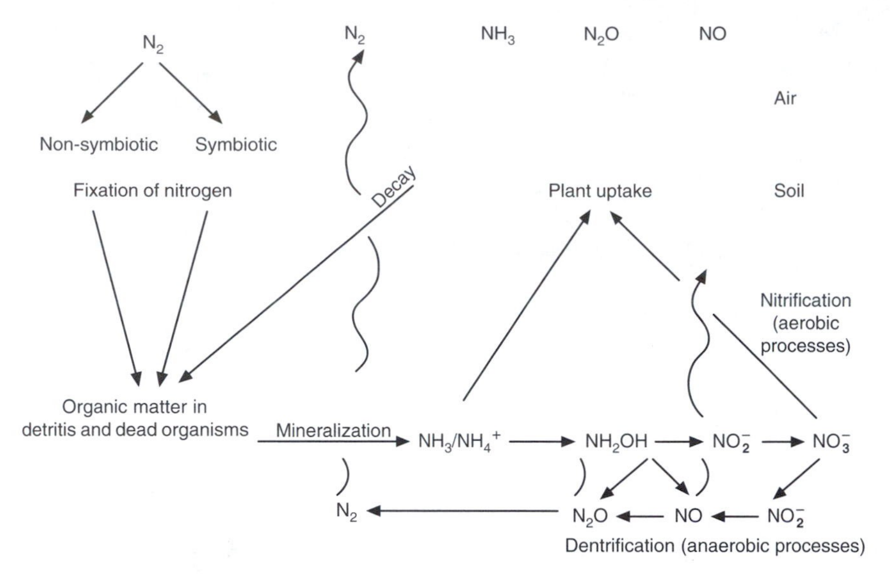

Fig. 11.1. Biochemical nitrogen cycle.

inorganic compounds. Once taken up by fauna (including humans), NO_3^- is converted to NO_2^-. Since NO_3^- is soluble and readily available as a nitrogen source for plants (e.g. to form plant tissue such as amino acids and proteins), farmers are the biggest users of NO_3^- compounds in commercial fertilizers (although even manure can contain high levels of NO_3^-).

Ingesting high concentrations of nitrates, e.g. in drinking water, can cause serious short-term illness and even death. The serious illness in infants is due to the conversion of nitrate to nitrite by the body, which can interfere with the oxygen-carrying capacity of the blood, known as methemoglobinemia. Especially in small children, when nitrates compete successfully against molecular oxygen, the blood carries methemoglobin (as opposed to healthy hemoglobin), giving rise to clinical symptoms. At 15–20% methemoglobin, children can experience shortness of breath and blueness of the skin (i.e. clinical cyanosis). At 20–40% methemoglobin, hypoxia will result. This acute condition can deteriorate a child's health rapidly over a period of days, especially if the water source continues to be used. Long-term, elevated exposures to nitrates and nitrites can cause an increase in the kidneys' production of urine (diuresis), increased starchy deposits and hemorrhaging of the spleen.[1]

Compounds of nitrogen and sulfur are important in every environmental medium. They are addressed throughout this book, as air pollutants. They are some of the best examples of the importance of balance. They are key nutrients. Nutrients are, by definition, essential to life processes, but in the wrong place under the wrong conditions, they become pollutants.

[1] US Environmental Protection Agency (EPA), *National Primary Drinking Water Regulations: Technical Fact Sheets*. Washington, DC: http://www.epa.gov/OGWDW/hfacts.html, 1999.

C. The Fluoride Cycle: Local Scale

The movement of fluoride through the atmosphere and into a food chain illustrates an air–water interaction at the local scale (<100 km) [3]. Industrial sources of fluoride include phosphate fertilizer, aluminum, and glass manufacturing plants. Domestic livestock in the vicinity of substantial fluoride sources are exposed to fluoride by ingestion of forage crops. Fluoride released into the air by industry is deposited and accumulated in vegetation. Its concentration is sufficient to cause damage to the teeth and bone structure of the animals that consume the crops.

The atmospheric movement of pollutants from sources to receptors is only one form of translocation. A second one involves our attempt to control air pollutants at the source. The control of particulate matter by wet or dry scrubbing techniques yields large quantities of waste materials—often toxic—which are subsequently taken to landfills. If these wastes are not properly stored, they can be released to soil or water systems. The prime examples involve the disposal of toxic materials in dump sites or landfills. The Resource Conservation and Recovery Act of 1976 and subsequent revisions are examples of legislation to ensure proper management of solid waste disposal and to minimize damage to areas near landfills [4].

II. TOTAL BODY BURDEN

The presence of air pollutants in the surrounding ambient air is only one aspect of determining the impact on human beings. An air pollution instrument can measure the ambient concentration of a pollutant gas, which may or may not be related to its interaction with individuals. More detailed information about where and for how long we are breathing an air pollutant provides additional information about our actual exposure. Finally, how an air pollutant interacts with the human body provides the most useful information about the dose to a target organ or bodily system.

The human body and other biological systems have a tremendous capacity to take in all types of chemicals and either utilize them to support some bodily function or eliminate them. As analytical capabilities have improved, lower and lower concentrations of chemicals have been observed in various parts of the body. Some of these chemicals enter the body by inhalation.

The concept of *total body burden* refers to the way a trace material accumulates in the human system. The components of the body that can store these materials are the blood, urine, soft tissue, hair, teeth, and bone. The blood and urine allow more rapid removal of trace materials than the soft tissue, hair, and bone [5]. Accumulation results when trace materials are stored more rapidly than they can be eliminated. It can be reversed when the source of the material is reduced. The body may eliminate the trace material over a period of a few hours to days, or may take much longer—often years.

Risk is an expression of the likelihood (statistical probability) that harm will occur when a receptor (e.g. human or a part of an ecosystem) is exposed to that hazard. An example of a toxic hazard is a carcinogen (a cancer-causing chemical). An example of a toxic risk is the likelihood that a certain population will have an incidence of a particular type of cancer after being exposed to that carcinogen. This is a way of describing the population risk; that is the risk of one person out of a million will develop lung cancer when exposed to a certain dose of a chemical carcinogen for a certain period of time.

Dose is the amount, often mass, of a contaminant administered to an organism (so-called "applied dose"), the amount of the contaminant that enters the organism ("internal dose"), the amount of the contaminant that is absorbed by an organism over a certain time interval ("absorbed dose"), or the amount of the contaminants or its metabolites that reach a particular "target" organ ("biologically effective dose" or "bio-effective dose"), such as the amount of a neurotoxin (a chemical that harms the nervous system) that reaches the nerve or other nervous system cells. Theoretically, as the organism increases its contact (exposure) to the hazardous substance, the adverse outcome (e.g. disease or damage) is also expected to increase. This relation is a biological gradient, which is known to toxicologists as the "dose–response" curve (Fig. 11.2).

The effect of accumulation in various systems depends greatly on the quantity of pollutants involved. Many pollutants can be detected at concentrations lower than those necessary to affect human health. For pollutants which are eliminated slowly, individuals can be monitored over long periods of time to detect trends in body burden; the results of these analyses can then be related

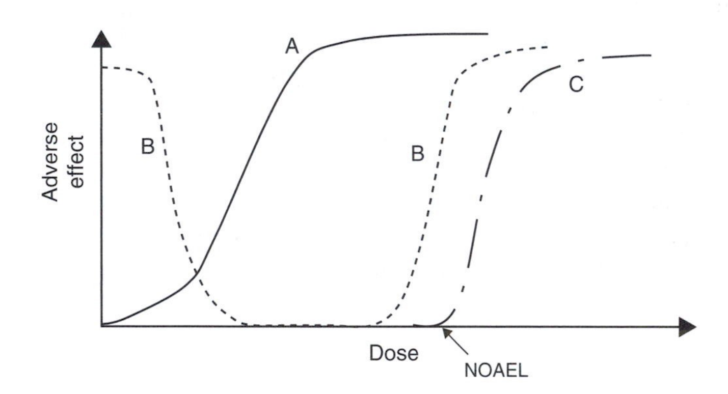

Fig. 11.2. Three prototypical dose–response curves. Curve A represents the no-threshold curve, which expects a response (e.g. cancer) even if exposed to a single molecule (this is the most conservative curve). Curve B represents the essential nutrient dose–response relationship, and includes essential metals, such as trivalent chromium or selenium, where an organism is harmed at the low dose due to a "deficiency" (left side) and at the high dose due to "toxicity" (right side). Curve C represents toxicity above a certain threshold (non-cancer). This threshold curve expects a dose at the low end where no disease is present. Just below this threshold is the NOAEL. *Sources*: US Environmental Protection Agency; and Vallero, D. A., *Environmental Contaminants: Assessment and Control*. Elsevier Academic Press, Burlington, MA, 2004.

to total pollutant exposure. Following are examples of air pollutants that contribute to the total body burden.

A. Lead and the Human Body

The major sources of airborne lead are, incineration of solid wastes and discarded oil, and certain manufacturing processes [6]. Although lead fuel additives have been banned in many countries (see Fig. 11.3), mobile sources continue to emit the metal to the air in those countries that still allow it. The populations most sensitive to lead exposure are unborn and young children. Lead can degrade renal function, impair hemoglobin synthesis, and alter the nervous system. The neuro behavioral impairment of children's intellectual development is a major concern for lead exposure. There are two principal routes for the entry of lead: inhalation and ingestion. The importance of each depends on the circumstances. As noted in Chapter 22, the US National Ambient Air Quality Standard (NAAQS) for lead is based on the ingestion route, which accounts for 80% of the allowed body burden, with only the remaining 20% permissible via inhalation. Inhalation results in primary exposure to airborne lead, whereas ingestion may result in secondary exposure via contamination of the ingested material by atmospheric lead. When lead is inhaled, some of it is absorbed directly into the bloodstream and a fraction into the gastrointestinal tract through lung clearance mechanisms that result in swallowing of mucus.

The absorption, distribution, and accumulation of lead in the human body may be represented by a three-part model [6]. The first part consists of red blood cells, which move the lead to the other two parts, i.e. soft tissue and bone. The blood cells and soft tissue, represented by the liver and kidney, constitute

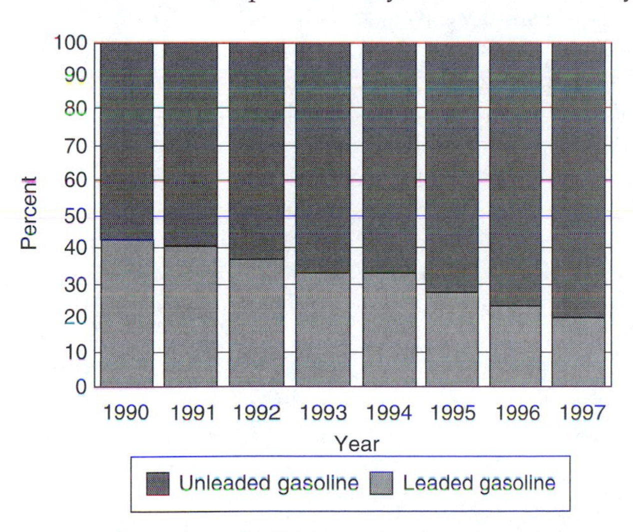

Fig. 11.3. Steady decline in the worldwide use of lead-based fuel additives. *Source*: United Nations Environmental Programme, Organization for Economic Cooperation and Development. *Phasing Lead Out of Gasoline: An Examination of Policy Approaches in Different Countries*, 1999.

the mobile part of the lead body burden, which can fluctuate depending on the length of exposure to the pollutant. Lead accumulation over a long period of time occurs in the bones, which store up to 95% of the total body burden. However, the lead in soft tissue represents a potentially greater toxicological hazard and is the more important component of the lead body burden. Lead measured in the urine has been found to be a good index of the amount of mobile lead in the body. The majority of lead is eliminated from the body in the urine and feces, with smaller amounts removed by sweat, hair, and nails.

Like several other heavy metals, lead interferes with physiological processes because, when ionized, divalent lead (Pb^{2+}) acts like divalent calcium (Ca^{2+}). Due its larger size and other chemical differences, however, Pb^{2+} induces biological responses that differ from those of Ca^{2+}. For example, during gestation and in early childhood, the developing brain is harmed when Pb^{2+}, competing with Ca^{2+}, induces the release of a neurotransmitter in elevated amounts and at the wrong time (e.g. during basal intervals, when a person is at rest). Thus, at high lead exposures, a person may have abnormally high amounts of brain activity (when it should be lower) and, conversely, when a neural response is expected, little or no increase in brain activity is observed. This may induce chronic effects when synaptic connections in the brain are truncated during early brain development.

Lead also adversely affects the release of the transmitter, glutamate, which is involved in brain activities associated with learning. The *N*-methyl-D-aspartate (NMDA) receptor seems to be selectively blocked when lead is present. Other lead effects include the activation of protein kinase C (known as "PKC") because PKC apparently has a greater affinity for lead than for the normal physiological activator, divalent calcium. This complicates and exacerbates the other neurotransmitter effects and harms the cell's chemical messaging (i.e. second-messenger systems), synthesis of proteins, and genetic expression.

All of these neurological effects, especially in the developing brain, began to be documented in earnest by the medical community only within the last half century. Herbert Needleman, a pediatrician at the University of Pittsburgh Medical Center, discovered that a correlation existed between the amount of lead in the teeth of infants and their intelligence as at age 16, as measured by their intelligence quotient (IQ) scores. His research has shown a dose–response between lead dose and IQ. That is, the higher the lead content, the lower the IQ in these teenagers. In a series of follow-up studies, Needleman determined that lead poisoning had long-term implications for a child's attentiveness, behavior, and school success.

Needleman and other scientists called for interventions even while the scientific evidence was still being gathered.[2] He was among the first to advocate

[2] At least on its face, this runs contrary to some of the measures calling for improved environmental risk-based science proposed in the 1980s, especially the separation of risk assessment and risk management. This advice actually was an attempt to make risk science more objective and empirical, so that the science does not become "contaminated" by vested interests in the findings (such as political and economic considerations).

the removal of tetraethyl lead from gasoline and to remove lead-based paints and to reduce exposure in houses where kids can chew on the paint chips.[3] The results have been dramatic, with average blood lead levels in the United States dropping an estimated 78% from 1976 to 1991.

Some important health thresholds are shown in Fig. 11.4. It is important to note that chronic lead thresholds (e.g. the NAAQS) are several orders of magnitude lower than acute thresholds (e.g. the lethal concentration 50 [LC_{50}], which is the concentration of lead at which 50% of the organisms in a specific test situation are killed). In other words, a person exposed for a long time will experience chronic effects even at very lower concentrations.

B. Carbon Monoxide and the Human Body

Another example of an air pollutant that affects the total body burden is carbon monoxide (CO). In addition to CO in ambient air, there are other sources for inhalation. People who smoke have an elevated CO body burden compared to nonsmokers. Individuals indoors may be exposed to elevated levels of CO from incomplete combustion in heating or cooking stoves. CO gas enters the human body by inhalation and is absorbed directly into the bloodstream; the total body burden resides in the circulatory system. The human body also produces CO by breakdown of hemoglobin. Hemoglobin breakdown gives every individual a baseline level of CO in the circulatory system. As the result of these factors, the body burden can fluctuate over a timescale of hours.

In the normal interaction between the respiratory and circulatory systems, O_2 is moved into the body for use in biochemical oxidation and CO_2, a waste product, is removed. Hemoglobin molecules in the blood play an important role in both processes. Hemoglobin combines with O_2 and CO_2 as these gases are moved between the lung and the blood cells. The stability of the hemoglobin–O_2 and hemoglobin–CO_2 complex is sufficiently strong to transport the gases in the circulatory system but not strong enough to prevent the release of CO_2 at the lung and O_2 where it is needed at the cellular level. CO interferes with this normal interaction by forming a much more stable complex with hemoglobin (COHb) [7]. This process reduces the number of hemoglobin molecules available to maintain the necessary transport of O_2 and CO_2.

The baseline level of COHb is ~0.5% for most individuals. Upon exposure to elevated levels of atmospheric CO, the percentage of COHb will increase in a very predictable manner. Analytical techniques are available to measure COHb from <0.1% to >80% in the bloodstream, providing a very rapid method for determining the total body burden. If elevated levels of CO are reduced, the percentage of COHb will decrease over a period of time.

[3] The phenomenon where children eat such non-food material as paint chips is known as *pica*. For young children, especially those in poorer homes with older residences, this type of ingestion was (and still is in some places) a major lead exposure pathway in children. Other pathways include soil ingestion (also pica), inhalation of lead on dust particles (which can be very high when older homes are renovated), and lead in food (such as lead leaching from glazes on cooking and dining ware into the food; common in some ethnic groups, e.g. Mexican and Mexican-American).

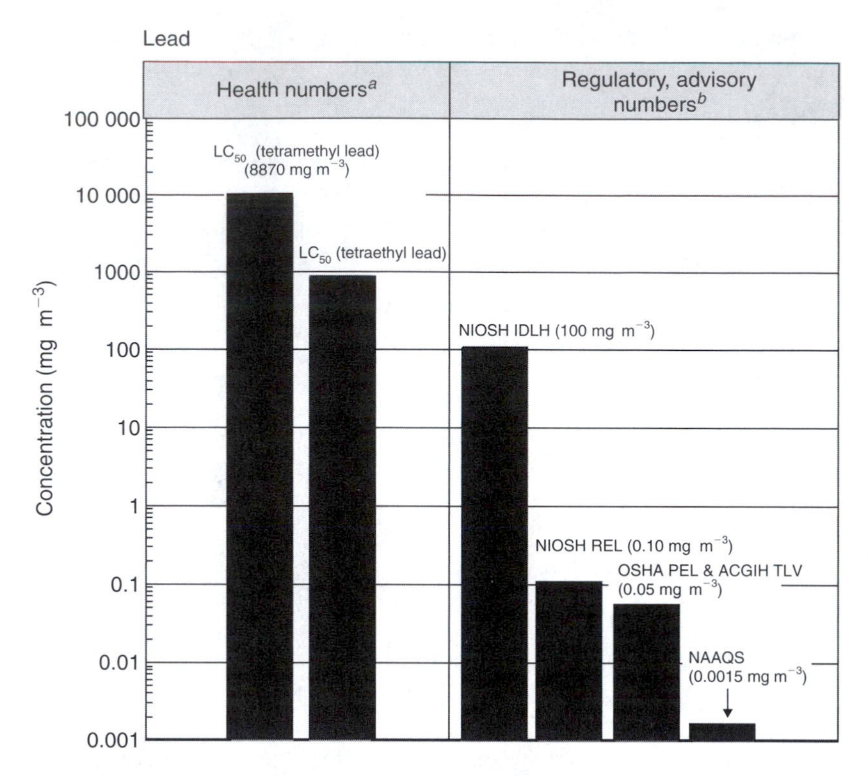

Fig. 11.4. Health related data for exposure to lead. *Source*: US Environmental Protection Agency, *Technology Transfer Network: Air Toxics: Lead Compounds*: http://www.epa.gov/ttn/atw/hlthef/lead.html; (accessed on January 18, 2007), 1999. ACGIH TLV is the American Conference of Governmental and Industrial Hygienists' threshold limit value expressed as a time-weighted average; the concentration of a substance to which most workers can be exposed without adverse effects. LC_{50} is a calculated concentration of a chemical in air to which exposure for a specific length of time is expected to cause death in 50% of a defined experimental animal population. NIOSH REL is the National Institute of Occupational Safety and Health's recommended exposure limit; NIOSH-recommended exposure limit for an 8- or 10-h time-weighted-average exposure and/or ceiling. NIOSH IDLH is NIOSH's immediately dangerous to life or health concentration; NIOSH recommended exposure limit to ensure that a worker can escape from an exposure condition that is likely to cause death or immediate or delayed permanent adverse health effects or prevent escape from the environment. NAAQS is the National Ambient Air Quality Standard set by EPA for pollutants that are considered to be harmful to public health and the environment; the NAAQS for lead is $1.5\,\mu g\,m^{-3}$, maximum arithmetic mean averaged over a calendar quarter. OSHA PEL is the Occupational Safety and Health Administration's permissible exposure limit expressed as a time-weighted average; the concentration of a substance to which most workers can be exposed without adverse effect averaged over a normal 8-h workday or a 40-h workweek. Table notes: [a]Health numbers are toxicological numbers from animal testing or risk assessment values developed by EPA. [b]Regulatory numbers are values that have been incorporated in Government regulations, while advisory numbers are nonregulatory values provided by the Government or other groups as advice. OSHA and NAAQS numbers are regulatory, whereas NIOSH and ACGIH numbers are advisory.

At low levels of COHb (0.5–2.0%) the body burden is measurable, but research has not shown any substantive effects at these low levels. When COHb increases to higher levels the body burden of CO is elevated, producing adverse effects on the cardiovascular system and reducing physical endurance.

C. Body Burden of Other Metals

Unlike lead and mercury (discussed in the next chapter), a number of elements are essential for normal physiologic functioning in all animal species. However, after a threshold of exposure is crossed, such toxic elements engender effects similar to those of lead and mercury. For example, manganese exposures have been associated with impairment of neural and behavioral problems. Thus, any quantitative risk assessment for manganese must take into account aspects of both the essentiality and the toxicity of manganese. That is, there is an optimal range of a number of metals, especially manganese, selenium and trivalent chromium, below which is deficiency and above which is toxicity (see curve B of Fig. 11.2). Also, the chemical form of the substance largely determines its essentiality and toxicity (e.g. all forms of hexavalent chromium appear to be toxic, but a number of trivalent chromium compounds are essential).

1. *Manganese and the Human Body*

The element manganese (Mn) is ubiquitous in the environment. The metal and its compounds have been an important constituent of numerous manufacturing processes, including[4]:

- Metallic manganese primarily in steel production to improve hardness, stiffness, and strength. It is also used in carbon steel, stainless steel, and high-temperature steel, along with cast iron and superalloys.
- Manganese dioxide (MnO_2) in the production of dry-cell batteries, matches, fireworks, and the production of other manganese compounds
- Catalyst manganese chloride ($MnCl_2 \cdot 4H_2O$) in the chlorination of organic compounds, in animal feed, and in dry-cell batteries.
- Manganese sulfate as a fertilizer, livestock nutritional supplement, in glazes and varnishes, and in ceramics.

The average manganese levels in various environmental media are[5]:

- levels in drinking water are approximately 0.004 ppm;
- average air levels are approximately $0.02\,\mu g\,m^{-3}$);
- levels in soil range from 40 to 900 ppm;
- average daily intake from food ranges from 1 to $5\,mg\,day^{-1}$).

Workers where manganese metal is produced from manganese ore or where manganese compounds are used to make steel or other products are

[4] Agency for Toxic Substances and Disease Registry (ATSDR), *Toxicological Profile for Manganese*. US Public Health Service, US Department of Health and Human Services, Atlanta, GA, September 2000.

[5] Ibid.

most likely to be exposed through inhalation to higher than normal levels of manganese.

Chronic inhalation exposure of humans to manganese results primarily in effects on the nervous system. Slower visual reaction time, poorer hand steadiness, and impaired eye-hand coordination were reported in several studies of workers occupationally exposed to manganese dust in air. Humans inhaling manganese at high levels may acquire the syndrome manganism that typically begins with feelings of weakness and lethargy and progresses to other symptoms such as gait disturbances, clumsiness, tremors, speech disturbances, a mask-like facial expression, and psychological disturbances. Other chronic effects reported in humans from inhalation exposure to manganese are respiratory effects such as an increased incidence of cough, bronchitis, dyspnea (difficult breathing) during exercise, and an increased susceptibility to infectious lung disease.[6] Several health-based thresholds are provided in Fig. 11.5.

Manganese is one of a few contaminants that can be eliminated before even being absorbed. This process is known as "presystemic elimination" and can take place while the contaminant is being transferred from the exposure site (e.g. the outer layer of the skin or the gastrointestinal (GI) tract. Manganese can be eliminated during uptake by the liver, even before it is absorbed into the bloodstream. Presystemic elimination, however, does not necessarily mean that an organism experiences no adverse effect. In fact, Mn exposure can damage the liver without ever being absorbed into the bloodstream. This is also one of the complications of biomarkers, since the body is protected against Mn toxicity by low rates of absorption or by the liver's presystemic Mn elimination.[7]

Arguably, one of the most important current issues associated with manganese is as a fuel additive, especially methylcyclopentadienyl manganese tricarbonyl (MMT). In 1994, the Environmental Protection Agency (EPA) issued an exposure assessment based on some key assumptions in Southern California:

- 100% of unleaded gasoline contains 1/32 g of Mn per gallon of gasoline (about 14% of the gasoline in the Riverside area used MMT in 1990).
- About 30% of the total MMT combusted is emitted from the tailpipe as manganese-containing particle matter with an aerodynamic diameter of $\leq 2.5\,\mu m$ ($PM_{2.5}$).
- 69% of the of Mn in larger particles, that is $\leq 4\,\mu m$ (PM_4) was derived from automotive sources.

The exposure assessment concluded that 5–10% of the general population in Riverside might be exposed via inhalation to manganese annual average levels of approximately $0.1\,\mu g\,m^{-3}$ PM_4 or higher, indicating that possibly hundreds of thousands of persons in the Los Angeles area could experience such

[6] Ibid, and US Environmental Protection Agency, *Integrated Risk Information System (IRIS) on Manganese*. National Center for Environmental Assessment, Office of Research and Development, Washington, DC, 1999.

[7] For example, see Greger, J., Dietary standards for manganese: overlap between nutritional and toxicological studies. *J. Nutr.*, **128**(2), 368S-371S, 1998.

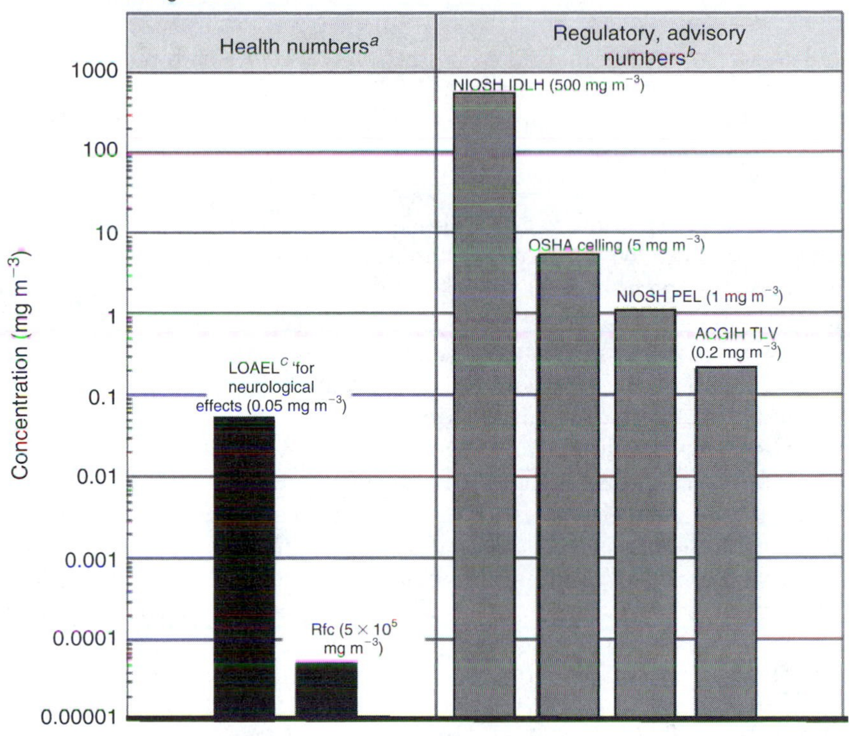

Fig. 11.5. Health data from inhalation exposure to manganese. *Sources*: US Environmental Protection Agency, *Integrated Risk Information System (IRIS) on Manganese*. National Center for Environmental Assessment, Office of Research and Development, Washington, DC, 1999; Occupational Safety and Health Administration (OSHA), Occupational Safety and Health Standards, Toxic and Hazardous Substances, *Code of Federal Regulations* 29 CFR 1910.1000, 1998; National Institute for Occupational Safety and Health (NIOSH), *Pocket Guide to Chemical Hazards*, US Department of Health and Human Services, Public Health Service, Centers for Disease Control and Prevention, Cincinnati, OH, 1997; and American Conference of Governmental Industrial Hygienists (ACGIH), *1999 TLVs and BEIs. Threshold Limit Values for Chemical Substances and Physical Agents, Biological Exposure Indices*, Cincinnati, OH, 1999. *Note*: ACGIH TLV is the American Conference of Governmental and Industrial Hygienists' threshold limit value expressed as a time-weighted average; the concentration of a substance to which most workers can be exposed without adverse effects. LOAEL is the lowest-observed-adverse-effect level. NIOSH REL is the National Institute of Occupational Safety and Health's recommended exposure limit; NIOSH-recommended exposure limit for an 8- or 10-h time-weighted-average exposure and/or ceiling. NIOSH IDLH is NIOSH's immediately dangerous to life or health concentration; NIOSH recommended exposure limit to ensure that a worker can escape from an exposure condition that is likely to cause death or immediate or delayed permanent adverse health effects or prevent escape from the environment. OSHA PEL is the Occupational Safety and Health Administration's permissible exposure limit expressed as a time-weighted average; the concentration of a substance to which most workers can be exposed without adverse effect averaged over a normal 8-h workday or a 40-h workweek. OSHA ceiling is OSHA's short-term exposure limit; 15-min time-weighted-average exposure that should not be exceeded at any time during a workday even if the 8-h time-weighted average is within the threshold limit value. [a]Health numbers are toxicological numbers from animal testing or risk assessment values developed by EPA. [b]Regulatory numbers are values that have been incorporated in government regulations, while advisory numbers are nonregulatory values provided by the government or other groups as advice. OSHA numbers are regulatory, whereas NIOSH and ACGIH numbers are advisory. [c]This LOAEL is from the critical study used as the basis for the EPA RfC.

exposures at or greater than EPA's current inhalation health benchmark[8] for manganese (reference concentration, RfC) of 0.05 $\mu g\, m^{-3}$. Decisions on the safety of Mn additive usage are still under consideration in the United States.

2. Selenium and the Human Body

The element selenium (Se) is found in food. In fact, food is the primary source of exposure to selenium, with an estimated selenium intake for the US population ranging from 0.071 to 0.152 mg day^{-1}. Inhalation of elevated Se concentrations in the ambient air is commonly not a major route of exposure, with an average selenium concentration estimated to be below 10 nanograms of Se per cubic meter (ng m^{-3}). Relatively high occupational, inhalation exposures are found in certain industrial categories, such as metal industries, selenium-recovery processes, painting, and special trades.[9]

Selenium is toxic at high concentrations but is also a nutritionally essential element. Hydrogen selenide (H_2Se) is the most acutely toxic selenium compound. Inhaling elemental selenium (Sn^0), H_2Se, and selenium dioxide (SeO_2) leads to acute respiratory effects, including irritation of the mucous membranes, pulmonary edema, severe bronchitis, and bronchial pneumonia. Epidemiological studies of humans chronically (long-term) exposed to high levels of selenium in food and water have reported discoloration of the skin, pathological deformation and loss of nails, loss of hair, excessive tooth decay and discoloration, lack of mental alertness, and listlessness.[10]

Epidemiological studies have reported an inverse association between selenium levels in the blood and cancer occurrence and animal studies have reported that selenium supplementation, as sodium selenate (Na_2SeO_4), sodium selenite (Na_2SeO_3), and organic forms of selenium, results in a reduced incidence of several tumor types. The only selenium compound that has been shown to be carcinogenic in animals is selenium sulfide (SeS), which resulted in an increase in liver tumors from oral exposure. The EPA has classified selenium sulfide as a probable human carcinogen (Group B2).[11] The health-based thresholds for selenium are provided in Fig. 11.6.

[8] IRIS, Integrated Risk Information System, Reference concentration (RfC) for chronic manganese exposure as revised December, 1993. Cincinnati, OH: US Environmental Protection Agency, National Center for Environmental Assessment. Available online: http://www.epa.gov/IRIS/subst/0373.htm (as of October 13, 2004)

[9] Agency for Toxic Substances and Disease Registry (ATSDR), *Toxicological Profile for Selenium*. Public Health Service, Department of Health and Human Services, Atlanta, GA, September 2003.

[10] Ibid.

[11] Ibid.

Fig. 11.6. Health Data from inhalation exposure to selenium. *Sources:* Agency for Toxic Substances and Disease Registry (ATSDR), *Toxicological Profile for Selenium* (Update). Public Health Service, Department of Health and Human Services, Atlanta, GA, 1996; US Department of Health and Human Services, *Registry of Toxic Effects of Chemical Substances (RTECS, online database)*. National Toxicology Information Program, National Library of Medicine, Bethesda, MD, 1993; US Environmental Protection Agency, *Integrated Risk Information System (IRIS) on Selenium and*

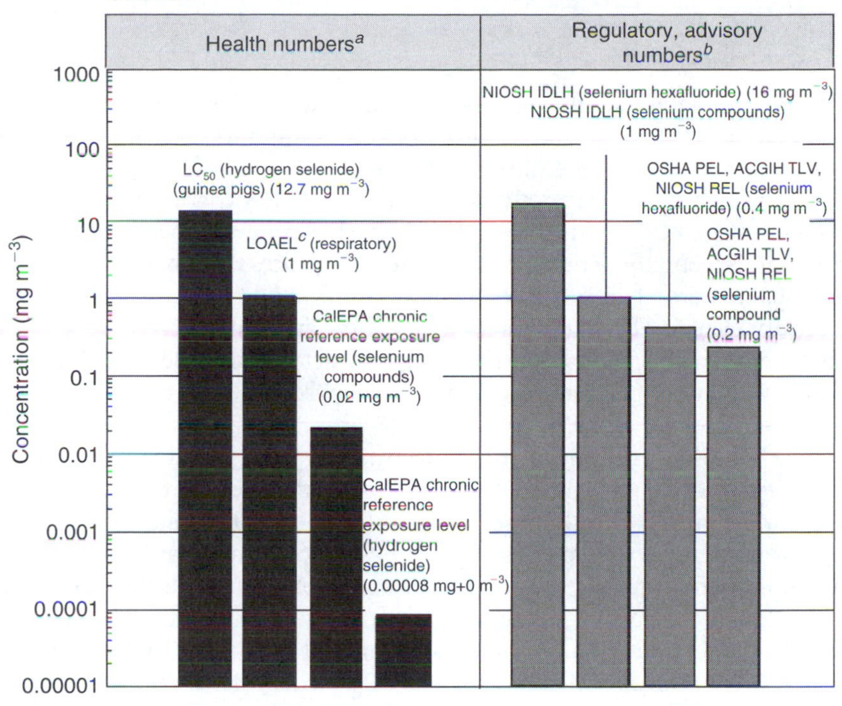

Compounds. National Center for Environmental Assessment, Office of Research and Development, Washington, DC, 1999. California Environmental Protection Agency (CalEPA), *Air Toxics Hot Spots Program Risk Assessment Guidelines: Part III. Technical Support Document for the Determination of Non-cancer Chronic Reference Exposure Levels*. SRP Draft. Office of Environmental Health Hazard Assessment, Berkeley, CA, 1999; American Conference Of Governmental Industrial Hygienists (ACGIH), *1999 TLVs and BEIs. Threshold Limit Values for Chemical Substances and Physical Agents. Biological Exposure Indices.* Cincinnati, OH, 1999; National Institute for Occupational Safety and Health (NIOSH), *Pocket Guide to Chemical Hazards.* US Department of Health and Human Services, Public Health Service, Centers for Disease Control and Prevention. Cincinnati, OH, 1997; and Occupational Safety and Health Administration (OSHA). Occupational Safety and Health Standards, Toxic and Hazardous Substances. *Code of Federal Regulations.* 29 CFR 1910.1000, 1998. *Note*: ACGIH TLV is the American Conference of Governmental and Industrial Hygienists' threshold limit value expressed as a time-weighted average; the concentration of a substance to which most workers can be exposed without adverse effects. LOAEL is the lowest-observed-adverse-effect level. NIOSH REL is the National Institute of Occupational Safety and Health's recommended exposure limit; NIOSH-recommended exposure limit for an 8- or 10-h time-weighted-average exposure and/or ceiling. NIOSH IDLH is NIOSH's immediately dangerous to life or health concentration; NIOSH recommended exposure limit to ensure that a worker can escape from an exposure condition that is likely to cause death or immediate or delayed permanent adverse health effects or prevent escape from the environment. OSHA PEL is the Occupational Safety and Health Administration's permissible exposure limit expressed as a time-weighted average; the concentration of a substance to which most workers can be exposed without adverse effect averaged over a normal 8-h workday or a 40-h workweek. OSHA ceiling is OSHA's short-term exposure limit; 15-min time-weighted-average exposure that should not be exceeded at any time during a workday even if the 8-h time-weighted average is within the threshold limit value. [a] Health numbers are toxicological numbers from animal testing or risk assessment values developed by EPA. [b] Regulatory numbers are values that have been incorporated in Government regulations, while advisory numbers are nonregulatory values provided by the government or other groups as advice. OSHA numbers are regulatory, whereas NIOSH and ACGIH numbers are advisory. [c] This LOAEL is from the critical study used as the basis for the EPA RfC.

3. Chromium and the Human Body

Chromium is perhaps the best example of how valence (electrons in outermost shell) affects toxicity and essentiality.[12] Chromium occurs in the environment predominantly in one of two valence states: trivalent chromium (Cr III or Cr^{3+}), which occurs naturally and is an essential nutrient, and hexavalent chromium (Cr VI Cr^{6+}), which, along with the less common metallic chromium (Cr 0), is most commonly produced by industrial processes. Trivalent chromium is essential to normal glucose, protein, and fat metabolism and is thus an essential dietary element. The body can detoxify some amount of Cr^6 via several systems for reducing it to $^+$ to Cr^{3+}. This $^+$ to Cr^{6+} detoxification leads to increased levels of Cr^{3+}, which is also toxic, but much less so and with different adverse effects than those of $^+$ to Cr^{6+}.

Air emissions of chromium are predominantly of trivalent chromium, and in the form of small particles or aerosols. Iron and chrome production are the most important industrial sources, but related industries such as refining, chemical and refractory processing, cement-producing plants, automobile brake lining and catalytic converters for automobiles, leather tanneries, and chrome-laden dyes also contribute to the atmospheric burden of chromium. The average US daily intake from air, water, and food is estimated to be less than 0.2–0.4 μg, 2.0 μg, and 60 μg, respectively. Occupational exposures can be much higher (For example, workers in chromate production, stainless-steel production, chrome plating, and tanning industries may be two orders of magnitude higher than exposure to the general population.[13] Persons living near these facilities or in sites receiving wastes from these industries would be expected to have elevated exposures to mixtures of Cr^{3+} and Cr^{6+} compounds.

The respiratory tract is the major target organ for Cr^{6+} toxicity, for acute (short-term) and chronic (long-term) inhalation exposures. Asthma-like symptoms, such as shortness of breath, coughing, and wheezing were reported from a case of acute exposure to Cr^{6+}, whereas damage to the septum, bronchitis, decreased pulmonary function, pneumonia, and other respiratory effects can result from chronic exposure. Human epidemiology and animal studies have clearly associated inhaled chromium (VI) with human cancers, especially an increased risk of lung cancer.[14] The health-based thresholds for chromium are shown in Fig. 11.7.

[12] Agency for Toxic Substances and Disease Registry (ATSDR). *Toxicological Profile for Chromium*. US Public Health Service, US Department of Health and Human Services, Atlanta, GA, 1998.

[13] Agency for Toxic Substances and Disease Registry (ATSDR), *Toxicological Profile for Chromium*. US Public Health Service, US Department of Health and Human Services, Atlanta, GA, 1998; and US Environmental Protection Agency, *Toxicological Review of Trivalent Chromium*. National Center for Environmental Assessment, Office of Research and Development, Washington, DC, 1998.

[14] Ibid.

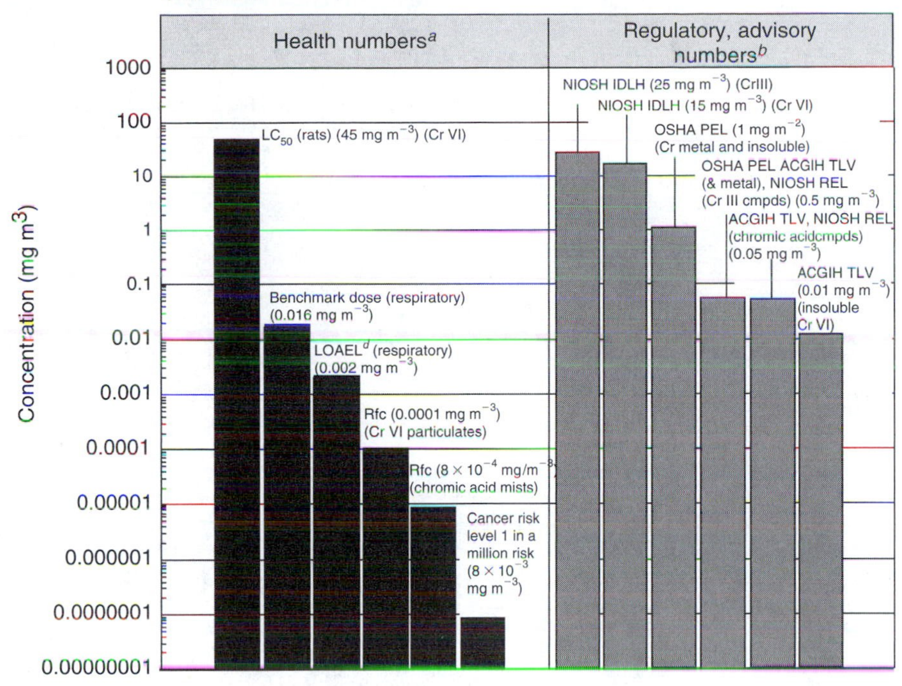

Fig. 11.7. Health data from inhalation exposure to chromium. *Sources*: Agency for Toxic Substances and Disease Registry (ATSDR), *Toxicological Profile for Chromium*. US Public Health Service, US Department of Health and Human Services, Atlanta, GA, 1998; US Environmental Protection Agency, *Integrated Risk Information System (IRIS) on Chromium VI*. National Center for Environmental Assessment, Office of Research and Development, Washington, DC, 1999; Occupational Safety and Health Administration (OSHA), Occupational Safety and Health Standards, Toxic and Hazardous Substances. *Code of Federal Regulations*. 29 CFR 1910.1000, 1998; American Conference of Governmental Industrial Hygienists (ACGIH), *1999 TLVs and BEIs. Threshold Limit Values for Chemical Substances and Physical Agents, Biological Exposure Indices*. Cincinnati, OH, 1999; and National Institute for Occupational Safety and Health (NIOSH), *Pocket Guide to Chemical Hazards*. US Department of Health and Human Services, Public Health Service, Centers for Disease Control and Prevention, Cincinnati, OH. 1997. *Note*: ACGIH TLV is the American Conference of Governmental and Industrial Hygienists' threshold limit value expressed as a time-weighted average; the concentration of a substance to which most workers can be exposed without adverse effects. LOAEL is the lowest-observed-adverse-effect level. NIOSH REL is the National Institute of Occupational Safety and Health's recommended exposure limit; NIOSH-recommended exposure limit for an 8- or 10-h time-weighted-average exposure and/or ceiling. NIOSH IDLH is NIOSH's immediately dangerous to life or health concentration; NIOSH recommended exposure limit to ensure that a worker can escape from an exposure condition that is likely to cause death or immediate or delayed permanent adverse health effects or prevent escape from the environment. OSHA PEL is the Occupational Safety and Health Administration's permissible exposure limit expressed as a time-weighted average; the concentration of a substance to which most workers can be exposed without adverse effect averaged over a normal 8-h workday or a 40-h workweek. OSHA ceiling is OSHA's short-term exposure limit; 15-min time-weighted-average exposure that should not be exceeded at any time during a workday even if the 8-h time-weighted average is within the threshold limit value. [a] Health numbers are toxicological numbers from animal testing or risk assessment values developed by EPA. [b] Regulatory numbers are values that have been incorporated in Government regulations, while advisory numbers are nonregulatory values provided by the government or other groups as advice. OSHA numbers are regulatory, whereas NIOSH and ACGIH numbers are advisory. [c] This LOAEL is from the critical study used as the basis for the EPA RfC.

TABLE 11.1

Physiological Unteractions Between Toxic and Essential Metals in the Human Body

Toxic metal	Essential metal	Effect
Cadmium	Zinc Iron	Nephrotoxicity (kidney dysfunction)
Lead	Calcium Iron Zinc	Neurotoxicity, including cognitive and behavioral effects in children
Mercury	Selenium	Neurotoxicity, including peripheral and central nervous system toxicity (*in utero* through adult)
Aluminum	Iron Calcium Magnesium Manganese	Central nervous system toxicity Bone diseases and dysfunction

Sources: US Environmental Protection Agency (EPA), *Mercury Study Report to Congress*. Office of Air Quality Planning and Standards and Office of Research and Development, Washington, DC; and Goyer, R. A., Toxic and essential metal interactions. *Annu. Rev. Nutr.* **17**, 37–50.

Further complicating the physiology of essential metals are the that numerous interactions that occur within an organism.[15] For example, toxic metals will react with essential metal interactions in the central nervous system (see Table 11.1). Thus, the mixtures and chemical speciation of essential compounds make estimations of the effects from exposure highly uncertain and complex, but from an air pollution control standpoint, emissions of essential metals almost always need to be decreased, since the inhalation route is not an ideal means for human intake of essential metals.

III. THE HUMAN RESPIRATORY SYSTEM

The primary function of the human respiratory system is to deliver O_2 to the bloodstream and to remove CO_2 from the body. These two processes occur concurrently as the breathing cycle is repeated. Air containing O_2 flows into the nose and/or mouth and down through the upper airway to the alveolar region, where O_2 diffuses across the lung wall to the blood-stream. The counterflow involves transfer of CO_2 from the blood to the alveolar region and then up the airways and out the nose. Because of the extensive interaction of the respiratory system with the surrounding atmosphere, air pollutants or trace gases can be delivered to the respiratory system.

The anatomy of the respiratory system is shown in Fig. 11.8. This system may be divided into three regions: the nasal, tracheobronchial, and pulmonary.

[15] This is actually a very common phenomenon in toxicology. Most exposures in the real world are mixtures. Effects can be additive, antagonistic (protective), or synergistic.

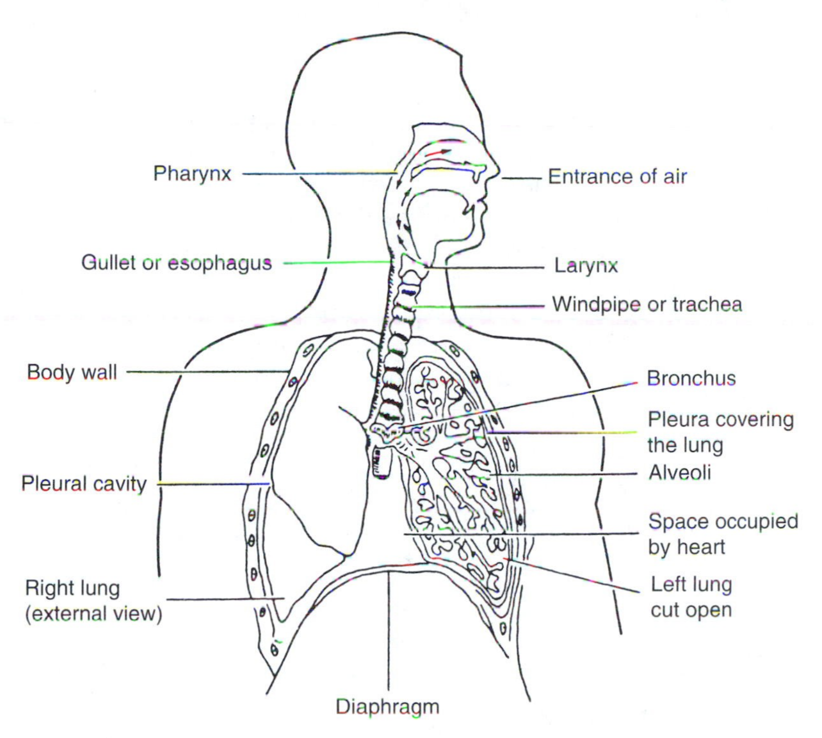

Fig. 11.8. Anatomy of the human respiratory system.

The nasal region is composed of the nose and mouth cavities and the throat. The tracheobronchial region begins with the trachea and extends through the bronchial tubes to the alveolar sacs. The pulmonary region is composed of the terminal bronchi and alveolar sacs, where gas exchange with the circulatory system occurs. Figure 11.8 illustrates the continued bifurcation of the trachea to form many branching pathways of increasingly smaller diameter by which air moves to the pulmonary region. The trachea branches into the right and left bronchi. Each bronchus divides and subdivides at least 20 times; the smallest units, bronchioles, are located deep in the lungs. The bronchioles end in about 3 million air sacs, the alveoli.

The behavior of particles and gases in the respiratory system is greatly influenced by the region of the lung in which they are located [8]. Air passes through the upper region and is humidified and brought to body temperature by gaining or losing heat. After the air is channeled through the trachea to the first bronchi, the flow is divided at each subsequent bronchial bifurcation until very little apparent flow is occurring within the alveolar sacs. Mass transfer is controlled by molecular diffusion in this final region. Because of the very different flows in the various sections of the respiratory region, particles suspended in air and gaseous air pollutants are treated differently in the lung.

A. Particle and Gas Behavior in the Lung

Particle behavior in the lung is dependent on the aerodynamic characteristics of particles in flow streams. In contrast, the major factor for gases is the solubility of the gaseous molecules in the linings of the different regions of the respiratory system. The aerodynamic properties of particles are related to their size, shape, and density. The behavior of a chain type or fiber may also be dependent on its orientation to the direction of flow. The deposition of particles in different regions of the respiratory system depends on their size. The nasal openings permit very large dust particles to enter the nasal region, along with much finer airborne particulate matter. Particles in the atmosphere can range from less than 0.01 μm to more than 50 μm in diameter.

The relationship between the aerodynamic size of particles and the regions where they are deposited is shown in Fig. 11.9 [9]. Larger particles are deposited in the nasal region by impaction on the hairs of the nose or at the bends of the nasal passages. Smaller particles pass through the nasal region and are deposited in the tracheobronchial and pulmonary regions. Particles are removed by impacts with the walls of the bronchi when they are unable to follow the gaseous streamline flow through subsequent bifurcations of the bronchial tree. As the airflow decreases near the terminal bronchi, the smallest particles are removed by Brownian motion, which pushes them to the alveolar membrane.

B. Removal of Deposited Particles from the Respiratory System

The respiratory system has several mechanisms for removing deposited particles [8]. The walls of the nasal and tracheobronchial regions are coated

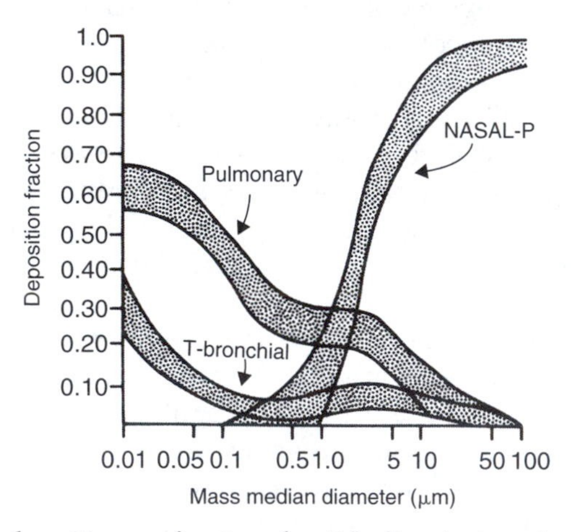

Fig. 11.9. Particle deposition as a function of particle diameter in various regions of the lung. The nasopharyngeal region consists of the nose and throat; the tracheobronchial (T-bronchial) region consists of the windpipe and large airways; and the pulmonary region consists of the small bronchi and the alveolar sacs. *Source*: Task Group on Lung Dynamics, *Health Phys.* **12**, 173 (1966).

with a mucous fluid. Nose blowing, sneezing, coughing, and swallowing help remove particles from the upper airways. The tracheobronchial walls have fiber cilia which sweep the mucous fluid upward, transporting particles to the top of the trachea, where they are swallowed. This mechanism is often referred to as the *mucociliary escalator*. In the pulmonary region of the respiratory system, foreign particles can move across the epithelial lining of the alveolar sac to the lymph or blood systems, or they may be engulfed by scavenger cells called *alveolar macrophages*. The macrophages can move to the mucociliary escalator for removal.

For gases, solubility controls removal from the airstream. Highly soluble gases such as SO_2 are absorbed in the upper airways, whereas less soluble gases such as NO_2 and O_3 may penetrate to the pulmonary region. Irritant gases are thought to stimulate neuroreceptors in the respiratory walls and cause a variety of responses, including sneezing, coughing, bronchoconstriction, and rapid, shallow breathing. The dissolved gas may be eliminated by biochemical processes or may diffuse to the circulatory system.

IV. IMPACT OF AIR POLLUTION ON HUMANS

Risk is a quantifiable engineering concept, and in its simplest form risk (R) is the product of the hazard (H) and the exposure (E) to that hazard:

$$R = H \times E \tag{11.12}$$

Environmental risk assessment consists of a number of steps.

A. Hazard Identification

A hazard is anything with the potential for causing harm. Air pollutants are hazards to health and the environment. The hazard is an intrinsic property of a substance, i.e. a concept of potential harm. For example, a chemical hazard is an absolute expression of a substance's properties, since all substances have unique physical and chemical properties. These properties can render the substance to be hazardous. Conversely, Eq. (11.12) shows risk can only occur with exposure. A person walking on a street in the summer has little likelihood of a person slipping on ice. Also, the total slipping risk is never necessarily zero (e.g. one could step on an oily surface any time of year). By analogy, if a person is a sufficient distance from an air pollution source, the health risk from that particular air pollutant is low. However, since certain air pollutants are persistent and can remain somewhere in the environment, the exposure is not zero. Also, even if the air pollutant exposure is near zero, that person's cancer risk is not zero, since he or she may be exposed to other cancer hazards.

Generally, increasing the amount of the dose means a greater incidence of the adverse outcome.

Dose–response assessment generally follows a sequence of five steps[16]:

1. Fitting the experimental dose–response data from animal and human studies with a mathematical model that fits the data reasonably well (see Fig. 11.10).
2. Expressing the upper confidence limit (e.g. 95%) line equation for the selected mathematical model.
3. Extrapolating the confidence limit line to a response point just below the lowest measured response in the experimental point (known as the "point of departure"), i.e. the beginning of the extrapolation to lower doses from actual measurements.

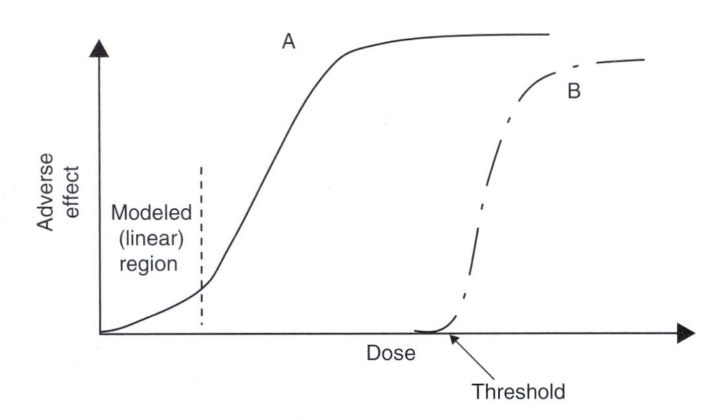

Fig. 11.10. Prototypical dose–response curves. Curve A represents the "no-threshold" curve, which predicts a response (e.g. cancer) even if exposed to a single molecule ("one-hit" model). As shown, the low end of the curve, i.e. below which experimental data are available, is linear. Thus, Curve A represents a linearized multistage model. Curve B represents toxicity above a certain threshold (NOAEL is the level below which no response is expected). Another threshold is the no observable effect concentration (NOEC), which is the highest concentration where no effect on survival is observed ($NOEC_{survival}$) or where no effect on growth or reproduction is observed ($NOEC_{growth}$). Note that both curves are sigmoidal in shape because of the saturation effect at high dose (i.e. less response with increasing dose). *Source*: Adapted from Vallero, D. A., *Environmental Contaminants: Assessment and Control*, Elsevier Academic Press, Burlington, MA, 2004.

[16] US Environmental Protection Agency, Guidelines for Carcinogen Risk Assessment, Report No. EPA/630/R-00/004, *Federal Register* **51**(185), 33992–34003, 1986, Washington, DC; and Larsen, R. A., An Air Quality Data Analysis System for Interrelating Effects, Standards, and Needed Source Reductions: Part 13—Applying the EPA Proposed Guidelines for Carcinogen Risk Assessment to a Set of Asbestos Lung Cancer Mortality Data. *J. Air Waste Manag. Assoc.*, **53**, 1326–1339.

4. Assuming the response is a linear function of dose from the point of departure to zero response at zero dose.
5. Calculating the dose on the line that is estimated to produce the response.

The curves in Fig. 11.10 represent those generally found for toxic chemicals.[17] Once a substance is suspected of being toxic, the extent and quantification of that hazard is assessed.[18] This step is frequently referred to as a dose–response evaluation because this is when researchers study the relationship between the mass or concentration (i.e. dose) and the damage caused (i.e. response). Many dose–response studies are ascertained from animal studies (*in vivo* toxicological studies), but they may also be inferred from studies of human populations (epidemiology). To some degree, "petri dish" (i.e. *in vitro*) studies, such as mutagenicity studies like the Ames test[19] of bacteria complement dose–response assessments, but are mainly used for screening and qualitative or, at best, semi-quantitative analysis of responses to substances. The actual name of the test is the "Ames *Salmonella*/microsome mutagenicity assay" which shows the short-term reverse mutation in histidine dependent *Salmonella* strains of bacteria. Its main use is to screen for a broad range of chemicals that induce genetic aberrations leading to genetic mutations. The process works by using a culture that only allows only those bacteria whose genes revert to histidine interdependence to form colonies. As a mutagenic chemical is added to the culture, a biological gradient can usually be determined. That is, the more chemical that is added, the greater the number and size of mutated colonies on the plate. The test is widely used to screen for mutagenicity of new or modified chemicals and mixtures. It is also a "red flag" for carcinogenicity, since cancer is a genetic disease and a manifestation of mutations.

The toxicity criteria include both acute and chronic effects, and include both human and ecosystem effects. These criteria can be quantitative. For example, a manufacturer of a new chemical may have to show that there are no toxic effects in rodents exposed to concentrations below $10\,mg\,m^{-3}$. If rodents show effects at $9\,mg\,m^{-3}$, the new chemical would be considered to be unacceptably toxic.

A contaminant is acutely toxic if it can cause damage with only a few doses. Chronic toxicity occurs when a person or ecosystem is exposed to a contaminant over a protracted period of time, with repeated exposures. The

[17] Duffus, J., and Worth, H., Training program: *The Science of Chemical Safety: Essential Toxicology—4; Hazard and Risk*, IUPAC Educators' Resource Material, International Union of Pure and Applied Chemistry.

[18] The optimal range as shown in Fig. 11.2 between deficiency and toxicity. Like the other curves, the safe levels of both effects would be calculated and appropriate factors of safety applied.

[19] For an excellent summary of the theory and practical applications of the Ames test, see: Mortelmans K., and Zeiger, E., The Ames *Salmonella*/Microsome mutagenicity assay. *Mutat. Res.* **455**, 29–60 (2000).

curves in Fig. 11.10 are sigmoidal because toxicity is often concentration dependent. As the doses increase, the response cannot mathematically stay linear (e.g. the toxic effect cannot double with each doubling of the dose). So, the toxic effect continues to increase, but at a decreasing rate (i.e. decreasing slope). Curve A is the classic cancer dose–response, i.e. any amount of exposure to a cancer-causing agent may result in an expression of cancer at the cellular level (i.e. no safe level of exposure). Thus, the curve intercepts the x-axis at 0.

Curve B is the classic non-cancer dose–response curve. The steepness of the three curves represents the potency or severity of the toxicity. For example, Curve B is steeper than Curve A, so the adverse outcome (disease) caused by chemical in Curve B is more potent than that of the chemical in Curve A. Obviously, potency is only one factor in the risk. For example, a chemical may be very potent in its ability to elicit a rather innocuous effect, like a headache, and another chemical may have a rather gentle slope (lower potency) for a dreaded disease like cancer.

With increasing potency, the range of response decreases. In other words, as shown in Fig. 11.11, a severe response represented by a steep curve will be manifested in greater mortality or morbidity over a smaller range of dose. For example, an acutely toxic contaminant's concentration that kills 50% of test animals (i.e. the LC_{50}) is closer to the concentration that kills only 5% (LC_5) and the concentration that kills 95% (LC_{95}) of the animals. The dose difference of a less acutely toxic contaminant will cover a broader range, with the differences between the LC_{50} and LC_5 and LC_{95} being more extended than that of the more acutely toxic substance.

The major differentiation of toxicity is between carcinogenic and non-cancer outcomes. The term "non-cancer" is commonly used to distinguish cancer outcomes (e.g. bladder cancer, leukemia, or adenocarcinoma of the lung) from other maladies, such as neurotoxicity, immune system disorders and endocrine disruption. The policies of many regulatory agencies and international organizations treat cancer differently than non-cancer effects, particularly in how the dose–response curves are drawn. As we saw in the dose–response curves, there is no safe dose for carcinogens. Cancer dose–response is almost always a non-threshold curve, i.e. no safe dose is expected while, theoretically at least, non-cancer outcomes can have a dose below which the adverse outcomes do not present themselves. So, for all other diseases safe doses of compounds can be established. These are known as reference doses (RfD), usually based on the oral exposure route. If the substance is an air pollutant, the safe dose is known as the RfC, which is calculated in the same manner as the RfD, using units that apply to air (e.g. $\mu g\, m^{-3}$). These references are calculated from thresholds below which no adverse effect is observed in animal and human studies. If the models and data were perfect, the safe level would be the threshold, known as the no observed adverse effect level (NOAEL).

The term "non-cancer" is very different from "anticancer" or "anticarcinogens." Anticancer procedures include radiation and drugs that are

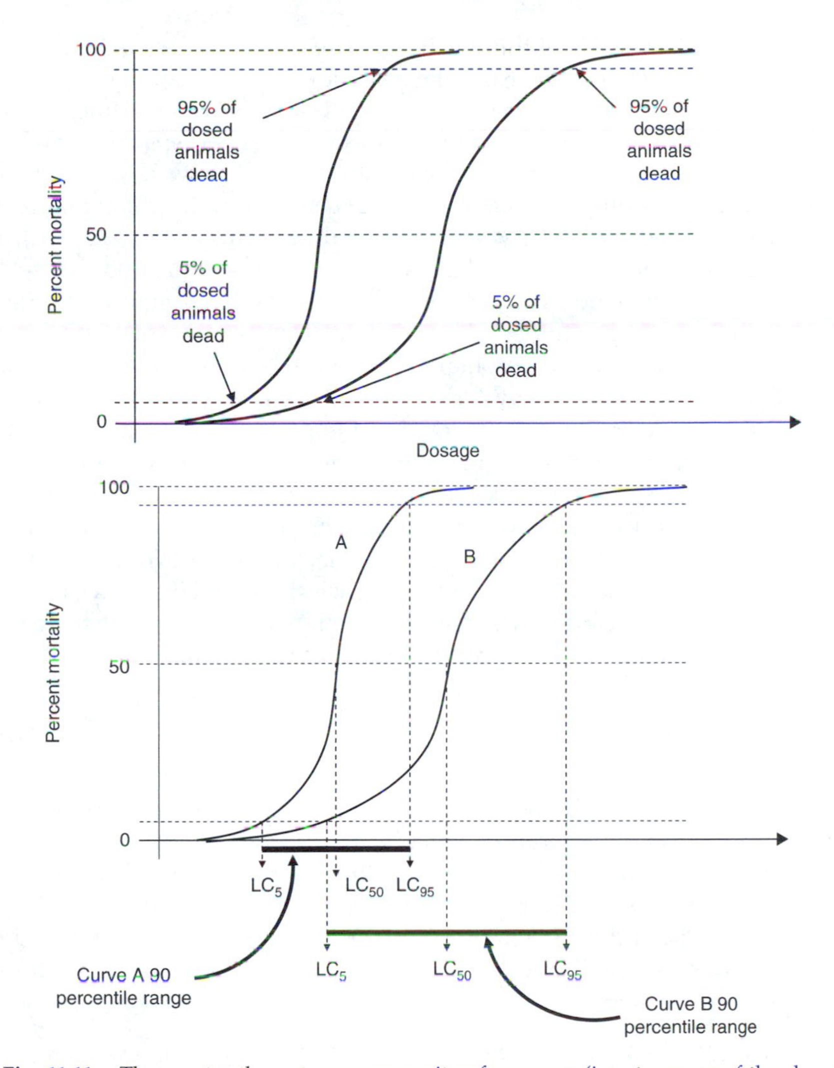

Fig. 11.11. The greater the potency or severity of response (i.e. steepness of the slope) of dose–response curve, the smaller the range of toxic response (90 percentile range shown in bottom graph). Also, note that both curves have thresholds and that curve B is less acutely toxic based on all three reported lethal doses (LC_5, LC_{50}, and LC_{95}). In fact, the LC_5 for curve B is nearly the same as the LC_{50} for curve A, meaning that about the same dose, contaminant A kills nearly half the test animals, but contaminant B has only killed 5%. Thus, contaminant A is much more acutely toxic. *Source*: Vallero, D. A., *Environmental Contaminants: Assessment and Control*, Elsevier Academic Press, Burlington, MA, 2004.

used to attack tumor cells. Anti-carcinogens are chemical substances that work against the processes that lead to cancer, such as antioxidants and essential substances that help the body's immune, hormonal and other systems to prevent carcinogenesis.

In reality, the hazard identification and dose–response research is inexact and often has much uncertainty. Chief reasons for this uncertainty include variability among the animals and people being tested, as well as differences in response to the compound by different species (e.g. one species may have decreased adrenal gland activity, while another may show thyroid effects). Sometimes, studies only indicate the lowest concentration of a contaminant that causes the effect, i.e. the lowest observed adverse effect level (LOAEL), but the NOAEL is unknown. If the LOAEL is used, one is less certain how close this is to a safe level where no effect is expected. Often, there is temporal incongruence, such as most of the studies taking place in a shorter timeframe than in the real world. Thus, acute or subchronic effects have to be used to estimate chronic diseases. Likewise, studies may have used different ways to administer the doses. For example, if the dose is oral, but the pollutant is more likely to be inhaled by humans, this route-to-route extrapolation adds uncertainty. Finally, the data themselves may be weak because the study may lack sufficient quality or the precision, accuracy, completeness and representativeness of the data are unknown. These are quantified as uncertainty factors (UFs). Modifying factors (MFs) address uncertainties that are less explicit than the UFs. Thus, any safe level must consider these uncertainties, so the RfD moves closer to zero, i.e. the threshold is divided by these factors (usually multiples of 10):

$$\mathrm{RfD} = \frac{\mathrm{NOAEL}}{\mathrm{UF} \times \mathrm{MF}} \tag{11.13}$$

For air pollutants, the RfC would use the no observable adverse effect concentration (NOAEC). Uncertainty can also come from error. Two errors can occur when information is interpreted in the absence of sound science. The first is the *false negative*, or reporting that there is no problem when one in fact exists. The need to address this problem is often at the core of the positions taken by environmental and public health agencies and advocacy groups. They ask questions like:

- What if the leak detector registers zero, but in fact toxic substances are being released from the tank?
- What if this substance really does cause cancer but the tests are unreliable?
- What if people are being exposed to a contaminant, but via a pathway other than the ones being studied?
- What if there is a relationship that is different from the laboratory when this substance is released into the "real world," such as the difference between how a chemical behaves in the human body by itself as opposed to when other chemicals are present (i.e. the problem of "complex mixtures")?

The other concern is, conversely, the *false positive*. This can be a major challenge for public health agencies with the mandate to protect people from exposures to environmental contaminants. For example, what if previous evidence shows that an agency had listed a compound as a potential endocrine disruptor, only to find that a wealth of new information is now showing that it has no such effect? This can happen if the conclusions were based on faulty models, or models that only work well for lower organisms, but subsequently developed models have taken into consideration the physical, chemical, and biological complexities of higher-level organisms, including humans. False positives may force public health officials to devote inordinate amounts of time and resources to deal with so-called "non-problems." False positives also erroneously scare people about potentially useful products. False positives, especially when they occur frequently, create credibility gaps between engineers and scientists and the decision makers. In turn the public, those whom we have been charged to protect, lose confidence in environmental professionals.

Both false negatives and false positives are rooted in science. Therefore, environmental risk assessment is in need of high quality, scientifically based information. Put in engineering language, the risk assessment process is a "critical path" in which any unacceptable error or uncertainty along the way will decrease the quality of the risk assessment and, quite likely, will lead to a bad environmental decision.

Intrinsic properties of compounds render them more or less toxic. For example, polycyclic aromatic hydrocarbons (PAHs) are a family of large, flat compounds with repeating benzene structures. This structure renders them highly hydrophobic, i.e. fat soluble, and difficult for an organism to eliminate (since most blood and cellular fluids are mainly water). This property also enhances the PAHs' ability to insert themselves into the deoxyribonucleic acid (DNA) molecule, interfering with transcription and replication. This is why some large organic molecules can be mutagenic and carcinogenic. One of the most toxic PAHs is benzo(a)pyrene, which is found in cigarette smoke, combustion of coal, coke oven emissions, and numerous other processes that use combustion.

After a compound is released into the environment, its chemical structure can substantially. Further, compounds change when taken up and metabolized by organisms. For example, methyl parathion, an insecticide used since 1954 and has been associated with numerous farm worker poisonings. It has also been associated with health problems in inner city communities. Methyl parathion can cause rapid, fatal poisoning through skin contact, inhalation, and eating or drinking. Due to its nature, it can linger in homes for years after its application. People living in low-income housing projects are disproportionately exposed to methyl parathion. Although methyl parathion is heavily restricted, residents and landlords have been able to obtain it since it is one of the most effective ways to deal with cockroaches. This has led to illnesses and even reports of death. In addition, the parent compound breaks down after the pesticide is applied. It may become less toxic, but it can also be transformed to more toxic metabolites, a process known as bioactivation. Figure 11.12 illustrates the forms that the pesticide methyl parathion can take after it is released

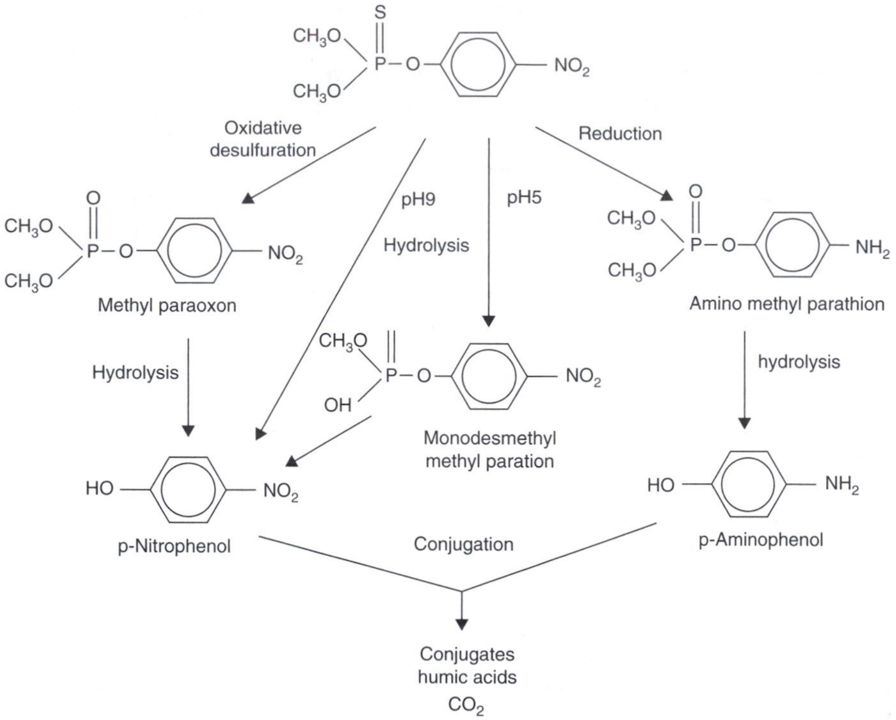

Oxidative desulfuration
Reduction
pH9
pH5
Hydrolysis
Methyl paraoxon
Amino methyl parathion
Hydrolysis
hydrolysis
Monodesmethyl methyl paration
p-Nitrophenol
Conjugation
p-Aminophenol
Conjugates humic acids CO_2

Fig. 11.12. Proposed pathway of methyl parathion in water. Environmental factors, including pH, available oxygen and water, determine the pathway. *Source*: World Health Organization, *International Programme on Chemical Safety*, Environmental Health Criteria 145: Methyl Parathion, Geneva, Switzerland, 1993; Bourquin, A. W., Garnas, R. L., Pritchard, P. H., Wilkes, F. G., Cripe, C. R., and Rubinstein, N. I., Interdependent microcosms for the assessment of pollutants in the marine environment. *Int.l J.Environ. Stud.*, **13**(2), 131–140, 1979; and Wilmes, R., *Parathion-methyl: Hydrolysis Studies*. Leverkusen, Germany, Bayer AG, Institute of Metabolism Research, 34 pp., 1987 (Unpublished report No. PF 2883, submitted to WHO by Bayer AG).

into the environment, and Fig. 11.13 shows the metabolism of methyl parathion in rodents.

Like the numerous air pollutants, catalysis plays a key role in its degradation and metabolism. Organic catalysts, such as hydrolases, are known as enzymes. Note that these reactions can generate byproducts that are either less toxic (i.e. detoxification) or more toxic (i.e. bioactivation) than the parent compound. For methyl parathion, the metabolic detoxification pathways are shown as 2 and 3 and the bioactivation pathway as 1 in Fig. 11.13. Methyl paroxon is more toxic than methyl parathion. Note that these reactions occur within and outside of an organism, so a person may be exposed to the more toxic byproduct some time after the pesticide has been applied. In other words, it is possible that the risk of health effects is increased with time until the less toxic byproducts (e.g. para-nitrophenol) replaces the more toxic substances (e.g. methyl paroxon).

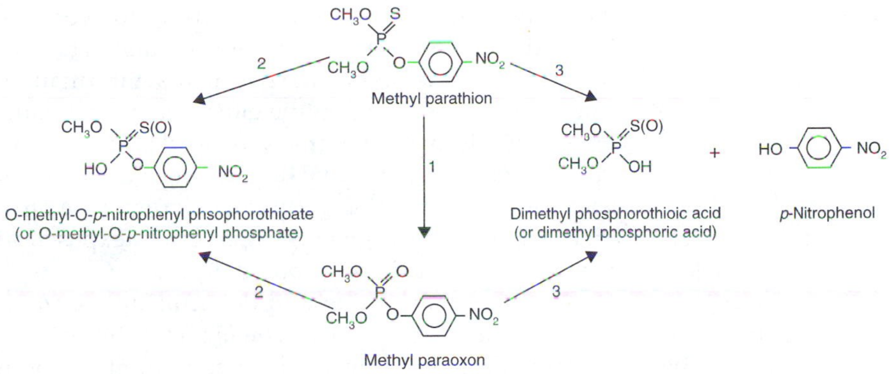

Methyl parathion

O-methyl-O-*p*-nitrophenyl phsophorothioate
(or O-methyl-O-*p*-nitrophenyl phosphate)

Dimethyl phosphorothioic acid
(or dimethyl phosphoric acid)

p-Nitrophenol

Methyl paraoxon

Fig. 11.13. Sometimes, chemicals become more toxic as a result of an organism's (in this instance, rodents) metabolism. For example, methyl parathion's toxicity changes according to the degradation pathway. During metabolism, the biological catalysts (enzymes) make the molecule more polar by hydrolysis, oxidation and other reactions. Bioactivation (pathway 1) renders the metabolites more toxic than the parent compound, while detoxification (pathways 2 and 3) produces less toxic metabolites. The degradation product, methyl paraoxon may be metabolized in the same pathways as those for methyl parathion. *Source*: International Agency for Research on Cancer, Methyl parathion, in *Miscellaneous Pesticides*, pp. 131–152. Lyon, France, 1983 (IARC Monographs on the evaluation of the carcinogenic risk of chemicals to humans, Volume 30).

The impact of air pollution on human beings has been the major force motivating efforts to control it. Most persons do not have the luxury of choosing the air they breathe. Working adults can make some choices in the selection of their occupation and the place where they live and work, but children and the nonworking elderly often cannot. The receptor population in an urban location includes a wide spectrum of demographic traits with respect to age, sex, and health status. Within this group, certain sensitive subpopulations have been identified: (1) very young children, whose respiratory and circulatory systems are still undergoing maturation; (2) the elderly, whose respiratory and circulatory systems function poorly; and (3) persons who have preexisting diseases such as asthma, emphysema, and heart disease. These subpopulations have been found to exhibit more adverse responses from exposure to air pollutants than the general population [10].

Air pollution principally affects the respiratory, circulatory, and olfactory systems. The respiratory system is the principal route of entry for air pollutants. However, any system can be affected. For example, although lead (Pb) and mercury (Hg) may enter via inhalation, their primary health effect is on the nervous system. Conversely, a pesticide may be an air pollutant that settles on surfaces and on food, so that exposure is through the skin and by ingestion, respectively.

Health effects data come from three types of studies: clinical, epidemiological, and toxicological. Clinical and epidemiological studies focus on human subjects, whereas toxicological studies are conducted on animals or simpler cellular systems. Ethical considerations limit human exposure to even low levels of air pollutants which do not have irreversible effects. Table 11.2 lists the advantages and disadvantages of each type of experimental information.

In general, clinical studies provide evidence on the effects of air pollutants under reproducible laboratory conditions. The exposure level may be accurately determined. The physiological effect may be quantified, and the health status of the subject is well known. This type of study can determine the presence or absence of various endpoints for a given sample group exposed to short-term, low-level concentrations of various air pollutants.

Epidemiology is the study of the distribution and determinants of states or events related to health in specific populations. The fact that the subjects are exposed to the actual pollutants existing in their community is both the greatest strength and the greatest weakness of epidemiological studies. Two key measures used to describe and analyze diseases in populations exposed to air pollutants are incidence, the number of newly reported cases in a population during a year, and prevalence, the total number of cases in a population.

Descriptive epidemiology might consider something like a specific hormonal dysfunction with an incidence of 150 per million and a prevalence of 150 per million. Analytical epidemiology would try to explain these numbers. For example, two possible explanations are that the cure rate could be equal to the number of new cases each year; or that the mortality rate could be 100% in every year studied, so that the numbers of new cases are the only

TABLE 11.2

Three Disciplinary Approaches for Obtaining Health Information

Discipline	Population	Strengths	Weaknesses
Epidemiology	Communities	Natural exposure	Difficulty of quantifying exposure
	Diseased groups	No extrapolations	Many covariates and confounders
		Susceptible groups	Minimal dose–response data
		Long-term, low-level effects	Association versus causation
Clinical studies	Experimental	Controlled exposure	Artificial exposure
	Diseased subjects	Few covariates	Acute effects only
		Vulnerable persons	Hazards
		Cause–effect	Public acceptance
Toxicology	Animals	Maximal dose–response data	Realistic models of human disease?
	Cells	Rapid acquisition of data	Threshold of human response?
	Biochemical systems	Cause–effect	Extrapolation
		Mechanism of response	

ones that show up in the data each year. Obviously, the former is infinitely better than the latter.

Another example might be that over a 10-year span, the incidence of a respiratory illness increased from 10 to 200 per million. There are at least three plausible explanations. If incidence could be increasing over 10 years, but prevalence is decreasing. The twentyfold increase could be the result of an actual increase in the number of new cases, possibly from an increase in the concentration of a stressor in the environment leading to increased exposures. Another explanation could be improving detection capabilities. For example, the 200 is closer to the actual number of new cases, but physicians have become better at recognizing the symptoms associated with the disease, improving the nosological data. A third possibility is misdiagnosis and erroneous reporting of health statistics. For example, physicians may increase their diagnoses of a new syndrome that was previously diagnosed as something else. The syndrome incidence may not have increased, but it has become popular to so designate. These interpretations point to the need to understand the data underlying health reports.

Two important study designs that have been important in air pollution epidemiology are the cohort and case–control studies. Investigators observe diseases and exposures over time.[20] Two types of cohort studies are life table studies and longitudinal. Life table cohort studies follow traditional life table methodologies, observing the ratio of general exposures and person-times to the incidences of diseases. Longitudinal studies follow populations and strata within these populations over time to link various types of exposures and diseases to specific changes experienced by the population and subpopulations over a specified time. Longitudinal studies include time-series and panel studies, as well as "ecologic" (between group differences) studies. Time-series studies collect observations sequentially to observe changes in exposures and health outcomes (e.g. changes in asthma incidence with changes in particulate matter concentrations over time). Panel studies involve measuring subjects continuously (e.g. daily) for symptoms and physiological functions and comparing these health metrics to possible exposures or ambient levels of contaminants. Longitudinal studies can be either prospective, where a group is identified and then followed for years after, or retrospective, where the group is identified and the investigators try to determine which risk factors and exposures appear to be associated with the group's present health status:

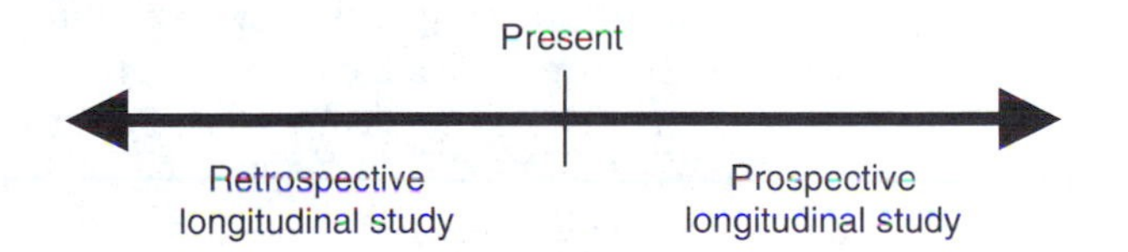

[20] Tager, I. B., Current view of epidemiologic study designs for occupational and environmental lung diseases. *Environ. Health Perspect.* **108** (Suppl. 4), 615–623 (August 2000).

Case–control studies are usually clinical, where investigators identify two groups: people who already have the health outcome (cases), and people who do not have the outcome (controls). The two groups are studied to determine the extent to which an exposure was more prevalent in the past history in comparison to the other group. A "nested" case–control study is one that is part of a cohort study. The advantage of a nested case–control study over a regular case–control study is that the exposure measurements are obtained before the health outcome has occurred, so bias is reduced.

Another type of study is the so-called "cluster." A particular group of people in a tightly defined area may develop a disease at a rate much higher than that of the general population. A cluster study is actually a type of retrospective longitudinal study in that it identifies the group to be studied because the members share a particular health outcome and researchers must investigate the myriad of exposures and risk factors that could explain the outcome.

No health data set is perfect. One of the weaknesses of epidemiological data is the inability to control for confounders, which are conditions or variables that are both a risk factor for disease and associated with an exposure of interest. An association between exposure and a confounder (a true risk factor for disease) will falsely indicate that the exposure is associated with disease. For example, if a person is exposed to chemical X at home and develops lung cancer, one must be sure that the chemical X is linked to the cancer, rather than a confounding condition, such as the fact that the person smoked two packs of cigarettes per day. This is why smoking is almost always a confounder in most epidemiological studies of air pollution. Confounding factors may not even be known at the time the epidemiological study was designed, so it was not controlled.

Similarly, not all populations respond in the same way to exposures. For example, exposure to an air pollutant can produce more severe effects in persons that are genetically predisposed to the disease than would be expected for the same exposure to the general population. Much variability exists among subpopulations' susceptibility to particular diseases. Another weakness of epidemiology has to do with the accuracy and representativeness of the data. For example, if physicians are inconsistent in disease taxonomy or in the ways that they report diseases, this will be reflected in the data. One physician may report pneumonia and another bronchitis, while a third may report acute asthma symptoms, all for an identical health episode. Spatial representation is difficult. For example, the address reported for a patient with a chronic disease may be near the health care facility where the patient has recently moved. However, the exposure or risk factors were encountered long ago and far away from the current address that is reported. The strength is the real-world condition of the exposure and the subjects; the weakness is the difficulty in quantifying the relationship between exposure and subsequent effects. In the future, the development of biomarkers may provide a better indication of target dose in epidemiological studies.

The effects attributed to air pollutants range from mild eye irritation to mortality. In many cases, the effect is to aggravate preexisting diseases or to

TABLE 11.3

Mechanisms by which Some Key Air Pollutants that May Increase Risk of Respiratory and Other Health Problems

Air pollutant	Mechanism	Potential health effects
$PM_{2.5}$	• Acute: bronchial irritation, inflammation and increased reactivity • Reduced mucociliary clearance • Reduced macrophage response and (?) reduced local immunity (?) • Fibrotic reaction • Autonomic imbalance, procoagulant activity, and oxidative stress	• Wheezing and exacerbation of asthma • Respiratory infections • Chronic bronchitis and chronic obstructive pulmonary disease (COPD) • Exacerbation of COPD • Excess mortality, including from • Cardiovascular disease
Carbon monoxide	• Binding with hemoglobin (Hb) to produce COHb which reduced O_2 delivery to key organs and the developing fetus	• Low birth weight (fetal COHb 2–10%, or higher) • Increase in perinatal deaths

Sources: US Environmental Protection Agency, Integrated Risk Information System, 2007; and Bruce, N., Perez-Padilla, R., and Albalak, R., *The Health Effects of Indoor Air Pollution Exposure in Developing Countries*. World Health Organization, Report No. WHO/SDE/OEH/02.05, 2002.

degrade the health status, making persons more susceptible to infection or development of a chronic respiratory disease. Some of the physiological mechanisms and effects associated with specific pollutants are listed in Tables 11.3 and 11.4, respectively. Further information is available in the US Environmental Protection Agency criteria documents summarized in Chapter 24.

TABLE 11.4

Specific Air Pollutants and Associated Health Effects

Pollutant	Effects
CO	Reduction in the ability of the circulatory system to transport O_2 Impairment of performance on tasks requiring vigilance Aggravation of cardiovascular disease.
NO_2	Increased susceptibility to respiratory pathogens
O_3	Decrement in pulmonary function Coughing and chest discomfort Increased asthma attacks
Lead	Neurocognitive and neuromotor impairment Heme synthesis and hematologic alterations
Peroxyacyl nitrates and aldehydes	Eye irritation
SO_2/particulate matter	Increased prevalence of chronic respiratory disease Increased risk of acute respiratory disease

V. IMPACT OF ODOR ON HUMANS

Odors are perceived via the olfactory system, which is composed of two organs in the nose: the olfactory epithelium, a very small area in the nasal system, and the trigeminal nerve endings, which are much more widely distributed in the nasal cavity [11]. The olfactory epithelium is extremely sensitive, and humans often sniff to bring more odorant in contact with this area. The trigeminal nerves initiate protective reflexes, such as sneezing or interruption of inhalation, with exposure to noxious odorants.

The health effects of odors are extremely hard to quantify, yet people have reported nausea, vomiting, and headache; induction of shallow breathing and coughing; upsetting of sleep, stomach, and appetite; irritation of the eyes, nose, and throat; destruction of the sense of well-being and the enjoyment of food, home, and the external environment; disturbance; annoyance; and depression [11]. Research under controlled conditions has qualitatively revealed changes in respiratory and cardiovascular systems. The difficulty has been in establishing the relationship between the intensity or duration of the exposure and the magnitude of the effects on these systems.

People living in the plume of industries like coke ovens experience the obnoxious smelling compounds, including metallic and sulfur compounds that volatilize during the conversion of coal to coke needed for steel manufacturing. While such areas continue to be industrialized, such ambient air quality as that in the 1960s is no longer tolerated in the West. But such conditions do persist in some lower socioeconomic communities. In junkyards, for example, the combination of fires, wet muck (comprised of soil, battery acid, radiator fluids, motor oil, corroded metal, and water), and oxidizing metals create a unique odor.

Odors have often been associated with public health nuisances. In addition to the link between memory and olfactory centers, however, the nasal–neural connection is important to environmental exposure. This goes beyond nuisance and is an indication of potential adverse health effects. For example, nitric oxide (NO) is a neurotoxic gas released from many sources, such as confined animal feeding operations, breakdown of fertilizers after they are applied to the soil and crops, and emissions from vehicles. Besides being inhaled into the lungs, NO can reach the brain directly. The gas can pass through a thin membrane via the nose to the brain.

The nasal exposure is a different paradigm from that usually used to calculate exposure. In fact, most risk assessment routines do not have a means for calculating exposures other than dermal, inhalation, and ingestion. People who live near swine and poultry facilities can be negatively affected when they smell odors from the facility. This is consistent with other research that has found that people experience adverse health symptoms more frequently when exposed to livestock odors. These symptoms include eye, nose, and throat irritation, headache, nausea, diarrhea, hoarseness, sore throat, cough, chest tightness, nasal congestion, palpitations, shortness of breath, stress, and drowsiness. There is quite a bit of diversity in response, with some people being highly sensitive to even low concentrations of odorant compounds while others are relatively unfazed even at much higher concentrations.

Actually, response to odors can be triggered by three different mechanisms. In the first mechanism, symptoms can be induced by exposure to odorant compounds at sufficiently high concentrations to cause irritation or other toxicological effects. The irritation, not the odor, evokes the health symptoms. The odor sensation is merely as an exposure indicator. In the second mechanism, symptoms of adverse effects result from odorants concentrations lower than those eliciting irritation. This can be owing to genetic predisposition or conditioned aversion. In the third mechanism, symptoms can result from a coexisting pollutant, e.g. an endotoxin, which is a component of the odorant mixture. The variety of health effects associated with air pollutants is vast. An understanding of the mechanisms and processes discussed in this chapter provides a foundation to risk assessment.

REFERENCES

1. Cowling, E. B., *Environ. Sci. Technol.* **16**, 110A–123A (1982).
2. Fantechi, R., and Ghazi, A. (eds.), *Carbon Dioxide and Other Greenhouse Gases: Climatic and Associated Impacts*. Kluwer Academic Publishers, Boston, MA, 1989.
3. International Symposium on Fluorides, *Fluorides: Effects on Vegetation, Animals and Humans*. Paragon Press, Salt Lake City, UT, 1983.
4. Code of Federal Regulations, *Protection of Environment Title 40, Subchapter I, Parts 240–272*. US Government Printing Office, Washington, DC, 1992.
5. Lee, D. H. K. (ed.), *Handbook of Physiology*, Vol. 9, *Reactions to Environmental Agents*. American Physiological Society, Bethesda, MD, 1977.
6. US Environmental Protection Agency, *Air Quality Criteria for Lead*, EPA 600/8-83-018F. Research Triangle Park, NC, June 1986.
7. National Research Council, *Carbon Monoxide*. National Academy of Sciences. Washington, DC, 1977.
8. American Lung Association, *Health Effects of Air Pollution*. American Lung Association, New York, 1978.
9. Task Group on Lung Dynamics, *Health Phys.* **12**, 173 (1966).
10. Shy, C., *Am. J. Epidemiol.* **110**, 661–671 (1979).
11. National Research Council, *Odors from Stationary and Mobile Sources*. National Academy of Sciences, Washington, DC, 1979.

SUGGESTED READING

Klaassen C. D., *Cassarett & Doull's Toxicology—The Basic Science of Poisons*. 6th ed. McGraw-Hill, New York, 2001.

Nagy, G. Z., The odor impact model. *J. Air Waste Manage. Assoc.* **42**, 1567 (1992).

National Research Council, Human Exposure Assessment for Airborne Pollutants: Advances and Opportunities, National Academy Press, Washington, DC, 1991.

Tomatis, L. (ed.), *Air Pollution and Human Cancer*. Springer-Verlag, New York, 1990.

US Environmental Protection Agency, *SI:300—Introduction to Air Pollution Toxicology, Air Pollution Training Institute Virtual Classroom*, 1994, http://yosemite.epa.gov/oaqps/EOGtrain.nsf/DisplayView/SI_300_0-5?OpenDocument.

Vallero, D. A., *Environmental Contaminants: Assessment and Control*. Elsevier Academic Press, Burlington, MA, 2004.

QUESTIONS

1. By extrapolation, what will be the concentration of CO_2 in the year 2050? How does this compare with the concentration in 1980?
2. What factors influence the accumulation of a chemical in the human body?
3. Describe normal lung function.
4. Explain why the inhalation route for lead is considered an important hazard when it accounts for only about 20% of the potential allowable body burden?
5. (a) Explain how CO interacts with the circulatory system, especially the relationship among CO, CO_2, and O_2 in blood cells and how exposure to CO influences normal oxygenation mechanisms. (b) Why are individuals with heart disease at greater risk when exposed to elevated CO levels?
6. From Figs. 11.8 and 11.9, form and defend a hypothesis of the types of particles and gases that may cause or exacerbate asthma.
7. How is particle deposition and removal from the lung influenced by the size of the particles?
8. How do exposure time and type of population influence the air quality standards established for the community and the workplace?
9. Compare the strengths and weaknesses of health effects information obtained from epidemiological, clinical, and toxicological studies.
10. Explain the role of valence in metal bioavailability and toxicity. Why is it unreasonable to try to "eliminate" metals?

12

Effects on Vegetation and Animals

I. INJURY VERSUS DAMAGE

The US Department of Agriculture makes a distinction between air pollution damage and air pollution injury. *Injury* is considered to be any observable alteration in the plant when exposed to air pollution. *Damage* is defined as an economic or aesthetic loss due to interference with the intended use of a plant. This distinction indicates that injury by air pollution does not necessarily result in damage because any given injury may not prevent the plant from being used as intended, e.g., marketed. Thus, damage is a value-laden concept.

Vegetation reacts with air pollution over a wide range of pollutant concentrations and environmental conditions. Many factors influence the outcome, including plant species, age, nutrient balance, soil conditions, temperature, humidity, and sunlight [1]. Any type of observable effect due to exposure can be termed plant injury. A schematic diagram of the potential levels of injury with increasing exposure to air pollution is presented in Fig. 12.1. At low levels of exposure for a given species and pollutant, no significant effects may be observed. However, as the exposure level increases, a series of potential injuries may occur, including biochemical alterations, physiological response, visible symptoms, and eventual death.

Air pollutants may enter plant systems by either a primary or a secondary pathway. The primary pathway is analogous to human inhalation. Figure 12.2

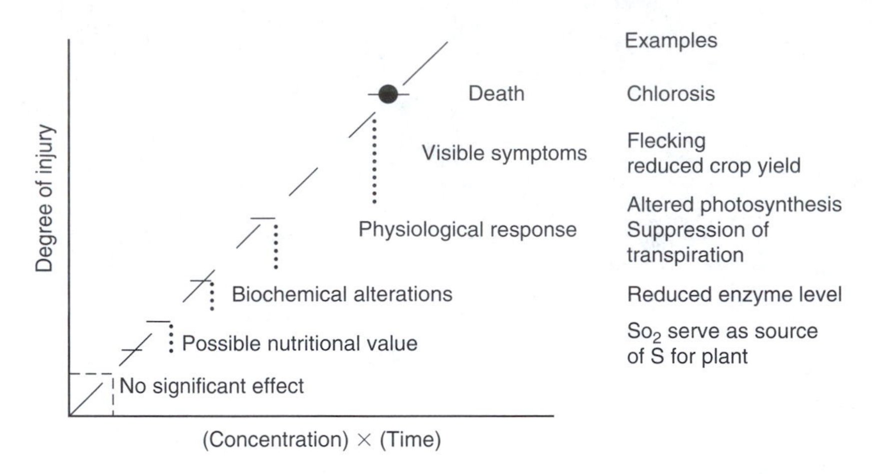

Fig. 12.1. Biological response spectrum for plants and air pollution. Note the similarities to the dose–response curves discussed in Chapter 11.

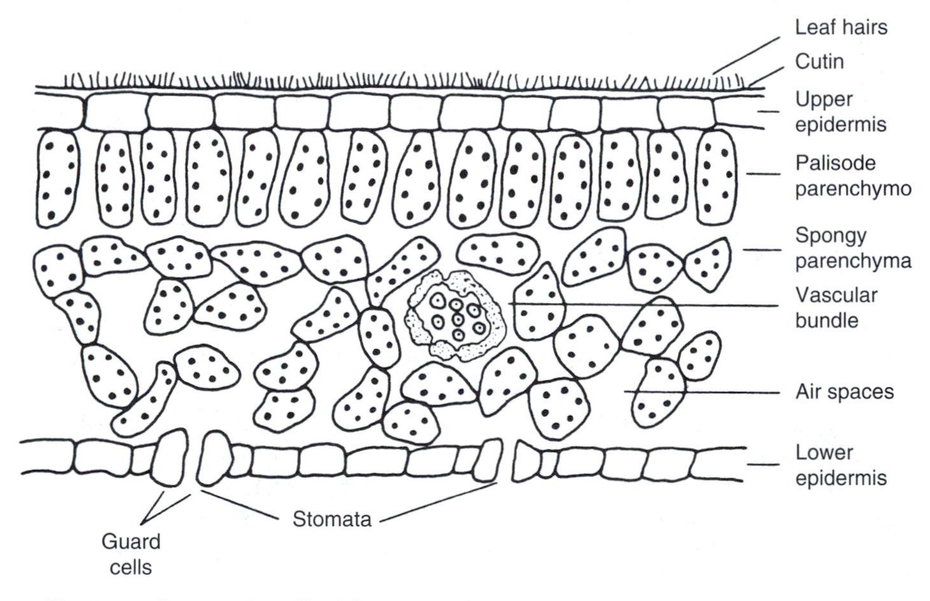

Fig. 12.2. Cross section of leaf showing various components.

shows the cross section of a leaf. Both of the outer surfaces are covered by a layer of epidermal cells, which help in moisture retention. Between the epidermal layers are the mesophyll cells–the spongy and palisade parenchyma. The leaf has a vascular bundle which carries water, minerals, and carbohydrates throughout the plant. Two important features shown in Fig. 12.2 are the openings in the epidermal layers called *stomates*, which are controlled by guard cells which can open and close, and air spaces in the interior of the leaf.

The leaf structure has several important functions, three of which are photosynthesis, transpiration, and respiration [2]. Photosynthesis is accomplished by chloroplasts in the leaf, which combine water and CO_2 in the presence of sunlight to form sugars and release O_2. A simplification of this process is:

$$6CO_2 + 6H_2O \rightarrow C_6H_{12}O_6 + 6O_2 \tag{12.1}$$

Transpiration is the physical movement of water from the root system up to the leaves and its subsequent evaporation to the atmosphere. This process moves nutrients throughout the plant and cools the plant. Respiration is a heat-producing process resulting from the oxidation of carbohydrates by O_2 to form CO_2 and H_2O:

$$C_6H_{12}O_6 + 6O_2 \rightarrow 6CO_2 + 6H_2O \tag{12.2}$$

These three functions involve the movement of O_2, CO_2, and H_2O through the epidermal layers of the leaf. The analogy to human inhalation is obvious. With the diffusion of gases into and out of the leaf, pollutant gases have a direct pathway to the cellular system of the leaf structure. Direct deposition of particulate matter also occurs on the outer surfaces of the leaves.

The indirect pathway by which air pollutants interact with plants is through the root system. Deposition of air pollutants on soils and surface waters can cause alteration of the nutrient content of the soil in the vicinity of the plant. This change in soil condition can lead to indirect or secondary effects of air pollutants on vegetation and plants.

Injury to plants and vegetation is caused by a variety of factors of which air pollution is only one. Drought, too much water, heat and cold, hail, insects, animals, disease, and poor soil conditions are some of the other causes of plant injury and possible plant damage [3]. Estimates suggest that less than 5% of total crop losses are related to air pollution. Air pollution has a much greater impact on some geographic areas and crops than others. Crop failure can be caused by fumigation from a local air pollution source or by more widespread and more frequent exposure to adverse levels of pollution.

The subtle interaction of air pollutants with these other stressors to plants and vegetation is the subject of ongoing research. For some plant systems, exposure to air pollutants may induce biochemical modifications which interfere with the water balance in plants, thereby reducing their ability to tolerate drought conditions.

II. EFFECTS ON VEGETATION AND CROPS

The effects of air pollution on plants range from subtle to catastrophic, as shown in Fig. 12.1. Historically, these effects have been classified as visible symptoms and non-visual or subtle effects [4]. Visible symptoms are deviations from the normal healthy appearance of the leaves. For broadleaf plants, a healthy leaf has good color, with a normal cell structure in the various layers. Deviations from this healthy appearance include tissue collapse and various degrees of loss of color. Extensive tissue collapse or necrosis results from injury to the spongy or palisade cells in the interior of the leaf. The leaf is severely discolored and loses structural integrity. Dead tissue may fall out of the leaf, leaving holes in the structure. Less dramatic discolorations are caused by a reduction in the number of chloroplasts, a symptom referred to as *chlorosis*. Injury to the outer or epidermal layer is referred to as *glazing* or *silvering* of the leaf surface. When the pattern is spotty, the terms *flecking* and *stippling* are used to describe the injury.

Other forms of visible injury are related to various physiological alterations. Air pollution injury can cause early senescence or leaf drop. Stems and leaf structure may be elongated or misshapen. Ornamentals and fruit trees can also show visible injury to the blooms of the fruit, which can result in decreased yield.

The nonvisual or subtle effects of air pollutants involve reduced plant growth and alteration of physiological and biochemical processes, as well as changes in the reproductive cycle. Reduction in crop yield can occur without the presence of visible symptoms. This type of injury is often related to low-level, long-term chronic exposure to air pollution. Studies have shown that field plantings exposed to filtered and unfiltered ambient air have produced different yields when no visible symptoms were present [5]. Reduction in total biomass can lead to economic loss for forage crops or hay.

Physiological or biochemical changes have been observed in plants exposed to air pollutants, including alterations in net photosynthesis, stomate response, and metabolic activity. Such exposure studies have been conducted under controlled laboratory conditions. An understanding of the processes involved will help to identify the cause of reduction in yield.

Laboratory studies have also investigated the interaction of air pollutants and the reproductive cycle of certain plants. Subtle changes in reproduction in a few susceptible species can render them unable to survive and prosper in a given ecosystem.

The major air pollutants which are phytotoxic to plants are ozone, sulfur dioxide, nitrogen dioxide, fluorides, and peroxyacetyl nitrate (PAN) [2]. Table 12.1 lists some of the types of plants injured by exposure to these pollutants. The effects range from slight reduction in yield to extensive visible injury, depending on the level and duration of exposure. Examples of the distinction between air pollution injury and damage are also given in Table 12.1. Visible markings on plants or crops such as lettuce, tobacco, and orchids caused by air pollution translate into direct economic loss (i.e. damage). In contrast, visible markings on the leaves of grapes, potatoes, or corn caused by air pollution will not result in a determination of damage if there is no loss in yield. Individual circumstances determine whether air pollution damage has occurred.

The costs of air pollution damage are difficult to estimate. However, estimates indicate crop losses of $1–$5 billion for the United States [6]. When compared to the crop losses due to all causes, this percentage is small. However, for particular crops in specific locations, the economic loss can be very high. Certain portions of the Los Angeles, California, basin are no longer suitable for lettuce crops because they are subject to photochemical smog. This forces producers either to move to other locations or to plant other crops that are less

TABLE 12.1

Examples of Types of Leaf Injury and Air Pollution

Pollutant	Symptoms	Maturity of leaf affected	Part of leaf affected	Injury threshold		
				ppm (vol)	µg m^{-3}	Sustained exposure
Sulfur dioxide	Bleached spots, bleached areas between veins, chlorosis; insect injury, winter and drought conditions may cause similar markings	Middle-aged leaves most sensitive; oldest least sensitive	Mesophyll cells	0.3	785	8 h
Ozone	Flecking, stippling, bleached spotting, pigmentation; conifer needle tips become brown and necrotic	Oldest leaves most sensitive; youngest least sensitive	Palisade or spongy parenchyma in leaves with no palisade	0.03	59	4 h
Peroxyacetyl nitrate (PAN)	Glazing, silvering, or bronzing on lower surface of leaves	Youngest leaves most sensitive	Spongy cells	0.01	50	6 h
Nitrogen dioxide	Irregular, white or brown collapsed lesions on intercostal tissue and near leaf margin	Middle-aged leaves most sensitive	Mesophyll cells	2.5	4700	4 h
Hydrogen fluoride	Tip and margin burns, dwarfing, leaf abscission; narrow brown-red band separates necrotic from green tissue; fungal disease, cold and high temperatures, drought, and wind may produce similar markings; suture red spot on peach fruit	Youngest leaves most sensitive	Epidermis and mesophyll cells	0.1 (ppb)	0.08	5 weeks
Ethylene	Sepal withering, leaf abnormalities; flower dropping, and failure of leaf to open properly; abscission; water stress may produce similar markings	Young leaves recover; older leaves do not recover fully	All	0.05	58	6 h
Chlorine	Bleaching between veins, tip and margin burn, leaf abscission; marking often similar to that of ozone	Mature leaves most sensitive	Epidermis and mesophyll cells	0.10	290	2 h

(*continued*)

TABLE 12.1 *(Continued)*

Examples of Types of Leaf Injury and Air Pollution

Pollutant	Symptoms	Maturity of leaf affected	Part of leaf affected	Injury threshold		
				ppm (vol)	$\mu g\ m^{-3}$	Sustained exposure
Ammonia	"Cooked" green appearance becoming brown or green on drying; overall blackening on some species	Mature leaves most sensitive	Complete tissue	~20	~14 000	4 h
Hydrogen chloride	Acid-type necrotic lesion; tip burn on fir needles; leaf margin necrosis on broad leaves	Oldest leaves most sensitive	Epidermis and mesophyll cells	~5–10	~11 200	2 h
Mercury	Chlorosis and abscission; brown spotting; yellowing of veins	Oldest leaves most sensitive	Epidermis and mesophyll cells	<1	<8200	1–2 days
Hydrogen sulfide	Basal and marginal scorching	Youngest leaves most affected		20	28 000	5 h
2,4-Dichlorophen-oxyacetic acid (2–4D)	Scalloped margins, swollen stems, yellow-green mottling or stippling, suture red spot (2,4,5–T); epinasty	Youngest leaves most affected	Epidermis	<1	<9050	2 h
Sulfuric acid	Necrotic spots on upper surface similar to those caused by caustic or acidic compounds; high humidity needed	All	All	—	—	—

susceptible to air pollution damage. Concern has been expressed regarding the future impact of air pollution on the much larger Imperial Valley of California, which produces up to 50% of certain vegetables for the entire United States.

III. EFFECTS ON FORESTS

Approximately 1.95×10^{10} km^2 of the earth's surface has at least 20% or more crown tree cover, representing about one-third of the total land area [7]. Several different types of forest ecosystems can be defined based on their location and the species present. The largest in area are tropical forest systems, followed by temperate forests, rain forests, and tidal zone systems. The temperate forest systems are located in the latitudes where the greatest industrialization is occurring and have the most opportunity to interact with pollutants in the atmosphere. The impact of air pollution on forest ecosystems ranges from beneficial to detrimental. Smith [8, 9] classified the relationship of air pollutants with forests into three categories: low dose (I), intermediate dose (II), and high dose (III). With this classification scheme, seemingly contradictory statements on the impact of air pollution on forests can be understood.

A. Low-Dose Levels

Under low-dose conditions, forest ecosystems act as sinks for atmospheric pollutants and in some instances as sources. As indicated in Chapter 11, the atmosphere, lithosphere, and oceans are involved in cycling carbon, nitrogen, oxygen, sulfur, and other elements through each subsystem with different timescales. Under low-dose conditions, forest and other biomass systems have been utilizing chemical compounds present in the atmosphere and releasing others to the atmosphere for thousands of years. Industrialization has increased the concentrations of NO_2, SO_2, and CO_2 in the "clean background" atmosphere, and certain types of interactions with forest systems can be defined.

Forests can act as sources of some of the trace gases in the atmosphere, such as hydrocarbons, hydrogen sulfide, NO_x, and NH_3. Forests have been identified as emitters of terpene hydrocarbons, which are a component of the photochemistry that leads to the formation of ozone.[1] In 1960, Went [10] estimated that hydrocarbon releases to the atmosphere were on the order of 108 tons per year. Later work by Rasmussen [11] suggested that the release of terpenes from forest systems is 2×10^8 tons of reactive materials per year on a global basis. This is several times the anthropogenic input. Yet, it is important to remember that forest emissions are much more widely dispersed and less concentrated than anthropogenic emissions. Table 12.2 shows terpene emissions from different types of forest systems in the United States.

[1] This harkens back to the 1980s when President Reagan postulated that trees accounted for much pollution, which spawned the phrase, "killer trees." This led to some hilarious running gags and skits on the show, Saturday Night Live, where such a tree (played by the late John Belushi) would attempt entry into homes with obvious mal intent. What made the skits so funny was, like most humor, they were based on some element of truth.

TABLE 12.2

Composition of US Forest-Type Groups by Foliar Terpene Emissions

	Percent of total US forest area	Percent α pinene emitters	Percent of isoprene emitters
Eastern type group			
Softwood types			
Loblolly-shortleaf pine	11	~100	Some from oak and sweetgum associates
Longleaf-slash pine	5	~100	Some from oak and sweetgum associates
Spruce-fir	4	~75	25% from spruce, which also emits (pinene
White-red-jack pine	2	~90	10% from aspen trees
Subtotal	22%	~91%	~9%
Hardwood types			
Oak-hickory	23	~10	70%, diluted by hickory, maple and black walnut
Oak-gum cypress	7	~50	50% from plurality of oak, cottonwood and willow
Oak-pine	5	~30	60%, diluted by black gum and hickory associates
Maple-beech-birch	6	~15	Terpene foliates are hemlock and white pine
Aspen-birch	5	~20	60%, diluted by birch, α pinene source balsam fir and balsam poplar
Elm-ash-cottonwood	4	–	30% from cottonwood, sycamore, and willow
	–	–	–
Subtotal	50%	~21%	~45%
Total	72%	–	–
Western type groups			
Softwoods			
Douglas fir	7	~100	–
Ponderosa pine	7	~100	5% from aspen associates
Lodgepole pine	3	~90	10% from Engelmann spruce and aspen
Fir-spruce	3	~100	40% from spruce trees
Hemlock-Sitka spruce	2	~100	25% from Sitka spruce
White pine	1	~100	5% from Engelmann spruce
Larch	1	~100	–
Redwood	0.5	~100	–
Subtotal	24.5%	~98%	~12%
Hardwoods	2	–	~100% from aspen trees
Total	26.5%		

Source: Rasmussen, R. A., *J. Air Pollut. Contrl. Assoc.* **22**, 537–543 (1972).

Forest systems also act as sources of CO_2 when controlled or uncontrolled burning and decay of litter occur. In addition, release of ethylene occurs during the flowering of various species. Thus, although trees are an important part of strategies to control global warming by storing (sequestering) carbon, especially in large root systems, they are also sources of CO_2. One additional form of emission to the atmosphere is the release of pollen grains. Pollen is essential to the reproductive cycle of most forest systems but becomes a human health hazard for individuals susceptible to hay fever. The contribution of sulfur from forests in the form of dimethyl sulfide is considered to be about 10–25% of the total amount released by soils and vegetation [12].

Trees and soils of forests act as sources of NH_3 and oxides of nitrogen. Ammonia is formed in the soil by several types of bacteria and fungi. The volatilization of ammonia and its subsequent release to the atmosphere are dependent on temperature and the pH of the soil. Fertilizers are used as a tool in forest management. The volatilization of applied fertilizers may become a source of ammonia to the atmosphere, especially from the use of urea.

Nitrogen oxides are formed at various stages of the biological denitrification process. This process starts with nitrate; as the nitrate is reduced through various steps, NO_2, NO, N_2O, and N_2 can be formed and, depending on the conditions, released to the atmosphere.

The interactions of air pollutants with forests at low-dose concentrations result in imperceptible effects on the natural biological cycles of these species. In some instances, these interactions may be beneficial to the forest ecosystem. Forests, as well as other natural systems, act as sinks for the removal of trace gases from the atmosphere.

B. Intermediate-Dose Levels

The second level of interaction, the intermediate-dose level, can result in measurable effects on forest ecosystems. These effects consist of a reduction in forest growth, change in forest species, and susceptibility to forest pests. Both laboratory investigations and field studies show SO_2 to be an inhibitor of forest growth. When various saplings have been exposed to SO_2 in the laboratory, they show reduction in growth compared with unexposed saplings [13]. Various field investigations of forest systems in the vicinity of large point sources show the effects of elevated SO_2 levels on the trees closer to the source. For example, Linzon [14] found that SO_2 from the Sudbury, Ontario (Canada), smelter caused a reduction in forest growth over a very large area, with the closer-in trees severely defoliated, damaged, and killed.

The effect of photochemical oxidants, mainly O_3 and PAN, on the forests located in the San Bernardino Mountains northeast of Los Angeles, California, has been to change the forest composition and to alter the susceptibility of forest species to pests. This area has been subjected to increasing levels of oxidant since the 1950s (Fig. 12.3). During the late 1960s and early 1970s, changes in the composition and aesthetic quality of the forest were observed [15].

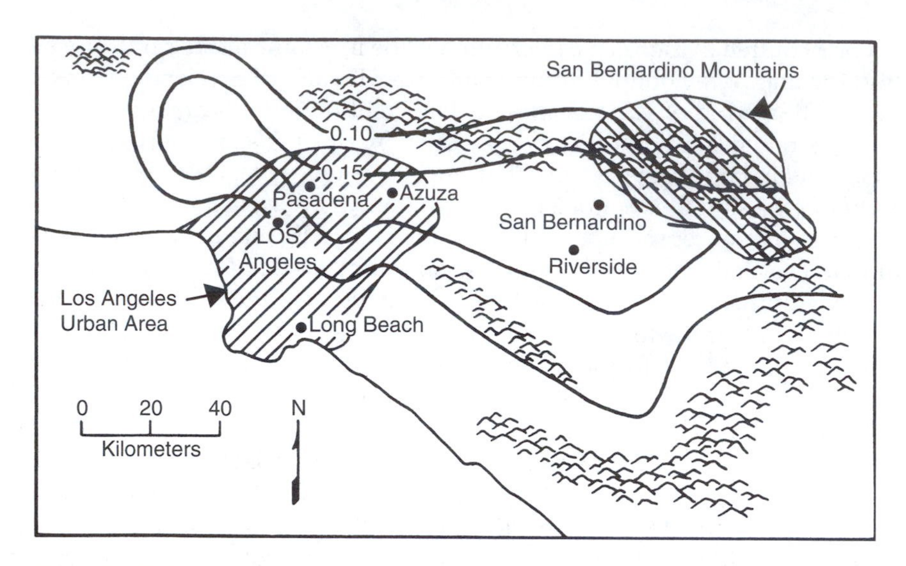

Fig. 12.3. Relationship between Los Angeles Basin's urban sources of photochemical smog and the San Bernardino Mountains, where ozone damage has occurred to the ponderosa pines. The solid lines are the average daily 1-h maximum does of ozone (ppm), July–September 1975–1977. *Source*: Adapted from Davidson, A., Ozone trends in the south coast air basin of California, in *Ozone/Oxidants: Interaction with the Total Environment*, pp. 433–450. Air Pollution Control Association, Pittsburgh, PA, 1979.

During this period, the photochemical problem was expanding to a wider geographical region; and photochemical oxidant was transported to the San Bernardino Mountains with increasing frequency and at higher concentrations. The receptor forest system has been described as a mixed conifer system containing ponderosa pine, Jeffrey pine, white fir, and cedar, along with deciduous black oak. The damage to the ponderosa pine ranged from no visible injury to death. As the trees came under increased stress due to exposure to oxidant, they became more susceptible to pine beetle, which ultimately caused their death. The ponderosa pine appears to be more susceptible than the other members of this forest system, and continued exposure to photochemical oxidant may very well shift it from the dominant species to a minor one.

The interactions in the intermediate-dose category may result in effects on the reproduction cycle of species, the utilization of nutrients, the production of biomass, and the susceptibility to disease.

C. High-Dose Levels

The third category for interactions is high dose (III). The effects produced by this level of interaction can be seen by the casual observer. The result of high-dose exposure is destruction or severe injury of the forest system. High-dose conditions are almost always associated with point source emissions. The pollutants most often involved are SO_2 and hydrogen fluoride. Historically, the most harmful sources of pollution for surrounding forest ecosystems have been smelters and aluminum reduction plants.

One example of high-dose interaction is the impact of a smelter on the surrounding area in Wawa, Ontario, Canada. This smelter began operating about 1940. Gordon and Gorham [16] documented the damage in the prevailing downwind northeast sector for a distance of 60 km. They analyzed vegetative plots and established four zones of impact in the downwind direction: Within 8 km of the plant, damage was classified as "very severe" where no trees or shrubs survived; "severe damage" occurred at ~17 km, where no tree canopy existed; "considerable damage" existed at ~25 km, where some tree canopy remained, but with high tree mortality; and "moderate damage" was found at ~35 km, where a tree canopy existed but was put under stress and where the number of ground flora species was still reduced.

This type of severe air pollution damage has occurred several times in the past. If care is not taken, additional examples will be documented in the future.

D. Acid Deposition

Acid deposition refers to the transport of acid constituents from the atmosphere to the earth's surface. This process includes dry deposition of SO_2, NO_2, HNO_3, and particulate sulfate matter and wet deposition ("acid rain") to surfaces. This process is widespread and alters distribution of plant and aquatic species, soil composition, pH of water, and nutrient content, depending on the circumstances.

The impact of acid deposition on forests depends on the quantity of acidic components received by the forest system, the species present, and the soil composition (notably ionic strength and depth). Numerous studies have shown that widespread areas in the eastern portion of North America and parts of Europe are being altered by acid deposition. Decreased pH in some lakes and streams in the affected areas was observed in the 1960s [17] and further evidence shows this trend.

When a forest system is subjected to acid deposition, the foliar canopy can initially provide some neutralizing capacity. If the quantity of acid components is too high, this limited neutralizing capacity is overcome. As the acid components reach the forest floor, the soil composition determines their impact. The soil composition may have sufficient buffering capacity to neutralize the acid components. However, alternation of soil pH can result in mobilization or leaching of important minerals in the soil. In some instances, trace metals such as calcium or magnesium may be removed from the soil, altering the aluminum tolerance for trees.

This interaction between airborne acid components and the tree-soil system may alter the ability of the trees to tolerate other environmental stressors such as drought, insects, and other air pollutants like ozone. In Germany, considerable attention is focused on the role of ozone and acid deposition as a cause of forest damage. Forest damage is a complex problem involving the interaction of acid deposition, other air pollutants, forestry practices, and naturally occurring soil conditions. As in all ecosystems, forests must always be understood as complex, integrated systems with myriad interrelationships that can be affected by air pollutants.

IV. EFFECTS ON ANIMALS

Acid deposition and the alteration of the pH of aquatic systems has led to the acidification of lakes and ponds in various locations in the world. Low-pH conditions result in lakes which contain no fish species.

Heavy metals on or in vegetation and water have been and continue to be toxic to animals and fish. Arsenic and lead from smelters, molybdenum from steel plants, and mercury from chlorine-caustic plants are major offenders. Poisoning of aquatic life by mercury is relatively new, whereas the toxic effects of the other metals have been largely eliminated by proper control of industrial emissions. Gaseous (and particulate) fluorides have caused injury and damage to a wide variety of animals-domestic and wild-as well as to fish. Accidental effects resulting from insecticides and nerve gas have been reported.

Autopsies of animals in the Meuse Valley, Donora, and London episodes described in Chapter 19, Section III, revealed evidence of pulmonary edema. Breathing toxic pollutants is not, however, the major form of pollutant intake for cattle; ingestion of pollution-contaminated feeds is the primary mode.

In the case of animals we are concerned primarily with a two-step process: accumulation of airborne contaminants on or in vegetation or forage that serves as their feed and subsequent effects of the ingested herbage on animals. In addition to pollution-affected vegetation, carnivores (humans included) consume small animals that may have ingested exotic chemicals including pesticides, herbicides, fungicides, and antibiotics. As in humans, the principal route of exposure to air pollutants can be indirect. After deposition, the animal ingests the contaminant. This reminds us that not every exposure to air pollution is by inhalation. Increasing environmental concern has pointed out the importance of the complete food chain for the physical and mental well-being of human beings.

A. Heavy Metal Effects

One of the earliest cattle problems involved widespread poisoning of cattle by arsenic at the turn of the century. Abnormal intake of arsenic results in severe colic (salivation, thirst, and vomiting), diarrhea, bloody feces, and a garlic-like odor on the breath; cirrhosis of the liver and spleen as well as reproductive effects may be noted. Arsenic trioxide in the feed must be approximately 10 mg kg^{-1} body weight for these effects to occur.

Cattle feeding on herbage containing 25–50 mg kg^{-1} (ppm wt.) lead develop excitable jerking of muscles, frothing at the mouth, grinding of teeth, and paralysis of the larynx muscles; a "roaring" noise is caused by the paralysis of the muscles in the throat and neck.

Symptoms of molybdenum poisoning in cattle include emaciation, diarrhea, anemia, stiffness, and fading of hair color. Vegetation containing 230 mg kg^{-1} of this substance affects cattle.

Mercury in fish has been found in waters in the United States and Canada. Mercury in the waters is converted into methyl mercury by aquatic vegetation and other processes. Small fish consume such vegetation and in turn are eaten by larger fish and eventually by humans; food with more than 0.5 ppm of mercury (0.5 mg kg^{-1}) cannot be sold in the United States for human consumption. The US Food and Drug Administration recommends that pregnant women and women of childbearing age should not exceed consuming 14 ounces of fish with 0.5 ppm mercury per week.

Mercury has a very complicated chemistry. It forms numerous chemical species with very different properties and behaviors in the environment. Three major types of mercury are important in air pollution: elemental mercury [Hg(0)], which is the uncombined chemical element; divalent mercury [Hg(II)], which is chemically reactive and under most environmental conditions is found to be combined with other substances into mercury salts (e.g. $HgCl_2$); and particulate-phase mercury [Hg(p)], most of which is actually the chemical species Hg(II) but is only slightly reactive because it is mixed with solid-phase material in the atmosphere. These three species have very different affinities for compartments and organisms (see Fig. 12.4). Thus, the mercury that is released from a stack can have a tortuous journey to and through the food chain.

B. Gaseous and Particulate Effects

Periodically, accidental emissions of a dangerous chemical affect animal well-being. During nerve gas experimentation in a desolate area in Utah, a high-speed airplane accidentally dropped several hundred gallons of nerve gas. As a result of the discharge, 6200 sheep were killed. Considering the

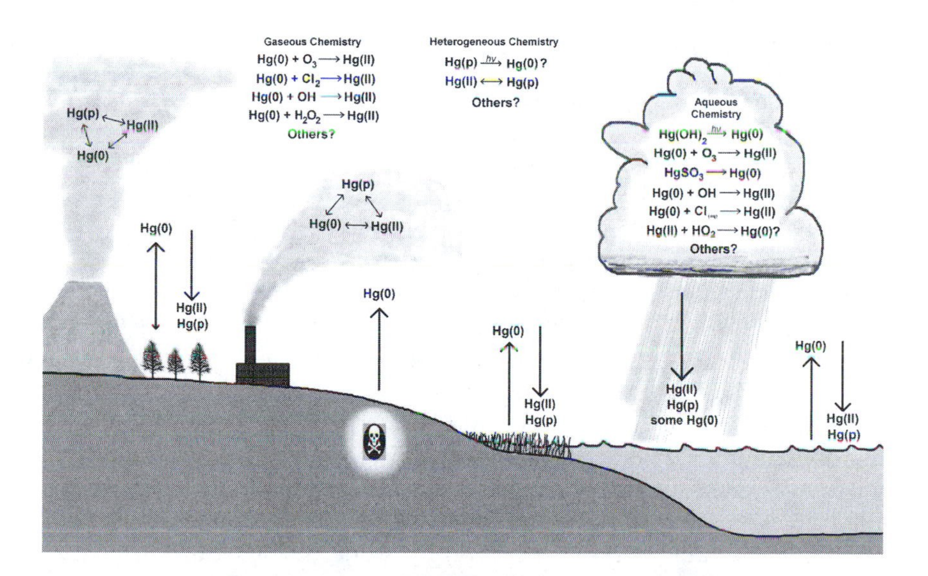

Fig. 12.4. Chemical speciation of mercury after it is released into the environment. Hg(0) is elemental, zero valence mercury; Hg(II) is divalent mercury (also known as reactive Hg); and Hg(p) is particulate-bound mercury. Drawing by Russell Bullock, National Oceanic and Atmospheric Administration, Research Triangle Park, North Carolina.

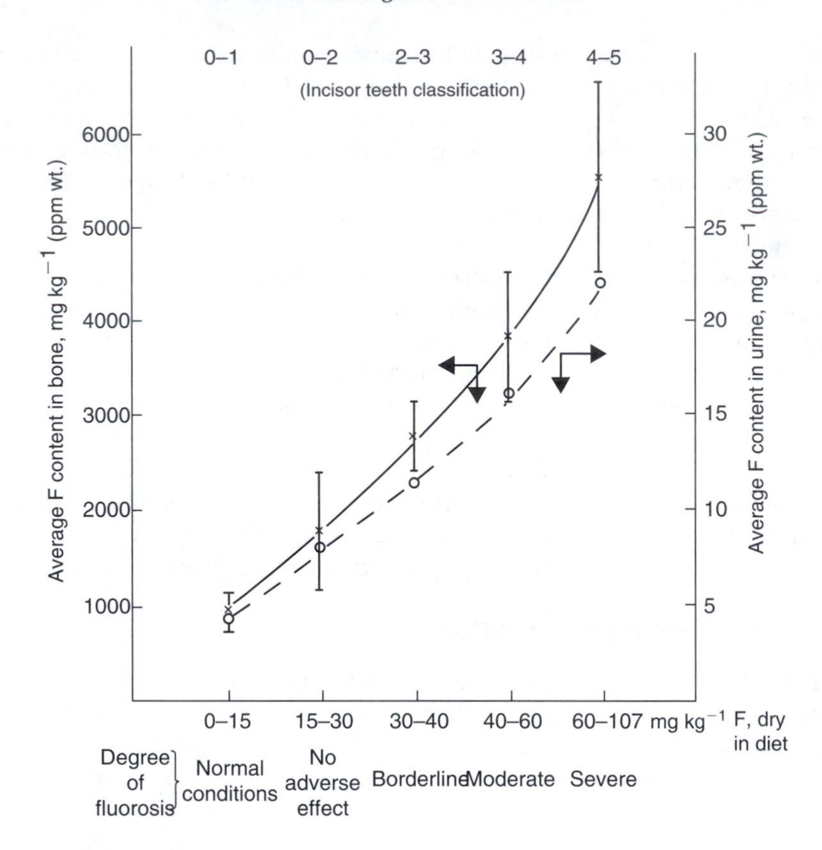

Fig. 12.5. Effects of fluoride on 4-year-old dairy cattle.

large number of exotic chemicals being manufactured, such unfortunate accidents may be anticipated in the future.

Fluoride emissions from industries producing phosphate fertilizers or phosphate derivatives have caused damage to cattle throughout the world; phosphate rock, the raw material, can contain up to 4% fluoride, part of which is discharged to air (and waters) during processing. In Polk and Hillsborough counties of Florida, the cattle population decreased by 30 000 between 1953 and 1960 as a result of fluoride emissions. Since 1950, research has greatly increased our knowledge of the effect of fluorides on animals; standards and guides for diagnosing and evaluating fluorosis in cattle have been compiled.

Chronic fluoride toxicity (fluorosis) is the type most frequently observed in cattle. The primary effects of fluorides in cattle are seen in the teeth and bones. Excessive intake weakens the enamel of developing teeth; the initially dulled erupted teeth can develop into soft teeth, with uneven wearing of molar teeth. Characteristic osteofluorotic bone lesions develop, causing intermittent lameness and stiffness in the animal. Fluoride content of the bone increases with dosage despite excretion in urine and feces. Secondary symptoms include reduced lactation, nonpliable skin, and dry, rough hair coat. As shown in Fig. 12.5, the fluoride ingestion level correlates with the fluoride content of bones and urine as well as incisor teeth classification [18].

TABLE 12.3

Fluoride Tolerance of Animals (ppm wt. in Ration, Dry)[a]

Species	Breeding or lactating animals (ppm)[b]	Finishing animals to be sold for slaughter with average feeding period (ppm)[b]
Dairy, beef heifers	30	100
Dairy cows	30	100
Beef cows	40	100
Steers	–	100
Sheep	50	160
Horse	60	–
Swine	70	–
Turkeys	–	100
Chickens	–	150

[a] Data based on soluble fluoride; increased values for insoluble fluoride compounds.
[b] 1 ppm wt. = 1 mg kg^{-1}.

Tolerance of animals for fluorides varies, dairy cattle being most sensitive and poultry least (Table 12.3). Fluorosis of animals in contaminated areas can be avoided by keeping the intake levels below those listed by incorporating clean feeds with those high in fluorides. It has also been determined that increased consumption of aluminum and calcium salts can reduce the toxicity of fluorides in animals. In addition, high-dose exposures can completely alter an ecosystem. For example, if the population of a species of animals or plants that is sensitive to an air pollutant falls dramatically, then the entire predator–prey interactions can change. Thus, non-human species are indicators of environmental quality.

REFERENCES

1. Levitt, J., *Responses of Plants to Environmental Stresses*. Academic Press, New York, 1972.
2. Hindawi, I. I., *Air Pollution Injury to Vegetation, AP-71*. United States Department of Health, Education, and Welfare, Raleigh, NC, 1970.
3. Treshow, M., *Environment and Plant Response*. McGraw-Hill, New York, 1970.
4. Heck, W. W., and Brandt, C. S., Effects on vegetation, in *Air Pollution*, 3rd ed., Vol. 111 (Stern, A. C., ed.), pp. 157–229. Academic Press, New York, 1977.
5. Hileman, B., *Environ. Sci. Technol.* **16**, 495A–499A (1982).
6. Heck, W. W., Heagle, A. S., and Shriner, D. S., Effects on vegetation: native, crops, forests, in *Air Pollution VI* (Stern, A. C., ed.), pp. 247–350. Academic Press, Orlando, FL, 1986.
7. *McGraw-Hill Encyclopedia of Science and Technology*, McGraw-Hill, New York; numerous air pollution articles available at http://www.accessscience.com/search/asearch, 2007.
8. Smith, W. H., *Environ. Pollut.* **6**, 111–129 (1974).
9. Smith, W. H., *Air Pollution and Forests: Interaction Between Air Contaminants and Forest Ecosystems*, 2nd ed. Springer-Verlag, New York, 1990.
10. Went, F. W., *Nature (London)* **187**, 641–643 (1960).
11. Rasmussen, R. A., *J. Air Pollut. Control Assoc.* **22**, 537–543 (1972).
12. Bremmer, J. M., and Steele, C. G., *Adv. Microbiol. Ecol.* **2**, 155–201 (1978).

13. Keller, T., *Environ. Pollut.* **16**, 243–247 (1978).
14. Linzon, S. N., *J. Air Pollut. Control Assoc.* **21**, 81–86 (1971).
15. Miller, P. R., and Elderman, M. J. (eds.), *Photochemical Oxidant Air Pollution Effects of a Mixed Conifer System*. EPA-600/3-77-104. US Environmental Protection Agency, Corvallis, OR, 1977.
16. Gordon, A. G., and Gorham, E., *Can. J. Bot.* **41**, 1063–1078 (1973).
17. Oden, S., *The Acidification of Air and Precipitation and Its Consequences in the Natural Environment*. Ecology Committee, Bulletin No. 1, Swedish National Science Research Council, Stockholm, 1967.
18. Shupe, J. L., *Am. Ind. Hyg. Assoc. J.* **31**, 240–247 (1970).

SUGGESTED READING

Adriano, D. C., Biological and ecological effects, in *Acidic Precipitation*, Vols. 1–5. Advances in Environmental Sciences, Springer-Verlag, London, 1989–1990.

Georgii, H. W., *Atmospheric Pollutants in Forest Areas: Their Deposition and Interception*. Kluwer Academic Publishers, Boston, MA, 1986.

Legge, A. H., and Krupa, S. V., *Air Pollutants and Their Effects on the Terrestrial Ecosystem*. Wiley, New York, 1986.

Mellanby, K. (ed.), *Air Pollution, Acid Rain, and the Environment*. Elsevier Science Publishers, Essex, England, 1988.

National Research Council, *Biologic Markers of Air Pollution Stress and Damage in Forests*. National Academy Press, Washington, DC, 1989.

Smith, W. H., *Air Pollution and Forests: Interactions Between Air Contaminants and Forest Ecosystems*, 2nd ed. Springer-Verlag, New York, 1990.

QUESTIONS

1. Distinguish between air pollution damage and injury.
2. Why is it difficult to prove that effects on plants in the field observed visually were cause by exposure to air pollution?
3. What functions do the stomates serve in gas exchange with the atmosphere?
4. Why is air pollution damage important when estimates suggest that it accounts for less than 5% of total crop losses in the United States?
5. List examples of air pollution effects on plants that cannot be detected by visual symptoms.
6. What types of trace gases are released to the atmosphere by forest ecosystems?
7. How have ozone and insects interacted to damage trees in the San Bernardino Mountain National Forest of California?
8. Why are animals used in research on air pollution effects?
9. Calculate the daily fluoride intake of a dairy animal from (a) air and (b) food and water, based on the conditions below and assuming 100% retention of the fluoride:
   - Animal breathing rate: 30 kg air per day containing 6 μg fluoride per cubic meter of air (STP).
   - Animal food and water intake:

     Herbage 10 kg containing 200 mg kg^{-1} of fluoride.

     Water 5 kg containing 1 mg kg^{-1} of fluoride.
10. If you were a forest manager for a woody grove in Atlanta, Georgia, how might you use the information in Table 12.2? What if our were doing the same in Los Angeles, California? What differences, if any, would come into play?

13

Effects on Materials and Structures

I. EFFECTS ON METALS

The principal effects of air pollutants on metals are corrosion of the surface, with eventual loss of material from the surface, and alteration in the electrical properties of the metals. Metals are divided into two categories: ferrous and nonferrous. Ferrous metals contain iron and include various types of steel. Nonferrous metals, such as zinc, aluminum, copper, and silver, do not contain iron.

Three factors influence the rate of corrosion of metals: moisture, type of pollutant, and temperature. A study by Hudson [1] confirms these three factors. Steel samples were exposed for 1 year at 20 locations throughout the world. Samples at dry or cold locations had the lowest rate of corrosion, samples in the tropics and marine environments were intermediate, and samples in polluted industrial locations had the highest rate of corrosion. Corrosion values at an industrial site in England were 100 times higher than those found in an arid African location.

The role of moisture in corrosion of metals and other surfaces is twofold: surface wetness acts as a solvent for containments and for metals is a medium for electrolysis. The presence of sulfate and chloride ions accelerates the corrosion of metals. Metal surfaces can by wetted repeatedly over a period of time as the humidity fluctuates.

Several studies have been conducted in urban areas to relate air pollution exposure and metal corrosion. In Tulsa, Oklahoma, wrought iron disks were exposed in various locations [2]. Using weight change as a measure of air pollution corrosion, the results indicated higher corrosion rates near industrial sectors containing an oil refinery and fertilizer and sulfuric acid manufacturing facilities. Upham [3] conducted a metal corrosion investigation in Chicago. Steel plates were exposed at 20 locations, and SO_2 concentrations were also measured. Figure 13.1 shows the relationship between weight loss during 3-, 6-, and 12-month exposure periods and the mean SO_2 concentration. Corrosion was also found to be higher in downtown locations than in suburban areas. Nonferrous metals are also subject to corrosion, but to a lesser degree than ferrous metals. Table 13.1 compares the weight loss of several nonferrous metals over a 20-year period [4]. The results vary depending on the type of exposure present.

Zinc is often used as a protective coating over iron to form galvanized iron. In industrial settings exposed to SO_2 and humidity, this zinc coating is subject to sufficient corrosion to destroy its protective capacity. Haynie and Upham [5] used their results from a zinc corrosion study to predict the useful life of a zinc-coated galvanized sheet in different environmental settings. Table 13.2 shows the predicted useful life as a function of SO_2 concentration.

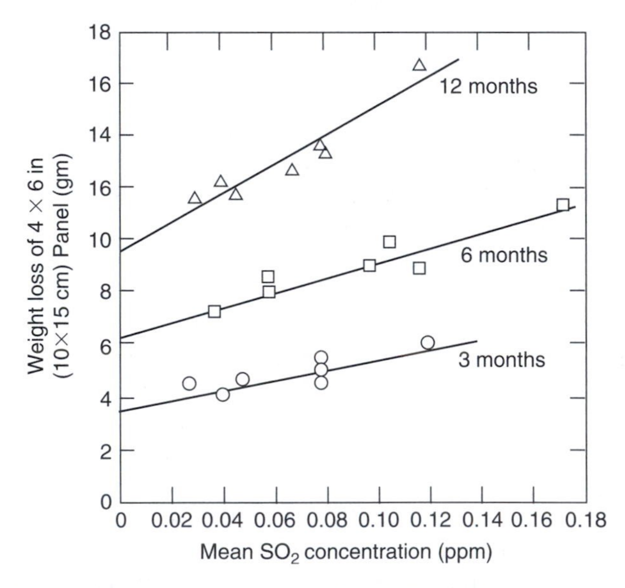

Fig. 13.1. Relationship between corrosion of mild steel and corresponding mean sulfur dioxide concentration at seven Chicago, Illinois sites. *Source*: Upham, J. B., *J. Air Pollut. Control Assoc.* **17**, 400 (1967).

TABLE 13.1

Weight Loss of Metal Panels[a] After 20 Years' Exposure in Various Atmospheres (*ca.* 1930–1954)[b]

City	Exposure classification	Average loss in weight (%)					
		Commercial copper (99.9% + Cu)	Commercial aluminum (99% + Al)	Brass (85% Cu, 15% Zn)	Nickel (99% + Ni)	Commercial lead (99.92% Pb, 0.06% Cu)	Commercial zinc (99% Zn, 0.85% Pb)
Altoona, PA	Industrial	6.1	–	8.5	25.2	1.8	30.7
New York, NY	Industrial	6.4	3.4	8.7	16.6	–	25.1
La Jolla, CA	Seacoast	5.4	2.6	1.3	0.6	2.1	6.9
Key West, FL	Seacoast	2.4	–	2.5	0.5	–	2.9
State College, PA	Rural	1.9	0.4	2.0	1.0	1.4	5.0
Phoenix, AZ	Rural	0.6	0.3	0.5	0.2	0.4	0.8

[a] Panels: 9 × 12 × 0.035 in. (22.86 × 30.48 × 0.089 cm).

[b] Data from H. R. Copson, Report of ASTM Subcommittee VI, of Committee B-3 on Atmospheric Corrosion, *Am. Soc. Test Mater.*, Special Technical Publication No. 175, 1955. Used by permission of the American Society for Testing and Materials, Philadelphia.

Source: Yocom, J. E., and Upham, J. B., Effects on economic materials and structures, in *Air Pollution*, 3rd ed., Vol. I (Stern, A. C., ed.), p. 80. Academic Press, New York, 1977.

TABLE 13.2

Predicted Useful Life of Galvanized Sheet Steel with a 53-μm Coating at an Average Relative Humidity of 65%[a]

		Useful life (years)		
SO_2 concentration ($\mu g\, m^{-2}$)	Type of environment	Predicted best estimate	Predicted range	Observed range
13	Rural	244	41.0	30–35
130	Urban	24	16.0–49.0	
260	Semi-industrial	12	10.0–16.0	15–20
520	Industrial	6	5.5–7.0	
1040	Heavy industrial	3	2.9–3.5	3–5

[a] *Source*: Yocom, J. E., and Upham, J. B., Effects of economic materials and structures, in *Air Pollution*, 3rd ed., Vol. I (Stern, A. C., ed.), p. 80. Academic Press, New York, 1977.

Aluminum appears to be resistant to corrosion from SO_2 at ambient air concentrations. Aluminum alloys tend to form a protective surface film that limits further corrosion upon exposure to SO_2. Laboratory studies at higher concentrations (280 ppm) show corrosion of aluminum at higher humidities (>70%), with the formation of a white powder of aluminum sulfate.

Copper and silver are used extensively in the electronics industry because of their excellent electrical conductivity. These metals tend to form a protective surface coating which inhibits further corrosion. When exposed to H_2S, a sulfide coating forms, increasing the resistance across contacts on electrical switches [6].

II. EFFECTS ON STONE

The primary concern in regard to air pollution is the soiling and deterioration of limestone, which is widely used as a building material and for marble statuary.[1] Figure 13.2 shows the long-term effects of urban air pollution on the appearance of stone masonry. Many buildings in older cities have been exposed to urban smoke, SO_2, and CO_2 for decades. The surfaces have become soiled and are subjected to chemical attack by acid gases. Exterior building surfaces are also subjected to a wet–dry cycle from rain and elevated humidity. SO_2 and moisture react with limestone ($CaCO_3$) to form calcium sulfate ($CaSO_4$) and gypsum ($CaSO_4 \cdot 2H_2O$). These two sulfates are fairly soluble in water, causing deterioration in blocks and in the mortar used to hold the blocks together. The soluble calcium sulfates can penetrate into the pores of the limestone and recrystallize and expand, causing further deterioration of the stone. CO_2 in the presence of moisture forms carbonic acid. This acid converts the limestone into bicarbonate, which is also water

[1] Marble is metamorphosed limestone.

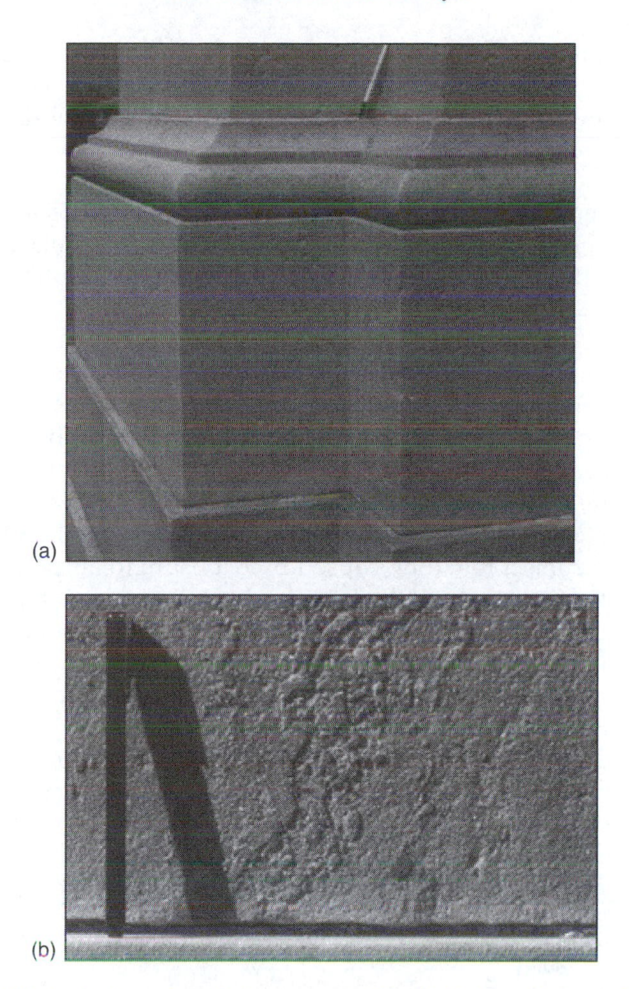

Fig. 13.2. Effects of urban air pollution on materials. (a) Pock marks in marble column on the south side of the U.S. Capitol building. Silicate mineral inclusions in the marble loosen and fall out when the calcite around them is dissolved by precipitation with pH that has been lowered (i.e. higher acidity) by urban air pollution. (b) Marble block that forms the northeast corner of the Capitol balustrade shows preferential erosion of the calcite around a silicate mineral inclusion. *Photos:* U.S. Geological Survey, http://pubs.usgs.gov/gip/acidrain/site1.html; accessed on June 30, 2007.

soluble and can be leached away by rain. This type of mechanism is present in the deterioration of marble statues.

III. EFFECTS ON FABRICS AND DYES

The major effects of air pollution on fabrics are soiling and loss of tensile strength. Sulfur oxides are considered to cause the greatest loss of tensile strength. The most widely publicized example of this type of problem has been damage to hosiery by air pollution, described in newspaper accounts. The mechanism is not understood, but it is postulated that fine droplets of sulfuric acid aerosol deposit on the very thin nylon fibers, causing them to fail under tension. Cellulose fibers are also weakened by sulfur dioxide. Cotton, linen, hemp, and rayon are subject to damage from SO_2 exposure.

Brysson and co-workers [7] conducted a study in St. Louis, Missouri, on the effects of urban air pollution on the tensile strength of cotton duck material. Samples were exposed at seven locations for up to 1 year. Figure 13.3 shows the relationship between tensile strength and pollutant exposure. For two levels of ambient air exposure, the materials exhibited less than one-half their initial tensile strength when exposed to air pollution for 1 year.

Particulate matter contributes to the soiling of fabrics. The increased frequency of washing to remove dirt results in more wear on the fabric, causing it to deteriorate in the cleaning process.

In addition to air pollution damage to fabrics, the dyes used to color fabrics have been subject to fading caused by exposure to air pollutants. Since the early 1900s, fading of textile dyes has been a continuing problem. The composition of dyes has been altered several times to meet demands of new fabrics and to "solve" the fading problem. Before World War I, dyes used on wool contained free or substituted amino acid groups, which were found to be susceptible to exposure to nitrogen dioxide.

When cellulose acetate rayon was introduced in the mid-1920s, old dye technology was replaced with new chemicals called dispersive dyes. Not

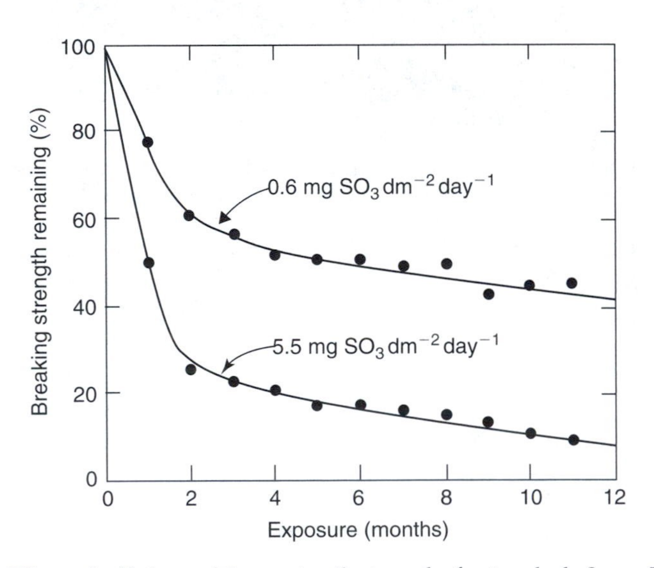

Fig. 13.3. Effects of sulfation and time on tensile strength of cotton duck. *Source*: Brysson, R. J., Trask, B. J., Upham, J. B., and Booras, S. A., *J. Air Pollut. Control Assoc.* **17**, 294 (1967).

long after their initial use, fading of blue, green, and violet shades began to be observed in material exposed to nitrogen oxides. The fabric was marked by a reddening discoloration. Laboratory studies duplicating ambient air levels of nitrogen dioxide and humidity reproduced these effects [8].

Ozone has also been found to cause fading of material. This was discovered when white fabrics developed a yellow discoloration [9, 10], leading researchers to investigate the effects of ozone on other chemicals added to the material, including optical brighteners, antistatic and soil-release finishes, and softeners. A very complex process was occurring where the dyes were migrating to the permanent-press-finish materials (e.g. softeners). Softeners have been found to be good absorbers of gases. Fading results from the combination of dye and absorbed nitrogen dioxide and ozone. This combination with high relative humidities has caused color fading in numerous types of material and dye combinations. However, dyes and pigments have improved in recent decades, so that fading from airborne pollutants has become less of a problem.

IV. EFFECTS ON LEATHER, PAPER, PAINT, AND GLASS

Sulfur dioxide affects the composition of leather and paper, causing significant deterioration. The major concern is the destruction of leatherbound books in the libraries of the world. SO_2 is absorbed by leather and converted to sulfuric acid, which attacks the structure of the leather. Initially, the edges of the exposed back of the book begin to crack at the hinges. As the cracks expand, more leather is exposed and the cracks widen, with the back eventually falling off the book. Preventive measures now include storage in sulfur dioxide free air.

The cellulose fiber in paper is attacked and weakened by sulfur dioxide. Paper made before about 1750 is not significantly affected by sulfur dioxide [11]. At about that time, the manufacture of paper changed to a chemical treatment process that broke down the wood fiber more rapidly. It is thought that this process introduces trace quantities of metals, which catalyze the conversion of sulfur dioxide to sulfuric acid. Sulfuric acid causes the paper to become brittle and more subject to cracking and tearing. New papers have become available to minimize the interaction with SO_2.

Paints are designed to decorate and protect surfaces. During normal wear, paint chalks moderately to clean the surface continuously. A hardened paint surface resists sorption by gases, although the presence of relatively high concentrations of 2620–5240 $\mu g\ m^{-3}$ SO_2 (1–2 ppm) increases the drying time of newly painted surfaces. Hydrogen sulfide reacts with lead base pigments:

$$Pb^{2+} + H_2S \rightarrow PbS + 2H^+ \quad (13.1)$$

to blacken white and light-tinted paints. Although most paints no longer contain Pb-based pigments, the reaction would expected to be similar for other metal pigments:

$$Me^{2+} + H_2Me \rightarrow MeS + 2H^+ \tag{13.2}$$

Wohlers and Feldstein [12] concluded that lead base paints could discolor surfaces in several hours at a concentration of $70\,\mu g\,m^{-3}$ H_2S (0.05 ppm). In time the black lead sulfide oxidizes to the original color. However, paints pigmented with titanium or zinc do not form a black precipitate. Alkyd or vinyl vehicles and pigments contain no heavy metal salts for reaction with H_2S. Painted surfaces are also dirtied by particulate matter. Contaminating dirt can readily become attached to wet or tacky paint, where it is held tenaciously and forms focal points for gaseous sorption for further attack. Dirt that collects on roofs or in gutters, blinds, screens, windowsills, or other protuberances is eventually washed over external surfaces to mar decorative effects.

Paints and coatings for automobiles have not been immune to damage by air pollution. Wolff and co-workers [13] found that damage to automobile finishes was the result of scarring by calcium sulfate crystals formed when sulfuric acid in rain or dew reacted with dry deposited calcium.

Glass is normally considered a very stable material. However, there is growing evidence that SO_2 air pollution may be accelerating the deterioration of medieval glass. A corrosion surface forms on these glass surfaces and the sulfate present helps prolonged surface wetness. This condition is conducive to further attack and degradation of the glass surface [14].

V. EFFECTS ON RUBBER

Although it was known for some time that ozone cracks rubber products under tension, the problem was not related to air pollution. During the early 1940s, it was discovered that rubber tires stored in warehouses in Los Angeles, California, developed serious cracks. Intensified research soon identified the causative agent as ozone that resulted from atmospheric reaction between sunlight (3000–4600 Å), oxides of nitrogen, and specific types of organic compounds, i.e., photochemical air pollution.

Natural rubber is composed of polymerized isoprene units. When rubber is under tension, ozone attacks the carbon–carbon double bond, breaking the bond. The broken bond leaves adjacent C=C bonds under additional stress, eventually breaking and placing still more stress on surrounding C=C bonds. This "domino" effect can be discerned from the structural formulas in Fig. 13.4. The number of cracks and the depth of the cracks in rubber under tension are related to ambient concentrations of ozone.

Rubber products may be protected against ozone attack by the use of a highly saturated rubber molecule, the use of a wax inhibitor which will "bloom" to the surface, and the use of paper or plastic wrappings to protect the surface. Despite these efforts, rubber products generally still crack more on the West Coast than on the East Coast of the United States.

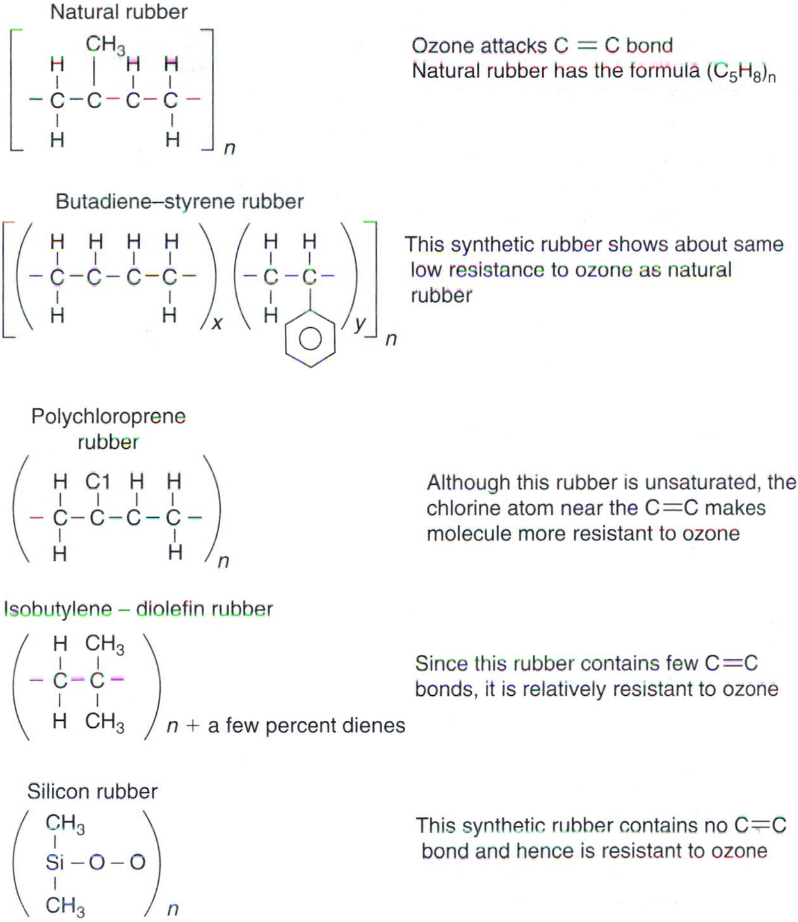

Natural rubber

$$\left[-\underset{\mathrm{H}}{\overset{\mathrm{H}}{\mathrm{C}}}-\overset{\mathrm{CH_3}}{\mathrm{C}}-\overset{\mathrm{H}}{\mathrm{C}}-\underset{\mathrm{H}}{\overset{\mathrm{H}}{\mathrm{C}}}- \right]_n$$

Ozone attacks C = C bond
Natural rubber has the formula $(C_5H_8)_n$

Butadiene–styrene rubber

$$\left[\left(-\underset{\mathrm{H}}{\overset{\mathrm{H}}{\mathrm{C}}}-\overset{\mathrm{H}}{\mathrm{C}}-\overset{\mathrm{H}}{\mathrm{C}}-\underset{\mathrm{H}}{\overset{\mathrm{H}}{\mathrm{C}}}- \right)_x \left(-\underset{\mathrm{H}}{\overset{\mathrm{H}}{\mathrm{C}}}-\underset{\mathrm{C_6H_5}}{\overset{\mathrm{H}}{\mathrm{C}}}- \right)_y \right]_n$$

This synthetic rubber shows about same low resistance to ozone as natural rubber

Polychloroprene rubber

$$\left(-\underset{\mathrm{H}}{\overset{\mathrm{H}}{\mathrm{C}}}-\overset{\mathrm{C1}}{\mathrm{C}}-\overset{\mathrm{H}}{\mathrm{C}}-\underset{\mathrm{H}}{\overset{\mathrm{H}}{\mathrm{C}}}- \right)_n$$

Although this rubber is unsaturated, the chlorine atom near the C=C makes molecule more resistant to ozone

Isobutylene – diolefin rubber

$$\left(-\underset{\mathrm{H}}{\overset{\mathrm{H}}{\mathrm{C}}}-\underset{\mathrm{CH_3}}{\overset{\mathrm{CH_3}}{\mathrm{C}}}- \right)_n + \text{a few percent dienes}$$

Since this rubber contains few C=C bonds, it is relatively resistant to ozone

Silicon rubber

$$\left(\underset{\mathrm{CH_3}}{\overset{\mathrm{CH_3}}{\mathrm{Si}}}-\mathrm{O}-\mathrm{O} \right)_n$$

This synthetic rubber contains no C=C bond and hence is resistant to ozone

Fig. 13.4. Susceptibility of natural and synthetic rubbers to attack by ozone.

REFERENCES

1. Hudson, J. D., *J. Iron Steel Inst. London* **148**, 161 (1943).
2. Galegar, W. C., and McCaldin, R. O., *Am. Ind. Hyg. Assoc. J.* **22**, 187 (1961).
3. Upham, J. B., *J. Air Pollut. Control Assoc.* **17**, 400 (1967).
4. Yocom, J. E., and Upham, J. B., Effects on economic materials and structures, in *Air Pollution*, 3rd ed., Vol. II (Stern, A. C., ed.), p. 80. Academic Press, New York, 1977.
5. Haynie, F. H., and Upham, J. B., *Mater. Prot. Perf.* **9**, 35–40 (1970).
6. Leach, R. H., Corrosion in liquid media, the atmosphere, and gases—silver and silver alloys, in *Corrosion Handbook*, Section 2 (Uhlig, H. H., ed.), p. 319. Wiley, New York, 1948.
7. Brysson, R. J., Trask, B. J., Upham, J. B., and Booras, S. A., *J. Air Pollut. Control Assoc.* **17**, 294 (1967).

8. Salvin, V. S., *Am. Dyest. Rep.* **53**, 33–41 (1964).
9. McLendon, V., and Richardson, F., *Am. Dyest. Rep.* **54**, 305–311 (1965).
10. Salvin, V. S., *Proceeding of Annual Conference, American Society for Quality Control, Textile and Needle Trades Division*, Vol. **16**, pp. 56–64, 1969.
11. Langwell, W. H., *Proc. R. Inst. Gr. Brit.* **37** (Part II, No. 166), 210 (1958).
12. Wohlers, H. C., and Feldstein, M., *J. Air Pollut. Control Assoc.* **16**, 19–21 (1966).
13. Wolff, G. T., Rodgers, W. R., Collins, D. C., Verma, M. H., and Wong, C. A., *J. Air Waste Manage. Assoc.* **40**, 1638–1648 (1990).
14. Newton, R. G., *The Deterioration and Conservation of Painted Glass: A Critical Bibliography*. Oxford University Press, Oxford, 1982.

SUGGESTED READING

Butlin, R. N., The effects of pollutants on materials, in *Energy and the Environment*. Royal Society of Chemistry, Cambridge, 1990.

Liu, B., and Yu, E. S., *Air Pollution Damage Functions and Regional Damage Estimates*. Technomic, Westport, CT, 1978.

Yocom, J. E., and Upham J. B., Effects on economic materials and structures, in *Air Pollution*, 3rd ed., Vol. II (Stern, A. C., ed.). Academic Press, New York, 1977.

QUESTIONS

1. Assuming that a relationship exists among corrosion, population, and sulfur dioxide, why might one expect this interdependence?
2. Compare the solubilities in water of calcium carbonate, calcium sulfite, calcium sulfate, magnesium sulfate, and dolomite.
3. Describe possible mechanisms for the deterioration of marble statuary.
4. Explain why soiling and corrosion are hidden costs of air pollution.
5. Describe possible preventive actions to limit deterioration of books and other print material in libraries.
6. Explain the role of moisture in corrosion of materials.
7. In your location, determine whether sulfur dioxide, ozone, or particulate matter contributes to soiling or corrosion problems.
8. Describe why some types of synthetic rubber are less susceptible to ozone attack than natural rubber.
9. Many people prefer cotton to synthetic fibers. What processes are available to increase cotton fabrics' resistance to the effects of air pollutants?

14

Effects on the Atmosphere, Soil, and Water Bodies

I. THE PHYSICS OF VISIBILITY

Impairment of visibility involves degradation of the ability to perceive the environment. Several factors are involved in determining visibility in the atmosphere (Fig. 14.1): the optical characteristics of the illumination source; the viewed targets; the intervening atmosphere; and the characteristics of the observer's eyesight [1].

To see an object, an observer must be able to detect the contrast between the object and its surroundings. If this contrast decreases, it is more difficult to observe the object. In the atmosphere, visibility can decrease for a number of reasons. For example, we may be farther away from the object (e.g. an airplane can move away from us); the sun's angle may change with the time of day; and if air pollution increases, the contrast may decrease, reducing our ability to see the object.

Objects close to us are easily perceived, but this diminishes as a function of distance. The lowest limit of contrast for human observers is called the *threshold contrast* and is important because this value influences the maximum distance at which we can see various objects. Thus, it is closely related

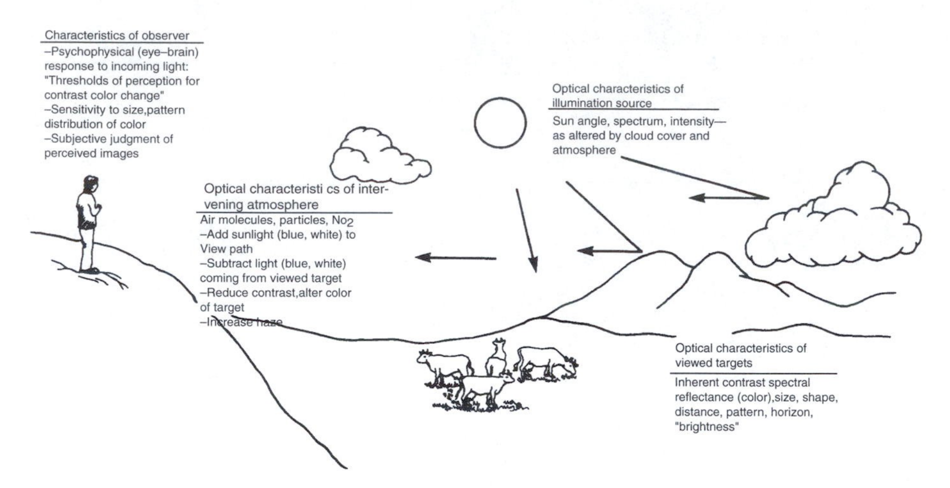

Fig. 14.1. Factors determining visibility in the atmosphere. *Source*: US Environmental Protection Agency, *Protecting Visibility*, EPA-450/5-79-008. Office of Air Quality Planning and Standards, Research Triangle Park, NC, 1979.

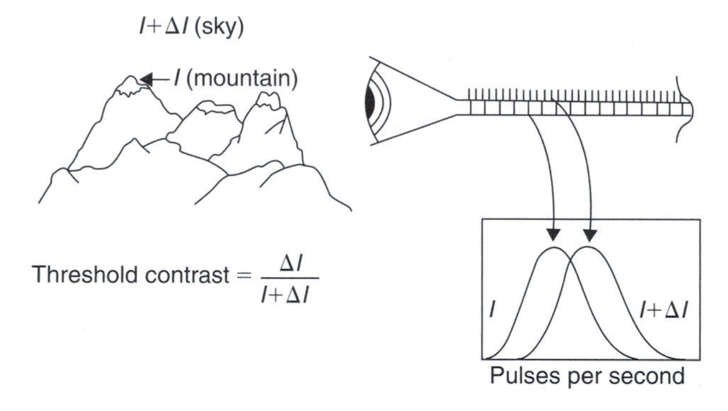

Fig. 14.2. Threshold contrast in distinguishing an object from its surroundings. The eye responds to an increment in light intensity by increasing the number of signals (pulses) sent to the brain. The detection of threshold contrast involves the ability to discriminate between the target (I) and the brighter background ($I + \Delta I$). *Source*: Gregory, R. L., *Eye and Brain: The Psychology of Seeing*. Weidenfeld and Nicolson, London, 1977.

to our understanding of good versus bad visibility for a particular set of environmental conditions.

Threshold contrast is illustrated in Fig. 14.2. I is the intensity of light received by the eye from the object, and $I + \Delta I$ represents the intensity coming from the surroundings. The threshold contrast can be as low as 0.018–0.03 and the object can still be perceptible. Other factors, such as the physical size of the visual image on the retina of the eye and the brain's response to the color of the object, influence the perception of contrast.

Let us consider the influence of gases and particles on the optical properties of the atmosphere. Reduction in visibility is caused by the following interactions in the atmosphere: light scattering by gaseous molecules and particles, and light absorption by gases and particles [2].

Light-scattering processes involve the interaction of light with gases or particles in such a manner that the direction or frequency of the light is altered. Absorption processes occur when the electromagnetic radiation interacts with gases or particles and is transferred internally to the gas or particle.

Light scattering by gaseous molecules is wavelength dependent and is the reason why the sky is blue. This process is dominant in atmospheres that are relatively free of aerosols or light-absorbing gases. Light scattering by particles is the most important cause of visibility reduction. This phenomenon is dependent on the size of the particles suspended in the atmosphere.

Light absorption by gases in the lower troposphere is limited to the absorption characteristics of nitrogen dioxide. This compound absorbs the shorter, or blue, wavelengths of visible light, causing us to observe the red wavelengths. We therefore perceive a yellow to reddish-brown tint in atmospheres containing quantities of NO_2. Light absorption by particles is related principally to carbonaceous or black soot in the atmosphere. Other types of fine particles such as sulfates, although not good light absorbers, are very efficient at scattering light.

A. Light Extinction in the Atmosphere

The interaction of light in the atmosphere is described mathematically in Eq. (14.1):

$$-\mathrm{d}I = b_{\mathrm{ext}} I \, \mathrm{d}x \tag{14.1}$$

where $-dI$ is the decrease in intensity, b_{ext} is the extinction coefficient, I is the original intensity of the beam of light, and $\mathrm{d}x$ is the length of the path traveled by the beam of light.

Figure 14.3(a) shows a beam of light transmitted through the atmosphere. The intensity of the beam $I(x)$ decreases with the distance from the illumination source as the light is absorbed or scattered out of the beam. For a short period, this decrease is proportional to the intensity of the beam and the length of the interval at that point. Here b_{ext} is the extinction or attenuation coefficient and is a function of the degree of scattering and absorption of the particles and gases which are present in the beam path.

Figure 14.3(b) illustrates a slightly more complicated case, but one more applicable to atmospheric visibility. In this example, the observer still depends on the ability to perceive light rays emanating from the target object and on the scattering and absorption of those rays out of the beam. In addition, however, the observer must contend with additional light scattered into the line of sight from other angles. This extraneous light is sometimes called

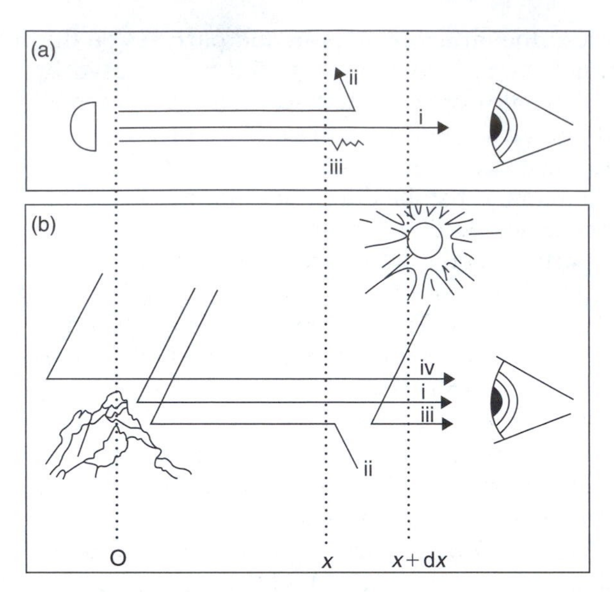

Fig. 14.3. (a) A diagram of extinction of light from a source such as an electric light in a reflector, illustrating (i) transmitted, (ii) scattered, and (iii) absorbed light. (b) A diagram of daylight visibility, illustrating (i) residual light from a target reaching an observer, (ii) light from a target scattered out of an observer's line of sight, (iii) air light from the intervening atmosphere, and (iv) air light constituting horizon sky. *Source*: US Environmental Protection Agency, *Protecting Visibility*, EPA-450/5-79-008. Office of Air Quality Planning and Standards, Research Triangle Park, NC, 1979.

air light. Equation (14.1) is modified to account for this phenomenon by adding a term to represent this background intensity.

$$-\mathrm{d}I = -\mathrm{d}I(\text{extinction}) + \mathrm{d}I(\text{air light}) \tag{14.2}$$

This air light term contributes to the reduced visibility we call *atmospheric haze*.

A simplified relationship developed by Koschmieder which relates the visual range and the extinction coefficient is given by

$$L_v = 3.92/b_{ext} \tag{14.3}$$

where L_v is the distance at which a black object is just barely visible [3]. Equation (14.3) is based on the following assumptions:

1. The background behind the target is uniform.
2. The object is black.
3. An observer can detect a contrast of 0.02.
4. The ratio of air light to extinction is constant over the path of sight.

While the Koschmieder relationship is useful as a first approximation for determining visual range, many situations exist in which the results are only qualitative.

The extinction coefficient b_{ext} is dependent on the presence of gases and molecules that scatter and absorb light in the atmosphere. The extinction coefficient may be considered as the sum of the air and pollutant scattering and absorption interactions, as shown in the following equation:

$$b_{ext} = b_{rg} + b_{ag} + b_{scat} + b_{ap} \tag{14.4}$$

where b_{rg} is scattering by gaseous molecules (Rayleigh scattering), b_{ag} is absorption by NO_2 gas, b_{scat} is scattering by particles, and b_{ap} is absorption by particles. These various extinction components are a function of wavelength. As extinction increases, visibility decreases.

The Rayleigh scattering extinction coefficient for particle-free air is $0.012\,km^{-1}$ for "green" light ($\gamma = 0.05\,\mu m$) at sea level [4]. This permits a visual range of ~320 km. The particle-free, or Rayleigh scattering, case represents the best visibility possible with the current atmosphere on earth.

The absorption spectrum of NO_2 shows significant absorption in the visible region (see Fig. 14.4) [5]. As a strong absorber in the blue region, NO_2 can color plumes red, brown, or yellow. Figure 14.5 shows a comparison of extinction coefficients of 0.1 ppm NO_2 and Rayleigh scattering by air [6]. In urban areas, some discoloration can be due to area-wide NO_2 pollution. In rural areas, the biggest problem with NO_2 is that in coherent plumes from power plants, it contributes to the discoloration of the plume.

Suspended particles are the most important factor in visibility reduction. In most instances, the visual quality of air is controlled by particle scattering and is characterized by the extinction coefficient b_{scat}. The size of particles plays a crucial role in their interaction with light. Other factors are the refractive index and shape of the particles, although their effect is harder to measure and is less well understood. If we could establish these properties, we could calculate the amount of light scattering and absorption. Alternatively, the extinction coefficient associated with an aerosol can be measured directly.

Light and suspended particles interact in the four basic ways shown in Fig. 14.6: refraction, diffraction, phase shift, and absorption. For particles with a diameter of 0.1–1.0 μm, scattering and absorption can be calculated by using the Mie equations [7]. Figure 14.7 shows the relative scattering and absorption efficiency per unit volume of particle for a typical aerosol containing some light-absorbing soot [8]. This clearly shows the importance of atmospheric particles in the diameter range 0.1–1.0 μm as efficient light-scattering centers. With

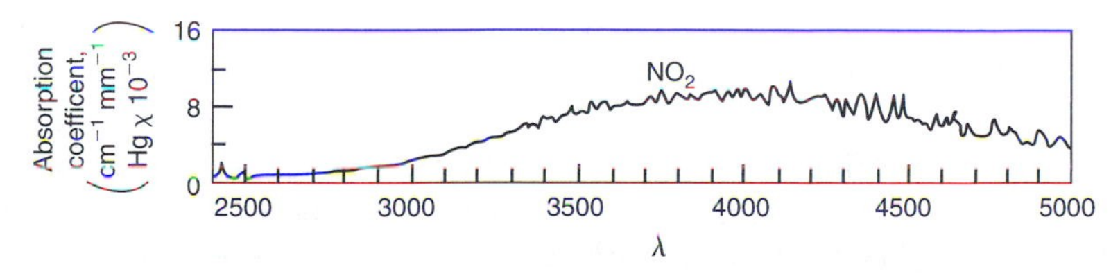

Fig. 14.4. Absorption spectrum of NO_2. *Source*: Hall Jr., T. C., and Blacet, F. E., *J. Chem. Phys.* **20**, 1745 (1952).

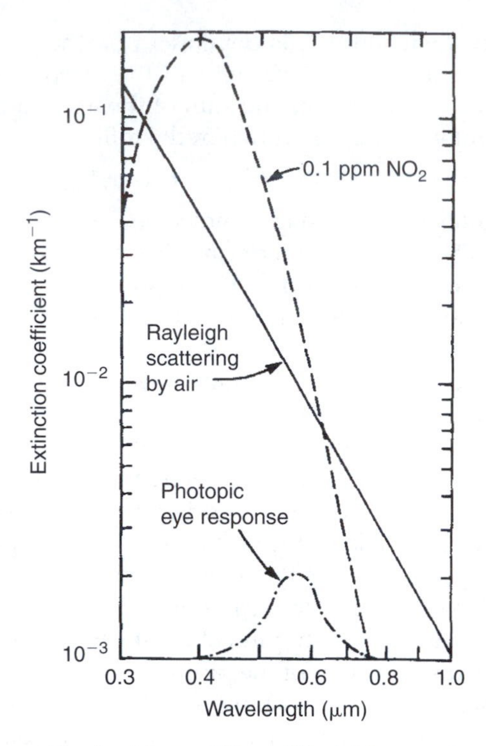

Fig. 14.5. Comparison of b_{ext} for 0.1 ppm NO_2 and Rayleigh scattering by air. The photopic eye response represents the range of wavelengths over which the eye detects light. *Source*: Husar, R., White, W. H., Paterson, D. E., and Trijonis, J., *Visibility Impairment in the Atmosphere*. Draft report prepared for the US Environmental Protection Agency under Contract No. 68022515, Task Order No. 28.

particles of larger and smaller diameters, scattering decreases. Absorption generally contributes less to the extinction coefficient than does the scattering processes. Atmospheric particles of different chemical composition have different refractive indices, resulting in different scattering efficiencies. Figure 14.8 shows the scattering-to-mass ratio for four different materials [9]. Clearly, carbon or soot aerosols, and aerosols of the same diameter with water content, scatter with different efficiencies at the same diameter.

Visibility is also affected by alteration of particle size due to hydroscopic particle growth, which is a function of relative humidity. In Los Angeles, California, the air, principally of marine origin, has numerous sea salt particles. Visibility is noticeably reduced when humidity exceeds about 67%. In a study of visibility related to both relative humidity and origin of air (maritime or continental), Buma [10] found that at a set relative humidity, continental air reduced visibility below 7 km more often than did air of maritime origin. This effect is presumably due to numerous hygroscopic aerosols from air pollution sources. Some materials, such as sulfuric acid mist, exhibit hygroscopic growth at humidity as low as 30%.

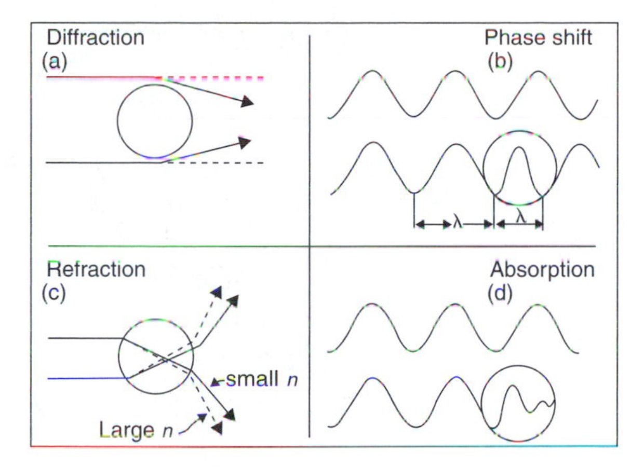

Fig. 14.6. Four forms of particle light interaction. Light scattering by coarse particles (>2 μm) is the combined effect of diffraction and refraction. (a) Diffraction is an edge effect whereby the light is bent to fill in the shadow behind the particle. (b) The speed of a wavefront entering a particle with refractive index $n > 1$ (for water, $n = 1.33$) is reduced. (c) Refraction produces a lens effect. The angular dispersion resulting from bending incoming rays increases with n. (d) For absorbing media, the refracted wave intensity decays within the particle. When the particle size is comparable to the wavelength of light (0.1–1.0 μm), these interactions (a)–(d) are complex and enhanced. *Source*: US Environmental Protection Agency, *Protecting Visibility*, EPA-450/5-79-008. Office of Air Quality Planning Standards, Research Triangle Park, NC, 1979.

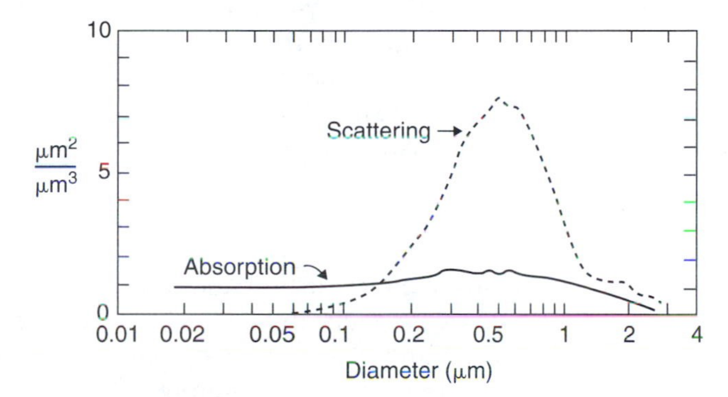

Fig. 14.7. Scattering and absorption cross section per unit volume as a function of particle diameter. *Source*: Charlson, R. J., Waggoner, A. P., and Thielke, H. F., *Visibility Protection for Class I Areas. The Technical Basis*. Report to the Council on Environmental Quality, Washington, DC, 1978.

B. Turbidity

The attenuation of solar radiation has been studied by McCormick and his associates [11, 12] using the Voltz sun photometer, which makes measurements at a wavelength of 0.5 μm. The ratio of the incident solar transmissivity to the extraterrestrial solar intensity can be as high as 0.5 in clean atmospheres but can drop to 0.2–0.3 in polluted areas, indicating a decrease

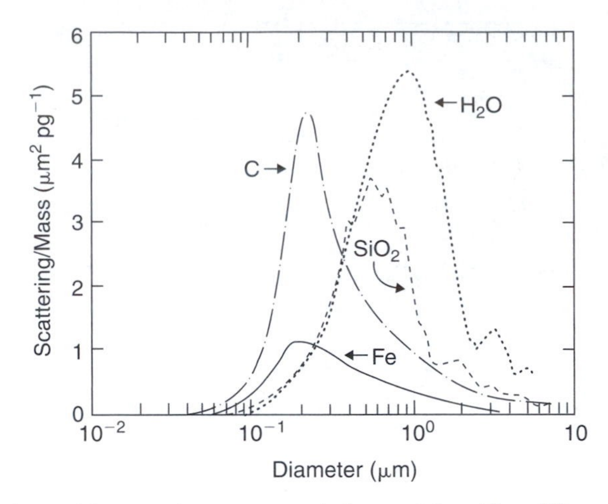

Fig. 14.8. Single particle scattering to mass ratio for particles of four different compositions. Carbon particles are also very efficient absorbers of light. *Source*: US Environmental Protection Agency, *Protecting Visibility*, EPA-450/5-79-008. Office of Air Quality Planning Standards, Research Triangle Park, NC, 1979.

of 50% in ground-level solar intensity. The turbidity coefficient can also be derived from these measurements and used to approximate the aerosol loading of the atmosphere. By assuming a particle size distribution in the size range 0.1–10.0 μm and a particle density, the total number of particles can be estimated. The mass loading per cubic meter can also be approximated. Because of the reasonable cost and simplicity of the sun photometer, it is useful for making comparative measurements around the world.

C. Precipitation

Pollution can cause opposite effects in relation to precipitation. Addition of a few particles that act as ice nuclei can cause ice particles to grow at the expense of supercooled water droplets, producing particles large enough to fall as precipitation. An example of this is commercial cloud seeding with silver iodide particles released from aircraft to induce rain. If too many particles are added, none of them grow sufficiently to cause precipitation. Therefore, the effects of pollution on precipitation are complex.

II. FORMATION OF ATMOSPHERIC HAZE

Atmospheric haze is the condition of reduced visibility caused by the presence of fine particles or NO_2 in the atmosphere. The particles must be 0.1–1.0 μm in diameter, the size range in which light scattering occurs. The source of these particles may be natural or anthropogenic.

Atmospheric haze has been observed in both the western and eastern portions of the United States. Typical visual ranges in the East are <15 miles and in the Southwest >50 miles. The desire to protect visual air quality in the United States is focused on the national parks in the West. The ability to see vistas over 50–100 km in these locations makes them particularly vulnerable to atmospheric haze. This phenomenon is generally associated with diffuse or widespread atmospheric degradation as opposed to individual plumes.

The major component of atmospheric haze is sulfate particulate matter (particularly ammonium sulfate), along with varying amounts of nitrate particulate matter, which in some areas can equal the sulfate. Other components include graphitic material, fine fly ash, and organic aerosols.

The sources of particulate matter in the atmosphere can be primary, directly injected into the atmosphere, or secondary, formed in the atmosphere by gas-to-particle conversion processes [13]. The primary sources of fine particles are combustion processes, e.g., power plants and diesel engines. Power plants with advanced control technology still emit substantial numbers and masses of fine particles with diameters <1.0 μm. The composition of these particles includes soot or carbonaceous material, trace metals, V_2O_5, and sulfates. In addition, large quantities of NO_2 and SO_2 are released to the atmosphere.

A. Particle Formation in the Atmosphere

The secondary source of fine particles in the atmosphere is gas-to-particle conversion processes, considered to be the more important source of particles contributing to atmospheric haze. In gas-to-particle conversion, gaseous molecules become transformed to liquid or solid particles. This phase transformation can occur by three processes: absorption, nucleation, and condensation. *Absorption* is the process by which a gas goes into solution in a liquid phase. Absorption of a specific gas is dependent on the solubility of the gas in a particular liquid, e.g., SO_2 in liquid H_2O droplets. *Nucleation* and *condensation* are terms associated with aerosol dynamics.

Nucleation is the growth of clusters of molecules that become a thermodynamically stable nucleus. This process is dependent on the vapor pressure of the condensable species. The molecular clusters undergo growth when the saturation ratio, S, is greater than 1, where *saturation ratio* is defined as the actual pressure of the gas divided by its equilibrium vapor pressure. $S > 1$ is referred to as a *supersaturated condition* [14].

The size at which a cluster may be thermodynamically stable is influenced by the Kelvin effect. The equilibrium vapor pressure of a component increases as the droplet size decreases. Vapor pressure is determined by the energy necessary to separate a single molecule from the surrounding molecules in the liquid. As the curvature of the droplet's surface increases, fewer neighboring molecules will be able to bind a particular molecule to the liquid phase, thus increasing the probability of a molecule escaping the liquid's surface. Thus, smaller droplets will have a higher equilibrium vapor pressure. This would affect the minimum size necessary for a thermodynamically stable cluster, suggesting that components with lower equilibrium saturation vapor pressures will form stable clusters at smaller diameters.

Condensation is the result of collisions between a gaseous molecule and an existing aerosol droplet when supersaturation exists. Condensation occurs at much lower values of supersaturation than nucleation. Thus, when particles already exist in sufficient quantities, condensation will be the dominant process occurring to relieve the supersaturated condition of the vapor-phase material.

A simple model for the formation and growth of an aerosol at ambient conditions involves the formation of a gas product by the appropriate chemical oxidation reactions in the gas phase. This product must have a sufficiently low vapor pressure for the gas-phase concentration of the oxidized product to exceed its saturation vapor pressure. When this condition occurs, nucleation and condensation may proceed, relieving supersaturation. These processes result in the transfer of mass to the condensed phase. Aerosol growth in size occurs while condensation is proceeding.

Coagulation, i.e., the process by which discrete particles come into contact with each other in the air and remain joined together by surface forces, represents another way in which aerosol diameter will increase. However, it does not alter the mass of material in the coagulated particle.

The clearest example of this working model of homogeneous gas-to-particle conversion is sulfuric acid aerosol formation. Sulfuric acid (H_2SO_4) has an extremely low saturation vapor pressure. Oxidation of relatively small amounts of sulfur dioxide (SO_2) can result in a gas-phase concentration of H_2SO_4 that exceeds its equilibrium vapor pressure in the ambient atmosphere, with the subsequent formation of sulfuric acid aerosol. In contrast, nitric acid (HNO_3) has a much higher saturation vapor pressure. Therefore, the gas-phase concentration of HNO_3 is not high enough to permit nucleation of nitric acid aerosol in typical atmospheric systems.

Atmospheric haze can occur over regions of several thousand square kilometers, caused by the oxidation of widespread SO_2 and NO_2 to sulfate and nitrate in relatively slow-moving air masses. In the eastern United States, large air masses associated with slow-moving or stagnating anticyclones have become sufficiently contaminated to be called *hazy blobs*. These blobs have been tracked by satellites as they develop and move across the country [15].

The evolution of regional hazy air masses has been documented in several case studies. The development of one such system is shown in Fig. 14.9. During a 10-day period in the summer of 1975, a large region of the eastern United States had decreased visibility associated with the presence of fine particles in the atmosphere. The phenomenon occurred in association with a slow-moving high-pressure system. Because it seldom rains during the passage of these systems, the fine particles may have stayed airborne for a longer period of time than usual.

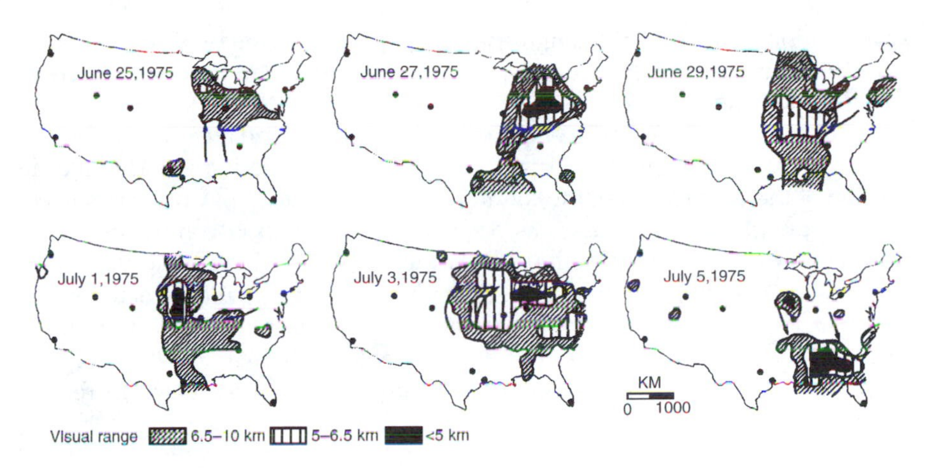

Fig. 14.9. The evolution and transport of a large hazy air mass. Contour maps of noon visibility for June 25–July 5, 1975. *Source*: Lyons, W. A., and Husar, R. B., *Mon. Weather Rev.* **104**, 1623–1626 (1976).

III. EFFECTS OF ATMOSPHERIC HAZE

The United States Clean Air Act of 1977 set as a national goal the prevention of any future degradation and the reduction of any existing impairment of visibility in mandatory class I federal areas caused by anthropogenic air pollution. The Clean Air Act Amendments of 1990 reinforce the support of these goals (see Chapter 24 for a discussion of federal classes of areas). These areas include most of the major national parks, such as the Grand Canyon, Yosemite, and Zion Park. This portion of the Clean Air Act addresses the problem of visibility degradation by atmospheric haze of anthropogenic origin. This legislation recognizes atmospheric haze as a cause of degradation in visual air quality.

All nations contain areas of exceptional scenic beauty. The value of these areas is largely determined by society. Many nations, determined to protect these areas, have established parks or preserves where only limited development can occur, in many instances limited to facilities such as food and lodging for visitors to the area.

The Grand Canyon National Park in the southwestern United States is a prime example of an area of natural beauty. This park is ~250 km long, varying in width up to ~45 km. The actual canyon is ~10 km at its widest point, with the Colorado River running in the bottom of the canyon, 1600 m below the edge of the outer rim. Visitors go to parks for many reasons, such as hiking, camping, wildlife, and the enjoyment of solitude, but the overwhelming majority visit the Grand Canyon to enjoy the magnificent views from its rim. These views have detail in the foreground (0–5 km), with colored layers of rock strata on canyon walls perhaps 5–25 km distant and in the far background (25–50 km) additional geologic features which contribute to the viewers' appreciation of the scene. To enjoy these views, one must have good visibility over the entire path length from the details in the foreground to the objects in the distant background.

A survey by national park personnel indicates that large areas of the United States are subject to varying degrees of visibility degradation [1]. The middle portion of the eastern half of the country and the Florida Gulf Coast are subject to widespread hazy air masses associated with stagnation conditions. Large portions of the western half of the country are subject to atmospheric haze problems associated with power plants, urban plumes, and agricultural activities.

Average airport visibilities over the eastern half of the United States have been determined over a period of approximately 25 years (1948–1974) [6]. Although seasonal variations occur, the long-term trend has been decreased visual air quality over the time period.

IV. VISIBILITY

Holzworth and Maga [16] developed a technique for examining the trend in visibility and analyzed data for several California airports. Bakersfield's visibility deteriorated over the period 1948–1957 and Sacramento's visibility decreased over the period 1935–1958. Los Angeles had decreasing visibility from 1932 to 1947, with little change over the period 1948–1959.

Holzworth [17] reported on the frequency of visibility of less than 7 miles for 28 cities. Two periods of records were compared for each city. There were increases in low visibility in only 26% of the comparisons from the early period (around 1930–1940) to a later period (around the mid-1950s).

Miller *et al.* [18], using analyses for Akron, Ohio; Lexington, Kentucky; and Memphis, Tennessee, concluded that "summer daytime visibilities were significantly lower during the period 1966–1969 than visibilities for the preceding 4-year period."

Faulkenberry and Craig [19], in examining the trends at three Oregon cities, utilized a modification of the Holzworth–Maga technique by which a single statistic can be calculated for each year, indicating the probability of observing better visibility at Salem, Oregon, with no trends at Portland and Eugene, Oregon over the period 1950–1971.

Arizona has traditionally been a large copper-producing state. SO_x emissions from copper smelters near Phoenix and Tucson are shown in Fig. 14.10 [1]. Phoenix is located 100 km from the nearest smelter, and Tucson is 60 km from the nearest smelter. The improvement in visibility in the 1967–1968 period was due to a decrease in SO_x emissions when there was a 9-month shutdown caused by a strike. Improvement in visibility in the mid-1970 was the result of better control technology and process changes.

Zannetti *et al.* [20] did an analysis of visual range in the eastern United States again showing the importance of humidity but also showing the importance of air mass type, which is usually related to its direction of origin.

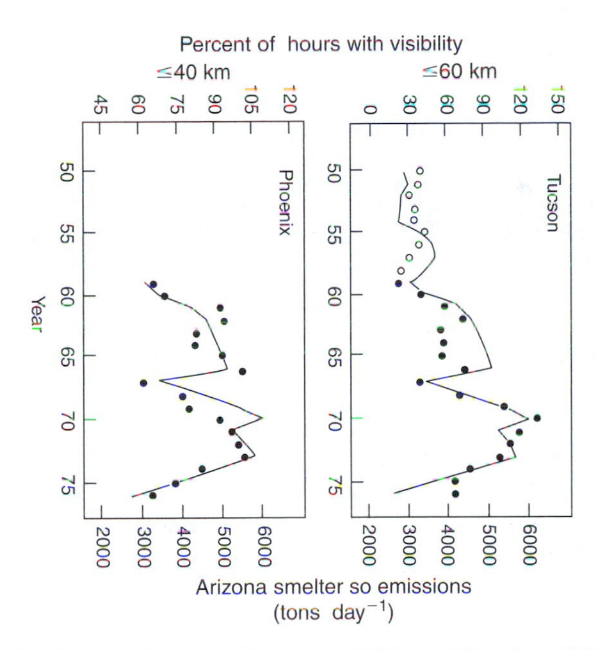

Fig. 14.10. Historic trends in hours of reduced visibility at Phoenix and Tucson, Arizona, compared to trends in SO_x emissions from Arizona smelters. The solid lines (———) represent yearly SO_x emissions. The dots (...) represent yearly percentages of hours of reduced visibility. Note the open dots used to represent different locations in Tucson before 1958. *Source*: US Environmental Protection Agency, *Protecting Visibility*, EPA-450/5-79-008. Office of Air Quality Planning Standards, Research Triangle Park, NC, 1979.

Mathai [21] summarized the specialty conference on atmospheric visibility. With the exception of water content of particles and the measurement of organic species, analytical laboratory techniques are readily available for particle analysis. Regulatory approaches to mitigate existing visibility impairment and to prevent further impairment are being formulated. A significant problem for regulation is the lack of proven techniques to quantify the contributions due to various sources.

V. ACIDIC DEPOSITION

Over the past 25 years, evidence has been accumulated on changes in aquatic life and soil pH in Scandinavia, Canada, and the northeastern United States. Many believe that these changes are caused by acidic deposition traceable to pollutant acid precursors that result from the burning of fossil fuels. Acid rain is only one component of *acidic deposition*, a more appropriate description of this phenomenon. Acidic deposition is the combined total of wet and dry deposition, with wet acidic deposition being commonly referred to as acid rain.

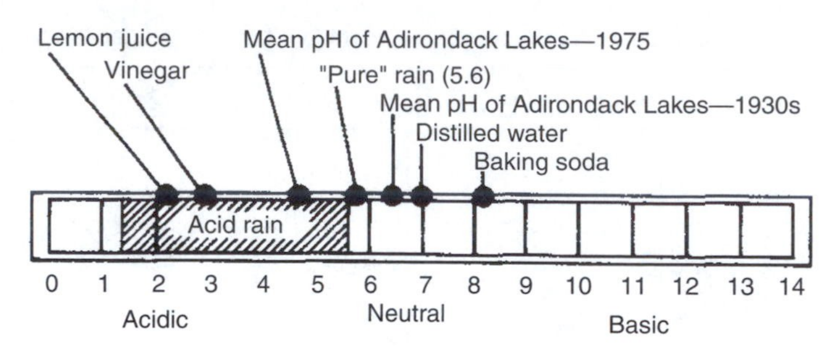

Fig. 14.11. The pH scale is a measure of hydrogen ion concentration. The pH of common substances is shown with various values along the scale. The Adirondack Lakes are located in the state of New York and are considered to be receptors of acidic deposition. *Source*: US Environmental Protection Agency, *Acid Rain—Research Summary*, EPA-600/8-79-028. Cincinnati, OH, 1979.

Acidity is defined in terms of the pH scale, where pH is the negative logarithm of the hydrogen ion $[H^+]$ concentration:

$$pH = -\log[H^+] \tag{14.5}$$

In the simplest case, CO_2 dissolves in raindrops, forming carbonic acid. At a temperature of 20°C, the raindrops have a pH of 5.6, the value often labeled as that of clean or natural rainwater. It represents the baseline for comparing the pH of rainwater which may be altered by SO_2 or NO_x oxidation products. Figure 14.11 illustrates the pH scale with the pH of common items and the pH range observed in rainwater. The pH of rainwater can vary from 5.6 due to the presence of H_2SO_4 and HNO_3 dissolved or formed in the droplets. These strong acids dissociate and release hydrogen ions, resulting in more acidic droplets. Basic compounds can also influence the pH. Calcium (Ca^{2+}), magnesium (Mg^{2+}), and ammonium (NH_4^+) ions help to neutralize the rain droplet and shift the overall H^+ toward the basic end of the scale. The overall pH of any given droplet is a combination of the effects of carbonic acid, sulfuric and nitric acids, and any neutralizers such as ammonia.

The principal elements of acidic deposition are shown in Fig. 14.12. Dry deposition occurs when it is not raining. Gaseous SO_2, NO_2, and HNO_3 and acid aerosols are deposited when they come into contact with and stick to the surfaces of water bodies, vegetation, soil, and other materials. If the surfaces are moist or liquid, the gases can go directly into solution; the acids formed are identical to those that fall in the form of acid rain. SO_2 and NO_2 can undergo oxidation, forming acids in the liquid surfaces if oxidizers are present. During cloud formation, when rain droplets are created, fine particles or acid droplets can act as seed nuclei for water to condense. This is one process by which sulfuric acid is incorporated in the droplets. While the droplets are in the cloud, additional gaseous SO_2 and NO_2 impinge on them and are absorbed. These absorbed gases can be oxidized by dissolved H_2O_2 or other oxidizers, lowering the pH of the raindrop. As the raindrop falls

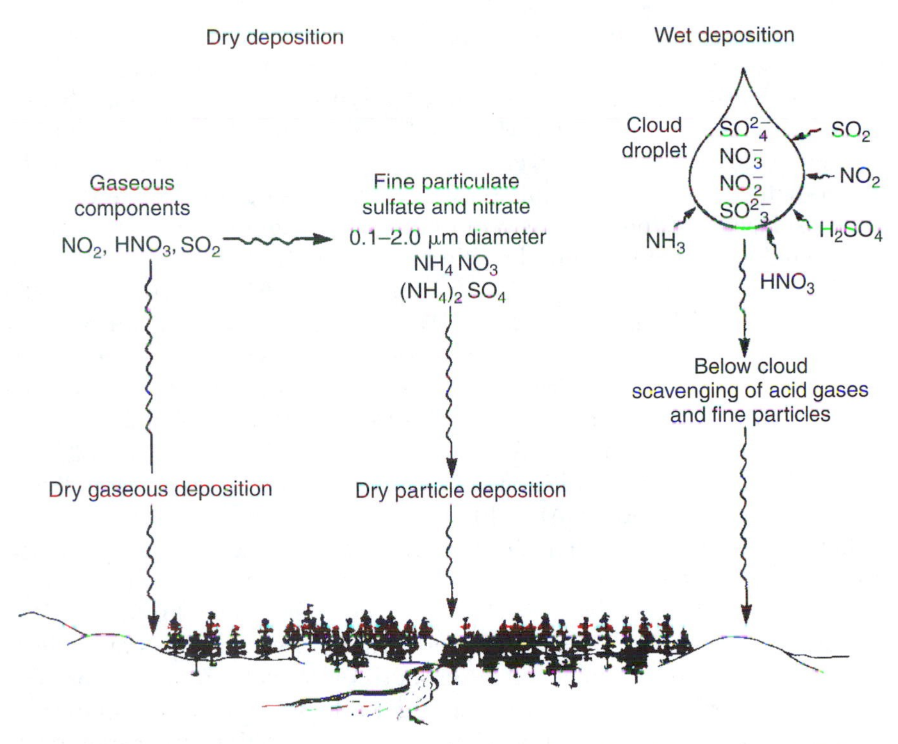

Fig. 14.12. Atmospheric processes involved in acidic deposition. The two principal deposition pathways are dry deposition (nonrain events) and wet deposition (rain events).

beneath the cloud, additional acidic gases and aerosol particles may be incorporated in it, also affecting its pH.

The United States has established a National Atmospheric Deposition Program (NADP), and Canada has established the CANSAP program, which consists of sampling networks and organizational and support structures to obtain quantitative information on the spatial and temporal distribution patterns of acid deposition [22, 23]. The lowest rainwater pH isopleths are associated with the regions of highest SO_2 emissions. Although there is considerable controversy over the quality and strength of the link between SO_2 and NO_x emissions from stationary sources and subsequent acid deposition hundreds of kilometers downwind, the National Research Council has concluded that a 50% reduction in the emissions of sulfur and nitrogen gases will produce about a 50% reduction in the acids deposited on the land and water downwind of the emission source. They also state that current meteorological models of atmospheric transport cannot identify specific sources with acid deposition at a particular downwind location [24].

A field study, the Eulerian Model Evaluation Field Study (EMEFS) [25], has been used to evaluate two models: the acidic deposition and oxidant model (ADOM) [26] and the regional acid deposition model (RADM) [27].

For both models the calculated values, such as air concentrations, are volume averages over grid cells which are 127 km on a side for the ADOM and 80 km for the RADM. These are compared with point measurements at a possible 97 locations. The ADOM tended to overestimate ground-level SO_2 and underestimate ground-level sulfate. Two factors not included in the model that may contribute to these results are consideration of conversion during surface fog and conversion in nonprecipitating clouds. The RADM also overestimated SO_2 and underestimated aerosol sulfate.

In the eastern United States, acid rain consists of ~65% sulfuric acid, ~30% nitric acid, and ~5% other acids. In the West, windblown alkaline dusts buffer the acidity in rains occurring over many rural areas, whereas in urban areas 80% of the acidity is due to nitric acid [28]. Average pH in rainfall over the eastern United States for the period April 1979–March 1980 was less than 5.0, with some areas less than 4.2 [29]. The lowest annual pH recorded was 3.78 at De Bilt, The Netherlands, in 1967, and the lowest in an individual rainfall was 2.4 at Pitlochry, Scotland, on April 10, 1974 [30].

One of the major effects of acidic deposition is felt by aquatic ecosystems in mountainous terrain, where considerable precipitation occurs due to orographic lifting. The maximum effect is felt where there is little buffering of the acid by soil or rock structures and where steep lakeshore slopes allow little time for precipitation to remain on the ground surface before entering the lake. Maximum fish kills occur in the early spring due to the "acid shock" of the first meltwater, which releases the pollution accumulated in the winter snowpack. This first melt may be 5–10 times more acidic than rainfall.

Although the same measurement techniques for rainfall acidity have not been used over a long period of time and sampling has been carried out at relatively few locations, the trend between 1955–1956 and 1975–1976 was for the area with a pH of less than 4.6 to expand greatly over the eastern United States. The largest increases occurred over the southeastern United States, where industrialization grew rapidly during the period. The last several decades have also seen an increased area of lower pH over northern Europe.

VI. EFFECTS OF ACIDIC DEPOSITION

Land, vegetation, and bodies of water are the surfaces on which acidic deposition accumulates. Bodies of freshwater represent the smallest proportion of the earth's surface area available for acidic deposition. Yet, the best-known effect is acidification of freshwater aquatic systems.

Consider a lake with a small watershed in a forest ecosystem. The forest and vegetation can be considered as an acid concentrator. SO_2, NO_2, and acid aerosol are deposited on vegetation surfaces during dry periods and rainfalls; they are washed to the soil floor by low-pH rainwater. Much of the acidity is neutralized by dissolving and mobilizing minerals in the soil. Aluminum, calcium, magnesium, sodium, and potassium are leached from the soil into surface waters. The ability of soils to tolerate acidic deposition is very dependent on the alkalinity of the soil. The soil structure in the northeastern United States and eastern Canada is quite varied, but much of the area is covered with thin soils with a relatively limited neutralizing capacity. In watersheds with this type of soil, lakes and streams are susceptible to low pH and elevated levels of aluminum. This combination has been found to be very toxic to some species of fish. When the pH drops to ~5, many species of fish are no longer to reproduce and survive. In Sweden, thousands of lakes are no longer able to support fish. In the United States the number of polluted lakes is much smaller, but many more may be pushed into that condition by continued acidic deposition. In Canada, damage to aquatic systems and forest ecosystems is a matter of considerable concern.

Aquatic systems in areas of large snowfall accumulation are subjected to a pH surge during the spring thaw. Acidic deposition is immobilized in the snowpack, and when warm springtime temperatures cause melting, the melted snow flows into streams and lakes, potentially overloading the buffering capacity of the aquatic system.

A second area of concern is reduced tree growth in forests, discussed in Chapter 13. As acidic deposition moves through forest soil, the leaching process removes nutrients. If the soil base is thin or contains barely adequate amounts of nutrients to support a particular mix of species, the continued loss of a portion of the soil minerals may cause a reduction in future tree growth rates or a change in the types of trees able to survive in a given location.

REFERENCES

1. US Environmental Protection Agency, *Protecting Visibility*, EPA-450/5-79-008. Office of Air Quality Planning and Standards, Research Triangle Park, NC, 1979.
2. Friedlander, S. K., *Smoke, Dust and Haze*. Wiley, New York, 1977.
3. Middleton, W. E. K., *Vision Through the Atmosphere*. University of Toronto Press, Toronto, Ont., 1952.
4. Van De Hulst, H. C., Scattering in the atmosphere of earth and planets, in *The Atmospheres of the Earth and Planets* (Kuiper, G. P., ed.), pp. 49–111. University of Chicago Press, Chicago, 1949.
5. Hall Jr., T. C., and Blacet, F. E., *J. Chem. Phys.* **20**, 1745 (1952).
6. Husar, R. B., Elkins, J. B., and Wilson, W. E., US Visibility Trends, 1906–1992, *Air and Waste Management Association 87th Annual Meeting and Exposition*, 1994.
7. Twomey, S., *Atmospheric Aerosols*. Elsevier, North-Holland, NY, 1977.
8. Charlson, R. J., Waggoner, A. P., and Thielke, H. F., *Visibility Protection for Class I Areas. The Technical Basis*. Report to the Council of Environmental Quality, Washington, DC, 1978.
9. Faxvog, F. R., *Appl. Opt.* **14**, 269–270 (1975).
10. Buma, T. J., *Bull. Am. Meteorol. Soc.* **41**, 357–360 (1960).
11. McCormick, R. A., and Baulch, D. M., *J. Air Pollut. Control Assoc.* **12**, 492–496 (1962).
12. McCormick, R. A., and Kurfis, K. R., *Quart. J. Roy. Meteorol. Soc.* **92**, 392–396 (1966).
13. National Research Council, *Airborne Particles*. University Park Press, Baltimore, MD, 1979.
14. Reiss, H., *Ind. Eng. Chem.* **44**, 1284–1288 (1952).

15. Lyons, W. A., and Husar, R. B., *Mon. Weather Rev.* **104**, 1623–1626 (1976).
16. Holzworth, G. C., and Maga, J. A., *J. Air Pollut. Control Assoc.* **10**, 430–435 (1960).
17. Holzworth, G. C., *Some Effects of Air Pollution on Visibility In and Near Cities.* Sanitary Engineering Center Technical Report A62-5, Department of Health, Education and Welfare. United States Public Health Service, Cincinnati, OH, 1962.
18. Miller, M. E., Canfield, N. L., Ritter, T. A., and Weaver, C. R., *Mon. Weather Rev.* **100**, 65–71 (1972).
19. Faulkenberry, D. G., and Craig, C. D., *Visibility Trends in the Willamette Valley, 1950–71*. Third Symposium on Atmospheric Turbulence, Diffusion, and Air Quality. American Meteorological Society, Boston, MA, 1976.
20. Zannetti, P., Tombach, I. H., and Cvencek, S. J., An analysis of visual range in the eastern United States under different meteorological regimes. *J. Air Pollut. Control Assoc.* **39**, 200–203 (1989).
21. Mathai, C. V., *J. Air Waste Manage. Assoc.* **40**, 1486–1494 (1990).
22. Interagency Task Force on Acid Precipitation. National Acid Precipitation Assessment Plan. NTIS, PB82-244 617, 1982.
23. Whelpdale, D. M., and Barrie, L. A., *J. Air Water Soil Pollut.* **14**, 133–157 (1982).
24. National Research Council, *Acid Deposition: Atmospheric Processes in Eastern North America.* National Academy Press, Washington, DC, 1983.
25. Hansen, D. A., Puckett, K. J., Jansen, J. J., Lusis, M., and Vickery, J. S., The Eulerian Model Evaluation Field Study (EMEFS). Paper 5.1, pp. 58–62, in *Preprints, Seventh Joint Conference on Applications of Air Pollution Meteorology with AWMA*, January 14–18, 1991, New Orleans. American Meteorological Society, Boston, MA, 1991.
26. Fung, C., Bloxam, R., Misra, P. K., and Wong, S., Understanding the performance of a comprehensive model. Paper N2.9, pp. 46–49, in *Preprints, Seventh Joint Conference on Applications of Air Pollution Meteorology with AWMA*, January 14–18, 1991, New Orleans. American Meteorological Society, Boston, MA, 1991.
27. Barchet, W. R., Dennis, R. L., and Seilkop, S. K., Evaluation of RADM using surface data from the Eulerian Model Evaluation Field Study. Paper 5.2, pp. 63–66, in *Preprints, Seventh Joint Conference on Applications of Air Pollution Meteorology with AWMA*, January 14–18, 1991, New Orleans. American Meteorological Society, Boston, MA, 1991.
28. Nicholas, G., and Boyd, R. R., *Dames and Moore Eng. Bull.* **58**, 4–12 (1981).
29. La Bastille, A., *Natl. Geogr.* **160**, 652–681 (1981).
30. Likens, G. E., Wright, R. F., Galloway, J. N., and Butler, T. J., *Sci. Am.* **241**, January 43–51 (1979).

SUGGESTED READING

Chang, J. S., Brost, R. A., Isaksen, I. S. A., Madronich, S., Middleton, P., Stockwell, W. R., and Walcek, C. J., A three-dimensional Eulerian acid deposition model: physical concepts and formulation. *J. Geophys. Res.* **92**, 14681–14700 (1987).

Hidy, G. M., Mueller, P. K., Grosjean, D., Appel, B. R., and Wesolowski, J. J. (eds.), *Advances in Environmental Science and Technology*, Vol. 9, *The Character and Origins of Smog Aerosols. A Digest of Results from the California Aerosol Characterization Experiment (ACHEX)*. Wiley, New York, 1980.

Keith, L. H., *Energy and Environmental Chemistry*, Vol. II, *Acid Rain*. Ann Arbor Science Publishers, Ann Arbor, MI, 1982.

Mathai, C. V. (ed.), *Visibility and Fine Particles*, TR-17. Air Waste Management Association, Pittsburgh, PA, 1990.

National Park Service, *Visibility Effects of Air Pollution*, 2007, accessible at http://www2.nature.nps.gov/air/AQBasics/visibility.cfm#types.

Suffet, I. H. (ed.), *Fate of Pollutants in the Air and Water Environments*, Part 2, *Chemical and Biological Fate of Pollutants in the Environment*. Wiley, New York, 1977.

US Environmental Protection Agency, *Protecting Visibility*, EPA-450/5-79-008. Office of Air Quality Planning and Standards, Research Triangle Park, NC, 1979.

QUESTIONS

1. Define threshold contrast.
2. List the four components which contribute to the extinction coefficient b_{ext}. Describe the circumstances in which each component would dominate extinction.
3. Derive the Koschmieder relationship from Eq. (14.1).
4. Compare visibility measurements at a nearby airport with those of particle-free clean air.
5. Explain why stringent emission standards for particulate matter based on mass/heat input will do little to improve visual air quality.
6. Explain the differences in visual air quality between the western and eastern portions of the United States.
7. Compare the wavelengths of visible light with the range of particle diameters which most efficiently scatter light.
8. Describe the impact of future visibility degradation on your area (e.g. on specific areas of scenic attraction).
9. In the 1980s and 1990s, where were the greatest increases in rainfall acidity in the United States? What is the suspected reason?

15

Long-Term Effects on the Planet

I. GLOBAL CLIMATE CHANGE

Warming on the global scale is expected to occur as a result of the increase of carbon dioxide (CO_2) and other greenhouse gases (those that absorb and reradiate portions of the infrared radiation from the earth). What is debatable is the amount of warming that will occur by a particular point in time. The CO_2 concentration has increased by about 25% since 1850 [1]. This is due to both combustion of fossil fuels and deforestation, which decreases the surface area available for photosynthesis and the resulting breakdown of CO_2 to oxygen and water vapor.

The average temperature of the earth is difficult to measure, but most measurements show a very small overall change that would not be detectable to humans due to short-term and regional variations. Overall, however, a majority of scientific evidence appears to indicate that the temperature of the earth is increasing. There have been wide fluctuations in mean global temperatures, such as the ice ages, but on balance the mean temperature has remained constant, prompting some scientists to speculate some whimsical causes for such consistency. Charles Keeling, an atmospheric scientist, measured CO_2 concentrations in the atmosphere using an infrared gas analyzer. Since 1958, these data have provided the single most important piece of information on global warming, and are now referred to as "Keeling curve" in honor of the scientist.

The Keeling curve shows that there has been more than 15% increase in CO_2 concentration, which is a substantial rise given that short time that the measurements have been taken. It is likely, if we extrapolate backward, that our present CO_2 levels are double what they were in pre-industrial revolution times, providing ample evidence that global warming is indeed occurring.[1]

Another hypothesis for this rise in temperature is that the presence of certain gases in the atmosphere is not allowing the earth to reflect enough of the heat energy from the sun back into space. The earth acts as a reflector to the sun's rays, receiving the radiation from the sun, reflecting some of it into space (called *albedo*) and adsorbing the rest, only to reradiate this into space as heat. In effect, the earth acts as a wave converter, receiving the high-energy, high-frequency radiation from the sun and converting most of it into low-energy, low-frequency heat to be radiated back into space. In this manner, the earth maintains a balance of temperature.

In order to better understand this balance, the light energy and the heat energy have to be defined in terms of their radiation patterns, as shown in Fig. 15.1. The incoming radiation (light) wavelength has a maximum at

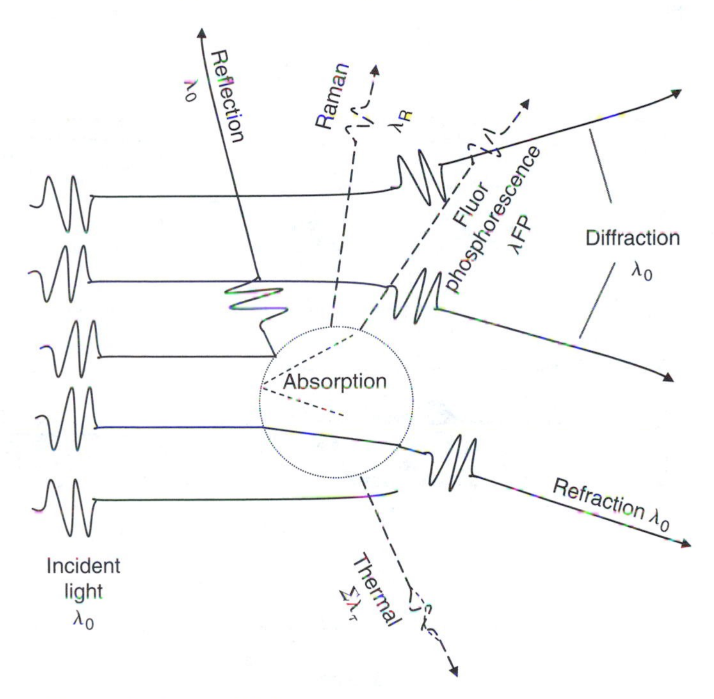

Fig. 15.1. Patterns for heat and light energy.

[1] The major source for this discussion is: Vallero, D.A., and Vesilind, P.A., *Socially Responsible Engineering: Justice in Risk Management*. John Wiley & Sons, Hoboken, NJ, 2006.

around 0.5 nm and almost all of it are lesser than 3 nm. The heat energy spectrum, or that energy reflected back into space, has the maximum at about 10 nm and almost all of it at a wavelength greater than 3 nm.

As both the light and heat energy pass through the earth's atmosphere, they encounter the aerosols and gases surrounding the earth. These can either allow the energy to pass through, or they can interrupt it by scattering or absorption. If the atoms in the gas molecules vibrate at the same frequency as the light energy, they will absorb the energy and not allow it to pass through. Aerosols will scatter the light and provide a "shade" for the earth.

The absorptive potential of several important gases is shown in Fig. 15.2, along with the spectra for the incoming light (short wavelength) radiation and the outgoing heat (long wavelength) radiation. The incoming radiation is impeded by water vapor and molecular oxygen and ozone. Most of the light energy comes through unimpeded.

The heat energy, however, encounters several potential impediments. As it is trying to reach outer space, it finds that water vapor, CO_2, CH_4, O_3, and N_2O all have absorptive wavelengths right in the middle of the heat spectrum. Quite obviously, an increase in the concentration of any of these will greatly limit the amount of heat transmitted into space. These gases are appropriately called *greenhouse gases* because their presence will limit the

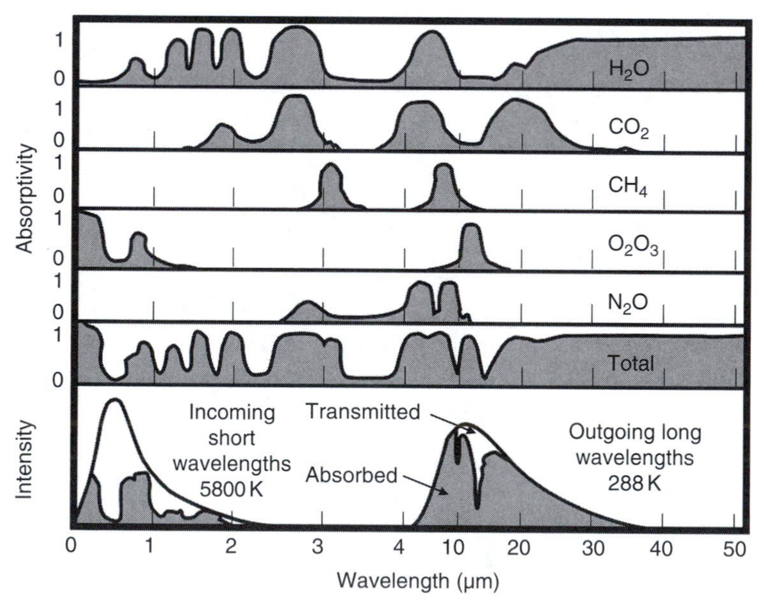

Fig. 15.2. Adsorptive potential of several important gases in the atmosphere. Also shown are the spectra for the incoming solar energy and the outgoing thermal energy from the earth. Note that the wavelength scale changes at 4 μm. Courtesy of Masters, G. M., *Introduction to Environmental Engineering and Science*. Prentice Hall, Englewood Cliffs, NJ, 1998.

heat escaping into space, much like the glass of a greenhouse or even the glass in a car limits the amount of heat that can escape, thus building up the temperature under the glass cover.

The effectiveness of a particular gas to promote global warming (or cooling, as is the case with aerosols) is known as *forcing*. The gases of most importance in forcing are listed in Table 15.1. Climate change results from natural internal processes and from external forcings. Both are affected by persistent changes in the composition of the atmosphere brought about by changes in land use, release of contaminants, and other human activities. Radiative forcing is the change in the net vertical irradiance within the atmosphere. Radiative forcing is often calculated after allowing for stratospheric temperatures to readjust to radiative equilibrium, while holding all tropospheric properties fixed at their unperturbed values. Commonly, radiative forcing is considered to be the extent to which injecting a unit of a greenhouse gas into the atmosphere changes global average temperature, but other factors can affect forcing, as shown in Figs. 15.3 and 15.4.

There is much uncertainty about the effects of the presence of these radiant gases (see Table 15.2), but the overall effect of the composite of gases is well understood. The effectiveness of CO_2 as a global warming gas has been known for over 100 years. However, the first useful measurements of atmospheric CO_2 were not taken until 1957. The data from Mauna Loa (Fig. 15.5) show that even in the 1950s the CO_2 concentration had increased from the baseline 280 to 315 ppm; and this has continued to climb over the last 50 years at a nearly constant rate of about 1.6 ppm year^{-1}. The most serious problem with CO_2 is that the effects on global temperature due to its greenhouse effect are delayed. Even in the completely impossible scenario of not emitting any new CO_2 into the atmosphere, CO_2 concentrations will continue to increase from our present 370 ppm to possibly higher than 600 ppm.

Methane is the product of anaerobic decomposition and human food production. One of the highest producers of methane in the world is New Zealand which boasts 80 million sheep. Methane is also emitted during the combustion of fossil fuels and cutting and clearing of forests. The concentration

TABLE 15.1

Relative Forcing of Increased Global Temperature

Gas	Percent of relative radiative forcing
Carbon dioxide, (CO_2)	64
Methane (CH_4)	19
Halocarbons (mostly CFCs)	11
Nitrous oxide (N_2O)	6

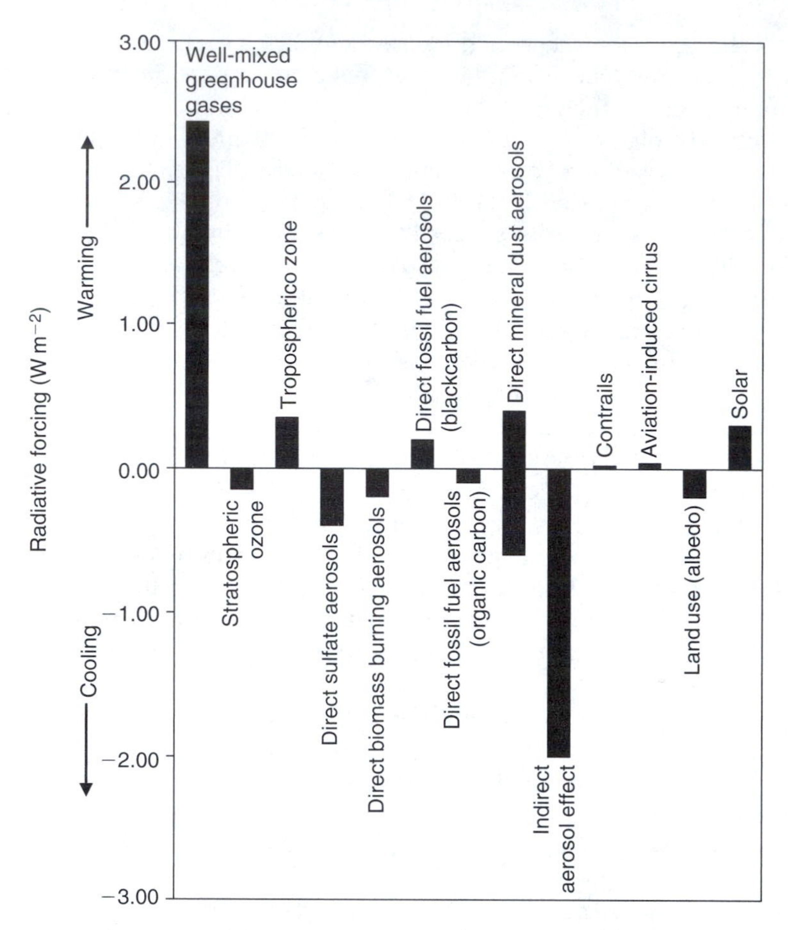

Fig. 15.3. The global mean radiative forcing (W m^{-2}) of the climate system for the year 2000, relative to 1750. The Intergovernmental Panel on Climate Change (IPCC) has applied a level of scientific understanding (LOSU) index in accorded to each forcing (see Table 15.2). This represents the Panel's subjective judgment about the reliability of the forcing estimate, involving factors such as the assumptions necessary to evaluate the forcing, the degree of knowledge of the physical/chemical mechanisms determining the forcing, and the uncertainties surrounding the quantitative estimate of the forcing. The relative contribution of the principal well-mixed greenhouse gases is shown in Fig. 15.4. *Data from:* IPCC, *Climate Change 2001: The Scientific Basis,* Chapter 6—Radiative Forcing of Climate Change, 2001.

of CH_4 in the atmosphere has been steady at about 0.75 ppm for over a thousand years, and then increased to 0.85 ppm in 1900. Since then, in the space of only a hundred years, it has skyrocketed to 1.7 ppm. Methane is removed from the atmosphere by reaction with the hydroxyl radical (OH) as:

$$CH_4 + OH + 9O_2 \rightarrow CO_2 + 0.5H_2 + 2H_2O + 5O_3 \tag{15.1}$$

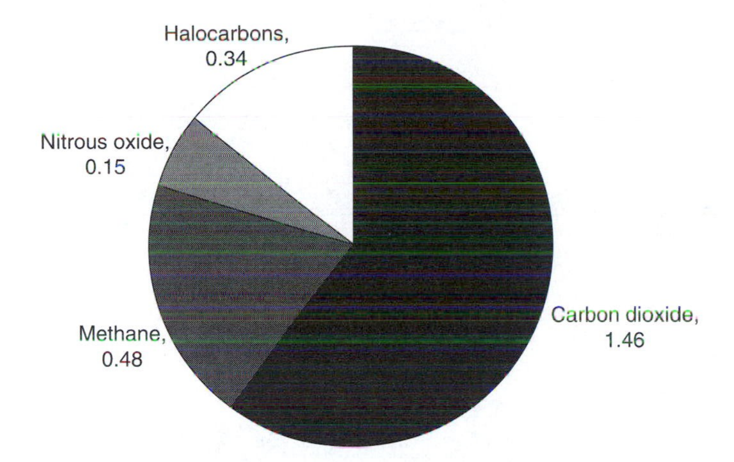

Fig. 15.4. Relative contribution of well-mixed greenhouse gases to the +2.43 W m^{-2} radiative forcing shown in Fig. 15.3. *Data from:* IPCC, *Climate Change 2001: The Scientific Basis*, Chapter 6—Radiative Forcing of Climate Change, 2001.

TABLE 15.2

Level of Scientific Understanding (LOSU) of Radiative Forcings

Forcing phenomenon	LOSU
Well-mixed greenhouse gases	High
Stratospheric O_3	Medium
Tropospheric O_3	Medium
Direct sulfate aerosols	Low
Direct biomass burning aerosols	Very low
Direct fossil fuel aerosols (black carbon)	Very low
Direct fossil fuel aerosols (organic carbon)	Very low
Direct mineral dust aerosols	Very low
Indirect aerosol effect	Very low
Contrails	Very low
Aviation-induced cirrus	Very low
Land use (albedo)	Very low
Solar	Very low

Source: IPCC, *Climate Change 2001: The Scientific Basis*, Chapter 6—Radiative Forcing of Climate Change, 2001.

This indicates that the reaction creates carbon dioxide, water vapor, and ozone, all of which are greenhouse gases, so the effect of one molecule of methane is devastating to the production of the greenhouse effect.

Halocarbons, or the same chemical class linked to the destruction of stratospheric ozone, are also radiant gases. The most effective global warming gases are CFC-11 and CFC-12, both of which are no longer manufactured, and the banning of these substances has shown a leveling off in the stratosphere. Nitrous oxide is also in the atmosphere mostly as a result of

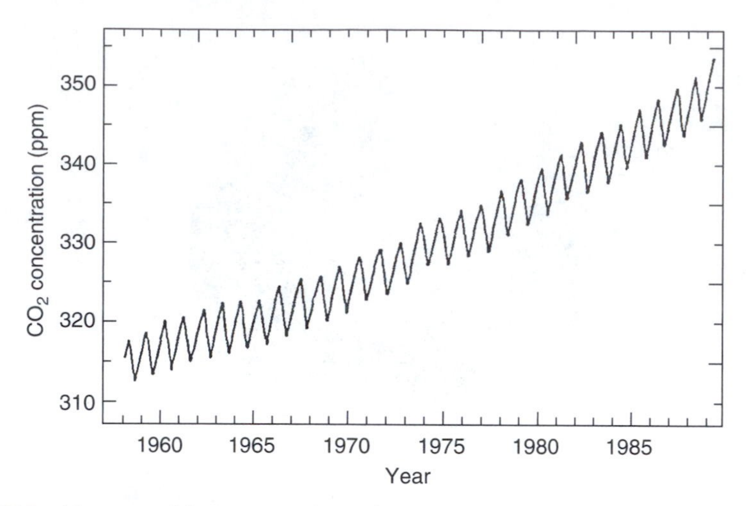

Fig. 15.5. Mean monthly concentrations of atmospheric CO_2 at Mauna Loa. The yearly oscillation is explained mainly by the annual cycle of photosynthesis and respiration of plants in the Northern Hemisphere. *Source*: Lindzen [2].

human activities, especially the cutting and clearing of tropical forests. The greatest problem with nitrous oxide is that there appears to be no natural removal processes for this gas and so its residence time in the stratosphere is quite long.

The net effect of these global pollutants is still being debated. Various atmospheric models used to predict temperature change over the next 100 years vary widely. They nevertheless agree that some positive change will occur, even if we do something drastic today (which does not seem likely). By the year 2100, even if we do not increase our production of greenhouse gases and if the United States signs the Kyoto Accord that encourages the reduction in greenhouse gas production, the global temperature is likely to be between 0.5°C and 1.5°C warmer.

Firm evidence for the amount of warming taking place in terms of actual temperature measurements has been complicated primarily by the magnitudes of natural climatic variations that occur. A summary of the available measurements shown by Kellogg [3] is given in Fig. 15.6. Other factors contributing to warming trends are the length of temperature records; the lack of representative measurements over large portions of the earth, primarily the oceans and polar regions; and the urban sprawl toward locations at which temperature measurements are made, such as airports.

Although there has been a warming trend over the past 100 years, it is not necessarily due to the greenhouse effect. The concern of the scientific community about accelerating changes in the next 40–50 years is based not only on the recent observations of temperature compared with past observations, but also on the physical principles related to the greenhouse effect.

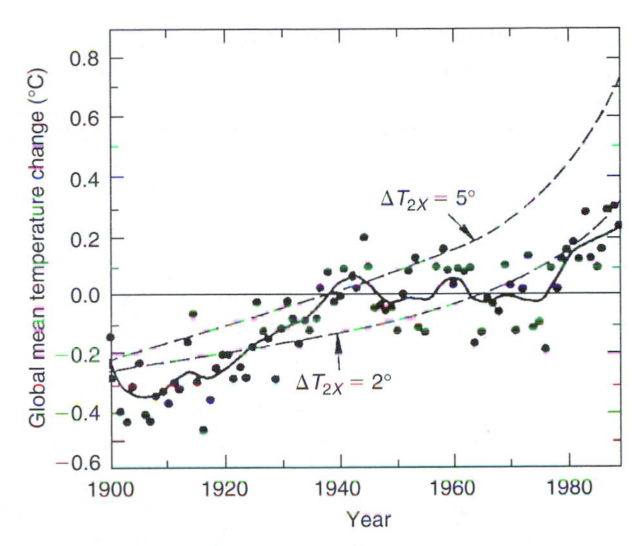

Fig. 15.6. Combined land–air and sea surface temperatures from 1900, relative to 1951–1980 average (solid line and dots), adapted from Folland *et al.* [4]. The land–air temperatures were derived by Jones [5] and the sea surface temperatures by the UK Meteorological Office and Farmer *et al.* [6]. The smoothed curve was obtained by a low-pass binomial filter operating on the annual data (shown by dots), passing fluctuations having a period of 290 years or more almost unattenuated. The dashed lines are calculated global temperature changes from the mean of the period 1861–1900, using the climate model of Wigley and Raper [7], and observed concentrations of greenhouse gases, adapted from Wigley and Barnett [8]. The upper curve assumes an equilibrium temperature increase for a doubling of greenhouse gases of 5 K, and the lower one assumes it to be 2 K; both curves are based on an ocean vertical diffusion coefficient (K) of 0.63 $cm^2 s^{-1}$ and temperature of the water in sinking regions (π), the same as the global mean. *Source*: Kellogg [3].

Global climate models have been used to estimate the effects in terms of temperature changes. Considerable difficulties are encountered in at least two areas. One is the difficulty in accounting properly for moisture changes including cloud formation. An important mechanism of heat transfer is through water vapor and water droplets. Of course, cloud cover alters the radiational heating at any given time. The second difficulty is accounting for solar radiation variability and the occasional injection of fine particulate matter into the atmosphere by volcanic activity, both of which alter the amount of solar radiation reaching the ground. Results from most model attempts suggest that global average surface temperatures will increase on the order of 1.5–4.5°C over the next century.

The greatest concern about global warming is the regional and seasonal effects that will result. Of considerable significance could be changes in the patterns of precipitation in the agricultural and forested regions during the growing seasons, in particular.

The Intergovernmental Panel on Climate Change (IPCC) was established in 1988 by the World Meteorological Organization (WMO) and United

Nations Environment Program (UNEP). This panel gave its official report in Geneva in November 1990 at the Second World Climate Conference. This group predicted that if no significant actions are taken to curtail consumption of fossil fuel worldwide, the global mean temperature will increase at a rate of 0.2–0.5°K per decade over the next century [9]. This is at a rate faster than seen over the past 10 000 years or longer.

Most recently, in *Climate Change 2007: The Physical Science Basis—Summary for Policymakers*, the IPCC has stated that "warming of the climate system is unequivocal, as is now evident from observations of increases in global average air and ocean temperatures, widespread melting of snow and ice, and rising global average sea level." They also assert:

- Average Arctic temperatures increased at almost twice the global average rate in the past 100 years. Satellite data since 1978 show that annual average Arctic sea ice extent has shrunk by 2.7% (2.1–3.3%) per decade, with larger decreases in summer of 7.4% (5.0–9.8%) per decade.
- Temperatures at the top of the permafrost layer have generally increased since the 1980s in the Arctic (by up to 3°C). The maximum area covered by seasonally frozen ground has decreased by about 7% in the Northern Hemisphere since 1900, with a decrease in spring of up to 15%.
- Long-term trends from 1900 to 2005 have been observed in precipitation amount over many large regions. Significantly increased precipitation has been observed in eastern parts of North and South America, northern Europe and northern and central Asia. Drying has been observed in the Sahel, the Mediterranean, southern Africa, and parts of southern Asia. Precipitation is highly variable spatially and temporally, and data are limited in some regions. Long-term trends have not been observed for the other large regions assessed.

Among the groups using advanced climate system models are the following five: National Center for Atmospheric Research (NCAR), NOAA Geophysical Fluid Dynamics Laboratory (GFDL), NASA Goddard Institute for Space Studies (GISS), United Kingdom Meteorological Office (UKMO), and Oregon State University (OSU). Proper simulation of cloudiness is difficult with the models. As part of a study of sensitivity to the inclusion of cloudiness, each of 14 models was run with clear skies and then with their simulation of cloudiness [10]. A climate sensitivity parameter (CSP) was determined for each model. If the ratio of the CSP with clouds included to the CSP with clear skies was 1.0, the clouds had no feedback effect on temperature, but if the ratio was greater than 1, the cloud feedback was positive (to increase temperature). For the foregoing five models the range of ratios was from unity (no feedback) to 1.55 (a fairly strong positive feedback). All of these models consider two cloud types, stratiform and convective, but there are differences in the way these are calculated. It can be concluded that considerable work will be needed before the treatment of clouds can be considered satisfactory [3]. Some arguments have been made that cloudiness may provide a negative feedback to temperature increases [2], i.e., cause a decrease in temperature. These results indicate that this is not likely.

Penner [11] has pointed out that short-lifetime constituents of the atmosphere such as nitric oxide, nitrogen dioxide, carbon monoxide, and non-methane hydrocarbons may also play roles related to global warming because of their chemical relations to the longer-lived greenhouse gases. Also, SO_2 with a very short life interacts with ozone and other constituents to be converted to particulate sulfate, which has effects on cloud droplet formation.

II. OZONE HOLES

During each September of the mid-1980s, scientists began to observe a decrease in ozone in the stratosphere over Antarctica. These observations are referred to as "ozone holes." To understand ozone holes, one needs to know how and why ozone is present in the earth's stratosphere.

Stratospheric ozone is in a dynamic equilibrium with a balance between the chemical processes of formation and destruction. The primary components in this balance are ultraviolet (UV) solar radiation, oxygen molecules (O_2), and oxygen atoms (O) and may be represented by the following reactions:

$$O_2 + hv \rightarrow O + O \tag{15.2}$$

$$O + O_2 + M \rightarrow O_3 + M \tag{15.3}$$

$$O_3 + hv \rightarrow O_2 + O \tag{15.4}$$

where hv represents a photon with energy dependent on the frequency of light, v, and M is a molecule of oxygen or nitrogen. The cycle starts with the photodissociation of O_2 to form atomic oxygen O (Eq. 15.2). O atoms react with O_2 in the presence of a third molecule (O_2 or N_2) to form O_3 (Eq. 15.3). Ozone absorbs UV radiation and can undergo photodissociation to complete the cycle of formation and destruction (Eq. 15.4). At a given altitude and latitude a dynamic equilibrium exists with a corresponding steady-state ozone concentration. This interaction of UV radiation with oxygen and ozone prevents the penetration of shortwave UV to the earth's surface. Stratospheric ozone thus provides a UV shield for human life and biological processes on the earth's surface.

In 1975, Rowland and Molina [12] postulated that chlorofluorocarbons (CFCs) could modify the steady-state concentrations of stratospheric ozone. CFCs are chemically very stable compounds and have been used for over 50 years as refrigerants, aerosol propellants, foam blowing agents, cleaning agents, and fire suppressants. The use has been curtailed since many of these compounds have been banned throughout the globe. The Montreal Protocol on Substances that Deplete the Ozone Layer is an international agreement designed to protect the stratospheric ozone layer. The treaty was originally signed in 1987 and substantially amended in 1990 and 1992 that stipulates that the production and consumption of compounds that deplete ozone in the stratosphere be phased out. Chlorofluorocarbons (CFCs), halons, carbon tetrachloride, and methyl chloroform were to be phased out by 2000 and methyl chloroform by 2005. Revisions to the Clean Air Act in 1998 induced the United States to limit the production and import of methyl bromide to 75% of the 1991 baseline. In 2001, production and import were further reduced to 50% of the 1991 baseline. In 2003, allowable production and import were again reduced, to 30% of the baseline, leading to a complete phaseout of production and import in 1995. Beyond 2005, continued production and import of methyl bromide are restricted to critical, emergency, and quarantine and pre-shipment uses. Because of their stability in the troposphere, CFCs remain in the troposphere for long periods of time, providing the opportunity for a portion of these chemicals to diffuse into the stratosphere. Rowland and Molina suggested that CFCs in the stratosphere would upset the balance represented by Eqs. (15.3) and (15.4). In the stratosphere, CFCs would be exposed to shortwave UV radiation with wavelengths $\lambda < 220\,\mathrm{nm}$ and undergo photodissociation, releasing chlorine atoms (Cl), and Cl would interfere with the ozone balance in the following manner:

$$CCl_3F + hv \rightarrow CCl_2F + Cl(\text{CFC role in formation of Cl}) \quad (15.5)$$

$$Cl + O_3 \rightarrow ClO + O_2 \quad (15.6)$$

$$ClO + O \rightarrow Cl + O_2 \quad (15.7)$$

$$O + O_3 \rightarrow O_2 + O_2$$

The chlorine atoms would provide another destruction pathway for ozone in addition to Eq. (15.4), shifting the steady-state ozone to a lower value. Because of the catalytic nature of Eqs. (15.6) and (15.7), one chlorine atom destroys many ozone molecules.

The discovery of ozone holes over Antarctica in the mid-1980s was strong observational evidence to support the Rowland and Molina hypothesis. The atmosphere over the South Pole is complex because of the long periods of total darkness and sunlight and the presence of a polar vortex and polar stratospheric clouds. However, researchers have found evidence to support the role of ClO in the rapid depletion of stratospheric ozone over the South Pole. Figure 15.7 shows the profile of ozone and ClO measured at an altitude of 18 km on an aircraft flight from southern Chile toward the South Pole on September 21, 1987. One month earlier the ozone levels were fairly uniform around 2 ppm (vol).

Ozone holes are considered by many as a harbinger of atmospheric modification. Investigators have found a similar but less intense annual decrease in ozone over the Arctic region of the globe. Additional studies are providing evidence for stratospheric ozone depletion over the northern temperate regions of the globe. These observations prompted a worldwide phaseout of

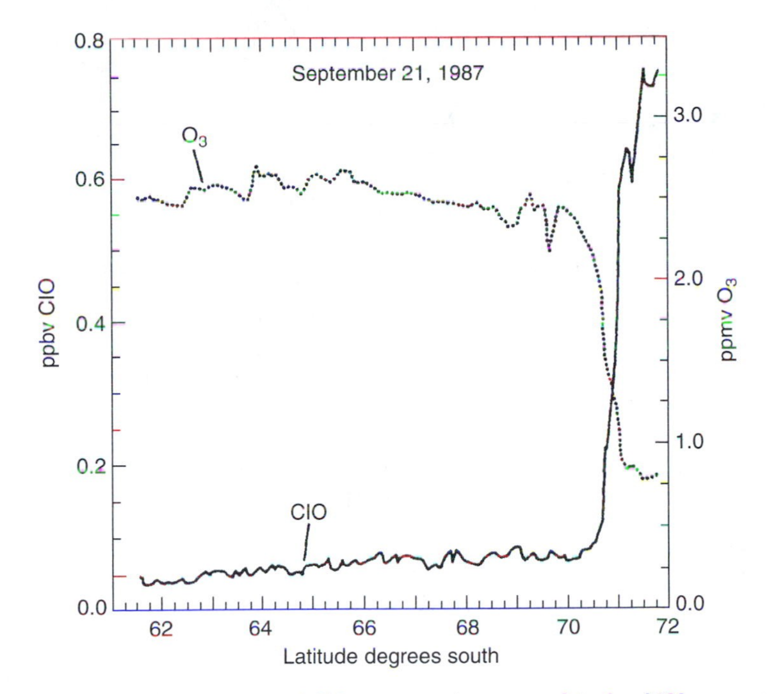

Fig. 15.7. Stratospheric ozone and ClO concentrations at an altitude of 18 km measured by aircraft flying south over Antarctica on September 27, 1987. The dramatic decrease in ozone at a latitude of 71° is attributed to the role of ClO in catalytic destruction of ozone. Adapted from Anderson *et al.* [13].

the manufacture and use of CFCs and halogens. These chemicals will be present at elevated levels for many years to come because of their stability.

CFCs represent only one class of chemicals being released to the atmosphere which have long-term effects. Replacement chemicals will be reviewed for potential adverse effects on the atmosphere. In addition, other radiant (greenhouse) gases will be subject to investigation and questioning of their role in global warming scenarios.

REFERENCES

1. Schneider, S. H., *Bull. Am. Meteorol. Soc.* **71**, 1292–1304 (1990).
2. Lindzen, R., *Bull. Am. Meteorol. Soc.* **71**, 288–299 (1990).
3. Kellogg, W. W., *Bull. Am. Meteorol. Soc.* **72**, 499–511 (1991).
4. Folland, C. K., Karl, T., and Vinnikov, K. Ya., Observed climate variations and change, in *Climate Change: The IPCC Scientific Assessment*, Working Group 1 Report (Houghton, J. T., Jenkins, G. J., and Ephramus, J. J., eds.), pp. 195–238. Cambridge University Press, Cambridge, 1990.
5. Jones, P. D., *J. Climatol.* **1**, 654–660 (1988).

6. Farmer, G., Wigley, T. M. L., Jones, P. D., and Salmon, M., *Documenting and Explaining Recent Global-Mean Temperature Changes*. Climatic Research Unit, Norwich, Final Report to NERC, UK, Contract No. GR3/6565, 1989.
7. Wigley, T. M. L., and Raper, S. C. B., *Nature* **330**, 127–131 (1987).
8. Wigley, T. M. L., and Barnett, T. P., Detection of the greenhouse effect in observations, in *Climate Change: The IPCC Scientific Assessment*, Working Group 1 Report (Houghton, J. T., Jenkins, G. J., and Ephramus, J. J., eds.), pp. 239–256. Cambridge University Press, Cambridge, 1990.
9. World Meteorological Organization, *WMO and Global Warming*, WMO No. 741, Geneva, Switzerland, 1990.
10. Cess, R. D., Potter, G. L., Blanchet, J. P., Boer, G. J., Ghan, S. J., Kiehl, J. T., Le Treut, H., Li, Z.-X., Liang, X.-Z., Mitchell, J. F. B., Morcrette, J.-J., Randall, D. A., Riches, M. R., Roeckner, E., Schlese, U., Slingo, A., Taylor, K. E., Washington, W. M., Wetherald, R. T., and Yagai, I., *Science* **245**, 513–516 (1989).
11. Penner, J. E., *J. Air Waste Manage. Assoc.* **40**, 456–461 (1990).
12. Rowland, F. S., and Molina, M. J., *Rev. Geophys. Space Phys.* **13**, 1–35 (1975).
13. Anderson, J. G., Brune, W. H., and Proffitt, M. H., *J. Geophys. Res.* **94**, 465–479 (1989).

SUGGESTED READING

American Meteorological Society, *Bull. Am. Meteorol. Soc.* **72**, 57–59 (1991).

Congressional Research Service, *Global Climate Change*. CRS Issue Brief IB89005, 2001.

Congressional Research Service, *Global Climate Change: US Greenhouse Gas Emissions—Status, Trends, and Projections*. Report No. 98-235 ENR, Updated March 12, 2002.

US Environmental Protection Agency, *Protection of Stratospheric Ozone: Incorporation of Montreal Protocol Adjustment for a 1999 Interim Reduction in Class I, Group VI Controlled Substances*. 40 *CFR* Part 82, 1999.

QUESTIONS

1. What are the difficulties in the use of climate models to estimate effects of increased CO_2?
2. What is the primary reason for the increase of CO_2 in the atmosphere?
3. What is the importance of having ozone in the stratosphere?
4. What is the effect of CFCs entering the stratosphere?
5. Why is there such strong disagreement over the models used to predict global climate change? What are the major sources of uncertainty?
6. Why do you think that the phaseout of methyl chloroform took longer than the other phased compounds in the Montreal Protocol?

Part IV

The Measurement and Monitoring of Air Pollution

16

Ambient Air Sampling

I. ELEMENTS OF A SAMPLING SYSTEM

The principal requirement of a sampling system is to obtain a sample that is representative of the atmosphere at a particular place and time and that can be evaluated as a mass or volume concentration.[1] The sampling system should not alter the chemical or physical characteristics of the sample in an undesirable manner. The major components of most sampling systems are an inlet manifold, an air mover, a collection medium, and a flow measurement device.

The inlet manifold transports material from the ambient atmosphere to the collection medium or analytical device, preferably in an unaltered condition. The inlet opening may be designed for a specific purpose. All inlets for ambient sampling must be rainproof. Inlet manifolds are made out of glass, Teflon, stainless steel, or other inert materials and permit the remaining components of the system to be located at a distance from the sample manifold inlet. The air mover provides the force to create a vacuum or lower pressure at the end of the sampling system. In most instances, air movers are pumps.

[1] Real-time, remote monitoring systems, discussed in Chapter 18, do not require subsequent analysis.

The collection medium for a sampling system may be a liquid or solid sorbent for dissolving gases, a filter surface for collecting particles, or a chamber to contain an aliquot of air for analysis. The flow device measures the volume of air associated with the sampling system. Examples of flow devices are mass flow meters, rotameters, and critical orifices.

Sampling systems can take several forms and may not necessarily have all four components (Fig. 16.1). Figure 16.1(a) is typical of many extractive sampling techniques in practice, e.g. SO_2 in liquid sorbents and polynuclear aromatic hydrocarbons on solid sorbents. Figure 16.1(b) is used for "open-face" filter collection, in which the filter is directly exposed to the atmosphere being sampled. Figure 16.1(c) is an evacuated container used to collect an aliquot of air or gas to be transported to the laboratory for chemical analysis; e.g., polished stainless steel canisters are used to collect ambient hydrocarbons for air toxic analysis. Figure 16.1(d) is the basis for many of the automated continuous analyzers, which combine the sampling and analytical processes in one piece of equipment, e.g. continuous ambient air monitors for SO_2, O_3, and NO_x.

Regardless of the configuration or the specific material sampled, several characteristics are important for all ambient air sampling systems. These are collection efficiency, sample stability, recovery, minimal interference, and an understanding of the mechanism of collection. Ideally, the first three would be 100% and there would be no interference or change in the material when collected.

One example is sampling for SO_2. Liquid sorbents for SO_2 depend on the solubility of SO_2 in the liquid collection medium. Certain liquids at the correct pH are capable of removing ambient concentrations of SO_2 with 100% efficiency until the characteristics of the solution are altered so that no more SO_2 may be dissolved in the volume of liquid provided. Under these circumstances, sampling is 100% efficient for a limited total mass of SO_2 transferred to the solution, and the technique is acceptable as long as sampling does not continue beyond the time that the sampling solution is saturated [1]. A second example is the use of solid sorbents such as Tenax for volatile hydrocarbons

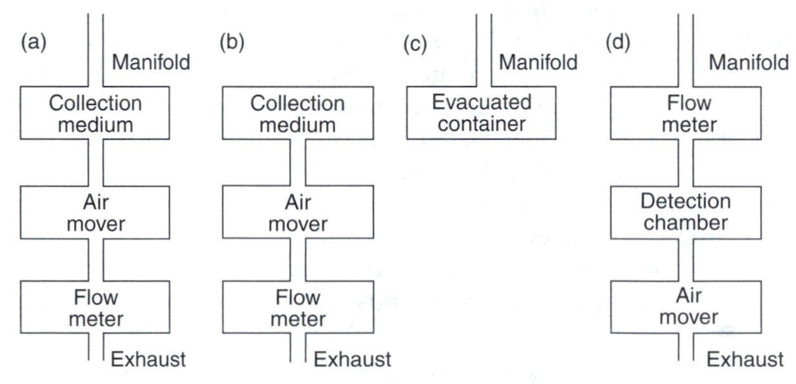

Fig. 16.1. Schematic diagram of various types of sampling systems.

by the physical adsorption of the individual hydrocarbon molecules on active sites of the sorbent [2]. Collection efficiency drops drastically when the active sites become saturated.

Sample stability becomes increasingly important as the time between sampling and analysis increases. Effects of temperature, trace contaminants, and chemical reactions can cause the collected species to be lost from the collection medium or to undergo a transformation that will prevent its recovery. Nearly 100% recovery is also required because a variable recovery rate will prevent quantification of the analysis. Interference should be minimal and, if present, well understood.

II. SAMPLING SYSTEMS FOR GASEOUS POLLUTANTS

Gaseous pollutants are generally collected by the sampling systems shown in Fig. 16.1(a)–(d). The sampling manifold's only function is to transport the gas from the manifold inlet to the collection medium in an unaltered state. The manifold must be made of nonreactive material. Tests of material for manifold construction can be made for specific gases to be sampled. In most cases, glass or Teflon will not adsorb or react with the gases. No condensation should be allowed to occur in the sampling manifold.

The volume of the manifold and the sampling flow rate determine the time required for the gas to move from the inlet to the collection medium. This residence time can be minimized to decrease the loss of reactive species in the manifold by keeping the manifold as short as possible.

The collection medium for gases can be liquid or solid sorbents, an evacuated flask, or a cryogenic trap. Liquid collection systems take the form of bubblers which are designed to maximize the gas–liquid interface. Each design is an attempt to optimize gas flow rate and collection efficiency. Higher flow rates permit shorter sampling times. However, excessive flow rates cause the collection efficiency to drop below 100%.

A. Extractive Sampling

When bubbler systems are used for collection, the gaseous species generally undergoes hydration or reaction with water to form anions or cations. For example, when SO_2 and NH_3 are absorbed in bubblers they form HSO_3^- and NHO_4^+, and the analytical techniques for measurement actually detect these ions. Table 16.1 gives examples of gases which may be sampled with bubbler systems.

Bubblers are more often utilized for sampling programs that do not require a large number of samples or frequent sampling. The advantages of these types of sampling systems are low cost and portability. The disadvantages are the high degree of skill and careful handling needed to ensure quality results. Solid sorbents such as Tenax, XAD, and activated carbon (charcoal)

TABLE 16.1

Collection of Gases by Absorption

Gas	Sampler	Sorption medium	Air flow ($L\,m^{-1}$)	Minimum sample (L)	Collection efficiency	Analysis	Interferences
Ammonia	Midget impinger	25 mL 0.1 N sulfuric acid	1–3	10		Nessler reagent	—
	Petri bubbler	10 mL of above	1–3	10	+95	Nessler reagent	—
Benzene	Glass bead column	5 mL nitrating acid	0.25	3–5	+95	Butanone method	Other aromatic hydrocarbons
Carbon dioxide	Fritted bubbler	10 mL 0.1 N barium hydroxide	1	10–15	60–80	Titration with 0.05 N oxalic acid	Other acids
Ethyl benzene	Fritted bubbler or midget impinger	15 mL spectrograde isooctane	1	20	+90	Alcohol extraction, ultraviolet analysis	Other aromatic hydrocarbons
Formaldehyde	Fritted bubbler	10 mL 1% sodium bisulfite	1–3	25	+95	Liberated sulfite titrated, 0.01 N iodine	Methyl ketones
Hydrochloric acid	Fritted bubbler	0.005 N sodium hydroxide	10	100	+95	Titration with 0.01 N silver nitrate	Other chlorides
Hydrogen sulfide	Midget impinger	15 mL 5% cadmium sulfate	1–2	20	+95	Add 0.05 N iodine, 6 N sulfuric acid, back-titrate 0.01 N sodium thiosulfate	Mercaptans, carbon disulfide, and organic sulfur compounds
Lead, tetraethyl, and tetramethyl	Dreschel-type scrubber	100 mL 0.1 M iodine monochloride in 0.3 N	1.8–2.9	50–75	100	Dithizone	Bismuth, thallium, and stannous tin
Mercury, diethyl, and dimethyl	Midget impinger	15 mL of above	1.9	50–75	91–95	Same as above	Same as above
	Midget impinger	10 mL 0.1 M iodine monochloride in .3 N hydrochloric acid	1–1.5	100	91–100	Dithizone	Copper
Nickel carbonyl	Midget impinger	15 mL 3% hydrochloric acid	2.8	50–90	+90	Complex with alpha-furil-dioxime	—

Nitrogen dioxide	Fritted bubbler (60–70 μm pore size)	20–30 mL Saltzman reagent[a]	0.4	Sample until color appears; probably 10 mL of air	94–99	Reacts with absorbing solution	Ozone in fivefold excess peroxyacyl nitrate
Ozone	Midget impinger	1% potassium iodide in 1 N potassium hydroxide	1	25	+95	Measures color of iodine liberated	Other oxidizing agents
Phosphine	Fritted bubbler	15 mL 0.5% silver diethyl dithiocarbamate in pyridine	0.5	5	86	Complexes with absorbing solution	Arsine, stibine, and hydrogen sulfide
Styrene	Fritted midget impinger	15 mL spectrograde isooctane	1	20	+90	Ultraviolet analysis	Other aromatic hydrocarbons
Sulfur dioxide	Midget impinger, fritted rubber	10 mL sodium tetrachloromercurate	2–3	2	99	Reaction of dichlorosulfito-mercurate and formaldehy-depararosaniline	Nitrogen dioxide,[b] hydrogen sulfide[c]
Toluene diisocyanate	Midget impinger	15 mL Marcali solution	1	25	95	Diazotization and coupling reaction	Materials containing reactive hydrogen attached to oxygen (phenol); certain other diamines
Vinyl acetate	Fritted midget impinger and simple midget impinger in series	Toluene	1.5	15	+99 (84 with fritted bubbler only)	Gas chromatography	Other substances with same retention time on column

[a] 5 g sulfanilic; 140 mL glacial acetic acid; 20 mL 0.1% aqueous *N*-(1-naphthyl) ethylene diamine.
[b] Add sulfamic acid after sampling.
[c] Filter or centrifuge any precipitate.

Source: Pagnotto, L. D., and Keenan, R. G., Sampling and analysis of gases and vapors, in *The Industrial Environment—Its Evaluation and Control*, pp. 167–179. US Department of Health, Education, and Welfare, US Government Printing Office, Washington, DC, 1973.

are used to sample hydrocarbon gases by trapping the species on the active sites of the surface of the sorbent. Figure 16.2 illustrates the loading of active sites with increasing sample time. It is critical that the breakthrough sampling volume, the amount of air passing through the tube that saturates its absorptive capacity, not be exceeded. The breakthrough volume is dependent on the concentration of the gas being sampled and the absorptive capacity of the sorbent. This means that the user must have an estimate of the upper limit of concentration for the gas being sampled.

Once the sample has been collected on the solid sorbent, the tube is sealed and transported to the analytical laboratory. To recover the sorbed gas, two techniques may be used. The tube may be heated while an inert gas is flowing through it. At a sufficiently high temperature, the absorbed molecules are desorbed and carried out of the tube with the inert gas stream. The gas stream may then be passed through a preconcentration trap for injection into a gas chromatograph for chemical analysis. The second technique is liquid extraction of the sorbent and subsequent liquid chromatography. Sometimes a derivatization step is necessary to convert the collected material chemically into compounds which will pass through the column more easily, e.g. conversion of carboxylic acids to methyl esters. Solid sorbents have increased our ability to measure hydrocarbon species under a variety of field conditions. However, this technique requires great skill and sophisticated equipment to obtain accurate results. Care must be taken to minimize problems of contamination of the collection medium, sample instability on the sorbent, and incomplete recovery of the sorbed gases.

Special techniques are employed to sample for gases and particulate matter simultaneously [3]. Sampling systems have been developed which permit the removal of gas-phase molecules from a moving airstream by diffusion to a coated surface and permit the passage of particulate matter downstream

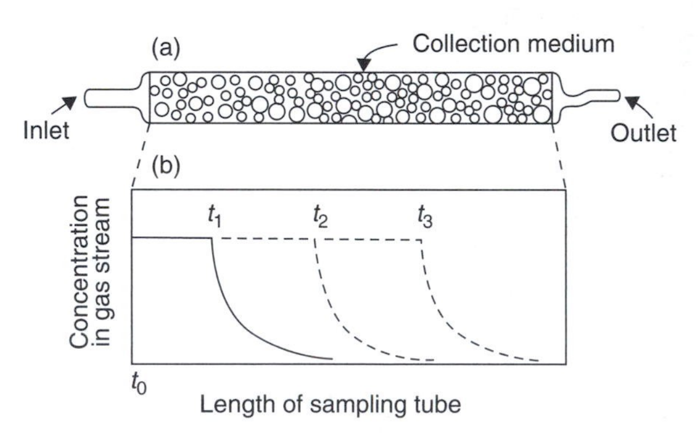

Fig. 16.2. Solid sorbent collection tube. (a) The tube is packed with a granular medium. (b) As the hydrocarbon-containing air is passed through the collection tube at t_1, t_2, and t_3, the collection medium becomes saturated at increasing lengths along the tube.

for collection on a filter or other medium. These diffusion denuders are used to sample for SO_2 or acid gases in the presence of particulate matter. This type of sampling has been developed to minimize the interference of gases in particulate sampling and vice versa.

The third technique, shown in Fig. 16.1(c), involves collection of an aliquot of air in its gaseous state for transport back to the analytical laboratory. Use of a preevacuated flask permits the collection of a gas sample in a specially polished stainless steel container. By use of pressure–volume relationships, it is possible to remove a known volume from the tank for subsequent chemical analysis. Another means of collecting gaseous samples is the collapsible bag. Bags made of polymer films can be used for collection and transport of samples. The air may be pumped into the bag by an inert pump such as one using flexible metal bellows, or the air may be sucked into the bag by placing the bag in an airtight container which is then evacuated. This forces the bag to expand, drawing in the ambient air sample.

B. *In Situ* Sampling and Analysis

The fourth sampling technique involves a combination of sampling and analysis. The analytical technique is incorporated in a continuous monitoring instrument placed at the sampling location. Most often, the monitoring equipment is located inside a shelter such as a trailer or a small building, with the ambient air drawn to the monitor through a sampling manifold. The monitor then extracts a small fraction of air from the manifold for analysis by an automated technique, which may be continuous or discrete. Instrument manufacturers have developed automated *in situ* monitors for several air pollutants, including SO_2, NO, NO_2, O_3, and CO.

This approach is also improving for organic pollutants. For example, real-time gas chromatograph–mass spectrometersand open Pater Fourier.

III. SAMPLING SYSTEMS FOR PARTICULATE POLLUTANTS

Sampling for particles in the atmosphere involves a different set of parameters from those used for gases. Particles are inherently larger than the molecules of N_2 and O_2 in the surrounding air and therefore behave differently with increasing diameter. When one is sampling for particulate matter in the atmosphere, three types of information are of interest: the mass concentration, size, and chemical composition of the particles. Particle size is important in determining adverse effects and atmospheric removal processes. The US Environmental Protection Agency has specified a $PM_{2.5}$ sampling method for compliance monitoring for the National Ambient Air Quality Standards (NAAQS) for particulate matter. This technique must be able to sample particulate matter with an aerodynamic diameter less than 10 μm with a prescribed efficiency.

Particles in the atmosphere come from different sources, e.g. combustion, windblown dust, and gas-to-particle conversion processes (see Chapter 10). Figure 2.2 illustrates the wide range of particle diameters potentially present in the ambient atmosphere. A typical size distribution of ambient particles is shown in Fig. 2.3. The distribution of number, surface, and mass can occur over different diameters for the same aerosol. Variation in chemical composition as a function of particle diameter has also been observed, as shown in Table 8.5.

The major purpose of ambient particulate sampling is to obtain mass concentration and chemical composition data, preferably as a function of particle diameter. This information is valuable for a variety of problems: effects on human health, identification of particulate matter sources, understanding of atmospheric haze, and particle removal processes.

The primary approach is to separate the particles from a known volume of air and subject them to weight determination and chemical analysis. The principal methods for extracting particles from an airstream are filtration and impaction. All sampling techniques must be concerned with the behavior of particles in a moving airstream. The difference between sampling for gases and sampling for particles begins at the inlet of the sampling manifold and is due to the discrete mass associated with individual particles.

A. Behavior of Particles at Sampling Inlets

Sampling errors may occur at the inlet, and particles may be lost in the sampling manifold while being transported to the collection surface. Figure 16.3 illustrates the flow patterns around a sampling inlet in a uniform flow

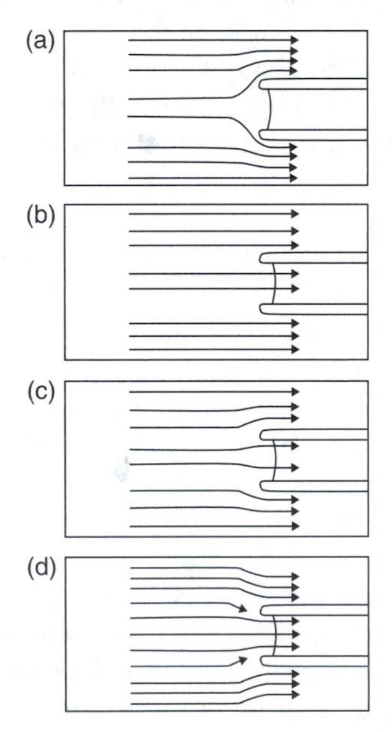

Fig. 16.3. The streamline flow patterns around a sampling inlet in a uniform flow field.

field. Figure 16.3(a) shows that when no air is permitted to flow into the inlet, the streamline flow moves around the edges of the inlet. As the flow rate through the inlet increases, more and more of the streamlines are attracted to the inlet. Figure 16.3(b) is called the *isokinetic condition*, in which the sampling flow rate is equal to the flow field rate. An example is an inlet with its opening into the wind pulling air at the wind speed. When one is sampling for gases, this is not a serious constraint because the composition of the gas will be the same under all inlet flow rates; i.e., there is no fractionation of the air sample by different gaseous molecules.

Transform infrared, differential optical absorption spectroscopy, and tuned lasers are being used with much success.

Particle-containing air streams present a different situation. Figure 16.3(b), the isokinetic case, is the ideal case. The ideal sample inlet would always face into the wind and sample at the same rate as the instantaneous wind velocity (an impossibility). Under isokinetic sampling conditions, parallel air streams flow into the sample inlet, carrying with them particles of all diameters capable of being carried by the stream flow. When the sampling rate is lower than the flow field (Fig. 16.3(c)), the streamlines start to diverge around the edges of the inlet and the larger particles with more inertia are unable to follow the streamlines and are captured by the sampling inlet. The opposite happens when the sampling rate is higher than the flow field. The inlet captures more streamlines, but the larger particles near the edges of the inlet may be unable to follow the streamline flow and escape collection by the inlet. The inlet may be designed for particle size fractionation; e.g., a $PM_{2.5}$ inlet will exclude particles larger than 2.5 μm aerodynamic diameter (see Fig. 2.4).

These inertial effects become less important for particles with diameters less than 5 μm and for low wind velocities, but for samplers attempting to collect particles above 5 μm, the inlet design and flow rates become important parameters. In addition, the wind speed has a much greater impact on sampling errors associated with particles more than 5 μm in diameter [4].

After the great effort taken to get a representative sample into the sampling manifold inlet, care must be taken to move the particles to the collection medium in an unaltered form. Potential problems arise from too long or too twisted manifold systems. Gravitational settling in the manifold will remove a fraction of the very large particles. Larger particles are also subject to loss by impaction on walls at bends in a manifold. Particles may also be subject to electrostatic forces which will cause them to migrate to the walls of nonconducting manifolds. Other problems include condensation or agglomeration during transit time in the manifold. These constraints require sampling manifolds for particles to be as short and have as few bends as possible.

The collection technique involves the removal of particles from the air stream. The two principal methods are filtration and impaction. Filtration consists of collecting particles on a filter surface by three processes: direct interception, inertial impaction, and diffusion [5]. Filtration attempts to remove a very high percentage of the mass and number of particles by these three processes. Any size classification is done by a preclassifier, such as an impactor, before the particle stream reaches the surface of the filter.

IV. PASSIVE SAMPLING SYSTEMS

Passive (or static) *sampling systems* are defined as those that do not have an active air-moving component, such as the pump, to pull a sample to the collection medium. This type of sampling system has been used for over 100 years. Examples include the lead peroxide candle used to detect the presence of SO_2 in the atmosphere and the dustfall bucket and trays or slides coated with a viscous material used to detect particulate matter. This type of system suffers from inability to quantify the amount of pollutant present over a short period of time, i.e., less than 1 week. The potentially desirable characteristics of a static sampling system have led to further developments in this type of technology to provide quantitative information on pollutant concentrations over a fixed period of time. Static sampling systems have been developed for use in the occupational environment and are also used to measure the exposure levels in the general community, e.g., radon gas in residences.

The advantages of static sampling systems are their portability, convenience, reliability, and low cost. The systems are lightweight and can be attached directly to individuals. Nonstatic sampling systems can, of course, also be attached to individuals, but are less convenient because the person must carry a battery-powered pump and its batteries. Static sampling systems are very reliable, and the materials used limit the costs to acceptable levels.

Two principles are utilized in the design of static samplers: diffusion and permeation [6,7]. Samplers based on the diffusion principle depend on the molecular interactions of N_2, O_2, and trace pollutant gases. If a concentration gradient can be established for the trace pollutant gas, under certain conditions the movement of the gas will be proportional to the concentration gradient (Fick's law of diffusion), and a sampler can be designed to take advantage of this technique. Figure 16.4 illustrates this principle. The sampler has a well-defined inlet, generally with a cylindrical shape, through which the pollutant gas must diffuse. At the end of the tube, a collection medium removes the pollutant gas for subsequent analysis and maintains a concentration gradient between the inlet of the tube and the collection medium. The mathematical relationship (Fick's law) describing this type of passive sampler is given by

$$R = -DA\left(\frac{dC}{dx}\right) \tag{16.1}$$

where R is the rate of transport by diffusion in moles per second, D is the diffusion coefficient in square centimeters per second, A is the cross-sectional area of the diffusion path in square centimeters, C is the concentration of species in moles per cubic centimeter, and x is the path length in centimeters.

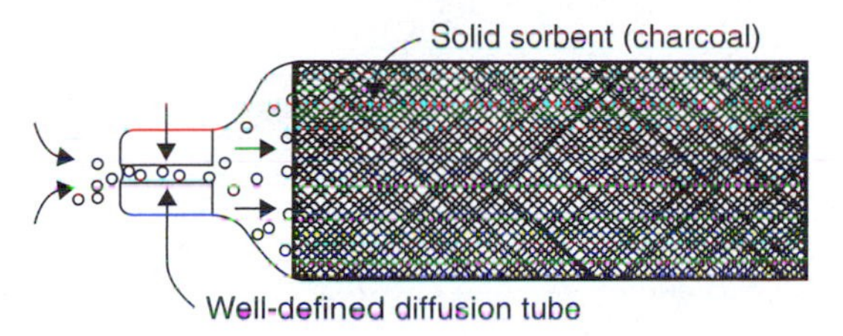

Fig. 16.4. Static sampler based on the diffusion principle.

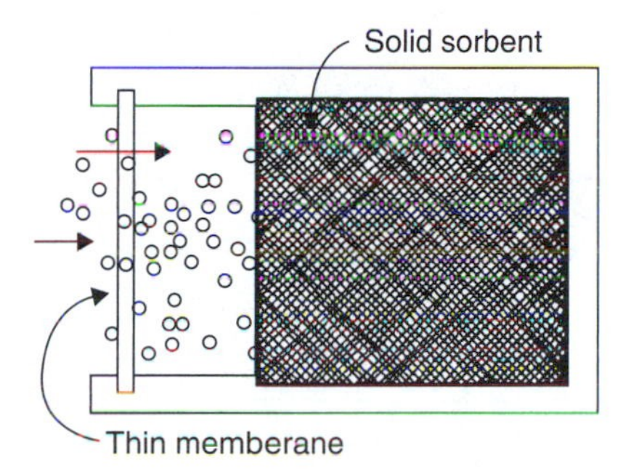

Fig. 16.5. Static sampler based on the permeation principle.

The ability of gases to permeate through various polymers at a fixed rate depending on a concentration gradient has been used to create static samplers. This principle was originally developed to provide a standard calibration source of trace gas by putting that gas in a polymer tube under pressure and letting the material diffuse or permeate through the wall to the open atmosphere. Permeation samplers operate in the reverse direction. Figure 16.5 illustrates this type of system. A thin film membrane is open to the atmosphere on one side and to a collection medium on the other. A pollutant gas in the atmosphere diffuses through the membrane and is collected in the medium. The mathematical relationship for a permeation sampler is given by

$$k = \frac{Ct}{m} \tag{16.2}$$

where k is the permeation constant, C is the concentration of gas in parts per million, t is the time of exposure, and m is the amount of gas absorbed in micrograms.

Permeation systems can be calibrated in the laboratory and then used in the field for sample collection for a fixed period of time, e.g., 8 h or 7 days. The sampler is returned to the laboratory for analysis. These systems can be made for specific compounds by selecting the appropriate collection medium and the polymer membrane (Table 16.2).

TABLE 16.2

Permeation Samplers for Selected Gases

Gas	Membrane	Sorber	Sensitivity
Chlorine	Dimethyl silicone (DMS) (single-backed)	Buffered (pH 7) fluorescein, 0.005% NaBr (0.31%)	0.013 ppm (8-h exposure)
Sulfur dioxide	DMS (single-backed)	Tetrachloromercurate (II)	0.01 ppm (8-h exposure)
Vinyl chloride	DMS (single-backed)	Activated charcoal (CS_2 desorption)	0.02 ppm (linear to 50 ppm +)
Alkyl lead	DMS (unbacked)	Silica gel (ICl desorption)	0.2 μg
Benzene	Silicon polycarbonate	Activated charcoal (CS_2 desorption)	0.02 ppm (8-h exposure)
Ammonia	Vinyl silicone	0.6% boric acid	0.4 ppm (8-h exposure)
Hydrogen sulfide	DMS (single-backed)	0.02 N NaOH, 0.003 M EDTA	0.01 ppm
Hydrogen cyanide	DMS (single-backed)	0.01 N NaOH	0.01 ppm (8-h exposure)

Source: West, P. W., *Am. Lab.* **12**, 35–39 (1980).

V. SAMPLER SITING REQUIREMENTS

Sampling site selection is dependent on the purpose or use of the results of the monitoring program. Sampling activities are typically undertaken to determine the ambient air quality for compliance with air quality standards, for evaluation of the impact of a new air pollution source during the preconstruction phase, for hazard evaluation associated with accidental spills of chemicals, for human exposure monitoring, and for research on atmospheric chemical and physical processes. The results of ambient air monitoring can be used to judge the effectiveness of the air quality management approach to air pollution problems. The fundamental reason for controlling air pollution sources is to limit the buildup of contaminants in the atmosphere so that adverse effects are not observed. This suggests that sampling sites should be selected to measure pollutant levels close to or representative of exposed populations of people, plants, trees, materials, structures, etc. Generally, sites in air quality networks are near ground level, typically 3 m aboveground, and are located so as not to be unduly dominated by a nearby source such as a roadway. Sampling sites require electrical power and adequate protection (which may be as simple as a fence). A shelter, such as a small building, may be necessary. Permanent sites require adequate heating and air conditioning to provide a stable operating environment for the sampling and monitoring equipment.

VI. SAMPLING FOR AIR TOXICS

Public awareness of the release of chemicals into the atmosphere has gone beyond the primary ambient pollutants (e.g. SO_2 or O_3) and governments require air toxics management plans. One component of this process is the characterization of the air quality via sampling.

Most of the airborne chemicals classified as "air toxics" are organic compounds with physical and chemical properties ranging from those similar to formaldehyde found in the gas phase to polycyclic aromatic hydrocarbons (PAHs) which may be absorbed on particle surfaces. Air toxics also include a number of metals and their compounds. This range of volatility and reactivity represented by air toxics requires a variety of sampling techniques—from grab sampling to filter techniques followed by extraction and detailed derivatization techniques. When these compounds are present in the atmosphere, the concentration level can be quite low, in the parts per billion (ppb) to sub-ppb range for gases and the picogram per cubic meter range for particulate components. Two concentrations must be calculated for particle-bound contaminants: the concentrations of particles in the atmosphere (mass of particles per volume of air) and the concentration of contaminants sorbed to the particle (mass of each chemical per mass of particles). This generally requires extended sampling times and very sensitive analytical techniques for laboratory analysis.

The US Environmental Protection Agency established a pilot Toxics Air Monitoring System network for sampling ambient volatile organic compounds (VOCs) at ppb levels in Boston, Chicago, and Houston for a 2-year period [8]. Evacuated stainless steel canisters were used to collect air at $3\,cm^3$ min for 24 h. The canisters were returned to a central laboratory and analyzed by cryogenic concentration of the VOCs, separation by gas chromatography, and mass-selective detection. This system provided information on 13 VOCs in three classes: chlorofluorocarbons, aromatics, and chlorinated alkanes.

A second sampling program in Southern California sampled for polychlorinated dioxins and polychlorinated dibenzofurans at seven locations [9]. Because of the semivolatile nature of these compounds, a tandem sampler was used with a glass fiber filter to collect the particulate-associated components followed by a polyurethane foam sorbent trap to collect the vapor-phase portion. These samples were returned to the laboratory, where they were extracted and analyzed with high-resolution gas chromatography and high-resolution mass spectrometry. The observed concentrations were in the picogram per cubic meter range. The techniques for these procedures are introduced in the next chapter.

More recently, in 2001 and 2002, the gases and dust released during and following the collapse of World Trade Center towers as well as that found in the aftermath of Hurricane Katrina in 2005 had to be analyzed.

Each of these examples suggest that air toxics sampling is complex and expensive and requires careful attention to quality assurance.

REFERENCES

1. Pagnotto, L. D., and Keenan, R. G., Sampling and analysis of gases and vapors, in *The Industrial Environment—Its Evaluation and Control*, pp. 167–179. US Department of Health, Education, and Welfare, US Government Printing Office, Washington, DC, 1973.
2. Tanaka, T., *J. Chromatogr.* **153**, 7–13 (1978).
3. Slanina, J., De Wild, P. J., and Wyers, G. P., *Adv. Environ. Sci. Technol.* **24**, 129–154 (1992).
4. Cadle, R. D., *The Measurement of Airborne Particles*. Academic Press, New York, 1976.
5. Liu, B. Y. H. (ed.), *Fine Particles*. Academic Press, New York, 1976.
6. Palmes, E. D., Gunnison, A. F., DiMatto, J., and Tomczyk, C., *Am. Ind. Hyg. Assoc. J.* **37**, 570–577 (1976).
7. West, P. W., *Am. Lab.* **12**, 35–39 (1980).
8. Evans, G. F., Lumpkin, T. A., Smith, D. L., and Somerville, M. C., *J. Air Waste Manage. Assoc.* **42**, 1319–1323 (1992).
9. Hunt, G. T., and Maisel, B. E., *J. Air Waste Manage. Assoc.* **42**, 672–680 (1992).

SUGGESTED READING

Friedlander, S. K., *Smoke, Dust and Haze*. Wiley, New York, 1977.

Hering, S. V., *Air Sampling Instruments for Evaluation of Air Contaminants*. ACGIH, Cincinnati, OH, 1989

Noll, K. E., and Miller, T. L., *Air Monitoring Survey Design*. Ann Arbor Science Publishers, Ann Arbor, MI, 1977.

Stanley-Wood, N. G., and Lines, R. W. (eds.), *Particle Size Analysis*. Royal Society of Chemistry, Cambridge, UK, 1992.

Willeke, K., and Baron, P. A., *Aerosol Measurement—Principles, Techniques, and Applications*. Van Nostrand Reinhold, New York, 1993.

QUESTIONS

1. Describe the four components of a sampling system.
2. List three examples of the four components, e.g., a metal bellows pump.
3. A solid sorbent Tenax cartridge has a capacity of 100 μg of toluene. If samples were collected at a rate of $5\,L\,min^{-1}$, calculate the maximum ambient concentration which can be determined by an hourly sample and a 15-min sample.
4. Describe the sampling approaches used for air pollutants by your state or local government.
5. List the possible sources of loss or error in sampling for particulate matter.
6. Why is sampling velocity not an important parameter when sampling for gases?
7. List the advantages of passive sampling systems.
8. Describe the precautions which should be considered when determining the location of the sampling manifold inlet for an ambient monitoring system.
9. What is the concentration of particle-phase mercury in the air if sample shows that the average (mean) particle concentration ($PM_{2.5}$) is $1.5\,\mu g\,m^{-3}$ and the mean concentration of mercury on the particles in the sample is $10\,ng\,g^{-1}$?
10. How important is the rest of the PM fraction (PM $>2.5\,\mu m$) in the above scenario? Explain.

11. How might a sample be analyzed in the above question's scenario (both the particle and the metals sorbed to the particle)? What are the advantages and disadvantages of each of the major types?
12. In the above examples, what might be some of the important differences in the physical and chemical characteristics between mercury and cadmium in sampling.
13. What if you wanted to know a particular species of mercury (e.g. mercury chloride or non-methyl mercury)? Can you use a technique like X-ray fluorescence? Why or why not? What might have to be done to be able to answer this question; i.e., what are the key features needed in your sampling and analysis plan?
14. What is the concentration of benzo(a)pyrene (B(a)p) in the scenario in Question 9 if its concentration on the particles (mean) is 10 ppb? How would you collect samples for gas phase B(a)p? How might this differ from the method used for mercury?

17

Ambient Air Pollutants: Analysis and Measurement

I. ANALYSIS AND MEASUREMENT OF GASEOUS POLLUTANTS

The samples gathered according to the protocols described in the previous chapter must be analyzed for their physical and chemical properties. The two major goals of testing for air pollutants are identification and quantification of a sample of ambient air. Air pollution measurement techniques generally pass through evolutionary stages. The first is the qualitative identification stage. This is followed by separate collection and quantification stages. The last stage is the concurrent collection and quantification of a given pollutant.

Gaseous SO_2 is an example. Very early procedures detected the presence of SO_2 in ambient air by exposing a lead peroxide candle for a period of time and then measuring the amount of lead sulfate formed. Because the volume of air in contact with the candle was not measured, the technique could not quantify the amount of SO_2 per unit volume of air.

The next stage involved passing a known volume of ambient air through an absorbing solution in a container in the field and then returning this container to the laboratory for a quantitative determination of the amount of absorbed SO_2. The United Nations Environmental Program–World Health Organization's worldwide air sampling and analysis network used this method for SO_2, the only gaseous pollutant measured by the network. The final evolutionary step has been the concurrent collection and quantification of SO_2. An example of this is the flame photometric SO_2 analyzer, in which SO_2-laden air is fed into an H_2 flame, and light emissions from electronically excited combustion products are detected by a photomultiplier tube. Prior calibration of the analyzer permits the rapid determination of SO_2. This is but one of the many methods available for the measurement of SO_2.

Hundreds of chemical species are present in urban atmospheres. The gaseous air pollutants most commonly monitored are CO, O_3, NO_2, SO_2, and nonmethane volatile organic compounds (NMVOCs). Measurement of specific hydrocarbon compounds is becoming routine in the United States for two reasons: (1) their potential role as air toxics and (2) the need for detailed hydrocarbon data for control of urban ozone concentrations. Hydrochloric acid (HCl), ammonia (NH_3), and hydrogen fluoride (HF) are occasionally measured. Calibration standards and procedures are available for all of these analytic techniques, ensuring the quality of the analytical results. See Table 17.1 for a summary of emission limits for one particular source class, incinerators.

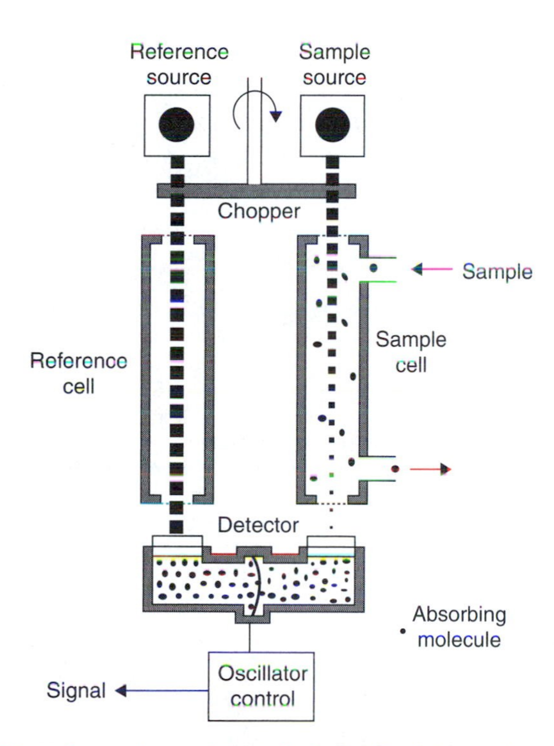

Fig. 17.1. NDIR analyzer. *Source*: Bryan, R. J., Ambient air quality surveillance, in *Air Pollution*, 3rd ed., Vol. III (Stern, A. C., ed.), p. 375. Academic Press, New York, 1976.

A. Carbon Monoxide

The primary reference method used for measuring carbon monoxide in the United States is based on nondispersive infrared (NDIR) photometry [1, 2]. The principle involved is the preferential absorption of infrared radiation by carbon monoxide. Figure 17.1 is a schematic representation of an NDIR analyzer. The analyzer has a hot filament source of infrared radiation, a chopper, a sample cell, reference cell, and a detector. The reference cell is filled with a non-infrared-absorbing gas, and the sample cell is continuously flushed with ambient air containing an unknown amount of CO. The detector cell is divided into two compartments by a flexible membrane, with each compartment filled with CO. Movement of the membrane causes a change in electrical capacitance in a control circuit whose signal is processed and fed to a recorder.

The chopper intermittently exposes the two cells to infrared radiation. The reference cell is exposed to a constant amount of infrared energy which is transmitted to one compartment of the detector cell. The sample cell, which contains varying amounts of infrared-absorbing CO, transmits to the detector cell a reduced amount of infrared energy that is inversely proportional to the CO concentration in the air sample. The unequal amounts of energy received by the two compartments in the detector cell cause the membrane to move, producing an alternating current (AC) electrical signal whose frequency is established by the chopper spacing and the speed of chopper rotation.

Water vapor is a serious interfering substance in this technique. A moisture trap such as a drying agent or a water vapor condenser is required to remove water vapor from the air to be analyzed.

Instruments based on other techniques are available which meet the performance specifications outlined in Table 17.1.

TABLE 17.1

Sampling Required to Demonstrate Compliance with Emission Limits for the New Source Performance Standards (NSPS) for Hospital/Medical/Infectious Waste Incinerators (HMIWI), Pursuant to the US Court of Appeals for the District of Columbia Circuit Ruling of March 2, 1999, Remanding the Rule to the US EPA for Further Explanation of the Agency's Reasoning In Determining the Minimum Regulatory "Floors" for New and Existing HMIWI

Pollutant (units)	Unit size[a]	Proposed remand limit for existing HMIWI[b]	Proposed remand limit for new HMIWI[b]
HCl (ppmv)	L, M, S	78 or 93% reduction[c]	15[c] or 99% reduction[c]
	SR	3100[c]	N/A[d]
CO (ppmv)	L, M, S	40[c]	32
	SR	40[c]	N/A[d]
Pb (mg/dscm)	L, M	0.78 or 71% reduction	0.060 or 98% reduction[c]
	S	0.78 or 71% reduction	0.78 or 71% reduction
	SR	8.9	N/A[d]

(*continued*)

TABLE 17.1 (*Continued*)

Pollutant (units)	Unit size[a]	Proposed remand limit for existing HMIWI[b]	Proposed remand limit for new HMIWI[b]
Cd (mg/dscm)	L, M	0.11 or 66% reduction[c]	0.030 or 93% reduction
	S	0.11 or 66% reduction[c]	0.11 or 66% reduction[c]
	SR	4[c]	N/A[d]
Hg (mg/dscm)	L, M	0.55[c] or 87% reduction	0.45 or 87% reduction
	S	0.55[c] or 87% reduction	0.47 or 87% reduction
	SR	6.6	N/A[d]
PM (gr/dscf)	L	0.015[c]	0.009
	M	0.030[c]	0.009
	S	0.050[c]	0.018
	SR	0.086[c]	N/A[d]
CDD/CDF, total ($ng\ dscm^{-1}$)	L, M	115	20
	S	115	111
	SR	800[c]	N/A[d]
CDD/CDF, TEQ ($ng\ dscm^{-1}$)	L, M	2.2	0.53
	S	2.2	2.1
	SR	15[c]	N/A[d]
NO_x (ppmv)	L, M, S	250[c]	225
	SR	250[c]	N/A[d]
SO_2 (ppmv)	L, M, S	55[c]	46
	SR	55[c]	N/A[d]

[a] L: large; M: medium; S: small; SR: small rural.
[b] All emission limits are measured at 7% oxygen.
[c] No change proposed.
[d] Not applicable.
Source: 40 *Code of Federal Regulations*, Part 60, Standards of Performance for New Stationary Sources and Emission Guidelines for Existing Sources: Hospital/Medical/Infectious Waste Incinerators; Proposed Rule, February 7, 2007.

B. Ozone

The principal method used for measuring ozone is based on chemiluminescence [3]. When ozone and ethylene react chemically, products are formed which are in an excited electronic state. These products fluoresce, releasing light. The principal components are a constant source of ethylene, an inlet sample line for ambient air, a reaction chamber, a photomultiplier tube, and signal-processing circuitry. The rate at which light is received by the photomultiplier tube is dependent on the concentrations of O_3 and ethylene. If the concentration of ethylene is made much higher than the ozone concentration to be measured, the light emitted is proportional only to the ozone concentration.

Instruments based on this principle may be calibrated by a two-step process shown in Fig. 17.2 [4]. A test atmosphere with a known source of ozone is produced by an ozone generator, a device capable of generating stable levels of O_3. Step 1 involves establishing the concentration of ozone in the

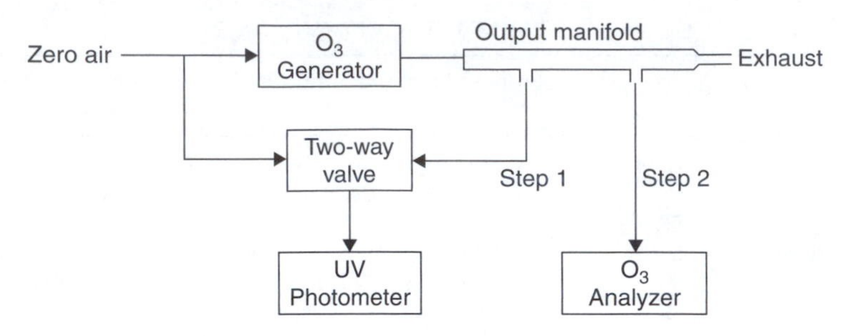

Fig. 17.2. Calibration apparatus for ozone analyzer (UV).

test atmosphere by ultraviolet (UV) photometry. This is followed by step 2, calibration of the instrument's response to the known concentration of ozone in the test atmosphere.

C. Nitrogen Dioxide

The principal method used for measuring NO_2 is also based on chemiluminescence (Fig. 17.3) [5]. NO_2 concentrations are determined indirectly from the difference between the NO and NO_x (NO + NO_2) concentrations in the atmosphere. These concentrations are determined by measuring the light emitted from the chemiluminescent reaction of NO with O_3 (similar to the reaction of O_3 with ethylene noted for the measurement of O_3), except that O_3 is supplied at a high constant concentration, and the light output is proportional to the concentration of NO present in the ambient air stream.

Figure 17.4 illustrates the analytical technique based on this principle. To determine the NO_2 concentration, the NO and NO_x (NO + NO_2) concentrations are measured. The block diagram shows a dual pathway through the instrument, one to measure NO and the other to measure NO_x. The NO pathway has an ambient air stream containing NO (as well as NO_2), an ozone stream from the ozone generator, a reaction chamber, a photomultiplier tube, and signal-processing circuitry. The NO_x pathway has the same components, plus a converter for quantitatively reducing NO_2 to NO. The instrument can also electronically subtract the NO from NO_x and yield as output the resultant NO_2.

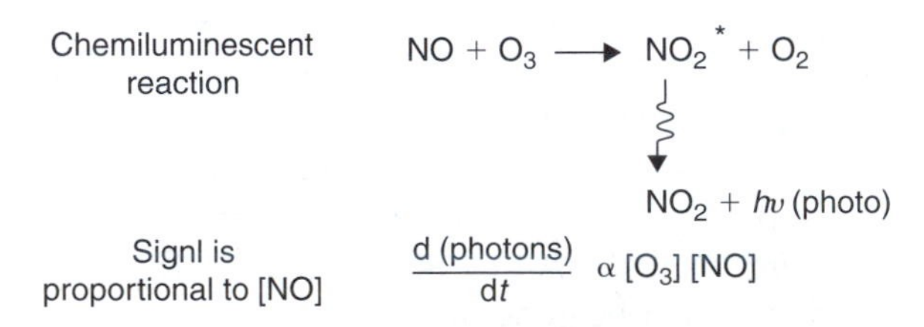

Fig. 17.3. NO_2 chemiluminescent detection principle based on the reaction of NO with O_3.

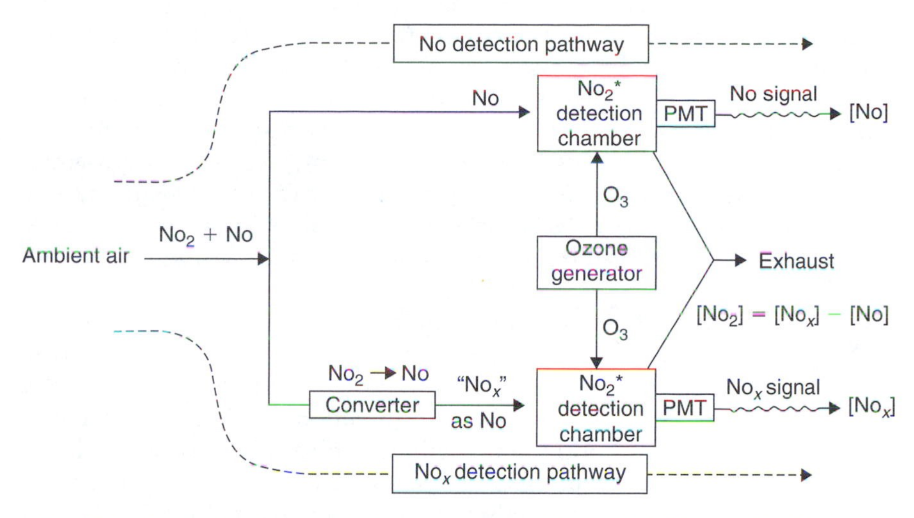

Fig. 17.4. Schematic diagram of chemiluminescent detector for NO_2 and NO. PMT: photomultiplier tube.

Air passing through the NO pathway enters the reaction chamber, where the NO present reacts with the ozone. The light produced is measured by the photomultiplier tube and converted to an NO concentration. The NO_2 in the air stream in this pathway is unchanged. In the NO_x pathway, the NO- and NO_2-laden air enters the converter, where the NO_2 is reduced to form NO; all of the NO_x exits the converter as NO and enters the reaction chamber. The NO reacts with O_3 and the output signal is the total NO_x concentration. The NO_2 concentration in the original air stream is the difference between NO_x and NO. Calibration techniques use gas-phase titration of an NO standard with O_3 or an NO_2 permeation device.

D. Sulfur Dioxide

Several manual and continuous analytical techniques are used to measure SO_2 in the atmosphere. The manual techniques involve two-stage sample collection and measurement. Samples are collected by bubbling a known volume of gas through a liquid collection medium. Collection efficiency is dependent on the gas–liquid contact time, bubble size, SO_2 concentration, and SO_2 solubility in the collection medium. The liquid medium contains chemicals which stabilize SO_2 in solution by either complexation or oxidation to a more stable form. Field samples must be handled carefully to prevent losses from exposure to high temperatures. Samples are analyzed at a central laboratory by an appropriate method.

The West–Gaeke manual method is the basis for the US Environmental Protection Agency (EPA) reference method for measurement of SO_2 [6]. The method uses the colorimetric principle; i.e., the amount of SO_2 collected is proportional to the amount of light absorbed by a solution. The collection medium is an aqueous solution of sodium or potassium tetrachloromercurate (TCM). Absorbed SO_2 forms a stable complex with TCM. This enhanced stability permits the collection, transport, and short-term storage of samples at a central laboratory. The analysis proceeds by adding bleached pararosaniline dye and formaldehyde to form red-purple pararosaniline methylsulfonic acid. Optical absorption at 548 nm is linearly proportional to the SO_2 concentration. Procedures are followed to minimize interference by O_3, oxides of nitrogen, and heavy metals.

The continuous methods combine sample collection and the measurement technique in one automated process. The measurement methods used for continuous analyzers include conductometric, colorimetric, coulometric, and amperometric techniques for the determination of SO_2 collected in a liquid medium [7]. Other continuous methods utilize physicochemical techniques for detection of SO_2 in a gas stream. These include flame photometric detection (described earlier) and fluorescence spectroscopy [8]. Instruments based on all of these principles are available which meet standard performance specifications.

E. Nonmethane Volatile Organic Compounds

The large number of individual hydrocarbons in the atmosphere and the many different hydrocarbon classes make ambient air monitoring a very difficult task. The ambient atmosphere contains a ubiquitous concentration of methane (CH_4) at approximately 1.6 ppm worldwide [9]. The concentration of all other hydrocarbons in ambient air can range from 100 times less to 10 times greater than the methane concentration for a rural versus an urban location. The terminology of the concentration of hydrocarbon compounds is potentially confusing. Hydrocarbon concentrations are referred to by two units—parts per million by volume (ppmV) and parts per million by carbon (ppmC). Thus, 1 μL of gas in 1 L of air is 1 ppmV, so the following is true:

Mixing ratio	ppmV	ppmC
$\frac{1\,\mu L \text{ of } O_3}{1\,L \text{ of air}} =$	1 ppm ozone	—
$\frac{1\,\mu L \text{ of } SO_2}{1\,L \text{ of air}} =$	1 ppmV SO_2	—
$\frac{1\,\mu L \text{ of } CH_4}{1\,L \text{ of air}} =$	1 ppmV CH_4	1 ppmC CH_4
$\frac{1\,\mu L \text{ of } C_2H_6}{1\,L \text{ of air}} =$	1 ppmV C_2H_6	2 ppmC C_2H_6

The unit ppmC takes into account the number of carbon atoms contained in a specific hydrocarbon and is the generally accepted way to report ambient hydrocarbons. This unit is used for three reasons: (1) the number of carbons atoms is a very crude indicator of the total reactivity of a group of hydrocarbon compounds; (2) historically, analytical techniques have expressed results in this unit; and (3) considerable information has been developed on the role of hydrocarbons in the atmosphere in terms of concentrations determined as ppmC.

Historically, measurements have classified ambient hydrocarbons into two classes: methane (CH_4) and all other NMVOCs. Analyzing hydrocarbons in the atmosphere involves a three-step process: collection, separation, and quantification. Collection involves obtaining an aliquot of air, e.g., with an evacuated canister. The principal separation process is gas chromatography (GC), and the principal quantification technique is with a calibrated flame ionization detector (FID). Mass spectroscopy (MS) is used along with GC to identify individual hydrocarbon compounds.

A simple schematic diagram of the GC/FID principle is shown in Fig. 17.5. Air containing CH_4 and other hydrocarbons classified as NMVOCs pass through a GC column and the air, CH_4, and NMVOC molecules are clustered into groups because of different absorption/desorption rates. As CH_4 and NMVOC groups exit the column, they are "counted" by the FID. The signal output of the detector is proportional to the two groups and may be quantified when compared with standard concentrations of gases. This simplified procedure has been used extensively to collect hydrocarbon concentration data for the ambient atmosphere. A major disadvantage of this technique is the grouping of all hydrocarbons other than CH_4 into one class. Hydrocarbon compounds with similar structures are detected by an FID in a proportional manner, but for compounds with significantly different structures the response may be different. This difference in sensitivity results in errors in measurements of NMVOC mixtures.

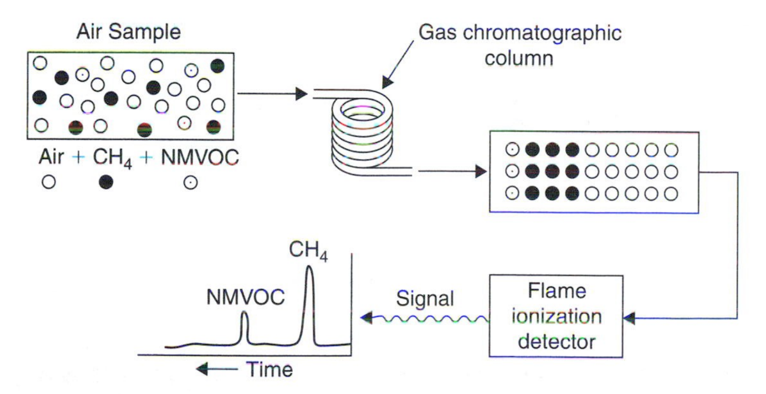

Fig. 17.5. Schematic diagram of hydrocarbon detection by GC. NMVOC: nonmethane volatile organic carbon.

More sophisticated GC columns and techniques perform more detailed separations of mixtures of hydrocarbons into discrete groups. Table 7.14 lists individual hydrocarbons measured in ambient air by advanced GC techniques.

Other types of detectors include the flame photometric detector (FPD) and the electron capture detector (ECD). The FID is composed of an H_2 flame through which the hydrocarbon gases are burned, forming charged carbon atoms, and an electrometer grid which generates a signal current proportional to the number of carbon atoms in the flame. The example of 1 ppmV methane (CH_4) and 1 ppmV (but 2 ppmC) ethane (C_2H_6) is related to this detection principle. One ppmV of CH_4 and 1 ppmV of C_2H_6 in air have the same number of molecules of hydrocarbon in a given volume of air, but if an aliquot of each mixture were run through an FID, the signal for ethane would be nearly twice the methane signal: 2 ppmC ethane compared to 1 ppmC methane.

The FPD is also used to measure sulfur-containing compounds and therefore is useful for measurement of sulfur-containing hydrocarbons such as dimethylsulfide or furan. The FPD has an H_2 flame in which sulfur-containing gases are burned. In the combustion process, electronically excited $S_2{}^*$ is formed. A photomultiplier tube detects light emitted from the excited sulfur at $\sim$395 nm. The ECD is preferred for measuring nitrogen-containing compounds such as PAN and other peroxyacyl nitrate compounds. The ECD contains a radioactive source which establishes a stable ion field. Nitrogen-containing compounds capture electrons in passing through the field. Alterations in the electronic signal are related to the concentration of the nitrogen species.

F. Laboratory Analysis of Air Pollutant Samples

When the sample arrives at the laboratory, the next step may be "extraction." The pollutant of concern on the environmental sample may be sorbed to particles on the filter or may be trapped on substrate and must be freed for analysis to take place. So, to analyze the sample, the chemicals must first be freed from the sorbant matrix. Dioxins provide an example. Under environmental conditions, dioxins are fat soluble and have low vapor pressures, so they may be found on particles, in the gas phase, or suspended to colloids. Therefore, to collect the gas-phase dioxins, the standard method calls for trapping it on polyurethane foam (PUF). These properties have influenced the design of the PS-1 monitor, which is used to collect semivolatile organic compounds (SVOCs) like the dioxins. It has both a filter and a polyurethane foam (PUF) trap to collect both particle and gas phases, respectively. Thus, to analyze dioxins in the air, the PUF and particle matter must first be extracted, and to analyze dioxins on filters, those particles that have been collected must also be extracted.

Extraction makes use of physics and chemistry. For example, many compounds can be simply extracted with solvents, usually at elevated temperatures. A common solvent extraction is the Soxhlet extractor, named after the German food chemist, Franz Soxhlet (1848–1913). The Soxhlet extractor (the US EPA Method 3540) removes sorbed chemicals by passing a boiling solvent through the media. Cooling water condenses the heated solvent and the extract is collected over an extended period, usually several hours. Other automated techniques apply some of the same principals as solvent extraction, but allow for more precise and consistent extraction, especially when large volumes of samples are involved. For example, supercritical fluid extraction (SFE) brings a solvent, usually carbon dioxide to the pressure and temperature near its critical point, where the solvent's properties are rapidly altered with very slight variations of pressure.[1] Solid phase extraction (SPE), which uses a solid and a liquid phase to isolate a chemical from a solution, is often used to clean up a sample before analysis. Combinations of various extraction methods can enhance the extraction efficiencies, depending on the chemical and the media in which it is found. Ultrasonic and microwave extractions may be used alone or in combination with solvent extraction. For example, the US EPA Method 3546 provides a procedure for extracting hydrophobic (that is, not soluble in water) or slightly water soluble organic compounds from particles such as soils, sediments, sludges, and solid wastes. In this method, microwave energy elevates the temperature and pressure conditions (i.e., 100–115°C and 50–175 psi) in a closed extraction vessel containing the sample and solvent(s). This combination can improve recoveries of chemical analytes and can reduce the time needed compared to the Soxhlet procedure alone.

Not every sample needs to be extracted. For example, air monitoring using canisters and bags allows the air to flow directly into the analyzer. Surface methods of particle matter, such as fluorescence, sputtering, and atomic absorption (AA), require only that the sample be mounted on specific media (e.g. filters). Also, continuous monitors like the chemiluminescent system mentioned earlier provide ongoing measurements.

Chromatography consists of separation and detection. Separation makes use of the chemicals' different affinities for certain surfaces under various temperature and pressure conditions. The first step, injection, introduces the extract to a "column." The term column is derived from the time when columns were packed with sorbents of varying characteristics, sometimes meters in length, and the extract was poured down the packed column to separate the various analytes. Today, columns are of two major types, gas and liquid. GC makes use of hollow tubes ("columns") coated inside with compounds that hold organic

[1] See Ekhtera, M., Mansoori, G., Mensinger, M., Rehmat, A., and Deville, B., Supercritical fluid extraction for remediation of contaminated soil, in *Supercritical Fluids: Extraction and Pollution Prevention* (Abraham, M., and Sunol, A., eds.), ACSSS, Vol. 670, pp. 280–298. American Chemical Society, Washington, DC, 1997.

chemicals. The columns are in an oven, so that after the extract is injected into the column, the temperature is increased, as well as the pressure, and the various organic compounds in the extract are released from the interior column surface differentially, whereupon they are collected by a carrier gas (e.g. helium) and transported to the detector. Generally, the more volatile compounds are released first (they have the shortest retention times), followed by the semi-volatile organic compounds. So, boiling point is often a very useful indicator as to when a compound will come off a column. This is not always the case, since other characteristics such as polarity can greatly influence a compound's resistance to be freed from the column surface. For this reason, numerous GC columns are available to the chromatographer (different coatings, interior diameters, and lengths). Rather than coated columns, liquid chromatography (LC) makes use of columns packed with different sorbing materials with differing affinities for compounds. Also, instead of a carrier gas, LC uses a solvent or blend of solvents to carry the compounds to the detector. In the high-performance LC (HPLC), pressures are also varied.

Detection is the final step for quantifying the chemicals in a sample. The type of detector needed depends on the kinds of pollutants of interest. Detection gives the "peaks" that are used to identify compounds (see Fig. 17.6). For example, if hydrocarbons are of concern, GC with FID may be used. GC–FID gives a count of the number of carbons, so for example, long chains can be distinguished from short chains. The short chains come off the column

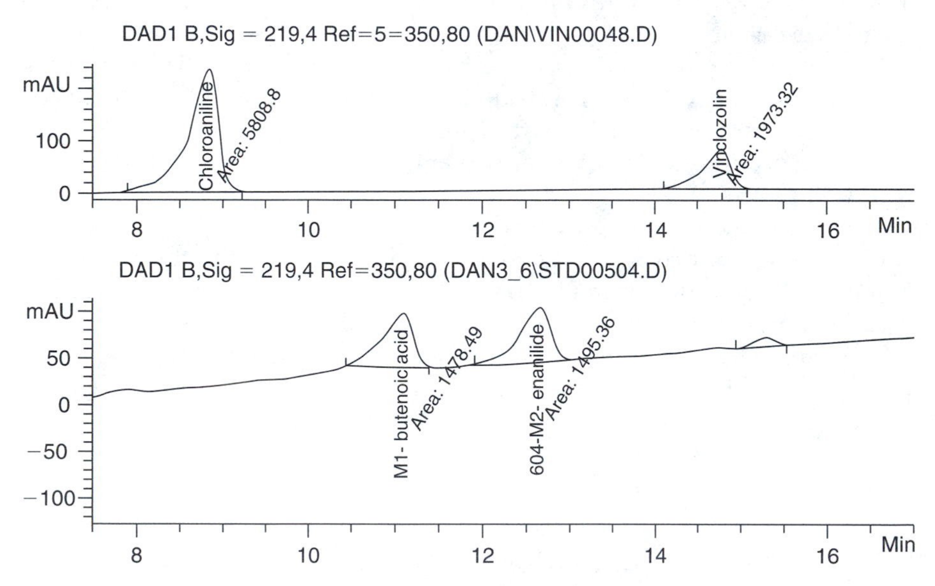

Fig. 17.6. High-performance liquid chromatograph/UV detection peaks for standard acetonitrile solutions: 9 mg L^{-1} 3,5-dichloroaniline and 8 mg L^{-1} the fungicide vinclozolin (top); and 7 mg L^{-1} M1 and 9 mg L^{-1}M2 (bottom). *Source*: Vallero, D., *Engineering the Risks of Hazardous Wastes*, Butterworth-Heinemann, Boston, MA, 2003.

first and have peaks that appear before the long-chain peaks. However, if pesticides or other halogenated compounds are of concern, ECD is a better choice.

A number of detection approaches are also available for LC. Probably the most common is absorption. Chemical compounds absorb energy at various levels, depending on their size, shape, bonds, and other structural characteristics. Chemicals also vary in whether they will absorb light or how much light they can absorb depending on wavelength. Some absorb very well in the UV range, while others do not. Diode arrays help to identify compounds by giving a number of absorption ranges in the same scan. Some molecules can be excited and will fluoresce. The Beer–Lambert law tells us that energy absorption is proportional to chemical concentration:

$$A = eb[C] \tag{17.1}$$

where, A is the absorbency of the molecule, e is the molar absorptivity (proportionality constant for the molecule), b is the light's path length, and [C] is the chemical concentration of the molecule. Thus, the concentration of the chemical can be ascertained by measuring the light absorbed.

One of the most popular detection methods is mass spectroscopy (MS), which can be used with either GC or LC separation. The MS detection is highly sensitive for organic compounds and works by using a stream of electrons to consistently break apart compounds into fragments. The positive ions resulting from the fragmentation are separated according to their masses. This is referred to as the "mass to charge ratio" or m/z. No matter which detection device is used, software is used to decipher the peaks and to perform the quantitation of the amount of each contaminant in the sample.

For inorganic substances and metals, the additional extraction step may not be necessary. The actual measured media (e.g. collected airborne particles) may be measured by surface techniques like AA, X-ray fluorescence (XRF), inductively coupled plasma (ICP), or sputtering. As for organic compounds, the detection approaches can vary. For example, ICP may be used with absorption or MS. If all one needs to know is elemental information, for example to determine total lead or nickel in a sample, AA or XRF may be sufficient. However, if it is speciation (i.e. knowing the various compounds of a metal), then significant sample preparation is needed, including a process known as "derivatization." Derivatizing a sample is performed by adding a chemical agent that transforms the compound in question into one that can be recognized by the detector. This is done for both organic and inorganic compounds, for example, when the compound in question is too polar to be recognized by MS.

The physical and chemical characteristics of the compounds being analyzed must be considered before visiting the field and throughout all the steps in the laboratory. Also, the quality of results generated about contamination depends on the sensitivity and selectivity of the analytical equipment. Table 17.2 defines some of the most important analytical chemistry threshold values.

TABLE 17.2

Expressions of Chemical Analytical Limits

Type of limit	Description
Limit of detection (LOD)	Lowest concentration or mass that can be differentiated from a blank with statistical confidence. This is a function of sample handling and preparation, sample extraction efficiencies, chemical separation efficiencies, and capacity and specifications of all analytical equipments being used (see IDL below).
Instrument detection limit (IDL)	The minimum signal greater than noise detectable by an instrument. The IDL is an expression of the piece of equipment, not the chemical of concern. It is expressed as a signal to noise (S:N) ratio. This is mainly important to the analytical chemists, but the engineer should be aware of the different IDLs for various instruments measuring the same compounds, so as to provide professional judgment in contracting or selecting laboratories and deciding on procuring for appropriate instrumentation for all phases of remediation.
Limit of quantitation (LOQ)	The concentration or mass above which the amount can be quantified with statistical confidence. This is an important limit because it goes beyond the "presence–absence" of the LOD and allows for calculating chemical concentration or mass gradients in the environmental media (air, water, soil, sediment, and biota).
Practical quantitation limit (PQL)	The combination of LOQ and the precision and accuracy limits of a specific laboratory, as expressed in the laboratory's quality assurance/quality control (QA/QC) plans and standard operating procedures (SOPs) for routine runs. The PQL is the concentration or mass that the engineer can consistently expect to have reported reliably.

Source: Vallero, D., *Engineering the Risks of Hazardous Wastes*. Butterworth-Heinemann, Boston, MA, 2003.

G. Semivolatile Organic Compounds

For sampling and analyzing SVOCs, a good place to start is US EPA "Method 1613," Tetra-through octa-chlorinated dioxins and furans by isotope dilution HRGC/HRMS (Rev. B), Office of Water, Engineering and Analysis Division, Washington, DC (1994), as well as "RCRA SW846 Method 8290," polychlorinated dibenzodioxins (PCDDs) and polychlorinated dibenzofurans (PCDFs) by high-resolution gas chromatograph/high-resolution mass spectrometry (HRGC/HRMS), Office of Solid Waste, US EPA (September 1994). For air, the best method is the PS-1 high-volume sampler system described in US EPA "Method TO-9A" in *Compendium of Methods for the Determination of Toxic Organic Compounds in Ambient Air* [24].

As mentioned, the extraction can be made by solvent extraction, by SFE, or by other techniques depending on the compound and the sorbant used to collect it. The procedure to analyze SVOCs begins with preparation of the sample for analysis by GC/MS using the appropriate sample preparation (e.g. EPA Method 3500) and, if necessary, sample cleanup procedures (i.e. EPA Method 3600). Next, the extract is introduced into the GC/MS by injecting the sample extract into a GC with a narrow-bore fused-silica capillary column. The GC column is temperature programmed to separate the analytes, which are then detected with MS. This is usually preferred. However, sometimes certain analytes cannot be detected directly with MS (e.g. highly polar compounds must first be derivatized). Thus, other detection systems, such as UV light, may need to be employed. The drawback is that the detection limits are often higher than that of MS.

Analytes eluted from the capillary column are introduced into the mass spectrometer using a jet separator or a direct connection. Identification of target analytes is accomplished by comparing their mass spectra with the electron impact (or electron impact-like) spectra of authentic standards (i.e. by the mass to charge [m/z] ratios of the molecular fragments). The column is selected based on the retention time (RT) of the particular SVOC. However, the most commonly used column for SVOCs is 30 m × 0.25 mm ID (or 0.32 mm ID) 1 μm film thickness silicone-coated fused-silica capillary column (J&W Scientific DB-5 or equivalent).

Quantitation is accomplished by comparing the response of a major (quantitation) ion relative to an internal standard using a five-point calibration curve that has been prepared in a solvent of known concentrations of the target SVOC.

Interference is a problem since SVOCs are ubiquitous in the environment (e.g. phthalates are used as platicizers even in laboratory settings). Thus, GC/MS data from all blanks, samples, and spikes must be evaluated for such interferences.

1. Air Pollution Chromatography Example

Consider the situation where an analytical laboratory has generated the following chromatogram and table from an HPLC/UV at 254 nm using a 5 μm, C_{18}, 4.6 × 250 mm column from a sample you submitted:

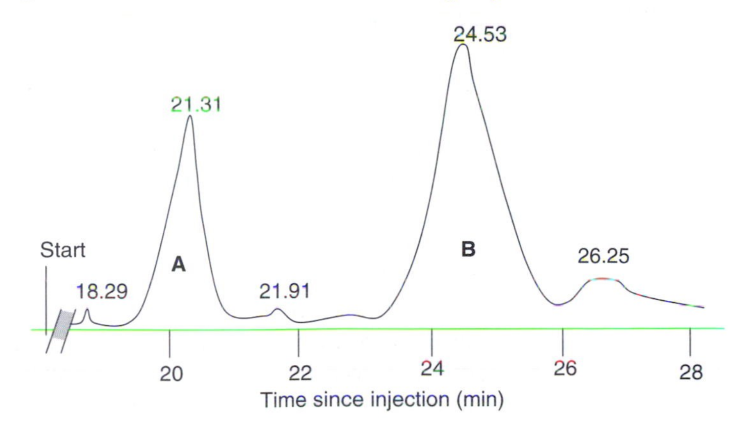

Retention time	Area	Type	Area/Height	Area (%)
18.29	NA	NA	NA	0.1
21.31	NA	NA	NA	31.4
21.91	NA	NA	NA	0.2
24.53	NA	NA	NA	67.2
26.25	NA	NA	NA	1.1

Even with the missing entries in the table, one can still ascertain certain information. What are the retention times of compounds A and B? Which compound is present in a larger amount? Which compound has the higher boiling point? What would happen to the retention times of compounds A and B if the column temperature were raised? You suspect that compound B is benzo(a)pyrene (B(a)p). How would you find out whether this is the case?

The retention time of compound A is 21.31 min, shown above of the peak and in the table's retention time column. The retention time of compound B is 24.53 min. You cannot tell from this table or chromatogram which compound is present in a larger amount, since the only way to do so is to have calibration curve from known concentrations of compound A and compound B (at least three, but preferably five). For example, you would run the HPLC successively with injections of pure solutions of 0.01, 0.1, 1, 10, and 100 $\mu g\ L^{-1}$ concentrations of compound A, and again with pure solutions of the same concentrations of compound B. These concentrations would give peak areas associated with each known concentration. Then you could calculate (actually the HPLC software will calculate) the calibration curve. So, for example, if a peak with an area of 200 is associated with 1 $\mu g\ L^{-1}$ of compound A and a peak with an area of 2000 is associated with 10 $\mu g\ L^{-1}$ of compound A (i.e. a linear calibration curve) at 21.31 min after the aliquot is injected into the HPLC, then when you run your unknown sample, a peak at 21.31 min with an area of 1000 would mean you have about 5 $\mu g\ L^{-1}$ concentration of compound A in your sample. The same procedure would be followed to draw a calibration curve for compound B at a retention time of 24.53 min.

The reason it is not sufficient to look at the percent area is that each compound is physically and chemically different, and recall from the Beer–Lambert law (Eq. 17.1) that the amount of energy absorbed (in this case, the UV light) is what gives us the peak. If a molecule of compound A absorbs UV at this wavelength (i.e. 254 nm) at only 25% as that of compound B, compound A's concentration would be higher than that of compound B (because even though compound B has twice the percent area, its absorbance is 4 times that of compound B).

Compound A likely has the lower boiling point since it comes off the column first. As mentioned, this is only true if other factors, especially polarity, are about the same. For example, if compound B has about the same polarity as the column being used, but compound A has a very different polarity, compound A will have a greater tendency to leave the column. Generally, however, retention time is a good indicator of boiling point; i.e., shorter retention times mean lower boiling points.

If the column temperature were raised, both compounds A and B would come off the column in shorter times. Thus, the retention times of both compounds A and B would be shorter than before the temperature was raised.

To determine whether the peak at 24.53 min is B(a)p, you must first obtain a true sample of pure B(a)p to place in a standard solution. This is the same process as you used to develop the calibration curve above. That is, you would inject this standard of known B(a)p into the same HPLC and the same volume of injection. If the standard gives a peak at a retention time of about 25 min, there is a good chance it is B(a)p. As it turns out, B(a)p absorbs UV at 254 nm and does come off an HPLC column at about 25 min.

The column type also affects retention time and peak area. The one used by the laboratory is commonly used for polycyclic aromatic hydrocarbons, includingB(a)p. However, numerous columns can be used for semivolatile organic compounds, so both the retention time and peak area will vary somewhat. Another concern is co-elution, i.e. two distinct compounds that have nearly the same retention times. One means of reducing the likelihood of co-elution is to target the wavelength of the UV detector. For example, the recommended wavelength for B(a)p is 254 nm, but 295 nm is preferred by environmental chromatographers because the interference peak in the B(a)p window is decreased at 295 nm. Another way to improve detection is to use a diode array detection system with the UV detector. This gives a number of different chromatograms simultaneously at various wavelengths. Finally, there are times when certain detectors do not work at all. For example, if a molecule does not absorb UV light (i.e. it lacks a group of atoms in a molecule responsible for absorbing the UV radiation, known as chromophores), there is no way to use any UV detector. In this case another detector, e.g. mass spectrometry, must be used.

H. General

The methods that have been discussed require specially designed instruments. Laboratories without such instruments can measure these gases using general-purpose chemical analytical equipment. A compendium of methods for these laboratories is the *Manual on Methods of Air Sampling and Analysis* [10].

II. ANALYSIS AND MEASUREMENT OF PARTICULATE POLLUTANTS

The three major characteristics of particulate pollutants in the ambient atmosphere are total mass concentration, size distribution, and chemical composition. In the United States, the $PM_{2.5}$ concentration, particulate matter with an aerodynamic diameter < 2.5 μm, is the quantity measured for an air quality standard to protect human health from effects caused by inhalation of suspended particulate matter. However, there remains a strong interest in the course fraction (PM_{10}) because it may be linked with certain diseases (e.g. asthma) and because it often has toxic components (e.g. sorbed metals and semivolatile organic compounds like dioxin). As shown in Chapter 11, the size distribution of particulate pollutants is very important in understanding the transport and removal of particles in the atmosphere and their deposition behavior in the human respiratory system. Their chemical composition may determine the type of effects caused by particulate matter on humans, vegetation, and materials.

Mass concentration units for ambient measurements are mass (μg) per unit volume (m^3). Size classification involves the use of specially designed inlet configurations, e.g., $PM_{2.5}$ sampling. To determine mass concentration, all the particles are removed from a known volume of air and their total mass is measured. This removal is accomplished by two techniques, filtration and impaction, described in Chapter 16. Mass measurements are made by pre- and postweighing of filters or impaction surfaces. To account for the absorption of water vapor, the filters are generally equilibrated at standard conditions ($T = 20°C$ and 50% relative humidity).

Size distributions are determined by classifying airborne particles by aerodynamic diameter, electrical mobility, or light-scattering properties. The most common technique is the use of multistage impactors, each stage of which removes particles of progressively smaller diameter. Figure 17.7

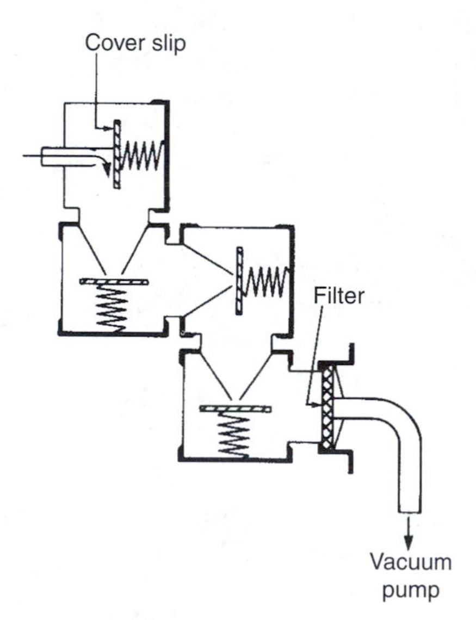

Fig. 17.7. Schematic diagram of a four-stage cascade impactor. *Source*: Giever, P. M., Particulate matter sampling and sizing, in *Air Pollution*, 3rd ed., Vol. III (Stern, A. C., ed.), p. 41. Academic Press, New York, 1976.

shows a four-stage impactor. The particulate matter collected on each stage is weighed to yield a mass size distribution or is subjected to chemical analysis to obtain data on its chemical size distribution. Impactors are used to determine size distributions for particle diameters of ~0.1 μm and larger.

Electrical mobility is utilized to obtain size distribution information in the 0.01–1.0 μm diameter range. This measurement method requires unipolar charging of particles and their separation by passage through an electrical field [11]. By incrementing the electrical field strength progressively, larger charged particles may be removed from a flowing air stream. The change in the amount of charge collected by an electrometer grid is then related to the number of particles present in a particular size increment. Instruments based on this principle yield a number size distribution.

Light-scattering properties of particles are also utilized to determine a number size distribution [12]. Individual particles interact with a light beam and scatter light at an angle to the original direction of the beam. The intensity of the scattered light is a function of the diameter and the refractive index of the particle. Inlet systems are designed to dilute a particle-laden air stream sufficiently to permit only one particle in the beam at a time. The intensity of the scattered light, as measured by a photomultiplier tube, is proportional to particle size. The number of electrical pulses of each magnitude is accumulated in a multichannel analyzer. By sampling at a known flow rate, the number of particles of different diameters are counted with this type of instrument.

The chemical composition of particulate pollutants is determined in two forms: specific elements, or specific compounds or ions. Knowledge of their chemical composition is useful in determining the sources of airborne particles and in understanding the fate of particles in the atmosphere. Elemental analysis yields results in terms of the individual elements present in a sample such as a given quantity of sulfur, S. From elemental analysis techniques we do not obtain direct information about the chemical form of S in a sample such as sulfate (SO_4^{2-}) or sulfide. Two nondestructive techniques used for direct elemental analysis of particulate samples are X-ray fluorescence (XRF) spectroscopy and neutron activation analysis (NAA).

XRF is a technique in which a sample is bombarded by X-rays [13]. Inner shell electrons are excited to higher energy levels. As these excited electrons return to their original state, energy with wavelengths characteristic of each element present in the sample is emitted. These high-energy photons are detected and analyzed to give the type and quantity of the elements present in the sample. The technique is applicable to all elements with an atomic number of 11 (sodium) or higher. In principle, complex mixtures may be analyzed with this technique. Difficulties arise from a matrix effect, so that care must be taken to use appropriate standards containing a similar matrix of elements. This technique requires relatively expensive equipment and highly trained personnel.

NAA involves the bombardment of the sample with neutrons, which interact with the sample to form different isotopes of the elements in the sample [14]. Many of these isotopes are radioactive and may be identified by comparing their radioactivity with standards. This technique is not quite as versatile as XRF and requires a neutron source.

Pretreatment of the collected particulate matter may be required for chemical analysis. Pretreatment generally involves extraction of the particulate matter into a liquid. The solution may be further treated to transform the material into a form suitable for analysis. Trace metals may be determined by AA spectroscopy, emission spectroscopy, polarography, and anodic stripping voltammetry. Analysis of anions is possible by colorimetric techniques and ion chromatography. Sulfate (SO_4^{2-}), sulfite (SO_3^{2-}), nitrate (NO_3^-), chloride (Cl^-), and fluoride (F^-) may be determined by ion chromatography [15].

Analytical methods available to laboratories with only general-purpose analytical equipment may be found in the *Methods of Air Sampling and Analysis* cited at the end of the previous section.

III. ANALYSIS AND MEASUREMENT OF ODORS

Odorants are chemical compounds such as H_2S, which smells like rotten eggs, and may be measured by chemical or organoleptic methods. Organoleptic methods are those which rely on the response to odor of the human nose. Although chemical methods may be useful in identifying and quantifying specific odorants, human response is the only way to assess the degree of acceptability of odorants in the atmosphere. This is due to several factors: the nonlinear relationship between odorant concentration and human response, the variability of individual responses to a given odorant concentration, and the sensory attributes of odor.

Four characteristics of odor are subject to measurement by sensory techniques: intensity, detectability, character (quality), and hedonic tone (pleasantness–unpleasantness) [16]. Odor intensity is the magnitude of the perceived sensation and is classified by a descriptive scale, e.g., faint–moderate–strong, or a 1–10 numerical scale. The detectability of an odor or threshold limit is not an absolute level but depends on how the odorant is present, e.g., alone or in a mixture. Odor character or quality is the characteristic which permits its description or classification by comparison to other odors, i.e., sweet or sour, or like that of a skunk. The last characteristic is the hedonic type, which refers to the acceptability of an odorant. For the infrequent visitor, the smell of a large commercial bread bakery may be of high intensity but pleasant. For the nearby resident, the smell may be less acceptable.

The sensory technique used for assessing human perception of odors is called *olfactometry.* The basic technique is to present odorants at different concentrations to a panel of subjects and assess their response. The process favored by the US National Academy of Sciences is dynamic olfactometry [16]. This technique involves a sample dilution method in which a flow of clean, nonodorous air is mixed with the odorant under dynamic or constant flow conditions. With this type of apparatus and standard operating conditions, it is possible to determine the detection threshold and the recognition threshold. At high dilution, the panel will be able to tell only whether an odorant is present or absent. Only at higher concentrations, typically by a factor of 2–10, will the subjects be able to identify the odorant.

The olfactometric procedure contains the following elements:

1. Dynamic dilution.
2. Delivery of diluted odorant for smelling through a mask or port.
3. Schedule of presentation of various dilutions and blanks.
4. Obtaining responses from the panelists.
5. Calculation of a panel threshold from experimental data.
6. Panelist selection criteria.

The first element, dynamic dilution, provides a reproducible sample for each panelist. The system must minimize the loss of the odorant to the walls of the delivery apparatus, provide clean dilution air of odor-free quality, maintain a constant dilution ratio for the duration of a given test, and have no memory effect when going from high to low concentrations or switching between odorants of different characters. The type of mask or port and the delivery flow rate have been found to influence the response of panelists in determining odor threshold and intensity.

The schedule of presentation may influence the results. The sensory effects are judgment criterion, anticipation, and adaptation. The judgment criterion determines how the panelist will respond when asked whether or not an odor is sensed. Individuals differ in their readiness to be positive or negative. The anticipation effect is a tendency to expect an odor over a given series of trials. Subjects show some positive response when no odorant is present. The adaptation effect is the temporary desensitization after smelling an odorant. This is also called olfactory fatigue and often occurs in occupational settings. Because of olfactory fatigue, investigators evaluating odor concentration in the field must breathe air deodorized by passage through an activated carbon canister before and after sniffing the ambient air being evaluated.

Individuals differ in their sensitivity to odor. Figure 17.8 shows a typical distribution of sensitivities to ethylsulfide vapor [17]. There are currently no guidelines on inclusion or exclusion of individuals with abnormally high or low sensitivity. This variability of response complicates the data treatment procedure. In many instances, the goal is to determine some mean value for the threshold representative of the panel as a whole. The small size of panels (generally fewer than 10 people) and the distribution of individual sensitivities require sophisticated statistical procedures to find the threshold from the responses.

Thresholds may also be determined by extrapolation of dose–response plots. In this approach, the perceived odor intensity is measured at several dilutions using some intensity rating method (Fig. 17.8). The threshold value may be selected at some value (e.g., zero intensity) and the concentration determined with the dilution ratio.

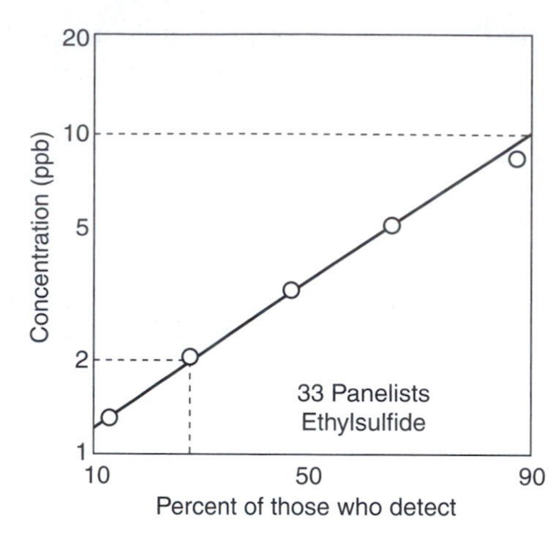

Fig. 17.8. Distribution of sensitivity to ethylene sulfide odor in 33 individuals. The abscissa is the percentage of the individuals who detected the presence of ethylene sulfide at various levels. *Source*: Dravnicks, A., and Jarke, F., *J. Air Pollut. Control Assoc.* **30**, 1284–1289 (1980).

IV. ANALYSIS AND MEASUREMENT OF VISIBILITY

Impairment of visibility is a degradation of our ability to perceive objects through the atmosphere. As discussed in Chapter 14, several components influence our concept of visibility: the characteristics of the source, the human observer, the object, and the degree of pollution in the atmosphere. Our attempts to measure visibility at a given location can take two approaches: human observations and optical measurements. In pristine locations such as national parks, use of human observers has permitted us to gain an understanding of the public's concept of visibility impairment. Although it is difficult to quantify the elements of human observations, this type of research, when coupled with optical measurements, provides a better measure of visibility at a given location [18].

Optical measurements permit the quantification of visibility degradation under different conditions. Several instruments are capable of measuring visual air quality, e.g., cameras, photometers, telephotometers, transmissometers, and scattering instruments.

Photography can provide a permanent record of visibility conditions at a particular place and time. This type of record can preserve a scene in a photograph in a form similar to the way it is seen. Photometers measure light intensity by converting brightness to representative electric signals with a photodetector. Different lenses and filters may be used to determine color and other optical properties. When used in combination with long-range lenses, photometers become telephotometers. This type of instrument may view distant objects with a much smaller viewing angle. The output of the photodetector is closely related to the perceived optical properties of distant targets. Telephotometers are often used to measure the contrast between a distant object and its surroundings, a measurement much closer to the human observer's perception of objects.

A transmissometer is similar to a telephotometer except that the target is a known light source. If we know the characteristics of the source, the average extinction coefficient over the path of the beam may be calculated. Transmissometers are not very portable in terms of looking at a scene from several directions. They are also very sensitive to atmospheric turbulence, which limits the length of the light beam.

Scattering instruments are also used to measure visibility degradation. The most common instrument is the integrating nephelometer, which measures the light scattered over a range of angles. The physical design of the instrument, as shown in Fig. 17.9, permits a point determination of the scattering coefficient of extinction, b_{ext} [19]. In clean areas, b_{ext} is dominated by scattering, so that the integrating nephelometer yields a measure of the extinction coefficient. As noted in Chapter 14, b_{ext} can be related to visual range through the Koschmieder relationship.

Other measurements important to visual air quality are pollutant related, i.e., the size distribution, mass concentration, and number concentration of airborne particles and their chemical composition. From the size distribution, the Mie theory of light scattering can be used to calculate the scattering coefficient [20]. Table 17.3 summarizes the different types of visual monitoring methods [21].

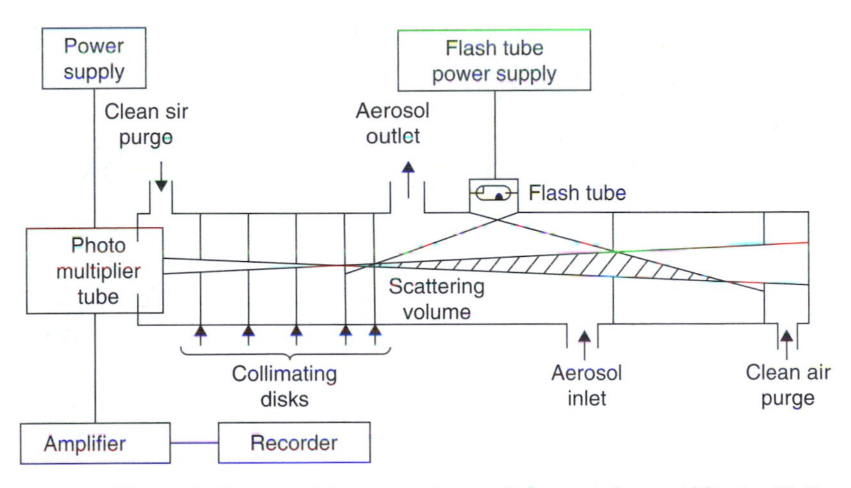

Fig. 17.9. Schematic diagram of the integrating nephelometer. *Source*: Ahliquist, N. C., and Charlson, R. J., *J. Air Pollut. Control Assoc.* **17**, 467 (1967).

TABLE 17.3

Visibility Monitoring Methods

Method	Parameters measured	Advantages	Limitations	Preferred use
Human observer	Perceived visual quality, atmospheric color, plume blight, visual range	Flexibility, judgment; large existing database (airport visual range)	Labor intensive; variability in observer perception; suitable targets for visual range not generally available	Complement to instrumental observations; areas with frequent plume blight, discoloration; visual ranges with available target distances
Integrating nephelometer	Scattering coefficient (b_{scat}) at site	Continuous readings; unaffected by clouds, night; b_{scat} directly relatable to fine aerosol concentration at a point; semiportable; used in a number of previous studies; sensitive models available; automated	Point measurement, requires assumption of homogeneous distribution of particles; neglects extinction from absorption, coarse particles (> 3–10 μm; must consider humidity effects at high relative humidity	Areas experiencing periodic, well-mixed general haze; medium to short viewing distances; small absorption coefficient (b_{abs}); relating to point composition measurements
Multiwavelength telephotometer	Sky and/or target radiance, contrast at various wavelengths	Measurement over long view path (up to 100 km) with suitable illumination and target, contrast transmittance, total extinction, and chromaticity over sight path can be determined; includes scattering and absorption from all sources; can detect plume blight; automated	Sensitive to illumination conditions; useful only in daylight; relationship to extinction, aerosol relationship possible only under cloudless skies; requires large, uniform targets	Areas experiencing mixed or inhomogeneous haze, significant fugitive dust; medium to long viewing distances (one-fourth of visual range); areas with frequent discoloration; horizontal sight path

Transmissometer	Long path extinction coefficient (b_{ext})	Measurement over medium view path (10–25 km); measures total extinction, scattering and absorption; unaffected by clouds, night	Calibration problems; single wavelength; equivalent to point measurement in areas with long view paths (50–100 km); limited applications to date still under development	Areas experiencing periodic mixed general haze, medium to short viewing distance areas with significant absorption (b_{abs})
Photography	Visual quality, plume blight, color, contrast (limited)	Related to perception of visual quality; documentation of vista conditions	Sensitive to lighting conditions; degradation in storage; contrast measurement from film subject to significant errors	Complement to human observation, instrumental methods; areas with frequent plume blight, discoloration
Particle samplers	Particles	Permit evaluation of causes of impairment	Not always relatable to visual air quality; point measurement	Complement to visibility measurements
Hi vol.	TSP	Large database, amenable to chemical analysis; coarse particle analysis	Does not separate sizes; sampling artifacts for nitrate, sulfate; not automated	Not useful for visibility sites
Cascade impactor	Size-segregated particles (more than two stages)	Detailed chemical, size evaluation	Particle bounce, wall losses; labor intensive	Detailed studies of scattering by particles <2 μm
Dichotomous and fine particle samplers (several fundamentally different types)	Fine particles (<2.5 μm) coarse particles (2.5–15 μm) inhalable particles (0–15 μm)	Size cut enhances resolution, optically important aerosol analysis, low artifact potential, particle bounce; amenable to automated compositional analysis; automated versions available; large networks under development	Some large-particle penetration; 24 h or longer sample required in clean areas for mass measurement; automated version relatively untested in remote locations	Complement to visibility measurement, source assessment for general haze, ground-level plumes

Source: US EPA. *Protecting Visibility*, EPA 450/5-79-008. Office of Air Quality Planning and Standards, Research Triangle Park, NC, 1979.

V. ANALYSIS AND MEASUREMENT OF ACIDIC DEPOSITION

The two components of acidic deposition described in Chapter 14 are wet deposition and dry deposition. The collection and subsequent analysis of wet deposition are intuitively straightforward. A sample collector opens to collect rainwater at the beginning of a rainstorm and closes when the rain stops. The water is then analyzed for pH, anions (negative ions), and cations (positive ions). The situation for dry deposition is much more difficult [22].

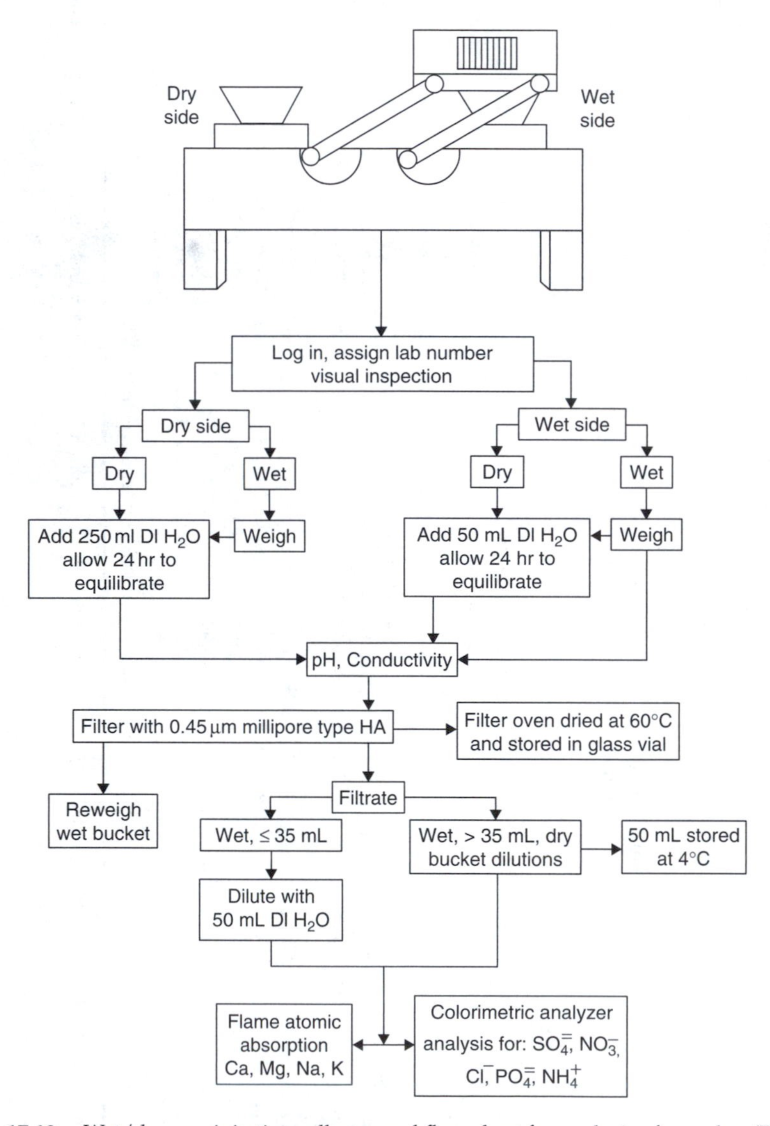

Fig. 17.10. Wet/dry precipitation collector and flow chart for analysis of samples. (DI H_2O: distilled water). *Source*: NADP Quality Assurance Report, Central Analytical Laboratory, Illinois Institute of Natural Resources, Champaign, III, March 1980.

Collection of particles settling from the air is very dependent on the surface material and configuration. The surfaces of trees, plants, and grasses are considerably different from that of the round, open-top canister often used to collect dry deposited particles. After collection, the material must be suspended or dissolved in pure water for subsequent analysis.

An overview of acid rain monitoring activities in North America shows several national and regional programs in operation in the United States, Canada, and Mexico [23]. The National Atmospheric Deposition Program has established the nationwide sampling network of ~100 stations in the United States. The sampler is shown in Fig. 17.10 with a wet collection container. The wet collection bucket is covered with a lid when it is not raining. A sensor for rain moves the lid to open the wet collector bucket and cover the dry bucket at the beginning of a rainstorm. This process is reversed when the rain stops.

The primary constituents to be measured are the pH of precipitation, sulfates, nitrates, ammonia, chloride ions, metal ions, phosphates, and specific conductivity. The pH measurements help to establish reliable long-term trends in patterns of acidic precipitation. The sulfate and nitrate information is related to anthropogenic sources where possible. The measurements of chloride ions, metal ions, and phosphates are related to sea spray and windblown dust sources. Specific conductivity is related to the level of dissolved salts in precipitation.

Figure 17.10 also shows a flowchart for analysis of wet and dry precipitation. The process involves weight determinations, followed by pH and conductivity measurements, and finally chemical analysis for anions and cations. The pH measurements are made with a well-calibrated pH meter, with extreme care taken to avoid contaminating the sample. The metal ions Ca^{2+}, Mg^{2+}, Na^{+}, and K^{+} are determined by flame photometry, which involves absorption of radiation by metal ions in a hot flame. Ammonia and the anions Cl^{-}, SO_4^{2-}, NO_3^{-}, and PO_4^{3-} are measured by automated colorimetric techniques.

Air pollution analytical methods continue to evolve and to improve. A good way to stay up-to-date on current methods is to visit the website for the US EPA's Ambient Monitoring Technology Information Center: http://www.epa.gov/ttn/amtic.

REFERENCES

1. Dailey, W. V., and Fertig, G. H., *Anal. Instrum.* **77**, 79–82 (1978).
2. US Environmental Protection Agency, **40** CFR, Part 50, Appendix C, July 1992.
3. Stevens, R. K., and Hodgeson, J. A., *Anal. Chem.* **45**, 443A–447A (1973).
4. US Environmental Protection Agency, Transfer Standards for Calibration of Air Monitoring Analyzers for Ozone, EPA-600/4-79-056. Office of Air Quality Planning and Standards, Research Triangle Park, NC, 1979.
5. US Environmental Protection Agency, *Fed. Regist.* **41**, 52686–52695 (1976).
6. US Environmental Protection Agency, **40** CFR, Part 50, Appendix A, July 1992.
7. Hollowell, C. D., Gee, G. Y., and McLaughlin, R. D., *Anal. Chem.* **45**, 63A–72A (1973).

8. Okake, H., Splitstone, P. L., and Ball, J. J., *J. Air Pollut. Control Assoc.* **23**, 514–516 (1973).
9. National Oceanic and Atmospheric Administration, *United States Standard Atmosphere*. US Government Printing Office, Washington, DC, 1976.
10. Lodge, J. P. (ed.), *Methods of Air Sampling and Analysis*, 3rd ed. American Public Health Association, CRC Press, Boca Raton, FL.
11. Liu, B. Y. H., Pui, D. Y. H., and Kapadia, A., Electrical aerosol analyzer, in *Aerosol Measurement* (Lundgren, D. A., Harris Jr., F. S., Marlow, W. H., Lippmann, M., Clark, W. E., and Durham, U. D., eds.), pp. 341–384. University Presses of Florida, Gainesville, FL, 1979.
12. Whitby, K. T., and Willeke, K., Single particle optical particle counters, in *Aerosol Measurement* (Lundgren, D. A., Harris Jr., F. S., Marlow, W. H., Lippmann, M., Clark, W. E., and Durham, U. D., eds.), pp. 241–284. University Presses of Florida, Gainesville, FL, 1979.
13. Dzubay, T. G., *X-ray Fluorescence Analysis of Environmental Samples*. Ann Arbor Science Publishers, Ann Arbor, MI, 1977.
14. Heindryckx, R., and Dams, R., *Prog. Nucl. Energy* **3**, 219–252 (1979).
15. Mulik, J. D., and Sawicki, E., *Ion Chromatographic Analysis of Environmental Pollutants*, Vol. 2. Ann Arbor Science Publishers, Ann Arbor, MI, 1979.
16. National Research Council, *Odors from Stationary and Mobile Sources*. National Academy of Sciences, Washington, DC, 1979.
17. Dravnicks, A., and Jarke, F., *J. Air Pollut. Control Assoc.* **30**, 1284–1289 (1980).
18. Malm, W., Kelley, K., Molenar, J., and Daniel, T., *Atmos. Environ.* **15**, 1875–1890 (1981).
19. Ahliquist, N. C., and Charlson, R. J., *J. Air Pollut. Control Assoc.* **17**, 467 (1967).
20. Twomey, S., *Atmospheric Aerosols*, pp. 200–216. Elsevier North-Holland, New York, 1977.
21. US Environmental Protection Agency, Protecting Visibility, EPA-450/5-79-008. Office of Air Quality Planning and Standards, Research Triangle Park, NC, 1979.
22. Hicks, B. B., Wesely, M. L., and Durham, J. L., Critique of Methods to Measure Dry Deposition—Workshop Summary, EPA-600/9-80-050. Environmental Sciences Research Laboratory, Research Triangle Park, NC, 1980.
23. Wisniewski, J., and Kinsman, J. D., *Bull. Am. Meteorol. Soc.* **63**, 598–618 (1982).
24. US EPA, Method TO-9A in Compendium of Methods for the Determination of Toxic Organic Compounds in Ambient Air, 2nd ed. EPA/625/R-96/010b, 1999.

SUGGESTED READING

Harrison, R. M., and Young, R. J. (eds.), *Handbook of Air Pollution Analysis*, 2nd ed. Chapman & Hall, London, 1986.

Lundgren, D. A., Harris Jr., F. S., Marlow, W. H., Lippmann, M., Clark, W. E., and Durham, U. D. (eds.), *Aerosol Measurement*. University Presses of Florida, Gainesville, FL, 1979.

Newman, L. (ed.), *Measurement Challenges in Atmospheric Chemistry*. American Chemical Society, Washington, DC, 1993.

Sickles II, J. E., *Adv. Environ. Sci. Technol.* **24**, 51–128 (1992).

Winegar, E. D., and Keith, L.H., *Sampling and Analysis of Airborne Pollutants*. Lewis Publishers, Boca Raton, FL, 1993.

QUESTIONS

1. Describe the rationale for the US EPA's establishment of a standard reference method for measurement of National Ambient Air Quality Standard air pollutants.

2. Under what conditions can another method be substituted for a standard reference method?
3. Describe the potential interferences (a) in the NDIR method for measuring CO and (b) in the chemiluminescent method for measuring NO_2.
4. The electrical aerosol analyzer and the optical counter are used to measure particle size distributions. Describe the size range and resolution characteristics of each of these instruments.
5. How can human observers, optical measurements along a line of sight, and point measurements by nephelometry provide conflicting information about visual air quality in the same location?
6. Using the Code of Federal Regulations, list the current reference methods for measuring NO_2, O_3, SO_2, CO, total suspended particulate matter, and lead.
7. List two types of calibration sources for gas analyzers.
8. Review the air pollution literature and describe the difficulties in establishing a standard reference method for measuring NO_2.
9. Describe the deficiencies of a total suspended particulate measurement for relating ambient concentrations to potential human health effects.
10. Explain how one chemical compound can have a larger peak on a chromatogram than another compound, yet have a much lower concentration.
11. Why is extraction important in analyzing air pollutants? When is it not necessary?

18

Air Pollution Monitoring and Surveillance

I. STATIONARY MONITORING NETWORKS

The US Environmental Protection Agency (EPA) has established National Ambient Air Quality Standards (NAAQS) for protection of human health and welfare. These standards are defined in terms of concentration and time span for a specific pollutant; for example, the NAAQS for carbon monoxide is 9 ppmV for 8 h, not to be exceeded more than once per year. For a state or local government to establish compliance with the NAAQS, measurements of the actual air quality must be made. To obtain these measurements, state and local governments have established stationary monitoring networks with instrumentation complying with federal specifications, as discussed in Chapter 17. The results of these measurements determine whether a given location is violating the air quality standard.

Stationary monitoring networks are also operated to determine the impact of new sources of emissions. As part of the environmental impact statement and prevention of significant deterioration processes, the projected impact of a new source on existing air quality must be assessed. Air quality monitoring is one means of making this type of assessment. A monitoring network is established at least 12 months before construction to determine prior air quality. Once the facility is completed and in operation, the network data determine the actual impact of the new source.

Long-term trends are measured by stationary air quality monitoring networks. Figure 8.7 shows the long-term decrease of SO_2 in the atmosphere resulting from the implementation of air pollution control technology. The trends in other atmospheric trace gases such as CH_4, NO, NO_2, and CO are similarly measured in rural as well as urban locations. Atmospheric budgets of various gases are developed to allow estimation of whether sources are anthropogenic or natural.

A stationary monitoring network should yield the following information: (1) background concentration levels, (2) highest concentration levels, (3) representative concentration levels in high-density areas, (4) the impact of local sources, (5) the impact of remote sources, and (6) the relative impact of natural and anthropogenic sources.

The spatial scale of an air monitoring network is determined by monitoring objectives. Spatial scales include personal (<1 m, on person), microscale (1–100 m), middle scale (100 m–0.5 km), neighborhood scale (0.5–4.0 km), urban scale (4–50 km), and regional scale (tens to hundreds of km). Table 18.1 shows the relationship between spatial scale and monitoring objectives [1].

Sampler siting within a network must meet the limitations of any individual sampling site and the relationship of sampling sites with each other [2]. The overall approach for selection of sampling sites is to (1) define the purpose of the collected data, (2) assemble site selection aids, (3) define the general areas for samplers based on chemical and meteorological constraints, and (4) determine the final sites based on sampling requirements and surrounding objects [3]. Several purposes of air quality monitoring were mentioned earlier, such as air quality standard compliance, long-term trends, and new facility siting.

The tools available for site selection include climatological data, topography, population data, emission inventory data, and diffusion modeling. Climatological data are useful in relating meteorology to emission patterns. For example, elevated levels of photochemical oxidant are generally related to stagnant meteorological conditions and warm temperatures. Seasonal climatological patterns of prevailing winds and frequency of inversions will influence the location of sampling stations. Various types of maps are useful for establishing topography, population density, and location of sources of

TABLE 18.1

Relationship of the Scale of Representativeness and Monitoring Objectives

Siting scales	Monitoring objectives
Personal	Personal cloud
Personal, micro, middle, neighborhood (sometimes urban)	Highest concentration affecting people
Neighborhood, urban	High-density population exposure
Micro, middle, neighborhood	Source impact
Neighborhood, region	General/background concentration

various pollutants. Wind roses overlaid with emission sources and population densities help to locate the general areas for location of samplers.

Various types of dispersion models are available which can use as input emission patterns, climatological data, and population data to rank sampling locations by concentration threshold, resolution of peak concentrations, and frequency of exposure [4] or to rank sampling locations for maximum sensitivity to source emission changes, to provide coverage of as many sources or to cover as large a geographic area as possible [5].

The last step in selecting specific sites is based on the following: availability of land and electrical power, security from vandalism, absence of nearby structures such as large buildings, probe height (inlet >3 m), and cost.

An example of matching scale and objective is the determination of CO exposure of pedestrians on sidewalks in urban street canyons. The location of a station to meet this objective would be an elevation of ~3 m on a street with heavy vehicular traffic and large numbers of pedestrians.

Figure 18.1 shows the Los Angeles, California, basin stationary air monitoring network, one of the most extensive in the United States [6]. At most of

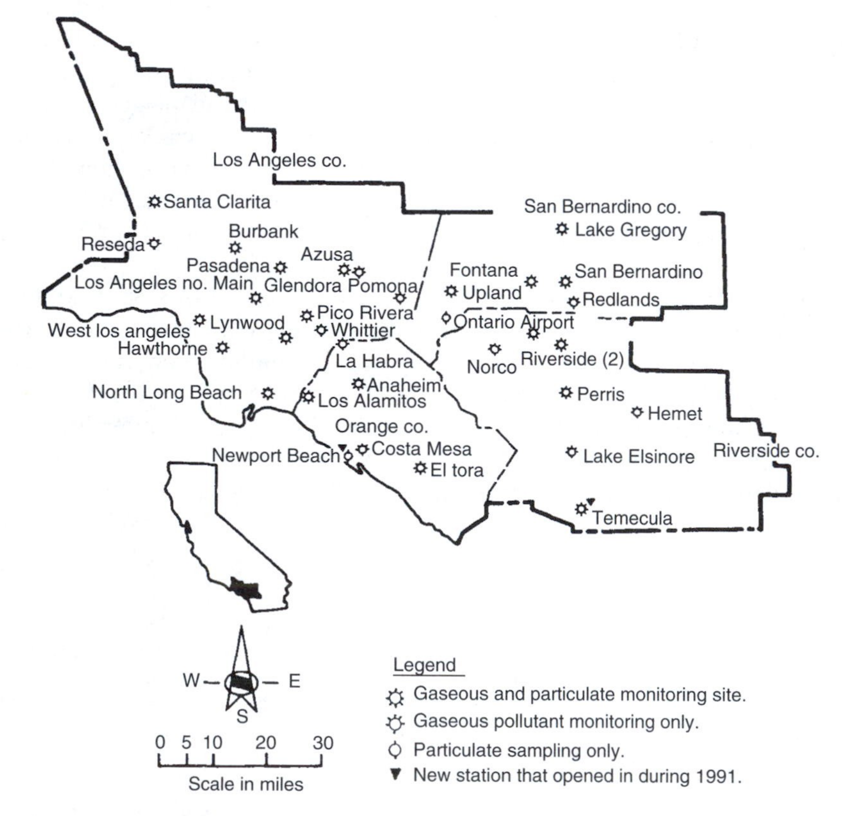

Fig. 18.1. California South Coast Air Basin stationary monitoring locations operating during 1991 (LA: Los Angeles). *Source*: California Air Resources Board, *Summary of 1991 Air Quality Data, Gaseous and Particulate Pollutants*, Vol. 23, 1991.

these locations, automated instruments collect air quality data continuously. Five pollutant gases are monitored, and particulate matter filter samples are collected periodically.

II. MOBILE MONITORING AND SURVEILLANCE

Mobile monitoring is accomplished from a movable platform, i.e., an aircraft or vehicle. Emissions are measured by source monitoring techniques (see Chapter 35). Atmospheric transport and chemical transformation processes occur in the region between the source and the receptor. By using mobile platforms containing air pollution instrumentation, one can obtain data to help understand the formation and transport of photochemical smog, acidic deposition, and the dispersion of air pollutants from sources. Mobile monitoring platforms may also be moved to *hot spots*, areas suspected of having high concentrations of specific air pollutants. These areas may be nearby locations downwind of a large source or a particular location that is an unfavorable receptor due to meteorological conditions. Vehicular and aircraft monitoring systems can also be moved to locations where hazardous chemical spills, nuclear and chemical plant accidents, or volcanoes or earthquakes have occurred.

The major advantage of a mobile monitoring system is its ability to obtain air quality information in the intermediate region between source monitors and stationary fixed monitors. The major disadvantage is the sparsity of suitable instrumentation that operates properly in the mobile platform environment. Limitations of existing instrumentation for the use on movable platforms are inadequate temperature and pressure compensation; incompatible power, size, and weight requirements; and excessive response time. Most movable platforms are helicopters, airplanes, trucks, or vans. These platforms do not provide the relatively constant-temperature environment required by most air quality instrumentation. Equipment mounted in aircraft is subject to large pressure variations with changing altitude. Most instrumentation is designed to operate with alternating current (AC) electrical power, whereas relatively low amounts of direct current (DC) power are available in aircraft or vans. Space is at a premium, and response times are often too slow to permit observation of rapid changes in concentration as aircraft move in and out of a plume.

Despite these limitations, mobile monitoring systems have been used to obtain useful information, such as the verification and tracking of the St. Louis, Missouri, urban plume. The measurement of a well-defined urban plume spreading northeastward from St. Louis is shown in Fig. 18.2 [7]. These data were collected by a combination of instrumented aircraft and mobile vans. Cross-sectional paths were flown by the aircraft at increasing distances downwind. Meteorological conditions of low wind speed in the same direction helped to maintain this urban plume in a well-defined condition for several hundred kilometers downwind. The presence of large point sources of NO and SO_2 was observed by the changes in the O_3 and b_{scat} profiles.

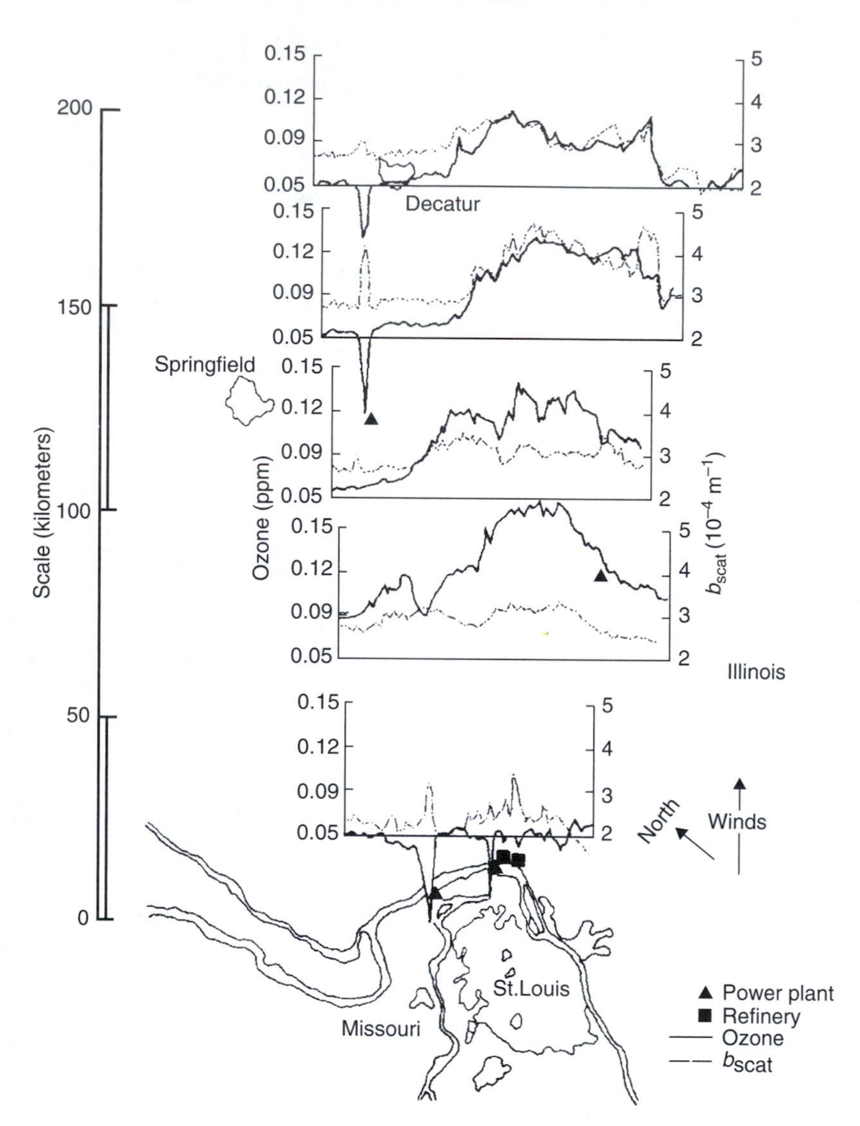

Fig. 18.2. The St. Louis, Missouri, urban plume. Ozone and b_{scat} profiles at four distances downwind of St. Louis track a detectable urban plume for 150 km. *Source*: Wilson Jr., W. E., *Atmos. Environ.* **12**, 537–547 (1978).

Sharp decreases in ozone concentration occurred when large amounts of NO were present, and rapid increases in b_{scat} were caused by the primary and secondary particulate matter from power plant plumes embedded in the larger urban plume. The overall increasing level of ozone and b_{scat} at greater downwind distances was caused by the photochemical reactions as the urban plume was transported farther away from St. Louis. This type of plume mapping can be accomplished only by mobile monitoring systems.

III. REMOTE SENSING

Remote sensing involves monitoring in which the analyzer is physically removed from the volume of air being analyzed. Satellites have been used to monitor light-scattering aerosol over large areas [8]. Large point sources such as volcanic activity and forest fires can be tracked by satellite. The development and movement of hazy air masses have been observed by satellite imagery (see Fig. 14.9). Remote sensing methods are available for determining the physical structure of the atmosphere with respect to turbulence and temperature profiles (see Chapter 5).

Differential absorption lidar (DIAL) is used for remote sensing of gases and particles in the atmosphere. *Lidar* is an acronym for *light detection and ranging*, and the technique is used for measuring physical characteristics of the atmosphere. DIAL measurements consist of probing the atmosphere with pulsed laser radiation at two wavelengths. One wavelength is efficiently absorbed by the trace gas, and the other wavelength is less efficiently absorbed. The radiation source projects packets of energy through the atmosphere, which interact with the trace gas. The optical receiver collects radiation backscattered from the target. By controlling the timing of source pulses and processing of the optical receiver signal, one can determine the concentration of the trace gas over various distances from the analyzer. This capability permits three-dimensional mapping of pollutant concentrations. Applications are plume dispersion patterns and three-dimensional gaseous pollutant profiles in urban areas.

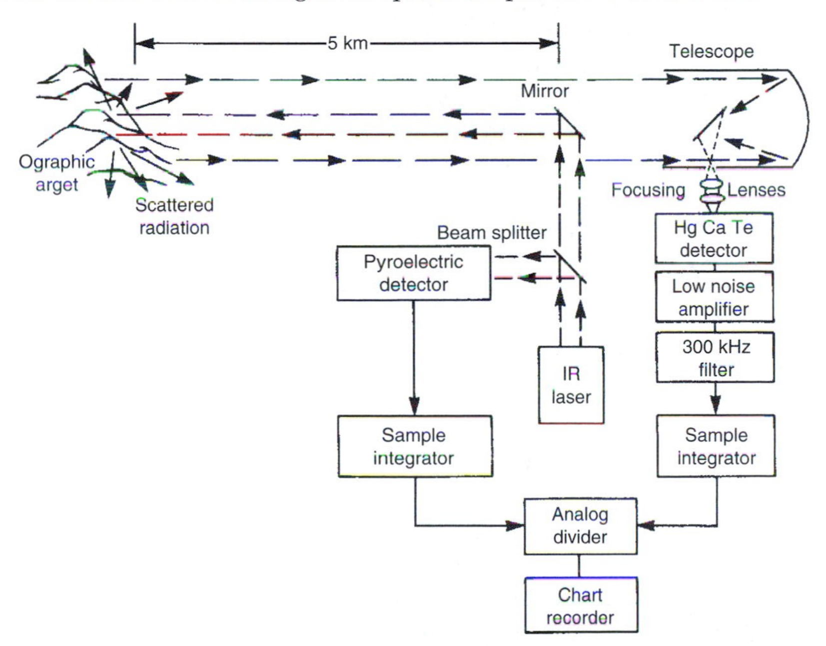

Fig. 18.3. An infrared DIAL system. *Source*: Murray, E. A., and Van der Laan, J. E., *Appl. Opt.* **17**, 814–817 (1978).

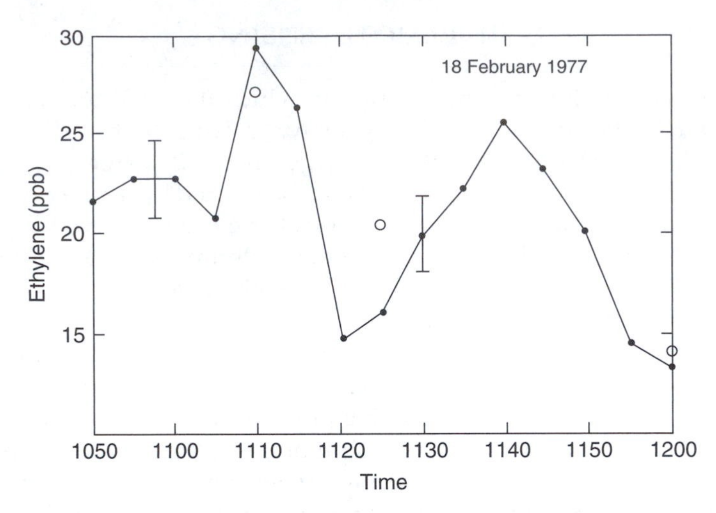

Fig. 18.4. Ethylene concentrations measured by the DIAL system along a 5-km path length near Menlo Park, California. The open circles are ethylene concentrations of samples taken at three ground-level locations near the line of sight and analyzed by gas chromatography. *Source*: Murray, E. A., and Van der Laan, J. E., *Appl. Opt.* **17**, 814–817 (1978).

SO_2 and O_3 are detected by an ultraviolet DIAL system operating at wavelengths near 300 nm [9]. Tunable infrared CO_2 lasers are used in applications of IR-DIAL systems which are capable of measuring SO_2, CO, HCl, CH_4, CO_2, H_2O, N_2O, NH_3, and H_2S [10]. The components of this type of system are shown in Fig. 18.3 [11]. The laser source is switched between the low-absorption and high-absorption frequencies for the trace gas to be detected. The system is pointed toward a target, and focusing lenses are used to collect the returning signal. The beam splitter diverts a portion of the transmitted beam to a detector. The backscattered and transmitted pulses are integrated to yield DC signals. Figure 18.4 shows ethylene concentrations measured on a 5-km path in Menlo Park, California, with this system.

Satellites are being used to detect global distributions of large areas of CO and O_3 [12].

IV. PERSONAL MONITORING

Ambient and even indoor measurements of air pollutants are not often a good indication of the levels of exposure to air pollutants that people actually receive. Their activities and locations often vary considerably from person to person. In fact, the phenomenon known as "personal cloud" has recently been characterized to help predict a person's exposures and risks to air pollutants. The cloud indicates an increased personal exposure to air pollutants compared

TABLE 18.2

Comparison of Ambient Concentrations of Air Pollutants to Personal Exposures

	Summer					Fall				
Pollutant[a]	n	ND	LOD	Mean ± SD	Maximum	n	ND	LOD	Mean ± SD	Maximum
Ambient concentrations										
Particles										
$PM_{2.5}$	65	0	0	20.1 ± 9.3	46.6	72	0	0	19.3 ± 12.2	50.7
SO_4^{2-}	61	0	0	7.7 ± 4.8	25.0	72	0	0	6.2 ± 4.7	22.4
EC	56	0	1	1.1 ± 0.5	2.9	71	0	0	1.1 ± 0.7	3.6
Gases										
O_3	62	0	4	29.3 ± 13.4	74.8	72	0	21	16.0 ± 8.1	42.4
NO_2	62	1	44	9.5 ± 7.4	37.9	71	0	16	11.3 ± 6.0	27.9
SO_2	63	23	53	2.7 ± 3.9	21.9	71	24	43	5.4 ± 9.6	63.6
Personal exposures										
Particles										
$PM_{2.5}$	169	0	0	19.9 ± 9.4	59.0	204	0	0	20.1 ± 11.6	66.0
SO_4^{2-}	165	0	2	5.9 ± 4.2	25.6	188	0	0	4.4 ± 3.3	16.3
EC	166	7	12	1.1 ± 0.6	4.6	197	1	1	1.2 ± 0.7	6.2
Gases										
O_3	183	2	168	5.3 ± 5.2	35.7	226	84	207	3.9 ± 4.4	21.3
NO_2	183	1	117	9.9 ± 6.0	38.9	228	1	32	12.1 ± 6.1	38.8
NO_2[b]	130	1	93	9.0 ± 5.2	38.9	139	1	28	9.9 ± 4.6	28.7
NO_2[c]	53	0	24	12.3 ± 7.1	33.5	89	0	4	15.7 ± 6.4	38.8
SO_2	185	99	173	1.5 ± 3.3	30.4	228	72	217	0.7 ± 1.9	14.2

[a] $PM_{2.5}$, SO_4^{2-}, and EC in units of micrograms per cubic meter; O_3, NO_2, and SO_2 in units of parts per billion.
[b] Samples from subjects without gas stoves in their homes.
[c] Samples from subjects with gas stoves in their homes.
ND: number of samples with values below the analytical LOD (i.e. not detected).
Source: Sarnat, S. E., Coull, B. A., Schwartz, J., Gold, D. R., and Suh, H. H., Factors affecting the association between ambient concentrations and personal exposures to particles and gases. *Environ. Health Persp.* **114** (5), 649–654, 2006.

to indoor concentrations, or compared to the time-weighted indoor and outdoor concentrations. Results from several recent studies suggest that a number of factors, such as relative humidity and ventilation may lead to very different exposures based on personal measurements versus ambient concentrations of air pollutants. For example, one recent study found that while ambient fine particle concentrations may somewhat represent exposures to fine particles, the ability of either ambient gases or ambient fine particles to represent exposure to gases is quite small (see Table 18.2).

With personal monitoring, the monitoring device is worn by individuals as they proceed through their normal activities. This approach is most common in workplaces. The radioactivity sensors worn by nuclear power plant workers are one example. Personal monitoring is increasingly being used, however, to estimate total human exposures, including exposures from the air people breathe, the water they drink, and the food they eat.[1] One advantage of personal monitoring is that the data provide valuable insights into the sources of the pollutants to which people are actually exposed. A challenge with personal monitoring is ensuring that sufficient sampling is done to be representative of the population being studied.

V. QUALITY ASSURANCE

Air quality monitoring for standards compliance, new facility siting, and long-term trend measurement has been going on for many years. Historically, a large number of federal, state, and local organizations, both governmental and nongovernmental, have been using a variety of technologies and approaches to obtain air quality data. This has resulted in multiple data sets of variable accuracy and precision. Questionable or conflicting air quality data are of little value in ascertaining compliance with air quality standards, determining whether air quality is improving or worsening in a given region over an extended period, or understanding the chemistry and physics of the atmosphere.

In order to minimize the collection of questionable air quality data, the US EPA has established and implemented stringent regulations requiring well-documented quality assurance programs for air quality monitoring activities [13].

Quality assurance programs are designed to serve two functions: (1) assessment of collected air quality data and (2) improvement of the data collection process. These two functions form a loop; as air quality data are collected,

[1] Examples of personal monitoring include EPA's Total Exposure Assessment Monitoring (TEAM) studies and its National Human Exposure Assessment Survey (NHEXAS), and the Relationship of Indoor, Outdoor, and Personal Air (RIOPA) study conducted by the Mickey Leland Center.

procedures are implemented to determine whether the data are of acceptable precision and accuracy. If they are not, increased quality control procedures are implemented to improve the data collection process.

The components of a quality assurance program are designed to serve the two functions just mentioned—control and assessment. Quality control operations are defined by operational procedures, specifications, calibration procedures, and standards and contain the following components:

1. Description of the methods used for sampling and analysis
2. Sampling manifold and instrument configuration
3. Appropriate multipoint calibration procedures
4. Zero/span checks and record of adjustments
5. Control specification checks and their frequency
6. Control limits for zero, span, and other control limits
7. The corrective actions to be taken when control limits are exceeded
8. Preventative maintenance
9. Recording and validation of data
10. Documentation of quality assurance activities

Table 18.3 contains a specific example of these components for ambient monitoring for ozone.

TABLE 18.3

Quality Control Components for Ambient Ozone Monitoring

Component	Description
Method	Chemiluminescent O_3 monitor Calibration method by certified ozone UV transfer method
Manifold/instrument configuration	Instrument connected to sampling manifold which draws ambient air at 3 m into instrument shelter
Calibration	Multipoint calibration on 0.5-ppm scale at 0.0, 0.1, 0.2, and 0.4 ppm weekly
Zero/span check	Zero check ±0.005 ppm Span check 0.08–0.10 on a 1.0-ppm full-scale daily
Control specification checks	Ethylene flow Sample flow, daily
Corrective limits	±0.005 ppm zero and span
Corrective action	Do multipoint calibration; invalidate data collection since last zero/span check within control limits
Preventive maintenance	Manufacturer's procedures to be followed
Recording and validating data	Data reported weekly to quality assurance coordinator, with invalid data flagged
Documentation	Data volume includes all quality control forms, e.g., zero/span control charts and multipoint calibration results

In addition to fulfilling the in-house requirements for quality control, state and local air monitoring networks which are collecting data for compliance purposes are required to have an external performance audit on an annual basis. Under this program, an independent organization supplies externally calibrated sources of air pollutant gases to be measured by the instrumentation undergoing audit. An audit report summarizes the performance of the instruments. If necessary, further action must be taken to eliminate any major discrepancies between the internal and external calibration results.

Data quality assessment requirements are related to precision and accuracy. Precision control limits are established, i.e., +10% of span value, as calculated from Eq. (18.1). The actual results of the may be used to calculate an average deviation (Eq. 18.3):

$$d_i = (y_i - x_i)/x_i \times 100 \tag{18.1}$$

where d_i is the percentage difference, y_i the analyzer's indicated concentration of the test gas for the ith precision check, and x_i the known concentration of the test gas for the ith precision check:

$$d_{\text{av}} = \frac{1}{n}\sum_{i=1}^{n} d_i \tag{18.2}$$

$$S_j = \sqrt{\frac{1}{n-1}\left[\sum_{i=1}^{n} d_i^2 - \frac{1}{n}\left(\sum_{i=1}^{n} d_i\right)^2\right]} \tag{18.3}$$

The external audit results are used to determine the accuracy of the measurements. Accuracy is calculated from percentage differences, d_i, for the audit concentrations and the instrument response.

VI. DATA ANALYSIS AND DISPLAY

In general, air quality data are classified as a function of time, location, and magnitude. Several statistical parameters may be used to characterize a group of air pollution concentrations, including the arithmetic mean, the median, and the geometric mean. These parameters may be determined over averaging times of up to 1 year. In addition to these three parameters, a measure of the variability of a data set, such as the standard deviation or the geometric standard deviation, indicates the range of data around the value selected to represent the data set.

Raw data must be analyzed and transformed into a format useful for specific purposes. Summary tables, graphs, and geographic distributions are some of the formats used for data display. Air quality information often consists of a large body of data collected at a variety of locations and over different seasons. Table 18.4 shows the tabular format used by a local air pollution authority to report the status of compliance with air quality standards [6].

TABLE 18.4

Summary of Air Quality Statistics in the California's South Coast Air Basin and the Desert Area of Coachella Valley in the Salton Sea Air Basin for December 2002.

			Maximum concentrations			
Pollutant averaging time	State standard	Federal standard	ppm $\mu g^{-1} m^{-3}$	% State standard	% Federal standard	Location
Ozone						
1 h	>0.09 ppm	>0.12 ppm	0.06	60	48	Several Locations
8 h		>0.08 ppm	0.055		65	Banning Airport
Carbon monoxide						
8 h	>9.0 ppm	>9 ppm	8.40	92	88	South Central Los Angeles County
Nitrogen dioxide						
1 h	>0.25 ppm		0.10	38		Southwest Coastal Los Angeles County
24 h			0.069			South San Gabriel Valley
Sulfur dioxide						
1 h	>0.25 ppm		0.02	8		South Coastal Los Angeles County
24 h	>0.04 ppm	>0.14 ppm	0.010	24	7	North Coastal Orange County
Particulate (PM_{10})						
24 h	>50 $\mu g\, m^{-3}$	>150 $\mu g\, m^{-3}$	95	186	63	Metropolitan Riverside County
Particulate ($PM_{2.5}$)						
24 h		>65 $\mu g\, m^{-3}$	55.4		85	South Coastal Los Angeles County
Sulfates						
24 h	≥25 $\mu g\, m^{-3}$		4.7	19		South Central Los Angeles County
Lead[a]						
30 Days	≥1.5 $\mu g\, m^{-3}$		0.03	2		Central Los Angeles
30 Days[a]			0.19	13		Several Locations

[a] Maximum monthly average concentration recorded at special monitoring sites in the immediate vicinity of major lead sources.

Source: South Coast Air Quality Management District Air Quality Standards Compliance Report, December 2002, and *Summary Statistics for 2002*, Vol. 15, No. 12.

The format has location, maximum values, annual means, and number of occurrences of hourly values above a given concentration as a function of the month of the year. One can quickly determine which areas are violating a standard, at what time of the year elevated concentrations are occurring, and the number of good data points collected.

Pollutant concentration maps may be constructed as shown in Fig. 18.5 [14]. In this example, elevated levels of ambient particulate matter are associated with population centers. For a given geographic area, isopleths, lines showing equal concentrations of a pollutant, are drawn on a map. Regions of high concentration are quickly identified. Further action may be taken to determine the cause, such as review of emission inventories of additional sampling.

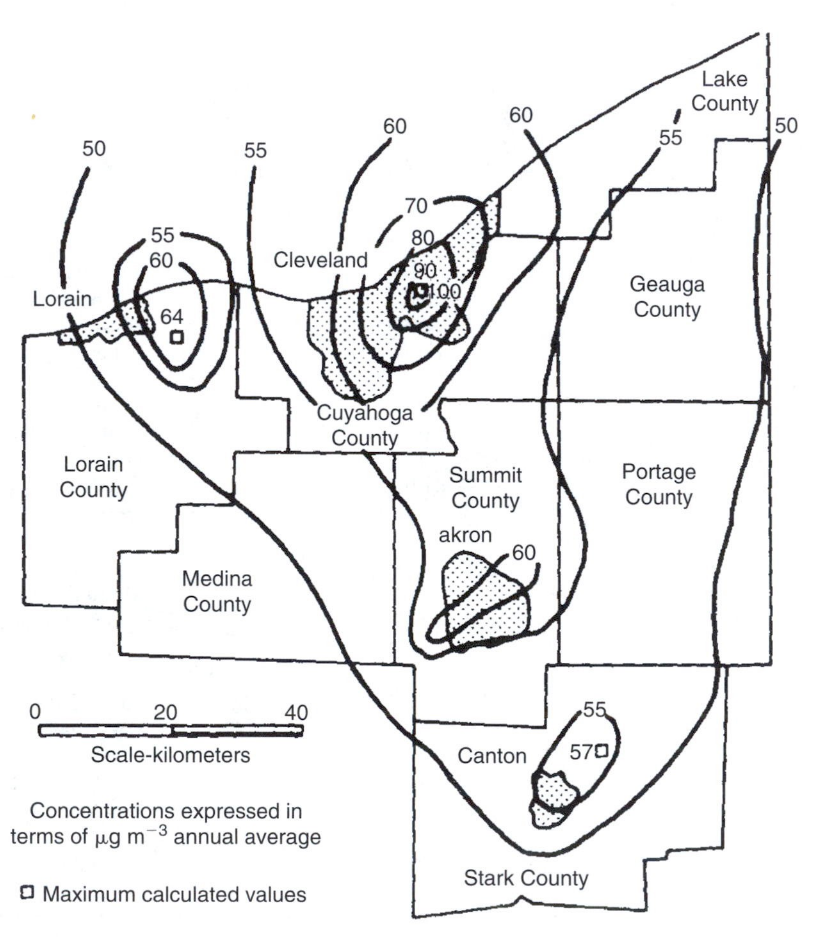

Fig. 18.5. Concentration isopleth diagram of ambient particulate matter calculated by a computer model. *Source*: Zimmer, C. E., *Air Pollution*, 3rd ed., Vol. 3 (Stern, A. C., ed.). Academic Press, New York, 1976.

REFERENCES

1. Code of Federal Regulations, Title 40, Part 58, *Ambient Air Quality Surveillance, Appendix D—Network Design for State and Local Air Monitoring Stations (SLAMS)*, pp. 158–172. US Government Printing Office, Washington, DC, July 1992.
2. Harrison, R. M., and Young, R. J. (eds.), *Handbook of Air Pollution Analysis*, 2nd ed. Chapman & Hall, London, 1986.
3. US Environmental Protection Agency, *Optimum Site Exposure Criteria for* SO_2 Monitoring, EPA 450/3-77-13. Office for Quality Planning and Standards, Research Triangle Park, NC, 1977.
4. Smith, D. G., and Egan, B. A., Design of monitoring networks to meet multiple criteria, in *Quality Assurance in Air Pollution Measurement* (Frederick, E. D., ed.), pp. 139–150. Air Pollution Control Association, Pittsburgh, 1979.
5. Houghland, E. S., Air quality monitoring network design by analytical techniques III, in *Quality Assurance in Air Pollution Measurement* (Frederick, E. D., ed.), pp. 181–187. Air Pollution Control Association, Pittsburgh, 1979.
6. *South Coast Air Quality Management District Air Quality Standards Compliance Report*, December 2002, and *Summary Statistics for 2002*, Vol. 15, No. 12.
7. Wilson Jr., W. E., *Atmos. Environ.* **12**, 537–547 (1978).
8. Alfodi, T. T., Satellite remote sensing for smoke plume definition, in *Proceedings of the 4th Joint Conference on Sensing of Environmental Pollutants*, pp. 258–261. American Chemical Society, Washington, DC, 1978.
9. Browell, E. V., Lidar remote sensing of tropospheric pollutants and trace gases, in *Proceedings of the 4th Joint Conference on Sensing of Environmental Pollutants*, pp. 395–402. American Chemical Society, Washington, DC, 1978.
10. Grant, W. B., *Appl. Opt.* **21**, 2390–2394 (1982).
11. Murray, E. A., and Van der Laan, J. E., *Appl. Opt.* **17**, 814–817 (1978).
12. Fishman, J., *Environ. Sci. Technol.* **25**, 613–621 (1991).
13. Code of Federal Regulations, Title 40, Part 58, Ambient Air Quality Surveillance, *Appendix A—Quality Assurance Requirements for State and Local Air Monitoring Stations (SLAMS)*, pp. 137–150. US Government Printing Office, Washington, DC, July 1992.
14. Zimmer, C. E., *Air Pollution*, 3rd ed., Vol. 111 (Stern, A. C., ed.), p. 476. Academic Press, New York, 1976.

SUGGESTED READING

Barrett, E. C., and Curtis, L. F., *Introduction to Environmental Remote Sensing*, 3rd ed. Chapman & Hall, London, 1992.

Beer, R., *Remote Sensing by Fourier Transform Spectroscopy*. Wiley, New York, 1992.

Cracknell, A. P., *Introduction to Remote Sensing*. Taylor & Francis, New York, 1991.

Keith, L. H., *Environmental Sampling and Analysis*. Lewis Publishers, Chelsea, MI, 1991.

QUESTIONS

1. List the two major functions of a quality assurance program and describe how they are interrelated.
2. List the advantages and disadvantages of remote sensing techniques by optical methods.

3. Determine which month and location have the greatest number of hours with ozone concentrations $\geqslant 0.01$ ppm, using Table 18.4.
4. Determine how many monitoring stations per million people are located in the four counties in southern California shown in Fig. 18.1.
5. What are the physical constraints in placing instrumentation in aircraft or motor vehicles?
6. What are appropriate uses of mobile platforms for monitoring?
7. Describe the chemical behavior of the b_{scat} and ozone concentration profiles of the St. Louis urban plume in Fig. 18.2. What is the reason for the sharp increase of b_{scat} and the sharp decrease of ozone in the vicinity of power plants?
8. List the reasons for establishing a stationary air monitoring network.
9. What factors can account for differences between ambient concentrations of a pollutant and actual, personal exposures to the same pollutant?

19

Air Pathways from Hazardous Waste Sites

I. INTRODUCTION

This chapter addresses the potential for hazardous air emissions from environmental remediation sites. These emissions can occur at hazardous spill locations, at undisturbed remediation sites, and during cleanup of remediation sites under the Comprehensive Environmental Response, Compensation, and Liability Act (CERCLA) or the Superfund Amendments and Reauthorization Act (SARA). Air emissions may pose a potential health risk at these sites.

The US Environmental Protection Agency (EPA) developed the Hazard Ranking System (HRS) [1] to determine priorities among releases, or threatened releases, from remediation sites. The HRS applies the appropriate consideration of each of the following site-specific characteristics of such facilities:

- The quantity, toxicity, and concentrations of hazardous constituents that are present in such waste and a comparison with other wastes.
- The extent of, and potential for, release of such hazardous constituents to the environment.
- The degree of risk to human health and the environment posed by such constituents.

II. MULTIMEDIA TRANSPORT

Air contaminant releases from hazardous waste sites can occur from wastes placed aboveground or belowground. The following are categories of air contaminant releases:

- Fugitive dust resulting from:
  - Wind erosion of contaminated soils
  - Vehicle travel over contaminated roadways
- Volatilization release from:
  - Covered landfills (with and without gas generation)
  - Spills, leaks, and landforming
  - Lagoons

The EPA has detailed procedures for conducting air pathway analysis for Superfund applications [2]. Decision network charts are given for all expected situations.

Figures 19.1 and 19.2 present the decision networks that guide contaminant release screening analysis. Figure 19.1 deals with contaminants in or under the soil and Fig. 19.2 addresses aboveground wastes. Any release mechanisms evident at the site will require a further screening evaluation to determine the likely environmental fate of the contaminants involved.

III. CONTAMINANT FATE ANALYSIS

Simplified environmental fate estimation procedures are based on the predominant mechanisms of transport within each medium, and they generally disregard intermedia transfer or transformation processes. In general, they produce conservative estimates (i.e. reasonable upper bounds) for final ambient concentrations and the extent of hazardous substance migration. However, caution should be taken to avoid using inappropriate analytical methods that underestimate or overlook significant pathways that affect human health.

When more in-depth analysis of environmental fate is required, the analyst must select the modeling procedure that is most appropriate to the circumstances. In general, the more sophisticated models are more data, time, and resource intensive.

Figures 19.3 through 19.5 present the decision network for screening contaminant fate in air, surface water, ground water, and biota. Pathways must be further evaluated to determine the likelihood of population exposure.

A. Atmospheric Fate

The following numbered paragraphs refer to particular numbered boxes in Fig. 19.3:

1. The atmospheric fate of contaminants must be assessed whenever it is determined that significant gaseous or airborne particulate contaminants

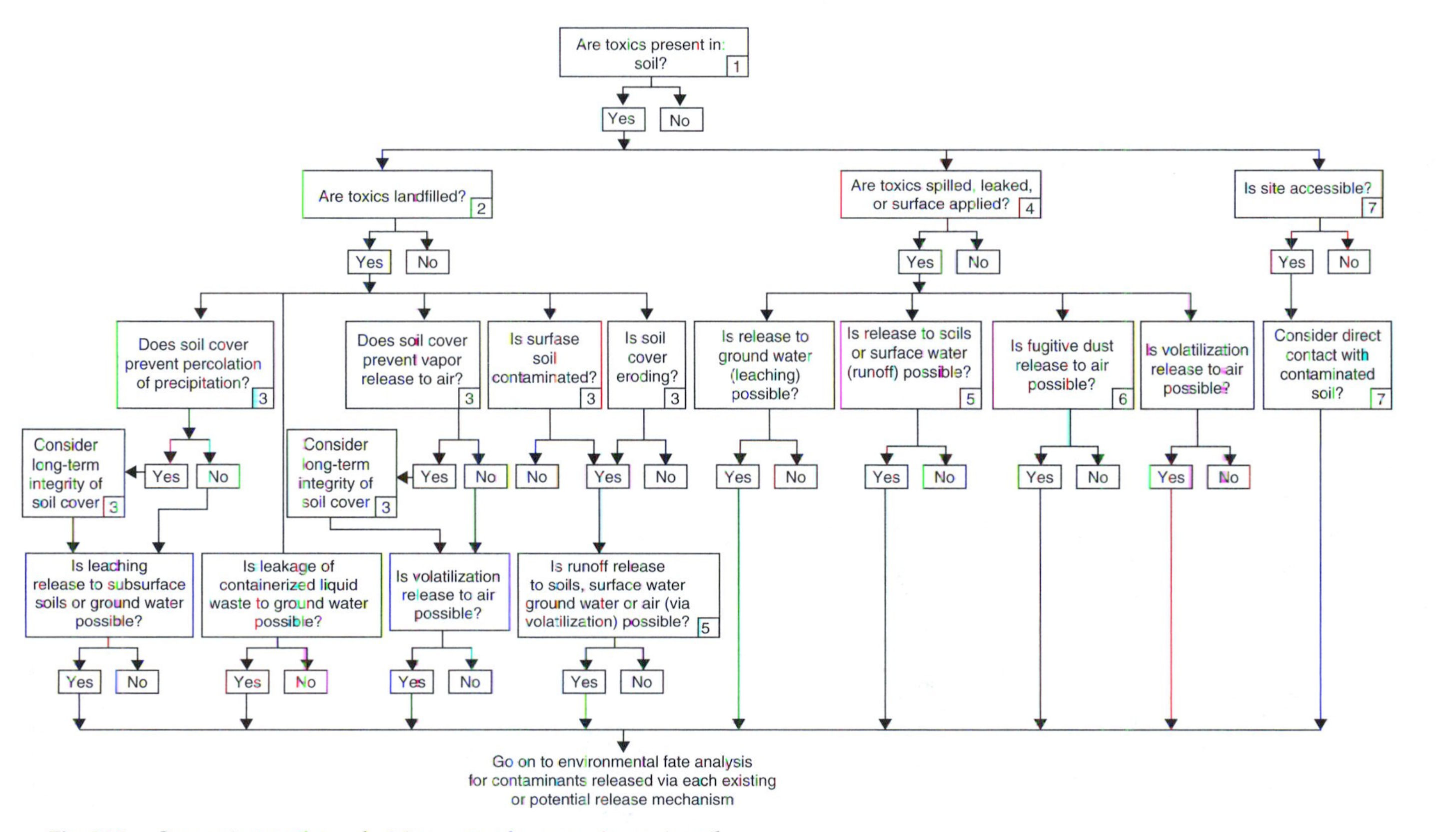

Fig. 19.1. Contaminant release decision network: contaminants in soil.

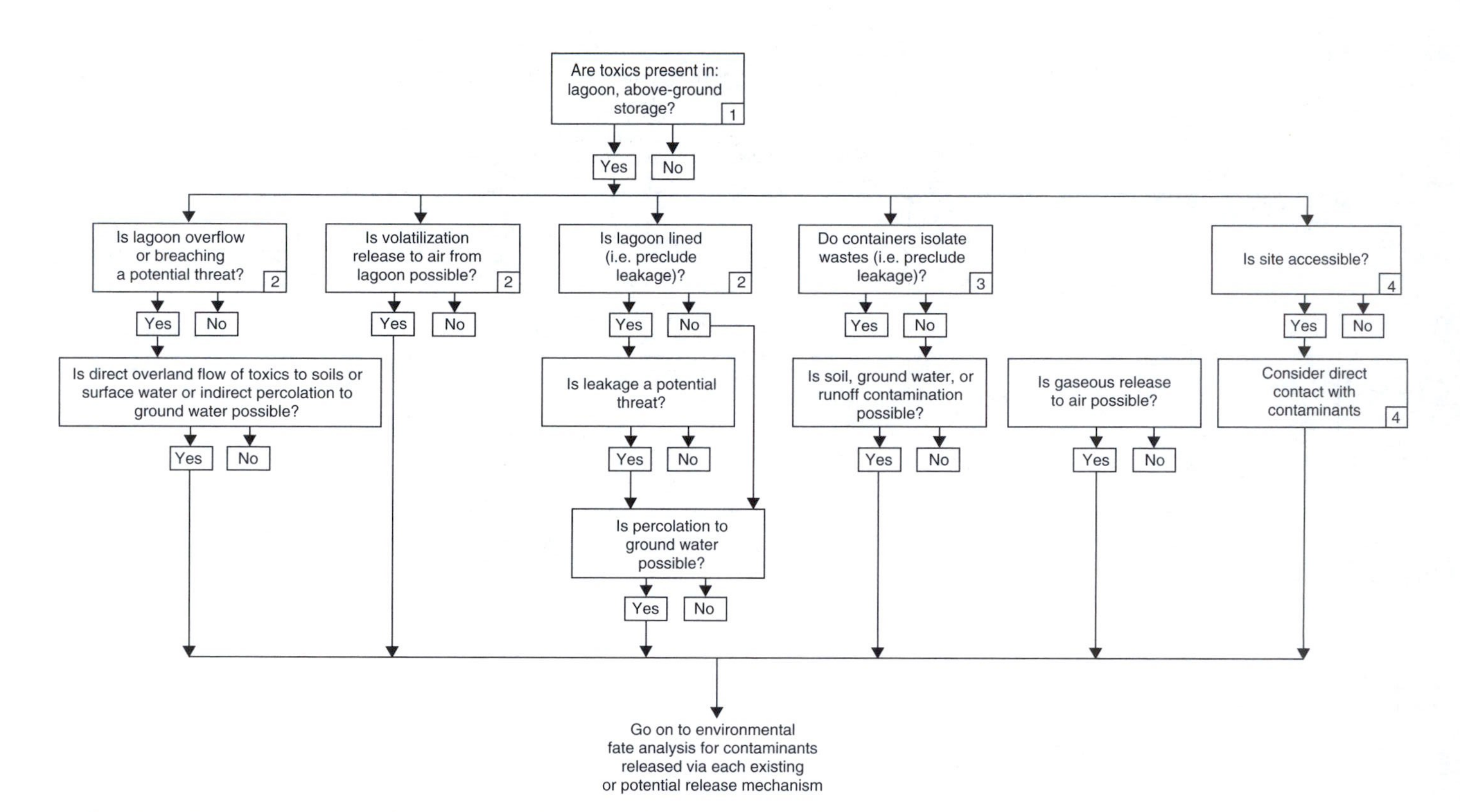

Fig. 19.2. Contaminant release decision network: contaminants above ground.

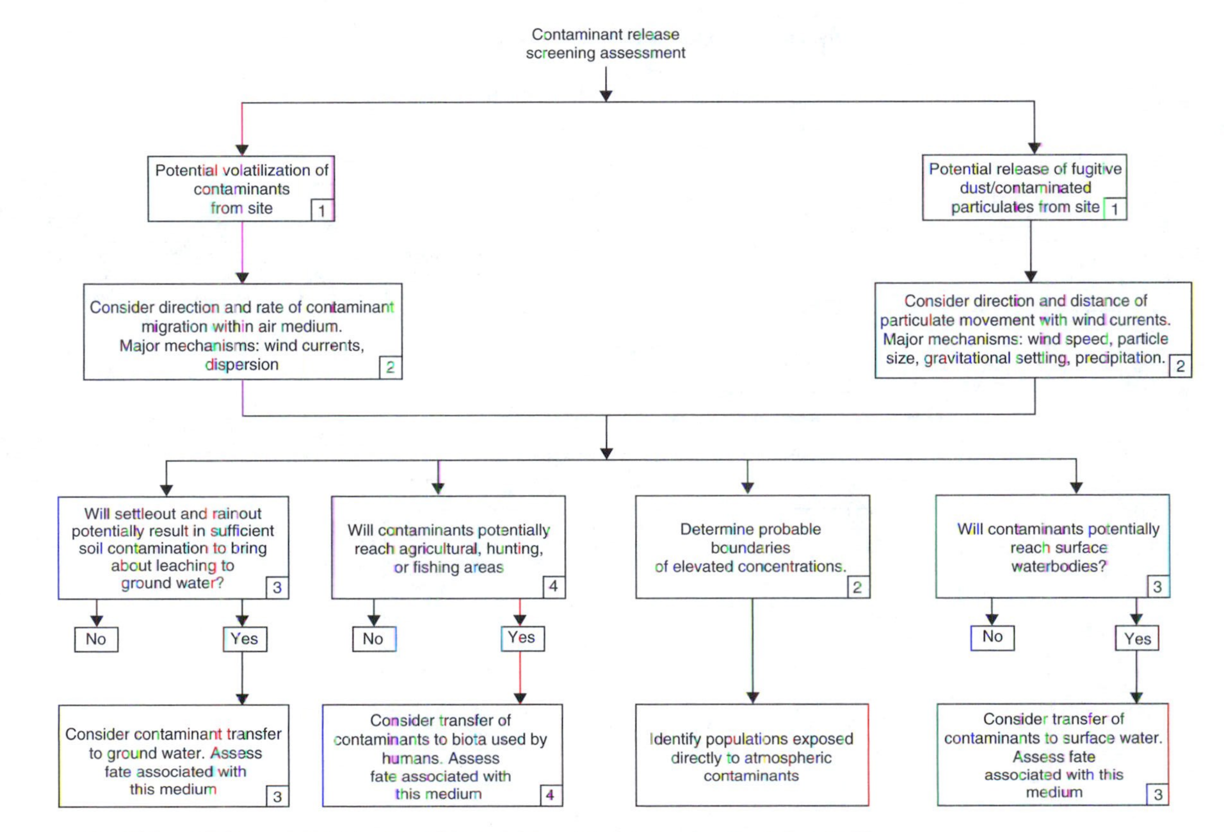

Fig. 19.3. Environmental fate screening assessment decision network: atmosphere.

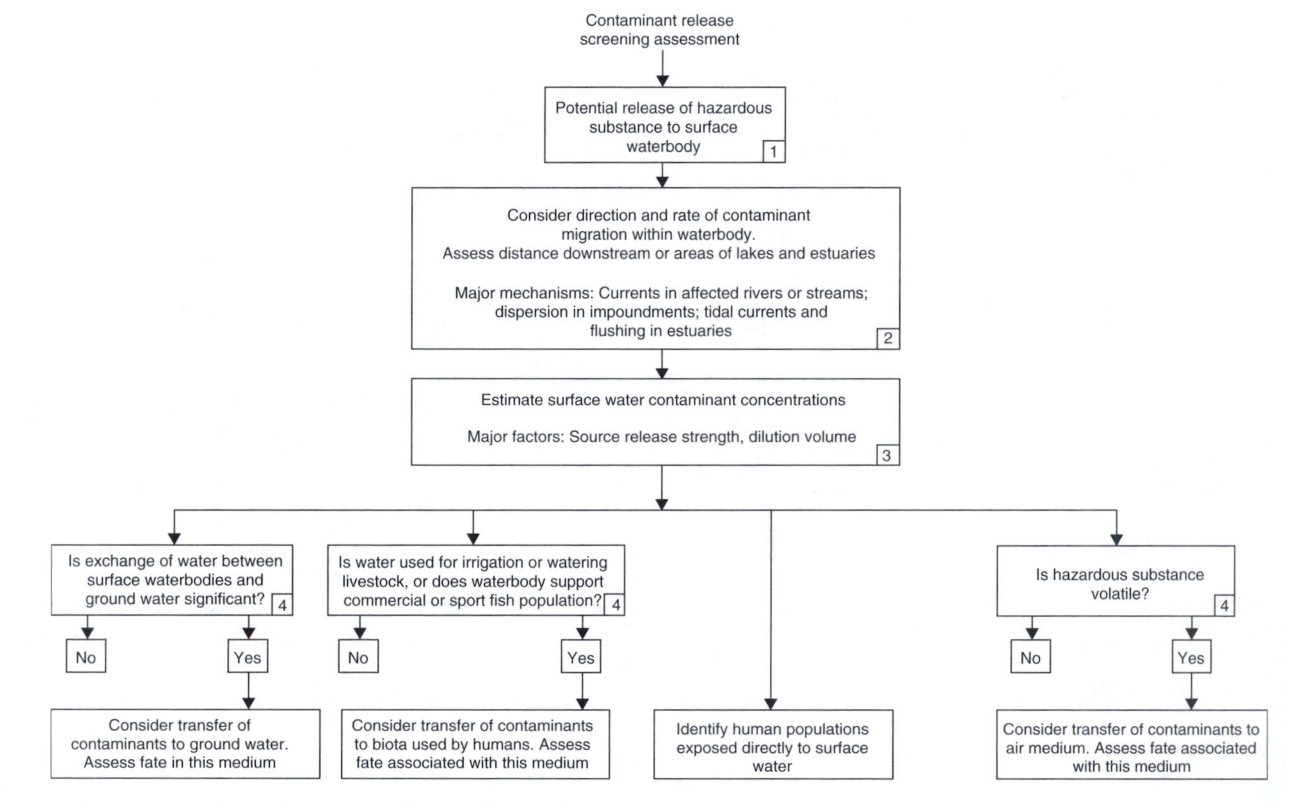

Fig. 19.4. Environmental fate screening assessment decision network: surface water.

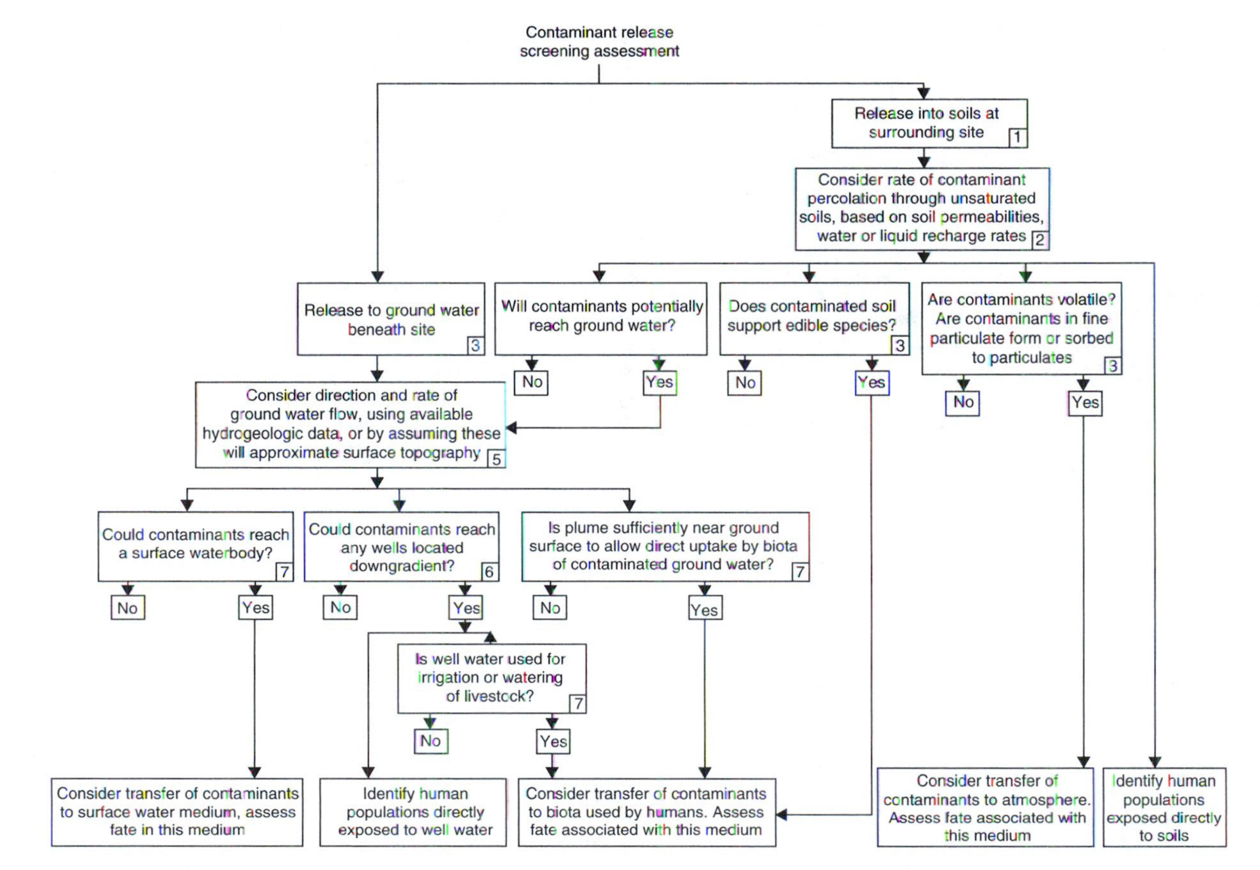

Fig. 19.5. Environmental fate screening assessment decision network: soils and ground water.

are released from the site. The atmospheric fate of contaminants released originally to other media, but eventually partitioned to the atmosphere beyond site boundaries, must also be assessed whenever this intermedia transfer is likely to be significant.

2. The predominant directions of contaminant movement will be determined by relative directional frequencies of wind over the site (as reflected in area-specific wind rose data). Atmospheric stability and wind speeds determine off-site areas affected by ambient concentrations of gaseous contaminants. Usually, high stability and low wind speed conditions result in higher atmospheric concentrations of gaseous contaminants close to the site. High stability and moderate wind speeds result in moderate concentrations over a larger downwind area. Low stability or high wind speed conditions cause greater dispersion and dilution of contaminants, resulting in lower concentrations over larger areas.

   For particulate contaminants (including those adsorbed to dust or soil particles), ambient concentrations in the atmosphere and areas affected by airborne contaminants are determined by wind speed and stability and also by particle size distribution. High winds result in greater dispersion and cause particles to remain airborne longer (which may also increase release rates). Low winds and high stability result in rapid settling of particles and in a more concentrated contaminant plume closer to the site. Larger particles settle rapidly, decreasing the atmospheric concentrations with distance from the site. Finer particles remain airborne longer, and their behavior more closely approximates that of gaseous contaminants, as described.

3. Settling and rainout are important mechanisms of contaminant transfer from the atmospheric media to both surface soils and surface waters. Rates of contaminant transfer caused by these mechanisms are difficult to assess qualitatively; however, they increase with increasing soil adsorption coefficients, solubility (for particulate contaminants or those adsorbed to particles), particle size, and precipitation frequency.

   Areas affected by significant atmospheric concentrations of contaminants exhibiting the foregoing physical and chemical properties should also be considered as potentially affected by contaminant rainout and settling to surface media. Contaminants dissolved in rainwater may percolate to ground water, run off or fall directly into surface waters, and adsorb to unsaturated soils. Contaminants settling to the surface through dry deposition may dissolve in or become suspended in surface waters or may be leached into unsaturated soils and ground water by subsequent rainfall. Dry deposition may also result in formation of a layer of relatively high contamination at the soil surface. When such intermedia transfers are likely, one should assess the fate of contaminants in the receiving media.

4. If areas identified as likely to receive significant atmospheric contaminant concentrations include areas supporting edible biota, the biouptake

of contaminants must be considered as a possible environmental fate pathway. Direct biouptake from the atmosphere is a potential fate mechanism for lipophilic contaminants. Biouptake from soil or water following transfer of contaminants to these media must also be considered as part of the screening assessments of these media.

B. Surface Water Fate

The following numbered paragraphs refer to particular numbered boxes in Fig. 19.4:

1. The aquatic fate of contaminants released from the CERCLA site as well as those transferred to surface water from other media beyond site boundaries must be considered.
2. Direction of contaminant movement is usually clear only for contaminants introduced into rivers and streams. Currents, thermal stratification or eddies, tidal pumping, and flushing in impoundments and estuaries render qualitative screening assessment of contaminant directional transport highly conjectural for these types of water bodies. In most cases, entire water bodies receiving contaminants must be considered potentially significant human exposure points. More in-depth analyses or survey data may subsequently identify contaminated and unaffected regions of these water bodies.
3. Similarly, contaminant concentrations in rivers or streams can be roughly assessed based on rate of contaminant introduction and dilution volumes. Estuary or impoundment concentration regimes are highly dependent on the transport mechanisms enumerated. Contaminants may be localized and remain concentrated or may disperse rapidly and become diluted to insignificant levels. The conservative approach is to conduct a more in-depth assessment and use model results or survey data as a basis for determining contaminant concentration levels.
4. Important intermedia transfer mechanisms that must be considered where significant surface water contamination is expected include transfers to ground water where hydrogeology of the area indicates significant surface water–ground water exchange, transfers to biota where waters contaminated with lipophilic substances support edible biotic species, and transfer to the atmosphere where surface water is contaminated by volatile substances. High temperatures, high surface area/volume ratios, high wind conditions, and turbulent stream flow also enhance volatilization rates.

   Contaminant transfer to bed sediments represents another significant transfer mechanism, especially in cases where contaminants are in the form of suspended solids or are dissolved hydrophobic substances that can become adsorbed by organic matter in bed sediments. For the purposes of this chapter, sediments and water are considered part of a single system because of their complex interassociation. Surface water–bed sediment transfer is reversible; bed sediments often act as temporary repositories for contaminants and gradually rerelease contaminants to surface waters. Sorbed or settled contaminants are frequently transported with bed sediment migration or flow. Transfer of sorbed contaminants to bottom-dwelling, edible biota represents a fate pathway potentially resulting in human exposure. Where this transfer mechanism appears likely, the biotic fate of contaminants should be assessed.

C. Soil and Ground Water Fate

The following numbered paragraphs refer to particular numbered boxes in Fig. 19.5:

1. The fate of contaminants in the soil medium is assessed whenever the contaminant release atmospheric, or fate screening, assessment results show that significant contamination of soils is likely.
2. The most significant contaminant movement in soils is a function of liquid movement. Dry, soluble contaminants dissolved in precipitation, run-on, or human applied water will migrate through percolation into the soil. Migration rates are a function of net water recharge rates and contaminant solubility.
3. Important intermedia transfer mechanisms affecting soil contaminants include volatilization or re-suspension to the atmosphere and biouptake by plants and soil organisms. These, in turn, introduce contaminants into the food chain.

IV. MODELING

An extremely difficult task is the estimation of emissions from hazardous waste sites. Frequently, both the amounts of materials existing within the site and the compounds and mixtures that are represented are not known. Even if both of these pieces of information are reasonably well known, the conditions of the containers holding these chemicals are not initially known.

Hazardous materials may enter the air pathway by evaporation from leaking containers and release of these gases through fissures and spaces between soil particles. Another pathway may release hazardous substances to the air if they are water soluble. Then ground water passing leaking containers may carry substances to or near the surface, where they may be released to the air near the original source or at locations at significant distances.

A. Estimates of Long-Term Impact

If the foregoing problems of emissions estimation can be overcome, or if it is possible to make estimates of maximum possible and minimum possible emissions, then it is quite easy to make estimates of resulting long-term impact on the surrounding area. The representation of the emissions may be through consideration of an area source or area sources; or if vent pipes are releasing material or flaring the gases, point sources should be used.

A single finite line source method is used to simulate area sources in the long-term (seasons to years) model ISCLT (Industrial Source Complex Long Term model) [3]. Although this method has been criticized as frequently underestimating concentrations for receptors that are quite close to the area source (within two or three side lengths away), this model is usually used for these estimates. In addition to the long-term estimate of emission rate for each constituent to be modeled, the ISCLT model requires meteorological data in the form of a joint frequency distribution of three parameters: wind direction (in 16 classes), wind speed (in 6 classes), and Pasquill stability class (in 6 classes). As long as the emissions can be considered relatively constant over the period of simulation, the long-term estimates will represent mean concentrations over the period represented by the meteorological data.

B. Estimates of Short-Term Impact during Remediation

If it is necessary to consider short-term (hours or days) impact, the point, area, and line (PAL) air quality model [4] will do a superior simulation of the area sources and a similar simulation of any point sources as done by the ISCLT model. PAL also includes a version with deposition and settling algorithms (PAL-DS).

In addition to short-term emission estimates, normally for hourly periods, the meteorological data include hourly wind direction, wind speed, and Pasquill stability class. Although of secondary importance, the hourly data also include temperature (only important if buoyant plume rise needs to be calculated from any sources) and mixing height.

The short-term model can then be used to estimate resulting concentrations during specific periods or to estimate concentrations for suspected adverse meteorological conditions, so that changes can be incorporated in the remediation process if concentrations are expected to be higher than desirable.

V. ASSESSMENT OF A HAZARDOUS WASTE SITE

The contaminant cleanup process is shown in Fig. 19.6. The first step of a contaminant cleanup is a preliminary assessment (PA). During the PA of a site, readily available information about a site and its surrounding area are collected to "distinguish between sites that pose little or no threat to human health and the environment and sites that may pose a threat and require further investigation."[1] Any possible emergency response actions may also be

[1] US EPA website (May 2003), http://www.epa.gov/superfund/whatissf/sfproces/pasi.htm.

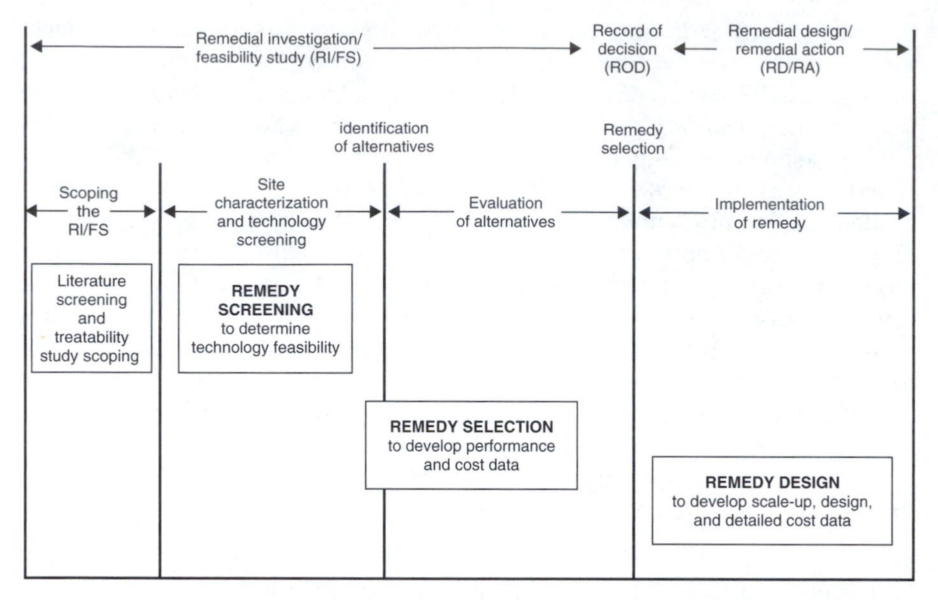

Fig. 19.6 Steps in a contaminated site cleanup, as mandated by Superfund. *Source*: US Environmental Protection Agency, 1992, Guide for Conducting Treatability Studies under CERCLA: Thermal Desorption, EPA/540/R-92/074 B.

identified. A site inspection or site investigation (SI) is performed if the preliminary assessment (PA), based on limited data, calls for one (that is why this step is often referred to as the PA/SI).[2]

In the US, certain hazardous waste sites are considered to be of sufficient concern to be "listed" on the National Priority List (NPL). The listing is actually a combination of the hazard (usually toxicity) of the contaminants found at the site and the likelihood that people or ecosystems will be exposed to these pollutants. Severely polluted sites and sites that contain very toxic compounds in measurable quantities are ranked higher than those with less toxic substances in lesser quantities.

A public disclosure of the condition following contaminant treatment must be made. This final record of decision (ROD) indicates that the specific engineering remedy has been selected for the site. Like any other aspects of hazardous waste cleanup, this decision is subject to later contests (legal, scientific, or otherwise). Since public officials are not exempt from personal tort liabilities in their decisions, the ROD is usually made as a collective, agency decision based on past and ongoing contaminant measurements, and includes provisions for monitoring for years to come to ensure that the engineered systems continue to perform according to plan. The ROD must also

[2] See the EPA publication Guidance for Performing Preliminary Assessments Under CERCLA, September 1991, PB92-963303, EPA 9345.0-01A) and the electronic scoring program "PA-Score" for additional information on how to conduct a PA.

ensure that a plan for operating and maintaining all systems is in place, including a plan for dealing with failures and other unexpected contingencies, such as improvements in measurement techniques that later identify previously undetected pollutants.

Scientists conduct exposure assessments to evaluate the kind and magnitude of exposure to contaminants. Such assessments are usually site specific for clearly identified contaminants of concern. For example, they may be conducted for an abandoned hazardous waste site or a planned industrial facility. For the former site, the list of contaminants of concern would be based on sampling and analysis of the various environmental compartments, while the latter would be based on the types of chemicals to be used or generated in the construction and operation of the industrial facility. Thus, the assessment considers sources of contaminants, pathways through which contaminants are moving or will be moving, and routes of exposure where the contaminants find their way to receptors (usually people, but also receptors in ecosystems, such as fish and wildlife). Table 19.1 includes some of the most important considerations in deciding on the quality of information needed to conduct an exposure assessment.

The necessary information to quantify is determined by both the characteristics of the contaminant and the route of exposure.

A. The Hazard Quotient

The hazard quotient (HQ) is the ratio of the potential exposure to a specific contaminant to the concentration at which no adverse effects are expected (known as the reference dose or RfD). The HQ is the ratio of a single contaminant exposure, over a specified time period, to a reference dose for that contaminant, derived from a similar exposure period:

$$HQ = \frac{\text{Exposure}}{\text{RfD}} \tag{19.1}$$

If the calculated HQ is less than 1, no adverse health effects are expected to occur at these contaminant concentrations. If the calculated HQ is greater than 1, there is a likelihood that adverse action can occur at these concentrations. For example, the chromic acid (Cr^{6+}) mists dermal chronic RfD of $6.00 \times 10^{-3}\,\text{mg}\,\text{kg}^{-1}\,\text{day}^{-1}$. If the actual dermal exposure of people living near a plant is calculated (e.g. by intake or lifetime average daily dose) to be $4.00 \times 10^{-3}\,\text{mg}\,\text{kg}^{-1}\,\text{day}^{-1}$, the HQ is 2/3 or 0.67. Since this is less than 1, one would not expect people chronically exposed at this level to show adverse effects from skin contact. However, at this same chronic exposure, i.e. $4.00 \times 10^{-3}\,\text{mg}\,\text{kg}^{-1}\,\text{day}^{-1}$, to hexavalent chromic acid mists via oral route, the RfD is $3.00 \times 10^{-3}\,\text{mg}\,\text{kg}^{-1}\,\text{day}^{-1}$, meaning the HQ = 4/3 or 1.3. The value is greater than 1, so we cannot rule out adverse non-cancer effects.

The calculated HQ value cannot be translated into a probability that adverse health effects will occur (i.e. it is not actually a metric of risk). The

TABLE 19.1

Questions to Be Asked When Determining the Adequacy of Information Needed to Conduct Exposure Assessments

Compartment	Question
Soil	If humans have access to contaminated soils, can ranges of contamination be provided on the basis of land use (i.e. restricted access, road/driveway/parking lot access, garden use, agriculture and feedlot use, residential use, playground and park use, etc.)?
	Have the soil depths been specified? Do soil data represent surface soil data (⩾3 in. in depth) or subsurface soil data (>3 in. in depth)? If soil depth is known, but does not meet surface or subsurface soil definitions, designate the data as *soil* and specify the depth (e.g. 0–6 in.). If the soil depth is unknown, the health assessor should designate the data as unspecified soil.
	Has soil been defined in the data? If not, the health assessor should assume soil includes any unconsolidated natural material or fill above bedrock that is not considered to be soil and excludes manmade materials such as slabs, pavements or driveways of asphalt, concrete, brick, rock, ash or gravel. A soil matrix may consist of pieces of each of these materials.
	Do soil data include uphill and downhill samples and upwind and downwind samples both on and off the site?
Sediment	Have the sediment samples been identified as grab samples or cores? Was the depth of the samples specified?
	Was the sampling program designed to collect sediment samples at regular intervals along a waterway or from depositional areas or both?
	Do the sediment data include results for upstream and downstream samples both on- and off-site?
	Has sediment been defined by the samplers? (To prevent confusion between sediment and soil, assume "sediment" is defined as any solid material, other than waste material or waste sludge that lies below a water surface, that has been naturally deposited in a waterway, water body, channel, ditch, wetland, or swale, or that lies on a bank, beach, or floodway land where solids are deposited.)
	Have any sediment removal activities (e.g. dredging, excavation, etc.) occurred that may have altered the degree of sediment contamination (leading to a false negative). This becomes important when the following occur:
	1. Sediment contamination in fishable waters is used to justify sampling and analyses of edible biota.
	2. Sediment data are used to justify additional downstream sampling, particularly at points of exposure and in areas not subject to past removal activities.
	3. The significance of past exposure is assessed.
Surface water	Do surface-water data include results for samples both upstream and downstream of the site?
	Was information obtained on the number of surface-water samples taken at each station, as well as the frequency, duration, and dates of sampling?
Groundwater	Were groundwater samples collected in the aquifer of concern?
	Did sampling occur both up-gradient and down-gradient of the site and the site's groundwater contamination plume?
All	Did the sampling design include selected hot spot locations and points of possible exposure?

Source: Agency for Toxic Substances and Disease Registry, 2003, ATSDR Public Health Assessment Guidance Manual.

HQ is merely a benchmark that can be used to estimate the likelihood of risk.[3] It is not even likely to be proportional to the risk. So, an HQ >1 does not necessarily mean that adverse effects will occur.

Non-cancer hazard estimates often have substantial uncertainties from a variety of sources. Scientific estimates of contaminant concentrations, exposures and risks, always incorporate assumptions to the application of available information and resources. Uncertainty analysis is the process used by scientists to characterize the just how good or bad the data are in making these estimates.

B. The Hazard Index

The HQ values are for individual contaminants. The hazard index (HI) is the sum of more than one HQ value to express the level of cumulative non-cancer hazard associated with inhalation of multiple pollutants (e.g. certain classes of compounds, such as solvents, pesticides, dioxins, fuels, etc.):

$$HI = \sum_{1}^{n} HQ \tag{19.2}$$

An HI can be developed for all pollutants measured, such as the 32 compounds measured in New Jersey as part of the National Air Toxics Assessment (Fig. 19.7). An HI can also be site specific. For example, if an environmental audit shows that only CCl_4 and Cr^{6+} were detected by sampling of soil. Recall that the previously calculated Cr^{6+} dermal HQ was 0.67. The dermal chronic RfD of CCl_4 is $4.55 \times 10^{-4}\,mg\,kg^{-1}\,day^{-1}$. If the exposure is $1.00 \times 10^{-4}\,mg\,kg^{-1}\,day^{-1}$, the HQ for chronic dermal exposure to tetrachloromethane is $1.00/4.55 = 0.22$.

Thus, the HI for this site is $0.67 + 0.22 = 0.89$. Since the HI is under 1, the non-cancer effect is not expected at these levels of exposure to the two compounds. However, if the chronic dermal exposure to CCl_4 had been $2.00 \times 10^{-4}\,mg\,kg^{-1}\,day^{-1}$, the HQ for CCl_4 would have been 0.44, and the HI would have been calculated as $0.67 + 0.44 = 1.11$. This is a benchmark that indicates that the cumulative exposures to the two contaminants may lead to non-cancer effects.

C. Comprehensive Risk Communication

The amount of data and information regarding contaminant concentrations, exposure and effects can be overwhelming when presented to the public and clientele. Thus, these data must be reduced into meaningful formats. A recent example of how the information discussed in this and the previous chapters can be presented is that of the Ohio Environmental Protection

[3] National Research Council, *Science and Judgment in Risk Assessment*. National Academy Press, Washington, DC, 1994.

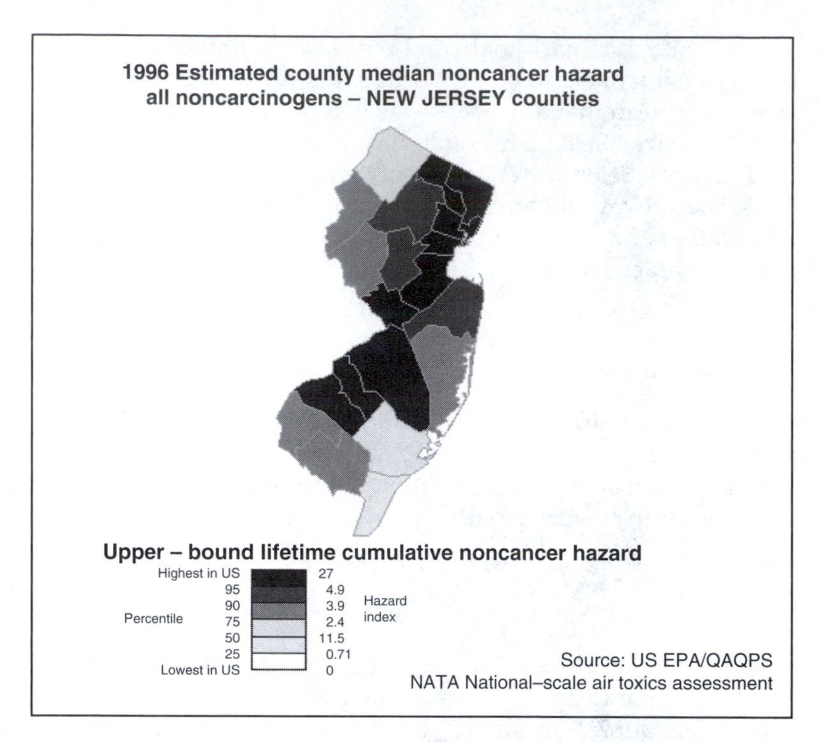

Fig. 19.7. Non-cancer hazard index for 32 air toxics included in the Clean Air Act, based on inhalation exposure data in New Jersey from political subdivisions. Estimates do not include indoor emissions and are based on exposure estimates for the median individual within each census tract, which EPA considers to be a "typical" exposure, meaning that individuals may have substantially higher or lower exposures based on their activities. *Source*: US Environmental Protection Agency, National Air Toxics Assessment.

Agency's Urban Air Toxic Monitoring Program,[4] that addresses potential risks in large urban areas with many industrial air pollution sources. Air quality samples were collected between 1989 and 1997 near a large industrial area in Cuyahoga County. The contaminant concentrations are typical of urban areas, the concentrations should be lower in the future. Pollution prevention activities by industry, vehicle emission tests by motorists and mandates in the Federal Clean Air Act will all help reduce toxics in the air. Samples were analyzed for volatile organic compounds (VOCs), heavy metals and polycyclic aromatic hydrocarbons (PAHs). The agency conducted a risk assessment based on both the cancer and non-cancer health risks, assuming that an individual is exposed constantly to the same concentration of the pollutant for a lifetime (i.e. exposure duration = total lifetime). The results of the cancer health risk assessment are provided in Table 19.2. Heavy metals contributed the majority of the cancer risk (about 66%).

[4] Ohio Environmental Protection Agency, *News Release*, August 17, 1999.

TABLE 19.2

Cumulative Cancer Risk Based on Air Sampling in Cuyahoga County, Ohio, 1989–1997

Source of cancer risk	Total estimated risk
VOCs	0.515×10^{-4}
Heavy metals	1.21×10^{-4}
PAHs	0.123×10^{-4}
Total carcinogenic risk	1.85×10^{-4}

Source: Ohio Environmental Protection Agency (1999).

Each category in the table shows the cumulative risks from exposure to all compounds detected under a specific contaminant class. The US EPA has defined acceptable exposure risks for individual compounds to range from 10^{-6} to 10^{-4}. Also, it is quite possible that one or a few contaminants are contributing the lion's share of risk to each contaminant class. For example, a particularly carcinogenic PAH, like benzo(a)pyrene or dibenz(a,h)anthracene (each with a inhalation cancer slope factor of 3.10), could account for most of the risk, even if its concentrations are about the same as other PAHs. In fact, this appears to be the case when looking at the individual chemical species listed in Table 19.3 that were used to derive the risks. Likewise, the VOC cancer risk was largely determined by the concentrations of benzene, while the heavy metals, although largely influenced by Cr^{6+}, were more evenly affected by arsenic and cadmium.

The cancer risk calculations are based on the unit risk estimate (URE), which is the upper-bound excess lifetime cancer risk that may result from continuous exposure to an agent at a defined concentration. For inhalation this concentration is $1\,\mu g\,m^{-3}$ in air. For example, if the URE $= 1.5 \times 10^{-6}$ per $\mu g\,m^{-3}$, then 1.5 excess tumors are expected to develop per million population being exposed daily for a lifetime to $1\,\mu g$ of the contaminant per cubic meter of air.

The cancer risk reported for each individual contaminant is below the level designated by federal health agencies as acceptable, and falls with the range of risks expected for large cities, with their numerous sources of toxic air contaminants (i.e. the so-called "urban soup").

The non-cancer hazard index calculations are provided in Table 19.4. Non-carcinogenic health effects include developmental, reproductive or cardiovascular health problems. Any total HI number below 100% is generally regarded as a safe level of exposure.

As was the case for cancer risk, a few compounds can drive the non-cancer hazard index. For example, the case above, 3-chloropropene and tetrachloromethane (account for an HI of 0.53, while all the other measured VOCs account for only 0.10. And, these two compounds account for almost 82% of the total non-carcinogenic risk estimates.

TABLE 19.3

Individual Chemical Species Used to Calculate Cancer Risks shown in Table 19.2

Compound	Carcinogenic unit risk $(\mu g\, m^{-3})^{-1}$	Source	Average concentration $(\mu g\, m^{-3})$	Carcinogenic risk
VOCs				
Methyl Chloride	1.8 E-06	HEAST	0.68	1.22 E-06
Dichloromethane	4.7 E-07	IRIS	2.06	9.70 E-07
Trichloromethane	2.3 E-05	IRIS	0.27	6.29 E-06
Benzene	8.3 E-06	IRIS	3.91	3.25 E-05
Carbon tetrachloride	1.5 E-05	IRIS	0.55	8.30 E-06
Trichloroethene	1.7 E-06	HEAST	0.55	9.42 E-07
Tetrachloroethene	9.5 E-07	HEAST	1.07	1.02 E-06
Styrene	5.7 E-07	HEAST	0.49	2.81 E-07
Sum				*5.15 E-05*
Heavy metals				
Arsenic	4.30 E-03	IRIS	0.00271	1.17 E-05
Cadmium	1.80 E-03	IRIS	0.00765	1.38 E-05
Chromium(total)[a]	1.20 E-02	IRIS	0.00800	9.60 E-05
Sum				*1.21 E-04*
PAHs				
Benzo(a)pyrene[b]	2.10 E-03	1	0.006	1.26 E-06
Benzo(a)anthracene	2.10 E-04	0.1	0.0048	1.01 E-06
Benzo(b)fluoranthene	2.10 E-04	0.1	0.0023	4.83 E-07
Benzo(k)fluoranthene	2.10 E-04	0.1	0.0007	1.47 E-07
Chrysene	2.10 E-05	0.01	0.0047	9.87 E-08
Dibenz(a,h)anthracene	2.10 E-03	1	0.0041	8.61 E-06
Indeno[1,2,3-cd]pyrene	2.10 E-03	0.1	0.0031	6.51 E-07
Sum				*1.23 E-05*
Total carcinogenic risk				*1.85 E-04*

[a] Estimation based on the slope factor of chromium (VI).
[b] Estimation based on the slope factor of oral route.
Source: Ohio Environmental Protection Agency, 1999, Cleveland Air Toxics Study Report.

TABLE 19.4

Cumulative Hazard Index Based on Air Sampling in Cuyahoga County, Ohio, 1989–1997

Source of non-carcinogenic risk	Hazard index (HI)
VOCs	0.63
Heavy metals	0.008
PAHs	0.012
Total non-carcinogenic risk	0.65

Source: Ohio Environmental Protection Agency (1999).

REFERENCES

1. Federal Register, Part II, Environmental Protection Agency, 40 CFR Part 300, *Hazard Ranking System: Final Rule*, Vol. 55, No. 241, December 14, 1990.
2. *Procedures for Conducting Air Pathway Analysis for Superfund Applications, Vol. I, Application of Air Pathway Analyses for Superfund Activities*, EPA-450/1-89-001, July 1989.
3. US EPA, *User's Guide for the Industrial Source Complex (ISC2) Dispersion Models*. EPA-450/4-92-008a (Vol. I—User Instructions), EPA-450/4-92-008b (Vol. II—Description of Model Algorithms), EPA-450/4-92-008c (Vol. III—Guide for Programmers). US Environmental Protection Agency, Research Triangle Park, NC, 1992.
4. Petersen, W. B., and Rumsey, E. D., *User's Guide for PAL 2.0—A Gaussian Plume Algorithm for Point, Area, and Line Sources*, EPA/600/8-87/009. US Environmental Protection Agency, Research Triangle Park, NC, 1987 (NTIS Accession No. PB87-168 787).

SUGGESTED READING

Air/Superfund National Technical Guidance Study Series, Vol. II: *Estimation of Baseline Air Emission at Superfund Sites*, EPA-450/1-89-002a, August 1990. Vol. III: *Estimation of Air Emissions from Clean-Up Activities at Superfund Sites*, EPA-450/1-89-003, January 1989. Vol. IV: *Procedures for Conducting Air Pathway Analyses for Superfund Applications*, EPA-450/1-89-004, July 1989.

An Act to Amend the Clean Air Act, Public Law No. 101–549, US Congress, November 15, 1990.

Summerhays, B. E., Procedures for estimating emissions from the cleanup of Superfund sites. *J. Air Waste Manage. Assoc.* **40**(1), 17–23, 1990.

Superfund Exposure Assessment Manual, US Environmental Protection Agency, EPA/540/1-88/001, OSWER Directive 9285.5-1, April 1988.

QUESTIONS

1. How would the release of a volatile gas from contaminated soil be affected by the soil temperature?
2. The EPA Hazardous Ranking System computes a numerical score for hazardous waste. If the score exceeds a predetermined value, the waste site is placed on the NPL for Superfund cleanup. Discuss the pros and cons of such a ranking system.
3. Describe a possible situation in which an air contaminant is controlled but the control system used transfers the contaminant problem to another medium, such as water or soil.
4. Explain the differences between a hazard quotient, hazard index, and cumulative cancer risk.

Part V

Air Pollution Modeling

20

The Meteorological Bases of Atmospheric Pollution

Air pollutants reach receptors by being transported and perhaps transformed in the atmosphere (Fig. 20.1). The location of receptors relative to sources and atmospheric influences affect pollutant concentrations, and the sensitivity of receptors to these concentrations determines the effects. The location, height, and duration of release, as well as the amount of pollutant released, are also of importance. Some of the influences of the atmosphere on the behavior of pollutants, primarily the large-scale effects, are discussed here, as well as several effects of pollutants on the atmosphere.

I. VENTILATION

If air movement past a continuous pollutant source is slow, pollutant concentrations in the plume moving downwind will be much higher than they would be if the air were moving rapidly past the source. If polluted air continues to have pollution added to it, the concentration will increase. Generally, a source emits into different volumes of air over time. However, there can be a buildup of concentration over time even with significant air motion if there are many sources.

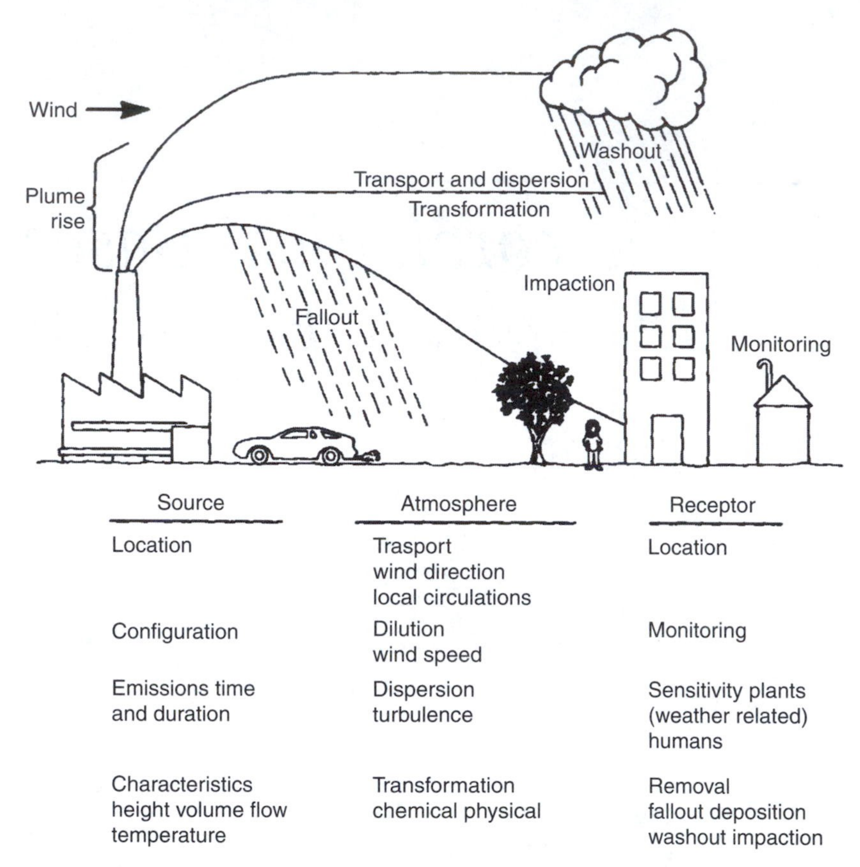

Source	Atmosphere	Receptor
Location	Trasport wind direction local circulations	Location
Configuration	Dilution wind speed	Monitoring
Emissions time and duration	Dispersion turbulence	Sensitivity plants (weather related) humans
Characteristics height volume flow temperature	Transformation chemical physical	Removal fallout deposition washout impaction

Fig. 20.1. The atmosphere's role in air pollution.

Low- and high-pressure systems have considerably different ventilation characteristics. Air generally moves toward the center of a low (Fig. 20.2) in the lower atmosphere, due in part to the frictional turning of the wind toward low pressure. This convergence causes upward vertical motion near the center of the low. Although the winds very near the center of the low are generally light, those away from the center are moderate, resulting in increased ventilation rates. Note the increased wind in the area to the west of the low in Fig. 20.2. Low-pressure systems generally cover relatively small areas (although the low-pressure system shown in Fig. 20.2 covers an extensive area) and are quite transient seldom remaining the same at a given area for a significant period of time. Lows are frequently accompanied by cloudy skies, which may cause precipitation. The cloudy skies minimize the variation in atmospheric stability from day to night. Primarily because of moderate horizontal wind speeds and upward vertical motion, ventilation (i.e. total air volume moving past a location) in the vicinity of low-pressure systems is quite good.

High-pressure systems are characteristically the opposite of lows. Since the winds flow outward from the high-pressure center, subsiding air from higher in the atmosphere compensates for the horizontal transport of mass.

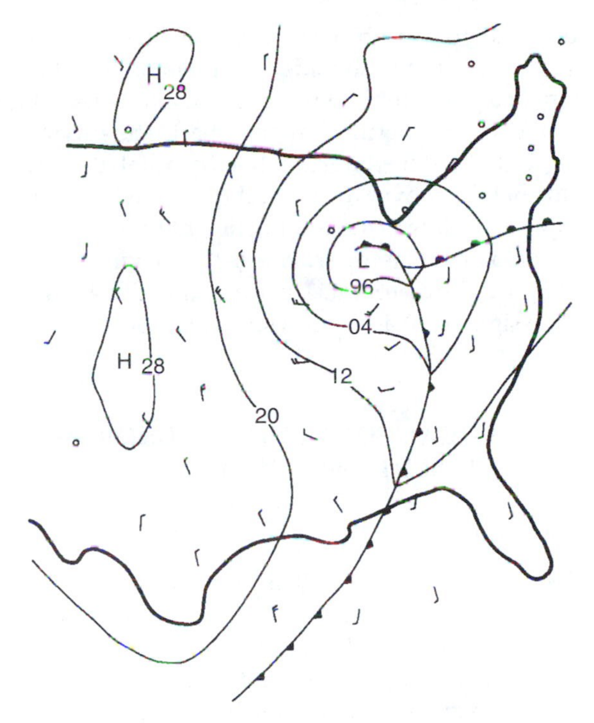

Fig. 20.2. Surface chart for 06Z Friday, November 20, 1981. Contours are isobars of atmospheric pressure; 12 is 1012 mb. Line with triangles, cold front; line with semicircles, warm front; line with both triangles and semicircles, an occluded front (a cold front that has caught up with a warm front). Wind direction is with the arrow; wind speed is 10 knots for 1 barb, 5 knots for one-half barb. Small station circles indicate calm. H, center of high pressure; L, center of low pressure.

This sinking air causes a subsidence inversion. Partially because of the subsiding vertical motion, the skies are usually clear, allowing maximum radiation—incoming during the day and outgoing at night—causing extremes of stability; there is instability during the day and stability at night, with frequent radiation inversions. Highs generally occupy large areas, and although they are transient, they are usually slow-moving. Winds over large areas are generally light; note the winds to the south of the high in the lower left corner of Fig. 20.2. Thus, the ventilation in the vicinity of high-pressure systems is generally much less than that of lows.

II. STAGNATION

At times the ventilation rate becomes very low. Such a lack of air motion usually occurs in the weak pressure gradient near the center of an anticyclone (i.e. of a high). If the high has a warm core, there is likely to be very little air movement near the center, i.e., stagnation. Under such circumstances, winds are very light. Skies are usually cloudless, contributing to the formation of surface-based radiation inversions at night. Although the clear skies contribute to instability in the daytime, the depth of the unstable layer (i.e. mixing height) may be severely limited due to the subsidence inversion over the high.

The mixing height at a given time may be estimated by use of the morning radiosonde ascent plotted on a thermodynamic chart. The surface temperature at the given time is plotted on the diagram. If a dry adiabat is drawn through this temperature, the height aboveground at the point where this dry adiabat intersects the morning sounding is the mixing height for that time. The mixing height for the time of maximum temperature is the maximum mixing height. Use of this sounding procedure provides an approximation because it assumes that there has been no significant advection since the time of the sounding.

III. METEOROLOGICAL CONDITIONS DURING HISTORIC POLLUTION EPISODES

A. Meuse Valley, Belgium

During the period December 1–5, 1930, an intense fog occupied the heavily industrialized Meuse Valley between Liege and Huy (about 24 km) in eastern Belgium [1]. Several hundred persons had respiratory attacks primarily beginning on the 4th and 63 persons died on the 4th and 5th after a few hours of sickness. On December 6 the fog dissipated; the respiratory difficulties improved and, in general, rapidly ceased.

The fog began on December 1 under anticyclonic conditions. What little air motion occurred was from the east, causing air to drift upvalley, moving smoke from the city of Liege and the large factories southwest of it into the narrow valley. The valley sides extend to about 100 m, and the width of the valley is about 1 km. A temperature inversion extended from the ground to a height of about 90 m, transforming the valley essentially into a tunnel deeper than the height of the stacks in the valley, which were generally around 60 m. Much of the particulate matter was in the 2–6 μm range. The fog was cooled by radiation from the top and warmed by contact with the ground. This caused a gentle convection in the "tunnel," mixing the pollutants uniformly and resulting in nearly uniform temperature with height.

The symptoms of the first patients began on the afternoon of December 3 and seemed to occur simultaneously along the entire valley. Deaths took place only on December 4 and 5, with the majority at the Liege end of the valley. Those affected were primarily elderly persons who had lung or heart problems. However, some previously healthy persons were among the seriously ill. There were no measurements of pollutants during the episode, but the five Liege University professors who participated in the subsequent inquiry indicated that part of the sulfur dioxide was probably oxidized to sulfuric acid.

Roholm [2], in discussing the episode, noted that 15 of the 27 factories in the area were capable of releasing gaseous fluorine compounds and suggested that the release of these compounds was of significance.

During the 30 years prior to the episode, fogs lasting for more than 3 days had occurred only five times, always in the winter, in 1901, 1911, 1917, 1919, and 1930. Some respiratory problems were also noted in 1911. Industrial activity was at a low level in 1917 and 1919.

It is prophetic that Firket [1], in speaking about public anxiety about potential catastrophes, said, "This apprehension was quite justified, when we think that proportionately, the public services of London, for example, might be faced with the responsibility of 3200 sudden deaths if such phenomenon occurred there". Indeed in 1952, such a catastrophe occurred (see Section III, C).

B. Donora, Pennsylvania

A severe episode of atmospheric pollution occurred in Donora, Pennsylvania, during the period October 25–31, 1948 [3]. Twenty persons died, 17 of them within 14 h on October 30.

During this period, a polar high-pressure area remained nearly stationary, with its center in the vicinity of northeastern Pennsylvania. This caused the regional winds, both at the ground and through the lowest layers, to be extremely light. Donora is southeast of Pittsburgh and is in the Monongahela River valley. Cold air accumulated in the bottom of the river valley and fog formed, which persisted past midday for 4 consecutive days. The top of the fog layer has a high albedo and reflects solar radiation, so that only part of the incoming radiation is available to heat the fog layer and eliminate it (Fig. 20.3(a)). During the night, longwave radiation leaves the top of the fog layer, further cooling and stabilizing the layer (Fig. 20.3(b)). Wind speeds at Donora were less than $3.1\,\mathrm{m\,s^{-1}}$ (7 miles $\mathrm{h^{-1}}$) from the surface up to 1524 m (5000 ft) for 3 consecutive days, so that pollutants emitted into the air within the valley were not transported far from their point of emission. Maximum temperatures at Donora at an elevation of 232 m (760 ft) mean sea level were considerably lower than those at the Pittsburgh airport, elevation 381 m (1250 ft), indicating the extreme vertical stability of the atmosphere. In the

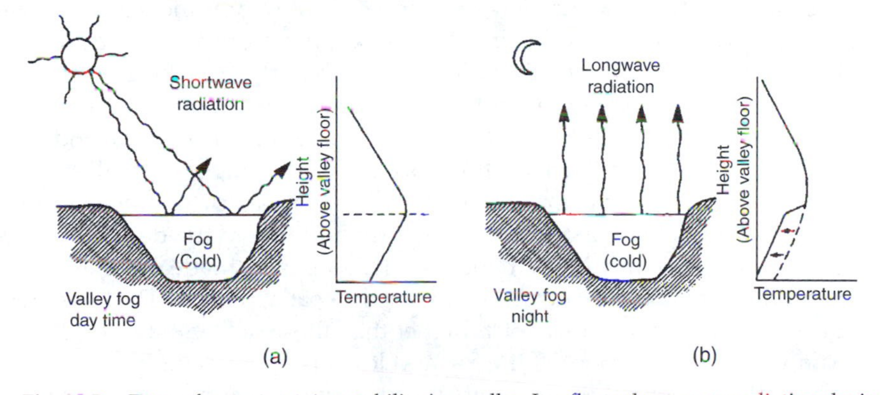

Fig. 20.3. Dense fog maintaining stability in a valley. It reflects short-wave radiation during the day and radiates heat from the top of the fog at night. *Source*: Adapted from Schrenk *et al.* [3].

vicinity of Donora there were sources of sulfur dioxide, particulate matter, and carbon monoxide. Previous recorded periods of stagnation had occurred in Donora in October 5–13, 1923 and October 7–18, 1938.

C. London, England

A dense 4-day fog occurred in London and its surroundings during December 5–9, 1952 [4, 5]. The fog began as the area came under the influence of an anticyclone approaching from the northwest early on December 5. This system became stationary, so that there was almost no wind until milder weather spread into the area from the west on December 9. Temperatures remained near freezing during the fog. The visibility was unusually restricted, with a 4-day average of less than 20 m over an area approximately 20 by 40 km and of less than 400 m over an area 100 by 60 km. The density of the fog was enhanced by the many small particles in the air available for condensation of fog droplets. The result was a very large number of very small fog droplets, more opaque and persistent than fog formed in cleaner air. The depth of the fog layer was somewhat variable, but was generally 100 m or less.

Measurements of particulate matter less than approximately 20 μm in diameter and of sulfur dioxide were made at 12 sites in the greater London area. The measurements were made by pumping air through a filter paper and then through a hydrogen peroxide solution. The smoke deposit on the filter was analyzed by reflectometer; the sulfur dioxide was determined by titrating the hydrogen peroxide with standard alkali, eliminating interference by carbon dioxide, Using the sampling procedure, sulfur dioxide existing as a gas and dissolved in fine fog droplets was measured. Any sulfur dioxide associated with larger fog droplets or adsorbed on particles collected on the filter would not be measured.

Smoke concentrations ranged from 0.3 to more than 4 mg m^{-3}. Daily means of the sampling stations are shown in Fig. 20.4. Sulfur dioxide measurements ranged from less than 0.1 ppm (260 μg m^{-3}) to 1.34 ppm (3484 μg m^{-3}). Also, 4 of the 11 stations had at least one daily value in excess of 1 ppm, and 9 of the 11 stations had at least one daily value in excess of 0.5 ppm. Compare these levels to the US primary standard for sulfur dioxide, which is a maximum 24-h concentration of 365 μg m^{-3} (0.14 ppm), not to be exceeded more than once per year. Daily means are shown in Fig. 20.4. The smoke and SO_2 means rose and later decreased in parallel. Daily concentrations of smoke averaged over all stations rose to about five times normal and sulfur dioxide to about six times normal, peaking on December 7 and 8, respectively. In addition to the daily measurements at 12 sites, monthly measurements at 117 sites were made using lead peroxide candles. This allowed determination of the spatial pattern. The December 1952 concentrations were about 50% higher than those of December 1951.

From the commencement of the fog and low visibility, many people experienced difficulty breathing, the effects occurring more or less simultaneously over a large area of hundreds of square kilometers. The rise in the number of deaths (Fig. 20.4) paralleled the mean daily smoke and sulfur

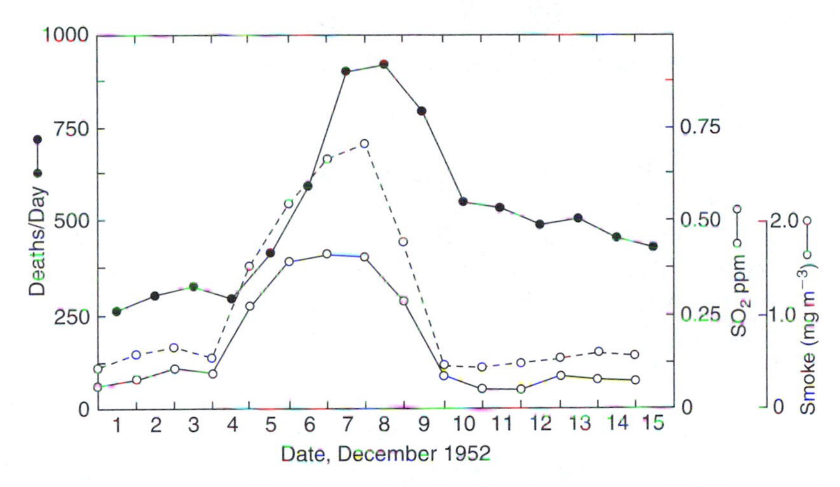

Fig. 20.4. Daily air pollution (SO_2 and smoke) and deaths during the 1952 London episode. *Source*: Adapted from Wilkins [4].

dioxide concentrations; daily deaths reached a peak on December 8 and 9, with many of them related to respiratory troubles. Although the deaths decreased when the concentrations decreased, the deaths per day remained considerably above the pre-episode level for some days. Would most of the persons who died have died soon afterward anyway? If this were the case, a below-normal death rate would have occurred following the episode. This situation did not seem to exist, but detailed analysis was complicated by increased deaths in January and February 1953 which were attributed primarily to an influenza outbreak.

Those who analyzed these excess deaths (the number of deaths above the normal number for each calendar day) believed that the level of sulfur dioxide was not near the toxic limit of 10 ppm necessary to affect healthy persons. They attributed the deaths to the synergistic effect of fine particles and sulfur dioxide combined. They believed that considerable sulfuric acid mist was formed from the oxidation of sulfur dioxide, but since no measurements were made, its amount was speculative.

D. Similarities of the Three Episodes

In the Meuse Valley, Donora, and London episodes, the areas were influenced by high pressure with nearly nonexistent surface air motion. Surface inversions caused the condensation of fog, which, once formed, persisted throughout the day, even during mid-afternoon. In each case the fog layer was relatively shallow, extending only about 100 m. The persistence of the fog past the third day and the lack of any air transport out of the region, as well as the existence of considerable emissions of pollutants, seem to separate these episodes from more common meteorological occurrences. Both the Meuse Valley and Donora had topography constraining the volume in which the pollutants were confined. This constraint apparently resulted from the lack of any transport wind in the London 1952 episode. Measurements of pollutant concentrations were made only in London.

E. Other Episodes

A number of somewhat less severe episodes are discussed in Goldsmith and Friberg [6]. Mention of the more important ones follows.

An air pollution episode responsible for approximately 300 excess deaths occurred in London between November 26 and December 1, 1948. Concentrations of smoke and sulfur dioxide were 50–70% of the values during the 1952 episode.

An accident complicated by fog, weak winds, and a surface inversion occurred in Poza Rica, Mexico, in the early morning of November 24, 1950, when hydrogen sulfide was released from a plant for the recovery of sulfur from natural gas. There were 22 deaths, and 320 persons were hospitalized.

In November and December 1962, a number of air pollution episodes occurred in the Northern Hemisphere. In London a fog occurred during the period December 3–7, with sulfur dioxide as high as during the 1952 episode, but with particulate concentrations considerably lower due to the partial implementation of the 1956 British Clean Air Act. Excess deaths numbered 340. High pollution levels were measured in the eastern United States between November 27 and December 5, 1962, notably in Washington, DC, Philadelphia, New York, and Cincinnati. Between December 2 and 7 elevated pollution levels were found in Rotterdam; Hamburg, Frankfurt, and the Ruhr area; Paris; and Prague. Pollution levels were high in Osaka between December 7 and 10, and mortality studies, which were under way, indicated 60 excess deaths.

F. Air Pollution Emergencies

Government authorities increasingly are facing emergencies that may require lifesaving decisions to be made rapidly by those on the scene. Of increasing frequency are transportation accidents involving the movement of volatile hazardous materials. A railroad derailment accident of a tank car of liquefied chlorine on February 26, 1978, at Youngtown, Florida, in which seven people died, and an accident in Houston, Texas, involving a truck carrying anhydrous ammonia on May 11, 1976, which also claimed seven lives, are examples. Two potentially dangerous situations involved barges with tanks of chlorine: one which sank in the lower Mississippi River and another which came adrift and came to rest on the Ohio River dam at Louisville, Kentucky; neither resulted in release of material.

Releases of radioactive materials from nuclear power plants have occurred, as at Three-Mile Island, Pennsylvania. In such situations, releases may be sufficient to require evacuation of residents.

Bhopal, India—On December 2, 1984 the contents of a methyl isocyanate (MIC) storage tank at the Union Carbide India plant in Bhopal became hot. Pressure in the tank became high. Nearly everything that could go wrong did. The refrigerator unit for the tank, which would have slowed the reactions, was turned off. After midnight, when the release valve blew, the vent gas scrubber that was to neutralize the gas with caustic soda failed to work. The flare tower, which would have burned the gas to harmless by-products, was down for repairs. As a result many tons of MIC were released from the tank. The gas spread as a fog-like cloud over a large, highly populated area to the south and east of the plant [7]. The number of fatalities was in excess of 2000 with thousands of others injured. Although little is available in the way of meteorological measurements, it is assumed that winds were quite light and that the atmosphere at this time of day was relatively stable.

Chernobyl, USSR—On April 26, 1986, shortly after midnight local time, a serious accident occurred at a nuclear power plant in Chernobyl in the Ukraine. It is estimated that 4% of the core inventory was released between April 26 and May 6. Quantities of Cs-137 (cesium) and I-131 (iodine) were released and transported, resulting in contamination, primarily by wet deposition of cesium, in Finland, northern Sweden and Norway, the Alps, and the northern parts of Greece. Because of temperatures of several thousand K during the explosion like release, the resulting pollutant cloud is assumed to have reached heights of 2000 m or more. The estimated southeast winds at plume level initially moved the plume toward Finland, northern Sweden, and northern Norway. As winds at plume level gradually turned more easterly and finally north and northwesterly, contaminated air affected the region of the Alps and northern Greece. A number of investigators, including Hass *et al.* [8], modeled the long-range transport including wet and dry removal processes. These attempts were considered quite successful, as radioactivity measurements provided some confirmation of the regions affected. Elevated levels of radioactivity were measured throughout the Northern Hemisphere. Because of the half-life of about 30 years for Cs-137, the contamination will endure.

World Trade Center, New York—On September 11, 2001, terrorists intentionally crashed fully fueled Boeing 767 jets into the twin towers of the heavily populated World Trade Center. The burning fuel, building materials and building contents, as well as the fibers and particles released during the collapse were the source of dangerous gas and particle phase pollutants released throughout the city. In addition to the immediate threat, the fire smoldered for months following the attack, which was a source of a variety of pollutants (see Fig. 10.19).

In such emergencies, it is most important to know the local wind direction at the accident site, so that the area that should be immediately evacuated can be determined. The next important factor is the wind speed, so that the travel time to various areas can be determined, again primarily for evacuation purposes. Both of these can be estimated on-site by simple means such as watching the drift of bubbles released by a bubble machine. It would be well to keep in mind that wind speeds are higher above ground and that wind direction is usually different. In fact, the World Trade Center episode dramatically demonstrated the importance of local meteorology. The wind directions at Ground Zero were almost always different than that measured at any of the three nearby, major airports. And, the three airports often have meteorological conditions different from one another.

As evacuation is taking place, it is important to determine whether meteorological events will cause a wind direction shift later on, requiring a change in the evacuation scenario. Particularly in coastal areas, or areas of significant terrain, authorities should be alert to a possible change in wind direction in going from night to day or vice versa. Useful advice may be obtained from the nearest weather forecaster, although accurate forecasting of wind direction for specific locations is not easy. Accurate air movement measurements and predictions are a matter of safety, even life and death. Portable meteorological stations have become accurate, precise and reasonably available. These should be deployed during any emergency involving airborne contaminants, or the potential for such.

If the situation is one of potential rather than current release, specific concentrations at various distances and localities may be estimated for various conditions.

IV. EFFECTS OF POLLUTION ON THE ATMOSPHERE

Pollutant effects on the atmosphere include increased particulate matter, which decreases visibility and inhibits incoming solar radiation, and increased gaseous pollutant concentrations, which absorb longwave radiation and increase surface temperatures. For a detailed discussion of visibility effects (see Chapter 14).

A. Turbidity

The attenuation of solar radiation has been studied by McCormick and his associates [9, 10] utilizing the Voltz sun photometer, which uses measurements at a wavelength of 0.5 μm. The ratio of ground-level solar intensity at 0.5 μm to extraterrestrial solar intensity can be as high as 0.5 in clean atmospheres but can drop to 0.2–0.3 in polluted areas, indicating that ground-level solar intensity can be decreased as much as 50% by pollution in the air. By making measurements using aircraft at various heights, the vertical extent of the polluted air can be determined. The turbidity coefficient can also be derived from the measurements and used to estimate the aerosol loading of the atmosphere. By assuming a particle size distribution in the size range 0.1–10 μm and a particle density, the total number of particles can be estimated. The mass loading per cubic meter can also be estimated. Because of the reasonable cost and simplicity of the sun photometer, it is useful for making comparative measurements around the world.

B. Precipitation

Depending on its concentration, pollution can have opposite effects on the precipitation process. Addition of a few particles that act as ice nuclei can cause ice particles to grow at the expense of supercooled water droplets, resulting in particles large enough to fall as precipitation. An example of this is commercial cloud seeding, with silver iodide particles released from aircraft to induce rain. If too many such particles are added, none of them will grow sufficiently to cause precipitation. Therefore, the effects of pollution on precipitation are not at all straightforward.

There have been some indications, although controversial, of increased precipitation downwind of major metropolitan areas. Urban addition of nuclei and moisture and urban enhancement of vertical motion due to increased roughness and the urban heat island effect have been suggested as possible causes.

C. Fogs

As mentioned in the previous section, the increased number of nuclei in polluted urban atmospheres can cause dense persistent fogs due to the many small droplets formed. Fog formation is very dependent on humidity and, in some situations, humidity is increased by release of moisture from industrial processes. Low atmospheric moisture content can also occur, especially in urban areas; two causes are lack of vegetation and rapid runoff of rainwater through storm sewers. Also, slightly higher temperatures in urban areas lower the relative humidity.

D. Solar Radiation

In the early part of this century, the loss of ultraviolet light in some metropolitan areas due to heavy coal smoke was of concern because of the resulting decrease in the production of natural vitamin D which causes the disease rickets. Recently, measurements in Los Angeles smog have revealed much greater decreases in ultraviolet than visible light. This is due to both absorption by ozone of wavelengths less than 0.32 μm and absorption by nitrogen dioxide in the 0.36–0.4 μm range. Heavy smog has decreased ultraviolet radiation by as much as 90%.

V. REMOVAL MECHANISMS

Except for fine particulate matter (0.2 μm or less), which may remain airborne for long periods of time, and gases such as carbon monoxide, which do not react readily, most airborne pollutants are eventually removed from the atmosphere by sedimentation, reaction, or dry or wet deposition.

A. Sedimentation (Settling by Gravity)

Particles less than about 20 μm are treated as dispersing as gases, and effects due to their fall velocity are generally ignored. Particles greater than about 20 μm have appreciable settling velocities. The fall velocity of smooth spheres as a function of particle size has been plotted (Fig. 20.5) by Hanna *et al.*

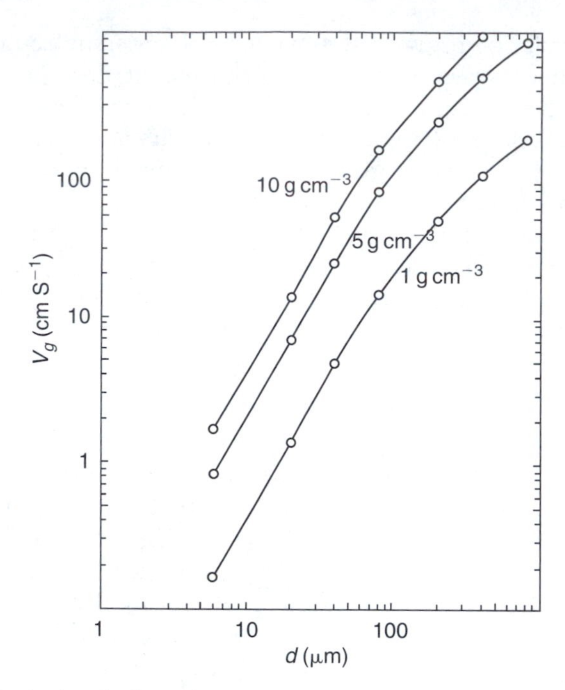

Fig. 20.5. Fall velocity of spherical particles as a function of particle diameter and density. *Source*: Adapted from Hanna *et al.* [11].

[11]. Particles in the range 20–100 μm are assumed to disperse approximately as gases, but with their centroid moving downward in the atmosphere according to the fall velocity. This can be accounted for by subtracting $v_g t$ from the effective height of release, where v_g is the gravitational fall velocity of the particles and t, in seconds, is x/u, where x is downwind distance from the source in m and u is wind speed. This is called the *tilted plume model*. The model may be modified to decrease the strength of the source with distance from the source to account for the particles removed by deposition.

For 20–100 μm particles, the deposition w on the ground is

$$w = \nu_g \chi(x,y,z) \tag{20.1}$$

where the air concentration χ is evaluated for a height above ground z of about 1 m.

Particles larger than 100 μm fall through the atmosphere so rapidly that turbulence has less chance to act upon and disperse them. The trajectories of such particles are treated by a ballistic approach.

B. Reaction (Transformation)

Transformations due to chemical reactions throughout the plume are frequently treated as exponential losses with time. The concentration $\chi(t)$ at travel time t when pollutant loss is considered compared to the concentration χ at the same position with no loss is

$$\frac{\chi(t)}{\chi} = \exp -(0.693t/L) \tag{20.2}$$

where L is the half-life of the pollutant in seconds. The half-life is the time required to lose 50% of the pollutant.

C. Dry Deposition

Although it does not physically explain the nature of the removal process, deposition velocity has been used to account for removal due to impaction with vegetation near the surface or for chemical reactions with the surface. McMahon and Denison [12] gave many deposition velocities in their review paper. Examples (in $\text{cm}\,\text{s}^{-1}$) are sulfur dioxide, 0.5–1.2; ozone, 0.1–2.0; iodine, 0.7–2.8; and carbon dioxide, negligible.

D. Wet Deposition

Scavenging of particles or gases may take place in clouds (rainout) by cloud droplets or below clouds (washout) by precipitation. A scavenging ratio or washout ratio W can be defined as

$$W = \frac{k\rho}{\chi} \tag{20.3}$$

where k is concentration of the contaminant in precipitation in $\mu\text{g}\,\text{g}^{-1}$; ρ is the density of air, approximately $1200\,\text{g}\,\text{m}^{-3}$; and χ is the concentration, $\mu\text{g}\,\text{m}^{-3}$, of

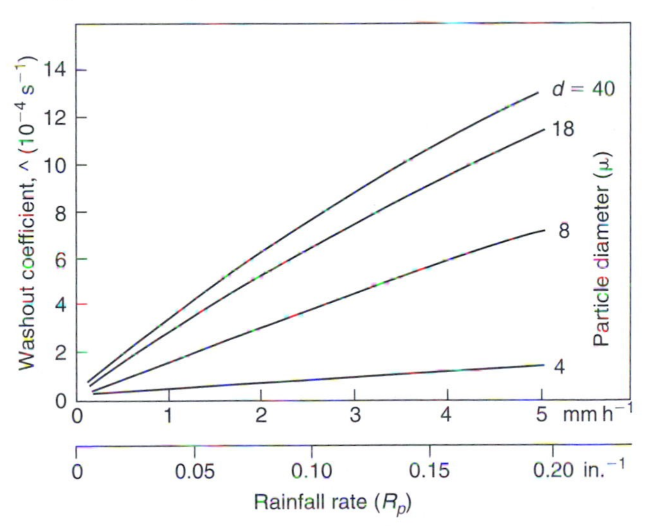

Fig. 20.6. Typical values of the washout coefficient as a function of rainfall rate and particle diameter. *Source*: After Engelmann [13].

the pollutant in the air prior to scavenging. McMahon and Denison [12] gave a table of field observations of washout ratios. The values for various pollutants range from less than 100 to more than 4000. These values are a function of particle size and rainfall intensity, generally decreasing with the latter and increasing with the former.

Scavenging may also be considered as an exponential decay process:

$$\chi(t) = \chi(0)e^{-\Lambda t} \tag{20.4}$$

where $\chi(t)$ is the concentration in $\mu g\, m^{-3}$ at time t in seconds, $\chi(0)$ is the concentration at time 0, and Λ is the scavenging or washout coefficient, s^{-1}. Figure 20.6, after Engelmann [13], gives the washout coefficient as a function of particle diameter and rainfall rate. McMahon and Denison [12] give a table of field measurements of scavenging coefficients. This same concept can be applied to gaseous pollutants. Fewer data are available for gases. Values ranging from 0.4×10^{-5} to 6×10^{-5} for SO_2 are given by McMahon and Denison [12] and compare reasonably well with an equation for SO_2 by Chamberlain [14]:

$$\Lambda = 10 \times 10^{-5} J^{0.53} \tag{20.5}$$

where J is rainfall intensity in $mm\, h^{-1}$.

REFERENCES

1. Firket, J., *Trans. Faraday Soc.* **32**, 1192–1197 (1936).
2. Roholm, K., *J. Ind. Hyg.* **19**, 126–137 (1937).
3. Schrenk, H. H., Heimann, H., Clayton, G. D., Gafafer, W. M., and Wexler, H., *Air Pollution in Donora, Pa, Public Health Bulletin 306*. US Public Health Service, Washington, DC, 1949.
4. Wilkins, E. T., *J. R. Sanit. Inst.* **74**, 1–15 (1954).
5. Wilkins, E. T., *Q. J. R. Meteorol. Soc.* **80**, 267–271 (1954).
6. Goldsmith, J. R., and Friberg, L. T., Effects of air pollution on human health, in *Air Pollution*, 3rd ed., Vol. II, *The Effects of Air Pollution* (Stern, A. C., ed.), pp. 457–610. Academic Press, New York, 1976.
7. Heylin, M., *Chem. Eng. News* (February 11, 1985).
8. Hass, H., Memmesheimer, M., Jakobs, H. J., Laube, M., and Ebel, A., *Atmos. Environ.* **24** (A), 673–692 (1990).
9. McCormick, R. A., and Baulch, D. M., *J. Air Pollut. Cont. Assoc.* **12**, 492–496 (1962).
10. McCormick, R. A., and Kurfis, K. R., *Q. J. R. Meteorol. Soc.* **92**, 392–396 (1966).
11. Hanna, S. R., Briggs, G. A., and Hosker Jr., R. P., *Handbook on Atmospheric Diffusion*, DOE/TIC-11223. Technical Information Center, US Department of Energy, Oak Ridge, TN, 1982.
12. McMahon, T. A., and Denison, P. J., *Atmos. Environ.* **13**, 571–585 (1979).
13. Engelmann, R. J., The calculation of precipitation scavenging, in *Meteorology and Atomic Energy—1968* (Slade, D., ed.), TID-24190, pp. 208–220.US Atomic Energy Commission, Oak Ridge, TN, 1968.
14. Chamberlain, A. C., *Aspects of Travel and Deposition ofAerosol and Vapour Clouds*, Atomic Energy Research Establishment HP/R-1261. Her Majesty's Stationery Office, London, 1953.

SUGGESTED READING

ApSimon, H. M., and Wilson, J. J. N., Modeling atmospheric dispersal of the Chernobyl release across Europe. *Boundary-Layer Meteorol.* **41**, 123–133 (1987).

Godish, T., *Air Quality*, 2nd ed. Lewis Publishers, Boca Raton, FL, 1991.

Hanna, S. R., and Drivas, P. J., *Guidelines for Use of Vapor Cloud Dispersion Models*. American Institute of Chemical Engineers, New York, 1987.

Knap, A. H., (ed.), *The Long-Range Atmospheric Transport of Natural and Contaminant Substances*. Kluwer Academic Press, Hingham, MA, 1989.

Kramer, M. L., and Porch, W. M., *Meteorological Aspects of Emergency Response*. American Meteorological Society, Boston, MA, 1990.

Puttock, J. S. (ed.), *Stably Stratified Flow and Dense Gas Dispersion*. Oxford University Press, New York, 1988.

Sandroni, S. (ed.), *Regional and Long-Range Transport of Air Pollution*. Elsevier Science Publishers, New York, 1987.

Scorer, R. S., *Air Pollution*. Pergamon, Oxford, 1968.

Seinfeld, J. H., *Air Pollution—Physical and Chemical Fundamentals*. McGraw-Hill, New York, 1975.

QUESTIONS

1. Characterize the conditions typical of low-pressure systems, particularly as they relate to ventilation.
2. Characterize the conditions typical of high-pressure systems, particularly as they relate to ventilation.
3. What atmospheric characteristics are usually associated with stagnating high-pressure systems?
4. What factors contribute to a high mixing height?
5. Discuss the simularities of the three major episodes of pollution (Meuse Valley, Donora, and London).
6. A railroad tank car has derailed and overturned, and some material is leaking out and apparently evaporating. The car is labeled "Toxic." In order to take appropriate emergency action, which meteorological factors would you consider and how would you assess them?
7. In addition to air pollutants, what meteorological factor has a profound effect on decreasing visibility, and what is the approximate threshold of its influence?
8. What pollution factors may affect precipitation?
9. What is the approximate lowering of the centroid of a dispersing cloud of particles at 2 km from the source whose mass medium diameter is 30 μm and whose particle density is $1\,g\,cm^{-3}$ in a $5\,m\,s^{-1}$ wind?
10. Prior to the onset of rain at the rate of $2.5\,mm\,h^{-1}$, the average concentration of 10-μm particles in a pollutant plume is $80\,\mu g\,m^{-3}$. What is the average concentration after 30 min of rain at this rate?

21

Transport and Dispersion of Air Pollutants

I. WIND VELOCITY

A. Wind Direction

The initial direction of transport of pollutants from their source is determined by the wind direction at the source. Air pollutant concentrations from point sources are probably more sensitive to wind direction than any other parameter. If the wind is blowing directly toward a receptor (a location receiving transported pollutants), a shift in direction of as little as 5° (the approximate accuracy of a wind direction measurement) causes concentrations at the receptor to drop about 10% under unstable conditions, about 50% under neutral conditions, and about 90% under stable conditions. The direction of plume transport is very important in source impact assessment where there are sensitive receptors or two or more sources and in trying to assess the performance of a model through comparison of measured air quality with model estimates.

There is normally considerable wind direction shear (change of direction) with height, especially near the ground. Although surface friction causes the wind to shift clockwise (veer) with height near the ground, the horizontal thermal structure of the atmosphere may exert a dominating influence at higher altitudes, such that the wind will shift counterclockwise (back) with additional height. Cold air advection in an air layer will cause the wind to back with height through that layer. Warm air advection will cause veering with height.

B. Wind Speed

Wind speed generally increases with height. A number of expressions describe the variation of wind speed in the surface boundary layer. A power law profile has frequently been used in air pollution work:

$$u(z) = u(z_a)(z/z_a)^p \tag{21.1}$$

where $u(z)$ is the wind speed at height z, $u(z_a)$ the wind speed at the anemometer measurement height z_a, and p an exponent varying from about 0.1 to 0.4. Figure 21.1 gives the measured wind speed variation with height for specific instances for five locations. The result of using the power law profile (Eq. 21.1) is also shown (open circles and dashed lines) using a value of p of 1/7. It should be noted that the power law wind profiles do not necessarily represent the data well. The exponent actually varies with atmospheric stability, surface roughness, and depth of the layer [1].

One of the effects of wind speed is to dilute continuously released pollutants at the point of emission. Whether a source is at the surface or elevated, this dilution takes place in the direction of plume transport. Figure 21.2 shows this effect of wind speed for an elevated source with an emission of

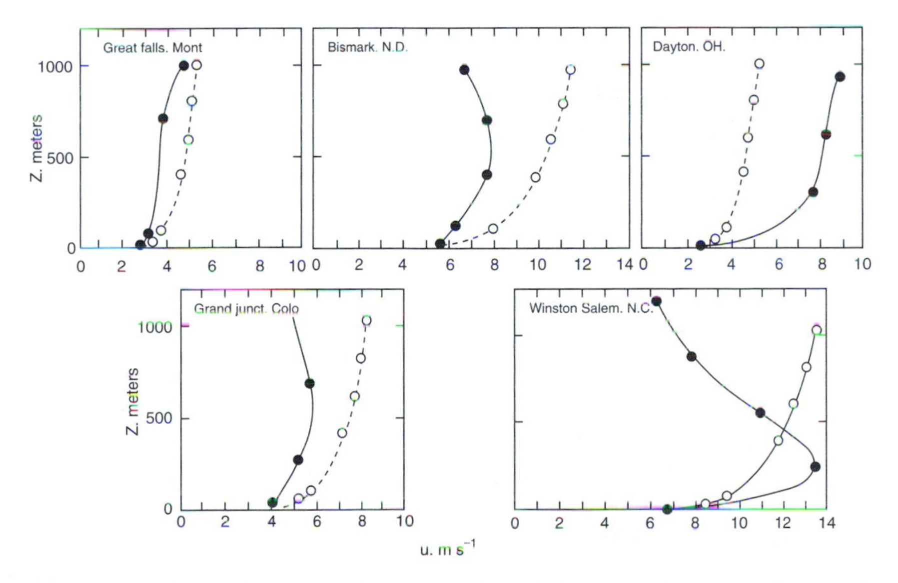

Fig. 21.1. Wind variation with height—measured (solid lines) and one-seventh power law (dashed lines).

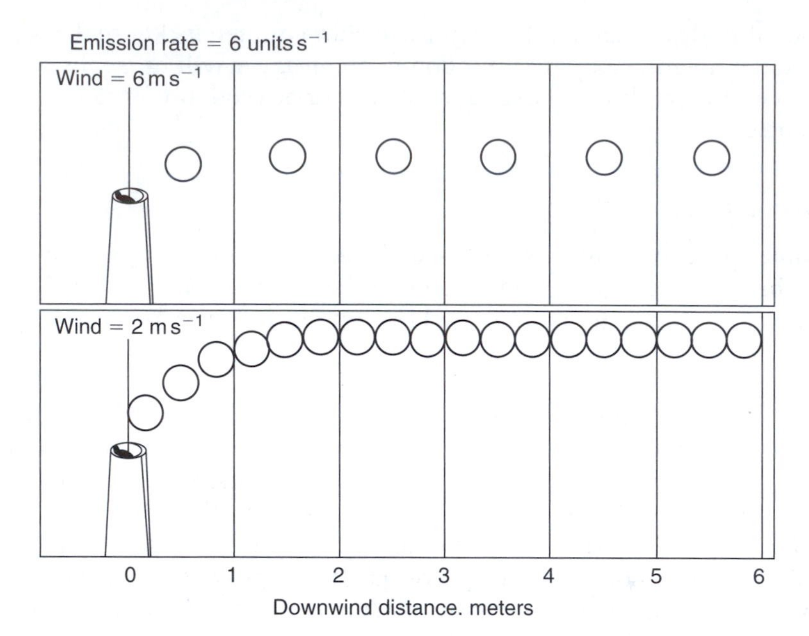

Fig. 21.2. Dilution by wind speed.

6 mass units per second. For a wind speed of $6\,\mathrm{m\ s^{-1}}$, there is 1 unit between the vertical parallel planes 1 m apart. When the wind is slowed to $2\,\mathrm{m\ s^{-1}}$, there are 3 units between those same vertical parallel planes 1 m apart. Note that this dilution by the wind takes place at the point of emission. Because of this, wind speeds used in estimating plume dispersion are generally estimated at stack top.

Wind speed also affects the travel time from source to receptor; halving of the wind speed will double the travel time. For buoyant sources, plume rise is affected by wind speed; the stronger the wind, the lower the plume. Specific equations for estimating plume rise are presented in Chapter 22.

II. TURBULENCE

Turbulence is highly irregular motion of the wind. The atmosphere does not flow smoothly but has seemingly random, rapidly varying erratic motions. This uneven flow superimposed on the mean flow has swirls or eddies in a wide range of sizes. The energy cascades through the eddy sizes, which are described by L. F. Richardson in verse:

> Big whirls have little whirls that feed on their velocity
> And little whirls have lesser whirls and so on to viscosity.

There are basically two different causes of turbulent eddies. Eddies set in motion by air moving past objects are the result of mechanical turbulence. Parcels of superheated air rising from the heated earth's surface, and the slower descent of a larger portion of the atmosphere surrounding these more rapidly rising parcels, result in thermal turbulence. The size and, hence, the scale of the eddies caused by thermal turbulence are larger than those of the eddies caused by mechanical turbulence.

The manifestation of turbulent eddies is gustiness and is displayed in the fluctuations seen on a continuous record of wind or temperature. Figure 21.3 displays wind direction traces during (a) mechanical and (b) thermal turbulence. Fluctuations due to mechanical turbulence tend to be quite regular; that is, eddies of nearly constant size are generated. The eddies generated by thermal turbulence are both larger and more variable in size than those due to mechanical turbulence.

The most important mixing process in the atmosphere which causes the dispersion of air pollutants is called *eddy diffusion*. The atmospheric eddies cause a breaking apart of atmospheric parcels which mixes polluted air with relatively unpolluted air, causing polluted air at lower and lower concentrations to occupy successively larger volumes of air. Eddy or turbulent dispersion is most efficient when the scale of the eddy is similar to that of the pollutant puff or plume being diluted. Smaller eddies are effective only at tearing at the edges of the pollutant mass. On the other hand, larger eddies will usually only transport the mass of polluted air as a whole.

The size and influence of eddies on the vertical expansion of continuous plumes have been related to vertical temperature structure [3]. Three appearances of instantaneous plumes related to specific lapse rates and three appearances of instantaneous plumes related to combinations of lapse rates are shown in Fig. 21.4. Strong lapse is decrease of temperature with height in excess of the adiabatic lapse rate. Weak lapse is decrease of temperature with height at a rate between the dry adiabatic rate and the isothermal condition (no change of temperature with height).

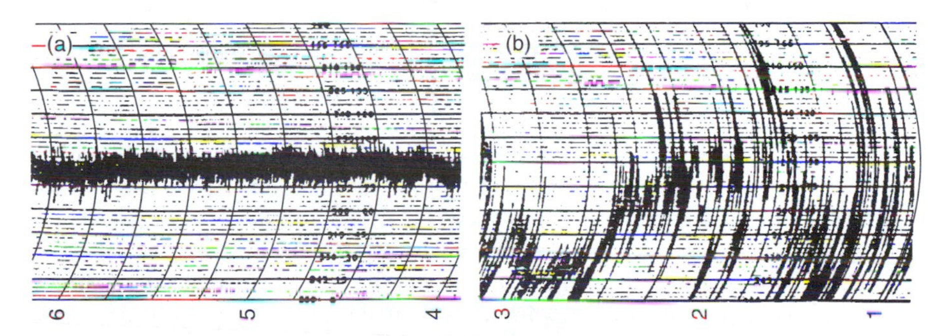

Fig. 21.3. Examples of turbulence on wind direction records: (a) mechanical and (b) thermal. *Source*: From Smith [2].

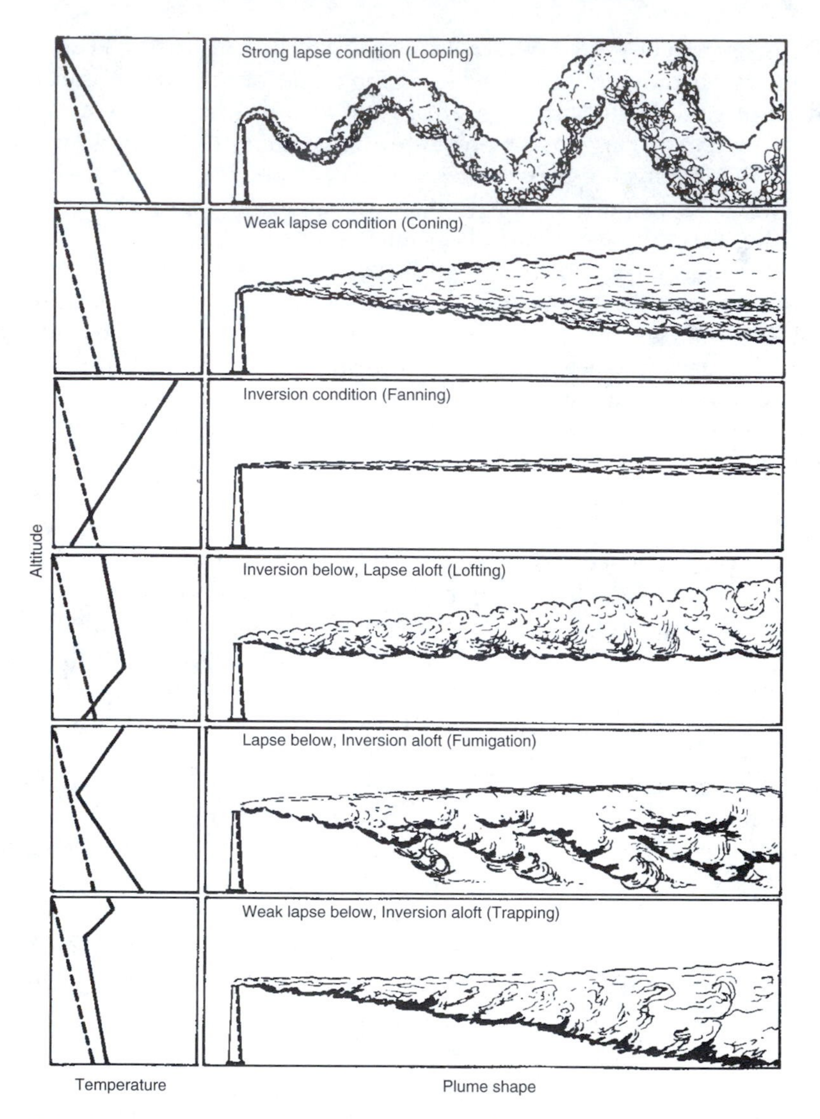

Fig. 21.4. Vertical expansion of continuous plumes related to vertical temperature structure. The dashed lines correspond to the dry adiabatic lapse rate for reference.

A number of methods have been used to measure or estimate the level of turbulence in the atmosphere and, in turn, its dispersive ability. These methods vary from direct measurement of wind fluctuations by sensitive wind measurement systems; to classification based on the appearance of the chart record of the wind direction trace; to classification of atmospheric stability indirectly by wind speed and estimates of insolation (incoming solar radiation) or outgoing longwave radiation. Details of these methods are given in the next section.

III. ESTIMATING CONCENTRATIONS FROM POINT SOURCES

The principal framework of empirical equations which form a basis for estimating concentrations from point sources is commonly referred to as the *Gaussian plume model*. Employing a three-dimensional axis system of downwind, crosswind, and vertical with the origin at the ground, it assumes that concentrations from a continuously emitting plume are proportional to the emission rate, that these concentrations are diluted by the wind at the point of emission at a rate inversely proportional to the wind speed, and that the time-averaged (about 1 h) pollutant concentrations crosswind and vertically near the source are well described by Gaussian or normal (bell-shaped) distributions. The standard deviations of plume concentration in these two directions are empirically related to the levels of turbulence in the atmosphere and increase with distance from the source.

In its simplest form, the Gaussian model assumes that the pollutant does not undergo chemical reactions or other removal processes in traveling away from the source and that pollutant material reaching the ground or the top of the mixing height as the plume grows is eddy-reflected back toward the plume centerline.

A. The Gaussian Equations

All three of the Gaussian equations (21.2 through 21.4) are based on a coordinate scheme with the origin at the ground, x downwind from the source, y crosswind, and z vertical. The normal vertical distribution near the source is modified at greater downwind distances by eddy reflection at the ground and, when the mixing height is low, by eddy reflection at the mixing height. *Eddy reflection* refers to the movement away ("reflection") of circular eddies of air from the earth's surface, since they cannot penetrate that surface. Cross sections in the horizontal and vertical at two downwind distances through a plume from a 20-m-high source with an additional 20 m of plume rise (to result in a 40-m effective height) are shown in Fig. 21.5. The following symbols are used:

- X concentration, g m^{-3}
- Q emission rate, g s^{-1}
- u wind speed, m s^{-1}
- σ_y standard deviation of horizontal distribution of plume concentration (evaluated at the downwind distance x and for the appropriate stability), m
- σ_z standard deviation of vertical distribution of plume concentration (evaluated at the downwind distance x and for the appropriate stability), m
- L mixing height, m
- h physical stack height, m

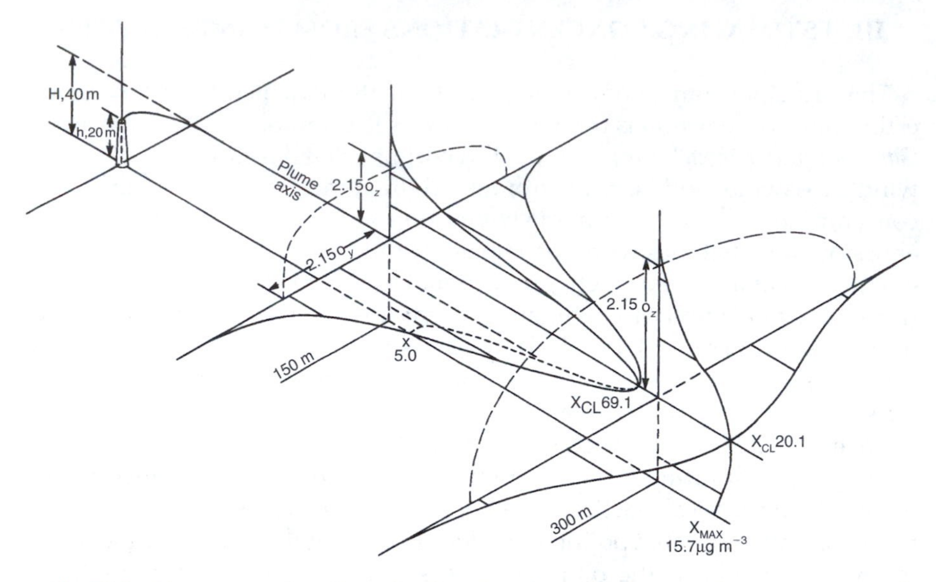

Fig. 21.5. Two cross sections through a Gaussian plume (total mass under curves conserved).

H effective height of emission, m
x downwind distance, m
y crosswind distance, m
z receptor height above ground, m

The concentration χ resulting at a receptor at (x, y, z) from a point source located at $(0, 0, H)$ is given by one of the three following equations. (Methods for obtaining values for the dispersion parameters σ_y and σ_z in the following equations are discussed later in this chapter.)

For stable conditions or unlimited vertical mixing (a very high mixing height), use

$$\chi = Q(1/u)\{g_1/[(2\pi)^{0.5}\,\sigma_y]\}\{g_2/[(2\pi)^{0.5}\,\sigma_z]\} \tag{21.2}$$

where

$$g_1 = \exp\left(-0.5y^2/\sigma_y^2\right)$$
$$g_2 = \exp\left[-0.5\,(H-z)^2/\sigma_z^2\right] + \exp\left[-0.5\,(H+z)^2/\sigma_z^2\right]$$

Note that if $y = 0$, or $z = 0$, or both z and H are 0, this equation is greatly simplified. For locations in the vertical plane containing the plume centerline, $y = 0$ and $g_1 = 1$.

For unstable or neutral conditions, where σ_z is greater than $1.6L$, use

$$\chi = Q(1/u)\{g_1/[(2\pi)^{0.5}\,\sigma_y]\}(1/L) \tag{21.3}$$

For these large σ_z values, eddy reflection has occurred repeatedly both at the ground and at the mixing height, so that the vertical expanse of the plume has been uniformly mixed through the mixing height, i.e., $1/L$.

For unstable or neutral conditions, where σ_z is less than $1.6L$, use the following equation provided that both H and z are less than L:

$$\chi = Q(1/u)\{g_1/[(2\pi)^{0.5}\,\sigma_y]\}\{g_3/[(2\pi)^{0.5}\,\sigma_z]\} \tag{21.4}$$

where

$$g_3 = \sum_{N=-\infty}^{\infty} \{\exp[-0.5(H - z + 2NL)^2/\sigma_z^2] + \exp[-0.5(H + z + 2NL)^2/\sigma_z^2]\}$$

This infinite series converges rapidly, and evaluation with N varying from -4 to $+4$ is usually sufficient. These equations are used when evaluating by computer, as the series g_3 can easily be evaluated.

When estimates are being made by hand calculations, Eq. (21.2) is frequently applied until $\sigma_z = 0.8L$. This will cause an inflection point in a plot of concentrations with distance.

By adding Eq. (21.4), which includes multiple eddy reflections, and changing the criteria for the use of Eq. (21.3) to situations in which σ_z is evaluated as being greater than $1.6L$, a smooth transition to uniform mixing, Eq. (21.3), is achieved regardless of source or receptor height. By differentiating Eq. (21.2) and setting it equal to zero, an equation for maximum concentration can be derived:

$$\chi_{\max} = \frac{2Q}{\pi u e H^2}\frac{\sigma_z}{\sigma_y} \tag{21.5}$$

and the distance to maximum concentration is at the distance where $\sigma_z = H/(2)^{0.5}$. This equation is strictly correct only if the σ_z/σ_y ratio is constant with distance.

B. Alternate Coordinate Systems for the Gaussian Equations

For estimating concentrations from more than one source, it is convenient to use map coordinates for locations. Gifford [4] has pointed out that the resulting calculated concentration is the same whether the preceding axis system is used or whether an origin is placed at the ground beneath the receptor, with the x-axis oriented upwind, the z-axis remaining vertical, and the y-axis crosswind.

This latter axis system is convenient in assessing the total concentration at a receptor from more than one source provided that the wind direction can be assumed to be the same over the area containing the receptor and the sources of interest.

Given an east–north coordinate system (R, S) the upward distance x and the crosswind distance y of a point source from a receptor are given by

$$x = (S_p - S_r)\cos\theta + (R_p - R_r)\sin\theta \quad (21.6)$$

$$y = (S_p - S_r)\sin\theta - (R_p - R_r)\cos\theta \quad (21.7)$$

where R_p, S_p are the coordinates of the point source; R_r, S_r are the coordinates of the receptor; and θ is the wind direction (the direction from which the wind blows). The units of x and y will be the same as those of the coordinate system R, S. In order to determine plume dispersion parameters, distances must be in kilometers or meters. A conversion may be required to convert x and y above to the appropriate units.

C. Determination of Dispersion Parameters

1. By Direct Measurements of Wind Fluctuations

Hay and Pasquill [5] and Cramer [6, 7] have suggested the use of fluctuation statistics from fixed wind systems to estimate the dispersion taking place within pollutant plumes over finite release times. The equation used for calculating the variance of the bearings (azimuth) from the point of release of the particles, σ_p^2, at a particular downwind location is

$$\sigma_p^2 = \sigma_a^2(\tau, s) \quad (21.8)$$

where σ_a^2 is the variance of the azimuth angles of a wind vane over the sampling period τ calculated from average wind directions averaged over averaging periods of duration s; s equals T/β, where T is the travel time to the downwind location; T is equivalent to x/u, where x is the downwind distance from the source and u is the transport wind speed. Here β is the ratio of the time scale of the turbulence moving with the air stream (Lagrangian) to the time scale of the turbulence at a fixed point (Eulerian). Although β has considerable variation (from about 1 to 9), a reasonable fit to field data has been found using a value of 4 for β.

A similar equation can be written for vertical spread from an elevated source. The standard deviation of the vertical distribution of pollutants at the downwind distance x is given by

$$\sigma_z = \sigma_e(\tau, s)x \quad (21.9)$$

where σ_z is in meters and σ_e is the standard deviation of the elevation angle, in radians, over the sampling period τ calculated from averaged elevation angles over averaging periods s. Here, as before, s equals T/β where T is travel time, and β can be approximated as equal to 4; x in Eq. (21.9) is in meters. In application, σ values can be calculated over several set averaging periods s. The distances to which each σ applies are then given by $x = \beta us$.

To calculate plume dispersion directly from fluctuation measurements, Draxler [8] used equations in the form:

$$\sigma_y = x\sigma_a f_y \tag{21.10}$$

$$\sigma_z = x\sigma_e f_z \tag{21.11}$$

He analyzed dispersion data from 11 field experiments in order to determine the form of the functions f_y and f_z, including release height effects. Irwin [9] has used simplified expressions for these functions where both f_y and f_z have the form:

$$f = 1/[1 + 0.9(T/T_0)^{0.5}] \tag{21.12}$$

where travel time T is x/u; T_0 is 1000 for f_y; T_0 is 500 for f_z for unstable (including daytime neutral) conditions; and T_0 is 50 for f_z for stable (including nighttime neutral) conditions.

2. *By Classification of Wind Direction Traces*

Where specialized fluctuation data are not available, estimates of horizontal spreading can be approximated from convential wind direction traces. A method suggested by Smith [2] and Singer and Smith [10] uses classification of the wind direction trace to determine the turbulence characteristics of the atmosphere, which are then used to infer the dispersion. Five turbulence classes are determined from inspection of the analog record of wind direction over a period of 1 h. These classes are defined in Table 21.1. The atmosphere is classified as A, B_2, B_1, C, or D. At Brookhaven National Laboratory, where the system was devised, the most unstable category, A, occurs infrequently enough that insufficient information is available to estimate its dispersion parameters. For the other four classes, the equations, coefficients, and exponents for the dispersion parameters are given in Table 21.2, where the source to receptor distance x is in meters.

TABLE 21.1

Brookhaven Gustiness Classes (Based on Variations of Horizontal Wind Direction over 1 h at the Height of Release)

A	Fluctuations of wind direction exceeding 90°
B_2	Fluctuations ranging from 45° to 90°
B_1	Similar to A and B_2, with fluctuations confined to a range of 15–45°
C	Distinguished by the unbroken solid core of the trace, through which a straight line can be drawn for the entire hour without touching "open space." The fluctuations must be 15°, but no upper limit is imposed
D	The trace approximates a line. Short-term fluctuations do not exceed 15°

Source: From Singer and Smith [10].

TABLE 21.2

Coefficients and Exponents for Brookhaven Gustiness Classes

Type	*a*	*b*	*c*	*d*
B_2	0.40	0.91	0.41	0.91
B_1	0.36	0.86	0.33	0.86
C	0.32	0.78	0.22	0.78
D	0.31	0.71	0.06	0.71

Note: $\sigma_y = ax^b$; $\sigma_z = cx^d$ (x is in meters).
Source: Adapted from Table 1 of Gifford [12].

3. *By Classification of Atmospheric Stability*

Pasquill [11] advocated the use of fluctuation measurements for dispersion estimates but provided a scheme "for use in the likely absence of special measurements of wind structure, there was clearly a need for broad estimates" of dispersion "in terms of routine meteorological data". The first element is a scheme which includes the important effects of thermal stratification to yield broad categories of stability. The necessary parameters for the scheme consist of wind speed, insolation, and cloudiness, which are basically obtainable from routine observations (Table 21.3).

TABLE 21.3

Pasquill Stability Categories

	Isolation			Night	
Surface wind speed (m s^{-1})	Strong	Moderate	Slight	Thinly overcast or ≥4/8 low cloud	≤3/8 cloud
<2	A	A–B	B	—	—
2–3	A–B	B	C	E	F
3–5	B	B–C	C	D	E
5–6	C	C–D	D	D	D
>6	C	D	D	D	D
	(for A–B, take the average of values for A and B, etc.)				

Notes:
1. Strong insolation corresponds to sunny midday in midsummer in England; slight insolation to similar conditions in midwinter.
2. Night refers to the period from 1 h before sunset to 1 h after sunrise.
3. The neutral category D should also be used, regardless of wind speed, for overcast conditions during day or night and for any sky conditions during the hour preceding or following night as defined above.

Source: From Pasquill [13].

Pasquill's dispersion parameters were restated in terms of σ_y and σ_z by Gifford [14, 15] to allow their use in the Gaussian plume equations. The parameters σ_y and σ_z are found by estimation from the graphs (Fig. 21.6), as a function of the distance between source and receptor, from the appropriate curve, one for each stability class [12]. Alternatively, σ_y and σ_z can be calculated using the equations given in Tables 21.4 and 21.5, which are used in the point source computer techniques PTDIS and PTMTP [16]. These parameter values are most applicable for releases near the ground (within about 50 m).

Other estimations of σ_y and σ_z by Briggs for two different situations, urban and rural, for each Pasquill stability class, as a function of distance between source and receptor, are given in Tables 21.6 and 21.7 [12].

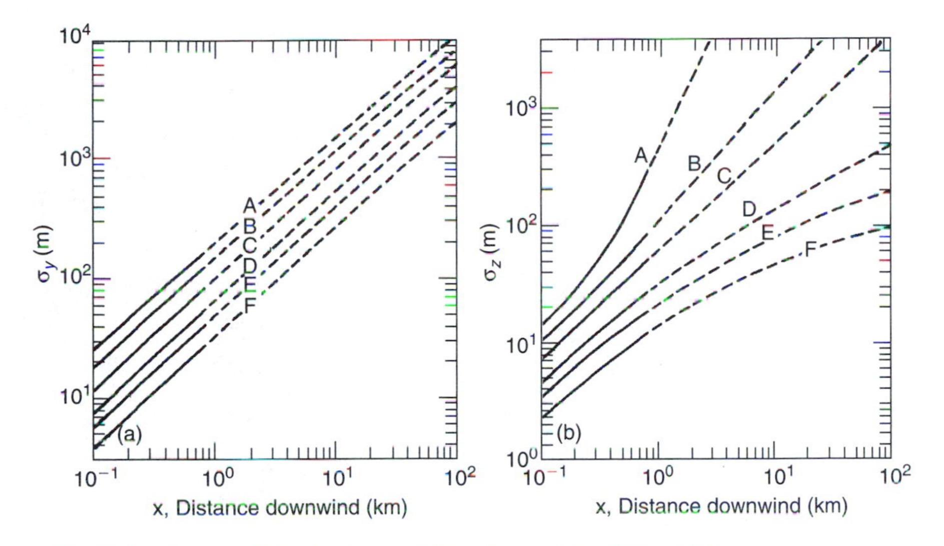

Fig. 21.6. Pasquill–Gifford: (a) σ_y and (b) σ_z. *Source*: From Gifford [12].

TABLE 21.4

Pasquill–Gifford Horizontal Dispersion Parameter

Stability	Parameter		
A	$T = 24.167$	-2.5334	$\ln x$
B	$T = 18.333$	-1.8096	$\ln x$
C	$T = 12.5$	-1.0857	$\ln x$
D	$T = 8.3333$	-0.72382	$\ln x$
E	$T = 6.25$	-0.54287	$\ln x$
F	$T = 4.1667$	-0.36191	$\ln x$

Note: σ_y (m) $= 465.116\,x \tan T$; x is downwind distance in km; T is one-half Pasquill's θ (degrees).

TABLE 21.5

Pasquill–Gifford Vertical Dispersion Parameter

Stability	Distance (km)	a	b	σ_z[a]
A	>3.11			5000 m
	0.5–3.11	453.85	2.1166	
	0.4–0.5	346.75	1.7283	104.7
	0.3–0.4	258.89	1.4094	71.2
	0.25–0.3	217.41	1.2644	47.4
	0.2–0.25	179.52	1.1262	37.7
	0.15–0.2	170.22	1.0932	29.3
	0.1–0.15	158.08	1.0542	21.4
	<0.1	122.8	0.9447	14.0
B	>35			5000 m
	0.4–35	109.30	1.0971	
	0.2–0.4	98.483	0.98332	40.0
	<0.2	90.673	0.93198	20.2
C	all x	61.141	0.91465	
D	>30	44.053	0.51179	
	10–30	36.650	0.56589	251.2
	3–10	33.504	0.60486	134.9
	1–3	32.093	0.64403	65.1
	0.3–1	32.093	0.81066	32.1
	<0.3	34.459	0.86974	12.1
E	>40	47.618	0.29592	
	20–40	35.420	0.37615	141.9
	10–20	26.970	0.46713	109.3
	4–10	24.703	0.50527	79.1
	2–4	22.534	0.57154	49.8
	1–2	21.628	0.63077	33.5
	0.3–1	21.628	0.75660	21.6
	0.1–0.3	23.331	0.81956	8.7
	<0.1	24.260	0.83660	3.5
F	>60	34.219	0.21716	
	30–60	27.074	0.27436	83.3
	15–30	22.651	0.32681	68.8
	7–15	17.836	0.4150	54.9
	3–7	16.187	0.4649	40.0
	2–3	14.823	0.54503	27.0
	1–2	13.953	0.63227	21.6
	0.7–1.0	13.953	0.68465	14.0
	0.2–0.7	14.457	0.78407	10.9
	<0.2	15.209	0.81558	4.1

[a] σ_z at boundary of distance range for all values except 5000 m.
Note: σ_z (m) $= ax^b$; x is downwind distance in km.

D. Example of a Dispersion Calculation

As an example of the use of the Gaussian plume equations using the Pasquill–Gifford dispersion parameters, assume that a source releases 0.37 g s^{-1} of a pollutant at an effective height of 40 m into the atmosphere with the wind blowing at 2 m s^{-1}. What is the approximate distance of the

TABLE 21.6

Urban Dispersion Parameters by Briggs (for Distances Between 100 and 10 000 m)

Pasquill type	σ_y, m	σ_z, m
A–B	$0.32x(1 + 0.0004x)^{-0.5}$	$0.24x(1 + 0.001x)^{0.5}$
C	$0.22x(1 + 0.0004x)^{-0.5}$	$0.20x$
D	$0.16x(1 + 0.0004x)^{-0.5}$	$0.14x(1 + 0.0003x)^{-0.5}$
E–F	$0.11x(1 + 0.0004x)^{-0.5}$	$0.08x(1 + 0.0015x)^{-0.5}$

Source: From Gifford [12].

TABLE 21.7

Rural Dispersion Parameters by Briggs (for Distances Between 100 and 10 000 m)

Pasquill type	σ_y, m	σ_z, m
A	$0.22x(1 + 0.0001x)^{-0.5}$	$0.20x$
B	$0.16x(1 + 0.0001x)^{-0.5}$	$0.12x$
C	$0.11x(1 + 0.0001x)^{-0.5}$	$0.08x(1 + 0.0002x)^{-0.5}$
D	$0.08x(1 + 0.0001x)^{-0.5}$	$0.06x(1 + 0.0015x)^{-0.5}$
E	$0.06x(1 + 0.0001x)^{-0.5}$	$0.03x(1 + 0.0003x)^{-1}$
F	$0.04x(1 + 0.0001x)^{-0.5}$	$0.016x(1 + 0.0003x)^{-1}$

Source: From Gifford [12].

maximum concentration, and what is the concentration at this point if the atmosphere is appropriately represented by Pasquill stability class B?

Solution: The maximum occurs approximately when $\sigma_z = H/(2)^{1/2} = 28.3\,\text{m}$. Under B stability, this occurs at $x = 0.28\,\text{km}$. At this point $\sigma_y = 49.0\,\text{m}$ (from Table 21. 4). First, the maximum can be estimated by Eq. (21.5):

$$\chi_{\text{max}} = \frac{2Q}{\pi u e H^2}\frac{\sigma_z}{\sigma_y} = \frac{2(0.37)}{\pi 2e40^2}\frac{28.3}{49.0} = 1.56 \times 10^{-5}\,\text{g m}^{-3}$$

To see if this is approximately the distance of the maximum, the equation

$$\chi = [Q/(\pi u \sigma_y \sigma_z)] \exp[-0.5(H/\sigma_z)^2] \tag{21.13}$$

which results from Eq. (21.2) with y and z equal to 0, is evaluated at three distances: 0.26, 0.28, and 0.30 km. The parameter values and the resulting concentrations are given in the following table:

x, km	σ_z, m	σ_y, m	χ, g m^{-3}
0.26	26.2	45.9	1.53×10^{-5}
0.28	28.2	49.0	1.56×10^{-5}
0.30	30.1	52.2	1.55×10^{-5}

σ_z is obtained from equations in Table 21.5.
σ_y is obtained from equations in Table 21.4.

which verifies that, to the nearest 20 m, the maximum is at 0.28 km. Note that the concentration obtained from this equation is the same as that obtained from the approximation equation for the maximum.

Buoyancy-induced dispersion, which is caused near the source due to the rapid expansion of the plume during the rapid rise of the thermally buoyant plume after its release from the point of discharge, should also be included for buoyant releases [15]. The effective vertical dispersion σ_{ze} is found from:

$$\sigma_{ze}^2 = (\Delta H/3.5)^2 + \sigma_z^2 \tag{21.14}$$

where ΔH, the plume rise, and σ_z are evaluated at the distance x from the source. Beyond the distance to the final rise, ΔH is a constant. At shorter distances, it is evaluated for the gradually rising plume (see Chapter 22).

Since in the initial growth phases of a buoyant plume the plume is nearly symmetrical about its centerline, the buoyancy-induced dispersion in the crosswind (horizontal) direction is assumed to be equal to that in the vertical. Thus, the effective horizontal dispersion σ_{ye} is found from

$$\sigma_{ye}^2 = (\Delta H/3.5)^2 + \sigma_y^2 \tag{21.15}$$

The Gaussian plume equations are then used by substituting the value of σ_{ye} for σ_y and σ_{ze} for σ_z.

IV. DISPERSION INSTRUMENTATION

A. Measurements near the Surface

Near-surface (within 10 m of the ground) meteorological instrumentation always includes wind measurements and should include turbulence measurements as well. Such measurements can be made at 10 m above ground by using a guyed tower. A cup anemometer and wind vane (Fig. 21. 7), or a vane with a propeller speed sensor mounted in front (Fig. 21.8), can be the basic wind system. The wind sensor should have a threshold starting speed of less than $0.5\,\mathrm{m\,s^{-1}}$, an accuracy of $0.2\,\mathrm{m\,s^{-1}}$ or 5%, and a distance constant of less than 5 m for proper response. The primary quantity needed is the hourly average wind speed. A representative value may be obtained from values taken each minute, although values taken at intervals of 1–5 s are better.

The vane can be used for both average wind direction and the fluctuation statistic σ_a, both determined over hourly intervals. The vane should have a distance constant of less than 5 m and a damping ratio greater than or equal to 0.4 to have a proper response. Relative accuracy should be 1° and absolute accuracy should be 5°. In order to estimate σ_a accurately, the direction should be sampled at intervals of 1–5 s. This can best be accomplished by microcircuitry (minicomputer) designed to sample properly the output from the vane and perform the calculations for both mean wind and σ_a, taking into account crossover shifts of the wind past the 360° and 0° point.

Fig. 21.7. Microvane and three-cup anemometer. *Source*: Photo courtesy of R. M. Young Co.

Fig. 21.8. Propeller vane wind system. *Source*: Photo courtesy of R. M. Young Co.

The elevation angle, and through appropriate data processing σ_e, can be measured with a bivane (a vane pivoted so as to move in the vertical as well as the horizontal). Bivanes require frequent maintenance and calibration and are affected by precipitation and formation of dew. A bivane is therefore more a research instrument than an operational one. Vertical fluctuations may be measured by sensing vertical velocity w and calculating σ_w from the output of a propeller anemometer mounted on a vertical shaft. The instrument should be placed away from other instrumentation and the propeller axis carefully aligned to be vertical. The specifications of this sensor are the same as those of the wind sensor. Because this instrument will frequently be operating near its lower threshold and because the elevation angle of the wind vector is small, such that the propeller will be operating at yaw angles where it has least accuracy, this method of measuring vertical velocity is not likely to be as accurate as the measurement of horizontal fluctuation.

Rather than using separate systems for horizontal and vertical wind measurements, a u–v–w anemometer system (Fig. 21.9) sensing wind along three orthogonal axes, with proper processing to give average wind direction and σ_a from the combination of the u and v components and w and σ_w from the w component may be used.

Fig. 21.9. *U, V, W* wind system. *Source*: Photo courtesy of R. M. Young Co.

Additional near-surface measurements may also be required to support calculated quantities such as the bulk Richardson number (a stability parameter):

$$\mathrm{Ri_B} = \frac{gh}{T}\frac{\theta_h - \theta_z}{u_h^2}$$

which requires a temperature gradient, a temperature, and wind speed at the height of the boundary layer h. For this purpose, in addition to the wind speed at 10 m from the instrumentation, a vertical temperature difference measurement is needed. This can be obtained for the interval of 2–10 m aboveground using two relatively slow response sensors wired to give the temperature difference directly. Again, hourly averages are of greatest interest. The specifications are response time of 1 min, accuracy of 0.1°C, and resolution of 0.02°C. Both sensors should use good-quality aspirated radiation shields to give representative values. Sensor sampling about every 30 s yields good hourly averages.

Radiation instruments are useful in determining stability such as F. B. Smith's [17] stability parameter P. Although somewhat similar to the Pasquill stability class (Table 21.3), P is continuous (rather than a discrete class) and is derived from wind speed and measurement of upward heat flux or, lacking this, incoming solar radiation (in daytime) and cloud amount at night. Pyranometers measure total sun and sky radiation. Net radiometers measure both incoming (mostly shortwave) radiation and outgoing (mostly longwave) radiation. Data from both are useful in turbulence characterization, and the values should be integrated over hourly periods. Care should be taken to avoid shadows on the sensors. The net radiometer is very sensitive to the condition of the ground surface over which it is exposed.

B. Measurements above the Surface

Measurements above the surface are also important to support pollutant impact evaluation. The radiosonde program of the National Weather Service (Fig. 21.10), established to support forecast and aviation weather activities, is a useful source of temperatures and data on winds aloft, although it has the disadvantage that measurements are made at 12-h intervals and the surface layer is inadequately sampled because of the fast rate of rise of the balloon. Mixing height, the height aboveground of the neutral or unstable layer, is calculated from the radiosonde information (see Chapter 5, Section II).

Measurements of wind, turbulence, and temperature aloft may also be made at various heights on meteorological towers taller than 10 m. Where possible, the sensors should be exposed on a boom at a distance from the tower equal to two times the diameter of the tower at that height.

Aircraft can take vertical temperature soundings and can measure air pollutant and tracer concentrations and turbulence intensity. Airborne lidar can

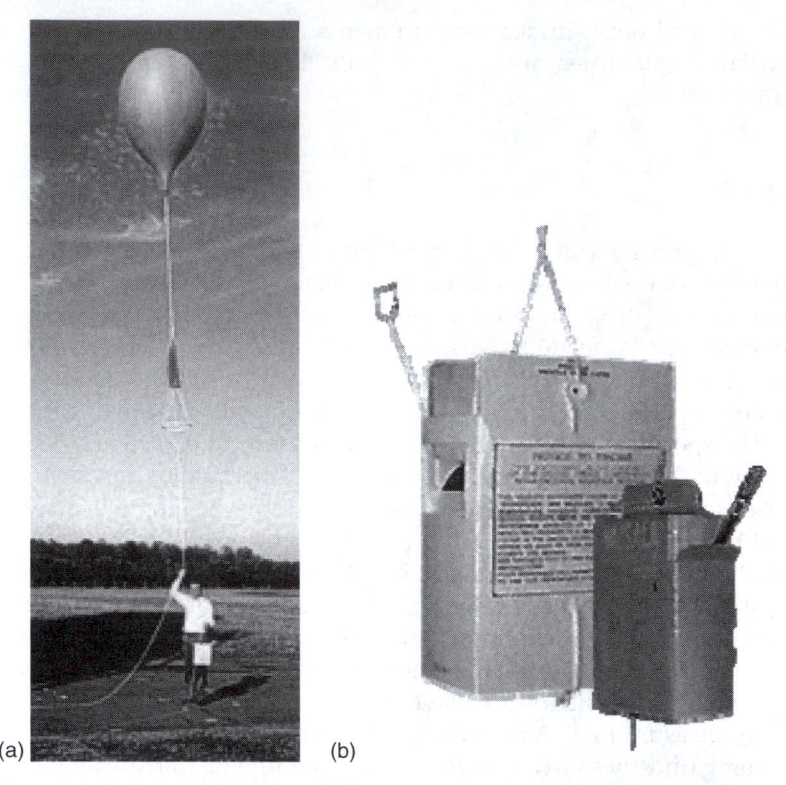

Fig. 21.10. (a) Radiosonde launch and (b) sensor transmitter. *Source*: Photos courtesy of National Oceanic and Atmospheric Administration.

measure plume heights, and integrating nephelometers can determine particle size distributions.

Since operating aircraft, building towers, and establishing instruments on towers are extremely expensive, considerable attention has focused on indirect upper-air sounding from the ground. Mixing height within the range of measurement (approximately 500–600 m aboveground) can be determined by either the Doppler or the monostatic version of sodar (sound direction and ranging) with a spatial resolution of about 30 m. Data on wind and turbulence can be determined by Doppler sodar, FM-CW radar, and lidar. Doppler sodar measurements of wind components are within approximately $0.5\,\mathrm{m\,s^{-1}}$ of tower measurements. Measurements represent 30-m volume averages in the vertical. A height of 500 m aboveground, and sometimes over 1000 m, can be reached routinely.

Some measurements that are completely impractical for routine measurement programs are useful during periods of intensive field programs. Winds and temperatures can be measured through frequent releases of balloon-carried sensors. Lidar is useful for determining plume dimensions. The particle lidar measures backscatter of laser radiation from particles in the plume and particles in the free air. The differential absorption lidar uses two wavelengths, one with strong absorption by sulfur dioxide and the other for weak absorption. The difference determines the amount of sulfur dioxide in the plume. Positioning of the lidar and its scanning mode determines whether vertical or horizontal dimensions of the plume are measured.

C. Data Reduction and Quality Assurance

A meteorological measurement program includes data reduction, calculation of quantities not directly measured, data logging, and archiving. Special-purpose minicomputers are used for sampling sensor output at frequent intervals (down to fractions of a second), calculating averages, and determining standard deviations. The output from the minicomputer should go to a data logger so that the appropriate information can be recorded on magnetic tape or disk or paper tape. If only hourly values must be archived, a considerable period of record for all data from a site can be contained on a single tape, disk, or cassette. Hard copy from a printer is usually also obtained. Immediate availability of this copy can aid in detecting system or sensor malfunctions. Sometimes analog charts are maintained for each sensor to provide backup data recovery (in case of reduction error or data logger malfunction) and to detect sensor malfunction.

An extremely important part of a measurement program is an adequate quality assurance program. Cost cutting in this part of the program can result in useless measurements. A good-quality assurance program includes calibration of individual components and of the entire system in the laboratory; calibration of the system upon installation in the field; scheduled maintenance and servicing; recalibration (perhaps quarterly); and daily examination of data output for unusual or unlikely values. More frequent servicing than that recommended by manufacturers may be required when sensors are placed in polluted atmospheres which may cause relatively rapid corrosion of instrument parts.

V. ATMOSPHERIC TRACERS

A. Technique

Tracer studies are extremely important in furthering our knowledge of atmospheric dispersion. These studies consist of release of a known quantity of a unique substance (the tracer), with measurements of that substance at one or more downwind sampling locations. Early experiments released uranine dye as a liquid spray; the water evaporated, leaving fine fluorescent particles to be sampled. Later, dry fluorescent particles (e.g. zinc–cadmium sulfide) having a relatively narrow range of particle sizes were used. Since the early 1970s, the gas sulfur hexafluoride has been used for most tracer studies, with collection in bags at sampling locations for later laboratory analysis using electron-capture gas chromatography.

Most recently, the US Departments of Homeland Security, Energy, and Commerce, and the Environmental Protection Agency (EPA) have used perfluoride tracers (PFTs) in their Urban Dispersion Program. The PFTs have the advantage over sulfur hexafluoride in that there are a very few sources that may interfere with the measurements. In addition, the various chemical forms of PFTs can be readily identified at very low concentrations (circa parts per quadrillion). Having a number of different chemical forms allows the tracer measurements to be linked to different sources. In fact, this was accomplished in tracer studies conducted in New York City in 2005 (see Fig. 21.11 and Fig. 21.12) [22].

Tracer studies are generally conducted by going into the field for a 2-week to 1-month intensive study period. The tracer is released, generally from a constant height, continuously at a constant rate for a set period (perhaps 2–3 h) on a day selected for its meteorological conditions with the wind forecast to blow toward the sampling network. Sampling equipment is arranged at ground level on constant-distance arcs usually at three or four distances. The samplers begin at a set time as switched on by the field crew or by radio control. More sophisticated samplers allow the unattended collection sequentially of several samples. Sampling time varies from around 20 min to several hours. This procedure measures horizontal dispersion at the height of the samplers.

Although it is highly desirable to determine vertical dispersion as well by direct measurement, it is seldom practical. Sampling in the vertical can be done by sampling on fixed towers or arranging samplers along the cables of captive balloons. Both of these methods are extremely expensive in terms of both equipment and personnel. Although it is possible to sample the tracer with aircraft, the pass through the pollutant plume occurs at such high speed that it is difficult to relate this instantaneous sample to what would occur over a longer sampling time of from 20 min to 1 h.

B. Computations

If the tracer concentration is χ_i measured at each sampling position that has its position at y_i on a scale along the arc (either in degrees or in meters), estimates of the mean position of the plume at ground level and the variance of the ground-level concentration distribution are given by:

$$y = \frac{\Sigma \chi_i y_i}{\Sigma \chi_i} \tag{21.16}$$

$$\sigma_y^2 = \frac{\Sigma \chi_i \; \Sigma \chi_i y_i^2 - (\Sigma \chi_i y_i)^2}{(\Sigma \chi_i)^2} \tag{21.17}$$

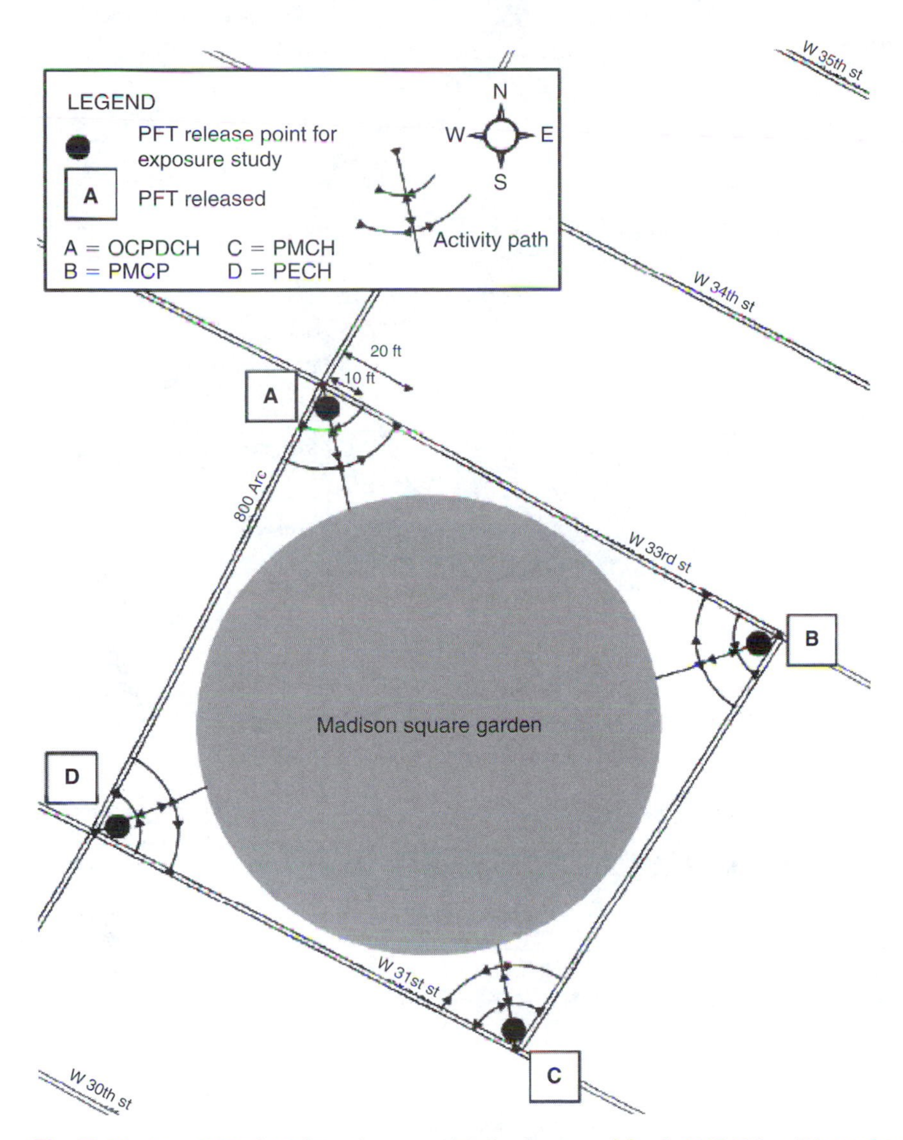

Fig. 21.11. Location of perfluoride tracer (PFT) releases in March 2005 Urban Dispersion Program study around Madison Square Garden in New York City. The tracers are: PMCP (perfluoromethylcyclopentane), oc-PDCH (perfluoro-1,2-dimethylcyclohexane), PMCH (perfluoromethylcyclohexane), and PECH (perfluoroethylcyclohexane). *Source*: Lioy, P., Vallero, D., Foley, G., Georgopoulos, P., Heiser, J., Watson, T., Reynolds, M., Daloia, J., Tong, S., and Isukapal, S., A personal exposure study employing scripted activities and paths in conjunction with atmospheric releases of perfluorocarbon tracers in Manhattan, New York. *J. Exposure Sci. Environ. Epidemiol.*, 17, 409–425 (2007).

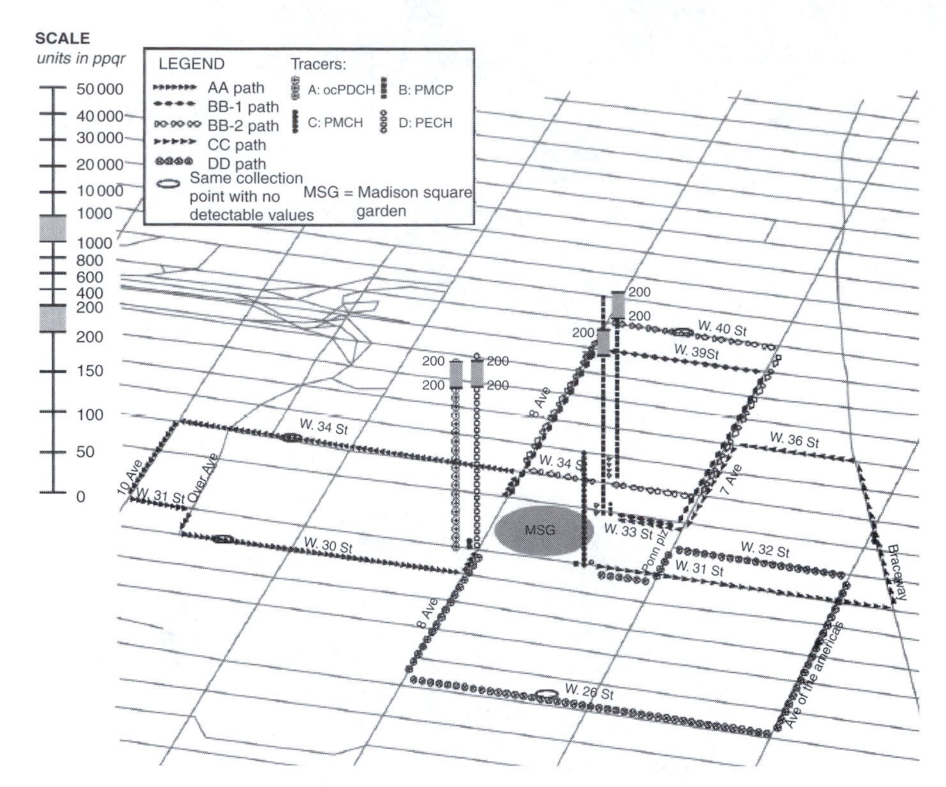

Fig. 21.12. Neighborhood scale personal exposure monitoring, 9:00–9:30 a.m., during release of PFTs on March 10, 2005 releases in the Urban Dispersion Program study around Madison Square Garden in New York City. The tracers are: PMCP (perfluoromethylcyclopentane), ocPDCH (perfluoro-1,2-dimethylcyclohexane), PMCH (perfluoromethylcyclohexane), and PECH (perfluoroethylcyclohexane). *Source*: Lioy, P., Vallero, D., Foley, G., Georgopoulos, P., Heiser, J., Watson, T., Reynolds, M., Daloia, J., Tong, S., and Isukapal, S. A personal exposure study employing scripted activities and paths in conjunction with atmospheric releases of perfluorocarbon tracers in Manhattan, New York. *J. Exposure Sci. Environ. Epidemiol.*, 17, 409–425 (2007).

In the example shown in Fig. 21.13, measurements were made every 2° on an arc 5 km from the source. The mean position of the plume is at an azimuth of 97.65° and the standard deviation is 4.806°.

$$\sigma_y(\text{meters}) = \sigma_y(\text{degrees})\frac{\pi}{180}x\ (\text{meters}) \tag{21.18}$$

In this case σ_y is 419 m. The peak concentration can be found from the measurements, or from the Gaussian distribution fitted to the data and the peak concentration obtained from the fitted distribution. Provided that the emission rate Q, the height of release H, and the mean wind speed u are known, the standard deviation of the vertical distribution of the pollutant

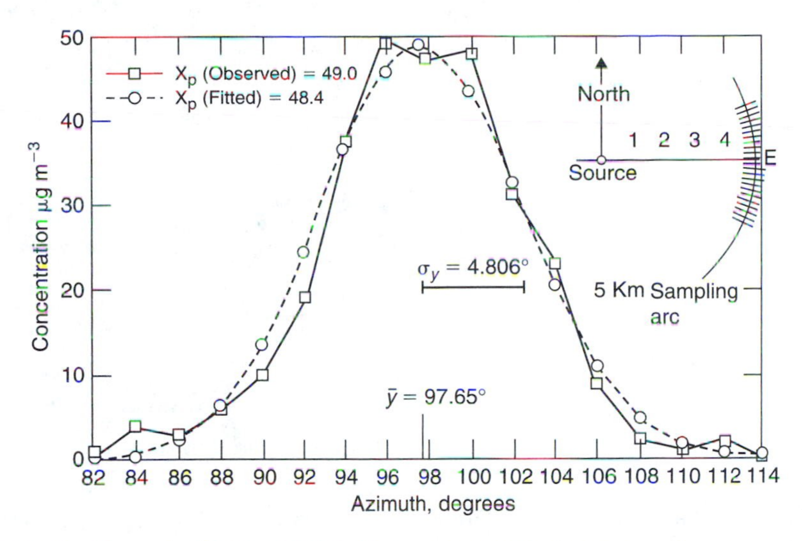

Fig. 21.13. Example of tracer concentration measurements along a sampling arc.

can be approximated from either the peak concentration (actual or fitted) or the crosswind integrated (CWI) concentration from one of the following equations:

$$\sigma_z \exp 0.5(H/\sigma_z)^2 = 2Q/[(2\pi)^{0.5}\, u\chi_{\mathrm{CWI}}] \tag{21.19}$$

$$\sigma_z \exp 0.5(H/\sigma_z)^2 = Q/[(\pi)u\sigma_y\, \chi_{\mathrm{peak}}] \tag{21.20}$$

The CWI concentration in g m^{-2} may be approximated from the tracer measurements from

$$\chi_{\mathrm{CWI}} = \text{sampler spacing (meters)}\ \Sigma\chi_i \tag{21.21}$$

Using the data from Fig. 21.14, the calculated σ_z from the CWI concentration is 239 m; from the observed peak concentration it is 232 m; and from the fitted peak concentration it is 235 m. Note that errors in any of the parameters H, Q, or u, will cause errors in the estimated σ_z.

Although extremely useful, tracer experiments require considerable capital expenditures and personnel. In addition to the difficulties and uncertainty in making estimates of various parameters, especially σ_z, one of the difficulties in interpreting tracer studies is relating the atmospheric conditions under which the study was conducted to the entire spectrum of atmospheric conditions. For example, trying to interpret a series of tracer experiments, even if conducted over a relatively large number of hours, in relation to the conditions that cause the second highest concentration once a year is extremely difficult, if not impossible.

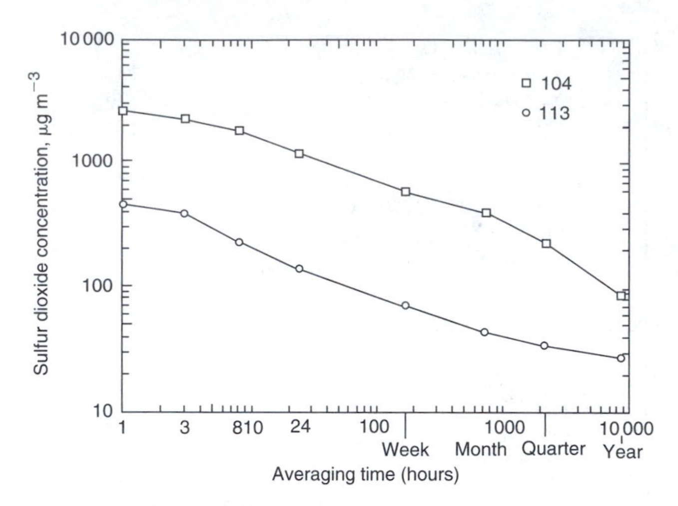

Fig. 21.14. Variation of St. Louis SO_2 maximum concentrations with sampling time for locations with highest (station 104) and lowest (station 113) maximum 1-h concentrations.

VI. CONCENTRATION VARIATION WITH AVERAGING TIME

If emission and meteorological conditions remained unchanged hour after hour, concentrations at various locations downwind would remain the same. However, since such conditions are ever-changing, concentrations vary with time. Even under fairly steady meteorological conditions, with the mean wind direction remaining nearly the same over a period of some hours, as the averaging time increases, greater departures in wind direction from the mean are experienced, thus spreading the time-averaging plume more and reducing the longer averaging time concentration compared with that experienced for shorter averaging times at the location of the highest concentrations. This effect is more pronounced for receptors influenced by single point sources than for those influenced by a number of point sources or by a combination of point and area sources, because there will be many hours when the wind is not blowing from the source to the receptor.

Figure 21.12 shows the maximum sulfur dioxide concentrations for eight averaging times over a 1-year period (1976) for two air monitoring stations in the Regional Air Monitoring (RAM) network in St. Louis. These two monitoring stations, 104 and 113, have the highest and lowest maximum 1-h concentrations of the 13 stations with sulfur dioxide measurements. These maximum concentrations deviate only slightly from a power law relation:

$$\chi_p = at_p^b \qquad (21.22)$$

where χ_p is the maximum concentration for the period p, t_p is the averaging time in hours, and a and b are appropriate constants. The power b is -0.28 for station 104 and -0.33 for station 113.

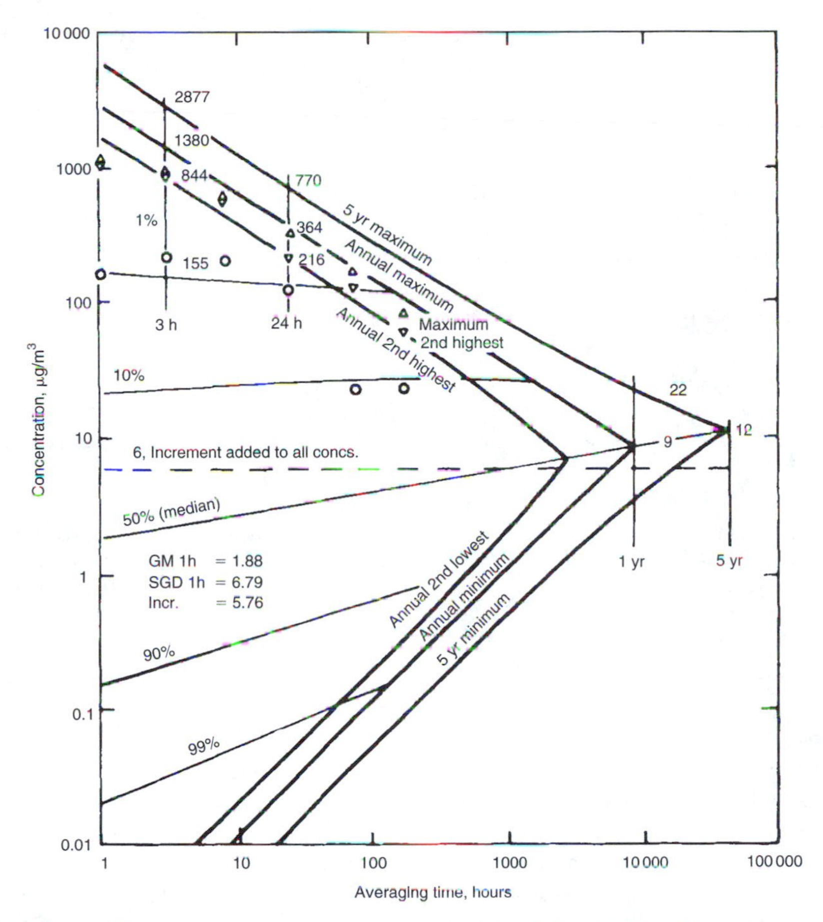

Fig. 21.15. Three-parameter averaging-time model fitted through the arithmetic mean and the second highest 3-h and 24-h SO_2 concentrations measured in 1972 a few miles from a coal-burning power plant. *Source*: From Larsen [21].

Larsen [18–21] has developed averaging time models for use in analysis and interpretation of air quality data. For urban areas where concentrations for a given averaging time tend to be lognormally distributed, that is, where a plot of the log of concentration versus the cumulative frequency of occurrence on a normal frequency distribution scale is nearly linear, the two-parameter averaging time model (Fig. 8.6) is adequate. The two parameters are the geometric mean and the standard geometric deviation. If these two parameters for a pollutant at a site can be determined for an averaging time, the model can calculate them and the annual maximum concentration expected for any other averaging time. For receptors in the vicinity of point sources, where for a given averaging time many concentrations will be zero, a three-parameter averaging time model is required. The third parameter is an increment (positive or negative) that is added to every observed concentration. In Fig. 21.15, showing the three-parameter model applied to data from the vicinity of a power plant, 6 μg m^{-3} have been added to each observed concentration.

REFERENCES

1. Irwin, J. S., *Atmos. Environ.* **13**, 191–194 (1979).
2. Smith, M. E., *Meteorol. Monogr.* No. 4, 50–55 (1951).
3. Church, P. E., *Ind. Eng. Chem.* **41**, 2753–2756 (1949).
4. Gifford, F. A., *Int. J. Air Pollut.* **2**, 109–110 (1959).
5. Hay, J. S., and Pasquill, F., Diffusion from a continuous source in relation to the spectrum and scale of turbulence, in *Advances in Geophysics*, Vol. 6, *Atmospheric Diffusion and Air Pollution* (Frenkie, F. N., and Sheppard, P. A., eds.), pp. 345–365. Academic Press, New York, 1959.
6. Cramer, H. E., *Am. Ind. Hyg. Assoc. J.* **20** (3), 183–189 (1959).
7. Cramer, H. E., Improved techniques for modeling the dispersion of tall stack plumes. *Proceedings of the Seventh International Technical Meeting on Air Pollution Modeling and Its Application*. North Atlantic Treaty Organization Committee on Challenges of Modern Society Publication No. 51, Brussels, 1976 (National Technical Information Service PB-270 799).
8. Draxler, R. R., *Atmos. Environ.* **10**, 99–105 (1976).
9. Irwin, J. S., *J. Clim. Appl. Meteorol.* **22** (1), 92–114 (1983).
10. Singer, I. A., and Smith, M. E., *J. Meteorol.* **10** (2), 121–126 (1953)
11. Pasquill, F., *Atmospheric Diffusion*, 2nd ed. Halstead Press, New York, 1974.
12. Gifford, F. A., *Nucl. Safety* **17** (1), 68–86 (1976).
13. Pasquill, F., *Meteorol. Mag.* **90** (1063), 33–49 (1961).
14. Gifford, F. A., *Nucl. Safety* **2**, 47–55 (1961).
15. Pasquill, F., *Atmospheric Dispersion Parameters in Gaussian Plume Modeling, Part II. Possible Requirements for Change in the Turner Workbook Values*, EPA-600/4-76-030b. US Environmental Protection Agency, Research Triangle Park, NC, 1976.
16. Turner, D. B., and Busse, A. D., *User's Guide to the Interactive Versions of Three Point Source Dispersion Programs: PTMAX, PTDIS, and PTMTP*. Meteorology Laboratory, US Environmental Protection Agency, Research Triangle Park, NC, 1973.
17. Smith, F. B., A scheme for estimating the vertical dispersion of a plume from a source near ground level, *Proceedings of the Third Meeting of the Expert Panel on Air Pollution Modeling*, October 2–3, 1972, Paris. North Atlantic Treaty Organization Committee on the Challenges of Modern Society Publication No. 14, Brussels, 1972 (National Technical Information Service, PB-240 574).
18. Larsen, R. I., *J. Air Pollut. Control Assoc.* **23**, 933–940 (1973).
19. Larsen, R. I., *J. Air Pollut. Control Assoc.* **24**, 551–558 (1974).
20. Larsen, R. I., and Heck, W. W., *J. Air Pollut. Control Assoc.* **26**, 325–333 (1976).
21. Larsen, R. I., *J. Air Pollut. Control Assoc.* **27**, 454–459 (1977).
22. Lioy, P., Vallero, D., Foley, G., Georgopoulos, P., Heiser, J., Watson, T., Reynolds, M., Daloia, J., Tong, S., and Isukapal, S., A personal exposure study employing scripted activities and paths in conjunction with atmospheric releases of perfluorocarbon tracers in Manhattan, New York. *J. Exposure Sci. and Environ. Epidemiol.*, 17, 409–425 (2007).

SUGGESTED READING

Cohen-Hubel, E. et al., Children's exposure assessment: a review of factors influencing children's exposure, and the data available to characterize and assess that exposure. *Environ. Health Persp.* **108**, 475–486 (2000).

Draxler, R. R., *A Summary of Recent Atmospheric Diffusion Experiments*. National Oceanic and Atmospheric Administration Technical Memorandum ERL ARL-78, Silver Spring, MD, 1979.

Electric Power Research Institute, *Preliminary Results from the EPRI Plume Model Validation Project—Plains Site*, EPRI EA-1788, RP 1616, Palo Alto, CA. Interim Report. Prepared by TRC Environmental Consultants, 1981.

Gryning, S. E., and Lyck, E., *Comparison between Dispersion Calculation Methods Based on In-Situ Meteorological Measurements and Results from Elevated-Source Tracer Experiments in an Urban Area*. National Agency of Environmental Protection, Air Pollution Laboratory, MST Luft—A40. Riso National Laboratory, Denmark, 1980.

Hanna, S. et al., *Description of the Madison 2005 (MSG 05) Tracer Experiment in New York City*. Harvard School of Public Health PO65, MSG05, November 1, 2004.

Haugen, D. A. (ed)., *Lectures on Air Pollution and Environmental Impact Analyses*. American Meteorological Society, Boston, MA, 1975.

Hewson, E. W., Meteorological measurements, in *Air Pollution*, 3rd ed., Vol. I, *Air Pollutants, Their Transformation and Transport* (Stern, A. C., ed.), pp. 563–642. Academic Press, New York, 1976.

Lenschow, D. H. (ed.), *Probing the Atmospheric Boundary Layer*. American Meteorological Society, Boston, MA, 1986.

Lioy, P. et al., A personal exposure study employing scripted activities and paths in conjunction with atmospheric releases of perfluorocarbon tracers in Manhattan, New York. *J. Exposure Sci. Environ. Epidemiol.* 17, 409–425 (2007).

Lockhart, T. J., *Quality Assurance Handbook for Air Pollution Measurement Systems*, Vol. IV, *Meteorological Measurements*. US Environmental Protection Agency, Research Triangle Park, NC, 1989.

Munn, R. E., *Boundary Layer Studies and Applications*. Kluwer Academic Press. Hingham, MA, 1989.

National Research Council, Human exposure assessment for airborne pollutants: advances and opportunities. *Committee on Advances in Assessing Human Exposure to Airborne Pollutants*. National Academy Press, Washington, DC, 1991.

Strimaitis, D., Hoffnagle, G., and Bass, A., *On-Site Meteorological Instrumentation Requirements to Characterize Diffusion from Point Sources—Workshop Report*, EPA-600/9-81-020. US Environmental Protection Agency, Research Triangle Park, NC, 1981.

Turner, D. B., The transport of Pollutants, in *Air Pollution*, 3rd ed., Vol. VI, *Supplement to Air Pollutants, Their Transformation, Transport, and Effects* (Stern, A. C., ed.), pp. 95–144. Academic Press, Orlando, FL, 1986.

Vaughan, R. A., *Remote Sensing Applications in Meteorology and Climatology*. Reidel, Norwell, MA, 1987.

Watson T. B. et al., *The New York City Urban Dispersion Program March 2005 Field Study: Tracer Methods and Results*, Brookhaven National Laboratory, BNL-75552-2006 (Report), January 2006.

QUESTIONS

1. In a situation under stable conditions with a wind speed of $4\ \mathrm{m\ s^{-1}}$ and $\sigma_a = 0.12$ radians, the wind is blowing directly toward a receptor 1 km from the source. How much must the wind

shift, in degrees, to reduce the concentration to 10% of its previous value? (At 2.15σ from the peak, the Gaussian distribution is 0.1 of the value at the peak.)

2. If the variation in wind speed with height is well approximated with a power law wind profile having an exponent equal to 0.15, how much stronger is the wind at 100 m above ground than at 10 m?
3. At a particular downwind distance, a dispersing plume is approximately 40 m wide. Which of the following three turbulent eddy diameters—5 m, 30 m, or 100 m—do you believe would be more effective in further dispersing this plume?
4. A pollutant is released from an effective height of 50 m and has a ground-level concentration of 300 $\mu g\ m^{-3}$ at a position directly downwind where the σ_z is 65 m. How does the concentration at 50 m above this point, that is, the plume centerline, compare with this ground-level concentration?
5. At a downwind distance of 800 m from a 75-m source having an 180-m plume rise, σ_y is estimated as 84 m and σ_z is estimated as 50 m. If one considers buoyancy-induced dispersion as suggested by Pasquill, by how much is the plume centerline concentration reduced at this distance?
6. Consider the parameters that need to be measured and the instrumentation needed to make the measurements for monitoring dispersion of effluent from a 30-m stack.
7. A tracer experiment includes sampling on an arc at 1000 m from the source. If the horizontal spread is expected to result in a σ_y between 120 and 150 m at this distance, and if the wind direction is within $\pm 15°$ azimuth of that forecast, how many samplers should be deployed and what should be that spacing? It is desirable to have above seven measurements within $\pm 2\sigma_y$ of the plume centerline and at least one sample on each side of the plume.
8. For a tracer release that can be considered to be at ground level, approximate the vertical dispersion σ_z at the downwind distance where measurements indicate that the concentration peak is 7×10^{-8} g m^{-3}, the horizontal σ_y is 190 m, and the crosswind integrated concentration is 3.16×10^{-5} g m^{-3}. The tracer release rate is 0.01 g s^{-1} and the wind speed is 3.7 m s^{-1}.
9. The maximum 1-h concentration at an urban monitoring station is 800 $\mu g\ m^{-3}$. If the concentration varies with averaging time with a power law relation when the power is –0.3, what is the expected maximum concentration for a 1-week averaging time?

22

Air Pollution Modeling and Prediction

To build new facilities or to expand existing ones without harming the environment, it is desirable to assess the air pollution impact of a facility prior to its construction, rather than construct and monitor to determine the impact and whether it is necessary to retrofit additional controls. Potential air pollution impact is usually estimated through the use of air quality simulation models. A wide variety of models is available. They are usually distinguished by type of source, pollutant, transformations and removal, distance of transport, and averaging time. No attempt will be made here to list all the models in existence at the time of this writing.

In its simplest form, a model requires two types of data inputs: information on the source or sources including pollutant emission rate, and meteorological data such as wind velocity and turbulence. The model then simulates mathematically the pollutant's transport and dispersion, and perhaps its chemical and physical transformations and removal processes. The model output is air pollutant concentration for a particular time period, usually at specific receptor locations.

Impact estimates by specific models are required to meet some regulatory requirements.

I. PLUME RISE

Gases leaving the tops of stacks rise higher than the stack top when they are either of lower density than the surrounding air (buoyancy rise) or ejected at a velocity high enough to give the exit gases upward kinetic energy (momentum rise). Buoyancy rise is sometimes called *thermal rise* because the most common cause of lower density is higher temperature. Exceptions are emissions of gases of higher density than the surrounding air and stack downwash, discussed next. To estimate effective plume height, the equations of Briggs [1–5] are used. The wind speed u in the following equations is the measured or estimated wind speed at the physical stack top.

A. Stack Downwash

The lowering below the stack top of pieces of the plume by the vortices shed downwind of the stack is simulated by using a value *h¢* in place of the physical stack height h. This is somewhat less than the physical height when the stack gas exit velocity v_s is less than 1.5 times the wind speed $u(\mathrm{m\,s^{-1}})$:

$$h' = h \qquad \text{for } v_s \geqslant 1.5u \tag{22.1}$$

$$h' = h + 2d[(v_s/u) - 1.5] \qquad \text{for } v_s < 1.5u \tag{22.2}$$

where d is the inside stack-top diameter, *m*. This *h¢* value is used with the buoyancy or momentum plume rise equations that follow. If stack downwash is not considered, h is substituted for *h¢* in the equations.

B. Buoyancy Flux Parameter

For most plume rise estimates, the value of the buoyancy flux parameter F in $\mathrm{m^4\,s^{-3}}$ is needed:

$$\begin{aligned} F &= g v_s d^2 (T_s - T)/(4T_s) \\ F &= 2.45 v_s d^2 (T_s - T)/T_s \end{aligned} \tag{22.3}$$

where g is the acceleration due to gravity, about $9.806\,\mathrm{m\,s^{-2}}$, T_s is the stack gas temperature in K, T is ambient air temperature in K, and the other parameters are as previously defined.

C. Unstable–Neutral Buoyancy Plume Rise

The final effective plume height H, in m, is stack height plus plume rise. Where buoyancy dominates, the horizontal distance x_f from the stack to where the final plume rise occurs is assumed to be at $3.5x^*$, where x^* is the horizontal distance, in km, at which atmospheric turbulence begins to dominate entrainment.

For unstable and neutral stability situations, and for F less than 55, H, in m, and x_f, in km, are

$$H = h' + 21.425F^{3/4}/u \quad x_f = 0.049F^{5/8} \qquad (22.4a,b)$$

For F equal to or greater than 55, H and x_f are

$$H = h' + 38.71F^{3/5}/u \qquad x_f = 0.119F^{2/5} \qquad (22.5a,b)$$

D. Stability Parameter

For stable situations, the stability parameter s is calculated by

$$s = g(\Delta\theta/\Delta z)/T$$

where $\Delta\theta/\Delta z$ is the change in potential temperature with height.

E. Stable Buoyancy Plume Rise

For stable conditions when there is wind, H and x_f are

$$H = h' + 2.6[F/(us)]^{1/3} \quad x_f = 0.00207us^{-1/2} \qquad (22.6a,b)$$

For calm conditions (i.e., no wind) the stable buoyancy rise is

$$H = h' + 4F^{1/4}S^{-3/8} \qquad (22.7)$$

Under stable conditions, the lowest value of Eq. (22.6a) or (22.7) is usually taken as the effective stack height.

The wind speed that yields the same rise from Eq. (22.6a) as that from Eq. (22.7) for calm conditions is

$$u = 0.2746F^{1/4}S^{1/8} \qquad (22.8)$$

F. Gradual Rise: Buoyancy Conditions

Plume rise for distances closer to the source than the distance to the final rise can be estimated from

$$H = h' + 160F^{1/3}x^{2/3}/u \qquad (22.9)$$

where x is the source-to-receptor distance, km. If this height exceeds the final effective plume height, that height should be substituted.

G. Unstable–Neutral Momentum Plume Rise

If the stack gas temperature is below or only slightly above the ambient temperature, the plume rise due to momentum will be greater than that due to buoyancy. For unstable and neutral situations,

$$H = h' + 3dv_s/u \qquad (22.10)$$

This equation is most applicable when v_s/u exceeds 4. Since momentum plume rise occurs quite close to the source, the horizontal distance to the final plume rise is considered to be zero.

H. Stable Momentum Plume Rise

For low-buoyancy plumes in stable conditions, plume height due to momentum is given by

$$H = h' + 1.5[(v^2{}_s d^2 T)/(4T_s u)]^{1/3} s^{-1/6} \tag{22.11}$$

Equation (22.10) should also be evaluated and the lower value is used.

I. Momentum–Buoyancy Crossover

There is a specific difference between stack gas temperature and ambient air temperature that gives the same result for buoyancy rise as for momentum rise. For unstable or neutral conditions this is as follows: For F less than 55,

$$(T_s - T)_c = 0.0297 T_s v_s^{1/3}/d^{2/3} \tag{22.12}$$

For F equal to or greater than 55,

$$(T_s - T)_c = 0.00575 T_s v_s^{2/3}/d^{1/3} \tag{22.13}$$

For stable conditions,

$$(T_s - T)_c = 0.01958 T v_s s^{1/2} \tag{22.14}$$

J. Maximum Concentrations as a Function of Wind Speed and Stability

Using the source in the example in Chapter 21 ($Q = 0.37$, $h = 20$, $d = 0.537$, $v_s = 20$, and $T_s = 350$), with plume rise calculated using the above equations, maximum ground-level concentrations are shown (Fig. 22.1) as functions of stability class and wind speed calculated using the Gaussian model with

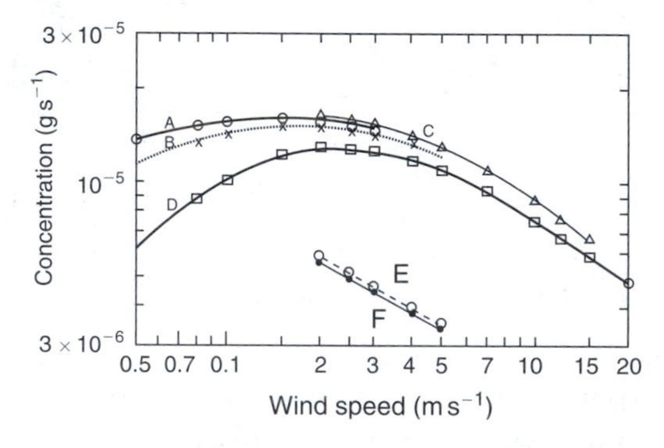

Fig. 22.1. Concentration of an air pollutant at the point of maximum ground-level concentration as a function of wind speed and Pasquill stability category (A–F).

Pasquill–Gifford dispersion parameters. Maximum concentrations are nearly the same for stabilities A, B, and C and occur at wind speeds of 1.5–2.0 $m\,s^{-1}$. The maximum for D stability occurs at around $u = 2.5\,m\,s^{-1}$. Because of the competing effects of dilution by wind and lower effective stack heights with higher wind speeds, concentrations do not change rapidly with wind speed. For E and F stabilities the concentrations are nearly the same (assuming that $\Delta\theta/\Delta z$ is 0.02 $K\,m^{-1}$ for E stability and 0.035 for F stability), but are considerably less than for the unstable and neutral cases.

II. MODELING TECHNIQUES

Gaussian techniques, discussed in Chapter 21, are reasonable for estimates of concentrations of nonreactive pollutants within 20 km of point sources. It is preferable to utilize on-site wind fluctuation measurements to estimate the horizontal and vertical spreading of a pollutant plume released from a point source.

In addition to the Gaussian modeling techniques already discussed, four other methods will be considered.

A. Box Model

Models which assume uniform mixing throughout the volume of a three-dimensional box are useful for estimating concentrations, especially for first approximations. For steady-state emission and atmospheric conditions, with no upwind background concentrations, the concentration is given by

$$\chi = \Delta x q_a/(z_i u) \tag{22.15}$$

where χ is the steady-state concentration, Δx is the distance over which the emissions take place, q_a is the area emission rate, z_i is the mixing height, and u is the mean wind speed through the vertical extent of the box.

When there is an upwind background concentration χ_b and the mixing height is rising with time into a layer aloft having an average concentration of χ_a, the equation of continuity is

$$\delta\chi/\delta t = [\Delta x q_a + u z_i(\chi_b - \chi) + \Delta x(\Delta z_i/\Delta t)(\chi_a - \chi)]/\Delta x z_i \tag{22.16}$$

This forms a basis for an urban photochemical box model (PBM) [6] discussed later in this chapter.

B. Narrow Plume Hypothesis

By assuming that the principal contributors to the concentration at a receptor are the sources directly upwind, especially those nearby, the concentration due to area sources can be calculated using the vertical growth rate rather than uniform vertical mixing and considering the specific area emission rate of each area upwind of the receptor. Area emission rate changes in the crosswind direction are neglected as being relatively unimportant. The expansion in the vertical is usually considered using the Gaussian vertical growth [7, 8].

C. Gradient Transport Models

The mean turbulent flux of concentration in the vertical direction z is $w'\chi'$. Assuming that this turbulent flux is proportional to the gradient of concentration, and in the direction from higher to lower concentrations, an overall diffusivity K can be defined as

$$\overline{w'\chi'} = -K(\delta\chi / \delta z) \tag{22.17}$$

The change in concentration with respect to time can be written as

$$\frac{\delta\chi}{\delta t} + \left(\overline{u}\frac{\delta\chi}{\delta x} + \overline{v}\frac{\delta\chi}{\delta y} + \overline{w}\frac{\delta\chi}{\delta z}\right) = \frac{\delta}{\delta x}K_x\frac{\delta\chi}{\delta x} + \frac{\delta}{\delta y}K_y\frac{\delta\chi}{\delta y} + \frac{\delta}{\delta z}K_z\frac{\delta\chi}{\delta z} + S \tag{22.18}$$

where the term in parentheses on the left accounts for advection. The terms on the right account for diffusivities in three directions, K_x, K_y, and K_z (where K_x is in the direction of the wind, K_y is horizontally crosswind, and K_z is vertically crosswind), and S represents emissions. This equation is the basis for the gradient transport model, which can handle varying wind and diffusivity fields. The vector speeds $\overline{u}, \overline{v}$, and $\overline{w}$ (where $\overline{u}$ is in the direction of the wind, $\overline{v}$ is horizontally crosswind, and $\overline{w}$ is vertically crosswind) and concentrations imply both time and space scales. Fluctuations over times and distances less than these scales are considered as turbulence and are included in the diffusivities.

The gradient transport model is most appropriate when the turbulence is confined to scales that are small relative to the pollutant volume. It is therefore most applicable to continuous line and area sources at ground level, such as automobile pollutants in urban areas, and to continuous or instantaneous ground-level area sources. It is not appropriate for elevated point source diffusion until the plume has grown larger than the space scale. Numerical rather than analytical solutions of Eq. (22.18) are used.

Errors in advection may completely overshadow diffusion. The amplification of random errors with each succeeding step causes numerical instability (or distortion). Higher-order differencing techniques are used to avoid this instability, but they may result in sharp gradients, which may cause negative concentrations to appear in the computations. Many of the numerical instability (distortion) problems can be overcome with a second-moment scheme [9] which advects the moments of the distributions instead of the pollutants alone. Six numerical techniques were investigated [10], including the second-moment scheme; three were found that limited numerical distortion: the second-moment, the cubic spline, and the chapeau function.

In the application of gradient transfer methods, horizontal diffusion is frequently ignored, but the variation in vertical diffusivity must be approximated [11–14].

D. Trajectory Models

In its most common form, a trajectory model moves in a vertical column, with a square cross section intersecting the ground, at the mean wind speed, with pollutants added to the bottom of the column as they are generated by each location over which the column passes. Treatment of vertical dispersion varies among models, from those which assume immediate vertical mixing throughout the column to those which assume vertical dispersion using a vertical coefficient K_z with a suitable profile [15].

Modeling a single parcel of air as it is being moved along allows the chemical reactions in the parcel to be modeled. A further advantage of trajectory models is that only one trajectory is required to estimate the concentration at a given endpoint. This minimizes calculation because concentrations at only a limited number of points are required, such as at stations where air quality is routinely monitored. Since wind speed and direction at the top and the bottom of the column are different, the column is skewed from the vertical. However, for computational purposes, the column is usually assumed to remain vertical and to be moved at the wind speed and direction near the surface. This is acceptable for urban application in the daytime, when winds are relatively uniform throughout the lower atmosphere.

Trajectory models of a different sort are used for long-range transport, because it is necessary to simulate transport throughout a diurnal cycle in which the considerable wind shear at night transports pollutants in different directions. Expanding Gaussian puffs can be used, with the expanded puff breaking, at the time of maximum vertical mixing, into a series of puffs initially arranged vertically but subsequently moving with the appropriate wind speed and direction for each height.

III. MODELING NONREACTIVE POLLUTANTS

A. Seasonal or Annual Concentrations

In estimating seasonal or annual concentrations from point or area sources, shortcuts can generally be taken rather than attempting to integrate over short intervals, such as hour-by-hour simulation. A frequent shortcut consists of arranging the meteorological data by joint frequency of wind direction, wind speed, and atmospheric stability class, referred to as a *STability ARray* (STAR). The ISCLT (Industrial Source Complex Long Term) [16] is a model of this type and is frequently used to satisfy regulatory requirements where concentrations averaged over 1 year (but not shorter averaging times) or longer are required. Further simplification may be achieved by determining a single effective wind speed for each stability–wind direction sector combination by weighting $1/u$ by the frequency of each wind speed class for each such wind direction–stability combination. Calculations for each sector are made, assuming that the frequency of wind direction is uniform across the sector.

B. Single Sources: Short-Term Impact

Gaussian plume techniques have been quite useful for determining the maximum impact of single sources, which over flat terrain, occurs within 10–20 km of the source. The ISCST model [16] is usually used to satisfy regulatory requirements. Because the combination of conditions that produces multihour high concentrations cannot be readily identified over the large range of source sizes, it has been a common practice to calculate the impact of a source for each hour of the year for a large number of receptors at specific radial distances from the source for 36 directions from the source, e.g., every 10°. Averaging and analysis can proceed as the calculations are made to yield, upon completion of a year's simulation, the highest and second-highest concentrations over suitable averaging times, such as 3 and 24 h. Frequently, airport surface wind data have been utilized as input for such modeling, extrapolating the surface wind speed to stack top using a power law profile, with the exponent dependent on stability class, which is also determined from the surface data. Although the average hourly wind direction at stack top and plume level is likely to be different from that at the surface, this has been ignored because hourly variations in wind direction at plume level closely parallel surface directional variations. Although the true maximum concentration may occur in a somewhat different direction from that calculated, its magnitude will be closely approximated.

Several point source algorithms, Hybrid Plume Dispersion Model (HPDM) [17, 18] and TUPOS [19, 20], incorporate the use of fluctuation statistics (the standard deviations of horizontal and vertical wind directions) and non-Gaussian algorithms for strongly convective conditions. Because during strong convective conditions, thermals with updrafts occupy about 30–35% of the area and slower descending downdrafts occupy 65–70% of the area, the resulting distribution of vertical motions are not Gaussian but have a smaller number of upward motions, but with higher velocity, and a larger number of downward motions, but with lower velocity. These skewed vertical motion distributions then cause non-Gaussian vertical distributions of pollutant concentrations.

C. Multiple Sources and Area Sources

The problem, already noted, of not having the appropriate plume transport direction takes on added importance when one is trying to determine the effects of two or more sources some distance apart, since an accurate estimate of plume transport direction is necessary to identify critical periods when plumes are superimposed, increasing concentrations.

In estimating concentrations from area sources, it is important to know whether there is one source surrounded by areas of no emissions or whether the source is just one element in an area of continuous but varying emissions.

To get an accurate estimate of the concentrations at all receptor positions from an isolated area source, an integration should be done over both the alongwind and crosswind dimensions of the source. This double integration is accomplished in the point, area, and line (PAL) source model [21] by approximating the area source using a number of finite crosswind line sources. The concentration due to the area source is determined using the calculated concentration from each line source and integrating numerically in the alongwind direction.

If the receptor is within an area source, or if emission rates do not vary markedly from one area source to another over most of the simulation area, the narrow plume hypothesis can be used to consider only the variation in emission rates from each area source in the alongwind direction. Calculations are made as if from a series of infinite crosswind line sources whose emission rate is assigned from the area source emission rate directly upwind of the receptor at the distance of the line source. The Atmospheric Turbulence and Diffusion Laboratory (ATDL) model [22] accomplishes this for ground-level area sources. The Regional Air Monitoring (RAM) model [8] does this for ground-level or elevated area sources.

Rather than examining the variation of emissions with distance upwind from the receptor as already described, one can simplify further by using the area emission rate of only the emission square in which the receptor resides [23]. The concentration χ is then given by

$$\chi = Cq_a/u \tag{22.19}$$

where q_a is the area emission rate, u is the mean wind speed over the simulation period, and the constant C is dependent on the stability, the effective height of emission of the sources, and the characteristics of the pollutant. For estimation of annual concentrations with this method, $C = 50$, 200, and 600 for SO_2, particulate matter (PM), and CO, respectively [24].

D. Pollutants that Deposit

The Fugitive Dust Model (FDM) [25] was formulated to estimate air concentrations as well as deposition from releases of airborne dust. It has a greatly improved deposition mechanism compared with that in previous models. It considers the mass removed from the plume through deposition as the plume is moved downwind. Up to 20 particle size fractions are available. The particle emissions caused as material is raised from the surface by stronger winds is built internally into this model. It has the capacity of making calculations for PAL sources. The area source algorithm has two options, simulation of the area source by five finite line sources perpendicular to wind flow, or a converging algorithm which provides greater accuracy. Currently, the model should be used for releases at or below 20 m above ground level.

E. Dispersion from Sources over Water

The Offshore and Coastal Dispersion (OCD) model [26] was developed to simulate plume dispersion and transport from offshore point sources to receptors on land or water. The model estimates the overwater dispersion by use of wind fluctuation statistics in the horizontal and the vertical measured at the overwater point of release. Lacking these measurements, the model can make overwater estimates of dispersion using the temperature difference between water and air. Changes taking place in the dispersion are considered at the shoreline and at any point where elevated terrain is encountered.

F. Dispersion over Complex Terrain

Development efforts in complex terrain by U.S. EPA researchers using physical modeling, both in the wind tunnel and in the towing tank, and field studies at three locations have resulted in the development of CTDMPLUS complex terrain dispersion model plus the calculation of concentrations for unstable conditions [27]. Complex terrain is the situation in which there are receptor locations above stack top. Using the meteorological conditions and the description of the nearby terrain feature, the model calculates the height of a dividing streamline. Releases that take place below the height of this streamline tend to seek a path around the terrain feature; releases above the streamline tend to rise over the terrain feature. Because the dispersion is calculated from fluctuation statistics, the meteorological measurements to provide data input for the model are quite stringent, requiring the use of tall instrumented towers. Evaluations of both the model [28] and a screening technique derived from the model [29] indicate that the model does a better job of estimating concentrations than previous complex terrain models.

IV. MODELING POLLUTANT TRANSFORMATIONS

A. Individual Plumes

Most pollutants react after release. However, many pollutants have such long half-lives in the atmosphere, it is safe to assume for most modeling purposes that they are nonreactive. However, numerous other pollutants undergo degradation within the atmosphere to such an extent that the model needs to consider their transformation to properly predict concentrations of the parent compound. The transformation is affected by numerous factors, including cloud chemistry and other aspects of water vapor, aerosol concentrations, catalysis, and photochemistry. In fact, many of the processes described in Table 6.1 occur in the atmosphere.

An understanding of the transformation of SO_2 and NO_x into other constituents no longer measurable as SO_2 and NO_x is needed to explain mass balance changes from one plume cross section to another. This loss of the primary pollutant SO_2 has been described as being exponential, and rates up to 1% per hour have been measured [30]. The secondary pollutants generated by transformation are primarily sulfates and nitrates.

The horizontal dispersion of a plume has been modeled by the use of expanding cells well mixed vertically, with the chemistry calculated for each cell [31]. The resulting simulation of transformation of NO to NO_2 in a power plant plume by infusion of atmospheric ozone is a peaked distribution of NO_2 that resembles a plume of the primary pollutants, SO_2 and NO. The ozone distribution shows depletion across the plume, with maximum depletion in the center at 20 min travel time from the source, but relatively uniform ozone concentrations back to initial levels at travel distances 1 h from the source.

B. Urban-Scale Transformations

Approaches used to model ozone formation include box, gradient transfer, and trajectory methods. Another method, the particle-in-cell method, advects centers of mass (that have a specific mass assigned) with an effective velocity that includes both transport and dispersion over each time step. Chemistry is calculated using the total mass within each grid cell at the end of each time step. This method has the advantage of avoiding both the numerical diffusion of some gradient transfer methods and the distortion due to wind shear of some trajectory methods.

It is not feasible to model the reaction of each hydrocarbon species with oxides of nitrogen. Therefore, hydrocarbon species with similar reactivities are lumped together, e.g., into four groups of reactive hydrocarbons: olefins, paraffins, aldehydes, and aromatics [32].

In addition to possible errors due to the steps in the kinetic mechanisms, there may be errors in the rate constants due to the smog chamber databases from which they were derived. A major shortcoming is the limited amount of quality smog chamber data available.

The emission inventory and the initial and boundary conditions of pollutant concentrations have a large impact on the ozone concentrations calculated by photochemical models.

To model a decrease in visibility, the chemical formation of aerosols from sulfur dioxide and oxidants must be simulated.

In a review of ozone air quality models, Seinfeld [33] indicates that the most uncertain part of the emission inventories is the hydrocarbons. The models are especially sensitive to the reactive organic gas levels, speciation, and the concentrations aloft of the various species. He points out the need for improvement in the three-dimensional wind fields and the need for hybrid models that can simulate sub-grid-scale reaction processes to incorporate properly effects of concentrated plumes. Schere [34] points out that we need to improve the way vertical exchange processes are included in the model. Also, although the current models estimate ozone quite well, the atmospheric chemistry needs improvement to better estimate the concentrations of other photochemical components such as peroxyacyl nitrate (PAN), the hydroxyl radical (OH), and volatile organic compounds (VOCs). In addition to the improvement of databases, including emissions, boundary concentrations, and meteorology, incorporation of the urban ozone with the levels at larger scales is needed.

C. Regional-Scale Transformations

In order to formulate appropriate control strategies for oxidants in urban areas, it is necessary to know the amount of oxidant already formed in the air reaching the upwind side of the urban area under various atmospheric conditions. Numerous physical and chemical processes are involved in modeling transformations [35, 36] on the regional scale (several days, 1000 km): (1) horizontal transport; (2) photochemistry, including very slow reactions; (3) nighttime chemistry of the products and precursors of photochemical reactions; (4) nighttime wind shear, stability stratification, and turbulence episodes associated with the nocturnal jet; (5) cumulus cloud effects—venting pollutants from the mixed layer, perturbing photochemical reaction rates in their shadows, providing sites for liquid-phase reactions, influencing changes in the mixed-layer depth, perturbing horizontal flow; (6) mesoscale vertical motion induced by terrain and horizontal divergence of the large-scale flow; (7) mesoscale eddy effects on urban plume trajectories and growth rates; (8) terrain effects on horizontal flows, removal, and diffusion; (9) sub-grid-scale chemistry processes resulting from emissions from sources smaller than the model's grid can resolve; (10) natural sources of hydrocarbons, NO_x, and stratospheric ozone; and (11) wet and dry removal processes, washout, and deposition.

Approaches to long-term (monthly, seasonal, annual) regional exchanges are EURMAP [37] for Europe and ENAMAP [38] for eastern North America. These two models can calculate SO_2 and sulfate air concentrations as well as dry and wet deposition rates for these constituents. The geographic region of interest (for Europe, about 2100 km N–S by 2250 km E–W) is divided into an emissions grid having approximately 50 by 50 km resolution. Calculations are performed by releasing a 12-h average emission increment or "puff" from each cell of the grid and tracking the trajectories of each puff by 3-h time steps according to the 850-mb winds interpolated objectively for the puff position from upper-air data.

Uniform mixing in the vertical to 1000 m and uniform concentrations across each puff as it expands with the square root of travel time are assumed. A $0.01\ h^{-1}$ transformation rate from SO_2 to sulfate and 0.029 and $0.007\ h^{-1}$ dry deposition rates for SO_2 and sulfate, respectively, are used. Wet deposition is dependent on the rainfall rate determined from the surface observation network every 6 h, with the rate assumed to be uniform over each 6-h period. Concentrations for each cell are determined by averaging the concentrations of each time step for the cell, and deposition is determined by totaling all depositions over the period.

The EURMAP model has been useful in estimating the contribution to the concentrations and deposition on every European nation from every other European nation. Contributions of a nation to itself range as follows: SO_2 wet deposition, 25–91%; SO_2 dry deposition, 31–91%; sulfate wet deposition, 2–46%; sulfate dry deposition, 4–57%.

In the application of the model to eastern North America, the mixing height is varied seasonally, and hourly precipitation data are used.

V. MODELING AIR POLLUTANTS

A. The Need for Models

Models provide a means for representing a real system in an understandable way.[1] They take many forms, beginning with "conceptual models" that explain the way a system works, such as delineation of all the factors and parameters of how a particle moves in the atmosphere after its release from a power plant. Conceptual models help to identify the major influences on where a chemical is likely to be found in the environment, and as such, need to be developed to help target sources of data needed to assess an environmental problem.

In general, developing an air pollution model requires two main steps. First, a model of the domain and the processes being studied must be defined. Then, at the model boundaries, a model of the boundary conditions is especially needed to represent the influential environment surrounding the study domain. The quality of the model study is related to the accuracy and representativeness to the actual study.

B. Physical Models

Research scientists often develop "physical" or "dynamic" models to estimate the location where a contaminant would be expected to move under controlled conditions, only on a much smaller scale. For example, the US Environmental Protection Agency's (EPA) wind tunnel facility in Research Triangle Part is sometimes used to support studies when local buildings and terrain have significant influences. For example, the wind tunnel housed a scaled model of the town of East Liverpool, Ohio and its surrounding terrain to estimate the movement of the plume from an incinerator. The plume could be observed under varying conditions, including wind direction and height of release. Like all models, the dynamic models' accuracy is dictated by the

[1] See Leete, J., Groundwater modeling in health risk assessment, in *A Practical Guide to Understanding, Managing and Reviewing Environmental Risk Assessment Reports* (Benjamin, S., and Belluck, D., eds.), Chapter 17. Lewis Publishers, Boca Raton, FL, 2001.

Fig. 22.2. Example of neutrally buoyant smoke released from the WTC physical model, showing flow from left to right is displayed; natural light is illuminating the smoke and a vertical laser sheet is additionally illuminating the plume near the centerline of source. *Source*: US Environmental Protection Agency. Perry, S., and Heath, D., *Fluid Modeling Facility*. Research Triangle Park, NC, 2006.

degree to which the actual conditions can be simulated and the quality of the information that is used. More recently, the wind tunnel was used to simulate pollutant transport and dispersion in Lower Manhattan to simulate possible plumes of pollutants from the collapse of the World Trade Center (WTC) towers on September 11, 2001. The 1:600 scale model was constructed on a turntable to test different wind directions (see Fig. 22.2). In addition to the building of Lower Manhattan and the rubble pile as it appeared approximately 1 week after the collapse, smoke for visualization and tracer gas for measurement dispersion patterns were released from positions throughout the simulated 16 acre site.

The study design includes smoke visualization for a qualitative examination of dispersion in this very complex urban landscape, known as an "urban street canyon." Detailed measurements were taken of flow velocities and turbulence, as well as concentration distribution, which suggested a number of flow phenomena that are not possible to consider using simple Gaussian dispersion algorithms (see Fig. 22.3). These include vertical venting behind large/tall buildings, channeling down street canyons, and both horizontal and vertical re-circulations associated with individual structures

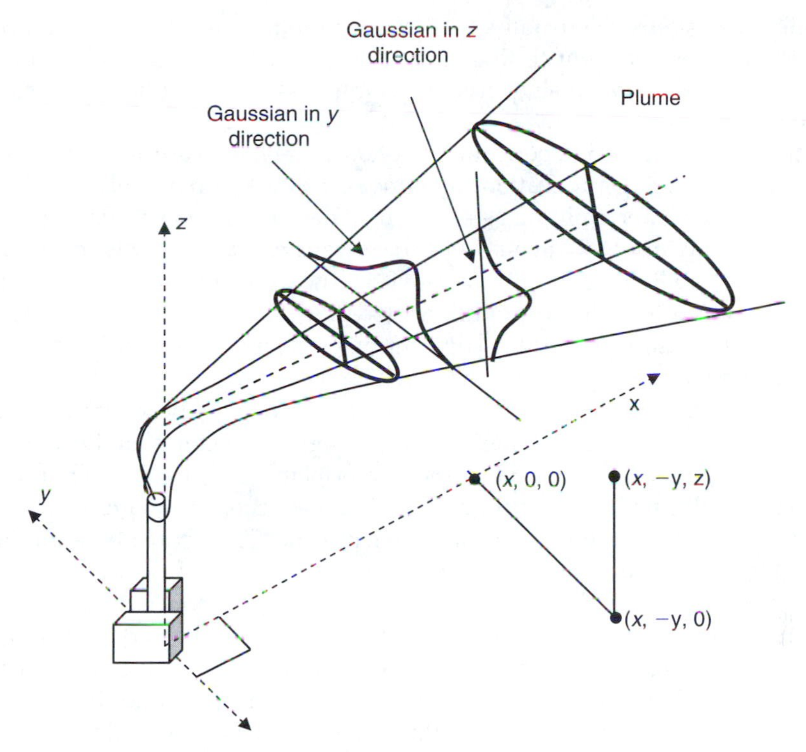

Fig. 22.3. Atmospheric plume model based on random distributions in the horizontal and vertical directions. *Source*: Turner, D., *Workbook of Atmospheric Dispersion Estimates*. Office of Air Programs Publication No. AP-26 (NTIS PB 191 482). US Environmental Protection Agency, Research Triangle Park, NC, 1970.

and groups of tall and tightly compacted buildings (such as the Wall Street area in the southeast edge of Manhattan, New York).

C. Numerical Simulation Models[2]

Numerical models apply mathematical expressions to approximate a system. System is a thermodynamic concept. Recall that we deal with three basic types of systems in air pollution sciences:

1. Isolated systems, in which no matter or energy crosses the boundaries of the system (i.e. no work can be done on the system).
2. Closed systems, in which energy can exchange with surroundings, but no matter crosses the boundary.
3. Open systems, in which both matter and energy freely exchange across system boundaries.

[2] Alan H. Huber contributed substantially to this discussion. The author appreciates these insights from one of the pioneers in the application of CFD to air pollutant transport phenomena.

Isolated systems are usually only encountered in highly controlled reactors, so are not pertinent to this discussion. In fact, most air pollution systems are open, but with simplifying assumptions, some subsystems can be treated as closed.

Rapid advances in high performance computing hardware and software are leading to increasing applications of numerical simulation models that characterize an atmospheric plume as a system. These methods simulate spatially and temporally resolved details of the pathways of air pollutants from source emissions to pollutant concentrations within the local "virtual" microenvironment. This approach to air pollution modeling applies numerical representation of fluid flow in a fluid continuum. Such an approach is known as computational fluid dynamics (CFD), which determines a numerical solution to the governing equations of fluid flow while advancing the solution through space or time to obtain a numerical description of the complete flow field of interest. CFD models are based on the first principles of physics; in particular, they start with single-phase flow based on Navier–Stokes equations.

The Navier–Stokes equations are deterministic. They consider simultaneously the conservation of energy, momentum, and mass. Practical CFD solutions require a sub-grid-scale model for turbulence. As computing capacities advance, the scale where turbulence is modeled can be reduced and the application of higher-order numerical methods can support increasingly representative and accurate turbulence models. The transport and dispersion of air pollutants is part of the fluid flow solution. CFD methods have been developed and routinely applied to aerospace and automotive industrial applications and are now being extended to environmental applications.

High-fidelity fine-scale CFD simulation of pollutant concentrations within microenvironments (e.g. near roadways or around buildings) is feasible now that high performance computing has become more accessible. Fine-scale CFD simulations have the added advantage of being able to account rigorously for topographical details such as terrain variations and building structures in urban areas as well as their local aerodynamics and turbulence. Thermal heat fluxes may be added to terrain and building surfaces to simulate the thermal atmospheric boundary layer, their influences on pollution transport and dispersion. The physics of particle flow and chemistry can be included in CFD simulations. The results of CFD simulations can be directly used to better understand specific case studies as well as to support the development of better-simplified algorithms for adoption into other modeling systems. For example, CFD simulations with fine-scale physics and chemistry can enhance and complement photochemical modeling with Community Multiscale Air Quality (CMAQ). Also, detailed CFD simulation for a complex site study can be used to develop reliable parameterizations to support simplified and rapid application air pollution model.

A few examples illustrate the utility of CFD modeling. Currently the main features of CFD application are the inclusion of site-specific geometry and dynamic processes affecting air pollution transport and dispersion. The

Fig. 22.4. A vertical slice of the resolved grid and sample (10%) of wind vectors in a CFD model of New York City. *Courtesy*: Huber, A. H., US Environmental Protection Agency, Research Triangle Park, NC (unpublished work, used with permission).

future will bring more refined spatial and temporal details, along with particle physics and chemistry. A vertical slice of the domain cells and a 10% sample of calculated wind vectors for these cells are shown in Fig. 22.4. The figure demonstrates a tendency for downward airflow on the windward faces of buildings and upward airflow on the leeward building faces. Figure 22.4 shows two different looking plume depictions for ground-level point emissions. In one case the emissions plume is caught in the leeward building updraft leading to significant vertical mixing, whereas the other case shows emissions remaining close to the ground as they move through the street canyons. Figure 22.5 shows vertical slices of concentration for roadway emissions represented as a box along the streets. There are significant differences for the case with a building on only one side of the roadway (a) relative to the street canyon (b), which induces a region of circulation (c).

CFD models have also found much use in estimating indoor and personal exposure to air pollutants. CFD modeling can also be used to simulate indoor air movement and pollution dispersion. For example, Furtaw[3] presents an example of this in a room where air movement vectors and pollutant concentration isopleths are around an emission source (represented by the small

[3] Furtaw., E. J., An overview of human exposure modeling activities at the USEPA's National Exposure Research Laboratory, *Toxicol. Ind. Health* **17** (5–10), 302–314 (2001).

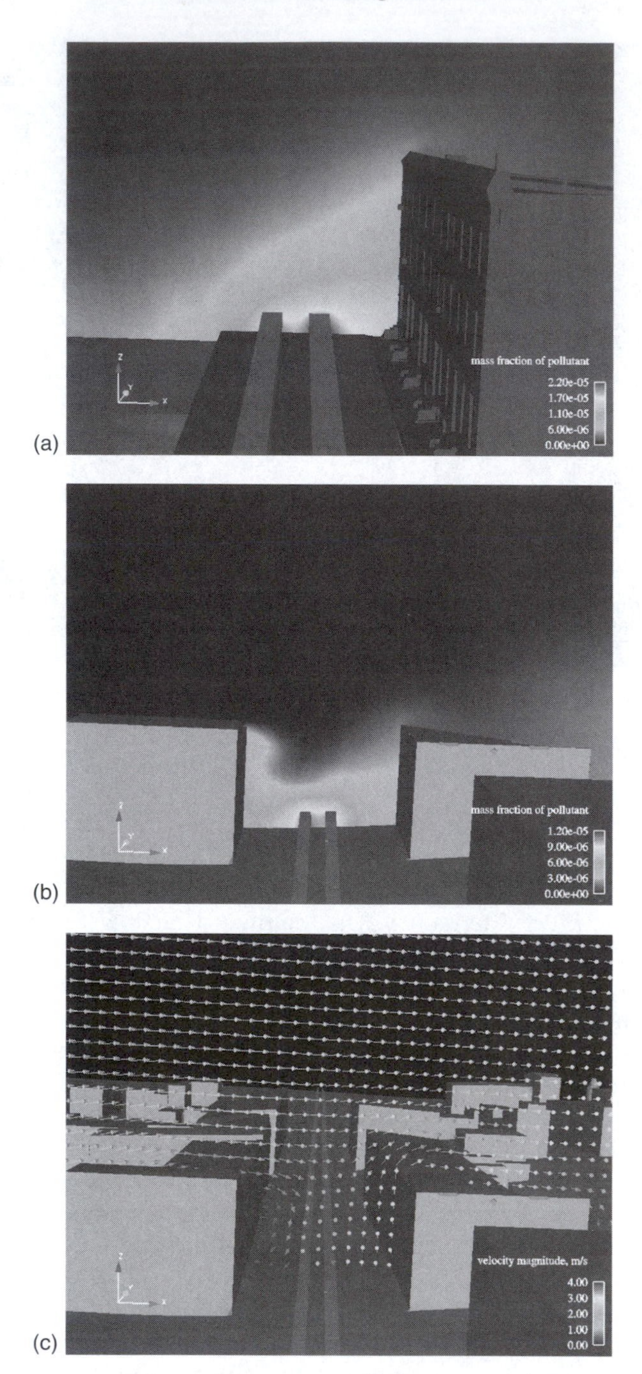

Fig. 22.5. Vertical slice view of a CFD model of roadway emissions represented as a source box along the roadway: (a) concentrations for street bounded by building on one side; (b) concentrations for street canyon; and (c) wind vectors for street canyon. *Courtesy*: Unpublished work by Huber, A. H., US Environmental Protection Agency, Research Triangle Park, NC.

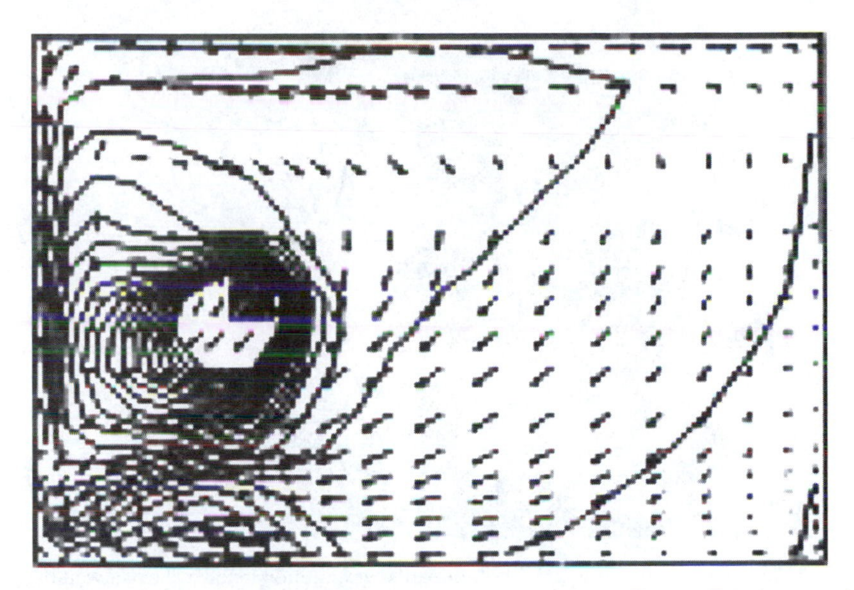

Fig. 22.6. CFD model simulation of indoor air in a room. *Source*: Furtaw, E. J., An overview of human exposure modeling activities at the USEPA's National Exposure Research Laboratory. *Toxicol.Ind. Health* **17** (5–10), 302–314 (2001).

square in Fig. 22.6). While CFD model simulation of every room in many buildings may not be a feasible, example case studies are possible for very critical situations or generic situations leading to the development of exposure factors for typical indoor room environments. Similarly, Settles[4] provides an excellent perspective on the increasing role for CFD model simulations for both ambient and indoor air concentrations as are critical to many homeland security issues involving potential human exposures. Figure 22.7 illustrates an application of CFD model simulations of trace concentrations behind a walking human. CFD model simulations at the human exposure scales can be very detailed and particularly useful in developing human exposure factors applicable for more general population-based human exposure models.

Some of atmospheric concentrations of pollutants consist of a regional background concentration due to long-range transport and regional-scale mixing as well as specific local microenvironmental concentrations. Concentrations within the local microenvironment often dominate a profile of total human exposure to the pollutant. Regional air quality applications are normally applied at grid resolutions larger than 10 km. Urban applications are applied at smaller grid scales but there is a meaningful limit due to the sub-grid-scale process models. In other words, there is no value in applying the model at fine scales without supporting details on the finer resolution. For example, the present framework of the CMAQ modeling system well supports environmental

[4] Settles, G. S., Fluid mechanics and homeland security. *Annu. Rev. Fluid Mech.* **38**, 87–110 (2006).

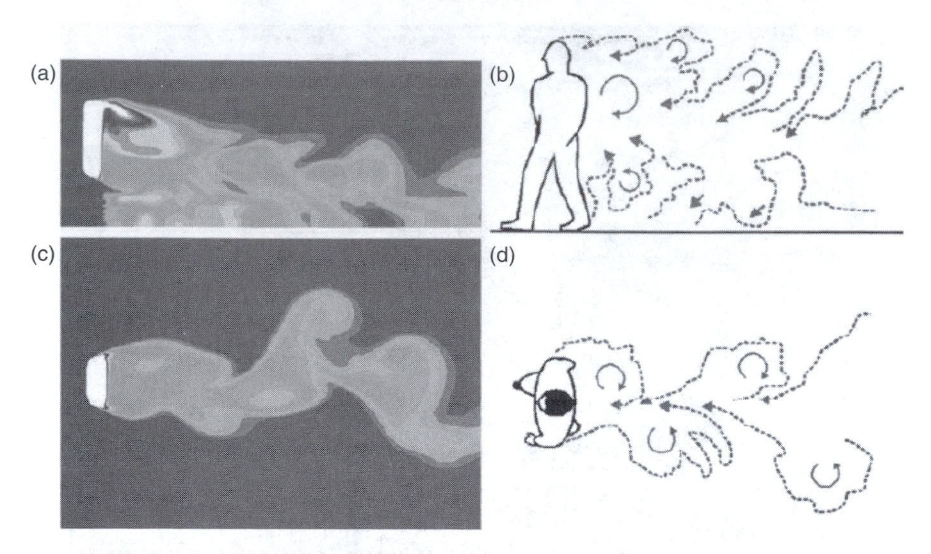

Fig. 22.7. The instantaneous trace contaminant concentration in the wake of a walking person, from Edge *et al.* (2005). Frames (a) and (c) are RANS solutions using a blended k–ϖ/k–ε 2-equation turbulence model and a simplified representation of the human body. Frames (b) and (d) are drawn from flow visualization experiments of an actual walking person. A side view is depicted in frames (a) and (b); frames (c) and (d) show the top view. *Source*: Settles, G. S., Fluid mechanics and homeland security. *Annu. Rev. Fluid Mech.* **38**, 87–110 (2006).

issues where large-sized grid-averaged concentrations are applicable. For example, regional emission control strategies are especially applicable while specific profiles of human exposure concentrations are not.

The large grid-sized modeling system may be used for estimating profiles of human exposure for pollutants having course temporal and spatial distributions. Human exposure concentrations on the sub-grid scales need to apply a sub-grid-scale model including human exposure factors to estimate exposed populations (such as that shown in Fig. 22.8). Such constraints related to local-scale air pollution and human exposure will be further discussed in the modeling sections below. Advances in numerical simulations models will continue to supplement and to improve the more simplified models presented in the following sections, particularly for simulations where building environments and complex terrain are significant.

Transport and fate models can be statistical and/or "deterministic." Statistical models include the pollutant dispersion models, such as the Lagrangian models, which assume Gaussian distributions of pollutants from a point of release (see Fig. 22.3). That is, the pollutant concentrations are normally distributed in both the vertical and horizontal directions from the source. The Lagrangian approach is common for atmospheric releases. "Stochastic" models are statistical models that assume that the events affecting the behavior of a chemical in the environment are random, so such models are based on probabilities.

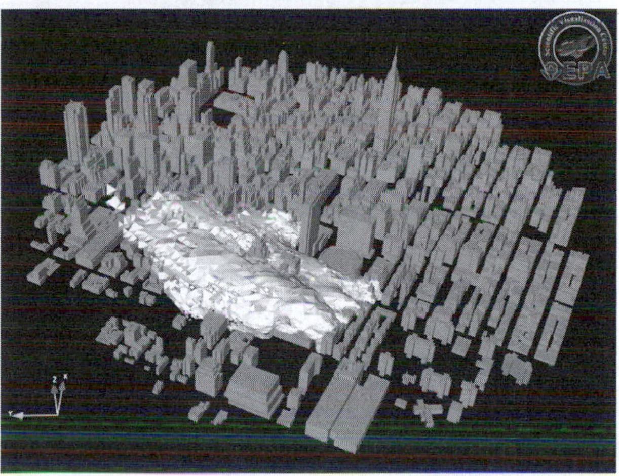

Fig. 22.8. Plumes from ground-level point source emissions in a CFD model of New York City (midtown area). *Courtesy*: Huber, A. H., US Environmental Protection Agency, Research Triangle Park, NC (unpublished work, used with permission).

Deterministic models are used when the physical, chemical, and other processes are sufficiently understood so as to be incorporated to reflect the movement and fate of chemicals. These are very difficult models to develop because each process must be represented by a set of algorithms in the model. Also, the relationship between and among the systems, such as the kinetics and mass balances, must also be represented.

Thus, the modeler must "parameterize" every important event following a chemical's release to the environment. Often, hybrid models using both statistical and deterministic approaches are used, for example, when one part of a system tends to be more random, while another has a very strong basis is physical principals. Numerous models are available to address the

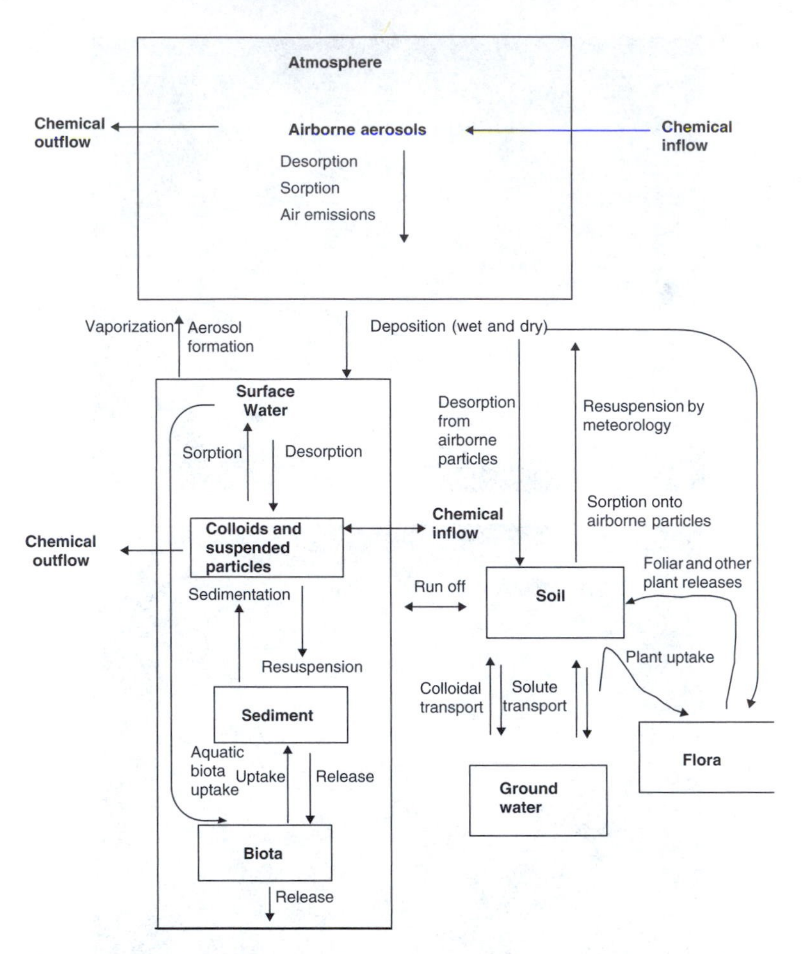

Fig. 22.9. Example framework and flow of a multimedia, compartmental chemical transport, and transformation model. Algorithms and quantities must be provided for each box. Equilibrium constants (e.g. partitioning coefficients) must be developed for each arrow and, steady-state conditions may not be assumed, reaction rates and other chemical kinetics must be developed for each arrow and box.

movement of chemicals through a single environmental media, but increasingly, environmental scientists and engineers have begun to develop "multimedia models," such as compartmental models that help to predict the behavior and changes to chemicals as they move within and among the soil, water, air, sediment, and biota (see Fig. 22.9).[5] Such models will likely see increased use in all environmental science and engineering.

[5] The compartmental or box models such as the one in Fig. 3.29 are being enhanced by environmental scientists and chemical engineers. Much of the information in this figure can be attributed to discussions with Yoram Cohen, a chemical engineering professor at UCLA, and Ellen Cooter, a National Oceanic and Atmospheric Administration modeler on assignment to the US EPA's National Exposure Research Laboratory in Research Triangle Park, NC.

Environmental chemodynamics describe the movement and change of chemicals in the environment (see Chapter 6). From an air pollution modeling perspective, the air is among one compartment among many. Most air pollution modeling is conducted as if the troposphere were the only compartment. This is usually a reasonable approach. However, we must keep in mind that the influences shown in Fig. 22.9 are always present. The air is a medium by which pollutants are transported. It is also the location of chemical reactions and physical changes, such as the agglomeration of smaller particles and the formation of particles from gas-phase pollutants. Fluid properties have a great bearing on the movement and distribution of contaminants. However, specific partitioning relationships that control the "leaving" and "gaining" of pollutants among particle, soil, and sediment surfaces, the atmosphere, organic tissues, and water can also be applied to model where a chemical ends up and the chemical form it assumes. This is known as environmental fate. These relationships are sorption, solubility, volatilization, and organic carbon–water partitioning, which are respectively expressed by coefficients of sorption (distribution coefficient (K_D) or solid–water partition coefficient (K_p)), dissolution or solubility coefficients, air–water partitioning (and the Henry's law (K_H) constant), and organic carbon–water (K_{oc}). These equilibrium coefficients are discussed in detail in Chapter 6.

D. Using Models to Predict Air Pollution

Air quality models mathematically simulate the physical and chemical processes that affect air pollutants as they disperse and react in the atmosphere. Meteorological data and source information (e.g. emission rates and stack height) are put into models to characterize pollutants that are emitted directly into the atmosphere. Air quality modelers generally refer to these as primary pollutants. Secondary pollutants, those that form as a result of complex chemical reactions within the atmosphere, can also be modeled. Models are a key component of air quality management at all scales. They are widely used by local, state and federal agencies charged with addressing air pollution, especially to identify source contributions to air quality problems and to help to design effective strategies aimed at reducing air pollutants. Models can be used to verify that before issuing a permit, a new source will not exceed ambient air quality standards or, if appropriate, to agree on appropriate additional control requirements. Models are useful as prediction tools, for example, to estimate the future pollutant concentrations from multiple sources after a new regulatory program is in place. This gives a measure of the expected effectiveness of various options in reducing harmful exposures to humans and the environment.

E. Dispersion Modeling

Dispersion modeling is used to predict concentrations at selected downwind receptor locations. According to the EPA, these air quality models help to determine compliance with National Ambient Air Quality Standards (NAAQS), and other regulatory requirements such as New Source Review (NSR) and Prevention of Significant Deterioration (PSD) regulations.[6] These guidelines are periodically revised to ensure that new model developments or expanded regulatory requirements are incorporated. The EPA recommends the AERMOD modeling system be used by states and local regulators to determine their progress in reducing criteria air pollutants under the Clean Air Act (CAA). AERMOD is a steady-state plume model that incorporates air dispersion based on planetary boundary-layer turbulence structure and scaling concepts, including treatment of both surface and elevated sources, and both simple and complex terrain.

AERMOD is better able to characterize plume dispersion than the model it replaced, the Industrial Source Complex (ISC3), which had been EPA's preferred model since 1980. In particular, AERMOD can simulate elevated air pollution sources as well as those at the surface, and can assimilate both simple and complex terrain (see Fig. 22.10). A substantial advance is that AERMOD makes use of updated physics algorithms. Thus, since November 9, 2006, AERMOD replaced ISC3 as the EPA's preferred general-purpose model. Regulators now use it to estimate local levels of ozone, VOCs, carbon monoxide, NO_2, SO_2, PM, and lead. Areas that have been unable to attain CAA standards for these criteria pollutants are required to use AERMOD to revise their State Implementation Plans (SIPs) for existing air pollution sources. States must also use AERMOD for their NSR and PSD determinations.

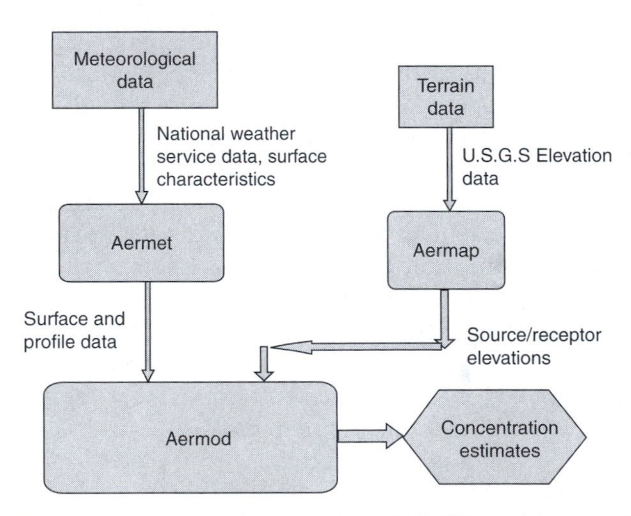

Fig. 22.10. Flow and processing of the complete AERMOD modeling system, which consists of the AERMOD dispersion model and two pre-processors. *Source*: US Environmental Protection Agency.

[6] These models are addressed in Appendix A of EPA's *Guideline on Air Quality Models* (also published as *Appendix W* of 40 CFR Part 51), which was originally published in April 1978 to provide consistency and equity in the use of modeling within the US air quality management system.

AERMOD modeling can be used to extrapolate from a limited number of data points. For example, it was recently used by the EPA in its regulatory impact analysis of the new standard for particulates by comparing a number of regulatory scenarios in terms of the costs and human health and welfare benefits. In this case, the comparison was among scenarios associated with attaining both the selected and one alternative standard for annual and daily $PM_{2.5}$ concentrations:

Combination of annual and daily values (annual/daily in $\mu g\,m^{-3}$)	Scenario
15/65	1997 Standards
15/35	New Standard
14/35	Alternative

One of the urban areas compared in analysis was the Detroit, Michigan. The primary $PM_{2.5}$ impacts at monitor locations in Wayne County, Michigan; the model shows locations where the annual ($15\,\mu g\,m^{-3}$) and daily ($35\,\mu g\,m^{-3}$) standards may well be exceeded.[7] Figure 22.11 provides the spatial gradient

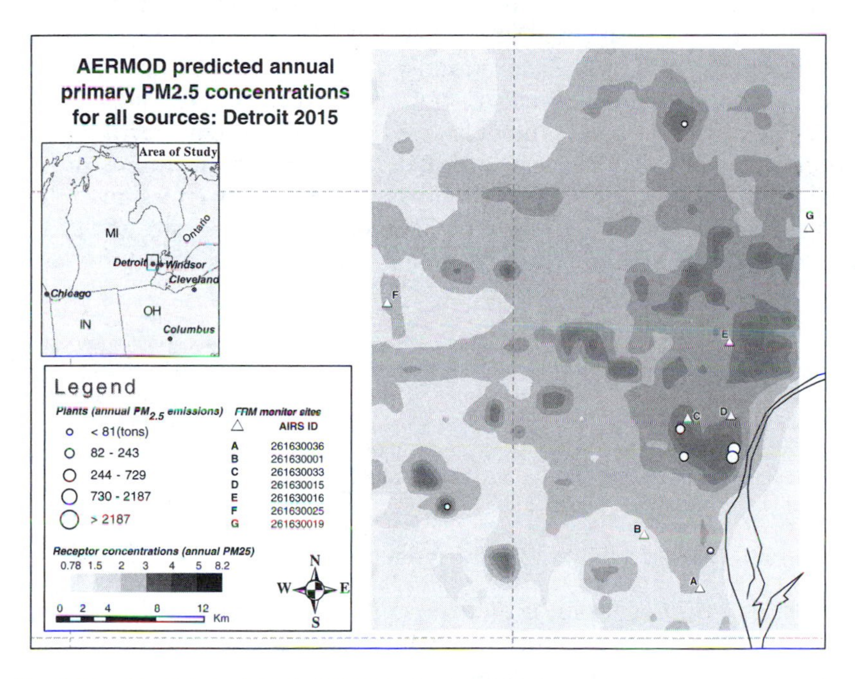

Fig. 22.11. Spatial gradient in Detroit, MI of AERMOD predicted annual primary $PM_{2.5}$ concentrations ($\mu g\,m^{-3}$) for all sources: 2015. *Note*: Dashed line reflect the 36-km grid cells from regional-scale modeling with CMAQ model.

[7] US Environmental Protection Agency, *Regulatory Impact Analysis for the Revised Particulate Matter*, National Ambient Air Quality Standards, 2006.

of primary $PM_{2.5}$ for the urban area associated with emissions from all sources. The model shows that this particular Wayne County monitor is expected to exceed $15\,\mu g\,m^{-3}$ by $2.4\,\mu g\,m^{-3}$ in the year 2015. This indicates that local sources of primary $PM_{2.5}$ contribute $3.3\,\mu g\,m^{-3}$ to this monitor location and that the application of controls for the 15/65 scenario would yield a $0.56\,\mu g\,m^{-3}$ reduction in $PM_{2.5}$ concentrations. The modeling results indicate little additional reductions at this monitor for the 15/35 scenario but an additional $0.46\,\mu g\,m^{-3}$ reduction in $PM_{2.5}$ concentrations for the 14/35 scenario. The AERMOD simulations of annual and daily exceedences are shown in Tables 22.1 and 22.2, respectively.

For non-steady-state conditions, the EPA recommends the CALPUFF modeling system, which is a non-steady-state puff dispersion model that simulates the effects of time- and space-varying meteorological conditions on pollution transport, transformation, and removal. CALPUFF is a multi-layer, multi-species non-steady-state puff dispersion model that simulates the effects of time- and space-varying meteorological conditions on pollution transport, transformation and removal. It can be applied at scales of tens to hundreds of kilometers. It includes algorithms for sub-grid-scale effects (such as terrain impingement), as well as, longer-range effects (such as pollutant removal due to wet scavenging and dry deposition, chemical transformation, and visibility effects of PM concentrations). Thus, CALPUFF is particularly useful for long-range pollutant transport and for complex terrain.

CALPUFF is one of the models used to characterize the plume from Ground Zero at WTC in New York City. The predicted plumes provided general insights into the likely pathway for pollutant emissions released from the collapse and the fire at the WTC site, both temporally and spatially (see Fig. 22.12). Huber *et al.*[8] give some perspective about the use of CALPUFF:

> A CALMET–CALPUFF type modeling system cannot provide as precise an estimate of pollution levels as fuller physics models might, is considered to be adequate for many applications. Even without the precise concentrations, knowing generally where the plume may have been can still help determine where to conduct more refined modeling of human exposures and where to study the population exposure in epidemiological studies. Having such results rapidly as a forecast or screening model could be very valuable to decision-makers.

Other models used by regulatory agencies include:

- BLP—A Gaussian plume dispersion model designed to handle unique modeling problems associated with aluminum reduction plants, and other industrial sources where plume rise and downwash effects from stationary line sources are important.

[8] Huber, A., Georgopoulos, P., Gilliam, R., Stenchikov, G., Wang, S-W., Kelly, B., and Feingersh, H., Modeling air pollution from the collapse of the World Trade Center and assessing the potential impacts on human exposures. *Environ. Manage.* 23–26 (February 2004).

TABLE 22.1

Summary of Modeled Source Contributions of Primary $PM_{2.5}$ to Monitors with Potential Annual Exceedences in Detroit, Michigan for the Year 2015

	Model predicted annual concentrations ($\mu g\ m^{-3}$)[a]				
Source sectors	Primary $PM_{2.5}$ emissions (ton $year^{-1}$)	Primary $PM_{2.5}$ contribution	15/65 Control scenario	15/35 Control scenario	14/35 Control scenario
Wayne County Monitor #261630033, Annual DV = 17.4					
Other industrial sources	1 375	0.712	0.171	0.000	0.222
CMV, aircraft, locomotive	638	0.540	0.191	0.000	0.000
Metal processing	852	0.484	0.037	0.000	0.000
Onroad (gasoline and diesel)	1 187	0.336	0.000	0.025	0.025
Commercial cooking	984	0.271	0.050	0.000	0.000
Area fugitive dust	10 270	0.237	0.000	0.000	0.000
Power sector	18 016	0.233	0.059	0.000	0.014
Other area	888	0.210	0.000	0.000	0.168
Nonroad (gasoline and diesel)	1 603	0.197	0.033	0.019	0.019
Natural gas combustion	119	0.034	0.000	0.000	0.000
Residential wood burning	703	0.026	0.005	0.000	0.000
Residential waste burning	1 741	0.015	0.000	0.000	0.007
Glass manufacturing	334	0.010	0.000	0.000	0.000
Cement manufacturing	700	0.009	0.009	0.000	0.000
Auto industry	413	0.005	0.000	0.000	0.000
Prescribed/open burning	444	0.004	0.000	0.000	0.003
Point fugitive dust	15	0.001	0.000	0.000	0.000
Wildfires	51	0.001	0.000	0.000	0.000
Total (all sources)	40 333	3.324	0.556	0.043	0.459
Wayne County Monitor #261630015, Annual DV = 15.69					
CMV, aircraft, locomotive	638	0.727	0.257	0.000	0.000
Metal processing	852	0.399	0.031	0.000	0.000
Other industrial sources	1 375	0.395	0.094	0.000	0.125
Commercial cooking	984	0.365	0.068	0.000	0.000
Power sector	18 016	0.311	0.064	0.000	0.031
Onroad (gasoline and diesel)	1 187	0.214	0.000	0.016	0.016
Area fugitive dust	10 270	0.183	0.000	0.000	0.000
Other area	888	0.154	0.000	0.000	0.123
Nonroad (gasoline and diesel)	1 603	0.147	0.025	0.014	0.014
Residential wood burning	703	0.024	0.005	0.000	0.000
Residential waste burning	1 741	0.013	0.000	0.000	0.007
Glass manufacturing	334	0.009	0.000	0.000	0.000
Cement manufacturing	700	0.008	0.008	0.000	0.000
Auto industry	413	0.005	0.000	0.000	0.000
Prescribed/open burning	444	0.003	0.000	0.000	0.003
Natural gas combustion	119	0.002	0.000	0.000	0.000
Point fugitive dust	15	0.001	0.000	0.000	0.000
Wildfires	51	0.000	0.000	0.000	0.000
Total (all sources)	40 333	2.962	0.550	0.030	0.319

[a] Natural gas combustion source category results are adjusted to reflect new emissions factor (94% reduction).

Source: US Environmental Protection Agency, Regulatory Impact Analysis for the Revised Particulate Matter National Ambient Air Quality Standards, 2006.

TABLE 22.2

Summary of Modeled Source Contributions of Primary $PM_{2.5}$ to Monitors with Potential Daily Exceedences in Detroit, Michigan for the Year 2015

	Model predicted annual concentrations ($\mu g\,m^{-3}$)[a]				
Source sectors	Primary $PM_{2.5}$ emissions (ton $year^{-1}$)	Primary $PM_{2.5}$ contribution	15/65 Control scenario	15/35 Control scenario	14/35 Control scenario
Wayne County Monitor #261630033, Daily DV = 39.06[b]					
Power sector	18016	0.896	0.344	0.000	0.021
Metal processing	852	0.623	0.048	0.000	0.000
Cement manufacturing	700	0.024	0.024	0.000	0.000
Glass manufacturing	334	0.025	0.000	0.000	0.000
Auto industry	413	0.014	0.000	0.000	0.000
Other industrial sources	1375	1.691	0.378	0.000	0.579
CMV, aircraft, locomotive	638	0.833	0.297	0.000	0.000
Nonroad (gasoline and diesel)	1603	0.296	0.041	0.030	0.030
Onroad (gasoline and diesel)	1187	0.475	0.000	0.035	0.035
Residential waste burning	1741	0.022	0.000	0.000	0.011
Residential wood burning	703	0.038	0.008	0.000	0.000
Commercial cooking	984	0.587	0.109	0.000	0.000
Prescribed/open burning	444	0.006	0.000	0.000	0.005
Wildfires	51	0.000	0.000	0.000	0.000
Area fugitive dust	10270	0.522	0.000	0.000	0.000
Point fugitive dust	15	0.002	0.000	0.000	0.000
Other area	888	0.362	0.000	0.000	0.289
Natural gas combustion	119	0.003	0.000	0.000	0.000
Total (all sources)	40333	6.418	1.250	0.065	0.970
Wayne County Monitor #261630015, Daily DV = 38.6[b]					
Power sector	18016	0.730	0.149	0.000	0.083
Metal processing	852	1.604	0.123	0.000	0.000
Cement manufacturing	700	0.003	0.003	0.000	0.000
Glass manufacturing	334	0.008	0.000	0.000	0.000
Auto industry	413	0.003	0.000	0.000	0.000
Other industrial sources	1375	0.844	0.264	0.000	0.153
CMV, aircraft, locomotive	638	2.082	0.725	0.000	0.000
Nonroad (gasoline and diesel)	1603	0.118	0.017	0.012	0.012
Onroad (gasoline and diesel)	1187	0.189	0.000	0.014	0.014
Residential waste burning	1741	0.021	0.000	0.000	0.011
Commercial cooking	984	0.534	0.099	0.000	0.000
Prescribed/open burning	444	0.005	0.000	0.000	0.004
Wildfires	51	0.000	0.000	0.000	0.000
Area fugitive dust	10270	0.159	0.000	0.000	0.000
Point fugitive dust	15	0.003	0.000	0.000	0.000
Other area	888	0.160	0.000	0.000	0.128
Natural gas combustion	119	0.024	0.000	0.000	0.000
Total (all sources)	40333	6.502	1.383	0.026	0.406

[a] Natural gas combustion source category results are adjusted to reflect new emissions factor (94% reduction).
[b] Each daily results reflects the 98th percentile day or the 3rd highest day modeled with AERMOD so for monitor #261630015, that day is November 18 and for monitor #261630033, that day is January 1.
Source: US Environmental Protection Agency, Regulatory Impact Analysis for the Revised Particulate Matter National Ambient Air Quality Standards, 2006.

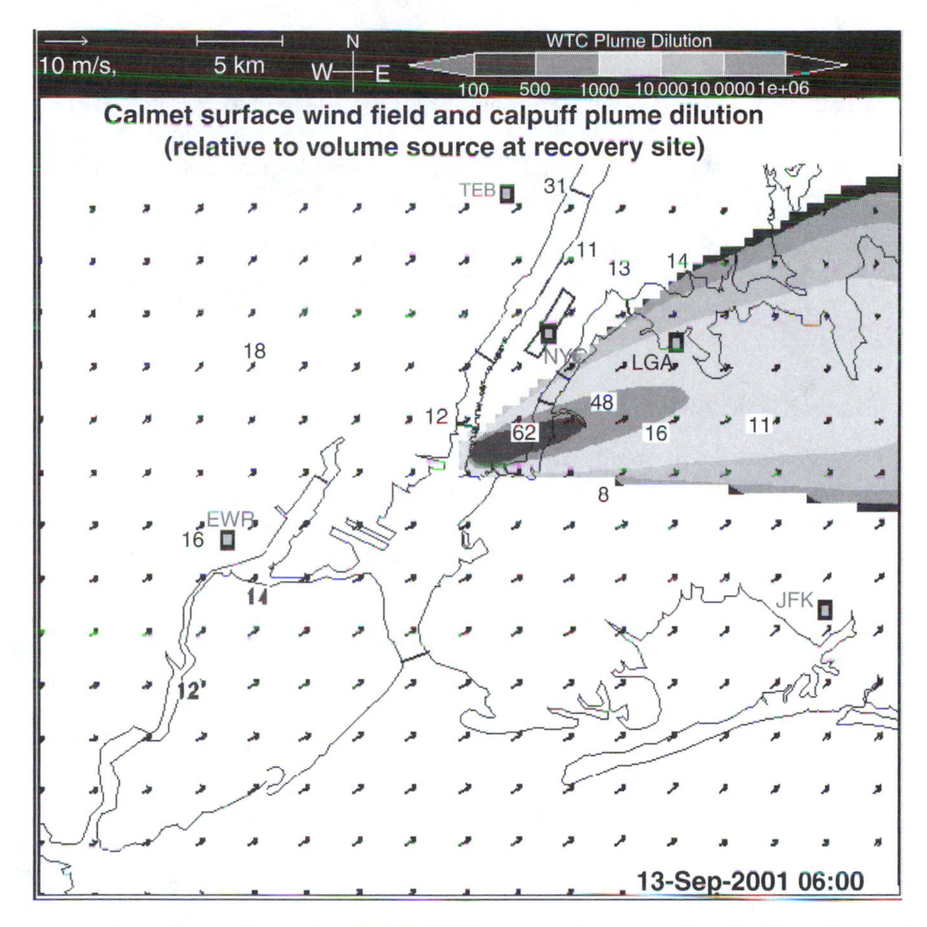

Fig. 22.12. Plume simulated with CALPUFF showing the averaged hourly $PM_{2.5}$ dilution of a volume source at the World Trade Center, New York City. *Source*: Huber, A., US Environmental Protection Agency.

- *CALINE3*: A steady-state Gaussian dispersion model esigned to determine air pollution concentrations at receptor locations downwind of roadways located in relatively uncomplicated terrain.
- *CAL3QHC*: A CALINE3-based CO model with queuing and hot spot calculations and with a traffic model to calculate delays and queues that occur at signalized intersections.
- *CAL3QHCR*: This is a more refined version based on CAL3QHC that requires local meteorological data.
- *Complex Terrain Dispersion Model Plus Algorithms for Unstable Situations (CTDMPLUS)*: A refined point source Gaussian air quality model for use in all stability conditions for complex terrain. The model contains, in its entirety, the technology of CTDM for stable and neutral conditions.

- *OCD Model Version 5*: A straight line Gaussian model developed to determine the impact of offshore emissions from point, area, or line sources on the air quality of coastal regions. OCD incorporates overwater plume transport and dispersion as well as changes that occur as the plume crosses the shoreline. Hourly meteorological data are needed from both offshore and onshore locations.

Documentation and access to these models are available at the EPA Technology Transfer Network website: http://www.epa.gov/scram001/dispersion_prefrec.htm#blp.

F. Photochemical Models

Photochemical air quality models have become widely recognized and routinely utilized tools for regulatory analysis and attainment demonstrations by assessing the effectiveness of control strategies. These models are large-scale air quality models that simulate the changes of pollutant concentrations in the atmosphere using a set of mathematical equations characterizing the chemical and physical processes in the atmosphere. They are applied at multiple spatial scales from local, regional, national, and global.

Photochemical models structure the atmosphere as a modeled volume with a three-dimensional grid with thousands of grid cells. Each grid cell can be envisioned as a box. The boxes are stacked one atop another and differ in height. Shorter boxes represent the parcels of air nearest to the ground surface. The photochemical calculates concentrations of pollutants, such as ozone, in each cell by simulating the movement of air into and out of cells by advection and dispersion. The model also includes algorithms to simulate mixing of pollutants vertically among layers, introduction of emissions from sources into each cell, as well as sets of chemical reactions, equations of pollution precursors and meteorology, especially incoming solar radiation (i.e. insolation) in each cell.

Photochemical air quality models are of two basic mathematical types: the Lagrangian trajectory model that employs a moving frame of reference, and the Eulerian grid model that uses a fixed coordinate system with respect to the ground. Early modeling efforts often used Lagrangian methods to simulate the formation of pollutants because of its computational simplicity. Describing physical processes, however, is incomplete, which is the major disadvantage of the Lagrangian method. Thus, most photochemical air quality models in operation today use three-dimensional Eulerian grid mathematics that allow for improved and more complete characterization of the physical processes in the atmosphere. This allows chemical-species concentrations to be predicted throughout the entire model domain.

Arguably, the principal photochemical model in use is the CMAQ modeling system, which has been developed to improve the environmental management community's ability to evaluate the impact of air quality management practices for multiple pollutants at multiple scales and enhance the scientist's ability to probe, to understand, and to simulate chemical and physical interactions in the atmosphere. The most recent CMAQ version is available for download from the Community Modeling and Analysis System website at: http://www.cmascenter.org/.

Several other photochemical models have been developed and are applied to various pollutants. These include:

- Comprehensive Air Quality Model with Extensions (CAMx)—simulates air quality over different geographic scales; treats a wide variety of inert and chemically active pollutants, including ozone, PM, inorganic and organic $PM_{2.5}/PM_{10}$, and mercury and other toxics; and has plume-in-grid and source apportionment capabilities.
- Regional Modeling System for Aerosols and Deposition (REMSAD)—calculates the concentrations of both inert and chemically reactive pollutants by simulating the physical and chemical processes in the atmosphere that affect pollutant concentrations over regional scales; and includes processes relevant to regional haze, PM, and other airborne pollutants, including soluble acidic components and mercury.
- Urban Airshed Variable Grid (UAM-V) Photochemical Modeling System—pioneering effort in photochemical air quality modeling in the early 1970s that has been used widely for air quality studies focusing on ozone. It is a three-dimensional photochemical grid model designed to calculate the concentrations of both inert and chemically reactive pollutants by simulating the physical and chemical processes in the atmosphere that affect pollutant concentrations. It is typically applied to model air quality episodes, i.e. periods during which adverse meteorological conditions result in elevated ozone pollutant concentrations.

G. Receptor Models

Receptor models are mathematical or statistical procedures for identifying and quantifying the sources of air pollutants at a receptor location. Unlike photochemical and dispersion air quality models, receptor models do not use pollutant emissions, meteorological data, and chemical transformation mechanisms to estimate the contribution of sources to receptor concentrations. Instead, receptor models use the chemical and physical characteristics of gases and particles measured at source and receptor to both identify the presence of and to quantify source contributions to receptor concentrations. These models are therefore a natural complement to other air quality models and are used as part of SIPs for identifying sources contributing to air quality problems. The EPA has developed the Chemical Mass Balance (CMB) and UNMIX models as well as the Positive Matrix Factorization (PMF) method for use in air quality management.

The CMB fully apportions receptor concentrations to chemically distinct source types depending on the source profile database, while UNMIX and

PMF internally generate source profiles from the ambient data. It is one of several receptor models that have been applied to air quality problems since the 1980s. Based on an effective-variance least-squares method (EVLS), it has supported numerous SIPs, when they include a source apportionment component. CMB requires speciated profiles of potentially contributing sources and the corresponding ambient data from analyzed samples collected at a single receptor site. CMB is ideal for localized nonattainment problems and has proven to be a useful tool in applications where steady-state Gaussian plume models are inappropriate, as well as for confirming or adjusting emissions inventories.

The EPA UNMIX model, as the name implies, disaggregates ("unmixes") the concentrations of chemical species measured in the ambient air to identify the contributing sources. Chemical profiles of the sources are not required, but instead are generated internally from the ambient data by UNMIX, using a mathematical formulation based on a form of factor analysis. For a given selection of chemical species, UNMIX estimates the number of sources, the source compositions, and source contributions to each sample.

The PMF technique is a form of factor analysis where the underlying covariability of many variables (e.g. sample to sample variation in PM species) is described by a smaller set of factors (e.g. PM sources) to which the original variables are related. The structure of PMF permits maximum use of available data and better treatment of missing and below-detection-limit values.

When the results of air pollution measurements are interpreted, one of the first questions asked by scientists, engineers, and policy makers is where did it come from? Sorting out the various sources of pollution is known as *source apportionment*. A number of tools are used to try to locate the sources of pollutants. A widely used approach is the "source–receptor model" or as it is more commonly known, the *receptor model*.

The distinction between receptor and dispersion models has to do with the direction of the prediction. Dispersion models start from the source and estimate where the plume and its contaminants are heading. Conversely, receptor models are based on measurements taken in the ambient environment and from these observations, make use of algorithms and functions to determine pollution sources. One common approach is the mathematical "back trajectory" model. Often, chemical co-occurrences are applied. So, it may be that a certain fuel is frequently contaminated with a conservative and, hopefully, unique element. Some fuel oils, for example, contain trace amounts of the element vanadium. Since there are few other sources of vanadium in most ambient atmospheric environments, its presence is a strong indication that the burning of fuel oil is a most likely source of the plume. The model, if constructed properly, can even quantify the contribution. So, if measurements show that sulfur dioxide (SO_2) concentrations are found to be $10\,\mu g\,m^{-3}$ in an urban area, and vanadium is also found at sufficient levels to indicate that home heating systems are contributing a certain amount of the SO_2 to the atmosphere, the model will correlate the amount of SO_2 coming from home heating systems. If other combustion sources, e.g. cars and power plants, also have unique trace elements associated with their SO_2 emissions, further SO_2 source apportionment can occur, so that the total may look something like Table 7.2.

For a discussion of source–receptor relationships and modeling, see Discussion Box: Source Apportionment, Receptor Models, and Carbon Dating in Chapter 7.

H. Personal Exposure Models

The actual exposures that people receive can also be modeled (see Fig. 22.13). According to the EPA, there are three techniques to estimate pollutant exposures quantitatively:

> Sometimes the approaches to assessing exposure are described in terms of "direct measures" and "indirect measures" of exposure (e.g. NRC, 1994). Measurements that actually involve sampling on or within a person, for example, use of personal monitors and biomarkers, are termed as "direct measures" of exposure. Use of models, microenvironmental measurements, and questionnaires, where measurements do not actually involve personal measurements, are termed as "indirect measures" of exposure. The direct/indirect nomenclature focuses on the type of measurements being made; the scenario evaluation/point-of-contact/reconstruction nomenclature focuses on how the data are used to develop the dose estimate. The three-term nomenclature is used in these guidelines to highlight the point that three independent estimates of dose can be developed.

There is seldom a sufficient amount of direct information from measurements to estimate and to predict exposures for a population. Thus, indirect

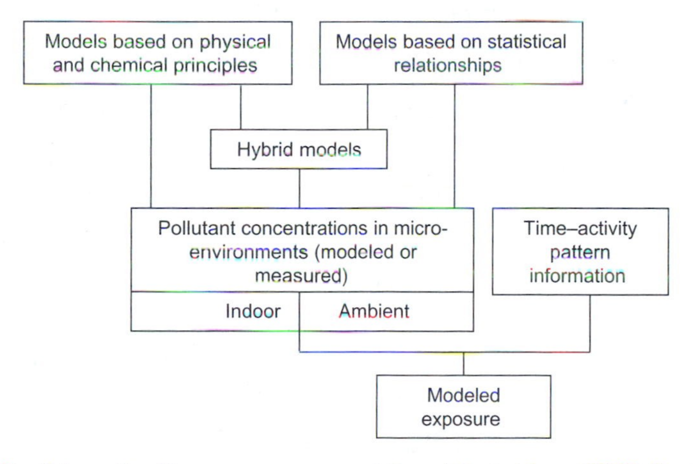

Fig. 22.13. Schematic of human exposure modeling. Adapted from: US National Research Council, Chapter 6: Models, in *Human Exposure Assessment for Airborne Pollutants: Advances and Applications*. Committee on Advances in Assessing Human Exposure to Airborne Pollutants and Committee on Geosciences, Environment, and Resources. National Academy Press, Washington, DC, 1990.

methods must be used. Personal exposure[9] can be expressed mathematically as a composite of the time in a microenvironment (e.g. in a garage, indoors at home, or in a car) and the concentration of the pollutant in that microenvironment:

$$E_i = \sum_{j=1}^{m} T_{ij} C_{ij} \tag{22.20}$$

where T_{ij} is the time spent in microenvironment j by person i with typical units of minutes, C_{ij} is the air pollutant concentration person i experiences in microenvironment j with typical units of $\mu g\, m^{-3}$, E_i represents the exposure for person i integrated T_{ij}, and m is the number of different microenvironments. Thus, the units of exposure are mass per volume-time ($\mu g\, m^{-3} min^{-1}$). Note, then, that exposure is the concentration per unit time. The calculation amounts to a weighted sum of concentrations with the weights being equal to the time spent experiencing a given concentration.

What people are doing is crucial to modeling their exposure to air pollutants. To support exposure, intake dose, and risk assessments, the EPA developed the Consolidated Human Activity Database (CHAD). CHAD is a relational database with a graphical user interface that facilitates queries and report generation.[10] It contains databases from previously existing human activity pattern studies, which were incorporated in two forms: (1) as the original raw data and (2) as data modified according to predefined format requirements. CHAD contains data obtained from pre-existing human activity studies that were collected at city, state, and national levels. People's diary information is input in CHAD. Figure 22.14 and Tables 22.3 and 22.4 provide the structure and data elements of CHAD. Most of the fields are answers from the questionnaire. Some fields vary from day to day for an individual and fields based on the diary entries will most likely be different for each day. The CHAD_DIARY table contains one record for each activity during a 24-h period. No record represents more than an hour or crosses the hour boundary. The minimal information in the diary data is the CHADID which links to the CHAD_DATA, a sequence number, which indicates the

[9] The principal reference for this discussion is: Klepeis, N. E., Modeling human exposure to air pollution, in *Exposure Analysis* (Ott, W. R., Wallace, L. A., and Steinemann, A., eds.). CRC Press, Boca Raton, 2006; and Klepeis, N. E., and Nazaroff, W. W., Modeling residential exposure to secondhand tobacco smoke. *Atmos. Environ.* **40** (23), 4393–4407 (2006). Other resources for all discussions about human exposure science have benefited from ongoing discussions with Wayne Ott and Lance Wallace, as well as the insights of Gerry Alkand, Larry Purdue, Judy Graham, Jerry Blancato, Paul Lioy, Tom McCurdy, Alan Huber, Alan Vette, Linda Sheldon, Dave Mage, Dale Pahl, Tom McKone, Edo Pellizzari, Ron Williams, and Curt Dary.

[10] The source of this discussion is: Stallings, C., Tippett, J. A., Glen, G., and Smith, L., *CHAD Users Guide: Extracting Human Activity Information from CHAD on the PC*. ManTech Environmental Services, modified by Systems Development Center Science Applications International Corporation, Research Triangle Park, NC, 2002.

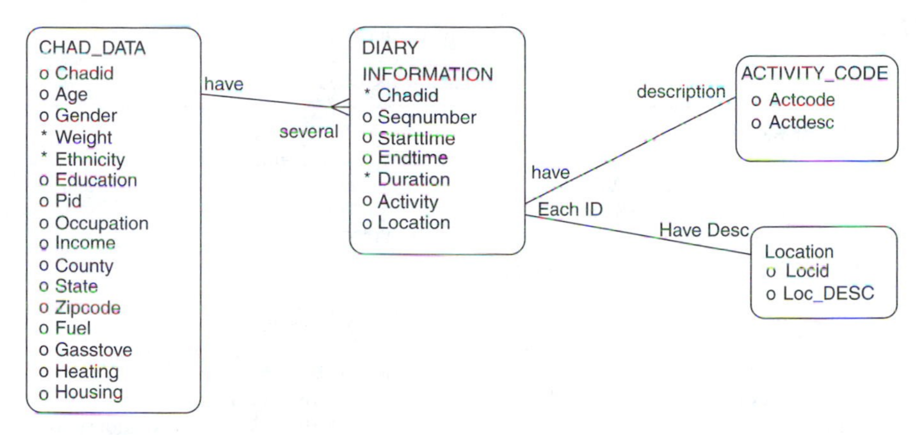

Fig. 22.14. Schematic of the tables and relationships among the tables in the combined CHAD. *Source:* Stallings, C., Tippett, J. A., Glen, G., and Smith, L., *CHAD Users Guide: Extracting Human Activity Information from CHAD on the PC*. ManTech Environmental Services, modified by Systems Development Center Science Applications International Corporation, Research Triangle Park, NC, 2002.

TABLE 22.3

CHAD Diary Location Codes

30000 Residence, general	31300 Waiting
30010 Your residence	31310 Wait for bus, train, ride (at stop)
30020 Other's residence	31320 Wait for travel, indoors
30100 Residence, indoor	31900 Other travel
30120 Your residence, indoor	31910 Travel by other vehicle
30121 Kitchen	32000 Other, indoor general
30122 Living room/family room	32100 Office building/bank/post office
30123 Dining room	32200 Industrial plant/factory/warehouse
30124 Bathroom	32300 Grocery store/convenience store
30125 Bedroom	32400 Shopping mall/non-grocery store
30126 Study/office	32500 Bar/night club/bowling alley
30127 Basement	32510 Bar/night Club
30128 Utility room/laundry room	32520 Bowling alley
30129 Other indoor	32600 Repair shop
30130 Other's residence, indoor	32610 Auto repair shop/gas station
30131 Other's kitchen	32620 Other repair shop
30132 Other's living room/family room	32700 Indoor gym/sports or health club
30133 Other's dining room	32800 Child care facility
30134 Other's bathroom	32810 Child care facility, house
30135 Other's bedroom	32820 Child care facility, commercial
30136 Other's study/office	32900 Public building/library/museum/theater
30137 Other's basement	32910 Auditorium, sport's arena/concert hall
30138 Other's utility room/laundry room	32920 Library/courtroom/museum/theater
30139 Other indoor	33100 Laundromat
30200 Residence, outdoor	33200 Hospital/health care facility/doctor's office
30210 Your residence, outdoor	33300 Beauty parlor/barber shop/hairdresser's

(*continued*)

TABLE 22.3 (*Continued*)

CHAD Diary Location Codes

30211 Your residence—pool, spa	33400 At work: no specific location, moving among locations
30219 Your residence—other outdoor	33500 At school
30220 Other's residence, outdoor	33600 At restaurant
30221 Other's residence—pool, spa	33700 At church
30229 Other's residence—other outdoor	33800 At hotel/motel
30300 Garage	33900 At dry cleaners
30310 Indoor garage	34100 Parking garage
30320 Outdoor garage	34200 Laboratory
30330 Your garage	34300 Other, indoor
30331 Your indoor garage	35000 Other outdoor, general
30332 Your outdoor garage	35100 Sidewalk/street/neighborhood
30340 Other's garage	35110 Within 10 yards of street
30341 Other's indoor garage	35200 Public garage/parking lot
30342 Other's outdoor garage	35210 Public garage
30400 Other, residence	35220 Parking lot
31000 Travel, general	35300 Service station/gas station
31100 Motorized travel	35400 Construction site
31110 Travel by car	35500 Amusement park
31120 Travel by truck	35600 School grounds/playgrounds
31121 Travel by truck (pick-up van)	35610 School grounds
31122 Travel by Truck (other than pick-up or van)	35620 Playground
31130 Travel by Motorcycle/moped/ motorized scooter	35700 Sports stadium and amphitheater
31140 Travel by bus	35800 Park/golf course
31150 Travel by train/subway/rapid transit	35810 Park
31160 Travel by airplane	35820 Golf course
31170 Travel by boat	35900 Pool, river, lake
31171 Travel by motorized boat	36100 Restaurant, picnic
31172 Travel by unmotorized boat	36200 Farm
31200 Non-motorized travel	36300 Other outdoor
31210 Travel by walk	U Uncertain
31220 Travel by bicycle/skateboard/roller-skates	X Missing
31230 Travel in a stroller or carried by an adult	

Source: Stallings, C., Tippett, J. A., Glen, G., and Smith, L., *CHAD Users Guide: Extracting Human Activity Information from CHAD on the PC*. ManTech Environmental Services, modified by Systems Development Center Science Applications International Corporation, Research Triangle Park, NC, 2002.

order of the records in a day, start, end, and duration times, and an activity and location.

Thus, if we measure concentrations with personal and indoor air pollution monitors and we apply activity patterns, such as those in CHAD, we can estimate and predict personal exposures to numerous pollutants.

Obviously the mean personal exposure can be expressed in concentration units ($\mu g\, m^{-3}$) by dividing E_i by the total time spent in each microenvironment and summing these exposures. Further, this can also be seen as a

TABLE 22.4

CHAD Diary Activity Code Descriptions

10000 Work and other income-producing activities, general
10100 Work, general
10110 Work, general, for organizational activities
10111 Work for professional/union organizations
10112 Work for special interest identity organizations
10113 Work for political party and civic participation
10114 Work for volunteer/ helping organizations
10115 Work of/for religious groups
10116 Work for fraternal organizations
10117 Work for child/youth/family organizations
10118 Work for other organizations
10120 Work, income-related only
10130 Work, secondary (income-related)
10200 Unemployment
10300 Breaks
11000 General household activities
11100 Prepare food
11110 Prepare and clean-up food
11200 Indoor chores
11210 Clean-up food
11220 Clean house
11300 Outdoor chores
11310 Clean outdoors
11400 Care of clothes
11410 Wash clothes
11500 Build a fire
11600 Repair, general
11610 Repair of boat
11620 Paint home/room
11630 Repair/maintain car
11640 Home repairs
11650 Other repairs
11700 Care of plants
11800 Care for pets/animals
11900 Other household

12000 Child care, general
12100 Care of baby
12200 Care of child
12300 Help/teach
12400 Talk/read

15200 Attend other classes
15300 Do homework
15400 Use library
15500 Other education
16000 General entertainment/social activities
16100 Attend sports events
16200 Participate in social, political, or religious activities
16210 Practice religion
16300 Watch movie
16400 Attend theater
16500 Visit museums
16600 Visit
16700 Attend a party
16800 Go to bar/lounge
16900 Other entertainment/social events
17000 Leisure, general
17100 Participate in sports and active leisure
17110 Participate in sports
17111 Hunting, fishing, hiking
17112 Golf
17113 Bowling/pool/ping pong/pinball
17114 Yoga
17120 Participate in outdoor leisure
17121 Play, unspecified
17122 Passive, sitting
17130 Exercise
17131 Walk, bike, or jog (not in transit)
17140 Create art, music, participate in hobbies
17141 Participate in hobbies
17142 Create domestic crafts
17143 Create art
17144 Perform music/drama/dance
17150 Play games
17160 Use of computers
17170 Participate in recess and physical education
17180 Other sports and active leisure
17200 Participate in passive leisure
17210 Watch
17211 Watch adult at work
17212 Watch someone provide child care

(continued)

TABLE 22.4 *(Continued)*

CHAD Diary Activity Code Descriptions

12500 Play indoors	17213 Watch personal care
12600 Play outdoors	17214 Watch education
12700 Medical care—child	17215 Watch organizational activities
12800 Other child care	17216 Watch recreation
13000 Obtain goods and services, general	17220 Listen to radio/listen to recorded music/watch TV
13100 Dry clean	17221 Listen to radio
13200 Shop/run errands	17222 Listen to recorded music
13210 Shop for food	17223 Watch TV
13220 Shop for clothes or household goods	17230 Read, general
13230 Run errands	17231 Read books
13300 Obtain personal care service	17232 Read magazines/not ascertained
13400 Obtain medical service	17233 Read newspaper
13500 Obtain government/financial services	17240 Converse/write
13600 Obtain car services	17241 Converse
13700 Other repairs	17242 Write for leisure/pleasure/paperwork
13800 Other services	17250 Think and relax
14000 Personal needs and care, general	17260 Other passive leisure
14100 Shower, bathe, personal hygiene	17300 Other leisure
14110 Shower, bathe	18000 Travel, general
14120 Personal hygiene	18100 Travel during work
14200 Medical care	18200 Travel to/from work
14300 Help and care	18300 Travel for child care
14400 Eat	18400 Travel for goods and services
14500 Sleep or nap	18500 Travel for personal care
14600 Dress, groom	18600 Travel for education
14700 Other personal needs	18700 Travel for organizational activity
15000 General education and professional training	18800 Travel for event/social activity
15100 Attend full-time school	18900 Travel for leisure
15110 Attend day-care	18910 Travel for active leisure
15120 Attend K-12	18920 Travel for passive leisure
15130 Attend college or trade school	U Unknown
15140 Attend adult education and special training	X Missing

Source: Stallings, C., Tippett, J. A., Glen, G., and Smith, L., *CHAD Users Guide: Extracting Human Activity Information from CHAD on the PC*. ManTech Environmental Services, modified by Systems Development Center Science Applications International Corporation, Research Triangle Park, NC, 2002.

continuum in which the personal exposure is assumed to be constant during each time comprising T_{ij}:

$$E_i = \int_1^2 C_{ij}(t;x;y;z)dt \qquad (22.21)$$

where $C_i(t, x, y, z)$ is the concentration occurring at a specific point occupied by person i at time t and spatial coordinates (x, y, z), and t_1 and t_2 are the respective starting and ending times of a given exposure episode. The concentrations can be measured using a real-time personal monitoring device. These data are not actually continuous, but can be quite frequent (e.g. every few seconds).

Other times, fully continuous space is not assumed, which means that discrete microenvionments are used. If this is the case the equation appears as

$$E_i = \left(\sum_{j=1}^{m} \int_{j1}^{j2} C_{ij}(t)dt \right) \quad (22.22)$$

where $C_{ij}(t)$ is the concentration to which the person is exposed in the discrete microenvironment j at a specific point in time t over the time interval defined by $[t_{j1}, t_{j2}]$, where t_{j1} is the starting time for the microenvironment and t_{j2} is the ending time. If we assume that every person in a population is exposed to the same microenvironment concentrations, a simplified population exposure can be derived to show the total time spent by all receptors in each microenvironment:

$$\tilde{E} = \sum_{j=1}^{m} C_j \tilde{T}_j \quad (22.23)$$

where m is the number of microenvironments encountered, C_j is the average pollutant concentration in microenvironment j assigned to every person i, $\tilde{E}$ is the integrated exposure over all members of the population. The total time spent by all persons in microenvironment j is

$$\tilde{T}_i = \sum_{i=1}^{n} T_{ij} \quad (22.24)$$

where n is the total number of people in the population being modeled. This can be further simplified if we assume that each person spends the same total amount of time across all microenvironments:

$$T = T_i = \sum_{j=1}^{m} T_{ij} \quad (22.25)$$

Further, if the time spent by certain individuals in particular microenvironments is zero, then the average personal exposure for the population would be

$$\bar{E}_c \frac{1}{nT} \sum_{j=1}^{m} C_j \tilde{T}_j \quad (22.26)$$

Therefore, time–activity relationships and concentrations can be used to model expected exposures to air pollutants. Indoor and personal exposures can dominate exposures to many pollutants. For example, Fig. 22.15 shows the modeled results of expected exposures to particulates ($PM_{2.5}$) for a simulated population based on available measurements of ambient, indoor, and personal measurements, combined with information about microenvironments (e.g. types and time in each) and activities. Note that ambient contribution is quite small compared to personal and indoor exposures.[11]

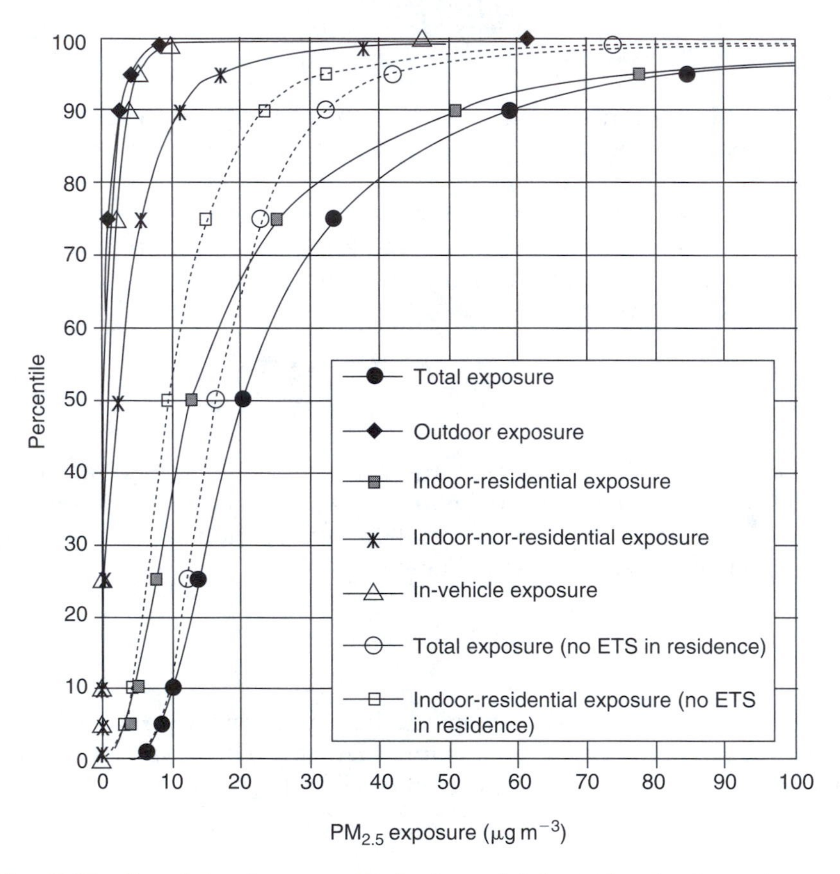

Fig. 22.15. Cumulative frequency distributions of daily total and microenvironmental $PM_{2.5}$ exposures ($\mu g\, m^{-3}$) for the simulated population of Philadelphia (solid lines). Distributions of total and indoor residential $PM_{2.5}$ exposures for the population without environmental tobacco smoke (ETS) exposure in the residence (dashed lines) are also shown. *Source*: Burke, J. M., Zufall, M. J., and Ozkaynak, A. H., *A Population Exposure Model for Particulate Matter: Case Study Results for $PM_{2.5}$* in Philadelphia, PA: http://www.epa.gov/heasd/pm/pdf/exposure-model-for-pm.pdf; accessed on February 23, 2007.

[11] Burke, J. M., Zufall, M. J., and Ozkaynak, A. H., A Population Exposure Model for Particulate Matter: Case Study Results for $PM_{2.5}$ in Philadelphia, PA: http://www.epa.gov/heasd/pm/pdf/exposure-model-for-pm.pdf; accessed on February 23, 2007.

I. Modeling Air Toxics

The Clean Air Act Amendments of 1990 (CAAA90) included a specific health outcome provision related to toxic air pollutants. The law included 190 substances that need to be addressed because they have been known or suspected to cause cancer or other serious health effects, such as reproductive effects or birth defects, or to cause adverse environmental effects. The 1990 amendments established a range of air toxics requirements for EPA to implement that generally fall into four categories:

1. Establishing emission standards based on existing pollution control technologies, called Maximum Achievable Control Technology (MACT), for an estimated 84 000 major stationary sources within 158 industries.
2. Examining the remaining health risk (called the "residual risk") from these sources 8 years after implementing each MACT standard and, if warranted, issuing additional standards to protect public health or the environment.
3. Regulating air toxics emissions from small stationary sources, such as dry cleaners.
4. Evaluating the need for and feasibility of regulation of air toxics emissions from mobile sources, such as cars, and regulating these sources based on this evaluation.

The amendments also required the EPA to periodically assess the costs and benefits of the entire CAA. This called for an assessment of the Act's costs and benefits prior to 1990 along with projections of future economic impacts resulting from the amendments (see the earlier discussion in this Chapter regarding the regulatory impact analysis as it pertains to PM).

Air toxics originate from a variety of sources (see Fig. 22.16). Of the numerous compounds that are classified as air toxics, however, about 60% are represented by just five pollutant classes (see Table 22.5). The EPA has completed MACT standards for major stationary sources. The EPA establishes the federal standard, but the state and local air pollution control agencies generally implement the EPA's emission standards. These agencies can choose to impose more stringent requirements than the federal standards, and some state and local agencies have developed innovative air toxics programs that go beyond the federal program, which is enabling them to address air toxics that would be insufficiently covered by EPA's standards.

The CAA residual risk program (i.e. the risk remaining after technology-based standards have been in place), must try to determine whether the most highly exposed individuals face excess cancer risk of more than 1 in 1 million (risk $= 1 \times 10^{-6}$). In cases where estimated risks exceed this threshold, the EPA must develop a new residual risk standard to provide an ample margin of safety for those potentially affected individuals. In fact, the EPA generally uses a lifetime cancer risk of 1 in 10 000 (risk $= 1 \times 10^{-4}$) as the upper boundary of acceptability. The ample margin of safety decision must

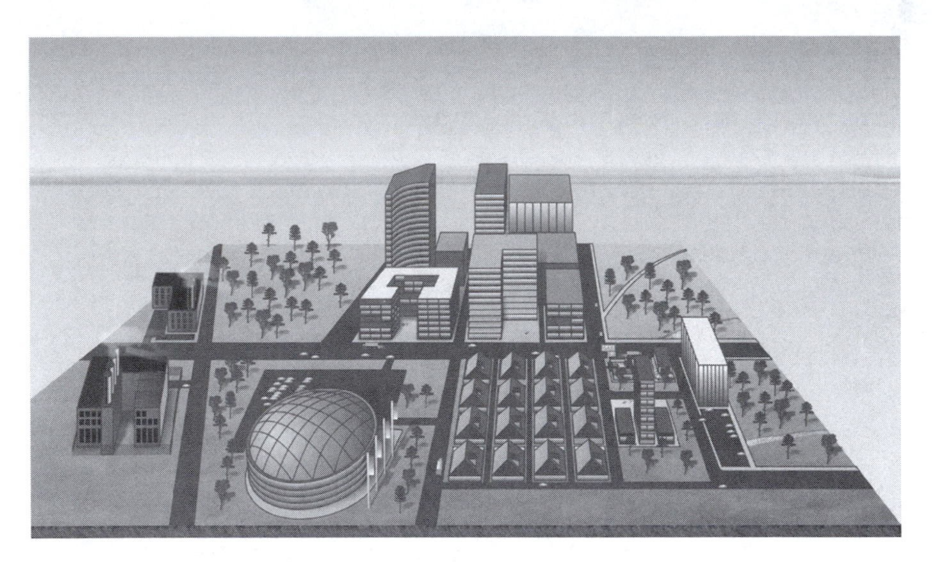

Fig. 22.16. Common sources of air toxics emissions. *Source*: US Government Accountability Office, *Clean Air Act: EPA Should Improve the Management of Its Air Toxics Program*. Report No. GAO-06-669, June 2006.

also consider costs, technological feasibility, uncertainties, and other relevant factors. In this feasibility step, EPA must also assess whether to adopt more stringent standards to prevent adverse effects to wildlife, aquatic life, or natural resources considering cost, energy, and other pertinent details.

EPA is not expected to complete the residual risk reviews until the year 2012, rather than by the 2008 date given in the amendment. These reviews are intended to provide information on any potential adverse health effects that may warrant further regulation. According to the US Government Accountability Office (GAO)[12]:

> EPA does not have reliable data on the degree to which its programs have reduced risks. Furthermore, the data that are available suggest that the agency still has substantial opportunities to control emissions from mobile and small stationary sources.
>
> EPA faces significant challenges in implementing the air toxics program, many of which stem from its relatively low priority within the agency. Importantly, the agency lacks a comprehensive strategy for managing its implementation of the remaining air toxics requirements. Senior EPA officials said that the program's agenda is largely set by external stakeholders who file litigation when the agency misses deadlines. For example, EPA currently faces a court order to issue emissions standards for small stationary sources. Previous reports by GAO identified inadequate funding for the air toxics program as a challenge, and key stakeholders—including senior EPA officials, environmental advocates, and state and local agency officials—said resource constraints continue to pose a major challenge. The percentage of funding for the air toxics program relative to all clean air programs ranged from 18% to 19% between 2000

[12] US Government Accountability Office, *Clean Air Act: EPA Should Improve the Management of Its Air Toxics Program*. Report No. GAO-06-669, June 2006.

TABLE 22.5

The Five Most Commonly Emitted Air Toxics, 2002

Pollutant	Percentage of total air toxics emissions	Primary sources of emissions	Health effects
Toluene	18	Mobile sources	Impairment of the nervous system with symptoms including tiredness, dizziness, sleepiness, confusion, weakness, memory loss, nausea, loss of appetite, and hearing and color vision loss; kidney problems; unconsciousness; and death.
Xylenes	13	Mobile sources, asphalt paving	Irritation of the skin, eyes, nose, and throat; headaches, dizziness, memory loss, and changes in sense of balance; lung problems; stomach discomfort; possible effects on the liver and kidneys; unconsciousness; and death.
Hydrochloric acid	12	Coal-fired utility and industrial boilers	Eye, nose, and respiratory tract irritation; corrosion of the skin, eyes, mucous membranes, esophagus, and stomach; severe burns; ulceration; scarring; inflammation of the stomach lining; chronic bronchitis; and inflammation of the skin.
Benzene	9	Mobile sources, open burning, pesticide application	Drowsiness, dizziness, vomiting, irritation of the stomach, sleepiness, convulsions, rapid heart rate, headaches, tremors, confusion, unconsciousness, anemia, excessive bleeding, weakened immune system, increased incidence of cancer (leukemia), and death.
Formaldehyde	7	Mobile sources, open burning	Irritation of the eyes, nose, throat, and skin; severe pain; vomiting; coma; limited evidence of cancer; and death.

US Environmental Protection Agency and Agency for Toxic Substances and Disease Registry

and 2003 and declined to 15% in 2004 and 12% in 2005. EPA has not estimated the level of resources necessary to comply with the remaining requirements of the 1990 amendments, according to a senior program official. We believe that such estimates would help inform congressional oversight and appropriations decisions. Senior EPA officials and other stakeholders also cited a lack of information on the benefits of regulating air toxics as a major challenge, which, in turn, reinforces the program's relative priority because the agency cannot demonstrate its effectiveness. The stakeholders identified a number of other challenges, but perceptions varied by stakeholder group. For example, EPA and industry stakeholders rated the large number of statutory requirements as a challenge, while environmental stakeholders rated a lack of reliable data on air toxics sources and their emissions as a challenge.

The EPA is required to conduct a National Air Toxics Assessment (NATA) as a part of the national air toxics program, which

- expands air toxics monitoring;
- develops emission inventories;
- conducts national and local-scale air quality analyses;
- characterizes risks associated with air toxics exposures;
- conducts research on health and environmental effects and exposures to both ambient and indoor sources;
- conducts exposure modeling.

This helps EPA set program priorities, characterize risks, and track progress toward meeting goals, especially providing trends in air quality as the United States transitions from technology-based standards to a greater number of risk-based standards (see Fig. 22.17).

As part of the NATA program, EPA conducted a national screening-level assessment using 1996 emissions inventory data to characterize air toxics risks nationwide. This helped to characterize the potential health risks associated with inhalation exposures to 33 air toxics and diesel PM. These 34 air toxics were identified as priority pollutants in EPA's Integrated Urban Air Toxics Strategy.[13]

The Assessment System for Population Exposure Nationwide (ASPEN) is the model used to estimate toxic air pollutant concentrations. It is based on

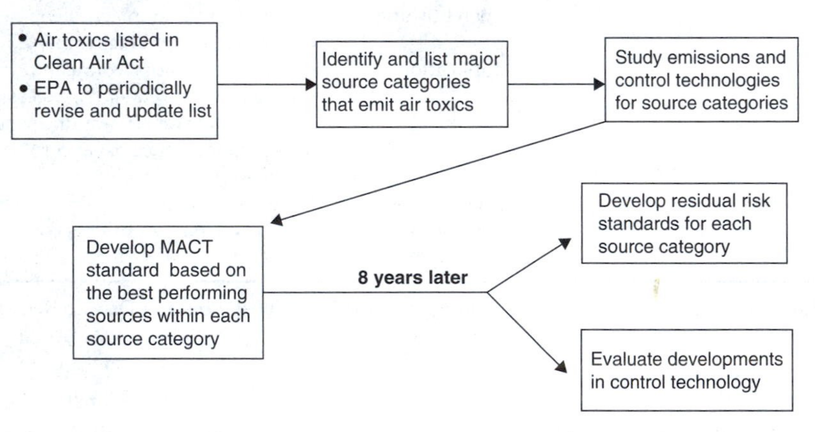

Fig. 22.17. Overview of the regulatory process for major stationary sources of air toxics, including MACT standards and 8-year technology and residual risk reviews. *Source*: US Government Accountability Office, *Clean Air Act: EPA Should Improve the Management of Its Air Toxics Program*. Report No. GAO-06-669, June 2006.

[13] A list of the 34 selected toxics can be found at: http://www.epa.gov/ttn/atw/nata/34poll.html.

the EPA's ISCLT, which simulates the behavior of the pollutants once they are emitted into the atmosphere. ASPEN uses estimates of toxic air pollutant emissions and meteorological data from National Weather Service stations to estimate air toxics concentrations nationwide.

ASPEN incorporates a number of critical pollutant concentration variables, including

- rate of release;
- location of release;
- the height from which the pollutants are released;
- wind speeds and directions from the meteorological stations nearest to the release;
- breakdown of the pollutants in the atmosphere after being released (i.e. reactive decay);
- settling of pollutants out of the atmosphere (i.e. deposition);
- transformation of one pollutant into another (i.e. secondary formation).

The model estimates toxic air pollutant concentrations for every census tract in contiguous United States, Puerto Rico, and the Virgin Islands. Census tracts are land areas defined by the US Bureau of the Census and typically contain about 4000 residents each. These census tracts are usually smaller than 2 square miles in size in cities, but much larger in rural areas.

NATA is intended to identify ambient concentrations of air toxics attributable to anthropogenic (human-generated) emissions, within 50 km of each source. However, for many pollutants, ambient concentrations have "background" components attributable to long-range transport, resuspension of previous emissions, and non-anthropogenic sources. To estimate ambient concentrations of air toxics accurately, these background concentrations must be addressed. In the 1996 assessment, with exception of diesel PM, background concentrations were based on measured values of 13 pollutants. The background concentrations used in the 1996 NATA are provided in Table 22.6.

The modeled estimates for most of the pollutants are generally lower than the measured ambient annual average concentrations when evaluated at the monitors. However, when the maximum modeled estimate for distances up to 10–20 km from the monitoring location are compared to the measured concentrations, the modeled estimates approach the monitored concentrations. This may be the result of spatial uncertainty of the underlying data (emissions and meteorological), as well as the tendency of existing air toxics monitoring networks to characterize the higher (if not maximum) air pollution impact areas in the ambient air. Also, the model estimates are more uncertain at the census tract scale, but are more reliable at more extensive geographic scales, such as a county or state. Further, ASPEN and underlying data are used to estimate exposures and risks (see Fig. 22.18).

TABLE 22.6

Background Concentration Estimates Used in the National-Scale Air Toxics Assessment

Pollutant	Background concentration ($\mu g\, m^{-3}$)
Benzene	0.48
Carbon tetrachloride	0.88
Chloroform	0.083
Ethylene dibromide	0.0077
Ethylene dichloride	0.061
Formaldehyde	0.25
Hexachlorobenzene	0.000093
Mercury compounds	0.0015
Methylene chloride	0.15
Polychlorinated biphenyls	0.00038
Perchloroethylene (tetrachloroethylene)	0.14
Trichloroethylene	0.081

Source: US Environmental Protection Agency.

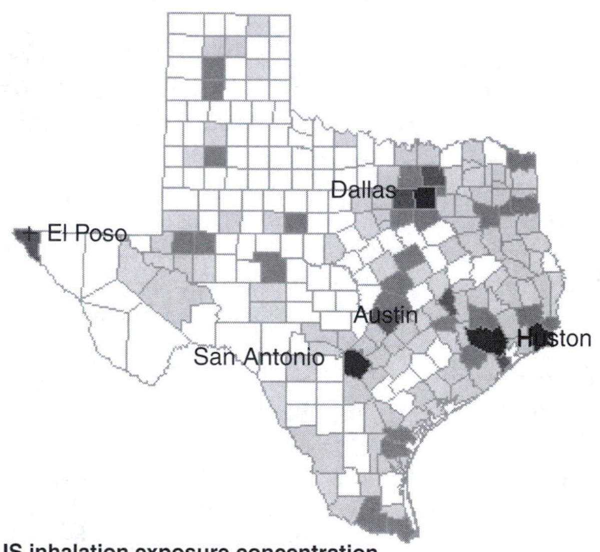

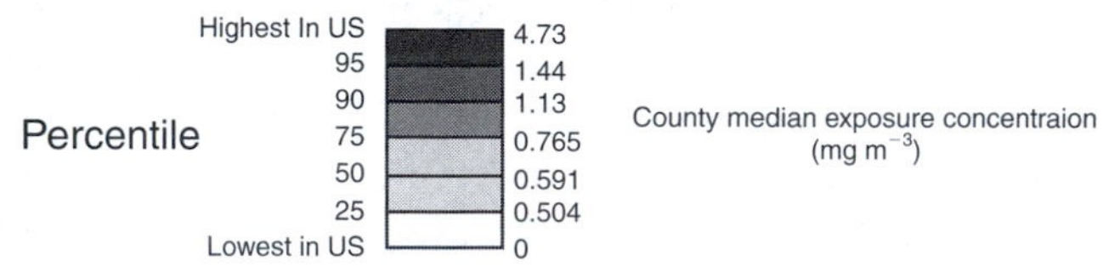

Fig. 22.18. Estimated county median exposure benzene concentrations in Texas in 1996. A county's shading indicates its exposure concentration compared with all other counties in the nation. The exposure concentration value displayed is the county median in micrograms of benzene per cubic meter of air ($\mu g\, m^{-3}$). Similar maps can be generated for other states and various air toxics at the US Environmental Protection Agency's National Air Toxics Assessment website: http://www.epa.gov/ttn/atw/nata/mapexpo.html.

VI. MODEL PERFORMANCE, ACCURACY, AND UTILIZATION

A. Methodology of Assessing Model Performance

A number of statistics have been suggested [39, 40] as measures of model performance. Different types of models and the use of models for different purposes may require different statistics to measure performance.

In time series of measurements of air quality and estimates of atmospheric concentration made by a model, residuals d can be computed for each location. The residual d is the difference between values paired timewise:

$$d = M - E \tag{22.27}$$

where M is the measured value and E is the value estimated by the model.

A measure of bias $\bar{d}$ is the first moment of the distribution of these differences, or the average difference:

$$\bar{d} = \frac{1}{N}\sum d \tag{22.28}$$

A measure of the variability of the differences is the variance S^2, which is the second moment of the distribution of these differences:

$$S^2 = \sum(d - \bar{d})^2/N = \frac{\sum d^2}{N} - \left(\frac{\sum d}{N}\right)^2 \tag{22.29}$$

The square root of the variance is the standard deviation.

The mean absolute error (MAE) is

$$\text{MAE} = \Sigma|d|/\text{N} \tag{22.30}$$

The mean square error (MSE) is

$$\text{MSE} = \Sigma d^2/N \tag{22.31}$$

The root-mean-square error is the square root of theMSE. Note that since the root-mean-square error involves the square of the differences, outliers have more influence on this statistic than on the MAE.

The fractional error (FE) [41] is

$$\text{FE} = (M - E)/0.5(M + E) \tag{22.32}$$

The mean fractional error (MFE) is

$$\text{MFE} = \Sigma\,\text{FE}/\text{N} \tag{22.33}$$

The FE is logarithmically unbiased; that is, an M which is k times E produces the same magnitude FE (but of opposite sign) as an M which is $1/k$ times E.

In addition to analyzing the residuals, it may be desirable to determine the degree of agreement between sets of paired measurements and estimates. The linear correlation coefficient r_{EM} is

$$r_{\mathrm{EM}} = \frac{N\Sigma E \cdot M - \Sigma E \Sigma M}{\{[N\Sigma E^2 - (\Sigma E)^2][N\Sigma M^2 - (\Sigma M)^2]\}^{0.5}} \tag{22.34}$$

The slope b and intercept a of the least-squares line of best fit of the relation $M = a + bE$ are

$$\text{Intercept:} \quad a = \frac{\Sigma E^2 \Sigma M - \Sigma E \Sigma E \cdot M}{N\Sigma E^2 - (\Sigma E)^2} \tag{22.35}$$

$$\text{Slope:} \quad b = \frac{N\Sigma E \cdot M - \Sigma E \Sigma M}{N\Sigma E^2 - (\Sigma E)^2} \tag{22.36}$$

The temporal correlation coefficient at each monitoring location can be calculated by analysis of the paired values over a time period of record. The spatial correlation coefficient at a given time can be calculated by analysis of the paired values from each station. For the spatial correlation coefficient to have much significance, there should be 20 or more monitoring locations.

Techniques to use for evaluations have been discussed by Cox and Tikvart [42], Hanna [43], and Weil *et al.* [44]. Hanna [45] shows how resampling of evaluation data will allow use of the bootstrap and jackknife techniques so that error bounds can be placed about estimates.

The use of various statistical techniques has been discussed [46] for two situations. For standard air quality networks with an extensive period of record, analysis of residuals, visual inspection of scatter diagrams, and comparison of cumulative frequency distributions are quite useful techniques for assessing model performance. For tracer studies the spatial coverage is better, so that identification of maximum measured concentrations during each test is more feasible. However, temporal coverage is more limited with a specific number of tests not continuous in time.

The evaluations cited in the following sections are examples of the use of various measures of performance.

B. Performance of Single-Source Models

Since wind direction changes with height above ground and plume rise will cause the height of the plume to vary in time, it is difficult to model accurately plume transport direction. Because of this transport wind direction error, analyses of residuals and correlations are not frequently utilized for assessment of single-source model performance. Instead, cumulative frequency distributions of estimated and measured concentrations for specific averaging times at each location are examined, as well as comparison of the highest concentrations such as the five highest for each averaging time of interest, e.g., 3 and 24 h.

The performance of one specific model, CRSTER [47], for the second-highest once-a-year 24-h concentrations, is summarized in Fig. 22.19 in terms of ratios of estimates to measurements. Overestimates by more than a factor of 2 occur for all receptors whose elevations are near or exceed stack top. The value ΔE is the elevation of the receptor minus the elevation of the stack base; h is the physical stack height. Over seven sites, having stacks varying from 81 to 335 m, for receptors having elevations above the stack base less than 0.7 of the stack height, the second-highest, once-a-year 24-h estimates were within a factor of 2 of the measurements for 25 of 35 monitoring stations.

Comparisons [49] of measured concentrations of SF_6 tracer released from a 36-m stack, and those estimated by the PTMPT model for 133 data pairs over Pasquill stabilities varying from B through F, had a linear correlation coefficient of 0.81. Here 89% of the estimated values were within a factor of 3 of the measured concentrations. The calculations were most sensitive to the selection of stability class. Changing the stability classification by 1 varies the concentration by a factor of 2–4.

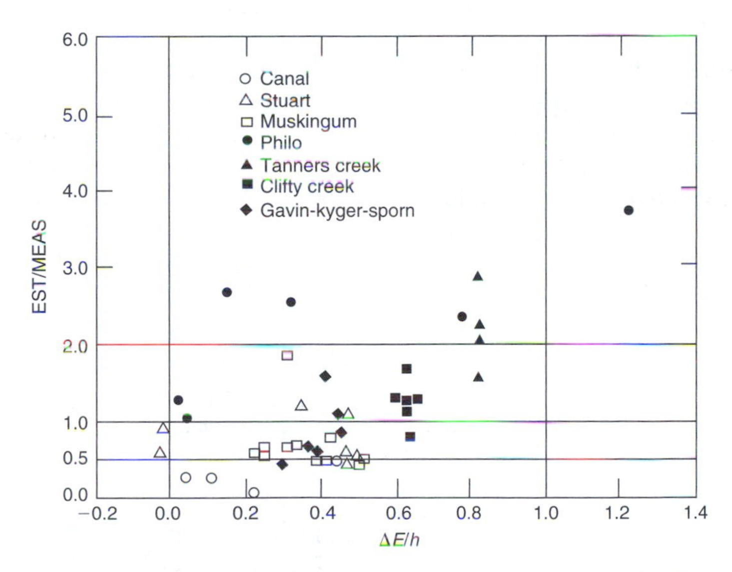

Fig. 22.19. Ratio of second-highest 24 h estimated concentrations from the CRSTER model [47] to measured concentrations as a function of the excess of receptor elevation over stack base evaluation ΔE relative to the stack height h. Names with each symbol are power plants. *Source*: From Turner and Irwin [48].

C. Performance of Urban Models for Nonreactive Pollutants

Because there are multiple sources of most pollutants in urban areas, and because the meteorology of urban areas is modified so that the extremes of stability are avoided, concentrations tend to vary much less in urban than in rural areas.

For sources having a large component of emissions from low-level sources, the simple Gifford–Hanna model given previously as Eq. (22.19), $\chi = Cq_a/u$, works well, especially for long-term concentrations, such as annual ones. Using the derived coefficients of 225 for PM and 50 for SO_2, an analysis of residuals (measured minus estimated) of the dependent data sets (those used to determine the values of the coefficient C) of 29 cities for PM and 20 cities for SO_2 and an independent data set of 15 cities for PM is summarized in Table 22.7. For the dependent data sets, overestimates result. The standard deviations of the residuals and the MAE are about equal for particulates and sulfur dioxide. For the independent data set the mean residual shows only slight overestimation, and the standard deviation of residuals and MAE are considerably smaller.

A version of the Gifford–Hanna model was evaluated [50] using 1969 data for 113 monitoring stations for PM and 75 stations for SO_2 in the New York metropolitan area. This version differed from Eq. (22.19) in considering major point source contributions and the stack height of emission release. This model produced results (Table 22.8) comparable to those of the much more complicated Climatological Dispersion Model (CDM) [51].

The urban RAM model was evaluated [52] using 1976 sulfur dioxide data from 13 monitoring locations in St. Louis on the basis of their second-highest once-a-year concentrations. The ratio of estimated to measured 3-h average concentrations was from 0.28 to 2.07, with a median of 0.74. Half of the values were between 0.61 and 1.11. For the 24-h average concentrations the rations ranged from 0.18 to 2.31, with a median of 0.70. Half of the values were between 0.66 and 1.21. Thus, the urban RAM model generally underestimates concentrations by about 25%.

TABLE 22.7

Residual Analysis for the Gifford–Hanna Model (Eq. (22.19))

Pollutant	N	$\overline{M}$	S_M	$\overline{E}$	S_E	$\overline{d}$	S_d	$\lvert\overline{d}\rvert$
Particulate	29	110.7	27.8	171.9	111.5	−61.2	103.7	81.3
SO_2	20	89.8	83.1	116.25	72.0	−26.5	101.2	82.5
Independent data set particulate	15	91.6	38.7	93.2	51.7	−1.6	24.9	21.3

Note: N is the number of cities, M is measured, E is estimated from the model, d is residual (measured − estimated) and S values are standard deviations.
Source: Data are from Gifford and Hanna [23].

TABLE 22.8

Residual Analysis Using 1969 New York Data[a]

Model	Pollutant	N	$\overline{M}$	S_M	$\overline{E}$	S_d	$\lvert\bar{d}\rvert$
CDM	Particulate	113	82	23	74	22	16
6B	Particulate	113	82	23	67	25	19
CDM	SO_2	75	135	72	138	52	37
6B	SO_2	75	135	72	127	56	38

[a] Adapted from Turner *et al.* [50].
Note: N is the number of monitoring stations, M is measured, E is estimated from the model, d is residual (measured − estimated), and S values are standard deviations.

TABLE 22.9

Residual Analyses of Three Photochemical Models

	Ozone (ppm)			
Model	N	$\bar{d}$	S_d	$\lvert\bar{d}\rvert$
Photochemical box model	10	–0.033	0.041	0.039
Lagrangian photochemical model	10	–0.004	0.110	0.080
Urban airshed model	10	0.074	0.033	0.074

Note: N is the number of days, d is residual (measured – estimated), and S is the standard deviation.
Source: Adapted from Shreffler and Schere [53].

D. Performance of Photochemical Models

The performance of photochemical models is evaluated by their ability to estimate the magnitude, time, and location of occurrence of the secondary pollutant oxidant (ozone). Several photochemical models were evaluated using 10 days' data from St. Louis (Table 22.9). Among those evaluated were the PBM [6], which features a single variable-volume reacting cell. This model is most appropriate for use with light winds, so that most of the emissions will remain in the cell and react, rather than being rapidly transported through the downwind side of the cell. The maximum measured concentrations for the 10 days ranged from 0.08 to 0.18 ppm; the maximum model estimates ranged from 0.07 to 0.22 ppm.

The second model evaluated was the Lagrangian Photochemical Model (LPM) [54], a trajectory model. Backward trajectories were first determined so that starting positions could be used which would allow trajectories to reach station locations at the times of measurement. Measured concentrations ranged from 0.20 to 0.26 ppm and estimated concentrations from 0.05 to 0.53 ppm.

The third model evaluated was the Urban Airshed Model (UAM) [55], a three-dimensional grid-type model. The area modeled was 60 by 80 km, consisting of cells 4 km on a side. There are four layers of cells in the vertical, the bottom simulating the mixing layer. The model provides both spatially and temporally resolved estimates. Hour-averaged estimates of each pollutant species at each monitoring site within the model domain are given. Measured concentrations ranged from 0.17 to 0.24 ppm and estimated concentrations ranged from 0.10 to 0.17 ppm. The model underestimates the concentration at the point of the observed maximum each day.

Seinfeld [33] indicates that photochemical models estimating peak ozone concentrations in urban areas are generally within 30% of measured peaks.

E. Utilization of Models

For the air quality manager to place model estimates in the proper perspective to aid in making decisions, it is becoming increasingly important to place error bounds about model estimates. In order to do this effectively, a history of model performance under circumstances similar to those of common model use must be established for the various models. It is anticipated that performance standards will eventually be set for models.

F. Modeling to Meet Regulatory Requirements

In the United States if the anticipated air pollution impact is sufficiently large, modeling has been a requirement for new sources in order to obtain a permit to construct. The modeling is conducted following guidance issued by the US EPA [56, 57]. The meeting of all requirements is examined on a pollutant-by-pollutant basis. Using the assumptions of a design that will meet all emission requirements, the impact of the new source, which includes all new sources and changes to existing sources at this facility, is modeled to determine pollutant impact. This is usually done using a screening-type model such as SCREEN [58]. The impacts are compared to the modeling significance levels for this pollutant for various averaging times. These levels are generally about 1/50 of the National Air Quality Standards. If the impact is less than the significance level, the permit can usually be obtained without additional modeling. If the impact is larger than the significance level, a radius is defined which is the greatest distance to the point at which the impact falls to the significance level. Using this radius, a circle is defined which is the area of significance for this new facility. All sources (not only this facility, but all others) emitting this pollutant are modeled to compare anticipated impact with the NAAQS and with the PSD increments.

The implementation of CAAA90 requires not only permission for new sources but also permits for existing facilities. There is also a requirement for reexamination of these permits at intervals not longer than 5 years. The permit programs are set up and administered by the states under the review and guidance of EPA. These programs include the collection of permit fees sufficient to support the program. Implementation of CAAA90 will require the application of MACT on an industry-by-industry basis as specified by EPA. Following the use of MACT, a calculation will be made of the residual risk due to the remaining pollutant emissions. This is accomplished using air quality dispersion modeling that is interpreted using risk factors derived from health information with the inclusion of appropriate safety factors.

Because of their inclusion of the effects of building downwash, the ISC2 (Industrial Source Complex) models [16], ISCST and ISCLT (short-term performing calculations by hourly periods and long-term using the joint frequencies of wind direction, wind speed, and Pasquill stability, commonly referred to as STAR data) are recommended. The ISC2 models differ only slightly from the earlier ISC models (several small errors that make little difference in the resulting calculations were corrected). The recoding of the model to result in ISC2 was requested by EPA to overcome a difficult-to-understand code that had resulted from a number of changes that had been incorporated. The ISC models will make calculations for three types of sources: point, area, and volume. There are recognized deficiencies in the area source algorithm incorporated in the ISC models.

Models are continuing to improve, but many challenges remain. Combined with better measurements and analytical procedures, the new models are crucial to air pollution control and planning decisions.

REFERENCES

1. Briggs, G. A., *Plume Rise*. United States Atomic Energy Commission Critical Review Series, TID-25075. National Technical Information Service, Springfield, VA, 1969.
2. Briggs, G. A., Some recent analyses of plume rise observation, in *Proceedings of the Second International Clean Air Congress* (Englund, H. M., and Beery, W. T., eds.), pp. 1029–1032. Academic Press, New York, 1971.
3. Briggs, G. A., *Atmos. Environ.* **6**, 507–510 (1972).
4. Briggs, G. A., *Diffusion Estimation for Small Emissions*. Atmospheric Turbulence and Diffusion Laboratory, Contribution File No. 79. (draft), Oak Ridge, TN, 1973.
5. Briggs, G. A., Plume rise predictions, in *Lectures on Air Pollution and Environmental Impact Analysis* (Haugen, D. A., ed.), Chapter 3, pp. 59–111. American Meteorological Society, Boston, MA, 1975.
6. Schere, K. L., and Demerjian, K. L., A photochemical box model for urban air quality simulation, in *Proceedings of the Fourth Joint Conference on Sensing of Environmental Pollutants*, pp. 427–433. American Chemical Society, Washington, DC, 1978.
7. Hanna, S. R., *J. Air Pollut. Control Assoc.* **21**, 774–777 (1971).
8. Novak, J. H., and Turner, D. B., *J. Air Pollut. Control Assoc.* **26**, 570–575 (1976).
9. Egan, B. A., and Mahoney, J., *J. Appl. Meteorol.* **11**, 312–322 (1972).
10. Long, P. E., and Pepper, D. W., A comparison of six numerical schemes for calculating the advection of atmospheric pollution, in *Proceedings of the Third Symposium on Atmospheric*

Turbulence, Diffusion and Air Quality, pp. 181–186. American Meteorological Society, Boston, MA, 1976.

11. Businger, J. A., Wyngaard, J. C., Izumi, Y., and Bradley, E. F., *J. Atmos. Sci.* **28**, 181–189 (1971).
12. Smith, F. B., A scheme for estimating the vertical dispersion of a plume from a source near ground level, in *Proceedings of the Third Meeting of the Expert Panel on Air Pollution Modeling*. North Atlantic Treaty Organization Committee on the Challenges of Modern Society Publication No. 14. Brussels, 1972 (National Technical Information Service PB 240–574).
13. Shir, C. C., *J. Atmos. Sci.* **30**, 1327–1339 (1973).
14. Hanna, S. R., Briggs, G. A., and Hosker Jr., R. P., *Handbook on Atmospheric Diffusion*, DOE/TIC-11223. Technical Information Center, US Department of Energy, Oak Ridge, TN, 1982.
15. Eschenroeder, A. Q., and Martinez, J. R., *Mathematical Modeling of Photochemical Smog*, No. IMR-1210. General Research Corp., Santa Barbara, CA, 1969.
16. US Environmental Protection Agency, *User's Guide for the Industrial Source Complex (ISC2) Dispersion Models*. Vol. I—*User Instructions*, Vol. II—*Description of Model Algorithms*, Vol. III *Guide to Programmers*. Technical Support Division, Office of Air Quality Planning and Standards, Research Triangle Park, NC, 1992.
17. Hanna, S. R., and Paine, R. J., *J. Appl. Meteorol.* **28** (3), 206–224 (1989).
18. Hanna, S. R., and Chang, J. C., *Modification of the Hybrid Plume Dispersion Model (HPDM) for Urban Conditions and its Evaluation Using the Indianapolis Data Set*. Report Number A089-1200.I, EPRI Project No. RP-02736-1. Prepared for the Electric Power Research Institute by Sigma Research Corporation, Westford, MA, 1990.
19. Turner, D. B., Chico, T., and Catalano, J. A., *TUPOS—A Multiple Source Gaussian Dispersion Algorithm using On-Site Turbulence Data*, EPA/600/8-86/010. US Environmental Protection Agency, Research Triangle Park, NC, 1986.
20. Turner, D. B., Bender, L. W., Paumier, J. O., and Boone, P. F., *Atmos. Environ.* **25A**, 2187–2201 (1991).
21. Petersen, W. B., *User's Guide for PAL—A Gaussian-Plume Algorithm for Point, Area, and Line Sources*, Publication No. EPA-600/4-78-013. US Environmental Protection Agency, Research Triangle Park, NC, 1978.
22. Gifford, F. A., and Hanna, S. R., Urban air pollution modeling, in *Proceedings of the Second International Clean Air Congress* (Englund, H. M., and Beery, W. T., eds.), pp. 1146–1151. Academic Press, New York, 1971.
23. Gifford, F. A., and Hanna, S. R., *Atmos. Environ.* **7**, 131–136 (1973).
24. Hanna, S. R., *J. Air Pollut. Control Assoc.* **28**, 147–150 (1978).
25. Winges, K. D., *User's Guide for the Fugitive Dust Model (FDM) (revised)*, EPA-910/9-88-202R. US Environmental Protection Agency, Region 10, Seattle, WA, 1990.
26. Hanna, S. R., Schulman, L. L., Paine, R. J., and Pleim, J. E., *User's Guide to the Offshore and Coastal Dispersion (OCD) Model*, DOI/SW/MT-88/007a. Environmental Research & Technology, Concord, MA for Minerals Management Service, Reston, VA, 1988 (NTIS Accession Number PB88-182 019).
27. Perry, S. G., Burns, D. J., Adams, L. A., Paine, R. J., Dennis, M. G., Mills, M. T., Strimaitis, D. G., Yamartino, R. J., and Insley, E. M., *User's Guide to the Complex Terrain Dispersion Model Plus Algorithms for Unstable Conditions (CTDMPLUS)*, Vol. I: *Model Description and User Instructions*, EPA/600/8-89/041. US Environmental Protection Agency, Research Triangle Park, NC, 1989.
28. Perry, S. G., Paumier, J. O., and Burns, D. J., Evaluation of the EPA Complex Terrain Dispersion Model (CTDMPLUS) with the Lovett Power Plant data base, in *Preprints of Seventh Joint Conference on Application of Air Pollution Meteorology with AWMA*, New Orleans, January 14–18, 1991, pp. 189–192. American Meteorological Society, Boston, MA, 1991.

29. Burns, D. J., Perry, S. G., and Cimorelli, A. J., An advanced screening model for complex terrain applications, in *Preprints of Seventh Joint Conference on Application of Air Pollution Meteorology with AWMA*, New Orleans, January 14–18, 1991, pp. 97–100. American Meteorological Society, Boston, MA, 1991.
30. Gartrell, F. E., Thomas, F. W., and Carpenter, S. B., *Am. Ind. Hyg. Assoc. J.* **24**, 113–120 (1963).
31. Liu, M. K., Stewart, D. A., and Roth, P. M., An improved version of the reactive plume model (RPM-II), in *Ninth International Technical Meeting on Air Pollution Modeling*. North Atlantic Treaty Organization Committee on the Challenges of Modern Society Publication No. 103, Umweltbundesamt, Berlin, 1978.
32. Demerjian, K. L., and Schere, K. L., Application of a photochemical box model for O_3 air quality in Houston, TX, in *Proceedings of Ozone/Oxidants: Interactions with the Total Environment II*, pp. 329–352. Air Pollution Control Association, Pittsburgh, PA, 1979.
33. Seinfeld, J. H., *J. Air Pollut. Control Assoc.* **38**, 616–645 (1988).
34. Schere, K. L., *J. Air Pollut. Control Assoc.* **38**, 1114–1119 (1988).
35. Viebrock, H. J., (ed.), *Fiscal Year 1980 Summary Report of NOAA Meteorology Laboratory Support to the Environmental Protection Agency*. National Oceanic and Atmospheric Administration Tech. Memo. ERL ARL-107. Air Resources Laboratories, Silver Spring, MD, 1981.
36. Lamb, R. G., *A Regional Scale (100 km Model of Photochemical Air Pollution. Part I: Theoretical Formulation*, EPA-600/3-83-035. US Environmental Protection Agency, Research Triangle Park, NC, 1983.
37. Johnson, W. B., Wolf, D. E., and Mancuso, R. L., *Atmos. Environ.* **12**, 511–527 (1978).
38. Bhumralkar, C. M., Johnson, W. B., Mancuso, R. L., Thuillier, R. A., Wolf, D. E., and Nitz, K. C., Interregional exchanges of airborne sulfur pollution and deposition in Eastern North America, in *Conference Papers, Second Joint Conference on Applications of Air Pollution Meteorology*, pp. 225–231. American Meteorological Society, Boston, MA, 1980.
39. National Research Council, *Human Exposure Assessment for Airborne Pollutants: Advances and Opportunities*. National Academy Press, Washington, DC, 1994.
40. Bencala, K. E., and Seinfeld, J. H., *Atmos. Environ.* **13**, 1181–1185 (1979).
41. Fox, D. G., *Bull. Am. Meteorol. Soc.* **62**, 599–609 (1981).
42. Gryning, S. E., and Lyck, E., *Comparison between Dispersion Calculation Methods Based on In-Situ Meteorological Measurements and Results from Elevated-Source Tracer Experiments in an Urban Area*. National Agency of Environmental Protection, Air Pollution Laboratory, MST Luft-A40. Riso National Laboratory, Denmark, 1980.
43. Cox, W. M., and Tikvart, J. A., *Atmos. Environ.* **24A**, 2387–2395 (1990).
44. Hanna, S. R., *J. Air Pollut. Control Assoc.* **38**, 406–412 (1988).
45. Weil, J. C., Sykes, R. I., and Venkatram, A., *J. Appl. Meteorol.* **31**, 1121–1145 (1992).
46. Hanna, S. R., *Atmos. Environ.* **23**, 1385–1398 (1989).
47. Bowne, N. E., Validation and performance criteria for air quality models, in *Conference Papers, Second Joint Conference on Applications of Air Pollution Meteorology*, pp. 614–626. American Meteorological Society, Boston, MA, 1980.
48. US Environmental Protection Agency, *User's Manual for the Single Source (CRSTER) Model*, EPA-450/2-77-013. Research Triangle Park, NC, 1977.
49. Turner, D. B., and Irwin, J. S., *Atmos. Environ.* **16**, 1907–1914 (1982).
50. Guzewich, D. C., and Pringle, W. J. B., *J. Air Pollut. Control Assoc.* **27**, 540–542 (1977).
51. Turner, D. B., Zimmerman, J. R., and Busse, A. D., An evaluation of some climatological dispersion models, in *Proceedings of the Third Meeting of the Expert Panel on Air Pollution Modeling*. North Atlantic Treaty Organization Committee on the Challenges of Modern Society Publication No. 14. Brussels, 1972 (National Technical Information Service PB 240–574).

52. Busse, A. D., and Zimmerman, J. R., *User's Guide for the Climatological Dispersion Model*. US Environmental Protection Agency Publication No. EPA-RA-73-024. Research Triangle Park, NC, 1973.
53. Turner, D. B., and Irwin, J. S., Comparison of sulfur dioxide estimates from the model RAM with St. Louis RAPS measurements, in *Air Pollution Modeling and Its Application II* (de Wispelaere, C., ed.). Plenum, New York, 1982.
54. Shreffler, J. H., and Schere, K. L., *Evaluation of Four Urban-Scale Photochemical Air Quality Simulation Models*. US Environmental Protection Agency Publication No. EPA-600/3-82-043. Research Triangle Park, NC, 1982.
55. Lurmann, F., Godden, D., Lloyd, A. C., and Nordsieck, R. A., *A Lagrangian Photochemical Air Quality Simulation Model*. Vol. I—*Model Formulation*; Vol. II—*User's Manual*. US Environmental Protection Agency Publication No. EPA-600/8-79-015a,b. Research Triangle Park, NC, 1979.
56. Killus, J. P., Meyer, J. P., Durran, D. R., Anderson, G. E., Jerskey, T. N., and Whitten, G. Z., *Continued Research in Mesoscale Air Pollution Simulation Modeling*, Vol. V, *Refinements in Numerical Analysis, Transport, Chemistry, and Pollutant Removal*, Report No. ES77-142. Systems Applications, Inc., San Rafael, CA, 1977.
57. US Environmental Protection Agency, *Guideline on Air Quality Models* (*Revised*), EPA-450/4-80-023R. Office of Air Quality Planning and Standards. Research Triangle Park, NC, 1986 (NTIS Accession Number PB86-245 248).
58. US Environmental Protection Agency, *Supplement A to the Guideline on Air Quality Models* (*Revised*), EPA-450/2-78-027R. Office of Air Quality Planning and Standards. Research Triangle Park, NC, 1987.
59. Brode, R. W., *Screening Procedures for Estimating the Air Quality Impact of Stationary Sources*, EPA-450/4-88-010. US Environmental Protection Agency, Research Triangle Park, NC, 1988.

SUGGESTED READING

Benarie, M. M., *Urban Air Pollution Modeling*. MIT Press, Cambridge, MA, 1980.

Beychok, M. R., Fundamentals of Stack Gas Dispersion, 4th ed., self-published, 2005.

De Wispelaere, C., (ed.), *Air Pollution Modeling and Its Application I*. Plenum, New York, 1980.

Electric Power Research Institute, *Survey of Plume Models for Atmospheric Application*. Report No. EPRI EA-2243. System Application, Inc., Palo Alto, CA, 1982.

Gryning, S.-E. and Schiermeier, F. A. (ed.), *Air Pollution Modeling and Its Application*, Vol. 14. Kluwer Publications, New York, 2001.

Huber, A. H., Incorporating building/terrain wake effects on stack effluents, in *Preprints, Joint Conference on Applications of Air Pollution Meteorology*, Salt Lake City, UT, November 29–December 2, 1977, pp. 353–356. American Meteorological Society, Boston, MA, 1977.

Huber, A. H., and Snyder, W. H., Building wake effects on short stack effluents, in *Preprints, Third Symposium on Atmospheric Turbulence, Diffusion and Air Quality*, Raleigh, NC, October 19–22, 1976, pp. 235–242. American Meteorological Society, Boston, MA, 1976.

Nieuwstadt, F. T. M., and Van Dop, H., *Atmospheric Turbulence and Air Pollution Modeling*. Reidel, Dordrecht, 1982.

Schulman, L. L., and Hanna, S. R., Evaluation of downwash modifications to the Industrial Source Complex model. *J. Air Pollut. Control Assoc.* **36** (3), 258–264 (1986).

Szepesi, D. J., *Compendium of Regulatory Air Quality Simulation Models*. Akadémiai Kiadó es Nyomda Vállalat, Budapest, 1989.

US Environmental Protection Agency, *Office of Air Quality Planning and Standards, Air Pollution Modeling*, 2007, http://www.epa.gov/oar/oaqps/modeling.html.

Venkatram, A., and Wyngaard, J. C., (eds.), *Lectures on Air Pollution Modeling*. American Meteorological Society, Boston, MA, 1988.

Watson, J. G., (ed.), *Receptor Models in Air Resources Management*. APCA Transaction Series, No. 14. Air and Waste Management Association, Pittsburgh, PA, 1989.

Wayne, R. P., *Principles and Applications of Photochemistry*. Oxford University Press, New York, 1988.

Zannetti, P., *Air Pollution Modeling: Theories, Computational Methods and Available Software*. Van Nostrand Reinhold, Florence, KY, 1990.

QUESTIONS

1. Assuming that the buoyancy flux parameter F is greater than 55 in both situations, what is the proportional final plume rise for stack A compared to stack B if A has an inside diameter three times that of B?
2. How much greater is the penetration of a plume through an inversion of 1°C per 100 m than through an inversion of 3°C per 100 m? Assume that the wind speed is $3\,\mathrm{m\,s^{-1}}$, ambient air temperature is 29 K and the stack characteristics are $T_s = 415\,\mathrm{K}$, $d = 3\,\mathrm{m}$, and $v_s = 20\,\mathrm{m\,s^{-1}}$.
3. What is the steady-state concentration derived from the box model for a 10 km city with average emissions of $2 \times 10^{-5}\mathrm{g\ m^{-2}s^{-1}}$ when the mixing height is 500 m and the wind speed is $4\,\mathrm{m\,s^{-1}}$?
4. In formulating and applying a gradient transfer model, what are two of the major difficulties?
5. What is the advantage in using trajectory models for estimating air pollutant concentrations at specific air monitoring stations?
6. What is the major difficulty in estimating the maximum short-term (hours) impact of two point sources 1 km apart?
7. Using simplified techniques for estimating the concentrations from area sources, what is the annual average PM concentration for a city with an average wind speed of $3.6\,\mathrm{m\ s^{-1}}$ and area emission rate of $8-10^{-7}\mathrm{g\,s^{-1}m^{-2}}$?
8. What are the major limitations in modeling pollutant transformations in urban areas?
9. From the results of the application of the EURMAP model to Europe, what pollutant and mechanism seem to cause the least pollution by a nation to itself?
10. Which measure of scatter is likely to be larger, the mean absolute error or the root-mean-square error?
11. Contrast the fractional error for a measurement of 20 and an estimate of 4 to the fractional error for a measurement of 4 and an estimate of 20.

23

Air Pollution Climatology

Climatology refers to averaged meteorology over a period of record, usually several years. Air pollution climatology involves meteorological variables that are important in air pollution. Alternatively, it is the interpretation of air pollution data from a meteorological perspective.

I. SOURCES OF DATA

There are numerous sources of meteorological data. Hourly observations, primarily to support forecast programs and aviation operations, are made 24 h a day. Observations throughout the world, including those of over 200 stations in the contiguous United States, are also made at other intervals, when the weather is changing significantly. Since January 1, 1966, when archiving of each hourly US observation in a computer-compatible form was discontinued as an economy move, only every third hour (00 GMT plus every 3 h) has been readily accessible. The other observations are available as reproductions of manually recorded observations and may be specially prepared on magnetic tape at cost for computer use. The US archive for such data is the National Oceanic and Atmospheric Administration's (NOAA)

National Climate Center in Asheville, North Carolina. The data available from the hourly observations are listed in Table 23.1.

Other data, gathered primarily once each day by cooperative observers, consist mostly of temperature and precipitation readings. These are of limited usefulness for air pollution analysis because wind data are generally lacking.

Other sources of data, especially wind data, may be routinely measured by industrial or commercial establishments. Availability of these data must be ascertained through contact with each data collector.

Many city and regional agencies responsible for air pollutant measurements also measure wind and temperature at some of their air pollutant sampling stations. Because exposure at air quality stations is generally considerably less ideal than at airport stations, the data may be representative of extremely local conditions.

Radiosonde balloons are released twice daily, approximately at 00:00 h and 12:00 h GMT. Measurements of temperature and humidity, alternated by a pressure switch, are transmitted by radio signals from the instrument package, which is also tracked by ground-based radio direction-finding equipment at the point of release. This allows computation of wind direction and wind speed at numerous heights above ground. Figure 23.1 shows the

TABLE 23.1

Hourly Surface Observation Variables

Station number, five digits[a]
Date—year, month, day, six digits[a]
Hour, two digits[a]
Ceiling height, hundreds of feet, three alphanumeric characters[a]
Sky condition, up to four layers, four alphanumeric characters
Visibility, miles, three digits (coded)
Weather and/or obstructions to vision, eight alphanumeric characters
Sea-level pressure, millibars, four digits
Dew point, °F, three digits
Wind direction, tens of degrees azimuth, two digits[a]
Wind speed, knots, two digits[a]
Station pressure, inches of mercury, four digits
Dry bulb temperature, °F, three digits[a]
Wet bulb temperature, °F, three digits
Relative humidity, percent, three digits
Clouds and obscuring phenomena
Total amount, tenths, one coded alphanumeric character[a]
Following for up to four layers:
Amount, tenths, one coded alphanumeric character
Type, one coded alphanumeric character
Height, hundreds of feet, three alphanumeric characters
Amount of opaque cloud cover, tenths, one alphanumeric character[a]

[a] Of particular interest in air pollution work.

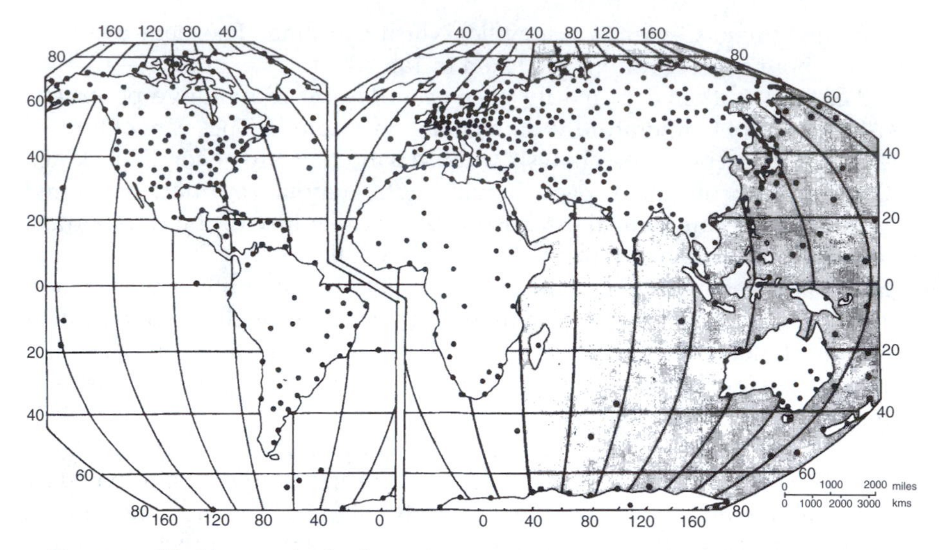

Fig. 23.1. World network of radiosonde stations. Ten stations, one in the Atlantic and nine in the Pacific, are not shown. Each dot represents a station at which an upper-air sounding is made each day at 00:00 h GMT, at 12:00 h GMT, or both. *Source*: The Secretary-General, World Meteorological Organization, Geneva.

locations of radiosonde stations throughout the world, including over 60 locations in the contiguous United States.

Numerous analyses of data routinely collected in the United States have been performed by the US National Climatic Center, results of these analyses are available at reasonable cost. The joint frequency of Pasquill stability class, wind direction class (primarily to 16 compass points), and wind speed class (in six classes) has been determined for various periods of record for over 200 observation stations in the United States from either hourly or 3-hourly data. A computer program called STAR (STability ARray) estimates the Pasquill class from the elevation of the sun (approximated from the hour and time of year), wind speed, cloud cover, and ceiling height. STAR output for seasons and the entire period of record can be obtained from the Center. Table 23.2 is similar in format to the standard output. This table gives the frequencies for D stability, based on a total of 100 for all stabilities.

Additional tables are furnished for the other stability classes. Note that calms have been distributed among the directions. Such joint frequency data can be used directly in climatological models such as the climatological dispersion model (CDM) [1]. The CDM calculates seasonal or annual concentrations at each receptor by considering sources in each wind sector (1/16 of the compass), performing a calculation for each wind speed–stability combination occurring for that sector, and weighting the calculation by the frequency for this combination. For annual concentrations, this saves a considerable number of calculations compared with simulating the period hour by hour.

TABLE 23.2

Relative Frequency of Winds for D Stability (O'Hare Airport, Chicago, 1965–1969)

	Speed (knots)						
Direction	0–3	4–6	7–10	11–16	17–21	>21	Total
N	0.0885	0.7123	1.3492	1.2670	0.1301	0.0411	3.5882
NNE	0.0646	0.5342	1.0547	1.1712	0.2123	0.0959	3.1328
NE	0.0605	0.4589	1.3972	1.0958	0.0959	0.0411	3.1494
ENE	0.0258	0.2123	0.9246	0.7260	0.0616	0.0068	1.9571
E	0.0521	0.3013	0.9109	0.8972	0.0342	0.0068	2.2027
ESE	0.0847	0.5068	0.9109	0.4794	0.0205	0.0068	2.0093
SE	0.0829	0.4726	0.6575	0.3150	0.0137	—	1.5417
SSE	0.0714	0.5274	0.9383	0.5890	0.0616	—	2.1877
S	0.1818	1.1095	2.7190	2.4245	0.2534	0.0479	6.7361
SSW	0.1495	0.7739	1.8423	2.2670	0.2397	0.0548	5.3272
SW	0.0985	0.6301	1.5889	1.4520	0.1781	0.0342	3.9818
WSW	0.1368	0.6712	1.2328	1.1712	0.2603	0.0822	3.5544
W	0.2485	1.0068	1.7191	2.0273	0.3698	0.0753	5.4467
WNW	0.1477	0.7397	1.4109	1.4794	0.2534	0.0274	4.0584
NW	0.1292	0.6643	1.3013	0.9999	0.0753	0.0068	3.1769
NNW	0.0349	0.3835	0.9109	1.0205	0.1781	0.0274	2.5553
Total	1.6574	9.7048	20.8684	19.3822	2.4382	0.5548	54.6058
Relative frequency of occurrence of D stability[a]							
Relative frequency of calms distributed above with D stability = 0.5753							

[a] Total frequency of all stability classes is 100.

Mixing heights for each day can be calculated from the radiosonce data. Such data for a 5-year period of record, 1960–1964, were calculated and used in a study by Holzworth [2]. Figure 23.2 shows the mean annual afternoon mixing height variation across the contiguous United States.

II. REPRESENTATIVENESS

The term *representativeness* in air pollution meteorology usually means the extent to which a particular parameter is measured by instrumentation sited in such a way and with sensitivity and accuracy such that it is useful for the designated purpose. For normal climatological purposes, wind measurements are made at locations relatively free from observation; thus, they are not influenced in different ways by winds coming from different directions and consequently present an unbiased record. A parameter such as wind, which varies with height above ground, must have the height of the measurement reported along with the data.

The use of a measurement generally dictates the circumstances of data collection. For example, to provide a best estimate of plume transport direction,

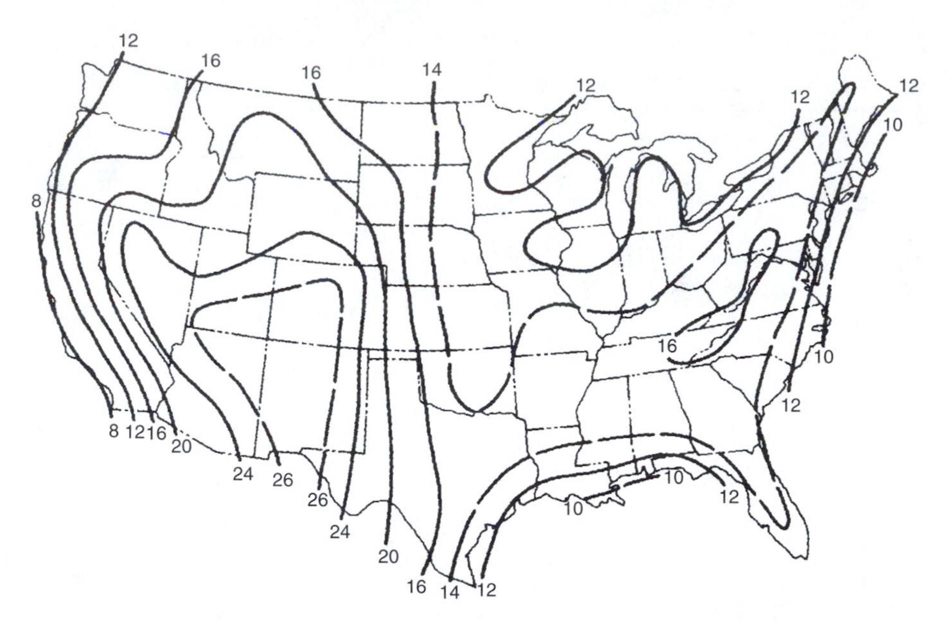

Fig. 23.2. Mean annual afternoon mixing height, in hundreds of meters. *Source*: Adapted from Holzworth [2].

hour by hour, of a release from a 75-m stack, a wind vane at the 100-m level of a tower will probably provide more representative wind direction measurements than a vane at 10 m above ground. If the release has buoyancy so that it rises appreciably before leveling off, even the 100-m measurement may not be totally adequate.

If one is studying the transport of material through the tree canopy of a forest, it is most desirable to disturb the natural environment as little as possible in making a wind measurement in the canopy. An extremely sensitive wind system is necessary because one would expect the winds to be extremely light. Also, it may be necessary to make supporting measurements both above and below the canopy, so that a wind speed profile is obtained.

Frequently, it is necessary to determine whether a record made at one location can be used to infer values of a particular parameter at another location. Many factors help to determine whether one location's measurements of a parameter are similar to those made at another location, such as the parameter in question, distance between sites, intervening topography, local topography, surface roughness, local vegetative cover, and the height above ground of the measurement. Generally, no measurements are made at both sites to check for comparability; if there were, there would be little need to use measurements at one site as a substitute for those at the other.

The foregoing factors are not independent. Two sites 50 km apart in Kansas, a state with large expanses of quite flat terrain largely planted in field crops, may be much more representative of one another than two sites

5 km apart in more rugged mountainous, wooded terrain, e.g. in the western part of North Carolina.

Data for one full year (1964) for Nashville, Tennessee, and Knoxville, Tennessee, 265 km (165 miles) apart, were compared to determine the extent to which the frequencies of various parameters were similar. Knoxville is located in an area with mountainous ridges oriented southwest-northeast; Nashville is situated in a comparatively flat area. The data available are the number of hours during which each of 36 wind directions (every 10° azimuth) occurred, the average wind speed for each direction, the number of hours of each Pasquill stability class for each direction, and the mean annual wind speed.

Table 23.3 indicates that average wind speeds are lower in Knoxville, where the average wind is 0.86 of that at Nashville. Considering the difference in topography, it is surprising that the difference is not greater. The percentage of calms, slightly greater at Knoxville, is entirely consistent with the wind speeds.

Because of the slight difference in average wind speed, one might expect this to cause a greater frequency of both stable and unstable conditions at Knoxville compared to Nashville. The stability comparisons are given in Table 23.4. The frequencies are very nearly the same, with only A and B stabilities being slightly greater at Knoxville at the expense of the D stability. Stability G is more stable than Pasquill's F.

TABLE 23.3

Average Wind Speed and Frequency of Calms

	Nashville	Knoxville
Average wind speed	$3.79\,\mathrm{m\,s^{-1}}$	$3.26\,\mathrm{m\,s^{-1}}$
Percentage of calms	9.21	9.77

TABLE 23.4

Occurrence of Pasquill Stability Classes

	Nashville		Knoxville	
Stability class	Percent	Cumulative	Percent	Cumulative
(G)	7.64	7.64	7.35	7.35
F	14.15	21.79	15.57	22.92
E	14.18	35.97	14.67	37.59
D	45.14	81.11	41.42	79.01
C	12.37	93.48	12.32	91.33
B	5.75	99.23	7.21	98.54
A	0.77	100.00	1.46	100.00

The maximum number of hours of each stability class in a single wind direction is given in Table 23.5. The total hours for A, C, and D stabilities are nearly the same. The maximum number of hours of B stability, with winds from a single direction, is about 50% higher at Knoxville. For all three stable cases, E, F, and (G), the maximum number of hours at Knoxville is about two-thirds that at Nashville.

The frequency of occurrence of winds from each of the 36 directions for the two sites is given in the second and third columns of Table 23.6. Knoxville has its maximum from 240°, with frequent winds also from adjacent points. There is a secondary maximum across the compass (in the vicinity of 60°). Nashville, on the other hand, seems to have only one principal maximum (from 180°). To explore the magnitude of the differences, the fourth column is the difference for each direction. Summing the absolute values of the differences and expressing them as a percentage of the total number of hours of observations shows that the differences constitute 62% of the total.

It is noted that backing the Knoxville frequencies by 60° would result in the maxima occurring together. This is tabulated in the fifth column of Table 23.6. The difference between Nashville and this artificial frequency is then obtained (the sixth column). The sum of the absolute values of the differences expressed as a percentage of the total is 30.2; the differences have been reduced by more than half. This does *not* prove that the frequencies are similar except that the winds at Knoxville are channeled by the major topographical features in the vicinity, but it does imply that this is an explanation.

Certainly the directional frequencies of these sites cannot be considered as representative of each other. However, this comparison would seem to indicate that the percentage of the stability categories may be more conservative over distance than the direction frequencies.

TABLE 23.5

Maximum Hours of Each Stability in One Direction

	Nashville			Knoxville		
Stability	Number of hours	Total for direction	Direction	Number of hours	Total for direction	Direction
A	7	184	22	8	{220 {364	{12 {20
B	25	252	16	38	616	6
C	79	736	36	81	616	6
D	324	736	36	332	616	6
E	123	736	36	82	431	22
F	118	736	36	79	342	25
(G)	66	736	36	45	174	31

TABLE 23.6

Number of Hours of Wind from Each Direction (Calms Are Distributed)

Direction (tens of degrees)	Hours from this direction: Nashville	Hours from this direction: Knoxville	Nashville–Knoxville	Knoxville backed 60°	Nashville–Knoxville backed 60°
1	140	279	−139	342	−202
2	201	364	−163	204	−3
3	158	295	−137	213	−55
4	184	431	−247	119	65
5	138	413	−275	104	34
6	165	426	−261	158	7
7	187	342	−155	174	13
8	134[a]	204	−70	157	−23
9	202	213	−11	81	121
10	146	119	27	86	60
11	167	104	63	67	100
12	170	158	12	137	33
13	155	174	−19	110	45
14	262	157	105	152	110
15	245	81	164	212	33
16	527	86	441	389	138
17	478	67[a]	411	410	68
18	736[b]	137	599	616	120
19	348	110	238	442	−94
20	303	152	151	302	1
21	226	212	14	343	−117
22	283	389	−106	209	74
23	243	410	−167	240	3
24	257	616[b]	−359	220	37
25	195	442	−247	174	21
26	182	302	−120	129	53
27	215	343	−128	129	86
28	174	209	−35	181	−7
29	231	240	−9	187	44
30	231	220	11	289	−58
31	338	174	164	279	59
32	300	129	171	364	−64
33	212	129	83	295	−83
34	252	181	71	431	−179
35	157	187	−30	413	−256
36	242	289	−47	426	−184
	8784	8784	Sum absolute 5450 = 62.0%		Sum absolute 2650 = 30.2%

[a] Minimum.
[b] Maximum.

III. FREQUENCY OF ATMOSPHERIC STAGNATIONS

At times when the surface pressure gradient is weak, resulting in light winds in the atmosphere's lowest layers, and there is a closed high-pressure system aloft, there is potential for the buildup of air pollutant concentrations. This is especially true if the system is slow-moving so that light winds remain in the same vicinity for several days. With light winds there will be little dilution of pollutants at the source and not much advection of the polluted air away from source areas.

Korshover [3] studied stagnating anticyclones in the eastern United States over two periods totaling 30 years. He found that for stagnation to occur for 4 days or longer, the high-pressure system had to have a warm core. Korshover's criteria included a wind speed of 15 knots or less, no frontal areas of precipitation, and persistence of these conditions for 4 or more days.

The numbers of occurrences of 4 days or more over the 30-year period (1936–1965) peak in October and September and reach a minimum in February and March. The total number of stagnation days for each part of the study area is shown in Fig. 23.3.

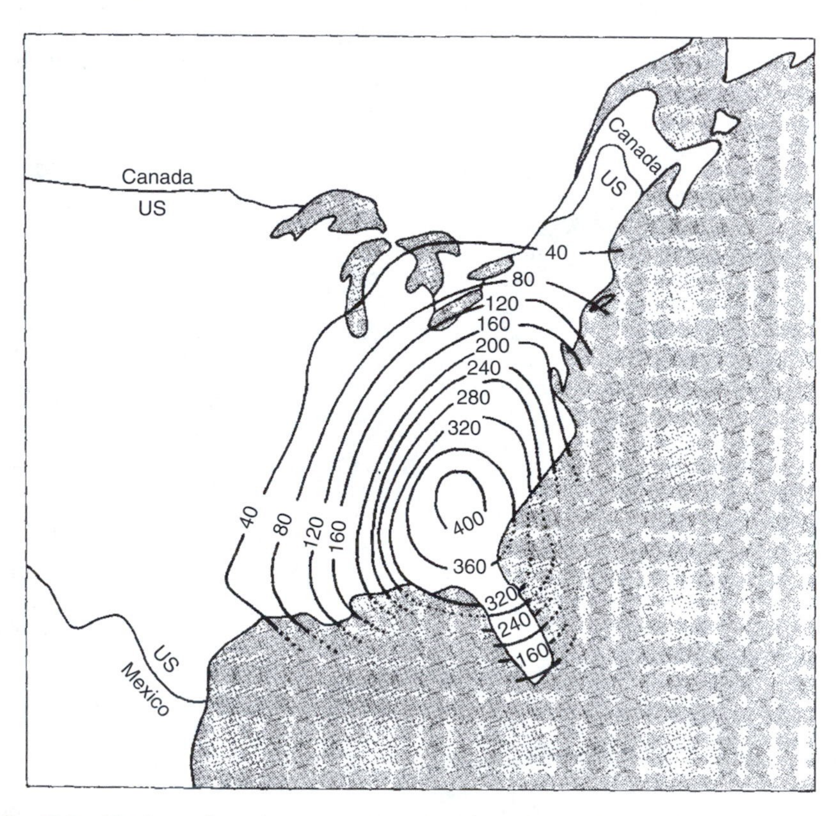

Fig. 23.3. Total number of extreme stagnation days during 1936–1965 east of the Rocky Mountains. *Source*: Korshover [3].

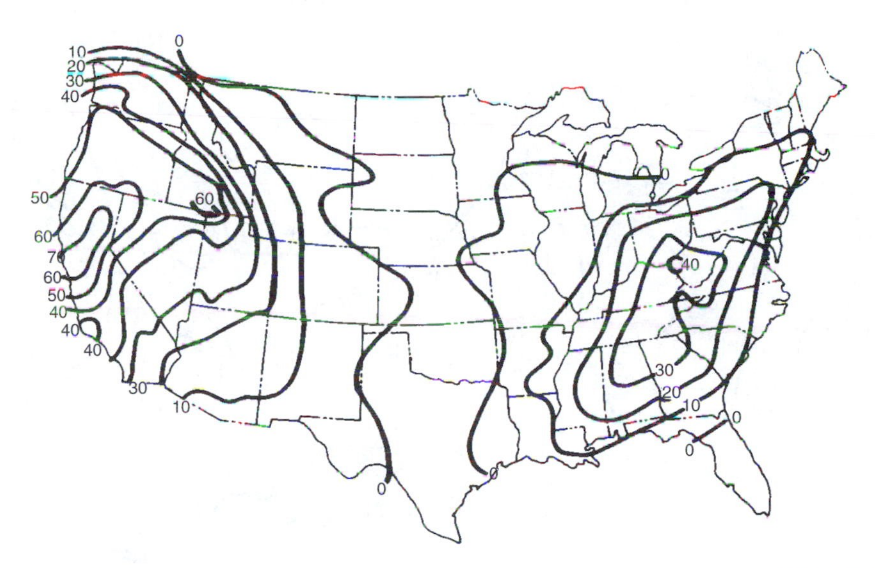

Fig. 23.4. Number of forecast days of high air pollution potential per 5 years (1960–1964). *Source*: Adapted from Holzworth [2].

Using criteria of mixing heights of 1500 m or less, with average speed through the mixing height of $4\,\text{m}\,\text{s}^{-1}$ or less, no precipitation, and persistence of these conditions for at least 2 days, Holzworth [2] tabulated high air pollution potential for the contiguous United States expressed as the number of days per 5 years (1960–1964) (Fig. 23.4). The pattern over the eastern United States is very similar to that of Korshover. It also shows no occurrences through the central plains. The number of days in the West exceed those in the East, with the maximum in central California.

IV. VENTILATION CLIMATOLOGY

Hosler [4], through study of radiosonde data for the lowest 500 ft at over 70 locations in the United States, determined inversion frequencies. The values were used to draw isolines of inversion frequency percentages on US maps for annual values and the four seasons. The percent of total hours of inversions for the annual period is shown in Fig. 23.5. Conditions frequently associated with radiation inversions—light winds and slight cloud cover at night—were also examined in terms of frequency. Both display maxima over the desert Southwest.

The study by Holzworth [2] also examined several other parameters in addition to mixing height. For example, because pollutants are diluted by the wind and mixing height limits the vertical dispersion of pollutants, Holzworth used the radiosonde data to determine the average wind speed through the mixing height for each season and annually. Figure 23.6 shows the distribution of mean annual wind speed averaged through the afternoon mixing layer.

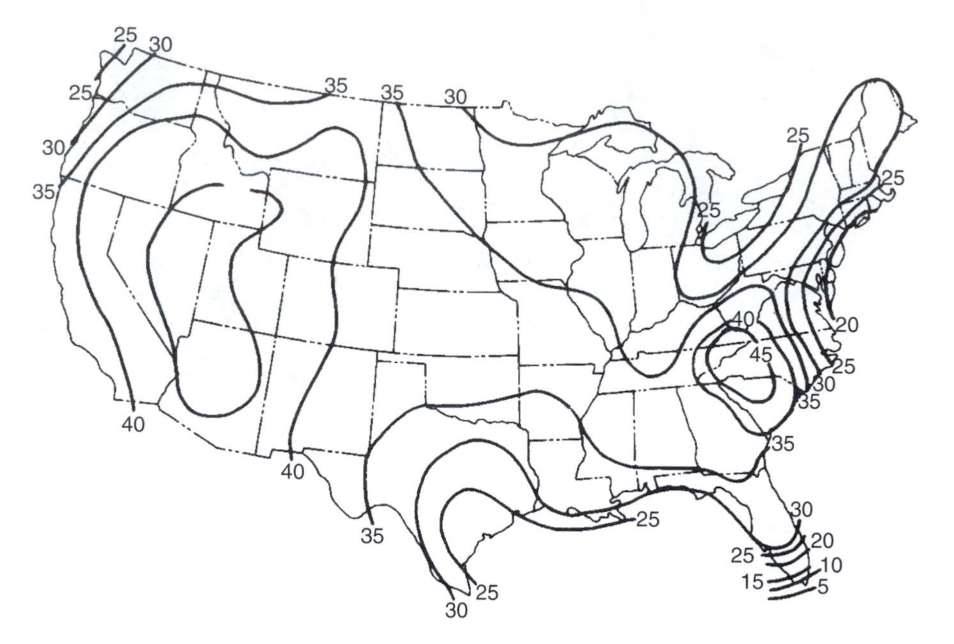

Fig. 23.5. Annual inversion frequency, percent of total hours. *Source*: Adapted from Hosler [4].

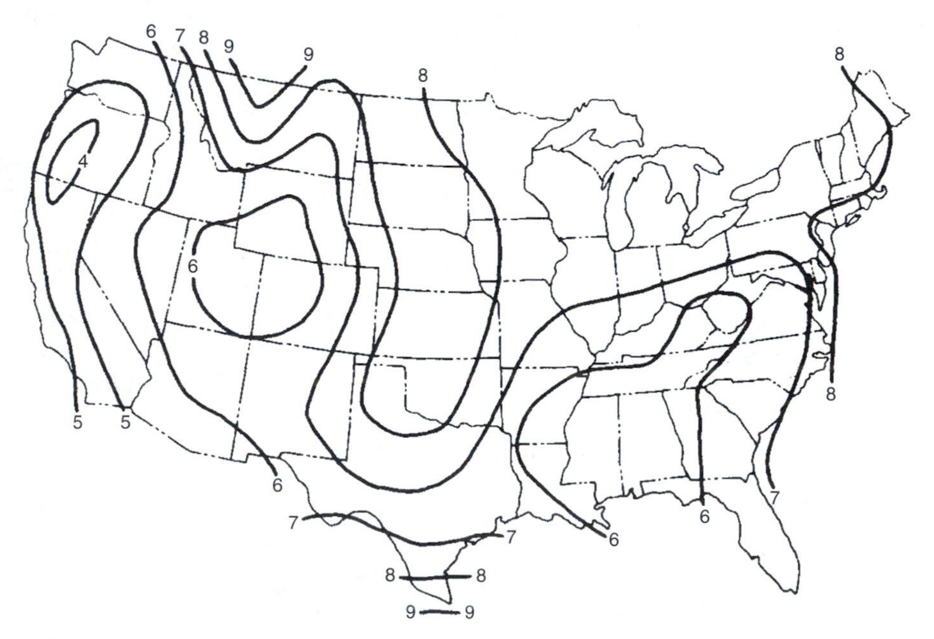

Fig. 23.6. Mean annual wind speed averaged through the afternoon mixing layer. Speeds are in meters per second. *Source*: Adapted from Holzworth [2].

Using the urban model of Miller and Holzworth [5], which requires wind speed and mixing height, Holzworth [2] used the mixing height and wind speed data to calculate concentrations for the median, upper quartile, and upper decile for hypothetical alongwind city lengths of 10 and 100 km. Results for the upper decile for the 10-km city for both the morning and the afternoon are shown in Fig. 23.7.

Another climatological study is of interest. Radiosonde observations for the 5-year period 1960–1964, used previously [2], were analyzed by Holzworth [6] to determine plume rise through the atmosphere's structure for two different stack heights, 50 and 400 m. This encompasses the range of stack heights normally encountered. The annual average effective height for the morning radiosonde ranged from less than 150 to 200 m for 50-m stacks and from less than 650 to greater than 750 m for 400-m stacks. The frequency of effective heights from 400-m stacks of the morning radiosonde that are exceeded by the afternoon mixing heights ranges from 50% to 60% near coastlines and from 70 to more than 90% throughout the rest of the contiguous United States.

Taylor and Marsh [7] investigated the long-term characteristics of temperature inversions and mixed layers in the lower atmosphere to produce an inversion climatology for the Los Angeles basin. In this area, the cooler ocean currents produce an elevated inversion that is nearly always present

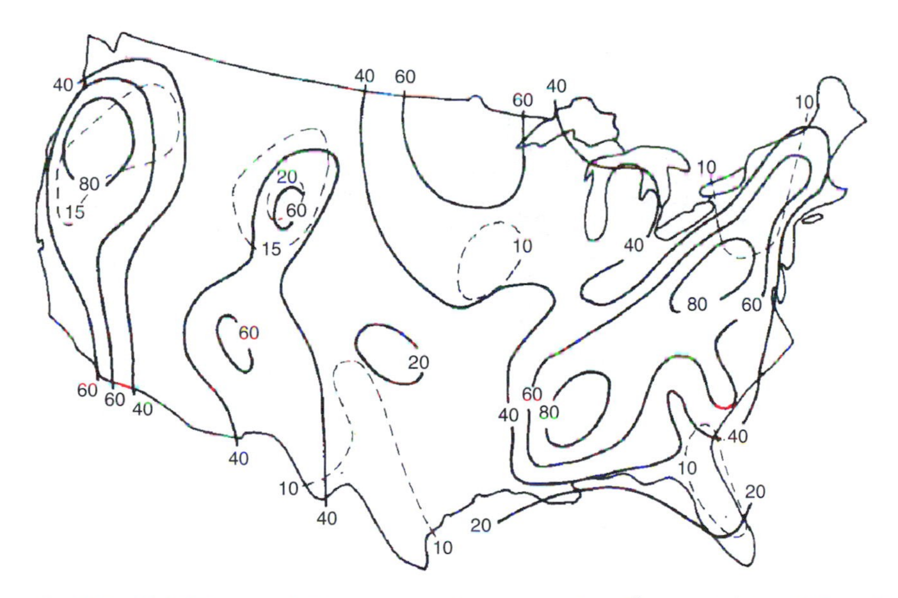

Fig. 23.7. Relative concentration in seconds per meter (s m^{-1}) exceeded in a 10-km city on 10% of all mornings (solid lines) and 10% of all afternoons (dashed lines). *Source*: After Holzworth [2].

and traps the pollutants released over the area within a layer seldom deeper than 1200 m and frequently much shallower.

V. WIND AND POLLUTION ROSES

Since wind is circular, it is frequently easier to interpret and visualize the frequency of wind flow subjectively by displaying a wind rose, that is, wind frequencies for each direction oriented according to the azimuth for that direction. Figure 23.8 is a wind rose showing both directional frequencies and wind speed frequencies by six classes from 3-hourly observations for a 5-year period (1965–1969) for O'Hare Airport, Chicago. The highest frequencies are from the south and west, the lowest from the southeast and east.

Figure 23.9 is a stability wind rose that indicates Pasquill stability class frequencies for each direction. For this location, the various stabilities seem to be nearly a set proportion of the frequency for that direction; the larger the total frequency for that direction, the greater the frequency for each stability.

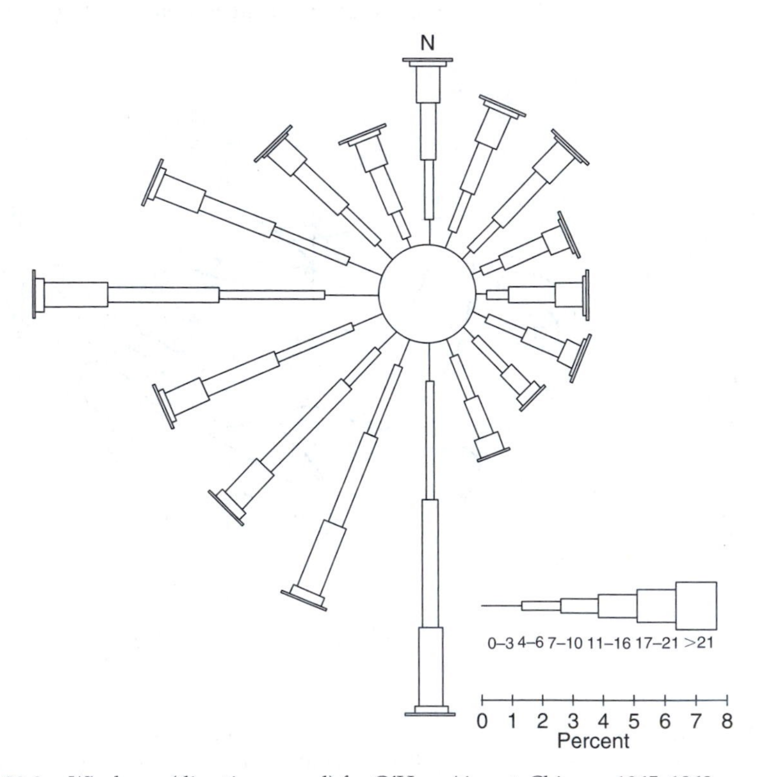

Fig. 23.8. Wind rose (direction–speed) for O'Hare Airport, Chicago, 1965–1969.

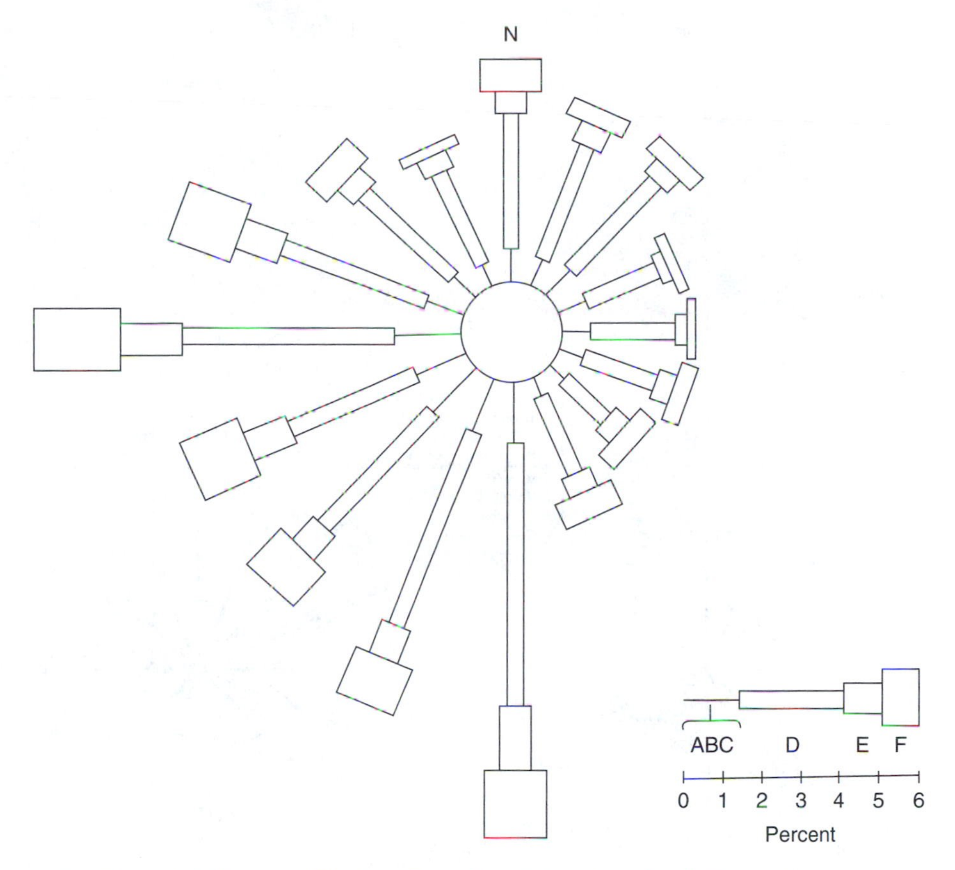

Fig. 23.9. Stability rose (direction–Pasquill stability class) for O'Hare Airport, Chicago, 1965–1969.

Since the frequencies of A and B stabilities are quite small (0.72% for A and 4.92% for B), all three unstable classes (A, B, and C) are added together and indicated by the single line.

Pollution roses are constructed by plotting either the average concentration for each direction or the frequency of concentrations above some particular concentration. Pollution roses for two pollutants at 2 times of the year are shown in Fig. 23.10, with wind frequencies by two speed classes (less than 7 miles h^{-1} and greater than 7 miles h^{-1}) superimposed. The average pollutant concentrations are connected so as to be depicted as areas rather than individual lines for each direction. Note that there is little seasonal change for hydrocarbon concentrations and only minor directional variation. SO_2, on the other hand, has very significant seasonal variation, with very low concentrations in the summer. SO_2 also has considerable directional variation.

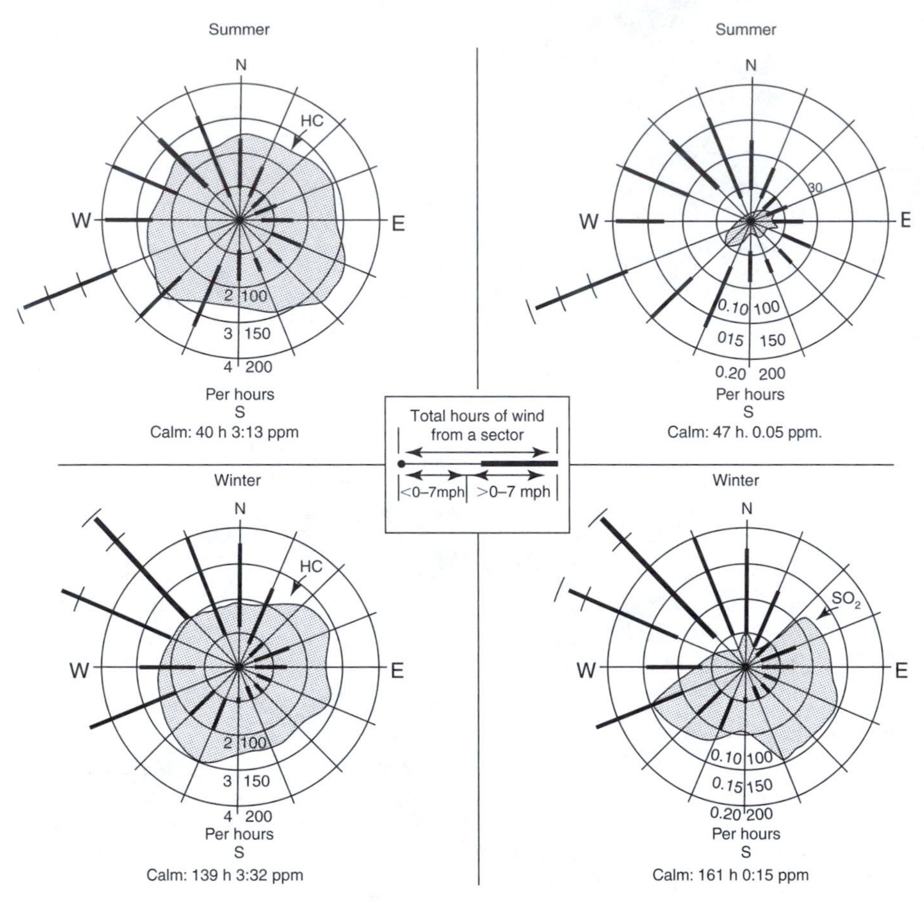

Fig. 23.10. Average concentrations of hydrocarbons and sulfur dioxide for each wind direction and wind direction frequency in two classes (0–7 miles h^{-1} and greater than 7 miles h^{-1}), Philadelphia, 1963. *Source*: US Department of Health, Education and Welfare [8].

The behavior of these pollution roses is intuitively plausible, because considerable hydrocarbon emissions come from motor vehicles which are operated in both winter and summer and travel throughout the urban area. On the other hand, sulfur dioxide is released largely from the burning of coal and fuel oil. Space heating emissions are high in winter and low in summer. The SO_2 emissions in summer are probably due to only a few point sources, such as power plants, and result in low average concentrations from each direction as well as large directional variability.

Concentrations resulting from dispersion models can also be depicted using a form of pollution rose. Figure 23.11 is a concentration rose for a measurement station in New York City for the SO_2 concentration estimates from the CDM [1]. The number in the circle is the total estimated annual SO_2 concentration from the model. The radial values represent the contribution to the annual concentration from each direction, with the length of the line proportional to the concentration resulting from area sources and the length of the rectangle proportional to the concentration resulting from the point sources. For the monitoring station in Fig. 23.11, the estimated annual concentration is 240 $\mu g\,m^{-3}$. The maximum annual contribution from area sources is from the south (39 $\mu g\,m^{-3}$); the maximum annual contribution from point sources is from the southwest (7.1 $\mu g\,m^{-3}$). The minimum concentration is from the east–southeast.

An example of frequencies of wind direction when the concentration exceeds a particular value is shown in Fig. 23.12. For this example, the concentration threshold is 0.1 ppm (262 $\mu g\,m^{-3}$). Although the maximum frequency from any one direction is only about 1%, this can be significant. Munn [9] is careful to point out that "The diagram suggests but, of course, does not prove that a major source of SO_2 is situated between the sampling stations" (p. 109).

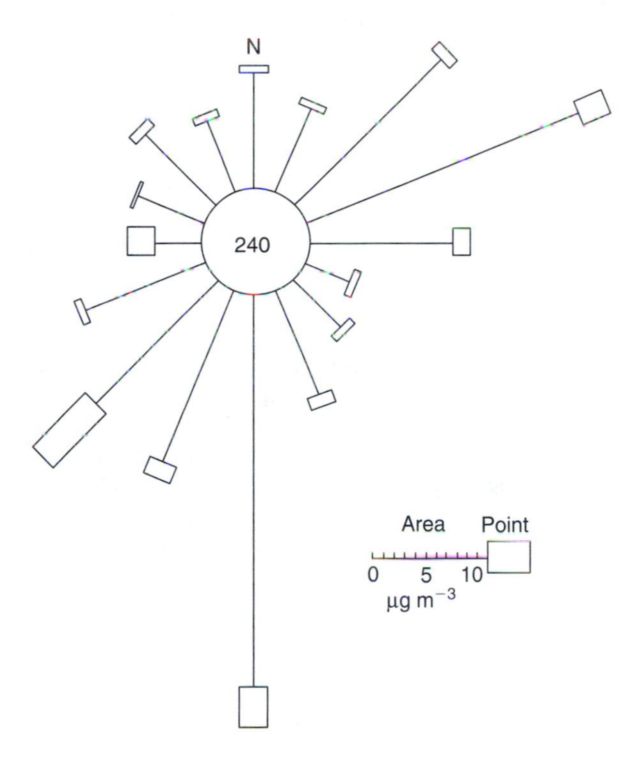

Fig. 23.11. Contributions to the annual sulfur dioxide concentration from each direction at a receptor in New York by area sources (lines) and point sources (rectangles) for 1969 using the CDM.

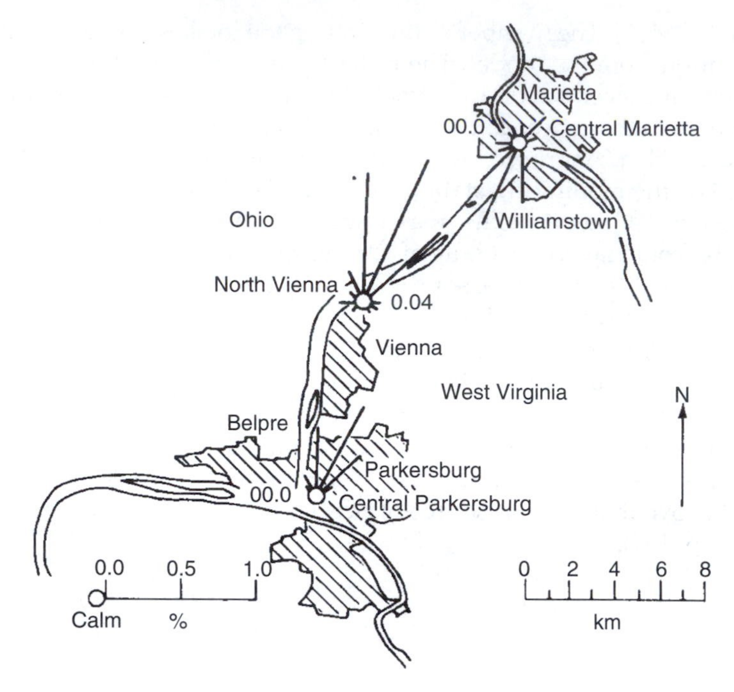

Fig. 23.12. Frequency of wind direction when sulfur dioxide exceeds 0.1 ppm near Parkersburg, West Virginia. *Source*: Munn [9].

REFERENCES

1. Busse, A. D., and Zimmerman, J. R., *User's Guide for the Climatological Dispersion Model*. Environmental Monitoring Series EPA-R4-73-024. US Environmental Protection Agency, Research Triangle Park, NC, 1973.
2. Holzworth, G. C., *Mixing Heights, Wind Speeds and Potential for Urban Air Pollution Throughout the Contiguous United States*. Office of Air Programs Publication No. *AP-101*. *US* Environmental Protection Agency, Research Triangle Park, NC, 1972.
3. Korshover, J., *Climatology of Stagnating Anticyclones East of the Rocky Mountains, 1936–1965*. Public Health Service Publication No. 999-AP-34. US Department of Health, Education and Welfare, Cincinnati, OH, 1967.
4. Hosler, C. R., *Mon. Weather Rev.* **89**, 319–339 (1961).
5. Miller, M. E., and Holzworth, G. C., *J. Air Pollut. Control Assoc.* **17**, 46–50 (1967).
6. Holzworth, G. C., Climatic data on estimated effective chimney heights in the United States, in *Preprints of Joint Conference on Application of Air Pollution Meteorology*, Salt Lake City, UT, November 29–December 2, 1977, pp. 80–87. American Meteorological Society, Boston, MA , 1977.
7. Taylor, G. H., and Marsh Jr., S. L., An inversion climatology for the Los Angeles basin, in *Preprints of Seventh Joint Conference on Application of Air Pollution Meteorology with AWMA*, New Orleans, LA, January 14–18, 1991, pp. 294–257. American Meteorological Society, Boston, MA, 1991.

8. US Department of Health, Education and Welfare, *Continuous Air Monitoring Projects in Philadelphia, 1962–1965*. National Air Pollution Control Administration Publication No. APTD 69-14. Cincinnati, OH, 1969.
9. Munn, R. E., *Biometeorological Methods*. Academic Press, New York, 1970.

SUGGESTED READING

Doty, S. R., Wallace, B. L., and Holzworth, G. C., *A Climatological Analysis of Pasquill Stability Categories Based on 'STAR' Summaries*. National Climatic Center, Environmental Data Service, National Oceanic and Atmospheric Administration, Asheville, NC, 1976.

Holzworth, G. C., and Fisher, R. W., *Climatological Summaries of the Lower Few Kilometers of Rawinsonde Observations*, EPA-600/4-79-026. US Environmental Protection Agency, Research Triangle Park, NC, 1979.

Lowry, W. P., *Atmospheric Ecology for Designers and Planners*. Peavine Publications, McMinnville, OR, 1988.

Nappo, C. J., Caneill, J. Y., Furman, R. W., Gifford, F. A., Kaimal, J. C., Kramer, M. L., Lockhart, T. J., Pendergast, M. M., Pielke, R. A., Randerson, D., Shreffler, J. H., and Wyngaard, J. C., *Bull. Am. Meteorol. Soc.* **63**(7), 761–764 (1982).

Oke, T. R., *Boundary Layer Climates*, 2nd ed. Methuen, New York, 1988.

QUESTIONS

1. What geographical or climatological conditions seem to be associated with the lowest mean annual mixing heights shown in Fig. 23.2? With the highest mixing heights?
2. In attempting to determine the air pollution impact of a small town containing several industrial facilities in a mountainous river valley about 1 km wide with sloping sides extending 400–500 m above the river, what meteorological measurement program would you recommend and what facts would you try to determine before finalizing your recommendation?
3. Over relatively flat terrain, which of the following measurements would be expected to represent conditions most closely at a second site 10 km away: wind velocity or total amount of cloud?
4. What areas of the contiguous United States have the least air pollution potential as defined by mixing heights of less than 1500 m with wind speeds of less than $4\,\mathrm{m\,s^{-1}}$ through the mixing height? Which areas have the greatest air pollution potential?
5. What is the variation in relative concentration exceeded 10% of the time in a 10-km city over the contiguous United States for mornings? For afternoons?
6. From Fig. 23.10, what wind directions are related to highest average winter sulfur dioxide concentrations at this sampling station?
7. If one considers a contribution to the annual concentration from point sources as significant when it exceeds $2\,\mu\mathrm{g\,m^{-3}}$, using Fig. 23.11, what wind directions account for significant point source contributions at this station?

Part VI

The Regulatory Control of Air Pollution

24

Air Quality Criteria and Standards

I. AIR QUALITY CRITERIA

For any pollutant, air quality criteria may refer to different types of effects. For example, Tables 24.1 through 24.6 list effects on humans, animals, vegetation, materials, and the atmosphere caused by various exposures to sulfur dioxide, particulate matter (PM), nitrogen dioxide, carbon monoxide, ozone, and lead. These data are from the Air Quality Criteria for these pollutants published by the US Environmental Protection Agency (EPA).

Criteria stipulate conditions of exposure and may refer to sensitive population groups or to the joint effects of several pollutants. Air quality criteria are descriptive. They describe effects that can be expected to occur wherever the ambient air level of a pollutant reaches or exceeds a specific concentration for a particular time period. Criteria will change as new information becomes available.

II. CONVERSION OF EFFECTS DATA AND CRITERIA TO STANDARDS

In developing air pollution cause–effect relationships, we must be constantly on guard lest we attribute to air pollution an effect caused by something else.

TABLE 24.1

US Ambient Air Quality Criteria for Carbon Monoxide

Percent of carboxyhemoglobin (CoHb) in blood	Human symptoms associated with this CoHb level
80	Death
60	Loss of consciousness; death if exposure is continued
40	Collapse on exercise; confusion
30	Headache, fatigue; judgment disturbed
20	Cardiovascular damage; electrocardiographic abnormalities
5	Decline (linear with increasing CoHb level) in maximal oxygen uptake of healthy young men undergoing strenuous exercise; decrements in visual perception, manual dexterity, and performance of complex sensorimotor tasks
4	Decrements in vigilance (i.e. ability to detect small changes in one's environment that occur at unpredictable times); decreased exercise performance in both healthy persons and those with chronic obstructive pulmonary disease
3–6	Aggravation of cardiovascular disease (i.e. decreased exercise capacity in patients with angina pectoris, intermittent claudication, or peripheral arteriosclerosis)

Source: Henderson, Y., and Haggard, H. W., *Noxious Gases*. Chemical Catalog Co., New York, 1927; and US Environmental Protection Agency, Air Quality Criteria for Carbon Monoxide, EPA/600/8-90/045F. Research Triangle Park, NC, December 1991.

Material damage due to pollution must be differentiated from that due to ultraviolet (UV) radiation, frost, moisture, bacteria, fungi, insects, and animals. Air pollution damage to vegetation has to be differentiated from quite similar damage attributable to bacterial and fungal diseases, insects, drought, frost, soil mineral deviations, hail, and cultural practices. In the principal animal disorder associated with air pollution, i.e. fluorosis, the route of animal intake of fluorine is by ingestion, the air being the means for transporting the substance from its source to the forage or hay used for animal feed. However, the water or feed supplements used may also have excess fluorine. Therefore, these sources and disease states, which may have symptoms similar to those of fluorosis, must be ruled out before a cause–effect relationship can be established between ambient air levels of fluorine and fluorosis in animals. Similarly, there are many instances of visibility reduction in the atmosphere by fog or mist for which air pollution is not a causative factor.

To study damage to materials, vegetation, and animals, we can set up laboratory experiments in which most confusing variables are eliminated and a direct cause–effect relationship is established between pollutant dosage and resulting effect. We are limited to low-level exposure experiments under controlled conditions with human beings for ethical reasons. Our cause–effect relationships

TABLE 24.2a

Summary of Lowest Observed Effect Levels for Key Lead-Induced Health Effects in Adults

Lowest observed effect level (PbB)[a] (μg dL^{-1})	Heme synthesis and hematological effects	Neurological effects	Effects on the kidney	Reproductive function effects	Cardiovascular effects
100–120		Encephalopathic signs and symptoms	Chronic nephropathy		
80	Frank anemia	↑	↓		
60				Female reproductive effects	
50	Reduced hemoglobin production	Overt subencephalopathic neurological symptoms	↓	Altered testicular function	
		↓		↓	
40	Increased urinary ALA and elevated coproporphyrins	Peripheral nerve dysfunction (slowed nerve conduction)			
30		↓			Elevated blood pressure (white males, aged 40–59)
25–30	Erythrocyte protoporphyrin (EP) elevation in males				↓
15–20	EP elevation in females				
<10	ALA-D[b] inhibition				?

[a]PbB = blood lead concentrations.

dL = deciliter = 0.1 L (a commonly used unit in medical literature).

ALA, ALA-D: aminolevulinic acid dehydrase.

Source: US Environmental Protection Agency, Air Quality Criteria for Lead, EPA-600-8-83/028aF, June 1986.

TABLE 24.2b

Summary of Lowest Observed Effect Levels for Key Lead-Induced Health Effects in Children

Lowest observed effect level (PbB)[a] ($\mu g\ dL^{-1}$)	Heme synthesis and hematological effects	Neurological effects	Renal system effects	Gastrointestinal effects
80–100		Encephalopathic signs and symptoms	Chronic nephropathy (aminoaciduria, etc.)	Colic, other overt gastrointestinal symptoms ↓
70	Frank anemia			
60		Peripheral neuropathies ↓		
50		?		
40	Reduced hemoglobin synthesis	Peripheral nerve dysfunction (slowed NCVs)[d]		
	Elevated coproporphyrin	CNS cognitive effects (IQ deficits, etc.)		
	Increased urinary ALA[c]	|		
30		?	Vitamin D metabolism interference ↓	
		|		
15	Erythrocyte protoporphyrin elevation	Altered CNS electro-physiological response ↓		
10	ALA-D[c] inhibition		?	
	Py-5-N activity inhibition[b] ↓	?		
	?			

[a] PbB: blood lead concentrations.
[b] Py-5-N: pyrimidine-5¢-nucleotidase.
[c] ALA, ALA-D: aminolevulinic acid dehydrase.
[d] NCV: nerve conduction velocity.
Source: US Environmental Protection Agency, Air Quality Criteria for Lead, EPA-600, EPA-600-8-83/028aF, June 1986.

TABLE 24.3

US Ambient Air Quality Criteria for Sulfur Dioxide

Concentration of sulfur dioxide in air (ppm)	Exposure time	Human symptoms and effects on vegetation
400	—	Lung edema; bronchial inflammation
20	—	Eye irritation; coughing in healthy adults
15	1 h	Decreased mucociliary activity
10	10 min	Bronchospasm
10	2 h	Visible foliar injury to vegetation in arid regions
8	—	Throat irritation in healthy adults
5	10 min	Increased airway resistance in healthy adults at rest
1	10 min	Increased airway resistance in asthmatics at rest and in healthy adults at exercise
1	5 min	Visible injury to sensitive vegetation in humid regions
0.5	10 min	Increased airway resistance in asthmatics at exercise
0.5	—	Odor threshold
0.5	1 h	Visible injury to sensitive vegetation in humid regions
0.5	3 h	United States National Secondary Ambient Air Quality Standard promulgated in 1973
0.2	3 h	Visible injury to sensitive vegetation in human regions
0.19	24 h[a]	Aggravation of chronic respiratory disease in adults
0.14	24 h	United States National Primary Ambient Air Quality Standard promulgated in 1971[b]
0.07	Annual[a]	Aggravation of chronic respiratory disease in children
0.03	Annual	United States National Primary Ambient Air Quality Standard promulgated in 1971[b]

[a] In the presence of high concentrations of PM.
[b] *Source*: Air Quality Criteria for Particulate Matter and Sulfur Oxides, final draft, US Environmental Protection Agency, Research Triangle Park, NC, December 1981; Review of the National Ambient Air Quality Standards for Sulfur Oxides: Assessment of Scientific and Technical Information, Draft OAQPS Staff Paper, US Environmental Protection Agency, Research Triangle Park, NC, April 1982.

for humans are based on (1) extrapolation from animal experimentation, (2) clinical observation of individual cases of persons exposed to the pollutant or toxicant (industrially, accidentally, suicidally, or under air pollution episode conditions), and (3) most important, epidemiological data relating population morbidity and mortality to air pollution. There are no human diseases uniquely caused by air pollution. In all air pollution-related diseases in which there is buildup of toxic material in the blood, tissue, bone, or teeth, part or all of the buildup could be from ingestion of food or water containing the material. Diseases which are respiratory can be caused by smoking or occupational exposure. They may be of a bacterial, viral, or fungal origin quite divorced from the inhalation of human-made pollutants in the ambient air. These causes in addition to the variety of congenital, degenerative, nutritional, and

TABLE 24.4

US Air Quality Criteria for Nitrogen Dioxide

Concentration of nitrogen dioxide in air (ppm)	Exposure time	Human symptoms and effects on vegetation, materials, and visibility
300	—	Rapid death
150	—	Death after 2 or 3 weeks by bronchiolitis fibrosa obliterans
50	—	Reversible, nonfatal bronchiolitis
10	—	Impairment of ability to detect odor of nitrogen dioxide
5	15 min	Impairment of normal transport of gases between the blood and lungs in healthy adults
2.5	2 h	Increased airway resistance in healthy adults
2	4 h	Foliar injury to vegetation
1.0	15 min	Increased airway resistance in bronchitics
1.0	48 h	Slight leaf spotting of pinto bean, endive, and cotton
0.3	—	Brownish color of target 1 km distant
0.25	Growing season	Decrease of growth and yield of tomatoes and oranges
0.2	8 h	Yellowing of white fabrics
0.12	—	Odor perception threshold of nitrogen dioxide
0.1	12 weeks	Fading of dyes on nylon
0.1	20 weeks	Reduction in growth of Kentucky bluegrass
0.05	12 weeks	Fading of dyes on cotton and rayon
0.03	—	Brownish color of target 10 km distant
0.003	—	Brownish color of target 100 km distant

Source: Draft Air Quality Criteria for Oxides of Nitrogen, US Environmental Protection Agency, Research Triangle Park, NC, 1981; Review of the National Ambient Air Quality Standard for Nitrogen Dioxide, Assessment of Scientific and Technical Information, EPA-450/5-82-002. US Environmental Protection Agency, Research Triangle Park, NC, March 1982.

psychosomatic causes of disease must all be ruled out before a disease can be attributed to air pollution. However, air pollution commonly exacerbates preexisting disease states. In human health, air pollution can be the "straw that breaks the camel's back."

Air quality standards prescribe pollutant levels that cannot legally be exceeded during a specific time period in a specific geographic area. Air quality standards are based on air quality criteria, with added safety factors as desired.

The main philosophical question that arises with respect to air quality standards is what to consider an adverse effect or a cost associated with air pollution. Let us examine several categories of receptors to see the judgmental problems that arise.

Two important pollutants that have undergone significant changes to air quality criteria are PM and ozone. Under the Clean Air Act, they are among the six principal (or "criteria") pollutants for which EPA has established National Ambient Air Quality Standards (NAAQS). Periodically, EPA reviews

TABLE 24.5

US Ambient Air Criteria for Ozone

Concentration of ozone in air (ppm)[a]	Human symptoms and vegetation injury threshold
10.0	Severe pulmonary edema; possible acute bronchiolitis; decreased blood pressure; rapid weak pulse
1.0	Coughing; extreme fatigue; lack of coordination; increased airway resistance; decreased forced expiratory volume
0.5	Chest constriction; impaired carbon monoxide diffusion capacity; decrease in lung function without exercise
0.3	Headache; chest discomfort sufficient to prevent completion of exercise; decrease in lung function in exercising subjects
0.25	Increase in incidence and severity of asthma attacks; moderate eye irritation
0.15	For sensitive individuals, reduction in pulmonary lung function; chest discomfort; irritation of the respiratory tract, coughing and wheezing. Threshold for injury to vegetation
0.12	United States National Primary and Secondary Ambient Air Quality Standard, attained when the expected number of days per calendar year with maximum hourly average concentrations above 0.12 ppm is equal to or less than 1, as determined in a specified manner

[a] 1 ppm: 1958 $\mu g\, m^{-3}$ ozone.
Source: Air Quality Criteria for Ozone and Other Photochemical Oxidants, EPA 600/8-78-004. US Environmental Protection Agency, Research Triangle Park, NC, April 1978; Revisions to National Ambient Air Quality Standards for Photochemical Oxidants. *Fed. Reg.* Part V, February 9, 8202–8237 (1979).
40 CFR § 50, July 1992.

the scientific basis for these standards by preparing an Air Quality Criteria Document (AQCD).[1]

The Clean Air Act requires an update and revision of the AQCD for PM every five years. The most recent started in 1998, following July 1997 promulgation of a new PM NAAQS. The PM AQCD is the scientific basis for the additional technical and policy assessments that form the basis for EPA decisions on the adequacy of the current PM NAAQS and the appropriateness of new or revised standards for PM. The original NAAQS for PM, issued in 1971 as "total suspended particulate" (TSP) standards, were revised in 1987 to focus on protecting against human health effects associated with exposure to ambient particles with aerodynamic diameters less than or equal to 10 microns ($\leq 10\,\mu m$). These are able to deposit in the thoracic (tracheobronchial and alveolar)

[1] The sources for the discussion in this section are the US EPA criteria documents for PM and ozone: US EPA, Air Quality Criteria for Particulate Matter, Report No. EPA/600/P-99/002aF, 2004; and US EPA,. Air Quality Criteria for Ozone and Related Photochemical Oxidants, Report No. EPA 600/R-05/004aF, 2006.

TABLE 24.6

US Ambient Air Quality Criteria for Particulate Matter

Concentration of PM in air ($\mu g\,m^{-3}$)				
Total suspended TSP > 25 μm	Thoracic TP > 10 μm	Fine FP > 2.5 μm	Exposure time	Human symptoms and effects on visibility
2000	—	—	2 h	Personal discomfort
1000	—	—	10 min	Direct respiratory mechanical changes
—	350	—		Aggravation of bronchitis
	150	—	24 h	United States Primary National Ambient Air Quality Standard as of September, 1987
180	90	—		Increased respiratory disease symptoms
	150	—	24 h	United States Primary National Ambient Air Quality Standard as of September, 1987
110	55	—	24 h	Increased respiratory disease risk
	50	—	Annual geometric mean	United States Primary National Air Quality Standard as of September, 1987
—	—	22	13 weeks	Usual summer visibility in eastern United States, nonurban sites

Source: Air Quality Criteria for Particulate Matter and Sulfur Oxides, Draft Final. US Environmental Protection Agency, Research Triangle Park, NC, December 1981; Review of the National Ambient Air Quality Standard for Particulate Matter: Assessment of Scientific and Technical Information, EPA-450/5-82-001. US Environmental Protection Agency, Research Triangle Park, NC, January 1982.
40 CFR § 50, July 1992.

portions of the lower respiratory tract. The PM_{10} NAAQS set in 1987 ($150\,\mu g\,m^{-3}$, 24 h; $50\,\mu g\,m^{-3}$, annual average) were retained in modified form and new standards ($65\,\mu g\,m^{-3}$, 24 h; $15\,\mu g\,m^{-3}$, annual average) for particles $\leqslant 2.5\,\mu m$ ($PM_{2.5}$) were promulgated in July 1997.

The current criteria document has pointed out a number of areas of uncertainty that need to be addressed. These are provided in Table 24.7. Though much has been learned about the hazards, exposure, effects, and risks of particulates, much still needs to be known.

The main focus of the PM document is the evaluation and interpretation of pertinent atmospheric science information, air quality data, human

TABLE 24.7

Key Areas of Uncertainty that Need to Be Addressed for PM as an Air Pollutant

RESEARCH TOPIC 1.	OUTDOOR MEASURES VERSUS ACTUAL HUMAN EXPOSURES
• What are the quantitative relationships between concentrations of particulate matter and gaseous co-pollutants measured at stationary outdoor air monitoring sites and the contributions of these concentrations to actual personal exposures, especially for subpopulations and individuals?	
RESEARCH TOPIC 2.	EXPOSURES OF SUSCEPTIBLE SUBPOPULATIONS TO TOXIC PARTICULATE MATTER COMPONENTS
• What are the exposure to biologically important constituents and specific characteristics of particulate matter that cause responses in potentially susceptible subpopulations and the general population?	
RESEARCH TOPIC 3.	CHARACTERIZATION OF EMISSION SOURCES
• What are the size distributions, chemical composition, and mass-emission rates of particulate matter emitted from the collection of primary-particle sources in the United States, and what are the emissions of reactive gases that lead to secondary particle formation through atmospheric chemical reactions?	
RESEARCH TOPIC 4.	AIR-QUALITY MODEL DEVELOPMENT AND TESTING
• What are the linkages between emission sources and ambient concentrations of the biologically important components of particulate matter?	
RESEARCH TOPIC 5.	ASSESSMENT OF HAZARDOUS PARTICULATE MATTER COMPONENTS
• What is the role of physicochemical characteristics of particulate matter in eliciting adverse health effects?	
RESEARCH TOPIC 6.	DOSIMETRY: DEPOSITION AND FATE OF PARTICLES IN THE RESPIRATORY TRACT
• What are the deposition patterns and fate of particles in the respiratory tract of individuals belonging to presumed susceptible subpopulations?	
RESEARCH TOPIC 7.	COMBINED EFFECTS OF PARTICULATE MATTER AND GASEOUS POLLUTANTS
• How can the effects of particulate matter be disentangled from the effects of other pollutants? How can the effects of long-term exposure to particulate matter and other pollutants be better understood?	
RESEARCH TOPIC 8.	SUSCEPTIBLE SUBPOPULATIONS
• What subpopulations are at increased risk of adverse health outcomes from particulate matter?	
RESEARCH TOPIC 9.	MECHANISMS OF INJURY
• What are the underlying mechanisms (local pulmonary and systemic) that can explain the epidemiologic findings of mortality/morbidity associated with exposure to ambient particulate matter?	
RESEARCH TOPIC 10.	ANALYSIS AND MEASUREMENT
• To what extent does the choice of statistical methods in the analysis of data from epidemiologic studies influence estimates of health risks from exposures to particulate matter? Can existing methods be improved? What is the effect of measurement error and misclassification on estimates of the association between air pollution and health?	

Source: National Research Council, *Research Priorities for Airborne Particulate Matter. III. Early Research Progress*. National Academy Press, Washington, DC, 2001.

exposure information, and health and welfare effects information published since what was assessed during the development of the 1996 PM AQCD. A number of draft versions of AQCD chapters have been evaluated via expert peer-review workshop discussions and peer reviews.

The document is a rich resource for anyone interested in PM and air pollution in general. It is divided into two volumes: Volume I (Chapters 1 through 5, EPA/600/P-99/002aD) and Volume II (Chapters 6 through 9, EPA/600/P-99/002bD). After the brief general introduction in Chapter 1, Chapters 2 and 3 provide background information on physical and chemical properties of PM and related compounds; sources and emissions; atmospheric transport; transformation and fate of PM; methods for the collection and measurement of PM; and ambient air concentrations; Chapter 4 describes PM environmental effects on vegetation and ecosystems, impacts on man-made materials and visibility, and relationships to global climate change processes; and Chapter 5 contains factors affecting exposure of the general population. Chapters 6 through 8 evaluate information concerning the health effects of PM. Chapter 6 discusses dosimetry of inhaled particles in the respiratory tract; and Chapter 7 assesses information on the toxicology of specific types of PM constituents, including laboratory animal studies and controlled human exposure studies. Chapter 8 discusses epidemiologic studies. Chapter 9 integrates key information on exposure, dosimetry, and critical health risk issues derived from studies reviewed in the prior chapters.

Tropospheric or "surface-level" ozone (O_3) is also the subject of a high profile AQCD. Following the review of criteria as contained in the EPA document, *Air Quality Criteria for Ozone and Other Photochemical Oxidants* published in 1978, the chemical designation of the standards was changed from photochemical oxidants to ozone (O_3) in 1979 and a 1-h O_3 NAAQS was set. The 1978 document focused mainly on the air quality criteria for O_3 and, to a lesser extent, on those for other photochemical oxidants (e.g. hydrogen peroxide and the peroxyacetal nitrates), as have subsequently revised versions of the document.

As is true for PM, to meet Clean Air Act requirements noted above for periodic review of criteria and NAAQS, the O_3 criteria document, *Air Quality Criteria for Ozone and Other Photochemical Oxidants*, was next revised and released in August 1986; and a supplement, *Summary of Selected New Information on Effects of Ozone on Health and Vegetation*, was issued in January 1992. These documents were the basis for a March 1993 decision by EPA that revision of the existing 1-h NAAQS for O_3 was not appropriate at that time. That decision, however, did not take into account newer scientific data that had become available after completion of the 1986 criteria document. Such literature was assessed in the next periodic revision of the O_3 AQCD which was completed in 1996 and provided scientific bases supporting the setting by EPA in 1997 of the current 8-h O_3 NAAQS. The purpose of this revised AQCD for O_3 and related photochemical oxidants is to critically evaluate and assess the latest scientific information published since that assessed in the 1996 O_3 AQCD, with the main focus being on pertinent new information useful in evaluating health and environmental effects data associated with ambient air O_3 exposures. However, other scientific data are also discussed in order to provide a better understanding of the nature, sources, distribution, measurement, and concentrations of O_3 and related photochemical oxidants and their precursors in the environment.

Like the PM document, the ozone AQCD has a wealth of information for the air pollution professional and student.

The Executive Summary summarizes key findings and conclusions as they pertain to background information on O_3-related atmospheric science and air quality, human exposure aspects, dosimetric considerations, health effect issues, and environmental effect issues.

Chapter 1 provides a general introduction, including an overview of legal requirements, the chronology of past revisions of O_3-related NAAQS, and orientation to the structure of this document.

Chapters 2 and 3 provide background information on atmospheric chemistry/physics of O_3 formation, air quality, and exposure aspects to help to place ensuing discussions of O_3 health and welfare effects into perspective.

Chapters 4 through 7 then assess dosimetry aspects, experimental (controlled human exposure and laboratory animal) studies, and epidemiologic (field/panel; other observational) studies. Chapter 8 then provides an integrative synthesis of key findings and conclusions derived from the preceding chapters with regard to ambient O_3 concentrations, human exposures, dosimetry, and health effects.

Chapter 9 deals with effects of O_3 on vegetation, crops, and natural ecosystems, whereas Chapter 10 evaluates tropospheric O_3 relationships to alterations in surface-level UVB flux and climate change and Chapter 11 assesses materials damage (these all being key types of welfare effects of relevance to decisions regarding secondary O_3 NAAQS review).

III. CONVERSION OF PHYSICAL DATA AND CRITERIA TO STANDARDS

Although air quality standards are based predominantly on biological criteria, certain physical criteria also deserve consideration.

Most materials will deteriorate even when exposed to an unpolluted atmosphere. Iron will rust, metals will corrode, and wood will rot. To prevent deterioration, protective coatings are applied. Their costs are part of the economic picture. Some materials, such as railroad rails, are used without protective coatings. There are costs associated with the decrease in their life in a polluted atmosphere as compared to an unpolluted one. One may argue that for materials on which protective coatings are used, only pollution levels that damage such coatings are of concern. One may further argue that some air pollution damage to protection coatings is tolerable, since by their very nature such coatings require periodic replenishment to maintain their protective integrity or appearance; therefore, only coatings that require more frequent replenishment than they would in an unpolluted atmosphere should enter into the establishment of deterioration costs and air quality standards. This argument certainly does hold with respect to the soiling of materials and structures. In fact, it is frequently the protective coatings themselves that require replacement because they become dirty long before their useful life as protectants has terminated. It can readily be shown that there are costs associated with soiling, including the cost of removing soil, the cost of protective coatings to facilitate the removal of soil, the premature disposal of material when it is no longer economical or practicable to remove soil, and the growth inhibition of vegetation due to leaf soiling. However, decision-making for air quality standards related to soiling is based less on economic evaluation than on esthetic considerations, i.e. on subjective evaluation of how much soiling the community will tolerate. This latter determination is judgmental and difficult to make. It may be facilitated by opinion surveys, but even when the limit of public tolerance for soiling is determined, it still has to be restated in terms of the pollution loading of the air that will result in this level of soiling.

An important effect of air pollution on the atmosphere is change in spectral transmission. The spectral regions of greatest concern are the UV and the visible. Changes in UV radiation have demonstrable adverse effects; e.g. a decrease in the stratospheric ozone layer permits harmful UV radiation to penetrate to the surface of the earth. Excessive exposure to UV radiation results in increases in skin cancer and cataracts. The worldwide effort to reduce the release of stratospheric ozone-depleting chemicals such as chlorofluorocarbons is directed toward reducing this increased risk of skin cancer and cataracts for future generations.

The fact that after a storm or the passage of a frontal system the air becomes crystal clear and one can see for many kilometers does not give a true measure of year-round visibility under unpolluted conditions. Between storms, even in unpolluted air, natural sources build up enough PM in the air so that on many days of the year there is less than ideal visibility. In many parts of the world, mountains are called "Smoky" or "Blue" or some other name to designate the prevalence of a natural haze, which gives them a smoky or bluish color and impedes visibility. When the Spanish first explored the area that is now Los Angeles, California, they gave it the name "Bay of the Smokes." The Los Angeles definition of air quality before the advent of smog was that "You could see Catalina Island on a clear day." The part of the definition that is lacking is some indication of how many clear days there were before the advent of smog.

There are costs associated with loss of visibility and solar energy. These include increased need for artificial illumination and heating; delays, disruptions, and accidents involving air, water, and land traffic; vegetation growth reduction associated with reduced photosynthesis; and commercial losses associated with the decreased attractiveness of a dingy community or one with restricted scenic views. However, these costs are less likely to be involved in deciding, for air quality standard-setting purposes, how much of the attainable visibility improvement to aim for than are esthetic considerations. Just as in the previously noted case of soiling, judgment on the limit of public tolerance for visibility reduction still has to be related to the pollutant loading of the atmosphere that will yield the desired visibility. Obviously, the pollutant level chosen for an air quality standards must be the lower of the values required for soiling or visibility, otherwise one will be achieved without the other. Whether the level chosen will not be lower than the atmospheric pollutant level required for prevention of health effects will depend on the esthetic standards of the jurisdiction.

IV. CONVERSION OF BIOLOGICAL DATA AND CRITERIA TO STANDARDS

There is considerable species variability with respect to damage to vegetation by any specific pollutant. There is also great geographic variability with respect to where these species grow naturally or are cultivated. Because of this, it is possible that in a jurisdiction none of the species particularly susceptible to damage by low levels of pollution may be among those indigenous or normally imported for local cultivation. As an example, the pollution level at which citrus trees are adversely affected, while meaningful in setting air quality standards in California and Florida, is meaningless for this purpose in Minnesota and Wisconsin. In like manner, a jurisdiction may take different viewpoints with respect to indigenous and imported species. It might set its air quality standards low enough to protect its indigenous vegetation even if this level is too high to allow satisfactory growth of imported species. Even if a particularly susceptible species is indigenous, it may be held in such low local esteem commercially or esthetically that the jurisdiction may be unwilling to let the damage level of that species be the air quality standard discriminator. In other words, the people would rather have that species damaged than assume the cost of cleaning up the air to prevent the damage. This same line of reasoning applies to effects on wild and domestic animals.

A jurisdiction may base part of its decision-making regarding vegetation and animal damage on esthetics. Its citizens may wish to grow certain ornamentals or raise certain species of pet birds or animals and allow these wishes to override the agricultural, forestry, and husbandry economics of the situation. Usually, however, economic considerations predominate in decision-making. Costs of air pollution effects on agriculture are the sum of the loss in income from the sale of crops or livestock and the added cost necessary to raise the crops or livestock for sale. To these costs must be added the loss in value of agricultural land as its income potential decreases and the loss suffered by the segments of local industry and commerce that are dependent on farm crops and the farmer for their existence. An interesting sidelight is that when such damage occurs on the periphery of an urban area, it is frequently a precursor to the breakup of such farmland into residential development, with a financial gain rather than a loss to the landowner. When the crop that disappears is an orchard, grove, or vineyard that took years to establish, and when usable farm buildings are torn down, society as a whole suffers a loss to the extent that it will take much time and money to establish a replacement for them at new locations. To some industries, air pollution costs include purchase of farm and ranch land to prevent litigation to recover damages, annual subsidy payments to farmers and ranchers in lieu of such litigation, and maintenance of air quality monitoring systems to protect themselves against unwarranted litigation for this purpose.

There is a range of ambiguity in our human health effects criteria data. In this range there is disagreement among experts as to its validity and interpretation. Thus, from the same body of health effects data, one could adopt an air quality standard on the high side of the range of ambiguity or one on the low side. Much soul searching is required before one accepts the results of questionable human health effects research and is accused of imposing large costs on the public by so doing, or of rejecting these results and being accused of subjecting the public to potential damage of human health.

V. AIR QUALITY STANDARDS

Since air pollution is controlled by air quality and emission standards, the principal philosophical discussions in the field of air pollution control focus on their development and application.

The US Clean Air Amendments of 1977 defined two kinds of air quality standards: primary standards, levels that will protect health but not necessarily prevent the other adverse effects of air pollution, and secondary standards, levels that will prevent all the other adverse effects of air pollution (Table 24.8). The amendments also define air quality levels that cannot be exceeded in specified geographic areas for "prevention of significant deterioration" (PSD) of the air of those areas. Although they are called "increments" over "baseline air quality" in the law, they are in effect tertiary standards, which are set at lower ambient levels than either the primary or secondary standards (Table 24.9). The PSD program applies to any "major emitting facility" in attainment areas. For 28 named categories a major emitting facility is one with a "potential to emit" 100 tons or more per year of any regulated air pollutant. However, any source is regulated under PSD if it has a potential to emit 250 or more tons per year. Emission potential assumes maximum design capacity (42 USC § 7479).

Increments are said to be "consumed" as new sources are given permits that allow pollution to be introduced into these areas. Jurisdictions with

TABLE 24.8

US Federal Primary and Secondary Ambient Air Quality Standards (NAAQS). ***Primary Standards*** **Set Limits to Protect Public Health, Including the Health of "Sensitive" Populations Such as Asthmatics, Children, and the Elderly.** ***Secondary Standards*** **Set Limits to Protect Public Welfare, Including Protection Against Decreased Visibility, Damage to Animals, Crops, Vegetation, and Buildings**

Pollutant	Primary standards	Averaging times	Secondary standards	Provisions
Carbon monoxide	9 ppm (10 mg m^{-3})	8 h	None	Not to be exceeded more than once per year.
	35 ppm (40 mg m^{-3})	1 h	None	Not to be exceeded more than once per year.
Lead	1.5 μg m^{-3}	Quarterly average	Same as primary	
Nitrogen dioxide	0.053 ppm (100 μg m^{-3})	Annual (arithmetic mean)	Same as primary	
Particulate matter (PM_{10})	Revoked	Annual (arithmetic mean)		Due to a lack of evidence linking health problems to long-term exposure to coarse particle pollution, the agency revoked the annual PM_{10} standard in 2006 (effective December 17, 2006).
	150 μg m^{-3}	24 h		Not to be exceeded more than once per year on average over 3 years.
Particulate matter ($PM_{2.5}$)	15.0 μg m^{-3}	Annual (arithmetic mean)	Same as primary	
	35 μg m^{-3}	24 h		To attain this standard, the 3-year average of the 98th percentile of 24-h concentrations at each population-oriented monitor within an area must not exceed 35 μg m^{-3}
Ozone	0.08 ppm	8 h	Same as primary	To attain this standard, the 3-year average of the fourth-highest daily maximum 8-h average ozone concentrations measured at each monitor within an area over each year must not exceed 0.08 ppm.

(continued)

TABLE 24.8 *(Continued)*

Pollutant	Primary standards	Averaging times	Secondary standards	Provisions
	0.12 ppm	1 h (applies only in limited areas)	Same as primary	(a) The standard is attained when the expected number of days per calendar year with maximum hourly average concentrations above 0.12 ppm is ≤1, as determined by appendix H. (b) As of June 15, 2005 EPA revoked the 1-h ozone standard in all areas except the fourteen 8-h ozone nonattainment Early Action Compact (EAC) Areas.
Sulfur oxides	0.03 ppm	Annual (arithmetic mean)	—	To attain this standard, the 3-year average of the weighted annual mean $PM_{2.5}$ concentrations from single or multiple community-oriented monitors must not exceed 15.0 $\mu g\,m^{-3}$.
	0.14 ppm	24 h	—	Not to be exceeded more than once per year.
	—	3 h	0.5 ppm (1300 $\mu g\,m^{-3}$)	Not to be exceeded more than once per year.

Source: 40 CFR Part 50, July 1999.

TABLE 24.9

US Federal Prevention of Significant Deterioration (PSD) Increments

Class I PSD increments[a]	
Sulfur dioxide	Increment ($\mu g\,m^{-3}$)
Annual arithmetic mean	2
24-h maximum	5
3-h maximum	25
Nitrogen dioxide	Increment ($\mu g\,m^{-3}$)
Annual arithmetic mean	25
Particulate matter (PM_{10})	Increment ($\mu g\,m^{-3}$)
Annual arithmetic mean	4
24-h maximum	8
Class II PSD increments[b]	
Sulfur dioxide	Increment ($\mu g\,m^{-3}$)
Annual arithmetic mean	20
24-h maximum	91
3-h maximum	512
Nitrogen dioxide	Increment ($\mu g\,m^{-3}$)
Annual arithmetic mean	25
Particulate matter (PM_{10})	Increment ($\mu g\,m^{-3}$)
Annual arithmetic mean	17
24-h maximum	30
Class III PSD increments[c]	
Sulfur dioxide	Increment ($\mu g\,m^{-3}$)
Annual arithmetic mean	40
24-h maximum	182
3-h maximum	700
Nitrogen dioxide	Increment ($\mu g\,m^{-3}$)
Annual arithmetic mean	50
Particulate matter (PM_{10})	Increment ($\mu g\,m^{-3}$)
Annual arithmetic mean	34
24-h maximum	60

[a] Class I areas are pristine, such as national parks, national seashores, and natural wilderness areas.
[b] Class II areas allow moderate deteriorations (unless otherwise designated, all areas are considered Class II).
[c] Class III are specifically designated for heavy industrial uses.
Source: 40 CFR § 51.166 (c).

authority to issue permits may choose to "allocate" portions of a PSD increment (or of the difference between actual air quality and the primary or secondary standard) for future consumption, rather than to allow its consumption on a first-come, first-served basis.

The states are required to submit to the federal EPA plans, known as State Implementation Plans (SIP), showing how they will achieve the standards in their jurisdictions within a specified time period. If after that time period there are areas within the states where these standards have not been attained, the states are required to submit and obtain EPA approval of revised plans to achieve the standards in these "nonattainment" areas. EPA also designates certain areas where the standards are being met, but which have the potential for future nonattainment, as Air Quality Maintenance Areas (AQMA). Such regions have stricter requirements than attainment areas for the granting of permits for new sources of the pollutant not in attainment status.

The Canadian Clean Air Act allows the provincial minister to formulate air quality objectives reflecting three ranges of ambient air quality for any contaminant. The *tolerable* range denotes a concentration that requires abatement without delay. The *acceptable* range provides adequate protection against adverse effects. The *desirable* range defines a long-term goal for air quality and provides the basis for a nondegradation policy for unpolluted parts of the country (Table 24.10). The Canadian ambient $PM_{2.5}$ and ozone standards shown in Table 24.11 are to be implemented by 2010.

TABLE 24.10

National Ambient Air Quality Standards for Canada

		Averaging time				
Pollutant		1 h	24 h	Annual	Published	Reviewed
Carbon monoxide ($mg\,m^{-3}$)	D	15	6 (8 h)		1974	1996
	A	35	15 (8 h)		1974	1996
	T		20 (8 h)		1978	1996
Hydrogen fluoride ($\mu g\,m^{-3}$)	RL		1.1	0.5(7 d)	1997	
Nitrogen dioxide ($\mu g\,m^{-3}$)	D			60	1975	1989
	A	400	200	100	1975	1989
	T	1000	300		1978	1989
Ozone ($\mu g\,m^{-3}$)	D	100	30		1974	1989
	A	160	50	30	1974	1989
	T	300			1978	1989
	RL				*	
PM < 10 ($\mu g\,m^{-3}$)	RL		25		1998	
PM < 2.5 ($\mu g\,m^{-3}$)	RL		15		1998	
Sulfur dioxide ($\mu g\,m^{-3}$)	D	450	150	30	1974	1989
	A	900	300	60	1974	1989
	T		800		1978	1989
Total reduced sulfur compounds	RL				*	
	AQO				*	
Total suspended particulates ($\mu g\,m^{-3}$)	D			60	1974	1989
	A		120	70	1974	1989
	T		400		1978	1989

* Reviews in progress.

Notes: D, desirable; A, acceptable; T, tolerable; RL, reference level; AQO, air quality objective.

Source: Canadian Council of Ministers of the Environment, Canadian national ambient air quality *Objectives*: process and status. In *Canadian Environmental Quality Guidelines*. Canadian Council of Ministers of the Environment, Winnipeg, Canada, 1999.

Some examples of air quality standards for other countries are given in Table 24.12.

TABLE 24.11

Canada Wide Standards (CWSs) Have Been Established for Ozone and PM Is to Minimize the Risk to Human Health and the Environment

Ozone	65 ppb (130 $\mu g\,m^{-3}$) averaged over an 8-h period. Achievement will be based on the 4th highest measurement annually as averaged over three consecutive years.
$PM_{2.5}$	30 $\mu g\,m^{-3}$ averaged over a 24-h period. Achievement will be based on the 98th percentile ambient measurement annually, averaged over three consecutive years.

Note: These standards represent a balance between the desire to achieve the best health and environmental protection possible in the near term with the feasibility and costs to reduce pollutant emissions contributing ozone and PM in the ambient air. These CWSs for ozone and $PM_{2.5}$ are to be implemented by the year 2010.

TABLE 24.12

Air Quality Standards for Selected Pollutants in Several Countries and for International Organizations Around the World

	Suspended particulate matter (SPM), $\mu g\,m^{-3}$		Sulphur dioxide (SO_2), $\mu g\,m^{-3}$		Oxides of Nitrogen (NO), $\mu g\,m^{-3}$	
Country	24 h	1 year	24 h	1 year	24 h	1 year
India						
Industrial	500	360	120	80	120	80
Residential	200	140	80	60	80	60
Sensitive	100	70	30	15	30	15
China						
Class III	500		250		150	
Class II	300		150		100	
Class I	150		50		50	
Australia		90		60		
Japan	100		100		100	
USA	260		365	80		100
EU	300	150	250	80		200
WHO	150–230	60–90	100–150	40–60	150	

Source: CONCAWE (Conservation of Clean Air and Water in Europe), *Potential for Exhaust After Treatment and Engine Technologies to Meet Future Emissions Limit*, Report No. 99/55. Brussels, Belgium, 1999.

SUGGESTED READING

Atkisson, A., and Gaines, R. S. (eds.), *Development of Air Quality Standards*. Merrill, Columbus, OH, 1970.

Cochran, L. S., Pielke, R. A., and Kovacs, E., Selected international receptor-based air quality standards. *J. Air Waste Manage. Assoc.* **42**, 1567–1572 (1992).

Kates, R. W., *Risk Assessment of Environmental Hazard*. Wiley, Chicester, England, 1978.

Schwing, R. C., and Albers, W. A. (eds.), *Society Risk Assessment*. Plenum, New York, 1980.

US Environmental Protection Agency, *Air Quality Criteria for Particulate Matter*, Report No. EPA/600/P-99/002aF, 2004.

US Environmental Protection Agency, *Air Quality Criteria for Ozone and Related Photochemical Oxidants*, Report No. EPA 600/R-05/004aF, 2006.

World Health Organization, *Air Quality Guidelines for Europe*. World Health Organization, Copenhagen, 1987.

QUESTIONS

1. Why are air quality criteria descriptive?
2. Why are air quality standards prescriptive?
3. Evaluate the use, effectiveness, and equity of local, state, provincial, or national air quality standards in your community.
4. Prepare a table similar in format to Tables 24.1 through 24.6 for another pollutant not yet required by the administrator of the US EPA to have a criteria document.
5. Discuss the relative merits of stating air quality standards as 1-h, 3-h, 8-h, 24-h, and annual averages.
6. Discuss the relative merits of national versus local air quality standards.
7. Discuss the differences in approach in using air quality standards (as in the United States), air quality objectives (as in Canada), and air quality goals (as in certain other countries).
8. Discuss the advantages and disadvantages of promulgating only one set of air quality standards (as in most countries) and of employing secondary and tertiary (PSD) standards, as in the United States.
9. Discuss the problem caused by cigarette smoking in the evaluation of epidemiological data on the effect of air pollution on respiratory disease.
10. Compare the NAAQS and NAAQO for US and Canada, respectively. What do they have in common and how do they differ?
11. How have air quality standards changed since those promulgated under the Clean Air Act of 1970?
12. Why have ozone and PM been the attention of intensive scrutiny since the mid-1990s? How have the AQCDs changed to address the research findings?

25

Indoor Air Quality

I. CHANGING TIMES

Societies' concern with air quality has evolved from medieval times, when breathing smelting fumes was a major hazard, to where we are today (see Chapter 1). In modern society, a parallel effort has been under way to improve air quality in the outside or ambient air, which is the principal focus of this book, and in the industrial occupational setting in manufacturing and other traditional jobs. A combination of events is moving many countries to consider the quality of air in other locations where we live parts of our lives. Attention is now being refocused on "indoor" air quality.

In developing countries, priorities have often differed from those of wealthier nations. Industrialization, water and food supply, sanitation, infrastructure improvements, and basic health care are often the focus of the leaders of a country. In some areas, the availability of a job is much more problematic than some consideration about the quality of the air in the workplace or the home. Many dwellings in developing countries do not have closable windows and doors, so the outdoor and indoor air quality issues are different. In some houses where cooking is done by firewood or charcoal, the air quality outdoors may be considerably better than that inside the smoky residence. The evolution of our modern society and the concomitant changes in lifestyle, workplace, and housing improvements place concerns about indoor air quality in a different category than for developing countries and from the times of our ancestors.

For many industrialized countries, efforts to improve the outdoor air quality have been under way for the majority of this century. In many locations around the world, significant improvements have taken place. Air quality in many major cities such as London, New York, and Chicago has improved from the conditions present in the first half of the twentieth century. Mechanisms and control programs are in place in the developed countries to continue the improvement of ambient air quality. Considerable effort and energy have been expended to characterize, evaluate, and control air pollution emissions to the atmosphere.

Buildings and their design have undergone major changes. Fifty years ago, central heating and windows which could be opened and closed depending on the season were the norm for commercial buildings. Now we have multistory buildings with central heating and air conditioning and sealed glass exterior walls. Residential housing has undergone similar design and structural changes, in some cases resulting in dwellings that may have poorer indoor air quality.

New residences and commercial buildings are designed and built with energy conservation as a major design criterion. New materials have been developed and are being used in construction. Although these modifications have helped to save energy, a consequence of some of these modifications is slower exchange of air with the outside. This helps considerably with the heating or cooling system because energy must condition this "new" air which is introduced into the structure. However, the decreased air exchange rates usually translate into dirtier air.

A second consideration is the change in lifestyle for individuals in industrialized societies. We are no longer a society dependent on occupations which require us to be outdoors for a significant part of our day. Over the past two decades, studies of daily activities have consistently shown for urban populations that, on average, we spend about 90% of our time indoors in our homes, cars, offices, factories, public buildings such as restaurants, malls, and others. Any given individual activity profile may differ significantly from this average.

Exposure assessment techniques now attempt to include as many as possible of the locations in which individuals now spend time. The concept involves identification of microenvironments which are important for potential exposure. For example, exposure to CO would include time spent in commuting, parking garages, in residences with gas stoves, as well as time spent outdoors. This approach classifies time spent in these microenvironments and the typical concentrations of CO in these locations.

II. FACTORS INFLUENCING INDOOR AIR QUALITY

Several factors influence the quality of air indoors: the rate of exchange of air with air from outdoors, the concentration of pollutants in outdoor air, the rate of emissions from sources indoors, the rate of infiltration from soil gases, and the rate of removal in the indoor environment (Fig. 25.1).

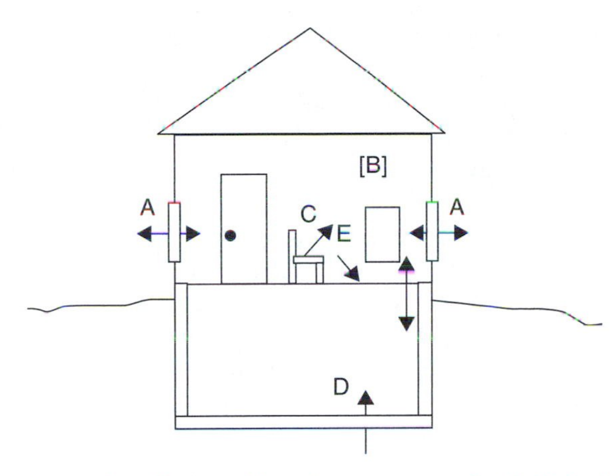

Fig. 25.1. Representation of home with various sources and sinks for indoor air pollutants: A, exchange; B, indoor concentration; C, outgassing of building and furniture materials; D, infiltration from soils; E, removal on interior surfaces.

The source of indoor air pollutants may be inside the building, or they may be transported into the interior space from the outside. Sources located indoors include building materials, combustion sources, furnishings, and pets. Emissions of organic gases are higher with increased temperature and humidity but usually decrease with age of the structure or furnishings. Construction materials and the composition of furnishings inside the building may give off or outgas pollutants into the interior airspace, e.g. glues or adhesives. Natural gas for cooking and kerosene space heaters release NO and CO_2 even when operating properly. Molds may grow in the ventilation ducts and be distributed throughout a building.

Radon from the soil can enter buildings through cracks in the foundation when the pressure inside is lower than in the soil. The rate of infiltration depends on the soil type, the building structure, and the pressure differential between the soil and the building.

Air is exchanged between indoors and outdoors by several ways: natural ventilation, mechanical ventilation, and infiltration or exfiltration. Natural ventilation involves movement of air through building openings like doors, windows, and vents. Mechanical ventilation involves fans and heating and air-conditioning systems. Infiltration and exfiltration represents undesirable movement of air in and out of the structure. Buildings are characterized as "tight" when infiltration rates are low.

The air exchange rate influences the concentration of indoor pollutants in two ways. At higher air exchange rates, the pollutants inside a structure are removed from the interior. As long as the ambient outside air has lower pollutant concentrations, high exchange rates help lower indoor air pollutant levels. However, if the pollutant concentration outside is elevated, then an increase in the air exchange rate will bring these materials into the building; e.g. an idling vehicle adjacent to an air intake will transfer exhaust fumes into the building. At lower exchange rates, pollutants released from sources inside the building can contribute to higher levels of indoor pollutants.

The concentration of indoor pollutants is a function of removal processes such as dilution, filtration, and destruction. Dilution is a function of the air exchange rate and the ambient air quality. Gases and particulate matter may also be removed from indoor air by deposition on surfaces. Filtration systems are part of many ventilation systems. As air is circulated by the air-conditioning system it passes through a filter which can remove some of the particulate matter. The removal efficiency depends on particle properties; including size and surface characteristics, such as electrostatics. In addition, some reactive gases like NO_2 and SO_2 are readily adsorbed on interior surfaces of a building or home.

III. INDOOR AIR POLLUTANTS

Table 25.1 shows the major categories of indoor air pollutants and sources. Table 25.2 shows a summary of reported indoor air pollutant concentrations compiled by the US Environment Protection Agency. Information in this table is not meant to be representative of typical indoor concentrations but only examples of measurements obtained by investigators and reported in the literature.

Airborne material affecting the quality of indoor air may be classified as gases or particulate matter. Gases which may be potential problems are radon, CO, NO_2, and hydrocarbons. Particulate matter may come from tobacco smoke, mold spores, animal dander, plant spores, and others as shown in Table 25.1. Other factors interact to influence our perception of indoor air quality, including humidity, temperature, lighting, and sound level.

TABLE 25.1

Indoor Air Pollutants and Typical Sources

Pollutant	Source
Combustion gases—CO and NO	Combustion—furnace, cooking stove, space heater, etc.
VOCs	Outgassing of building materials, coatings, wall and floor coverings, and furnishings
Formaldehyde and other aldehydes and carbonyls	Outgassing of pressed wood and insulation foam
Pesticides	Household products
Particulate matter	Combustion
Biological agents—molds, spores, and dander	Contaminated ventilation systems, ceiling tile and walboard, outdoor and indoor plant materials, and pets
Environmental tobacco smoke (ETS)	Smoking in building
Radon	Infiltration from soil beneath structure
Asbestos	Construction coatings, tile, and insulation

TABLE 25.2

Summary of Reported Indoor Air Pollutant Concentrations

Pollutant	Measured concentration			Types of building	Reference[d]
	Minimum	Maximum	Mean		
Radon			$0.8\ pCi\ L^{-1}$	Residences	EPA (1987d)
	$0.5\ pCi\ L^{-1}$	$2000\ pCi\ L^{-1}$		Residences	EPA (1987b)
	$0.14\ pCi\ L^{-1}$	$4.11\ pCi\ L^{-1}$	—	New public buildings	Sheldon *et al.* (1988)
	$0.3\ pCi\ L^{-1}$	$1.68\ pCi\ L^{-1}$	—	Old public buildings	Sheldon *et al.* (1988)
	—	—	$1.7–2.4\ pCi\ L^{-1}$	Three office buildings	Bayer and Black (1988a)
ETS (as respirable suspended particulate or RSP)	—	—	$28\ \mu g\ m^{-3(1)a}$	Residences	NRC (1986b)
	—	—	$74\ \mu g\ m^{-3(2)}$	Residences	NRC (1986b)
	—	—	$32\ \mu g\ m^{-3(3)}$	Residences	DHHS (1986)
	—	—	$50\ \mu g\ m^{-3(4)}$	Residences	DHHS (1986)
(as nicotine)	—	—	0.7–3.2 ppb	Three office buildings	Bayer and Black (1988a)
Biological contaminants	—	$564–5360\ CFU\ m^{-3b}$		Three office buildings	Bayer and Black (1988a)
Formaldehyde	—	$131–319\ \mu g\ m^{-3}$	$78–144\ \mu g\ m^{-3}$	Residences	Hawthorne *et al.* (1984)
	ND[c]	192 ppb	—	New public buildings	Sheldon *et al.* (1987)
	ND	103 ppb	—	Old public buildings	Sheldon *et al.* (1987)
	—	—	25–39 ppb	Three office buildings	Bayer and Black (1988a)
Benzene	—	$120\ \mu g\ m^{-3}$	$20\ \mu g\ m^{-3}$	Various	Wallace *et al.* (1983)
Carbon tetrachloride	—	$14\ \mu g\ m^{-3}$	$2.5\ \mu g\ m^{-3}$	Various	Wallace *et al.* (1983)
Trichloroethylene	—	$47\ \mu g\ m^{-3}$	$3.6\ \mu g\ m^{-3}$	Various	Wallace *et al.* (1983)
Tetrachloroethylene	—	$250\ \mu g\ m^{-3}$	$10\ \mu g\ m^{-3}$	Various	Wallace *et al.* (1983)

(continued)

TABLE 25.2 *(Continued)*

Pollutant	Measured concentration			Types of building	Reference[d]
	Minimum	Maximum	Mean		
Chloroform	—	$200\,\mu g\,m^{-3}$	$8\,\mu g\,m^{-3}$	Various	Wallace *et al.* (1983)
Dichlorobenzenes	—	$1200\,\mu g\,m^{-3}$	$41\,\mu g\,m^{-3}$	Various	Wallace *et al.* (1983)
Pesticides					
Diazionon	ND	$8.9\,\mu g\,m^{-3}$	$1\,4\,\mu g\,m^{-3}$	Residences	Lewis *et al.* (1986)
Chlordane	ND	$1.7\,\mu g\,m^{-3}$	$0.51\,\mu g\,m^{-3}$	Residences	Lewis *et al.* (1986)

[a] (1) 73 residences without smokers; (2) 73 residences with smokers; (3) nonsmokers not exposed to ETS; and (4) nonsmokers exposed to ETS.
[b] Summation of mesophilic bacteria, fungi, and thermophilic bacteria. CFU: colony-forming units.
[c] ND: not detected.
[d] *References*

Bayer, C. W., and Black, M. S., *Indoor Air Quality Evaluations of Three Office Buildings: Two of Conventional Construction Designs and One of a Special Design to Reduce Indoor Air Contaminants*. Georgia Institute of Technology, Athens, GA, 1988a.

Department of Health and Human Services (DHHS), *The Health Consequences of Involuntary Smoking—A Report to the Surgeon General.* US Department of Health and Human Services, Rockville, MD, 1986.

EPA, *Compendium of Methods for the Determination of Air Pollutants in Indoor Air*. Environmental Monitoring Systems Laboratory, US Environmental Protection Agency, Cincinnati, OH, 1987b.

EPA, *Radon Reference Manual*, EPA 520/1-87-20. Office of Radiation Programs, US Environmental Protection Agency, Washington, DC, 1987d.

Hawthorne, A., *et al.*, Models for estimating organic emissions from building materials: formaldehyde example. *Atmos. Environ.* **21** (2), 419–424 (1987).

Lewis, R. G., *et al.*, Monitoring for non-occupational exposure to pesticides in indoor and personal respiratory air. *Presented at the 79th Annual Meeting of the Air Pollution Control Association*, Minneapolis, MN, 1986.

National Research Council (NRC), *Environmental Tobacco Smoke: Measuring Exposures and Assessing Health Effects*. National Academy Press, Washington, DC, 1986b.

Sheldon, L., Zelon, H., Sickles, J., Eaton, C., and Hartwell, T., *Indoor Air Quality in Public Buildings*, Vol II. Environmental Monitoring Systems Laboratory, Office of Research and Development, US Environmental Protection Agency, Washington, DC, 1988.

Wallace, L. A., *et al.*, *Personal Exposure Assessment Methodology (TEAM) Study: Summary and Analysis: Vol. I.* Office of Research and Development, US Environmental Protection Agency, Washington, DC, 1983.

Source: US Environmental Protection Agency Report to Congress on Indoor Air Quality, EPA/400/1-89/001c, August 1989.

IV. EFFECTS OF INDOOR AIR POLLUTANTS

Effects of indoor air pollutants on humans are essentially the same as those described in Chapter 11. However, there can be some additional pollutant exposures in the indoor environment that are not common in the ambient setting. From the list in Table 25.1, radon exposure indoors present a radiation hazard for the development of lung cancer. Environmental tobacco smoke (ETS) has been found to cause lung cancer and other respiratory diseases. Biological agents such as molds and other toxins may be a more likely exposure hazard indoors than outside.

When present, arguably the most hazardous indoor pollutant is asbestos. Chronic exposures are directly associated with mesothelioma, lung cancer, and asbestosis. That is why governments at all levels have aggressively implemented programs to protect children and others in schools and other indoor microenvironments from exposures to airborne asbestos fibers. Asbestos is discussed in detail in Chapter 2.

Mold is ubiquitous, but indoor exposures need not be high. Mold's growth is usually increased with increasing temperature and humidity under environmental conditions, but this does not mean molds cannot grow in colder conditions. Species may be of wide range of colors and often elicit particles and gases that render odors, often referred to as "musty." Like other fungi, molds reproduce by producing spores that are emitted into the atmosphere. Living spores are disseminated to colonize growth wherever conditions allow. Most ambient air contains large amounts of so-called "bioaerosols," i.e. particles that are part of living or once living organisms. In the instance, the bioaerosols are live mold spores, meaning that inhalation is a major route of exposure.

Indoor sources of molds include leaking pipes and fixtures, damp spaces such as those in basements and crawl spaces, heating, air conditioning and ventilation (HVAC) systems, especially those that allow for condensation from temperature differentials between surfaces and ambient air, kitchens, and showers.

Some molds produce toxic substances called mycotoxins. There is much uncertainty related to possible health effects associated with inhaling mycotoxins over a long-time periods. Extensive mold growth may cause nuisance odors and health problems for some people. It can damage building materials, finishes, and furnishings and, in some cases, cause structural damage to wood.

Sensitive persons may experience allergic reactions, similar to common pollen or animal allergies, flu-like symptoms, and skin rash. Molds may also aggravate asthma. Rarely, fungal infections from building-associated molds may occur in people with serious immune disease. Most symptoms are temporary and eliminated by correcting the mold problem, although much variability exists on how people are affected by mold exposure. Particularly sensitive subpopulations include:

- Infants and children
- Elderly people

- Pregnant women
- Individuals with respiratory conditions or allergies and asthma
- Persons with weakened immune systems (e.g. chemotherapy patients, organ or bone marrow transplant recipients, and people with HIV infections or autoimmune diseases)

Persons cleaning mold should wear gloves, eye protection, and a dust mask or respirator to protect against breathing airborne spores (Fig. 8.18). A professional experienced in mold evaluation and remediation, such as an industrial hygienist, may need to be consulted to address extensive mold growth in structures. It is important to correct large mold problems as soon as possible by first eliminating the source of the moisture and removing contaminated materials, cleaning the surfaces, and finally drying the area completely.

If visible mold is present, then it should be remediated, regardless of what species are present and whether samples are taken. In specific instances, such as cases where health concerns are an issue, litigation is involved, or the source(s) of contamination is unclear, sampling may be considered as part of a building evaluation. Sampling is needed in situations where visible mold is present and there is a need to have the mold identified. A listing of accredited laboratories can be found at www.aiha.org/LaboratoryServices/html/lists.htm.

Radon gas is formed in the process of radioactive decay of uranium. The distribution of naturally occurring radon follows the distribution of uranium in geological formations. Elevated levels have been observed in certain granite-type minerals. Residences built in these areas have the potential for elevated indoor concentrations of radon from radon gas entering through cracks and crevices and from outgassing from well water.

Radon gas is radioactive and emits alpha particles in the decay process. The elements resulting from radon decay are called radon daughters or progeny. These radon daughters can attach to airborne particles, which can deposit in the lung. The evidence supporting the radon risk of lung cancer comes from studies of uranium mine workers, in whom elevated rates of lung cancer have been observed. When an analysis of the potential exposure to radon inside homes was conducted by the US Environmental Protection Agency, an estimate of 5000–20 000 excess lung cancer deaths was projected annually [1]. The risk is associated directly with increased lifetime doses; i.e. the longer the time spent living in a residence with elevated levels of radon, the higher the risk. Indoor levels of radon range from less than 1–200 pCi L^{-1}. Levels as high as 12 000 pCi L^{-1} have been observed but most levels are much lower. The US Environmental Protection Agency has established an action level of 4 pCi L^{-1} for indoor radon, and if a home screening test shows concentrations below 4 pCi L^{-1}, no remedial action is suggested.

Environmental tobacco smoke (ETS), sometimes referred to as "second-hand smoke" or "passive smoking" represents the exposure to tobacco smoke by individuals other than the smoker. For decades, the US Surgeon General has indicated that smoking is a cause of lung cancer and cardiovascular disease for individuals who smoke. The US Environmental Protection Agency has also concluded that ETS is a lung carcinogen for others breathing it.

The presence of biological contaminants gained widespread recognition with the outbreak of Legionnaires' disease in Philadelphia, Pennsylvania in 1976. In that year, 221 persons attending a convention of Legionnaires developed pneumonia symptoms and 34 subsequently died. The agent, a bacterium later named *Legionella pneumophila,* was found in the cooling tower of the hotel's air-conditioning system. This bacterium has subsequently been responsible for other outbreaks of Legionnaires' disease. The bacteria in water supplies may be eliminated by suitable treatment procedures. *Legionnella* represents one of many types of biological agents which can cause allergic reactions and illness in the indoor environment.

One of the more difficult challenges remaining is the characterization of "sick building" syndrome. On numerous occasions, some employees in certain office buildings or other workplaces have developed a combination of symptoms including respiratory problems, dryness of the eyes, nose, and throat, headaches, and other nonspecific complaints. In such situations a substantial portion of the workers may exhibit these symptoms, which decrease in severity or stop when the worker is away from the building over the weekend or for longer periods. Investigations into the cause of these symptoms sometimes provide explanations, uncover ventilation problems, or identify an irritant gas. But many of these problem buildings are very difficult to understand and additional research is necessary to understand the cause-and-effect relationships.

V. CONTROL OF INDOOR AIR POLLUTANTS

The control and regulation of indoor air quality are influenced by individual property rights and a complicated mosaic of federal, state, and local government jurisdiction with conflicts, overlaps, and gaps in addressing these issues. Table 25.3 shows a large number of agencies and departments involved in indoor air quality control efforts at the federal level.

Government can institute certain laws and regulations for the citizens' well-being. Environmental and occupational examples abound, such as clean water and air legislation and workplace safety and health regulations. As the extension of this role into the home occurs, implementation and enforcement become more problematic. Examples of proactive regulatory approaches are building codes, zoning, consumer product standards, and safety requirements. Table 25.4 shows what various parties, from the individual to the federal government, can do to improve indoor air quality. Many of these efforts focus on education, improved materials, and better design of products and structures.

The technological control strategies are related back to Fig. 25.1. If the hazard is the result of elevated concentrations, then the technological solution is to reduce or remove the sources or dilute or remove the agent.

TABLE 25.3

Indoor Air Responsibility of Federal Agencies

Point of impact	Agency/Activity	Comments
Direct control of indoor concentrations and/or exposures	OSHA air standards	Limited to industrial environments
	BPA radon action level	Limited to residents in BRA's weatherization program
	NASA air standards	Adopted OSHA standards
Control of emissions by restricting activities or product composition	EPA drinking water MCLs for radon and VOCs	Indoor air exposures considered in determining drinking water levels
	EPA pesticide restrictions	Restricts use and sales of pesticides which may cause indoor air pollution
	CPSC consumer product bans	Bans on use of some potential indoor pollutants in consumer products
	Smoking restrictions imposed by DOD, DOT, and GSA	Restricts smoking in specified indoor environments
	VA restrictions on asbestos use	Restricts use of asbestos in VA buildings
Control through assessment and mitigation procedures	EPA asbestos rules	Provides for the assessment and mitigation of asbestos hazards in schools
	GSA building assessments	Investigates GSA-controlled buildings for indoor air problems
	NIOSH building assessments	Responds to air quality health complaints
	DOD/USAF chlordane assessments	Investigates USAF facilities for chlordane problems
	NASA HVAC system maintenance	Assesses and corrects HVAC operation to optimize indoor air quality
Effort to increase knowledge of indoor air quality problems and controls	Research efforts by EPA, CPSC, DOE, HHS, BPA, DOT, NASA, NIST, NSF, TVA, HUD, and GSA	
	Information dissemination by EPA, CPSC, DOE, HHS, BPA, HUD, TVA, FTC, NASA, NIST, and NIBS	

OSHA, Occupational Safety and Health Administration; BPA, Bonneville Power Administration; NASA, National Aeronautics and Space Administration; MCL, Maximum Contaminant Levels; VOC, Volatile Organic Compounds; CPSC, Consumer Products Safety Commission; DOD, Department of Defense; DOT, Department of Transportation; CSA, General Services Administration; VA, Veterans Administration; NIOSH, National Institute of Occupational Safety and Health; USAF, United States Air Force; HVAC, Heating Ventilation and Air Conditioning; DOE, Department of Energy; HHS, Health and Human Services; NIST, National Institute of Standards and Technology; NSF, National Science Foundation; TVA, Tennessee Valley Authority; HUD, Housing and Urban Development; FTC, Federal Trade Commission; NIBS, National Institute of Building Sciences.
Source: US Environmental Protection Agency Report to Congress on Indoor Air Quality, EPA/400/1-89/001c, August 1989.

TABLE 25.4

Stakeholder Interests in Improving Indoor Air Quality

Individuals	Consumer and health professionals	Manufacturers	Building owners and managers	Builders and architects	State and local governments	Federal government
Find low-emission products in purchasing decisions	Be knowledgeable of symptoms, effects, and mitigation and advise clients	Adopt test procedures and standards to minimize product and material emissions	Adopt ventilation maintenance procedures to eliminate and prevent contamination and ensure and adequate supply of clean air to building occupants	Adopt indoor air quality as a design objective	Conduct studies of specific problems in state or local area and adopt mitigation strategies	Conduct research and technology transfer programs
Maintain and use products to minimize emissions	Develop information and education programs for constituent publics	Adequately label products as to emission level and proper use and maintenance of products	Use zone ventilation or local exhaust for indoor sources	Ensure compliance with indoor air quality ventilation standards	Establish building codes for design, construction, and ventilation requirements to ensure adequate indoor air quality	Coordinate actions of other sectors

(continued)

TABLE 25.4 (*Continued*)

Individuals	Consumer and health professionals	Manufacturers	Building owners and managers	Builders and architects	State and local governments	Federal government
Exercise discretionary control over ventilation to ensure clean air supply	Substitute materials to minimize emissions from products manufactured	Develop specific procedures for use of cleaning solvents, paints, herbicides, insecticides, and other contaminants to protect occupants	Adopt low emission requirements in procurement specifications for building materials from manufacturers	Enforce and monitor code compliance	Coordinate actions of other sectors, encourage, or require specific sectors to take actions toward mitigation	
Be knowledgeable of indoor air quality problems and take actions to avoid personal exposure	Develop training programs for commercial users to ensure low emissions Conduct research to advance mitigation technology	Adopt investigatory protocols to respond to occupant complaints	Contain or ventilate known sources	Educate and inform building community, health community, and public about problems and solutions		

Source: US Environmental Protection Agency Report to Congress on Indoor Air Quality, EPA/400/1-89/001c, August 1989.

Control techniques are discussed for agents mentioned earlier—radon, ETS, and biological agents—and also for volatile organic compounds (VOCs). Radon enters the residence by two principal routes: infiltration from soil beneath the structure and outgassing from well water during showers. Elevated levels of radon are generally observed in basements or first-floor rooms. The mitigation techniques available include increased ventilation of the crawl space beneath the first floor, soil gas venting from beneath a basement floor, and sealing of all openings to the subsurface soil. These steps reduce the entry of radon into the home. Elevated levels of radon in well water can be removed by aeration or filtration by absorbent-filled columns.

Control of ETS is complicated because of the personal behavior of individuals. For public buildings and facilities like offices, restaurants, and malls, many governmental bodies are placing restrictions on smoking in these areas, which can range from complete bans to requiring a restaurant to have a portion of a dining area for smokers and the remainder for nonsmokers. The difficulty for the restaurant owner is ensuring that the nonsmoking section is free of ETS. Education is the primary approach to "control" in the home. Information about the effects of ETS on family members, especially the health of pregnant women, unborn and young children, has modified the behavior of some smokers.

Control of biological agents is multifaceted. In the case of *Legionnella*, cleaning and maintenance of heating and air-conditioning systems are generally sufficient to reduce the risk of this disease. In home heating and air-conditioning systems, mold and bacteria may be present and controlled with maintenance procedures and increasingly available technologies (e.g. UV disinfection). Growth may be inhibited by lower humidity levels. Keeping a house clean lowers the presence of dust mites, pollen, dander, and other allergens.

For VOCs, control options are multiple. Source reduction or removal includes product substitution or reformulation. Particleboard or pressed wood has been developed and used extensively in building materials for cabinet bases and subflooring and in furniture manufacturing for frames. If the product is not properly manufactured and cured prior to use as a building material, VOCs can outgas into the interior of the residence or building. Other sources of VOCs may be paints, cleaning solutions, fabrics, binders, and adhesives. Proper use of household products will lower volatile emissions.

In many of the industries associated with building or household products, efforts are under way to reduce the potential for subsequent VOC release to the interior of residences or commercial buildings. Modification of the manufacturing process, solvent substitution, product reformulation, and altering installation procedures are a few of the approaches available.

When VOCs are present indoors at elevated concentrations, modification of ventilation rates is a control option for diluting and reducing these concentrations. The American Society of Heating, Refrigerating and Air-Conditioning Engineers (ASHRAE) has established standards for ventilation rates for outside air per individual. The guideline is 15 cfm per person [2]. This guideline is designed to bring sufficient fresh air into a building to minimize the buildup of contaminants and odors.

REFERENCES

1. *A Citizen's Guide to Radon: What It Is and What to Do About It*. US Environmental Protection Agency and US Department of Health and Human Services, OPA-8-004. Government Printing Office, Washington, DC, 1986.
2. *Ventilation for Acceptable Indoor Air Quality*. American Society of Heating, Refrigerating and Air-Conditioning Engineers, Atlanta, GA, 1981.

SUGGESTED READING

Godish, T., *Indoor Air Pollution Control*. Lewis Publishers, Chelsea, MI, 1989.

Nazaroff, W. W., and Teichmann, K., Indoor radon. *Environ. Sci. Technol.* **24**, 774–782 (1990).

US Environmental Protection Agency, *Indoor Air Quality*, 2007, http://www.epa.gov/iaq/index.html 2007. Includes separate links to asbestos, mold and other contaminants.

US Environmental Protection Agency and US Consumer Product Safety Commission, *The Inside Story: A Guide to Indoor Air Quality*. EPA/400/1-88/004, September 1988.

QUESTIONS

1. Define "sick building" syndrome.
2. What controls are available to control indoor air quality?
3. Why will the command and control approach not work for residential indoor air quality?
4. Describe an implication of high radon levels during the sale of a home.
5. The mass balance can be determined for any control volume. So, let us consider a house to be such a control volume. We can construct a simple *box model* for this control volume:

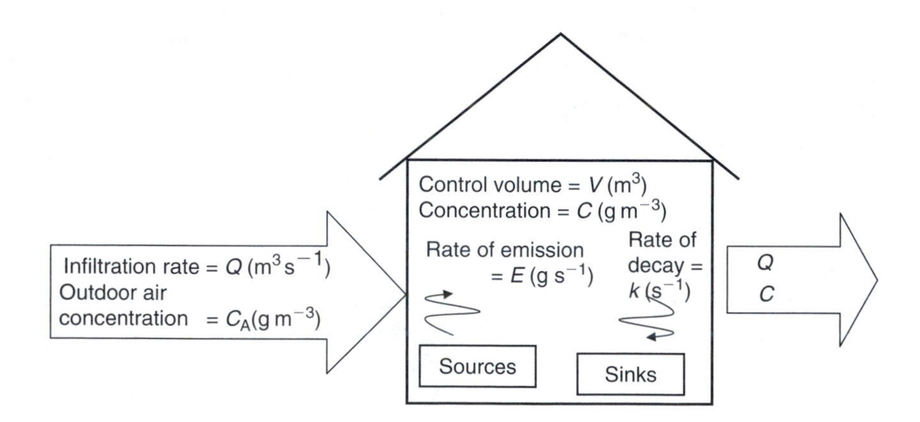

Recall that the mass balance equation is:

Rate of pollutant increase in control volume	=	Rate of pollutant entering control from outdoors	+	Rate of pollutant entering control volume from indoor source emissions	−	Rate of pollutant leaving control volume by leakage to outdoors	−	Rate of pollutant leaving control volume by chemical decay

Stated mathematically, the equation is:

$$V\frac{dC}{dt} = QC_A + E - QC - kCV$$

Some k values in inverse seconds for some important indoor pollutants (from Lawrence Berkeley National Laboratory) are:

Formaldehyde (HCHO) $= 1.11 \times 10^{-4}$
Nitric acid (NO) $= 0.0$
Aerosols (fine particles $<0.5\,\mu m$ aerodynamic diameter) $= 1.33 \times 10^{-4}$
Rn $= 2.11 \times 10^{-6}$
Sulfur dioxide (SO_2) $= 6.39 \times 10^{-5}$
Carbon monoxide $= 0.0$

To find the time-dependent concentration (i.e. the general solution), use:

$$C_t = \frac{(E/V) + C_A(Q/V)}{(Q/V) + k}\left[1 - \exp\left(-\left(\frac{Q}{V} + k\right)t\right)\right] + C_0 \exp\left[-\left(\frac{Q}{V} + k\right)t\right]$$

And, for steady state conditions (i.e. $(dC/dt) = 0$), the indoor concentrations are simplified to:

$$C = \frac{QC_A + E}{Q + kV}$$

Further, if the pollutant does not decay appreciably (i.e. $k \approx 0$); it is considered to be a *conservative pollutant*. Many indoor studies use conservative tracers (e.g. perfluorides). So, when outdoor concentrations of a conservative compound are negligible, and the initial indoor concentration is zero, we can further simplify the mass balance to:

$$C_t = \frac{E}{Q}\left[1 - \exp\left(-\left(\frac{Q}{V}\right)t\right)\right]$$

So, let us consider an example. You are heating your apartment with an unvented kerosene heater for 1 h. The air volume of your apartment is $200\,m^3$. If the heater emits SO_2 at a rate of $50\,\mu g\,s^{-1}$, and the ambient (outdoor air) and your initial concentrations in your apartment are both $100\,\mu g\,m^{-3}$, the ventilation (air exchange rate) is $50\,L\,s^{-1}$, and your ceiling fan is sufficiently mixing the air (i.e. the control volume is assumed to be well-mixed), what is the indoor concentration of SO_2 in your apartment after 1 h?

These conditions allow us to use the time-dependent concentration (i.e. the general solution) equation:

$$C_t = \frac{(E/V) + C_A(Q/V)}{(Q/V) + k}\left[1 - \exp\left(-\left(\frac{Q}{V} + k\right)t\right)\right] + C_0 \exp\left[-\left(\frac{Q}{V} + k\right)t\right]$$

$$C_t = \frac{(50\,\mu g\,s^{-1}/200\,m^3) + 100\,\mu g\,m^{-3}(0.050\,m^3\,s^{-1}/200\,m^3)}{(0.050\,m^3\,s^{-1}/200\,m^3) + 6.39 \times 10^{-5}\,s^{-1}}$$

$$\times\left[1 - \exp\left(-\left(\frac{0.050\,m^3\,s^{-1}}{200\,m^3} + 6.39 \times 10^{-5}\,s^{-1}\right)3600\,s\right)\right]$$

$$+ \left(100\,\mu g\,m^{-3}\right)\exp\left(-\left(\frac{0.050\,m^3\,s^{-1}}{200\,m^3} + 6.39 \times 10^{-5}\,s^{-1}\right)3600\,s\right)$$

$$= 876.08(1 - \exp(-1.13)) + 100\ \exp(-1.13)$$

$$= 876.08(1 - 0.323) + 100(0.323)$$

$$= 593.09 + 32.3 = 625.39\,\mu g\,m^{-3}$$

Question: How long will it take after shutting off the heater (i.e. $E = 0$), to reach the same concentrations of SO_2 as that found outside the house?

6. Assuming that you been working on your car's engine in a garage with the same dimensions and air exchanges as the apartment above, and that the concentrations of carbon monoxide were also 625.39 $\mu g\,m^{-3}$? How long will it take to reach 0 $\mu g\,m^{-3}$ CO, assuming the outdoor concentration is 0 $\mu g\,m^{-3}$ CO? In other words, how long will it take to flush the CO?
   What could you do to increase the flushing rate?
   You may have noticed some interesting things about the mathematics of the time-dependent equation when applied to a conservative pollutant ($k = 0$) at an ambient concentration = 0. So, find the time needed to reach a relatively safe (low, but not zero) indoor CO concentration of 300 ppb, with $k = 0.0000000002/s$. After all, even conservative gases like CO break down (e.g. by reactions with hydroxyl radicals).
   Compare the results and try to explain any differences in theory *versus* practical applications. What is going on here? Be creative and bold in thinking about the differences. This is one of the interesting differences between engineering and basic science.
7. You are the city engineer sitting in a discussion about a program to give kerosene space heaters to needy people. Recent inspections of target homes show that the air exchange rates (QV^{-1}) at present to be an average 4 per hour. Before the heaters are given away, the city, with the help of a team of volunteers will fix windows, caulk and add insulation so that any home receiving the heater will be at most 1.5 air exchanges per hour. The manufacturer specs show that the heaters will produce 23 mg of carbon monoxide per second. If the average home will use one heater in the living room (500 m^3) and one in each of the three bedrooms (each 400 m^3), do you foresee any problems with this plan? Would you want to live and sleep in a home with these CO concentrations (e.g. based on health standards)? Support your recommendation in a memo to the mayor and city council.
8. Later in the same meeting, you are asked to consider the use of building materials for the new library. The architects call for library panel constructed of fiberboard. It is much more aesthetically appealing than the alternate material (drywall). The fiberboard emits formaldehyde (HCHO) at 360 $\mu g\,m^{-2}h^{-1}$ and the library walls to be covered represent a surface area of 20 000 m^2. The emission rate is expected to drop by 40 $\mu g\,m^{-2}h^{-1}$ each year as the product off gases and the formaldehyde decomposes. The drywall emits no HCHO, but in a few places the drywall will be reinforced with plywood which emits 4 $\mu g\,m^{-2}h^{-1}$ (but only from 1500 m^2 of wall surface). However, the grout used if the

drywall is selected emits radon at a rate of $1 \times 10^{-5} \mu g\,m^{-2} h^{-1}$. (For some additional information about conversion of radiation units to concentrations, see the following website: http://www.hps.org/publicinformation/ate/q2548.html. Remember that the volume being used here is liters, not cubic meters, and that the density of water is much higher than that of air.)

The volume of the library will be $250\,000\,m^3$. Assume that the library air will be well mixed. The library is designed to have 1.2 air exchanges per hour. Recent measurements in the area around the building site have detected $5 \mu g\,m^{-3}$ HCHO in the ambient air. Write a memo to the library planning commission communicating your calculations and estimates, as well as your recommendations about the choice of wall materials. Also, make some recommendations on what corrective actions will need to be taken if they still wish to install the panel as the architects wish, or if they go with the alterative.

In these problems, can you draw any conclusions about the relationship of indoor air quality and energy efficiency (e.g. the role that air exchange rates play in both)? Can you think of some approaches that could ameliorate both air quality and energy efficiency in the same building?

26

Regulating Air Pollution

I. INTRODUCTION

The US Clean Air Act Amendments of 1990 (CAAA90) [1] were a revision of the original US Clean Air Act passed in 1963, amended in 1970, and amended again in 1977 [2]. It is one of the most significant pieces of environmental legislation ever enacted. Estimates of the yearly economic impacts of CAAA90 range from US $12 to US $53 billion in 1995 and from US $25 to US $90 billion by 2005.

CAAA90 is a technology-based program rather than the health-based program used in the original Clean Air Act (CAA). The standards and emission limits are based on Maximum Achievable Control Technology (MACT). The final emission limits are set forth in permits issued by the individual states.

Stenvaag, in his book, *Clean Air Act 1990 Amendments Law and Practices* [3], has written:

> By far the most important interpretive documents have not yet been written: the countless EPA regulations, decisions, and explanatory preambles that must be published in the Federal Register to carry out the Act. As these rulings are systematically promulgated over the coming decades, flesh and muscle will be added to the massive skeleton of the 1990 amendments and we will all gain a better understanding of what the 101st Congress has wrought.

II. TITLES

CAAA90 contains 11 "Titles." Some of these are new, and some are greatly expanded from similar titles in the original CAA.

A. Title I: Provisions for Attainment and Maintenance of National Ambient Air Quality Standards

Title I changed the existing nonattainment program of the original Clean Air Act Title I, Part D, by establishing new nonattainment requirements for ozone, carbon monoxide (CO), and particulate matter (PM_{10}). Also, the new nonattainment requirements vary from location to location and with the severity of nonattainment. The changes are accomplished through revisions of the State Implementation Plan (SIP). The SIP must designate "milestones" according to specified timetables. Failure to meet milestones results in mandatory sanctions.

The most widespread and persistent urban pollution problem is ozone. The causes of this and the lesser problem of CO and PM_{10} pollution in our urban areas are largely due to the diversity and number of urban air pollution sources. One component of urban smog, hydrocarbons, comes from automobile emissions, petroleum refineries, chemical plants, dry cleaners, gasoline stations, house painting, and printing shops. Another key component, nitrogen oxides, comes from the combustion of fuel for transportation, utilities, and industries.

Although there are other reasons for continued high levels of ozone pollution, such as growth in the number of stationary sources of hydrocarbons and continued growth in automobile travel, the remaining sources of hydrocarbons are the most difficult to control. These are the small sources, those that emit less than 100 tons of hydrocarbons per year. These sources, such as auto shops and dry cleaners, may individually emit less than 10 tons per year but collectively emit many hundreds of tons of pollution.

Title I allows the EPA to define the boundaries of "nonattainment" areas for ozone, CO, and PM_{10}. Emission standards for these areas are based on a new set of "nonattainment categories." EPA has established a classification system for ozone design values (goals) and attainment deadlines. Table 26.1 lists these parameters.

TABLE 26.1

Classification and Attainment Dates for Ozone Nonattainment Areas

Classification	Ozone design values	Attainment deadline (from enactment)
Marginal	0.121 up to 0.138 ppm	1993 (3 years)
Moderate	0.138 up to 0.160 ppm	1996 (6 years)
Serious	0.160 up to 0.180 ppm	1999 (9 years)
Severe	0.180 up to 0.280 ppm	2005 (15 years)
Extreme	0.280 ppm and above	2010 (20 years)

TABLE 26.2

Title I Emission Sources Requiring Control

EPA ozone nonattainment classification	Allowable emissions of NO_x and VOC combined (tons per year)
Extreme and severe	10
Serious	50
Moderate and marginal	100

If a nonattainment area is classified as serious, based on ambient ozone measurements, then the state must modify its SIP to bring the area into compliance in 9 years. The CAAA90 also specify the size and, therefore, the number of sources subject to regulatory control as a function of nonattainment classification. Table 26.2 illustrates these requirements for ozone nonattainment classifications of extreme and severe; the state must include sources with combined NO_x and VOC emissions of 10 tons per year in their control plans.

As mentioned, nonattainment areas must implement different control measures, depending on their classification. Marginal areas, for example, are the closest to meeting the standard. They are required to conduct an inventory of their ozone-causing emissions and institute a permit program. Nonattainment areas with more serious air quality problems must implement various control measures. The worse the air quality, the more control areas will have to implement.

The new law also establishes similar programs for areas that do not meet the federal health standards for the pollutants carbon monoxide and particulate matter. Areas exceeding the standards for these pollutants are divided into "moderate" and "serious" classifications. Depending on the degree to which they exceed the carbon monoxide standard, areas are required to implement programs introducing oxygenated fuels and/or enhanced emission inspection programs, among other measures. Depending on their classification, areas exceeding the particulate matter standard will have to implement either Reasonably Available Control Technology (RACT) or Best Available Control Technology (BACT), among other requirements.

B. Title II: Provisions Related to Mobile Sources

Title II of the CAAA90 is related mainly to vehicles that operate on roads and highways. Off-road, or nonroad, engines and vehicles used for site drilling, remediation, or related construction may be regulated if the administrator of EPA determines that some degree of emission reduction is necessary.

The EPA has summarized the provisions related to mobile sources [4] as follows:

> While motor vehicles built today emit fewer pollutants (60% to 80% less, depending on the pollutant) than those built in the 1960s, cars and trucks still account for almost half the emissions of the ozone precursors VOCs and NO_x, and up to 90% of the CO emissions in urban areas. The principal reason for this problem is the rapid growth in the number of vehicles on the roadways and total miles driven. This growth has offset a large portion of the emission reductions gained from motor vehicle controls.
>
> In view of the unforeseen growth in automobile emissions in urban areas combined with the serious air pollution problems in many urban areas, the Congress has made significant changes to the motor vehicle provisions on the 1977 Clean Air Act.
>
> The Clean Air Act of 1990 establishes tighter pollution standards for emissions from automobiles and trucks. These standards will reduce tailpipe emissions of hydrocarbons, carbon monoxide, and nitrogen oxides on a phased-in basis beginning in model year 1994. Automobile manufacturers will also be required to reduce vehicle emissions resulting from the evaporation of gasoline during refueling.
>
> Fuel quality will also be controlled. Scheduled reductions in gasoline volatility and sulfur content of diesel fuel, for example, will be required. New programs requiring cleaner (so-called "reformulated" gasoline) will be initiated in 1995 for the nine cities with the worst ozone problems. Other cities can "opt in" to the reformulated gasoline program. Higher levels (2.7%) of alcohol-based oxygenated fuels will be produced and sold in 41 areas during the winter months that exceed the federal standard for carbon monoxide.
>
> The new law also establishes a clean fuel car pilot program in California, requiring the phase-in of tighter emission limits for 150 000 vehicles in model year 1996 and 300 000 by the model year 1999. These standards can be met with any combination of vehicle technology and cleaner fuels. The standards become even stricter in 2001. Other states can 'opt in' to this program, though only through incentives, not sales or production mandates.
>
> Further, 26 of the dirtiest areas of the country will have to adopt a program limiting emissions from centrally fueled fleets of 10 or more vehicles beginning as early as 1998.

C. Title III: Hazardous Air Pollutants

Toxic air pollutants are pollutants which are hazardous to human health or the environment but which are not specifically regulated by the CAA. These pollutants are typically carcinogens, mutagens, and teratogens. The CAAA of 1977 failed to result in substantial reductions in the emissions of these harmful substances.

The toxic air pollution problem is widespread. Information generated from the Superfund "Right to Know" rule from the Superfund Authorization and Recovery Act (SARA Section 313) indicates that more than 2.7 billion pounds of toxic air pollutants are emitted annually in the United States. EPA studies indicate that exposure to such quantities of air toxics may result in 1000–3000 cancer deaths each year.

The CAAA90 offers a comprehensive plan for achieving significant reductions in emissions of hazardous air pollutants from major sources. The law has improved EPA's ability to address this problem effectively and it has accelerated progress in controlling major toxic air pollutants.

EPA issued MACT standards for each listed source category according to a prescribed schedule. These standards are based on the best demonstrated control technology or practices within the regulated industry, and EPA was required to issue the standards for 40 source categories within 2 years of passage of the new law. The remaining source categories are controlled according to a schedule which ensures that all controls will have been achieved within 10 years of enactment. Companies that voluntarily reduce emissions according to certain conditions can get a 6-year extension to meet the MACT requirements.

Eight years after MACT is installed on a source, EPA must examine the risk levels remaining at the regulated facilities and determine whether additional controls are necessary to reduce unacceptable residual risk.

The EPA developed [4] a one-page summary of the key points of Title III. The following is this summary.

1. *Title III: Air Toxics, Key Points*

- *List of Pollutants and Source Categories*: The law lists 189 hazardous air pollutants. One year after enactment EPA lists source categories (industries) which emit one or more of the 189 pollutants. In 2 years, EPA must publish a schedule for regulation of the listed source categories.
- *MACT*: MACT regulations are emission standards based on the best demonstrated control technology and practices in the regulated industry. MACT for existing sources must be as stringent as the average control efficiency or the best controlled 12% of similar sources excluding sources which have achieved the lowest achievable emission rate (LAER) within 18 months prior to proposal or 30 months prior to promulgation. MACT for new sources must be as stringent as the best controlled similar source. For all listed major point sources, EPA must promulgate MACT standards—40 source categories plus coke ovens within 2 years and 25% of the remainder of the list within 4 years, an additional 25% in 7 years, and the final 50% in 10 years.
- *Residual Risk*: 8 years after MACT standards are established (except for those established 8 years after enactment), standards to protect against the residual health and environmental risks remaining must be promulgated, if necessary. The standards would be triggered if more than one source in a category exceeds a maximum individual risk of cancer of 1 in 1 million. These residual risk regulations would be based on current CAA language that specifies that standards must achieve an "ample margin of safety."
- *Accidental Releases*: Standards to prevent against accidental release of toxic chemicals are required. EPA must establish a list of at least 100 chemicals and threshold quantities. All facilities with these chemicals on site in excess of the threshold quantities would be subject to the regulations which would include hazard assessments and risk management

plans. An independent chemical safety board is established to investigate major accidents, conduct research, and promulgate regulations for accidental release reporting.

D. Title IV: Acid Deposition Control

The EPA summary [4] of Title IV states the basics of the acid deposition control amendments:

> Acid deposition occurs when sulfur dioxide and nitrogen oxide emissions are transformed in the atmosphere and return to the earth in rain, fog or snow. Approximately, 20 million tons of SO_2 are emitted annually in the United States, mostly from the burning of fossil fuels by electric utilities. Acid rain damages lakes, harms forests and buildings, contributes to reduced visibility, and is suspected of damaging health.
>
> The new Clean Air Act will result in a permanent 10 million ton reduction in sulfur dioxide (SO_2) emissions from 1980 levels. To achieve this, EPA will allocate allowances of 1 ton of sulfur dioxide in two phases. The first phase, effective January 1, 1995, requires 110 powerplants to reduce their emissions to a level equivalent to the product of an emissions rate = (2.5 lbs of SO_2/mm Btu) × (the average mm Btu of their 1985–1987 fuel use). Plants that use certain control technologies to meet their Phase I reduction requirements may receive a 2-year extension of compliance until 1997. The new law also allows for a special allocation of 200 000 annual allowances per year each of the 5 years of Phase I to powerplants in Illinois, Indiana and Ohio.
>
> The second phase, becoming effective January 1, 2000, will require approximately 2000 utilities to reduce their emissions to a level equivalent to the product of an emissions rate of (1.2 lbs of SO_2/mm Btu) × (the average mm Btu of their 1985–1987 fuel use). In both phases, affected sources will be required to install systems that continuously monitor emissions in order to track progress and assure compliance.
>
> The new law allows utilities to trade allowances within their systems and/or buy or sell allowances to and from other affected sources. Each source must have sufficient allowances to cover its annual emissions. If not, the source is subject to a $2,000/ton excess emissions fee and a requirement to offset the excess emissions in the following year.
>
> Nationwide, plants that emit SO_2 at a rate below 1.2 lbs mm^{-1} Btu will be able to increase emissions by 20% between a baseline year and 2000. Bonus allowances will be distributed to accommodate growth by units in states with a statewide average below 0.8 lbs mm Btu. Plants experiencing increases in their utilization in the last 5 years also receive bonus allowances. 50 000 bonus allowances per year are allocated to plants in 10 midwestern states that make reductions in Phase I. Plants that repower with a qualifying clean coal technology may receive a 4 year extension of the compliance date for Phase II emission limitations.
>
> The new law also includes specific requirements for reducing emissions of nitrogen oxides, based on EPA regulations to be issued not later than mid-1992 for certain boilers and 1997 for all remaining boilers.

Title IV represents legislation designed to reduce total SO_2 emissions by approximately 50% over a 10-year period. Provisions of the title are designed to introduce economic flexibility for the electric power industry, to recognize controls already implemented by progressive utilities and to reduce the economic impact on high-sulfur coal regions of the United States.

E. Title V: Permits

The 1990 law introduced an operating permits program modeled after a similar program under the Clean Water Act's National Pollution Elimination Discharge System (NPDES) law. The purpose of the operating permits program is to ensure compliance with all applicable requirements of the CAAA90 and to enhance EPA's ability to enforce the Act. Air pollution sources subject to the program must obtain an operating permit; states must develop and implement the program; and EPA must issue permit program regulations, review each state's proposed program, and oversee the state's efforts to implement any approved program. EPA must also develop and implement a federal permit program when a state fails to adopt and implement its own program. The final rulemaking for this Title V program was published on July 21, 1992 as Part 70 of Chapter I of Title 40 of the Code of Federal Regulations (57FR32250).

The program clarifies and makes more enforceable a source's pollution control requirements. Currently, a source's pollution control obligations may be scattered throughout numerous hard-to-find provisions of state and federal regulations, and in many cases, the source is not required, under the applicable SIP, to submit periodic compliance reports to EPA or the states. The permit program ensures that all of a source's obligations with respect to its pollutants are contained in one permit document and that the source will file periodic reports identifying the extent to which it has complied with those obligations. Both of these requirements greatly enhance the ability of federal and state agencies to evaluate a source's air quality situation.

In addition, the program provides a ready vehicle for states to assume administration, subject to federal oversight, of significant parts of the air toxics program and the acid rain program. Through the permit fee provisions discussed later, the program greatly augments a state's resources to administer pollution control programs by requiring sources of pollution to pay their fair share of the costs of a state's air pollution program.

Under the 1990 law, EPA was required to issue program regulations by November 15, 1991. By November 15, 1993, each state must submit to EPA a permit program meeting these regulatory requirements. After receiving the state submittal, EPA has 1 year to accept or reject the program. EPA must level sanctions against a state that does not submit or enforce a permit program.

If a state fails to comply, EPA will promulgate and administer the state's program. That could mean lengthy permitting delays and additional paperwork. By supporting the state's permitting authority, a facility can simplify the permitting process and avoid EPA intervention.

Each permit issued to a facility is for a fixed term of up to 5 years. The new law establishes a permit fee whereby the state collects a fee from the permitted facility to cover reasonable direct and indirect costs of the permitting program.

All sources subject to the permit program must submit a complete permit application within 12 months of the effective date of the program. The state permitting authority must determine whether or not to approve an application within 18 months of the date it receives the application.

To ensure compliance with the standards, the permit also must contain provisions for the inspection, entry, monitoring, certification, and reporting of compliances with the permit conditions.

EPA has 45 days to review each permit and to object to permits that violate the CAAA. If EPA fails to object to a permit that violates the Act or the implementation plan, any person may petition EPA to object within 60 days following EPA's 45-day review period, and EPA must grant or deny the permit within 60 days. Judicial review of EPA's decision on a citizen's petition can occur in the federal court of appeals. The public is guaranteed the right to inspect and review all permit applications and documents. There are provisions for three kinds of permit revisions: administrative amendment, minor permit modification, and significant modification.

These regulations will apply to an estimated 34 000 "major" industrial sources. "Major" sources are defined according to their "potential to emit" and the cutoff levels vary depending on both the pollutant and the local areas' compliance status with the National Ambient Air Quality Standard (NAAQS) for that pollutant. For the present, the EPA has exempted all "non-major" sources, of which there are estimated to be about 350 000, from this permitting, until they have studied further the feasibility of permitting them. However, the states can require permitting of some of these sources.

The regulations provide for the collection of fees from permit seekers and the states must require fees sufficient to cover the cost of administering the program. If a state does not submit, properly administer, or enforce a permit program, federal sanctions must be levied including the withholding of highway funds and requiring offset pollution for new sources by reducing emissions by 2 tons from existing sources for every 1 tons of emissions from a proposed new source.

All permits must include a cap on emissions which cannot be exceeded without an approved revision of the permit. Permitted sources must periodically test and monitor their emissions and report on these activities every 6 months. Civil penalties include fines of not less than $10 000 per day for permit violations and criminal penalties for deliberate false statements or representations, or for rendering any inaccurate monitoring device or method required in the permit.

F. Title VI: Stratospheric Ozone Protection

The EPA summary [4] for stratospheric ozone and global climate protection lists the basics of the title:

> The new law builds on the market-based structure and requirements currently contained in EPA's regulations to phase out the production of substances that deplete the ozone layer. The law requires a complete phase-out of CFCs and halons with

interim reductions and some related changes to the existing Montreal Protocol, revised in June 1990.

Under these provisions, EPA must list all regulated substances along with their ozone-depletion potential, atmospheric lifetimes and global warming potentials within 60 days of enactment.

In addition, EPA must ensure that Class I chemicals be phased out on a schedule similar to that specified in the Montreal Protocol—CFCs, halons, and carbon tetrachloride by 2000; methyl chloroform by 2002—but with more stringent interim reductions. Class II chemicals (HCFCs) will be phased out by 2030. Regulations for Class I chemicals will be required within 10 months, and Class II chemical regulations will be required by December 31, 1999.

The law also requires EPA to publish a list of safe and unsafe substitutes for Class I and II chemicals and to ban the use of unsafe substitutes.

The law requires nonessential products releasing Class I chemicals to be banned within 2 years of enactment. In 1994 a ban will go into effect for aerosols and non-insulating foam using Class II chemicals, with exemptions for flammability and safety. Regulations for this purpose will be required within 1 year of enactment, to become effective 2 years afterwards.

G. Title VII: Provisions Relating to Enforcement

The CAAA90 contains a broad array of authorities to make the law more readily enforceable, thus bringing it up to date with the other major environmental statutes. EPA has new authorities to issue administrative penalty orders up to US $200 000 and field citations up to US $5000 for lesser infractions. Civil judicial penalties are enhanced. Criminal penalties for knowing violations are upgraded from misdemeanors to felonies, and new criminal authorities for knowing and negligent endangerment are established.

In addition, sources must certify their compliance, and EPA has authority to issue administrative subpoenas for compliance data. EPA will also be authorized to issue compliance orders with compliance schedules of up to 1 year.

The citizen suit provisions have also been revised to allow citizens to seek penalties against violators, with the penalties going to a US Treasury fund for use by EPA for compliance and enforcement activities. The US Government's right to intervene is clarified and citizen plaintiffs will be required to provide the US Government with copies of pleadings and draft settlements.

Any atmospheric emissions for which EPA does not develop standards may be regulated by state or regional authorities.

A review of the enforcement and liability provisions of CAAA90 [5] recommends that because of the new enforcement tools available to the federal government, the regulated community should implement effective self-auditing and compliance programs at facilities to reduce the risk of criminal liability. Stenvaag [3] covers the provisions relating to enforcement from a legal standpoint. He states the new language of this title to be "quite confusing, particularly in specifying when criminal sanctions are appropriate."

H. Title VIII: Miscellaneous Provisions

Section 130, Emission Factors requires revising emission inventory factors every 3 years:

> Within 6 months after enactment of the Clean Air Act Amendments of 1990, and at least every 3 years thereafter, the Administrator shall review and, if necessary, revise, the methods ('emission factors') used for purposes of this Act to estimate the quantity of emissions of carbon monoxide, volatile organic compounds, and oxides of nitrogen from sources of such air pollutants (including area sources and mobile sources). In addition, the Administrator shall permit any person to demonstrate improved emissions estimating techniques, and following approval of such techniques, the Administrator shall authorize the use of such techniques. Any such technique may be approved only after appropriate public participation. Until the Administrator has completed the revision required by this section, nothing in this section shall be construed to affect the validity of emission factors established by the Administrator before the date of the enactment of the Clean Air Act Amendments of 1990.

I. Title IX: Clean Air Research

Title IX of the CAAA90 addresses air pollution research areas including monitoring and modeling, health effects, ecological effects, accidental releases, pollution prevention and emissions control, acid rain, and alternative motor vehicle fuels. The provisions require ecosystem studies on the effects of air pollutants on water quality, forests, biological diversity, and other terrestrial and aquatic systems exposed to air pollutants; mandate the development of technologies and strategies for air pollution prevention from stationary and area sources; and call for several major studies. The EPA must improve methods and techniques for measuring individual air pollutants and complex mixtures and conduct research on long- and short-term health effects, including the requirement for a new interagency task force to coordinate these research programs. Finally, the Agency must develop improved monitoring and modeling methods to increase the understanding of tropospheric ozone formation and control.

To implement the research provisions, the EPA plans to conduct research in emissions inventories, atmospheric modeling, source/ambient monitoring, control technologies, health, and ecological monitoring. Both ecological monitoring and ambient monitoring are to be done jointly with other agencies who also need these data to meet their mission. Other proposed work includes developing improved risk assessment methods, maintaining existing networks or establishing new ones for aquatic and terrestrial effects monitoring, and continuing work on deposition chemistry. Again, these efforts in particular are to be supported, in part, by other agencies.

J. Title X: Disadvantaged Business Concerns and Title XI: Clean Air Employment Transition Assistance

These two final titles were added to cover the subject areas of concern. They relate to procedural matters and direct the appropriate federal agencies to implement and oversee the necessary compliance action.

REFERENCES

1. Public Law No. 101–549; 101st Congress, November 15, 1990, An Act to amend the Clean Air Act to provide for attainment and maintenance of health protective national ambient air quality standards, and for other purposes.
2. US Public Law No. 88–206, 77 Stat. 392 (1963), US Public Law No. 91–604, 84 Stat. 1676 (1970), US Public Law No. 95–95, 91 Stat. 686 (1977).
3. Stenvaag, J. M., *Clean Air Act 1990 Amendments Law and Practices*. Wiley, New York, 1991.
4. Office of Air and Radiation, US Environmental Protection Agency, *The Clean Air Act Amendments of 1990, Summary Materials*, November 1990.
5. Elliott, E. D., Schwartz, R. M., Goldman, A. V., Horowitz, A. B., and Laznow, J., The Clean Air Act: new enforcement and liability provisions. *J. Air Waste Manage. Assoc.* **42** (11), 1261–1270 (1992).

SUGGESTED READING

Burke, R. L., *Permitting for Clean Air—A Guide to Permitting under Title V of the Clean Air Act Amendments of 1990*. Air and Waste Management Association, Pittsburgh, PA, 1992.

Gas Research Institute briefing on Clean Air Act Amendments of 1990. Presentation by J. G. Holmes and R. W. Crawford, Energy and Environmental Analysis, Inc., December 1990.

Lee, B., Highlights of the Clean Air Act Amendments of 1990. *J. Air Waste Manage. Assoc.* **41** (1), 48–55 (1991).

Quarles, J., and Lewis Jr., W. H., *The New Clean Air Act: A Guide to the Clean Air Program as Amended in 1990*. Lewis and Brockius, Washington, DC, 1990.

US Environmental Protection Agency (EPA), *Implementation Strategy for the Clean Air Act Amendments of 1990*. EPA Office of Air and Radiation, Washington, DC, January 15, 1991.

QUESTIONS

1. Give an example of a health-based air pollution standard with the justification for the standard.
2. Give an example of a technology-based air pollution standard with the justification for the standard.
3. Choose a specific metropolitan area and determine its classification as an ozone nonattainment area. Find the alternative deadline and allowable emissions of NO_x and VOC combined.
4. For the specific metropolitan area in question 3, discuss how the attainable deadline and allowable emissions can be met.
5. List the alternatives that are possible to replace present automobiles with vehicles, or systems, that will reduce emissions of VOCs, NO_x, and CO.
6. List five categories (such as hazardous waste incineration) that may be considered as "major" or "area" sources of hazardous air pollutants.
7. Develop an outline of a permit for a new plastic molding company. Include a schedule of costs and reporting requirements.

27

Emission Standards

I. SUBJECTIVE STANDARDS

Limits on emissions are both subjective and objective. Subjective limits are based on the visual appearance or smell of an emission. Objective limits are based on physical or chemical measurement of the emission. The most common form of subjective limit is that which regulates the optical density of a stack plume, measured by comparison with a Ringelmann chart (Fig. 27.1). This form of chart has been in use for over 90 years and is widely accepted for grading the blackness of black or gray smoke emissions. Within the past four decades, it has been used as the basis for "equivalent opacity" regulations for grading the optical density of emissions of colors other than black or gray.

The original Ringelmann chart was a reflectance chart; the observer viewed light reflected from the chart. More recently, light transmittance charts have been developed for both black [1] and white [2] gradations of optical density, which correlate with the Ringelmann chart scale. It is now common practice in the United States to send air pollution inspectors to a "smoke school" where they are trained and certified as being able to read the density of black and white plumes with an accuracy that is acceptable for court testimony.

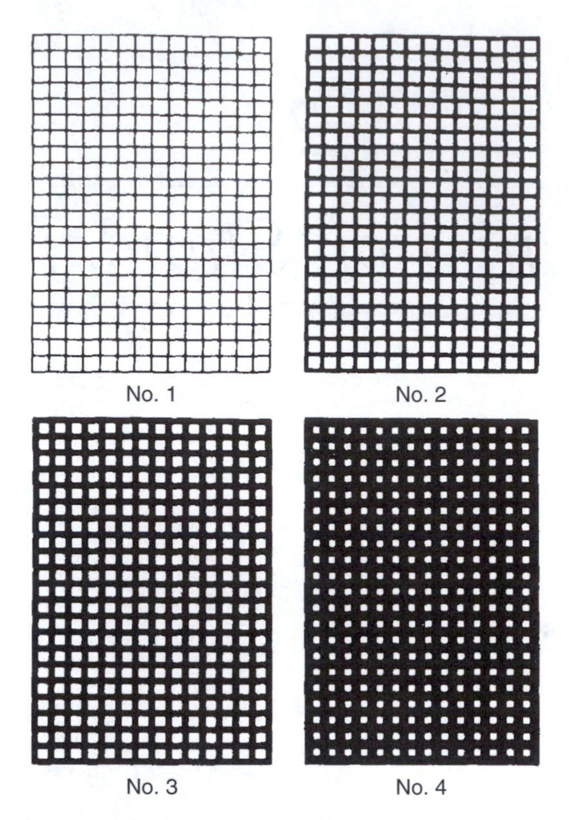

Fig. 27.1. Ringelmann smoke chart.

Before the widespread acceptance of the Ringelmann scale, smoke was regulated by prohibiting the emission of black smoke. Now regulatory practice prohibits the emission of smoke whose density is greater than a specified Ringelmann number. Over recent decades, there has been a progressive decrease in the Ringelmann number thus specified, culminating in the specification of number 1 for large new steam power plants in the United States. In addition to subjective observation of smoke density, continuous emission monitoring systems (CEMs) have been developed to measure objectively, by means of a photocell, the decrease in intensity of a beam of light projected through the plume prior to its emission (Fig. 27.2). Such systems are discussed in Chapter 32.

Subjective evaluation of odor emission is made difficult by the phenomenon of odor fatigue, which means that after persons have been initially subjected to an odor, they lose the ability to perceive the continued presence of low concentrations of that odor. Therefore, all systems of subjective odor evaluation rely on preventing olfactory fatigue by letting the observer breathe odor-free air for a sufficient time prior to breathing the odorous air

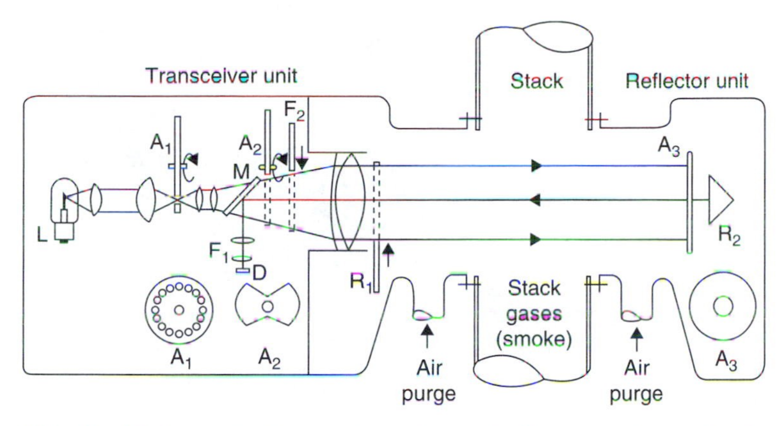

Fig. 27.2. Double-beam, double-pass transmissometer for measuring smoke density in stacks. A_1, chopper wheel; A_2, beam gating wheel; A_3, aperture; D, detector; F_1, spectral filter; F_2, solenoid-activated neutral density filter; L, lamp; M, half-mirror/beam splitter; R_1, solenoid-activated zero calibration reflector; R_2, retroreflector (alignment bullseye not shown). Design patented. *Source*: Drawing courtesy of Lear Siegler, Inc.

and evaluating its odor content. Usually an activated charcoal bed is used to clean up the air in order to provide the odor-free air required by the observer.

II. OBJECTIVE STANDARDS

There are two major categories of objective emission limits: those which limit the emission of a specific pollutant, regardless of the process or equipment from which it is emitted, and those which limit the emissions of a specific pollutant from a specific process or type of equipment. Regulations may require the same emission limit for all sources, regardless of size or capacity; or they may vary the allowable emission with the size or capacity of the source. Limits may be stated in absolute terms, i.e. not more than a specified mass of pollutant per unit of time, or in relative terms, i.e. not more than a specified mass of pollutant per unit mass of fuel burned, material being processed, or product produced or per unit of heat released in a furnace. In the case of gaseous pollutants, limits may be stated in volumetric rather than gravimetric terms. Emission limits are sometimes stated as mass of pollutant per unit volume of effluent. Effluent volume varies with gas temperature and pressure, the presence or absence of diluting air, and the amount thereof. Therefore, volumes must be reduced to a specified temperature, pressure, and percent of diluting air. In the case of flue gases from fuel combustion, dilution is usually expressed as percentage of excess air or percentage of carbon dioxide or oxygen in the flue gas.

The pollutants, source categories, and affected facilities for which the United States has established New Source Performance Standards (NSPS) are listed in Table 27.1. Certain categories listed in Table 27.1 are subject to US

TABLE 27.1

US National New Source Performance Standards

Source category (in sequence of promulgation)	Affected facility	Pollutants regulated									
		Particulate matter	Opacity	SO_2	NO_x	H_2S	CO	TRS	VOC	Pb	Total F
Fossil fuel and electric utility steam generator[a] [>73 MW (>250 million BTU h^{-1}) input] (264 KJ h^{-1})	Solid fossil fuel and wood residue-fired	x	x	x	x	—	—	—	—	—	—
	Oil- or gas-fired boilers	x	x	x	x	—	—	—	—	—	—
Incinerators[a]	>45 metric tons day^{-1}	x	—	—	—	—	—	—	—	—	—
Portland cement plants[a]	Kiln and clinker cooler	x	x	—	—	—	—	—	—	—	—
	Fugitive emission points	—	x	—	—	—	—	—	—	—	—
Nitric acid plants[a]	Process equipment	—	x	—	x	—	—	—	—	—	—
Sulfuric acid plants	Process equipment	x[b]	x	x	—	—	—	—	—	—	—
Asphalt concrete plants	Dryers, etc.[c]	x	x	—	—	—	—	—	—	—	—
Petroleum refineries[a]	Fluid catalyst regenerator	x	x	—	—	—	x	—	—	—	—
	Fuel gas combustion	—	—	x	—	x	—	—	—	—	—
	Claus sulfur recovery TRS X[d]	—	—	x	—	—	—	x	—	—	—
Storage vessels for petroleum liquids[a]	Storage tanks	—	—	—	—	—	—	—	x	—	—
Secondary lead smelters[a]	Reverberatory and blast furnaces	x	x	—	—	—	—	—	—	—	—
	Pot furnaces	—	x	—	—	—	—	—	—	—	—
Secondary brass and bronze plants[a]	Reverberatory furnaces	x	x	—	—	—	—	—	—	—	—
	Blast and electric furnaces	—	x	—	—	—	—	—	—	—	—
Iron and steel plants[a]	Basic O_2 and electric arc furnaces	x	x	—	—	—	—	—	—	—	—
	Dust-handling equipment	—	x	—	—	—	—	—	—	—	—

Sewage treatment plants	Sludge incineration	x	x	—	—	—	—	—	—	—	—
Primary copper smelters	Dryer	x	x	—	—	—	—	—	—	—	—
	Roaster, furnace, convertor	—	x	x	—	—	—	—	—	—	—
Primary lead smelters[a]	Blast and reverberatory furnaces[e]	x	x	—	—	—	—	—	—	—	—
	Electric furnace converter[f]	—	x	x	—	—	—	—	—	—	—
Primary zinc smelters	Sinter machine	x	x	—	—	—	—	—	—	—	—
	Roaster	—	x	x	—	—	—	—	—	—	—
Primary aluminum reduction plants[a]	Potrooms and anode baking	—	x	—	—	—	—	—	—	—	x
Phosphate fertilizer plants[a]	Wet process phosphoric acid[g]	—	—	—	—	—	—	—	x		
Coal preparation plants[a]	Coal cleaning and dryer	x	x	—	—	—	—	—	—	—	—
	Transfer, loading[h]	—	x	—	—	—	—	—	—	—	—
Ferroalloy production facilities	Electric submerged-arc furnaces	x	x	—	—	—	x	—	—	—	—
	Dust-handling equipment	—	x	—	—	—	—	—	—	—	—
Kraft pulp mills[a]	Smelt tanks and lime kilns	x	—	—	—	—	—	x	—	—	—
	Recovery furnaces	x	x	—	—	—	—	x	—	—	—
	Digester, washer, etc.[i]	—	—	—	—	—	—	x	—	—	—
Glass manufacturing plants	Melting furnaces	x	—	—	—	—	—	—	—	—	—
Grain dryers	Dryers	x	x	—	—	—	—	—	—	—	—
Grain elevators	Truck-loading stations[j]	x	x	—	—	—	—	—	—	—	—
Gas turbines > 1000 hp ($10^7\,GJ\,h^{-1}$)	Simple, regen., and combined	—	—	x	x	—	—	—	—	—	—
	Lime hydrators	x	—	—	—	—	—	—	—	—	—
Automobile and truck surface coating	Prime, guide, and topcoat	—	—	—	—	—	—	—	x	—	—
Ammonium sulfate manufacturing	Dryers	x	x	—	—	—	—	—	—	—	—
Lead acid battery manufacturing	Casting, mixing reclamation and oxide manufacture	—	—	—	—	—	—	—	—	x	—
Phosphate rock plants[a]	Drying, calcining, grinding	x	x	—	—	—	—	—	—	—	—

(*continued*)

TABLE 27.1 (*Continued*)

Source category (in sequence of promulgation)	Affected facility	Pollutants regulated									
		Particulate matter	Opacity	SO_2	NO_x	H_2S	CO	TRS	VOC	Pb	Total F
Rotogravure printing	Each press	—	—	—	—	—	—	—	x	—	—
Large appliance surface coating	Each coating operation	—	—	—	—	—	—	—	x	—	—
Metal coil surface coating	Each coating operation	—	—	—	—	—	—	—	x	—	—
Asphalt processing and roofing manufacture	Saturator	x	x	—	—	—	—	—	—	—	—
	Blowing still	x	—	—	—	—	—	—	—	—	—
	Storage and handling	—	x	—	—	—	—	—	—	—	—
Beverage can surface coating	Each coating operation	—	—	—	—	—	—	—	—	x	—
Bulk gasoline terminals	All loading racks	—	—	—	—	—	—	—	x	—	—

Note: TRS, total reduced sulfur; VOC, volatile organic carbon; F, fluorides.

[a] Subject to PSD review if emission potential exceeds 90 metric tons $year^{-1}$.

[b] Acid mist.

[c] Screening and weighing systems; storage, transfer, and loading systems; and dust handling equipment.

[d] Reduced sulfur compounds plus H_2S.

[e] Sintering machine discharge.

[f] Sintering machine.

[g] Superphosphoric acid, diammonium phosphate, triple superphosphate, and granular triple superphosphate.

[h] Processing and conveying equipment and storage systems.

[i] Evaporator, condensate stripper, and black liquor oxidation systems.

[j] Truck unloading stations, barge, ship, or railcar loading and unloading stations; grain and rack dryers and handling operations.

Source: Pahl, D., *J. Air Pollut. Control Assoc.* **33**, 468–482 (1983).

Prevention of Significant Deterioration (of air quality) (PSD) review if their emission potential of a regulated pollutant exceeds 100 tons per year. In addition, PSD review is required for the following sources, whose emission potential exceeds 100 tons per day: hydrofluoric acid plants, coke oven batteries, furnace process carbon black plants, fuel conversion plants, taconite ore processing plants, gas fiber processing plants, and charcoal production plants.

III. TYPES OF EMISSION STANDARDS

The most common rationale for developing emission limits for stationary sources is the application of the best practicable means for control. Under this rationale, the degree of emission limitation achievable at the best designed and operated installation in a category sets the emission limits for all other installations of that category. As new technology is developed, what was the best practicable means in 1993 may be short of the best attainable in 2003. Because there is thus a moving target, means must be provided administratively to allow the plant which complied with a best practicable means limit to be considered in compliance for a reasonable number of years thereafter. It is important to note that best practicable means limits are set without regard to present or background air quality or the air quality standards for the pollutants involved, the number and location of sources affected by the limit, and the meteorology or topography of the area in which they are located. However, an additional limit on minimum stack height, or buffer zone, based on these noted factors may be coupled with a best practicable means limit on mass or emission.

The other major rationales for developing emission limits have been based on some or all of the noted factors—air quality, air quality standards, number and location of sources, meteorology, and topography. These include the rollback approach, which involves all these factors except source location, meteorology, and topography; and the single-source mathematical modeling approach, which considers only the air quality standard and meteorology, ignoring the other factors listed.

IV. VARIANT FORMS OF EMISSION STANDARDS

Some variants of best practicable means were spelled out in the US Clean Air Act of 1977. One is the requirement that Best Available Control Technology (BACT) for a specific pollutant be employed on new "major sources" that are to be located in an area that has attained the National Ambient Air Quality Standard (NAAQS) for that pollutant. BACT is also required for pollutants for which there is no NAAQS [e.g. total reduced sulfur (TRS), for which emission limits are specified by a Federal NSPS]. BACT must be at least as stringent as NSPS but is determined on a case-by-case basis.

Another variant is the lowest achievable emission rate (LAER) for a specific pollutant required for a new source of that pollutant to be located in a nonattainment area (i.e. one which has not attained the NAAQS for that pollutant). LAER is the lowest emission rate allowed or achieved anywhere without regard to cost or energy usage. LAER is intended to be more stringent than BACT or NSPS and is also determined on a case-by-case basis.

This changed with the passage of the Clean Air Act Amendments of 1990 and the subsequent regulations. The Maximum Achievable Control Technology (MACT) standards continued for some years, but incrementally have been phased-out in favor of standards to address the margin of risk that remained from the technology-based standards. This is known as residual risk. Section 112(f) of the Act directs the Environmental Protection Agency (EPA) to prepare a report to Congress on the methods to be used to address this residual risk from sources that emit hazardous air pollutants (HAPs) have been promulgated and applied. The EPA presents a discussion of how it intends to provide the Act's mandate to "provide an ample margin of safety to protect public health" or to set more stringent standards, if necessary, "to prevent, taking into consideration costs, energy, safety, and other relevant factors, an adverse environmental effect." This includes both human and ecological risk.

All of the 96 MACT regulations for 174 industry source categories are being reviewed to eliminate 1.5 million tons per year of 188 toxic air pollutants listed in the Clean Air Act. The 1989 National Emission Standard for Hazardous Air Pollutants (NESHAP) for benzene[1] presented the following risk management framework for cancer risk, which reflects the two-step approach suggested by the court. The benzene rule preamble states that in determining acceptable risk:

> The Administrator believes that an MIR [maximum individual risk] of approximately 1 in 10 thousand should ordinarily be the upper-end of the range of acceptability. As risks increase above this benchmark, they become presumptively less acceptable under Section 112, and would be weighed with the other health risk measures and information in making an overall judgment on acceptability. Or, the Agency may find, in a particular case, that a risk that includes MIR less than the presumptively acceptable level is unacceptable in light of the other health risk factors.

The MIR, the distribution of risks in the exposed population, incidence, the science policy assumptions and uncertainties associated with risk measures, and the weight of evidence that a pollutant is harmful to health are all key determinants used to judge safety and risk. To ensure protection of public health and the environment, the Section 112(f) requires a human health risk and adverse environmental effects based "needs test" in the second regulatory phase of the air toxics program. This residual risk standard setting considers the need for additional national standards on stationary emission sources following regulation under Section 112(d) to "provide an ample

[1] US Environmental Protection Agency, 1989, National emission standards for hazardous air pollutants; benzene. *Fed. Regist.* 54 (177), 38044–38072, Rule and Proposed Rule, September 14.

margin of safety to protect public health." The section also mandates the EPA to determine whether residual risk standards are necessary to prevent adverse environmental effects, taking into consideration "costs, energy, safety, and other relevant factors" in deciding what level is protective. Adverse environmental effect is defined in Section 112(a)(7) as "any significant and widespread adverse effect, which may reasonably be anticipated, to wildlife, aquatic life, or other natural resources, including adverse impacts on populations of endangered or threatened species or significant degradation of environmental quality over broad areas."

On March 31, 2005 the EPA issued final amendments to reduce emissions of toxic air pollutants, or air toxics, from coke oven batteries; these were the first of a series of emissions reductions under the new residual risk standards. Across the industry, this rule will require reductions of allowable emissions from 11.3 tons per year under the 1993 MACT to 9.8 tons per year under the amendments. This rule was also designed to ensure that the affected batteries do not increase their emissions back to the levels allowed in 1993. The final amendments include more stringent requirements for certain processes at several coke oven batteries to address health risks remaining after implementing EPA's October 1993 air toxics emission MACT standards.

The final amendments also include requirements for new or reconstructed coke oven batteries that reflect improvements in emission control practices that have occurred in the years since the 1993 MACT standards. This applies to coke oven emissions from nine batteries at five coke plants for which the 1993 MACT standard applied. There are approximately 14 other coke oven plants in the country that are not affected by this rule because they chose to install more stringent controls than the MACT the LAER, beginning in 1993. Because these LAER batteries opted to reduce their emissions by more than the 1993 MACT required, they do not have to comply with residual risk rule until the year 2020. The final amendments apply to charging, leaks, and bypass stacks at coke oven batteries. Emissions from these processes occur at the start of the process of turning coal into coke, primarily as the coke ovens are heating up. Thus, residual risk measures focus in a number of areas, beyond technology.

V. MEANS FOR IMPLEMENTING EMISSION STANDARDS

When owners wish to build a new source which will add a certain amount of a specific pollutant to an area that is in nonattainment with respect to that pollutant, they must, under US federal regulations, document a reduction of at least that amount of the pollutant from another source in the area. They can effect this reduction, or "offset," as it is called, in another plant they own in the area or can shut down that plant. However, if they do not own another such plant or do not wish to shut down or effect such reduction in a plant they own, they can seek the required reduction or offset from another owner.

Thus, such offsets are marketable credits that can be bought, sold, traded, or stockpiled ("banked") as long as the state or local regulatory agency legitimizes, records, and certificates these transactions. The new source will still have to meet NSPS, BACT or MACT, and/or LAER standards, whichever are applicable.

A. Bubble Concept

When a new source has been added to a group of existing sources under the same ownership in the same industrial complex, the usual practice has been to require the new source independently to meet the offset, NSPS, BACT, and/or LAER, disregarding the other sources in the complex. Under the more recent "bubble concept" (Fig. 27.3) adopted by some states with the approval of the EPA, the addition of the new source is allowed, whether or not it meets NSPS, BACT, or LAER, provided the total emission of the relevant regulated pollutants from the total complex is decreased. This can be accomplished by obtaining the required offset from one or more of the other sources within the complex, by shutdown, or by improvement in control efficiency. The bubble concept has been subject to litigation and may require a ruling or challenge to make it acceptable.

B. Fugitive and Secondary Sources

In computing the potential for emission of a new source or source complex, it is necessary to consider two other source categories, "fugitive" and "secondary" sources. Fugitive emissions are those from other than point sources, such as unpaved plant roads, outdoor storage piles swept by the wind, and surface mining. Secondary sources are those small sources with emissions of a different character from those of a major source, necessary for the operation of the major source, or source complex.

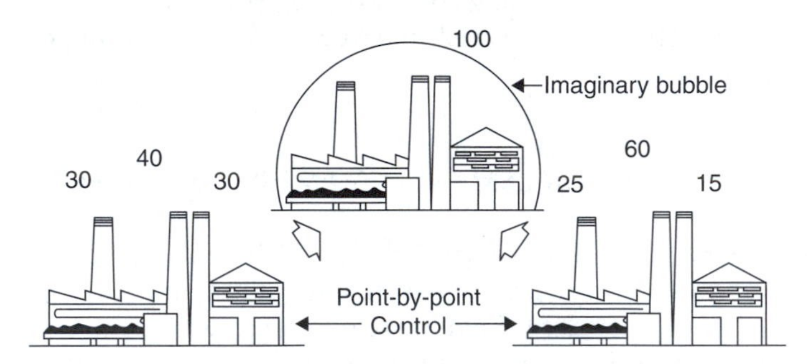

Fig. 27.3. Bubble concept. This pollution control concept places an imaginary bubble over an entire industrial plant, evaluating emissions from the facility as a whole instead of requiring control point-by-point on emission sources. Numbers represent emissions from individual sources, some of which can be fugitive sources, and from the entire industrial plant. *Source*: Drawing courtesy of the Chemical Manufacturers Association.

C. Indirect Sources

The term *indirect sources* is used to describe the type of source created by the parking areas of shopping malls, sports arenas, etc. which attract large numbers of motor vehicles, frequently arriving and leaving over relatively short periods of time. Of somewhat similar character are traffic interchanges, as at an intersection of major highways, each highway being a line source in its own right.

D. Rollback

The rollback approach assumes that emissions and atmospheric concentrations are linearly related, i.e. that a given percentage reduction in emission will result in a similar percentage reduction in atmospheric concentrations. This is most likely a valid assumption for a nonreactive gas such as carbon monoxide, whose principal source is the automobile. The model is

$$R = \frac{g(P) - D}{g(P) - B} \times 100 \tag{27.1}$$

where R is the required percentage reduction in emission, P is the present air quality, D is the air quality standard, B is the background concentration, and g is the growth factor in emissions (g is projected to a year in the future when emissions are expected to apply to all vehicles on the road).

An example of a set of emission limits based on the rollback approach is the limits adopted by the United States for carbon monoxide, hydrocarbons, and oxides of nitrogen emissions from new automobiles (Table 27.2).

E. Standard for Hazardous Pollutants

Before the US Clean Air Act Amendments of 1990 [3], HAPs were regulated through federal promulgation of the NESHAPS. The EPA listed only eight HAPs under NESHAPS.

Title III of the 1990 amendments completely changed the US standards for the hazardous air pollutant control program. Stensvaag [6] has summarized these changes as follows:

- Massively expand the list of HAPs. This initial list contained 189 HAPs.
- Direct the EPA to list source categories of major sources and area sources.
- Promulgate technology-based emission standards for listed pollutants and sources.
- Set up programs to control accidental releases.
- Establish structures for future risk assessment.
- Require regulation of solid waste combustion facilities.

TABLE 27.2

Emission limits for light-duty vehicles. The Tier 2 regulations were adopted on December 21, 1999, with a phase-in implementation schedule from 2004 to 2009. They introduced more stringent numerical emission limits relative to the previous Tier 1 requirements. The regulations are structured into eight permanent and three temporary certification levels of different stringency, called "certification bins," and an average fleet standard for oxides of nitrogen (NO_x) emissions. The same emission standards apply to all vehicle weight categories, i.e. cars, minivans, light-duty trucks, and SUVs have the same emission limit

Engine type and pollutant	Prior to control[d]	1968–1969	1970–1971	1972	1973–1974	1975–1976	1977–1979	1980	1981	1982–1986	1987–1993	Tier 1[i] 1994–2003[b]		Interim Tier 2[i] 2004–2006		Tier 2[i] 2007+	
Gasoline																	
HC (total)	11	*g*	2.2	3.4	3.4	1.5	1.5	0.41	0.41	0.41	0.41	0.41	(*h*)	*h*	*h*	*h*	*h*
NMHC	*e*	*h*	*h*	*h*	*h*	*h*	*h*	*h*	*h*	*h*	*h*	0.25	(0.31)	*h*	*h*	*h*	*h*
NMOG	*e*	*h*	*h*	*h*	*h*	*h*	*h*	*h*	*h*	*h*	*h*	*h*	*h*	0.125	(0.156)	0.100	(0.125)
CO	80	*g*	23	39	39	15	15	7.0	3.4	3.4	3.4	3.4	(4.2)	3.4	(4.2)	3.4	(4.2)
Cold-temp. CO[c]	*e*	*h*	*h*	*h*	*h*	*h*	*h*	*h*	*h*	*h*	*h*	10	(*h*)	10	(*h*)	10	(*h*)
NO_x	4	*h*	*h*	*h*	3.0	3.1	2.0	2.0	1.0	1.0	1.0	0.4	(0.6)	0.4	(0.6)	0.14	(0.20)
Particulates	*e*	*h*	*h*	*h*	*h*	*h*	*h*	*h*	*h*	*h*	*h*	0.08	(0.10)	0.08	(0.08)	0.02	(0.02)
Formaldehyde	*e*	*h*	*h*	*h*	*h*	*h*	*h*	*h*	*h*	*h*	*h*	*h*	*h*	0.015	(0.018)	0.015	(0.018)
Diesel																	
HC (total)	11	*h*	*h*	*h*	*h*	1.5	1.5	0.41	0.41	0.41	0.41	0.41	(*h*)	*h*	*h*	*h*	*h*
NMHC	*e*	*h*	*h*	*h*	*h*	*h*	*h*	*h*	*h*	*h*	*h*	0.25	(0.31)	*h*	*h*	*h*	*h*
NMOG	*e*	*h*	*h*	*h*	*h*	*h*	*h*	*h*	*h*	*h*	*h*	*h*	*h*	*h*	(0.156)	0.100	(0.125)
CO	80	*h*	*h*	*h*	*h*	15	15	7.0	3.4	3.4	3.4	3.4	(4.2)	*h*	(4.2)	3.4	(4.2)
NO_x	4	*h*	*h*	*h*	*h*	3.1	2.0	2.0	1.0	1.0	1.0	1.0	(1.25)	*h*	(0.6)	0.14	(0.20)
Particulates	*e*	*h*	*h*	*h*	*h*	*h*	*h*	*h*	*h*	0.60	0.20	0.08	(0.10)	*h*	(0.10)	0.02	(0.02)
Formaldehyde	*e*	*h*	*h*	*h*	*h*	*h*	*h*	*h*	*h*	*h*	*h*	*h*	*h*	*h*	(0.018)	0.015	(0.018)
Test procedure[a]		7-mode		CVS-72		CVS-75	CVS-75	CVS-75	CVS-75	CVS-75	CVS-75	CVS-75	CVS-75	CVS-75	CVS-75	CVS-75	CVS-75
Useful life[f], intermediate[b,d] (full)		*h*	*h*	*h*	*h*	*h*	*h*	*h*	*h*	*h*	*h*	5 years/ 50 000 miles	5 years/ 50 000 miles	5 years/ 50 000 miles	5 years/ 50 000 miles	5 years/ 50 000 miles	5 years / 50 000 miles

Useful life, intermediate[b, d] (full)	5 years/ 50 000 miles	5 years/ 50 000 miles	5 years/ 50 000 miles	5 years/ 50 000 miles	5 years/ 50 000 miles	5 years/ 50 000 miles	5 years/ 50 000 miles	5 years/ 50 000 miles	5 years/ 50 000 miles	5 years/ 50 000 miles	10 years/ 100 000 miles	10 years/ 100 000 miles	10 years/ 100 000 miles	10 years/ 100 000 miles	10 years/ 120 000 miles	10 years/ 120 000 miles

[a] The test procedure for measuring exhaust emissions has changed several times over the course of vehicle emissions regulations. The 7-mode procedure was used through model year 1971 and was replaced by the CVS-72 procedure beginning in model year 1972. The CVS-75 procedure became the test procedure as of model year 1975. While it may appear that the total HC and CO standards were relaxed in 1972–74, these standards were actually more stringent due to the more stringent nature of the CVS-72 test procedure. Additional standards for CO and composite standards for NMHC and NO_x tested under the new Supplemental Federal Test Procedure will be phased-in beginning with model year 2000; these standards are not shown in this table.

[b] All emissions standards must be met for a useful life of 5 years/50 000 miles. Beginning with model year 1994, a second set of emissions standards must also be met for a full useful life of 10 years/100 000 miles; these standards are shown in parentheses. Tier 1 exhaust standards were phased-in during 1994–96 at a rate of 40%, 80%, and 100%, respectively.

[c] The cold CO emissions standard is measured at 20 F (rather than 75 F) and is applicable for a 5 year/50 000 mile useful life.

[d] The "Prior to control" column reports emissions estimates of a typical newly manufactured car in the years before exhaust emissions certification standards were implemented.

[e] No estimate available.

[f] Manufacturers can opt to certify vehicles for a full useful life of 15 years/150 000 miles and have either (1) intermediate useful life standards waived or (2) receive additional NO_x credits.

[g] In 1968–1969, exhaust emissions standards were issued in parts per million rather than grams per mile and are, therefore, incompatible with this table.

[h] No standard has been set.

[i] The term "tier" refers to a level of standards and is associated with specific years. Interim Tier 2 refers to an intermediate level of standards that move manufacturers toward compliance with Tier 2 standards. Interim Tier 2 and Tier 2 standards are established as "bins." Each bin is a set of standards for NO_x, CO, NMOG, formaldehyde, and particulate matter; HC and NMHC standards are dropped for Tier 2 and Interim Tier 2. Manufacturers may certify any given vehicle family to any of the bins available for that vehicle class as long as the resulting sales-weight corporate average NO_x standard is met for the full useful life of the vehicle. The Tier 2 corporate average NO_x standard is 0.07 g $mile^{-1}$. Interim corporate-based average NO_x standards are based on vehicle type. The interim sales-weighted average for light-duty vehicles (LDVs) is 9.3 g $mile^{-1}$. For LDVs, Tier 2 standards will be phased-in at a rate of 25% in 2004, 50% in 2005, 75% in 2006, and 100% in 2007. During this period, all LDVs not meeting the Tier 2 standards must meet Interim Tier 2 standards.

CO, carbon monoxide; CVS, constant volume sampler; HC, hydrocarbons; NMHC, non-methane hydrocarbons; NMOG, non-methane organic gases; NO_x, nitrogen oxides.

Source: 40 CFR 86, Subpart A (July 1, 2000).

Federal Register, Vol. 65, No. 28, pp. 6851–6858.

F. Emission Standards for Existing Installations in the United States

The states and cities of the United States sometimes have emission standards for existing installations, which are usually less restrictive than those federally required for new installations, i.e. NSPS, BACT, MACT, LAER. The Clean Air Act Amendments of 1990 will change most of these state and local emission standards. The amendments instruct the state air pollution control agencies to submit, by enumerated deadlines, highly detailed revisions to their existing implementation plans. The term "revision" appears almost 250 times in the 1990 amendments [6].

G. Emission Standards for Industrialized Countries of the World

The most recent compilation of emission standards for processes and substances emitted from processes in the industrialized countries of the world was the companion Volume II of the source of Table 24.10 (see Jarrault in Suggested Reading).

H. Fuel Standards

To reduce emissions from fuel-burning sources, one can limit the sulfur, ash, or volatile content of fuels. A listing of such limitations as they existed in 1974 is given in Martin and Stern (see Suggested Reading).

The US EPA required major gasoline retailers to begin to sell one grade of unleaded gasoline by July 1, 1974. This mandate was primarily focused on protecting emissions control systems (e.g. catalytic converters). It was at this time that the working definition of "unleaded" gasoline was to mean "gasoline containing not more than 0.05 gram of lead per gallon and not more than 0.005 gram of phosphorus per gallon" [38FR1255, January 10, 1973]. The agency also called for the gradual phase-out of leaded gasoline. The schedule for reduction of lead content in automobile gasoline was 1.7 grams per gallon ($g\,gal^{-1}$) in 1975, to 1.4 $g\,gal^{-1}$ in 1976, 1.0 $g\,gal^{-1}$ in 1977, 0.8 $g\,gal^{-1}$ in 1978, and 0.5 $g\,gal^{-1}$ in 1979 [38FR33741, December 6, 1973]. Subsequent regulations reduced the allowable lead content to 0.1 $g\,gal^{-1}$ in 1986 [50FR9397, March 7, 1985], and prohibited leaded gas use after 1995 [61FR3837, February 2, 1996]. The EPA established a limit of 1.1 grams per gallon for the content of leaded gasoline beginning on July 1, 1985 and announced its intent to further reduce lead in gasoline to 0.5 $g\,gal^{-1}$ after July 1, 1985, and 0.1 $g\,gal^{-1}$ after January 1, 1986. An EPA program to allow trading in lead credits among refiners facilitated this schedule of lead reduction. Without trading in lead credits, two alternatives were likely: (1) the phase down would be protracted or (2) short-term contractions may occur in the supply of gasoline and possible supply disruptions in some areas.

On July 1, 1983 EPA allowed refiners and importers of gasoline to trade lead reduction credits to meet the limit for the average lead content of gasoline.

Refiners and importers that reduced the average lead content of their gasoline below the EPA limit generated credits that could be sold to refiners or importers that exceeded the limit. Once the limit for the average content of leaded gasoline reached 0.1 g gal^{-1}, trading would not be allowed because of concern that gasoline with less than 0.1 grams of lead per gallon could cause excessive valve seat wear in older vehicles. In 1985 EPA allowed refiners to bank lead credits for subsequent use before the end of 1987 [5].

The Council of the European Economic Community, established under the 1957 Treaty of Rome, in 1973 issued a declaration on the environment [4], which the European Commission in Brussels has interpreted as giving it authority to issue directives on matters related to the emission of air pollutants, such as one limiting the sulfur content of fuel oil [5].

REFERENCES

1. Rose Jr., A. H., and Nader, J. S., *J. Air Pollut. Control Assoc.* **8**, 117–119 (1958).
2. Connor, W. D., Smith, C. F., and Nader, J. S., *J. Air Pollut. Control Assoc.* **18**, 748–750 (1968).
3. Public Law No. 101–549, 101st US Congress, *Clean Air Act Amendments of 1990*, November 1990.
4. Declaration of Council of the European Communities ... on the environment. *Official J.* **16**, C-112 (December 1973).
5. US Environmental Protection Agency, *The United States Experience with Economic Incentives to Control Environmental Pollution*, 230-R-92-001, Washington, DC, July 1992.
6. Proposal for a council directive on the use of fuel oils, with the aim of decreasing sulfurous emissions. *Official J.* **19**, C-54 (March 1976).
7. Stensvaag, J. M., *Clean Air Act 1990 Amendments Law and Practice*. Wiley, New York, 1991.

SUGGESTED READING

Jarrault, P. Limitation des Émission des Pollutants et Qualité de l'Aire-Valeurs Réglemantaires en 1980 dans les principaux Pays Industrielles, Vol. II, Limitation des Émissions des Pollutants. Institut Francais de l'Energie Publ. I.F.E. No. 66, Paris, 1980.

Marin, W., and Stern, A. C., *The World's Air Quality Management Standards*, Vol. I—*Air Quality Management Standards of the World*, Vol. II—*Air Quality Management Standards of the United States*, Publication No. EPA 65019-75-001/002, Miscellaneous Series. US Environmental Protection Agency, Washington, DC, 1974.

Murley, L. (ed.), *Clean Air Around the World. National and International Approaches to Air Pollution Control*. International Union of Air Pollution Prevention Associations, Air & Waste Management Association, Pittsburgh, PA, 1991.

National Research Council, *On Prevention of Significant Deterioration of Air Quality*. National Academy Press, Washington, DC, 1981.

National Research Council, *Board on Environmental Studies and Toxicology, Interim Report of the Committee on Changes in New Source Review Programs for Stationary Sources of Air Pollutants*. National Academy Press, Washington, DC, 2005.

Stern, A. C., *J. Air Pollut. Control Assoc.* **27**, 440 (1977).

QUESTIONS

1. Discuss remote sensing equipment for measurement of emissions after they have left the stack and have become part of the plume.
2. Discuss the means for determining the strength of an odorous emission from a source and of definitively relating the odor to the presumed source.
3. Discuss the relative merits of the prototype testing of automotive vehicles for certification and of certification based on production line testing of each vehicle produced.
4. Select one source category and affected facility from Table 27.1 and determine the detailed performance standards for the pollutants regulated.
5. Discuss some of the variants of the rollback equation (Eq. 27.1) that have been proposed, and evaluate them.
6. In some countries, the enforcement agency is allowed to exercise judgment on how much emission to allow rather than to adhere to a rigid emission limit. Discuss the advantages and disadvantages of this system.
7. Discuss the availability and reliability of in-stack continuous emission monitors when they are required by US NSPS.
8. Discuss the extent to which cost–benefit analysis should be considered in setting emission standards.
9. Trace the history of fuel standards for control of air pollution.

28

The Elements of Regulatory Control

Regulatory control is governmental imposition of limits on emission from sources. In addition to quantitative limits on emissions from chimneys, vents, and stacks, regulations may limit the quantity or quality of fuel or raw material permitted to be used; the design or size of the equipment or process in which it may be used; the height of chimneys, vents, or stacks; the location of sites from which emissions are or are not permitted; or the times when emissions are or are not permitted. Regulations usually also specify acceptable methods of testing or measurement.

One instance of an international air pollution control regulation was the cessation of atmospheric testing of nuclear weapons by the United States, the USSR, and other powers signatory to the cessation agreement. Another was the Trail Smelter Arbitration [1] in which Canada agreed to a regulatory protocol affecting the smelter to prevent flow of its stack emissions into the United States. Another example is the Montreal Protocol to Protect stratospheric ozone, developed by the United Nations, that has greatly reduced and which may eventually eliminate the release of chlorofluorocarbons (CFCs) and other ozone-depleting chemicals.

National air pollution control regulations in some instances are preemptive in that they do not allow subsidiary jurisdictions (states, provinces, counties, towns, boroughs, cities, or villages) to adopt different regulations. In other instances, they are not preemptive in that they allow subsidiary jurisdictions to adopt regulations that are not less stringent than the national regulations. In the United States, the regulations for mobile sources are an example of the latter (there is a provision of a waiver to allow more stringent automobile regulations in California); those for stationary sources are an example of the former.

In many countries, provinces or states have enacted air pollution control regulations. Unless or until superseded by national enactment, these regulations are the ones currently in force. In some cases, municipal air pollution control regulatory enactments are the ones currently in force and will remain so until superseded by state, provincial, or national laws or regulations.

A regulation may apply to all installations, to new installations only, or to existing installations only. Frequently, new installations are required to meet more restrictive limits than existing installations. Regulations which exempt from their application installations made before a specified date are called *grandfather clauses*, in that they apply to newer installations but not to older ones. Regulatory enactments sometimes contain time schedules for achieving progressively more restrictive levels of control. To make regulatory control effective, the regulatory agency must have the right to enter premises for inspection and testing, to require the owner to monitor and report noncompliance, and, where necessary, to do the testing.

I. CONTROL OF NEW STATIONARY SOURCES

In theory, if starting at any date, all new sources of air pollution are adequately limited, all sources installed prior to that date eventually will disappear, leaving only those adequately controlled. The weakness of relying solely on new installation control to achieve community air pollution control is that installations deteriorate in control performance with age and use and that the number of new installations may increase to the extent that what was considered adequate limitation at the time of earlier installations may not prove adequate in the light of the cumulative impact of these increased numbers. Although in theory old sources will disappear, in practice they take such a long time to disappear that it may be difficult to achieve satisfactory air quality solely by new installation control.

One way to achieve new installation control is to build and test prototype installations and allow the use of only replicates of an approved prototype. This is the method used in the United States for the control of emissions from new automobiles (Table 27.2).

Another path is followed where installations are not replicates of a prototype but rather are each unique, e.g., cement plants. Here two approaches are possible. In one, the owner assumes full responsibility for compliance with regulatory emission limits. If at testing the installation does not comply, it is the owner's responsibility to rebuild or modify it until it does comply. In the alternative approach, the owner makes such changes to the design of the proposed installation as the regulatory agency requires after inspection of the plans and specifications. The installation is then deemed in compliance if, when completed, it conforms to the approved plans and specifications. Most regulatory agencies require both testing and plan approval. In this case, the testing is definitive and the plan filing is intended to prevent less sophisticated owners from investing in installations they would later have to rebuild or modify.

In following the approved replicate route, the regulatory agency has the responsibility to sample randomly and test replicates to ensure conformance with the prototype. In following the owner responsibility route, the agency has the responsibility for locating owners, particularly those who are not aware of the regulatory requirements applicable to them, and then of testing their installations. Locating owners is a formidable task requiring good communication between governmental agencies concerned with air pollution control and those which perform inspections of buildings, factories, and commercial establishments for other purposes. Testing installations for emissions is also a formidable task and is discussed in detail in Chapter 36. Plan examination requires a staff of well-qualified plan examiners.

Because of the foregoing requirements, a sophisticated organization is needed to do an effective job of new installation control. Prototype testing is best handled at the national level; installation testing and plan examinations at the state, provincial, or regional level. Location of owners is a local operation with respect to residential and commercial structures but may be a state or provincial operation for industrial sources.

II. CONTROL OF EXISTING STATIONARY SOURCES

Existing installations may be controlled on a retrospective, a present, or a prospective basis. The retrospective basis relies on the receipt of complaints from the public and their subsequent investigation and control. It follows the theory that "the squeaking wheel gets the grease". Complaint investigation and control can become the sole function of a small air pollution control organization to the detriment of its overall air pollution control effectiveness. Activities of this type may improve an agency's short-term image by appeasing the more vocal elements of the public, but this may be accomplished at the expense of the agency's ability to achieve long-range goals.

On a present basis, an agency's staff can be used to investigate and bring under control selected source categories, selected geographic areas, or both. The selection is done, not on the basis of the number or intensity of complaints, but rather on an analysis of air quality data, emission inventory data, or both. An agency's field activity may be directed to the enforcement of existing regulatory limits or to the development of data from which new regulatory limits may be set and later enforced.

On a prospective basis, an agency can project its source composition and location and their emissions into the future and by the use of mathematical models and statistical techniques to determine what control steps have to be taken now to establish future air quality levels. Since the future involves a mix of existing and new sources, decisions must be made about the control levels required for both categories and whether these levels should be the same or different.

Regulatory control on a complaint basis requires the least sophisticated staffing and is well within the capability of a local agency. Operation on present basis requires more planning expertise and a larger organization and therefore is better adapted to a regional agency. To operate on a prospective basis, an agency needs a still higher level of planning expertise, such as may be available only at the state or provincial level. To be most effective, an air pollution agency must operate on all three bases, retrospective, present, and prospective, because an agency must address complaints, attend to special source or area situations, and conduct long-range planning.

III. CONTROL OF MOBILE SOURCES

Mobile sources include railroad locomotives, marine vessels, aircraft, and automotive vehicles. Over the past 100 years, we have gained much experience in regulating smoke and odor emission from locomotives and marine craft. Methods of combustion equipment improvement, firefighter training, and smoke inspection for these purposes are well documented. This type of control is best at the local level.

Regulation of aircraft engine emissions has been made a national responsibility by law in the United States. The Administrator of the Environmental Protection Agency is responsible for establishing emission limits of aircraft engines, and the Secretary of Transportation is required to prescribe regulations to ensure compliance with these limits.

In the United States, regulation of emissions from new automotive vehicles has followed the prototype-replicate route. The argument for routine annual automobile inspection is that cars should be regularly inspected for safety (brakes, lights, steering, and tires) and that the additional time and cost required to check the car's emission control system during the same inspection will be minimal. Such an inspection certainly pinpoints cars whose emission control system has been removed, altered, damaged, or deteriorated and force such defects to be remedied. The question has been whether the improvement in air quality that results from correcting these defects is worth the cost to the public of maintaining the inspection system. Another way of putting the question is whether the same money would be better invested in making the prototype test requirements more rigid with respect to the durability of the emission control system (with the extra cost added to the cost of the new car) than in setting up and operating an inspection system for automotive emissions from used cars. A final question in this regard is whether the factor of safety included in the new car emission standards is sufficient to allow a percentage of all cars on the road to exceed the emission standards without jeopardizing the attainment of the air quality standard.

To date, inspection and maintenance has been a successful part of the overall strategy to improve air quality. Advances in diagnostics and improved emission reduction technologies have also been important.

IV. AIR QUALITY CONTROL REGIONS

Workers in the field of water resources are accustomed to thinking in terms of watersheds and watershed management. It was these people who introduced the term *airshed* to describe the geographic area requiring unified management for achieving air pollution control. The term airshed was not well received for regulatory purposes because its physical connotation is wrong. It was followed by the term *air basin*, which was closer to the mark but still had the wrong physical connotation, since unified air pollution control management is needed in flat land devoid of valleys and basins. The term that finally evolved was *air quality control region*, meaning the geographic area including the sources significant to production of air pollution in an urbanized area and the receptors significantly affected thereby. If long averaging time isopleths (i.e. lines of equal pollution concentration) of a pollutant such as suspended particulate matter are drawn on the map of an area, there will be an isopleth that is at essentially the same concentration as the background concentration. The area within this isopleth meets the description of an air quality control region.

For administrative purposes, it is desirable that the boundaries of an air quality control region be the same as those of major political jurisdictions. Therefore, when the first air quality control regions were officially designated in the United States by publication of their boundaries in the *Federal Register*, the boundaries given were those of the counties all or part of which were within the background concentration isopleth.

When about 100 such regions were designated in the United States, it was apparent that only a small portion of the land area of the country was in officially designated regions. For uniformity of administration of national air pollution legislation, it became desirable to include all the land area of the nation in designated air quality control regions. The Clean Air Amendments of 1970 therefore gave the states the option of having each state considered an air quality control region or of breaking the state into smaller air quality control regions mutually agreeable to the state and the US Environmental Protection Agency. The regions thus created need bear no relation to concentration isopleths, but rather represent contiguous counties which form convenient administrative units. Therefore, for purposes of air pollution control, the United States is now a mosaic of multicounty units, all called air quality control regions, some of which were formed by drawing background concentration isopleths and others of which were formed for administrative convenience.

Some of the former group are interstate, in that they include counties in more than one state. All of the latter are intrastate.

None of the interstate air quality control regions operates as a unified air pollution control agency. Their control functions are all exercised by their separate intrastate components.

V. TALL STACKS AND INTERMITTENT AND SUPPLEMENTARY CONTROL SYSTEMS

For years, it was an item of faith that the higher the stack from which pollution was emitted, the better the pollution would be dispersed and the lower the ground-level concentrations resulting form it. As a result, many tall stacks were built. Subsequently, this practice came under attack on the basis that discharge from tall stacks is more likely to result in long-range transport with the associated problems of interregional and international transport, fumigation, and acidic deposition. In retrospect, tall stacks seem to be an example of doing the wrong thing for the right reasons. It is argued that, to prevent these problems, pollutants must be removed from the effluent gases so that, even if plumes are transported for long distances, they will not cause harm when their constituents eventually reach ground level, and that, if such cleanup of the emission takes place, the need for such tall stacks disappears. Despite these arguments, the taller the stack, the better the closer-in receptors surrounding the source are protected from its effluents.

The ability of a tall stack to inject its plume into the upper air and disperse its pollutants widely depends on prevalent meteorological conditions. Some conditions aid dispersion; others retard it. Since these conditions are both measurable and predictable, intermittent and supplementary control systems (ICS/SCS) have been developed to utilize these measurements and predictions to determine how much pollution can be released from a stack before ground-level limits are exceeded. If such systems are used primarily to protect close-in receptors by decreasing emissions when local meteorological dispersion conditions are poor, they tend to allow relatively unrestricted release when the upper-air meteorology transports the plume for a long distance from the source. This leads to the same objections to ICS/SCS as were noted to the use of tall stacks without a high level of pollutant removal before emission. Therefore, here again it is argued that if a high level of cleanup is used, the need to project the effluent into the upper air is reduced and the benefits of operating an ICS/SCS (which is costly) in lieu of installing more efficient pollutant removal equipment disappear. The US 1990 Clean Air Act Amendment [2] does not allow ICS/SCS to be substituted for pollutant removal from the effluent from fossil fuel-fired steam-powered plants. However, ICS/SCS may still be useful for other applications.

REFERENCES

1. Dean, R. S., and Swain, R. E., Report submitted to the Trail Smelter Arbitral Tribunal, *US Department of the Interior, Bureau of Mines Bulletin 453*. US Government Printing Office, Washington, DC, 1944.
2. Public Law No. 101–549, US 101st Congress, *Clean Air Act Amendments of 1990*, November 1990.

SUGGESTED READING

Eisenbud, M., *Environment, Technology and Health—Human Ecology in Historical Perspective*. New York University Press, New York, 1978.

Friedlaender, A. F. (ed.), *Approaches to Controlling Air Pollution*. MIT Press, Cambridge, MA, 1978.

Gangoiti, G., Sancho, J., Ibarra, G., Alonso, L., García, J. A., Navazo, M., Durana, N., and Ilardia, J. L. Rise of moist plumes from tall stacks in turbulent and stratified atmosphere. *Atmos. Environ.*, **31**(2), 253–269 (1997).

Hauchman, F. S., The role of risk assessment in Title III of the 1990 Clean Air Act Amendments, in *Proceedings of the Conference on Air Toxics Issues in the 1990s: Policies, Strategies, and Compliance*, pp. 73–80. Air and Waste Management Association, Pittsburgh, PA, 1991.

Powell, R. J., and Wharton, L. M., *J. Air Pollut. Cont. Assoc.* **32**, 62 (1982).

Stern, A. C., *J. Air Pollut. Control. Assoc.* **32**, 44 (1982).

US Environmental Protection Agency Civil Enforcement: CAA National Enforcement Programs, 2007, http://www.epa.gov/compliance/civil/caa/caaenfprog.html.

QUESTIONS

1. What are the geographic boundaries of the air quality region (or its equivalent in countries other than the United States) in which you reside?
2. Are there regulatory limits in the jurisdiction where you reside which are different from your state, provincial, or national air quality or emission limits? If so, what are they?
3. Discuss the application of the prototype testing–replicate approval approach to stationary air pollution sources.
4. One form of air pollution control regulation limits the pollution concentration at the owner's "fence line." Find an example of this type of regulation and discuss its pros and cons.
5. How are pollutant emissions to the air from used automobiles classified where you reside? Discuss the merits and the extent of such regulation.
6. Limitation of visible emission was the original form of control of air pollution a century ago. Has this concept outlived its usefulness? Discuss this question.
7. Discuss the use of data telemetered to the office of the air pollution control agency from automatic instruments measuring ambient air quality and automatic instruments measuring pollutant emissions to the atmosphere as air pollution control regulatory means.
8. The ICS/SCS control system has been used for control of emissions from nonferrous smelters. Discuss at least one such active system in terms of its success or failure.
9. Discuss the reasons why the interstate air pollution control region concept has failed in the United States.

29

Organization for Air Pollution Control

The best organizational pattern for an air pollution control agency is that which most effectively and efficiently performs all its functions. There are many functions a control agency or industrial organization could conceivably perform. The desired budget and staff for the agency or organization are determined by listing the costs of performing all desired functions. The actual functions performed by the agency or organization are determined by limitations on staff, facilities, and services imposed by its budget.

I. FUNCTIONS

The most elementary function of an air pollution control agency is its *control* function, which breaks down into two subsidiary functions: *enforcement* of the jurisdiction's air pollution control laws, ordinances, and regulations and *evaluation* of the effectiveness of existing regulations and regulatory practices and the need for new ones.

The enforcement function may be subdivided in several ways, one of which is control of *new sources* and *existing sources*. New-source control can involve all or some of the following functions:

1. *Registration* of new sources.
2. *Filing* of applicants' plans, specifications, air quality monitoring data, and mathematical model predictions.
3. *Review* of applicant's plans, specifications, air quality monitoring data, and mathematical model predictions for compliance with emission and air quality limitations.
4. *Issuance* of certificates of approval for construction.
5. *Inspection* of construction.
6. *Testing* of installation.
7. *Issuance* of certificates of approval for operation.
8. *Receipt* of required fees for the foregoing services.
9. *Appeal* and *variance* hearings and actions.
10. *Prosecution* of violations.

Control of existing sources can involve all or some of the following functions:

1. Visible emission *inspection*.
2. Complaint *investigation*.
3. Periodic and special industrial category or geographic area *inspections*.
4. *Bookkeeping* of offsets and offset trades.
5. Fuel and fuel dealer *inspection and testing*.
6. *Testing* of installations.
7. *Renewal* of certificates of operation.
8. *Receipt* of required fees for the foregoing services.
9. Appeal and variance *hearings and actions*.
10. *Prosecution* of violations.

Note that items 6–10 of the proceeding two lists are the same.

The evaluation function may be subdivided into *retrospective evaluation* of existing regulations and practices and *prospective planning* for new ones. Retrospective evaluation involves the following functions:

1. *Air quality monitoring* and surveillance.
2. *Emission inventory* and monitoring.
3. *Statistical analysis* of air quality and emission data and of agency activities.
4. *Analytic evaluation*.
5. *Recommendation* of required regulations and regulatory practices.

Prospective planning involves the following activities:

1. *Prediction* of future trends.
2. *Mathematical modeling*.
3. *Analytical evaluation*.
4. *Recommendation* of required regulations and regulatory practices.

Meteorological services are closely related to both retrospective evaluation and prospective planning. The last two items on the proceeding two lists are the same.

Functions such as those already noted require extensive technical and administrative support. The *technical* support functions required include the following:

1. *Technical* information services—libraries, technical publications, etc.
2. *Training* services—technical.
3. *Laboratory* services—analytical instrumentation, etc.
4. *Computer* services.
5. *Field* support—equipment calibration.
6. *Shop* services.

An agency requires, either within its own organization or readily available from other organizations, provision of the following *administrative* support functions:

1. *Personnel.*
2. *Procurement.*
3. *Budget*—finance and accounting.
4. *Information technology* (IT).
5. *General* services—secretarial, clerical, reproduction, mail, telecommunications, building maintenance, etc.

Since an air pollution control agency must maintain its *extramural* relationships, the following functions must be provided:

1. *Public relations* and information.
2. *Public education.*
3. *Liaison* with other agencies.
4. *Publication* distribution (including website updates and management).

An agency needs *legal* services for prosecution of violations, appeal and variance hearings, and the drafting of regulations. In most public and private organizations, these services are provided by lawyers based in organizational entities outside the agency. However, organizationally, it is preferable that the legal function be provided within the agency.

One category that has been excluded from all of the proceeding lists is research and development (R & D). R & D is not a necessary function for an air pollution control agency at the state, provincial, regional, or municipal level. It is usually sufficient if the national agency, e.g. the US. Environmental Protection Agency, maintains an R & D program that addresses the nation's needs and encourages each industry to undertake the R & D required to solve its particular problems. National agencies tend to give highest priority to problems common to a number of areas in the country. However, where major problems of a state, province, region, or municipality are unique to its locality and not likely to have high national priority, the area may have to undertake the required R & D. In such instances, the State universities, especially the land grant colleges, are an excellent resource.

It is apparent that some of the functions listed overlap. An example is the overlap of the laboratory function in technical support and the testing and air quality monitoring functions in control. It is because of such overlaps that different organizational structures arise. Many of the functions listed for a control agency are not required in an industrial air pollution control organization. A control agency must emphasize its enforcement function, sometimes at the expense of its evaluation function. The converse is the case for an industrial organization. Also, in an industrial organization, the technical and administrative support, legal, and R & D functions are likely to be supplied by the parent organization.

II. ORGANIZATION

In the foregoing section functions were grouped into categories. The most logical way to organize an air pollution control or industrial organization is along these categorical lines (Figs 29.1 and 29.2), deleting from the organizational structure those functions and categories with which the agency or organization is presently not concerned. When budget and staff are small, one person is required to cover all the agency's or organization's activities in more than one function and, in very small agencies or organizations, in more than one category.

Agencies and organizations which cover a large geographic area must decide to what extent they will either centralize or decentralize functions and categories. Centralization consolidates the agency's or organization's technical expertise, facilitating the resolution of technical matters particularly in relation to large or complex sources. Decentralization facilitates the agency's ability to deal with a larger number of smaller sources. The ultimate decentralization is delegation of certain of an agency's functions to lesser jurisdictions. This can lead to a three-tiered structure, with certain functions fully centralized, certain ones delegated, and certain ones decentralized to regional offices or laboratories.

In an industrial organization the choice is between centralization in corporate or company headquarters and decentralization to the operating organizations or individual plans. The usual organization is a combination of headquarters centralization and company decentralization.

A major organizational consideration is where to place an air pollution control agency in the hierarchy of government. As state, provincial, and municipal governmental structure evolved during the nineteenth century, smoke abatement became a function of the departments concerned with buildings and with boilers. In the twentieth century, until the 1960s, air pollution control shifted strongly to national, state, provincial, and municipal health agencies. However, since the mid-1960s there has been an increasing

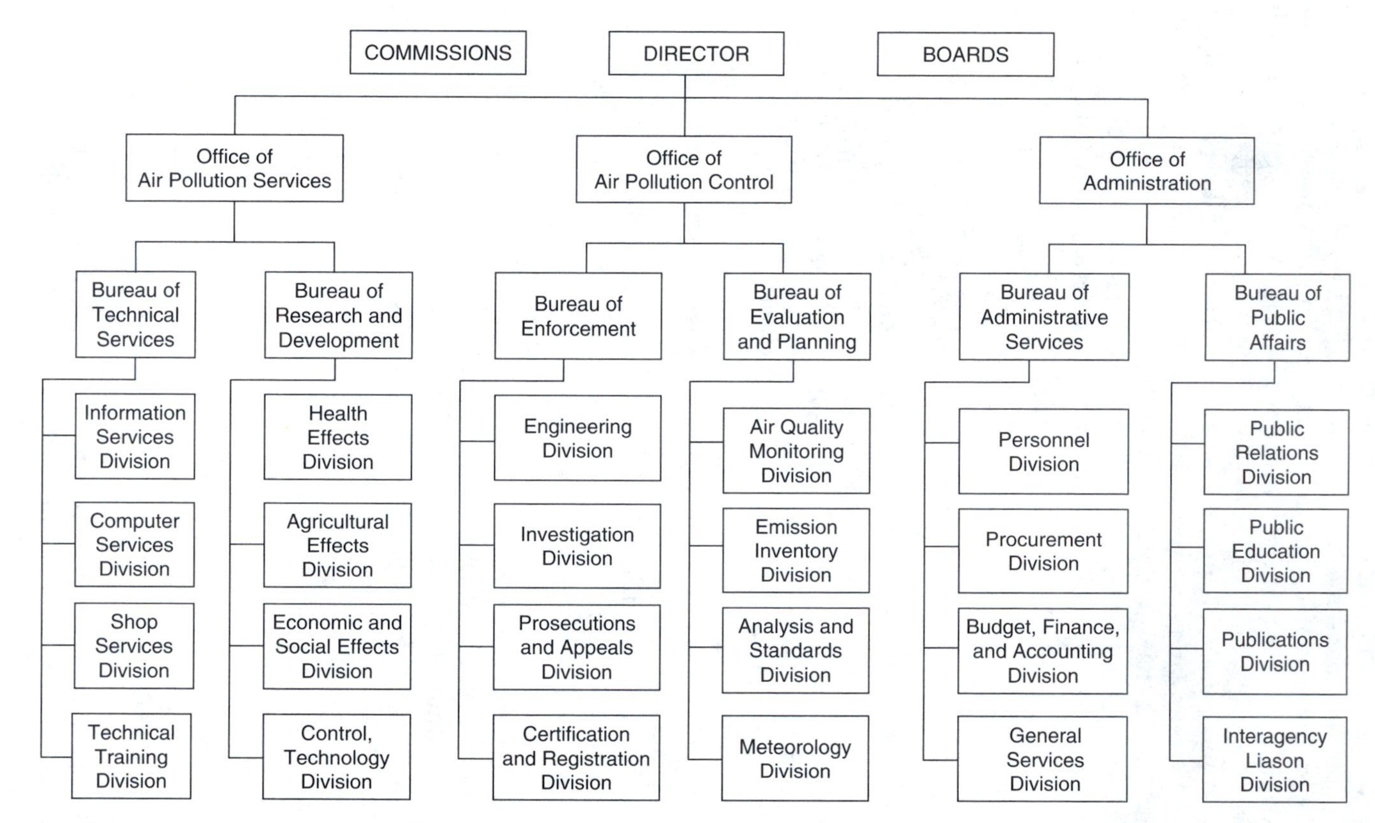

Fig. 29.1. Organization plan which encompasses all functions likely to be required of a governmental air pollution control agency.

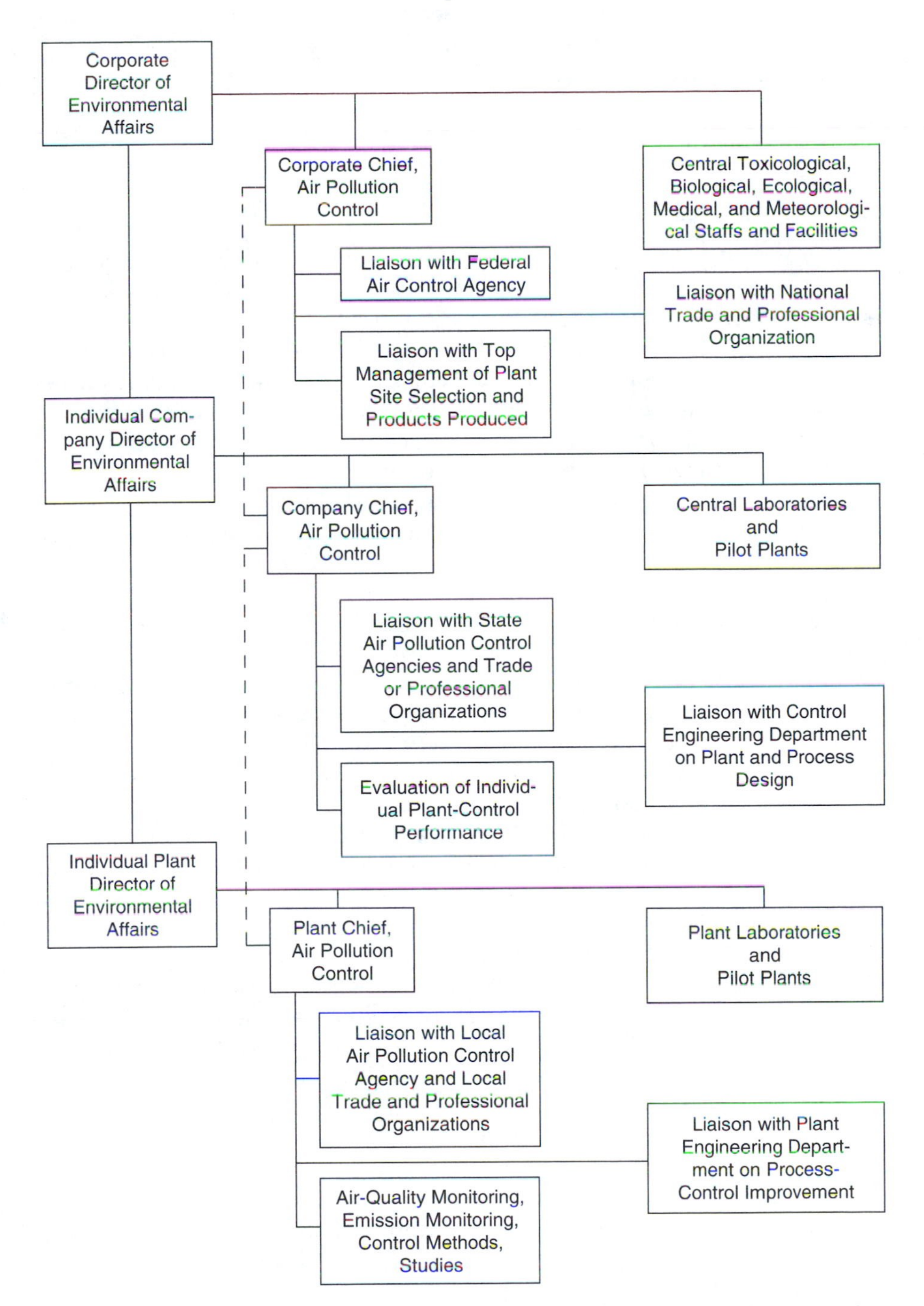

Fig. 29.2. Organization plan which encompasses all functions likely to be required of an industrial air pollution organization.

tendency to locate air pollution control in agencies concerned with natural resources and the environment.

III. FINANCE

A. Fines and Fees

Governmental air pollution control agencies are primarily tax supported. Most agencies charge fees for certain services, such as plan filing, certificate issuance, inspection, and tests. In some agencies, these fees provide a substantial fraction of the agency's budget. In general, this is not a desirable situation because when the continuity of employment of an agency's staff depends on the amount of fees collected, there is an understandable tendency to concentrate on fee-producing activities and resist their deemphasis. This makes it difficult for an agency to change direction with changes in program priorities. Even when these fees go into the general treasury, rather than being retained by the agency, such collection gives some leverage to the agency in its budget request and has the same effect as it would if the fees were retained by the agency.

In most jurisdictions, violators of air pollution control laws and regulations are subject to fines. In the past, it has been considered good administrative practice for these fines to go into the general treasury in order to discourage imposition of fines as a means of supporting the agency's program, which might result if the agency retained the fines. Collection of fines could become an end in itself if the continued employment of the agency staff were dependent on it. Moreover, as in the case of fees, an agency can use its fine collection to give it leverage in budget allocation.

There has been extensive recent rethinking of the role of fees and fines as means of influencing industrial decision-making with regard to investment in pollution control equipment and pollution-free processes. In their new roles, fees and fines take the form of tax write-offs and credits for pollution control investment; taxes on the sulfur and lead content of fuels; continuing fines based on the pollution emission rate; and effluent fees on the same basis. Tax write-offs and credits tend to be resisted by treasury officials because they diminish tax income. Air pollution control agencies tend to look with favor on such write-offs and credits because they result in air pollution control with minimal effort on the part of their staffs and with minimal effect on their budget.

One problem with fuel taxes, continuing fines, and effluent fees is how to use the funds collected most effectively. Among the ideas which have been developed are to use the funds as the basis for loans or subsidies for air pollution control installations by industrial or domestic sources and for financing air pollution control agency needs of jurisdictions subsidiary to those which collect and administer the fines or fees. By ensuring that the continuity of employment of agency personnel is divorced from the fine and fee setting, collection, and administrative process, it is argued that the use of fines and fees can be a constructive rather than a stultifying process.

There is a discussion in Chapter 9, Section IV, of financial incentives to supplement or replace regulations.

B. Budget

An agency's budget consists of the following categories:

- Personnel
    - Salaries
    - Fringe benefits
    - Consultant fees
- Space, furniture, and office equipment
    - Acquisition or rental
    - Operation
    - Maintenance
- Technical and laboratory and other related, capitalized equipment
    - Acquisition or rental
    - Operation
    - Maintenance (including maintenance agreements and contracts)
- Transportation equipment
    - Acquisition
    - Operation
    - Maintenance
- Travel and sustenance
- Supplies and expendables
- Reproduction services
- Communication services
- Contractual services
- Other—miscellaneous

The principal precaution is to avoid allocating too high a percentage of an agency's budget to the personnel category. When this happens, the ability of the agency staff to perform its functions is unduly restricted.

An industrial organization must decide whether to charge off air pollution control costs as a corporate charge, so that the plant manager does not include them in the accounts; or the reverse, so that the plant manager must show a profit for the plant as a cost center, after including costs of air pollution control.

IV. ADVISORY GROUPS

It is common practice, at least in the United States, for air pollution control agencies to be associated with nonsalaried groups who meet from time to time in capacities ranging from official to advisory. The members of such groups may be paid fees or expenses for their days of service, or they may contribute their time and expenses without cost to the agency. The members may be appointed by elected officials, legislative bodies, department heads, program directors, or directors of program categories. They may have to be sworn into office, and the records of their meetings are usually official documents, even if they are informal.

In general, when such bodies are official, laws or ordinances set a statutory base for their creation. This base may be as specific as the statutory requirement that a specified group be appointed in a particular manner with specified duties and authorities, or as broad as a general statutory statement that advisory groups may be created, without enumerating any details of their creation or duties. Such groups may provide their own secretariat and have their own budget authority, or be completely dependent on the agency for services and costs.

The highest levels of such groups are commissions or boards, which can promulgate standards and regulations, establish policy, issue variances, and award funds. The next highest level consists of the hearings or appeals boards, which can issue variances but which neither promulgate standards or regulations nor set policy. Next in rank are the boards, panels, or committees which formally recommend standards or regulations but do not have authority to promulgate them. Next in line are formally organized groups which review requests for funds and make recommendations but lack final award authority. These groups are particularly prevalent in federal programs, such as that of the United States, where large amounts of money are awarded as grants, fellowships, or contracts. These review groups tend to be specialized, e.g. a group of cardiologists to review requests for funds to do cardiovascular research. Although such groups do not have final award authority, the award process tends to follow their recommendations, thereby making their real authority greater than their statutory authority.

At the lowest level are groups that review program content and make recommendations. The impact of groups in this category depends on whether their reports and recommendations are or are not published and the ground rules for determining whether or not they shall be published. If the groups are creations of the agency whose programs they review and the decision on whether or not to publish rests with the agency, minimal impact can be expected. Where review groups are relatively independent of the agency whose programs they review and have their own authority and means for publication, maximum impact can be expected. An example of the latter is the statutory requirement in Section 202 of the US Clean Air Act Amendments of 1977 that automobile exhaust standards be reviewed by the National Academy of Sciences.

SUGGESTED READING

1993 Government Agencies Directory, Section 4, Directory and Resource Book. Air and Waste Management Association, Pittsburgh, 1993.

Jones, C. O., *Clean Air*. University of Pittsburgh Press, Pittsburgh, 1975.

Kokoszka, L. C., and Flood, J. W., *Environmental Management Handbook*. Marcel Decker, New York, 1989.

Mallette, F., Minutes of the Proceedings, *Institution of Mechanical Engineers (GB)*, **N68** (2), 595–625 (1954).

Marsh, A., *Smoke: The Problem of Coal and the Atmosphere*. Faber and Faber, London, 1947.

Portney, P. R. (ed.), *Public Policies for Environmental Protection*. Resources for the Future, Washington, DC, 1990.

QUESTIONS

1. How is the air pollution control agency in your state or province organized? Where is its principal office? Who is its head?
2. Is there a local air pollution control agency in your city, county, or region? How is it organized? Where is its principal office? Who is its head?
3. Draw an organization chart for a governmental air pollution control organization which is limited by budget to 10 professional persons.
4. Write a job specification for the chief of the Bureau of Enforcement in Fig. 25.1.
5. Discuss the relative roles of the staff of an air pollution control agency, its advisory board, and its chief executives in the development and promulgation of air quality and emission standards.
6. Discuss the alternatives mentioned in last paragraph of Section III, (B) regarding allocation of costs of air pollution control in an industrial organization.
7. How does a governmental air pollution control agency or an industrial air pollution control organization organize to ensure that its registration of new sources does not miss significant new sources?
8. Draw an organizational chart for an air pollution organization with a number of plants limited by budget to three professional persons.
9. How does the job market for air pollution control personnel respond to changes in regulatory requirements and to the state of the economy?

Part VII

Preventing and Controlling Air Pollution

30

Preventing Air Pollution

I. INTRODUCTION

The *green* professional is no longer a term for a neophyte to the profession (opposite of a "grey beard"). It is now more likely to mean an environmentally oriented engineer, scientist, planner, or other environmental decision maker. In fact, green engineering is simply systematic engineering done well. If a process is considered comprehensively and systematically, many of the expenses and social costs of air pollution can be prevented. This is always preferable to treating and controlling the problems, which is the subject of the next few chapters.

We can begin understanding air pollution prevention by considering the concept of *sustainability*.

II. SUSTAINABILITY

Their recognition of an impending global threat of environmental degradation led the World Commission on Environment and Development, sponsored by the United Nations, to conduct a study of the world's resources. Also known as the Brundtland Commission, their 1987 report, *Our Common Future,* introduced the term *sustainable development* and defined it as "development that meets the needs of the present without compromising the ability of future generations to meet their own needs" [1]. The United Nations Conference on Environment and Development (UNCED), i.e. the Earth Summit held in Rio de Janeiro in 1992 communicated the idea that sustainable development is both a scientific concept and a philosophical ideal. The document, *Agenda 21,* was endorsed by 178 governments (not including the United States) and hailed as a blueprint for sustainable development. In 2002, the World Summit on Sustainable Development (WSSD) identified five major areas that are considered key for moving sustainable development plans forward.

The underlying purpose of sustainable development is to help developing nations manage their resources, such as rain forests, without depleting these resources and making them unusable for future generations. In short, the objective is to prevent the collapse of the global ecosystems. The Brundtland report presumes that we have a core ethic of intergenerational equity, and that future generations should have an equal opportunity to achieve a high quality of life. The report is silent, however, on just why we should embrace the ideal of intergenerational equity, or why one should be concerned about the survival of the human species. The goal is a sustainable global ecologic and economic system, achieved in part by the wise use of available resources.

From a thermodynamics standpoint, a sustainable system is one that is in equilibrium or changing at a tolerably slow rate. In the food chain, for example, plants are fed by sunlight, moisture and nutrients, and then become food themselves for insects and herbivores, which in turn act as food for larger animals. The waste from these animals relenishes the soil, which nourishes plants, and the cycle begins again [2].

"Sustainable design" is a systematic approach. At the largest scale, manufacturing, transportation, commerce and other human activities that promote high consumption and wastefulness of finite resources cannot be sustained. At the individual designer scale, the products and processes must be considered for their entire lifetimes and beyond.

III. GREEN ENGINEERING AND SUSTAINABILITY

To attain sustainability, people need to adopt new and better means of using materials and energy. The operationalizing of the quest for sustainability is defined as *green engineering,* a term that recognizes that engineers are central to the practical application of the principles sustainability to everyday life [3].

The relationship between sustainable development, sustainability, and green engineering can be depicted as:

Sustainable development → Green engineering → Sustainability

Sustainable development is an ideal that can lead to sustainability, but this can only be done through green engineering.

Green engineering [4] treats environmental quality as an end in itself. The US EPA has defined green engineering as:

> ... the design, commercialization, and use of processes and products, which are feasible and economical while minimizing (1) generation of pollution at the source and (2) risk to human health and the environment. The discipline embraces the concept that decisions to protect human health and the environment can have the greatest impact and cost effectiveness when applied early to the design and development phase of a process or product [5].

Green engineering approaches are being linked to improved computational abilities (see Table 30.1) and other tools that were not available at the outset of the environmental movement. Increasingly, companies have come to recognize that improved efficiencies save time, money, and other resources in the long run. Hence, companies are thinking systematically about the entire product stream in numerous ways:

- applying sustainable development concepts, including the framework and foundations of "green" design and engineering models;
- applying the design process within the context of a sustainable framework: including considerations of commercial and institutional influences;
- considering practical problems and solutions from a comprehensive standpoint to achieve sustainable products and processes;
- characterizing waste streams resulting from designs;
- understanding how first principles of science, including thermodynamics, must be integral to sustainable designs in terms of mass and energy relationships, including reactors, heat exchangers, and separation processes;
- applying creativity and originality in group product and building design projects.

There are numerous industrial, commercial, and governmental green initiatives, including design for the environment (DFE), design for disassembly (DFD), and design for recycling (DFR) [6]. These are replacing or at least changing pollution control paradigms. For example, concept of a "cap and trade" has been tested and works well for some pollutants. This is a system where companies are allowed to place a "bubble" over a whole manufacturing complex or trade pollution credits with other companies in their industry

TABLE 30.1

Principles of Green Programs. First Two Columns, Except "Nano-materials"

Principle	Description	Example	Role of computational toxicology
Waste prevention	Design chemical syntheses and select processes to prevent waste, leaving no waste to treat or clean up.	Use a water-based process instead of an organic solvent-based process.	Informatics and data mining can provide candidate syntheses and processes.
Safe Design	Design products to be fully effective, yet have little or no toxicity.	Using microstructures, instead of toxic pigments, to give color to products. Microstructures bend, reflect and absorb light in ways that allow for a full range of colors.	Systems biology and "omics" (genomics, proteomics, and metabonomics) technologies can support predictions of cumulative risk from products used in various scenarios.
Low hazard chemical synthesis	Design syntheses to use and generate substances with little or no toxicity to humans and the environment.	Select chemical synthesis with toxicity of the reagents in mind upfront. If a reagent ordinarily required in the synthesis is acutely or chronically toxic, find another reagent or new reaction with less toxic reagents.	Computational chemistry can help predict unintended product formation and reaction rates of optional reactions.
Renewable material use	Use raw materials and feedstocks that are renewable rather than those that deplete nonrenewable natural resources. Renewable feedstocks are often made from agricultural products or are the wastes of other processes; depleting feedstocks are made from fossil fuels (petroleum, natural gas, or coal) or that must be extracted by mining.	Construction materials can be from renewable and depleting sources. Linoleum flooring, for example, is highly durable, can be maintained with nontoxic cleaning products, and is manufactured from renewable resources amenable to being recycled. Upon demolition or re-flooring, the linoleum can be composted.	Systems biology, informatics, and "omics" technologies can provide insights into the possible chemical reactions and toxicity of the compounds produced when switching from depleting to renewable materials.
Catalysis	Minimize waste by using catalytic reactions. Catalysts are used in small amounts and can carry out a single reaction many times. They are preferable	The Brookhaven National Laboratory recently reported that it has found a "green catalyst" that works by removing one stage of the reaction, eliminating the need to use solvents	Computation chemistry can help to compare rates of chemical reactions using various catalysts.

TABLE 30.1 (*Continued*)

Principle	Description	Example	Role of computational toxicology
	to stoichiometric reagents, which are used in excess and work only once.	in the process by which many organic compounds are synthesized. The catalyst dissolves into the reactants. Also, the catalyst has the unique ability of being easily removed and recycled because, at the end of the reaction, the catalyst precipitates out of products as a solid material, allowing it to be separated from the products without using additional chemical solvents.[a]	
Avoiding chemical derivatives	Avoid using blocking or protecting groups or any temporary modifications if possible. Derivatives use additional reagents and generate waste.	Derivativization is a common analytical method in environmental chemistry, i.e. forming new compounds that can be detected by chromatography. However, chemists must be aware of possible toxic compounds formed, including left over reagents that are inherently dangerous.	Computational methods and natural products chemistry can help scientists start with a better synthetic framework.
Atom economy	Design syntheses so that the final product contains the maximum proportion of the starting materials. There should be few, if any, wasted atoms.	Single atomic and molecular scale logic used to develop electronic devices that incorporate DFD, DFR, and design for safe and environmentally optimized use.	The same amount of value, e.g. information storage and application, is available on a much smaller scale. Thus, devices are smarter and smaller, and more economical in the long-term. Computational toxicology enhances the ability to make product decisions with better predictions of possible adverse effects, based on the logic.
Nano-materials	Tailor made materials and processes for	Emissions, effluent, and other environmental	Improved, systematic catalysis in emission

(*Continued*)

TABLE 30.1 (*Continued*)

Principle	Description	Example	Role of computational toxicology
	specific designs and intent at the nanometer scale (≤100 nm).	controls; design for extremely long life cycles. Limits and provides better control of production and avoids over-production (i.e. "throwaway economy").	reductions, e.g. large sources like power plants and small sources like automobile exhaust systems. Zeolite and other sorbing materials used in air pollution treatment and emergency response situations can be better designed by taking advantage of surface effects; this decreases the volume of material used.
Selection of safer solvents and reaction conditions	Avoid using solvents, separation agents, or other auxiliary chemicals. If these chemicals are necessary, use innocuous chemicals.	Supercritical chemistry and physics, especially that of carbon dioxide and other safer alternatives to halogenated solvents are finding their way into the more mainstream processes, most notably dry cleaning.	To date, most of the progress has been the result of wet chemistry and bench research. Computational methods will streamline the process, including quicker "scale-up."
Improved energy efficiencies	Run chemical reactions and other processes at ambient temperature and pressure whenever possible.	To date, chemical engineering and other reactor-based systems have relied on "cheap" fuels and, thus, have optimized on the basis of thermodynamics. Other factors, e.g. pressure, catalysis, photovoltaics and fusion, should also be emphasized in reactor optimization protocols.	Heat will always be important in reactions, but computational methods can help with relative economies of scale. Computational models can test feasibility of new energy efficient systems, including intrinsic and extrinsic hazards, e.g. to test certain scale-ups of hydrogen and other economies. Energy behaviors are scale-dependent. For example, recent measurements of H_2SO_4 bubbles when reacting with water have temperatures in range of those found the surface of the sun.[b]
Design for degradation	Design chemical products to break down to innocuous substances after use	Biopolymers, e.g. starch-based polymers can replace styrene and other halogen-based	Computation approaches can simulate the degradation of substances as they enter

TABLE 30.1 (*Continued*)

Principle	Description	Example	Role of computational toxicology
	so that they do not accumulate in the environment.	polymers in many uses. Geopolymers, e.g. silane-based polymers, can provide inorganic alternatives to organic polymers in pigments, paints, etc. These substances, when returned to the environment, become their original parent form.	various components of the environment. Computational science can be used to calculate the interplanar spaces within the polymer framework. This will help to predict persistence and to build environmentally friendly products, e.g. those where space is adequate for microbes to fit and biodegrade the substances.
Real-time analysis to prevent pollution and concurrent engineering	Include in-process real-time monitoring and control during syntheses to minimize or eliminate the formation of byproducts.	Remote sensing and satellite techniques can provide be linked to real-time data repositories to determine problems. The application to terrorism using nano-scale sensors is promising.	Real-time environmental mass spectrometry can be used to analyze whole products, obviating the need for any further sample preparation and analytical steps. Transgenic species, while controversial, can also serve as biological sentries, e.g. fish that change colors in the presence of toxic substances.
Accident prevention	Design processes using chemicals and their forms (solid, liquid, or gas) to minimize the potential for chemical accidents including explosions, fires, and releases to the environment.	Scenarios that increase probability of accidents can be tested.	Rather than waiting for an accident to occur and conducting failure analyses, computational methods can be applied in prospective and predictive mode; that is, the conditions conducive to an accident can be characterized computationally.

[a] US Department of Energy, *Research News*, http://www.eurekalert.org/features/doe/2004–05/dnl-brc050604.php; accessed on March 22, 2005.
[b] Flannigan, D. J., and Suslick, K. S., Plasma formation and temperature measurement during single-bubble cavitation. *Nature* **434**, 52–55 (2005).
Source: Adapted from US Environmental Protection Agency, 2005, *Green Chemistry*: http://www.epa.gov/greenchemistry/principles.html; accessed on April 12, 2005. Other information from discussions with Michael Hays, US EPA, National Risk Management Research Laboratory, April 28, 2005.

instead of a "stack-by-stack" and "pipe-by-pipe" approach, i.e. the so-called "command and control" approach. Such policy and regulatory innovations call for some improved technology-based approaches as well as better quality-based approaches, such as leveling out the pollutant loadings and using less expensive technologies to remove the first large bulk of pollutants, followed by higher operation and maintenance (O&M) technologies for the more difficult to treat stacks and pipes. But, the net effect can be a greater reduction of pollutant emissions and effluents than treating each stack or pipe as an independent entity. This is a foundation for most sustainable design approaches, i.e. conducting a life cycle analysis, prioritizing the most important problems, and matching the technologies and operations to address them. The problems will vary by size (e.g. pollutant emission rates), difficulty in treating, and feasibility. The easiest ones are the big ones that are easy to treat (so-called "low hanging fruit"). These improvements are relatively easy to bring about in a short time period. However, the most intractable problems are often those that are small but very expensive and difficult to control, i.e. less feasible. The expectations of the client, the regulators, and those of the individual engineer must be realistic in how rapidly the new approaches can be incorporated.

Historically, air pollution considerations have been approached by engineers as constraints on their designs. For example, hazardous substances generated by a manufacturing process were dealt with as a waste stream (including releases and emissions from vents and stacks) that must be contained and treated. The pollutant generation had to be constrained by selecting certain manufacturing types, increasing waste handling facilities, and if these did not entirely do the job, limiting rates of production. Green engineering emphasizes that these processes are often inefficient financially and environmentally, calling for a comprehensive, systematic life cycle approach. Green engineering attempts to achieve four goals:

1. Waste reduction
2. Materials management
3. Pollution prevention
4. Product enhancement

Waste reduction involves finding efficient material uses. It is compatible with other engineering efficiency improvement programs, such as total quality management and real-time or just-in-time manufacturing. The overall rationale for waste reduction is that if materials and processes are chosen intelligently at the beginning, less waste will result. In fact, a relatively new approach to engineering is to design and manufacture a product simultaneously rather than sequentially, known as *concurrent engineering*. Combined with DFE and life cycle analysis, concurrent engineering approaches may allow air quality improvements under real-life, manufacturing conditions. However, changes made in any step must consider possible effects on the rest of the design and implementation.

CASE STUDY BOX: INDOOR AIR POLLUTION AND CONCURRENT ENGINEERING FAILURE

One of the most perplexing and tragic medical mysteries of the past 50 years has been sudden infant death syndrome (SIDS). The syndrome was first identified in the early 1950s.

Numerous etiologies have been proposed for SIDS, including a number of environmental causes. Air pollution is one of the likely suspects. A recent study, for example, found a statistically significant link between exposure of newborn infants to fine aerosols and SIDS [7]. The study found that approximately 500 of the 3800 SIDS cases in 1994 were associated with elevated concentrations of particle matter with aerodynamic diameters less than 10 μm (PM_{10}) in the United States. This estimate is based only on metropolitan areas in counties with standard PM_{10} monitors. Based on the metropolitan area with the lowest particle concentrations, there appears to be a threshold, that is, particulate-related infant deaths occurred when PM_{10} levels below 11.9 $\mu g\ m^{-3}$.

Extrapolations from these data show that almost 20% of all SIDS cases each year in the top twelve most polluted metro areas in the United States are associated with PM_{10} pollution. The number of annual SIDS cases associated with PM_{10} in Los Angeles, New York, Chicago, Philadelphia, and Detroit metropolitan areas range from 20 to 44. The study found that 10 states accounted for more than 60% of the particle-related SIDS cases, with 93 in California, 37 in Texas, and 32 in Illinois.

Since particle matter has been linked to SIDS cases, a logical extension would be to suspect the role of environmental tobacco smoke (i.e. "side stream" exposure) in some cases, since this smoke contains both particulate and gas phase contaminants that are released into the infant's breathing zone. Also, *in utero* exposures to toxic substances when a pregnant woman smokes (e.g. nicotine and other organic and inorganic toxins) may make the baby more vulnerable.

Another suspected etiology for SIDS is the exposure to pollutants via consumer products. For example, polyvinyl chloride (PVC) products have been indirectly linked to SIDS. The most interesting link is not the PVC itself, but the result of an engineering "solution."

Plastics came into their own in the 1950s, replacing many other substances, because of their lightweight and durability. However, being a polymer, physical and chemical conditions affect the ability of PVC to stay "hooked together." This can be a big problem for plastics used for protection, such as waterproofing. One such use was as a tent material.

Serendipity often plays a role in linking harmful effects to possible causes. In 1988, Barry Richardson was in the process of renting a tent for his daughter's wedding. Richardson, an expert in material science and deterioration, while renting a tent from proprietor Peter Mitchell inquired about its durability and found that PVC tents tend to break down. Richardson surmised that the rapid degradation was microbial and in fact due to fungi. The tent manufacturers decided to correct the PVC durability by changing the manufacturing process, that is, by concurrent engineering. In this case, they decided to increase the amount of fungicide, 10-10′-oxybis(phenoxarsine) (OBPA):

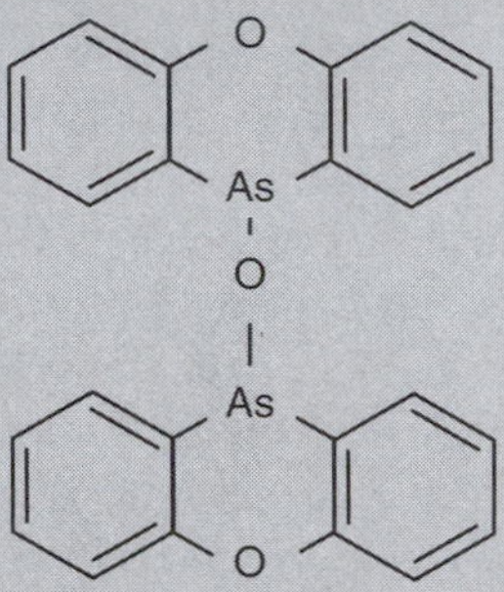

A quick glance at the OBPA structure shows that when it breaks down it is likely to release arsenic compounds. In this case, it is arsine (AsH_3), a toxic gas (vapor pressure = 11 mmHg at 20 °C). It is rapidly absorbed when inhaled, and easily crosses the alveolo-capillary membrane and enters red blood cells. Arsine depletes the reduced glutathione content of red blood cells, leading to the oxidation of sulfhydryl groups in hemoglobin and, possibly, red cell membranes. These effects produce membrane instability with rapid and massive intravascular hemolysis. It also binds to hemoglobin, forming a metalloid-hemaglobin complex [8]. These can lead to acute cardiovascular, neurotoxic, and respiratory effects.

Increasing the OBPA to address the problem of PVC disintegration is an example of the problem of ignoring the life cycle and systematic aspects of most engineering problems. In this case, production and marketing would greatly benefit from a type of PVC that does not break down readily under ambient conditions. In fact, if that problem cannot be solved, the entire camping market might be lost, since fungi are ubiquitous in the places where these products are used.

Had the engineers and planners considered the chemical structure and the possible uses, however, they at least might have restricted the PVC treated with high concentrations of OBPA to certain uses, such as only on tent materials, and not in materials that come in contact or near humans (bedding materials, toys, etc.). To the contrary, the PVC manufacturers blatantly disregarded the science. Richardson, the expert, from the outset had warned that increasing the amount of fungicide would not only increase the hazard and risk, but also would make the product less efficacious (even more vulnerable to fungal attack). He stated, "The biocide won't kill this fungus—instead, the fungus will consume the biocide as well as the plasticizer. Since the biocide contains arsenic, the fungus will generate a very poisonous gas which would be harmful to your staff working with the marquees." Plasticizers are semivolatile organic compounds (e.g. phthalates) that can serve as a food source for microbes, once they become acclimated. The engineers should have known this, since it is one of the biological principles upon which much wastewater treatment is based. But, the manufacturers wanted to approach the situation as a linear problem with a simple solution, that is, increase fungicide and decrease fungus. The PVC manufacturer even argued that the fungicide was even approved for use in baby mattresses.

The extent to which arsine gas released by the degradation of OBPA was a causative agent in SIDS cases is a matter of debate. But, the physics and chemistry certainly indicate that a toxic gas *could* be released leading to exposures of a highly susceptible population (babies) is not debatable.

Pollution and consumer products are only some of the possible causes of SIDS. Others include breathing position (probably increased carbon dioxide inhalation), poor nutrition, and physiological stress (e.g. overheating) [9].

The overall lesson is that are many advantages to concurrent engineering, such as real-time feedback between design and build stages, adaptive approaches, and continuous improvement. However, concurrent engineering works best when the entire life cycle is considered. The designer must ask how even a small change to improve one element in the process can affect other steps and systems within the design and build process.

IV. LIFE CYCLE ANALYSIS

One means of understanding questions of material and product use and waste production is to conduct what has become known as a *life cycle assessment*. Such an assessment is a comprehensive approach to pollution prevention by analyzing the entire life of a product, process or activity, encompassing raw materials, manufacturing, transportation, distribution, use, maintenance, recycling, and final disposal. In other words, assessing its *life cycle* should yield a complete picture of the environmental impact of a product.

The first step in a life cycle assessment is to gather data on the flow of a material. Once the quantities of various components of such a flow are known, the environmental effect of each step in the production, manufacture, use, and recovery/disposal is estimated.

Life cycle analyses are performed for several reasons, including the comparison of products for purchasing and a comparison of products by industry. In the former case, the total environmental effect of glass returnable bottles, for example, could be compared to the environmental effect of non-recyclable plastic bottles. If all of the factors going into the manufacture, distribution, and disposal of both types of bottles are considered, one container might be shown to be clearly superior.

Life cycle analyses often suffer from a dearth of data. Some of the information critical to the calculations is virtually impossible to obtain. For example, something as simple as the tonnage of solid waste collected in the United States is not readily calculable or measurable. And even if the data *were* there, the procedure suffers from the unavailability of a single accounting system. Is there an optimal level of pollution, or must all pollutants be removed 100% (a virtual impossibility)? If there is air pollution and water pollution, how must these be compared?

A recent study supported by the US EPA developed complex models using principles of life cycle analysis to estimate the cost of materials recycling. The models were able to calculate the dollar cost, as well as the cost in environmental damage caused at various levels of recycling. Contrary to intuition, and the stated public policy of the US EPA, it seems that there is a breakpoint at about 25% diversion. That is, as shown in Fig. 30.1, the cost in dollars and adverse environmental impact start to increase at an exponential rate at about 25% diversion. Should we therefore even strive for greater

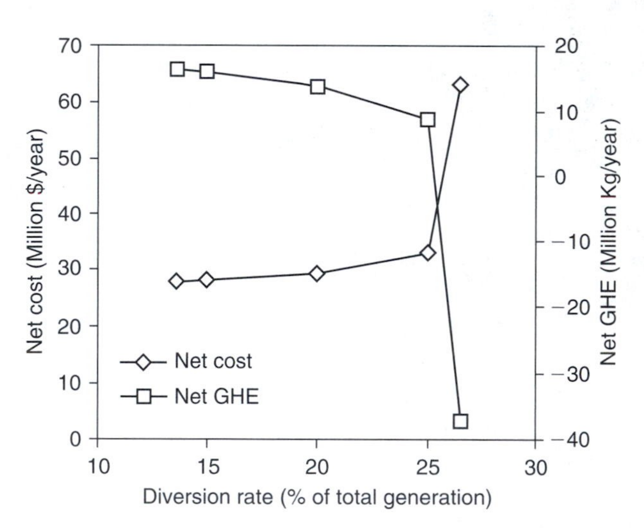

Fig. 30.1. The cost in dollars and adverse environmental impact increases dramatically when the fraction of solid waste recycled exceeds 25%. *Source*: Solano. E., Dumas, R. D., Harrison, K. W., Ranjithan, S., Barlaz, M. A., and Brill, E. D., *Integrated Sold Waste Management Using a Life-Cycle Methodology for Considering cost, Energy, and Environmental Emissions -2. Illustrative Applications.* Department of Civil Engineering, North Carolina State University, Raleigh, NC, 1999.

DISCUSSION BOX: THE MORNING CONTAINER

A simple example of the difficulties in life cycle analysis would be in finding the solution to the great coffee cup debate—whether to use paper coffee cups or polystyrene coffee cups. The answer most people would give is not to use either, but instead to rely on the permanent mug. But there nevertheless are times when disposable cups are necessary (e.g. in hospitals), and a decision must be made as to which type to choose [10]. So let us use life cycle analysis to make a decision.

The paper cup comes from trees, but the act of cutting trees results in environmental degradation. The foam cup comes from hydrocarbons such as oil and gas and this also results in adverse environmental impact, including the use of nonrenewable resources. The production of the paper cup results in significant water pollution while the production of the foam cup contributes essentially no water pollution. The production of the paper cup results in the emission of chlorine, chlorine dioxide, reduced sulfides and particulate, while the production of the foam cup results in none of these. The paper cup does not require chlorofluorocarbons (CFCs), but neither do the newer foam cups ever since the CFCs in polystyrene were phased out. The foam cups however results in the emission of pentane while the paper cup contributes none. From a materials separation perspective, the ability to recycle the foam cup is much higher than that of the paper cup since the latter is made from several materials, including the plastic coating on the paper. They both burn well, although the foam cup produces 17 200 Btu lb^{-1} (40 000 kJ kg^{-1}) while the paper cup produces only 8600 Btu lb^{-1} (20 000 kJ kg^{-1}). In the landfill, the paper cup degrades into CO_2 and CH_4, both greenhouse gases, while the foam cup is inert. Since it is inert, it will remain in the landfill for a very long time, while the paper cup will eventually (but very slowly) decompose. If the landfill is considered a waste storage receptacle, then the foam cup is superior, since it does not participate in the reaction, while the paper cup produces gases and probably leachate. If on the other hand the landfill is thought of as a treatment facility, then the foam cup is highly detrimental since it does not biodegrade.

So, then, which cup is better for the environment? If you wanted to do the right thing, which cup should you use? Private individuals can of course practice pollution prevention by such a simple expedient as not using either plastic or paper disposable coffee cups but by using a refillable mug instead. The argument as to which kind of cup, plastic or paper, is better is then moot. It is better not to produce the waste in the first place. In addition, the coffee tastes better from a mug! Of course, the life cycle of the mug must also be factored into the decision (e.g. do the glazes contain toxic metals or were toxic materials and fossil fuels used?) This is an example of considering the function, not simply selecting a preferred device or product.

diversion rates, if this results in unreasonable cost in dollars and actually does harm to the environment?

Once the life cycle of a material or product has been analyzed, the next engineering step is to manage the life cycle. If the objective is to use the least energy and to cause the least detrimental effect on the environment, then it is clear that much of the onus is on the manufacturers of these products. The users of the products can have the best intentions for reducing adverse environmental effects, but if the products are manufactured in such a way as to make this impossible, then the fault is with the manufacturers. On the other hand, if the manufactured materials are easy to separate and recycle, then most likely energy is saved and the environment is protected. This process has become known as *pollution prevention* in industry, and there are numerous examples of how industrial firms have reduced emissions or the production of other wastes, or have made it easy to recover waste products, and in the process saved money. Some automobile manufacturers, for example, are modularizing the engines so that junked parts can be easily reconditioned and reused. Printer cartridge manufacturers have found that refilling cartridges with ink or toner is far cheaper than remanufacturing them, and now offer trade-ins. All of the efforts by industry to reduce waste (and save money in the process) will influence the solid waste stream in the future.

V. POLLUTION PREVENTION

The EPA defines "pollution prevention" as the following:

> The use of materials, processes, or practices that reduce or eliminate the creation of pollutants or wastes at the source. It includes practices that reduce the use of hazardous materials, energy, water or other resources and practices that protect natural resources through conservation or more efficient use [11].

In the widest sense, pollution prevention is the idea of eliminating waste, regardless of how this might be done.

Originally, pollution prevention was applied to industrial operations with the idea of reducing either the amount of the wastes being produced or to change their characteristics in order to make them more readily disposable. Many industries changed to water-soluble paints, for example, thereby eliminating organic solvents, clean up time, etc. and often in the process saving considerable money. In fact, the concept was first introduced as "pollution prevention pays," emphasizing that many of the changes would actually save the companies money. In addition, the elimination or reduction of hazardous and otherwise difficult wastes also has a long-term effect—it reduces the liability the company carries as a consequence of its disposal operations.

With the passage of the Pollution Prevention Act of 1990, the EPA was directed to encourage pollution prevention by setting appropriate standards for pollution prevention activities, to assist federal agencies in reducing wastes generated, to work with industry and to promote the elimination of wastes by creating waste exchanges and other programs, seek out and eliminate barriers to the efficient transfer of potential wastes, and to do this with the cooperation of the individual states.

In general, the procedure for the implementation of pollution prevention activities is to:

1. recognize a need,
2. assess the problem,
3. evaluate the alterative,
4. implement the solutions.

Contrary to most pollution control activities, industries generally have welcomed this governmental action, recognizing that pollution prevention can and often does result in the reduction of costs to the industry. Thus, recognition of the need quite often is internal and the company seeks to initiate the pollution prevention procedure.

During the assessment phase, a common procedure is to perform a "waste audit," which is the black box mass balance, using the company as the black box.

EXAMPLE BOX: WASTE AUDIT[1]

A manufacturing company is concerned about the air emissions of volatile organic carbons. These chemicals can volatilize during the manufacturing process, but the company is not able to estimate accurately the rate of volatilization, or even which chemicals are going to the vapor phase. The company conducts an audit of three of their most widely used volatile organic chemicals, with the following results:

Purchasing department records

Material	*Purchase quantity (barrels)*
Carbon tetrachloride[2] (CCl_4)	48
Methyl chloride[3] (CH_2Cl_3)	228
Trichloroethylene (C_2HCl_3)	505

Wastewater treatment plant influent

Material	*Average concentration ($mg\,L^{-1}$)*
Carbon tetrachloride	0.343
Methylene chloride	4.04
Trichloroethylene	3.23

The average influent flow rate to the treatment plant is $0.076\,m^3\,s^{-1}$.

[1] This example is taken from: Vallero, D. A., and Vesilind, P. A., *Socially Responsible Engineering: Justice in Risk Management*. Wiley, Hoboken, NJ, 2006.

[2] The correct name is tetrachloromethane, but the compound was in such common use throughout the twentieth century and was referred to as carbon tetrachloride that the name is still frequently used in the engineering and environmental professions.

[3] Also known as chloromethane.

Hazardous waste manifests (what leaves the company by truck, headed to a hazardous waste treatment facility)

Material	*Barrels*	*concentration (%)*
Carbon tetrachloride	48	80
Methyl chloride	228	25
Trichloroethylene	505	80

Unused barrels at the end of the year

Material	*Barrels*
Carbon tetrachloride	1
Methyl chloride	8
Trichloroethylene	13

How much VOC is escaping?

Conduct a black box mass balance, as

$$[A_{\underline{acc}}] = [A_{in}] - [A_{out}] + [A_{prod}] - [A_{cons}]$$

where $A_{\underline{acc}}$ = mass of A per unit time accumulated

A_{in} = mass of A per unit time in
A_{out} = mass of A per unit time out
A_{prod} = mass of A per unit time produced
A_{cons} = mass of A per unit time consumed

The materials A are, of course, the three VOC's.

Barrels must be converted to cubic meters and the density of each chemical must be known. Each barrel is 0.12 m^2, and the density of the three chemicals is 1548, 1326, and 1476 kg m^{-3}. The mass per year of carbon tetrachloride accumulated is

$$[A_{\underline{acc}}] = 1 \text{ barrel year}^{-1} \times 0.12 \text{ m}^3 \text{ barrel}^{-1} \times 1548 \text{ kg m}^{-3}$$
$$= 186 \text{ kg year}^{-1}$$

Similarly,

$$[A_{in}] = 48 \times 0.12 \times 1548 = 8916 \text{ kg year}^{-1}$$

The mass out is in three parts; mass discharge to the wastewater treatment plant, mass leaving on the trucks to the hazardous waste disposal facility, and the mass volatilizing. So the equation,

$$[A_{out}] = [0.343 \text{ g m}^{-3} \times 0.076 \text{ m}^3\text{s}^{-1} \times 86\,400 \text{ s day}^{-1} \times 365 \text{ day year}^{-1} \times 10^{-3} \text{kg g}^{-1}] + [48 \times 0.12 \times 1548 \times 0.80] + A_{air}$$
$$= 822.1 + 7133 + A_{air}$$

where A_{air} is the mass per unit time emitted to the air.

Since there is no carbon tetrachloride consumed or produced,

$$186 = 8916 - [822.1 + 7133 + A_{air}] + 0 - 0$$

and

$$A_{air} = 775 \text{ kg year}^{-1}.$$

If a similar balance is done on the other chemicals, it appears that the losses to air of methyl chloride is about 16 000 kg $year^{-1}$ and the trichloroethylene is about 7800 kg $year^{-1}$.

If the intent is to cut total VOC emissions, it is clear that the first target should be the methyl chloride, at least in terms of the mass released. But, another important consideration in preventing pollution is *relative risk*.

Although methyl chloride is two orders of magnitude more volatile than the other pollutants, all three compounds are likely to be found in the atmosphere. Thus, inhalation is a likely exposure pathway.

Since risk is the product of exposure times hazard ($R = E \times H$), we can compare the risks by applying a hazard value (e.g. cancer potency). We can use the air emissions calculated above as a reasonable approximation of exposure via the inhalation pathway[4] and the inhalation cancer slope factors to represent the hazard. These slope factors are published by the US EPA and are found to be:

Carbon tetrachloride = 0.053 kg day mg^{-1}
Methyl chloride = 0.0035 kg day mg^{-1}
Trichloroethylene = 0.0063 kg day mg^{-1}

The relative risk for the three compounds can be estimated by removing the units (i.e. we are not actually calculating the risk, only comparing the three compounds against each other, so we do not need units.) If we were calculating risks, the units for exposure would be mass of contaminant per body mass per time, e.g. mg kg^{-1} day^{-1}, whereas the slope factor unit is the inverse of this kg day mg^{-1} so risk itself is a unitless probability:

Carbon tetrachloride = $0.053 \times 775 = 41$
Methyl chloride = $0.0035 \times 16\,000 = 56$
Trichloroethylene = $0.0063 \times 7800 = 49$

Thus, in terms of relative risk, methyl chloride is again the most important target chemical, but the other two a much closer. In fact, given the uncertainties and assumptions, from a relative risk perspective, the importance of the removing the three compounds is nearly identical, owing to the much higher cancer potency of CCl_4. However, it is important to keep in mind that numerous compounds are regulated individually (including several VOCs). Thus, an action plan that addresses overall VOC reductions also needs to ensure that individual compounds do not exceed emission limits.

[4] Even without calculating the releases, is probably reasonable to assume that the exposures will be similar since the three compounds have high vapor pressures (more likely to be inhaled):

Carbon tetrachloride = 115 mmHg
Methyl chloride = 4300 mmHg
Trichloroethylene = 69 mmHg

After identifying and characterizing the environmental problems, the next step is to discover useful options. These options fall generally into three categories:

1. Operational changes
2. Materials changes
3. Process modifications

Operational changes might consist simply of better housekeeping; plugging up leaks, eliminating spills, etc. A better schedule for cleaning, and segregating the water might similarly yield large return on a minor investment.

Materials changes often involve the substitution of one chemical for another which is less toxic or requires less hazardous materials for clean-up. The use of trivalent chromium (Cr^{3+}) for chrome plating instead of the much more toxic hexavalent chrome has found favor, as has the use of water-soluble dyes and paints. In some instances, ultraviolet radiation has been substituted for biocides in cooling water, resulting in better quality water and no waste cooling water disposal problems. In one North Carolina textile plant, biocides were used in air washes to control algal growth. Periodic "blow down" and cleaning fluids were discharged to the stream but this discharge proved toxic to the stream and the State of North Carolina revoked the plant's discharge permit. The town would not accept the waste into its sewers, rightly arguing that this may have serious adverse effects on its biological wastewater treatment operations. The industry was about to shut down when it decided to try ultraviolet radiation as a disinfectant in its air wash system. Fortunately, they found that the ultraviolet radiation effectively disinfected the cooling water and that the biocide was no longer needed. This not only eliminated the discharge but it eliminated the use of biocides all together, thus saving the company money. The payback was 1.77 years [12].

Process modifications usually involve the greatest investments, and can result in the most rewards. For example, a countercurrent wash water use instead of a once-through batch operation can significantly reduce the amount of wash water needing treatment, but such a change requires pipes, valves and a new process protocol. In industries where materials are dipped into solutions, such as in metal plating, the use of drag out recovery tanks, an intermediate step, has resulted in the savings of the plating solution and reduction in waste generated.

Pollution prevention has the distinct advantage over stack controls that most of the time the company not only eliminates or greatly reduces the release of hazardous materials, but it also saves money. Such savings are in several forms including of course the direct savings in processing costs as with the ultraviolet disinfection example above. The most obvious costs are those normally documented in company records, such as direct labor, raw materials, energy use, capital equipment, site preparation, tie-ins, employee training, and regulatory recordkeeping (e.g. permits) [13]. In addition, there are other savings, including those resulting from not having to spend time on submitting

TABLE 30.2

Pollution Cost Categories

Cost category	Typical cost components
Usual/normal	Direct labor Raw materials Energy and fuel Capital equipment and supplies Site preparation Tie-ins Training Permits: administrative and scientific
Hidden or direct	Monitoring Permitting fees Environmental transformation Environmental impact analyses and assessments Health and safety assessments Service agreements and contracts Legal Control instrumentation Reporting and recordkeeping Quality assurance planning and oversight
Future liabilities	Environmental cleanup, removal and remedial actions Personal injury Health risks and public insults More stringent compliance requirements Inflation
Less tangible	Consumer reaction and loss of investor confidence Employee relations Lines of credit (establishing and extending) Property values Insurance premiums and insurability Greater regulatory oversight (frequency, intensiveness, onus) Penalties Rapport and leverage with regulators

Source: Adapted from Chreremisinoff, N. P., *Handbook of Solid Waste Management and Waste Minimization Technologies*. Butterworth-Heinemann, Burlington, MA, 2003.

compliance permits and suffering potential fines for noncompliance. Future liabilities weigh heavily where hazardous wastes have to be buried or injected. Additionally, there are the intangible benefits of employee relations and safety (see Table 30.2).

VI. MOTIVATIONS FOR PRACTICING GREEN ENGINEERING

In order to understand the reasons why humans behave as they do, one must identify the driving forces that lead to particular activities [14]. The concept of the driving force can also be used to explain engineering processes. For example, in gas transfer the driving force is the difference in concentrations of a particular gas on either side of an interface. We express the rate of this transfer mathematically as ($dM/dt = k(\Delta C)$) where M is mass, t is time, k is a proportionality constant, and ΔC is the difference in concentrations on either side of the interface. The rate at which the gas moves across the interface is thus directly proportional to the difference in concentrations. If ΔC approaches zero, the rate drops until no net transfer occurs. The driving force is therefore ΔC, the difference in concentrations.

Analogously in engineering, driving forces spur the adoption of new technologies or practices. The objective here is to understand what these motivational forces are for adopting green engineering practices. We that the three diving forces supporting green engineering seem to be legal considerations, financial considerations, and finally ethical considerations.

A. Legal Considerations

At the simplest and most basic level, green engineering is practiced in order to comply with the law. For example, a supermarket recycles corrugated cardboard because it is the law—either a state law such as in North Carolina or a local ordinance as in Pennsylvania. Engineers and managers comply with the law because of the threat of punishment for noncompliance. So in this situation, managers and engineers choose to do "the right thing," not because it is the right thing to do—but simply because they feel it is their only choice.

History has shown that the vast majority of firms will comply with the law regardless of the financial consequences. It will not even bother to conduct a cost-benefit analysis because it assumes breaking the law is not worth the cost.

Occasionally, however, firms may prioritize financial concerns over legal concerns and the managers may determine that by adopting an illegal practice (or failing to adopt a practice codified in law) they can enhance profitability. In such cases they argue that either the chances of getting caught are low, or that the potential for profit is large enough to override the penalty if they do get caught.

For example, in November 1999 the US Environmental Protection Agency sued seven electric utility companies—American Electric Power (AEP), Cinergy, FirstEnergy, Illinois Power, Southern Indiana Gas & Electric Company, Southern Company, Tampa Electric Company—for violating "the Clean Air Act by making major modifications to many of their coal burning plants without installing the equipment required to control smog, acid rain and soot" [15]. On August 7, 2003, "Judge Edmund Sargus of the US District Court for the Southern District of Ohio found that Ohio Edison, an affiliate of FirstEnergy Corp., violated the Clean Air Act's NSR provisions by undertaking 11 construction projects at one of its coal-fired plants from 1984 to 1998 without obtaining necessary air pollution permits and installing modern pollution controls on the facility" [16]. Given the number of violations, it seems obvious that the companies had calculated that breaking the law and possibly getting caught was the least cost solution and thus behaving illegally was "the right answer."

In some cases private firms can take advantage of loopholes in tax laws that inadvertently allow companies to pretend to be environmentally green while in reality doing nothing but gouging the taxpayer. An example of this is the great synfuel scam [17]. In the 1970s, the US Congress decided to promote the use of cleaner fuels in order to take advantage of both the huge coal reserves in the United States and the environmental benefits derived from burning a clean gaseous fossil fuel made from coal. Producing such synfuel from coal had already been successfully implemented in Canada and the United States Government wanted to encourage our power companies to get into the synfuel business. In order to promote this industry Congress wrote in huge tax credits for companies that would produce synfuel and defined a synfuel as chemically altered coal, anticipating that the conversion would be to a combustible gas that could be used much as natural gas is used today.

Unfortunately, the synfuel industry in the United States did not develop as expected because cheaper natural gas supplies became available. The synfuel tax credit idea remained dormant until the 1990s when a number of corporations (including seemingly unrelated businesses like a hotel chain) found the tax break and went into the synfuel business. Since the only requirement was to change the chemical nature of the fuel, it became evident that even spraying the coal with diesel oil or pine tar would alter the fuel chemistry and that this fuel would then be legally classified as a synfuel. The product of these synfuel plants was still coal, and more expensive coal than raw coal at that, but the tax credits were enormous. Companies formed specifically to take advantage of the tax break, often with environmentally attractive names like Earthco, and made huge profits by selling their tax credits to other corporations that needed them. The synfuels industry presently is receiving a gift from the US taxpayer of over $1 billion annually, while doing nothing illegal, but also while doing nothing to benefit the environment.

B. Financial Considerations

Decisions about the adoption of green practices are also driven by financial concerns. This level of involvement with "greening" is at the level promoted by the economist Milton Friedman, who stated famously, "The one and only social responsibility of business [is] to use its resources and engage in activities designed to increase its profits so long as it … engages in open and free competition, without deception or fraud" [18]. In line with this stance, the firm calculates the financial costs and benefits of adopting a particular practice and makes its decision based on whether the benefits outweigh the costs or vice versa.

Many companies seek out green engineering opportunities solely on the basis of their providing a means of lowering expenses, thereby increasing profitability. Here are some examples [19]:

- In one of its facilities at Deepwater, New Jersey, Dupont uses phosgene, an extremely hazardous gas, and used to ship the gas to the plant. In an effort to reduce the chance of accidents, DuPont redesigned the plant to produce phosgene on site and to use almost all of it in the manufacturing process, avoiding and costs associated with hazardous gas transport and disposal.
- Polaroid did a study of all of the materials it used in manufacturing and grouped them into five categories based on risk and toxicity. Managers are encouraged to alter product lines in order to reduce the amount of material in the most toxic groups. In the first five years, the program resulted in a reduction of 37% of the most toxic chemicals, and saved over $19 million in money not spent on waste disposal.
- Dow Chemical challenged its subsidiaries in Louisiana to reduce energy use, and sought ideas on how this should be done. Following up on the best ideas, Dow invested $1.7 million, and received a 173% return on its investment.

Other firms may believe that adopting a particular green engineering technology will provide them with public relations opportunities: green engineering is a useful tool for enhancing the company's reputation and community standing. If the result is likely to be an increase in sales for the business, and if sales are projected to rise *more* than expenses, so that profits rise, the firm is likely to adopt such a technology. The same is true if the public relations opportunities can be exploited to provide the firm with expense reductions, such as decreased enforcement penalties or tax liabilities. Similarly, green technologies that not only yield increased sales, but also at the same time decrease expenses are the perfect recipes for the adoption of green practices by a company whose primary driving forces are financial concerns. For instance:

- Dupont's well-publicized decision to discontinue its $750 million a year business producing CFCs was a public relations bonanza. Not only did DuPont make it politically possible for the United States to become a signatory to the Montreal Protocol on ozone depletion, but it already had alternative refrigerants in the production stage and were able to smoothly transition to these. In 1990, the US Environmental Protection Agency gave DuPont the Stratospheric Protection Award in recognition of their decision to get out of CFC manufacturing [20].
- The seven electric companies sued by the EPA in November 1999 for Clean Air Act violations heavily publicized their efforts to reduce greenhouse gas emissions. For example, AEP issued news releases on May 8, June 11, and November 21, 2002 regarding emissions reduction efforts at various plants.

These examples clearly demonstrate bottom-line thinking: cases in which managers were simply trying to practice "good business," seeking ways to increase the difference between revenues and expenses so that profits would rise. These decisions apparently were not influenced by the desire to "do the right thing" for the environment. Businesses are organized around the idea that they will either make money or cease to survive; in the "financial concerns" illustrations so far provided, green practices were adopted as a means of making more money.

On occasion, though, managers are *forced* into considering the adoption of greener practices by the threat that not doing so will cause expenses to rise and/or revenues to fall. For example, in October 1998, Earth Liberation Front (ELF)) targeted Vail Ski Resort, burning a $12 million expansion project to the ground [21]. In the wake of this damage, the National Ski Areas Association (NSAA) began developing its Environmental Charter in 1999 with "input from stakeholders, including ... environmental groups" [22] and officially adopted the charter in June 2000 [23]. In accordance with the charter, NSAA has produced its Sustainable Slopes Annual Report each year since 2001 [24]. NSAA was spurred to create the Environmental Charter by concerns about member companies' bottom lines: further "ecoterrorist" activity could occur, thereby causing expenses to rise; and the ELF action may have sufficiently highlighted the environmental consequences of resort development to the point that environmentally minded skiers might pause before deciding to patronize resorts where development was occurring, thereby causing revenues to fall.

Similarly, for firms trying to do business in Europe, adopting ISO 14000 (environmental management) is close to a required management practice. The ISO network has penetrated so deeply into business practices that firms are nearly locked out if they do not gain ISO 14000 certification.

There is ample evidence that one of the reasons businesses participate in the quest for sustainability is because it is good for business. The leaders of eight leading firms that adopted an environmentally proactive stance on sustainability were asked in one study to justify the firms' adoption of such a strategy [25]. All companies reported that they were motivated first by regulations such as the control of air emissions, pretreatment of wastewater, and the disposal of hazardous materials. One engineer in the study admitted that: "The [waste disposal] requirements became so onerous that many firms recognized that benefits of altering their production processes to generate less waste."

The second motivator identified in this study was competitive advantage. Lawrence and Morell quote one director of a microprocessor company, who noted that "by reducing pollution, we can cut costs and improve our operating efficiencies." The company recognized the advantage of cutting costs by reducing its hazardous waste stream [26].

Another study, conducted by PriceWaterhouseCoopers, confirmed these findings [27]. When companies were asked to self-report on their stance on sustainable principles, the top two reasons for adopting sustainable development were found to be:

1. enhanced reputation (90%),
2. competitive advantage (cost savings) (75%).

It is not clear if the respondents were given the option of responding that they practiced sustainable operations because this was mandated by law. If it had, there is no doubt that all companies would have publicly stated that they are, indeed, law abiding.

So it seems likely that the two primary driving forces behind the adoption of green business and engineering practices are (1) legal concerns and (2) financial concerns. According to Sethi [28], one can argue instead that actions undertaken in response to legal and financial concerns are actually *obligatory*, in that society essentially demands that businesses make their decisions within legal and financial constraints. For an action to be morally admirable, however, the motivation has to be far different in character.

C. Ethical Considerations

The first indication that some engineers and business leaders are making decisions where the driving force may not be due to legal or financial concerns comes from several cases in American business. Although most business or engineering decisions are made on the basis of legal or financial concerns, some companies believe that behaving more environmentally responsible is simply the right thing to do. They believe that saving resources for the generations that will follow is an important part of their job. When making decisions, they are guided by the "triple bottom line." Their goal is to balance the financial, social, and environmental impacts of each decision.

A prime example of this sort of thinking is the case of Interface Carpet Company [29]. Founded in 1973, its founder and CEO until 2001 was Ray Anderson, now Chairman of the Board. By the mid-1990s, Interface had grown to nearly $1.3 billion in sales, employed some 6600 people, manufactured on four continents, and sold its products in over 100 countries worldwide. In 1994, several members of Interface's research group asked Anderson to give a kick-off speech for a task force meeting on sustainability: they wanted him to provide Interface's environmental vision. Despite his reluctance to do so—Anderson had no "environmental vision" for the company except to comply with the law—he agreed. Fortuitously, as Anderson struggled to determine what to say, someone sent him a copy of Paul Hawken's *The Ecology of Commerce* [30]; Anderson read it, and it completely changed not only his view of the natural environment, not only his vision for Interface Carpet Company, but also his entire conception of business. In the coming years, he held meetings with employees throughout the Interface organization explaining to them his desire to see the company spearhead a sustainability revolution. No longer would they be content to keep pollutant emissions at or below regulatory levels: Instead, they were going to strive to be a company that created zero waste and did not emit any pollutants *at all*. The company began to employ "The Natural Step" [31] and notions of "Natural Capitalism" [32] as part of its efforts to become truly sustainable. The program continues today, and although the company has saved many millions of dollars as a result of adopting green engineering technologies and practices, the reason for adopting these principles was not to earn more money but rather to do the right thing.

Yet another example is that of Herman Miller, an office furniture manufacturing company located in western Michigan. Its pledge in 1993 to stop sending any materials to landfills by 2003 has resulted in the company's adoption of numerous progressive, but sometimes expensive, practices. For example, the company ceased taking scrap fabric to the landfill and began shredding it and trucking it to a firm in North Carolina that processes it into automobile insulation. This environmentally friendly process costs Herman Miller $50 000 each year, but the company leaders agree that a decision that is right for the environment is the right decision. Similarly, the company's new waste-to-energy plant has increased costs, but again company leaders feel it is worth the cost, as employees and managers are proud of the company's leadership in preserving the natural environment in their state [33].

The decisions made by the leadership of Interface Carpet Company and Herman Miller were not morally admirable simply because they enabled these companies to reduce toxic emissions (among many other positive outcomes for the environment); they were morally admirable because the *driving force* behind those decisions was the desire to protect it the environment so that future generations would be able to enjoy it as much as, or even more than, we do today. Conversely, in the cases of Dupont, Polaroid, and Dow Chemical cited earlier, the *driving force* behind their decisions to adopt green technologies was appeared to be a desire to save the company money; the benefits to the Earth were simply a fortunate byproduct of those decisions.

VII. FUTURE PEOPLE

One of the unique characteristics of humans is that we have self-awareness. We can see ourselves in the world today, and we know that humans existed in days gone by, and our species will (we hope) exist tomorrow. We thus are able to plan for the future and accept delayed gratification.

But there will come a time in the future where we individually are long dead and we can no longer personally benefit from any actions we might have taken on our own behalf. Or, for that matter, there will come a time, after our death, when we are not longer burdened by the ill-considered actions that might have led to unhappiness. Why, then worry about the future?

We can, based on empirical evidence, assume that there will be a future, of some sort, and we have some confidence that this future will be inhabited by human beings. It is this future—the future without you and me, that we now address.

While the "client" for engineers is almost always an existing person or organization, the work in which engineers engage can have far-reaching consequences for persons who are not yet even born, or future people. It is easy to argue that engineers have a moral responsibility by virtue of their position in society to existing people, but does this extend as well to these future persons, those as yet unborn, who may or may not even exist?

We believe there are two reasons why the engineer has moral responsibilities to future people:

- Many engineering works, be they small gadgets or huge buildings, will certainly last for more than one generation and will be used by people who were not yet born when the product or facility was constructed.
- Engineers can and do appreciably alter the environment, and the health, safety, and welfare of future people will depend on maintaining a sustainable environment.

Engineers conceive, design, and construct products and facilities that last for generations. Indeed, many engineering decisions have no effects until decades later. For example, suppose engineers choose to dispose of some hazardous waste in steel containers buried underground. It may take generations for waste containers to corrode, for their contents to leach, for the leachate to migrate and pollute groundwater, and for toxic effects to occur in people coming in contact with the water. Such a problem is not of concern for present people since it will be decades before the effects are felt. The only persons to be adversely affected by such an engineering decision are future people, and they are the only ones who have no say in the decision.

Some would argue that we owe no moral obligations to future generations because they do not exist and the alleged obligation has no basis because we do not form a moral community with them. This is a fallacious argument, however. Even if future generations do not yet exist (by definition); we can still have obligations to them. If we agree that we have moral obligations to distant peoples who we do not know, then it would be reasonable to argue that we have similar moral obligations to people who are yet unborn.

Vesilind and Gunn [34] use an analogy to illustrate this point. Consider a terrorist who plants a bomb in a primary school. Plainly the act is wrong, and in breach of a general obligation not to cause (or recklessly risk) harm to fellow citizens. Even though the terrorist may not know the identities of the children we would all agree that this is an evil act. And the same would be true if the terrorist bomb had a very long fuse, say 20 years. This would be equally heinous, even though the children, at the time the bomb was placed, had yet not been born. Some engineering works, such as the hazardous waste disposal alluded to above, have very long fuses, and there is no doubt that future people can be harmed by irresponsible engineering activities. The act of burying wastes in the ground where they will not find their way to drinking water supplies for some decades is no different from the act pouring the wastes down a well, except in terms of time.

The second way that engineers have responsibility to future generations is by consciously working to maintain a sustaining environment. Global warming is one instance where the damage done to date is so severe that the effects will not be felt until many years from now. Most models predict that by building up greenhouse gasses at the present time, the temperature of the earth will be slowly getting warmer even if and when we begin to reduce the emission of such gases. This is analogous to heating a pot of water on an electric stove. The burner is turned on and the water begins to heat. When the burner is turned off the temperature of the water does not drop immediately to room temperature. The burner is still warm and heat continues to be transmitted to the pot and the temperature of the water continues to rise even after the burner is turned off. This effect will also occur with global warming (although rather than heat being buffered, the decrease in the concentrations of greenhouse gases will resist change even after the sources are removed). We therefore may have already exceeded the level of sustainability with regard to the earth's temperature but we will not know about it until decades from now [35].

Some argue that we have no obligation to maintain a quality environment for future generations because we cannot know what kind of an environment they will want. Our sole responsibility to future generations is therefore not to plan for them [36].

We know very well that future generations will *not* want contaminated air or water, dramatically reduced number of species, or global warming. Certainly there will be changes in style and fashion, and future generations will no doubt have different views on many of our present moral issues, but they will want a sustainable environment for themselves and their children. Irreparable global warming, or large-scale radiation, or the destruction of the ozone layer are not, under any circumstances, what our progeny would want. Parents do not know what careers their children will choose when they grow up, whom they will marry, or what their life style will be like, but the parents *do* know that their children will want to be healthy, and thus the parents are morally obligated to provide heath care for their children. Also, future generations most likely will not want to suffer genetic damage or to produce babies with severe birth defects, and thus our obligation to them is to control chemical pollution. The argument that because we do not know the desires of future people our only obligation is to not plan for them is therefore wrong.

The engineers' responsibilities to society are the control and prevention of pollution, and they are therefore entrusted to help maintain a healthy environment. Because this responsibility extends into the future, the "public" in the first canon in the codes of ethics should refer to all people, present and future.

Future, therefore, is the future beyond the careers of present engineers. But unlike some laborers or trades-people, the effect of their work will last long after then are no longer around. Is it important to you, today, to know that what you do will have a positive effect on future people?

The profession and practice of engineering is changing, but we will always be required to have strong analytical skills. The engineer of the future will increasingly need "practical ingenuity," as well as the ability to find new ways of doing things (i.e. creativity) built on a framework of high ethical standards, professionalism, and lifelong learning [37]. These are the qualities of a *good* engineer.

REFERENCES

1. World Commission on Environment and Development, United Nations, *Our Common Future.* Oxford Paperbacks, Oxford, UK, 1987.
2. *American Society of Mechanical Engineers, 2004, Professional Practice Curriculum: Sustainability,* http://www.professionalpractice.asme.org/communications/sustainability/index.htm; accessed on November 2, 2004.
3. This section is based on a paper originally authored by P. A. Vesilind, L. Heine, J. R. Hendry, and S. A. Hamill.
4. The source for this discussion is: Billatos, S. B., and Basaly, N. A., *Green Technology and Design for the Environment.* Taylor & Francis, Bristol, PA, 1997.
5. US Environmental Protection Agency, *What is Green Engineering?* Accessed at: http://www.epa.gov/oppt/greenengineering/whats_ge.html, November 2, 2004.
6. See: Billatos S. B., *Green Technology And Design For The Environment.* Taylor & Francis, Washington, DC, 1997; and Allada, V., Preparing engineering students to meet the ecological challenges through sustainable product design, *Proceedings of the 2000 International Conference on Engineering Education,* Taipei, Taiwan, 2000.
7. Woodruff, T. J., Grillo, J., and Schoendorf, K. C., The relationship between selected causes of postneonatal infant mortality and particulate air pollution in the United States. *Environ. Health Persp.* **105** (6) (June 1997).
8. Gosselin R. E., Smith R. P., and Hodge H. C., *Clinical Toxicology of Commercial Products*, 5th ed. Williams and Wilkins, Baltimore, MD, 1984.
9. Since we brought it up, the SIDS Alliance recommends a number of risk reduction measures that should be taken to protect infants from SIDS:

   - *Place your baby on his or her back to sleep*: The American Academy of Pediatrics recommends that healthy infants sleep on their backs or sides to reduce the risk for SIDS. This is considered to be most important during the first six months of age, when baby's risk of SIDS is greatest.
   - *Stop smoking around the baby*: SIDS is long associated with women who smoke during pregnancy. A new study at Duke University warns against use of nicotine patches during pregnancy as well. Findings from the National Center for Health Statistics now demonstrate that women who quit smoking during pregnancy, but resume after delivery, put their babies at risk for SIDS, too.
   - *Use firm bedding materials*: The US Consumer Product Safety Commission has issued a series of advisories for parents regarding hazards posed to infants sleeping on top of beanbag cushions, sheepskins, sofa cushions, adult pillows, and fluffy comforters. Waterbeds have also been identified as unsafe sleep surfaces for infants. Parents are advised to use a firm, flat mattress in a safety-approved crib for their baby's sleep.

- *Avoid overheating, especially when your baby is ill*: SIDS is associated with the presence of colds and infections, although colds are not more common among babies who die of SIDS than babies in general. Now, research findings indicate that overheating too much clothing, too heavy bedding, and too warm a room may greatly increase the risk of SIDS for a baby who is ill.
- *If possible, breastfeed*: Studies by the National Institute of Health show that babies who died of SIDS were less likely to be breastfed. In fact, a more recent study at the University of California, San Diego found breast milk to be protective against SIDS among non-smokers but not among smokers. Parents should be advised to provide nicotine-free breast milk, if breastfeeding, and to stop smoking around your baby particularly while breastfeeding.
- *Mother and baby need care*: Maintaining good prenatal care and constant communication with your health care professional about changes in your baby's behavior and health are of the utmost importance.

10. Pritchard, M. S., *On Being Responsible*. University Press of Kansas, Lawrence, KS, 1991.
11. Environmental Protection Agency Pollution Prevention Directive, US EPA, May 13, 1990, quoted in *Industrial Pollution Prevention: A Critical Review* by H. Freeman *et al.*, presented at the Air and Waste Management Association Meeting, Kansas City, MO, 1992.
12. Richardson, S., Pollution prevention in textile wet processing: an approach and case studies, *Proceedings Environmental Challenges of the 1990s*, US EPA 66/9-90/039, September 1990
13. Chreremisinoff, N. P., *Handbook of Solid Waste Management and Waste Minimization Technologies*. Butterworth-Heinemann, Burlington, MA, 2003.
14. Much of this discussion appeared in an earlier paper, *Ethics of Green Engineering* by P. A. Vesilind, L. Heine, J. R. Herndry, and S. A. Hamill.
15. Lazaroff, C., *US Government Sues Power Plants to Clear Dirty Air*, Environment News Service, http://ens.lycos.com/ens/nov99/1999L-11-03-06.html, 1999.
16. Fowler, D., *Bush Administration, Environmentalists Battle Over 'New Source Review' Air Rules*. Group Against Smog and Pollution Hotline, Fall 2003, http://www.gasp-pgh.org/hotline/fall03_4.html; accessed on January 3, 2004.
17. Barlett, D. L., and Steele, J. B.,The great energy scam. *Time* **162** (15), 60–70 (October 13, 2003).
18. Friedman, M. *Capitalism and Freedom*, p. 133. University of Chicago Press, Chicago, IL, 1962.
19. Gibney, K.,Sustainable development: a new way of doing business, *Prism*. American Society of Engineering Education, Washington, DC, January 2003.
20. Billatos, S. B. and Basaly, N. A., *Green Technology and Design for the Environment*. Taylor & Francis, Washington, DC, 1997.
21. Faust, J.,*Earth Liberation Who?* ABCNEWS.com website: http://more.abcnews.go.com/sections/us/DailyNews/elf981022.html; retrieved on January 3, 2004.
22. National Ski Areas Association (NSAA), Environmental Charter, http://www.nsaa.org/nsaa2002/_environmental_charter.asp, 2002.
23. Jesitus, J., Charter promotes environmental responsibility in ski areas. *Hotel Motel Manage.* **215** (19), 64 (2000).
24. National Ski Areas Association (NSAA), *Sustainable Slopes Annual Report*, http://www.nsaa.org/nsaa2002/_environmental_charter.asp?mode=s, 2002.
25. Lawrence, A. T. and Morell, D., Leading-edge environmental management: motivation, opportunity, resources and processes in *Research in Corporate Social Performance and Policy* (Post, J. E., Collins, D., and Starik, M., eds.), Suppl. 1. JAI Press, Greenwich, CT, 1995.
26. Ibid.
27. PriceWaterhouseCooper, *Sustainability Survey Report*. New York, August, 2002.
28. Sethi, S. P., Dimensions of corporate social performance: an analytical framework. *California Manage. Rev.* **17** (3), Spring, 58–64 (1975).

29. Anderson, R. C., *Mid-Course Correction—Toward a Sustainable Enterprise: The Interface Model.* The Peregrinzilla Press, Atlanta, GA, 1998.
30. Hawken, P., *The Ecology of Commerce: A Declaration of Sustainability.* Harper Business, New York, 1994.
31. See Robért, K.-H., Educating a nation: The natural step, *In Context,* No. 28, Spring 1991. According to their website (http://www.naturalstep.org/): Since 1988, The Natural Step has worked to accelerate global sustainability by guiding companies, communities and governments onto an ecologically, socially and economically sustainable path. More than 70 people in twelve countries work with an international network of sustainability experts, scientists, universities, and businesses to create solutions, innovative models and tools that will lead the transition to a sustainable future.
32. Lovins, A. B., Lovins, L. H., and Hawken, P. A road map for natural capitalism. *Harvard Business Rev.* **7** (3), 145—158 (1999).
33. An earlier version of this discussion appeared as *Ethical Motivations for Green Business and Engineering*" by P. A. Vesilind and J. R. Hendry. *Clean Technol. Environ. Policy.*
34. Vesilind, P. A. and Gunn, A. S., *Engineering, Ethics, and the Environment*, p. 39. Cambridge University Press, New York, 1998.
35. This is one of the hottest topics of debate in environmental circles. It is not our purpose here to take one side or another, but *if* the predictions are correct, the most dramatic and harmful effects would be in the coming decades. Our point is that, even with the gaps in knowledge, these effects would only be avoided by prudent decisions now. After the warming has reached severity, several decades of recovery may well be needed to reach a new chemical and energy atmospheric equilibrium.
36. Golding, M., Obligations to future generations. *Monist* **56**, 85–99 (1972).
37. National Academy of Engineering, *The Engineer of 2020: Visions of Engineering in the New Century*. The National Academies Press, Washington, DC, 2004.

SUGGESTED READING

Allen, D. T., and D. R. Shonnard, *Green Engineering: Environmentally. Conscious Design of Chemical Processes, Prentice Hall*, Englewood, Cliffs, NJ, 2001.

Becktold, R., *Alternative Fuels Guidebook: Properties, Storage, Dispensing and Vehicle Facility Modifications*. Society of Automotive Engineers, Warrendale, PA, 1997.

Flannigan, D. J., and Suslick, K. S., Plasma formation and temperature measurement during single-bubble cavitation. *Nature* **434**, 52–55 (2005).

Harrison, K. W., Dumas, R. D., Nishtala, S. R. and Morton, A. Barlaz, 2000, "A Life-Cycle Inventory Model of Municipal Solid Waste Combustion," *Journal of Air and Waste Management Association*, **50**, 993–1003.

US Environmental Protection Agency, 2005, Green Chemistry: http://www.epa.gov/greenchemistry/principles.html; accessed on April 12, 2005.

QUESTIONS

1. Consider the black box balance example. Calculate the amount of vapors escaping from 500 barrels of carbon tetrachloride (85%), with 10 barrels remaining; 1000 barrels of methyl chloride (30%) with 100 barrels remaining; and, 2000 barrels of TCE (85%) with 200 barrels

remaining. Which of the three VOCs presents the greatest risks? What are the relative risks of from the three compounds?

2. Consider a facility in your hometown. What three steps could be taken in the life cycle to improve air pollution emissions?
3. Give an example of a company in your home state that has turned an environmental problem into a profit. What were the major drivers? What were the obstacles that had to be overcome?

31

Engineering Control Concepts

I. INTRODUCTION

The application of control technology to air pollution problems assumes that a source can be reduced to a predetermined level to meet a regulation or some other target value. Control technology cannot be applied to an uncontrollable source, such as a volcano, nor can it be expected to control a source completely to reduce emissions to zero. The cost of controlling any given air pollution source is usually an exponential function of the percentage of control and therefore becomes an important consideration in the level of control required [1]. Figure 31.1 shows a typical cost curve for control equipment.

If the material recovered has some economic value, the picture is different. Figure 31.2 shows the previous cost of control with the value recovered curve superimposed on it. The plant manager looking at such a curve would want to be operating in the area to the left of the intersection of the two curves, whereas the local air pollution forces would insist on operation as far to the right of the graph as the best available control technology would allow.

Control of any air pollution source requires a complete knowledge of the contaminant and the source. The engineers controlling the source must be thoroughly familiar with all available physical and chemical data on the effluent

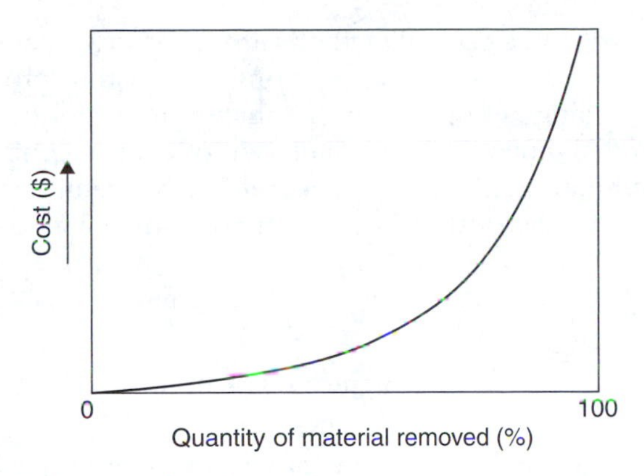

Fig. 31.1. Air pollution control equipment cost.

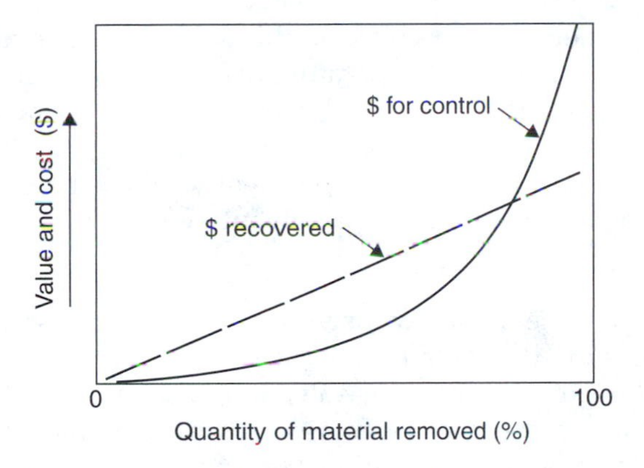

Fig. 31.2. Control equipment cost with value recovered.

from the source. They must know the rules and regulations of the control agencies involved, including not only the Air Pollution Control Agency but also any agencies, which may have jurisdiction over the construction, operation, and final disposal of the waste from the source [2].

In many cases, heating or cooling of the gaseous effluent will be required before it enters the control device. The engineer must be thoroughly aware of the gas laws, thermodynamic properties, and reactions involved to secure a satisfactory design. For example, if a gas is cooled there will be condensation when the temperature drops below the dewpoint. If water is sprayed into the hot gas for cooling, it adds greatly to the specific volume of the mixture. As the gases pass through hoods, ducts, fans, and control equipment, temperatures and pressures change and hence, also, specific volumes and velocities [3].

The control of atmospheric emissions from a process will generally take one of three forms depending on the process, fuel, types, availability of control equipment, etc. The three general methods are (1) process change to a less polluting process or to lowered emission from the existing process through a modification or change in operation, (2) change to a fuel that will give the desired level of emissions, and (3) installation of control equipment between the point of pollutant generation and its release to the atmosphere. Control may consist of either removal of the pollutant or conversion to a less polluting form [3].

II. PROCESS CHANGE

A process change can be either a change in operating procedures for an existing process or the substitution of a completely different process. In recent years, this has been labeled "pollution prevention." Consider a plant manager who for years has been using solvent A for a degreasing operation. By past experimentation, it has been found that with the conveyor speed at 100 units per hour, with a solvent temperature of 80°C, one gets maximum cleaning with solvent a loss that results in the lowest overall operating cost for the process.

A new regulation is passed requiring greatly reduced atmospheric emissions of organic solvents, including solvent A. The manager has several alternatives:

1. Change to another more expensive solvent, which by virtue of its lower vapor pressure would emit less organic matter.
2. Reduce the temperature of the solvent and slow down the conveyor to get the same amount of cleaning. This may require the addition of another line or another 8 h shift.
3. Put in the necessary hooding, ducting, and equipment for a solvent recovery system which will decrease the atmospheric pollution and also result in some economic solvent recovery.
4. Put in the necessary hooding, ducting, and equipment for an afterburner system which will burn the organic solvent vapors to a less polluting emission, but with no solvent recovery.

In some cases, the least expensive control is achieved by abandoning the old process and replacing it with a new, less polluting one. Any increased production and/or recovery of material may help offset a portion of the cost. It has proved to be cheaper to abandon old steel mills and to replace them with completely new furnaces of a different type than to modify the old systems to meet pollution regulations. Kraft pulp mills found that the least costly method of meeting stringent regulations was to replace the old, high-emission recovery furnaces with a new furnace of completely different design. The kraft mills have generally asked for, and received, additional plant capacity to offset partially the cost of the new furnace type.

III. FUEL CHANGE

In the past, for many air pollution control situations, a change to a less polluting fuel offered the ideal solution to the problem. If a power plant was emitting large quantities of SO_2 and fly ash, conversion to natural gas was cheaper than installing the necessary control equipment to reduce the pollutant emissions to the permitted values. If the drier at an asphalt plant was emitting 350 mg of particulate matter per standard cubic meter of effluent when fired with heavy oil of 4% ash, it was probable that a switch to either oil of a lower ash content or natural gas would allow the operation to meet an emission standard of 250 mg per standard cubic meter.

Fuel switching based on meteorological or air pollution forecasts was, in the past, a common practice to reduce the air pollution burden at critical times. Some control agencies allowed power plants to operate on residual oil during certain periods of the year when pollution potential was low. Some large utilities for years have followed a policy of switching from their regular coal to a more expensive but lower-sulfur coal when stagnation conditions were forecast.

Caution should be exercised when considering any change in fuels to reduce emissions. This is particularly true considering today's fuel costs. Specific considerations might be the following:

1. What are current and potential fuel supplies? In many areas natural gas is already in short supply. It may not be possible to convert a large plant with current allocations or pipeline capacity.
2. Most large boilers use a separate fuel for auxiliary or standby purposes. One actual example was a boiler fired with wood residue as the primary fuel and residual oil as the standby. A change was made to natural gas as the primary fuel, with residual oil kept for standby. This change was made to lower particulate emissions and to achieve a predicted slightly lower cost. Because of gas shortages, the plant now operates on residual oil during most of the cold season, and the resulting particulate emission greatly exceeds that of the previously burned wood fuel. In addition, an SO_2 emission problem exists with the oil fuel that never occurred with the wood residue. Overall costs have not been lowered because natural gas rates have increased since the conversion.
3. Charts or tables listing supplies or reserves of low-sulfur fuel may not tell the entire story. For example, a large percentage of low-sulfur coal is owned by steel companies and is therefore not generally available for use in power generating stations even though it is listed in tables published by various agencies.
4. Strong competition exists for low-pollution fuels. While one area may be drawing up regulations to require use of natural gas or low-sulfur fuels, it is probable that other neighboring areas are doing the same. Although there may have been sufficient premium fuel for one or two

Fig. 31.3. Trojan nuclear power plant. *Source*: Portland General Electric Company.

areas, if the entire region changes, not enough exists. Such a situation has resulted in extreme fuel shortages during cold spells in some large cities. The supply of low-sulfur fuels has been exhausted during period of extensive use.

Nuclear reactors (Fig. 31.3) used for power generation has been questioned from several environmental points of view. These appear to be relatively pollution free compared to the more familiar fossil fuel-fired plant, which emits carbon monoxide and carbon dioxide, oxides of nitrogen and sulfur, hydrocarbons, and fly ash. However, waste and spent-fuel disposal problems may offset the apparent advantages. These problems (along with steam generator leaks) caused the plant shown in Fig. 31.3 to close permanently in 1993.

IV. POLLUTION REMOVAL

In many situations, sufficient control over emissions cannot be obtained by fuel or process change. In cases such as these, the levels of the pollutants of concern in the exhaust gases or process stream must be reduced to allowable values before they are released to the atmosphere.

The equipment for the pollutant removal system includes all hoods, ducting, controls, fans, and disposal or recovery systems that might be necessary. The entire system should be engineered as a unit for maximum efficiency and economy. Many systems operate at less than maximum efficiency because a portion of the system was designed or adapted without consideration of the other portions [4].

Efficiency of the control equipment is normally specified before the equipment is purchased. If a plant is emitting a pollutant at $500\,\mathrm{kg\,h^{-1}}$ and the regulations allow an emission of only $25\,\mathrm{kg\,h^{-1}}$, it is obvious that at least 95% efficiency is required of the pollution control system. This situation requires the regulation to state "at least 95% removal on a weight basis." The regulation should further specify how the test would be made to determine the efficiency. Figure 31.4 shows the situation as it exists.

The efficiency for the device shown in Fig. 31.4 may be calculated in several ways:

$$\text{Efficiency, \%} = 100\left(\frac{C}{A}\right), \quad \text{but since } A = B + C \tag{31.1}$$

$$\text{Efficiency, \%} = 100\left(\frac{C}{B+C}\right) \quad \text{or} \quad 100\left(\frac{A-B}{A}\right) \quad \text{or} \quad 100\left(\frac{A-B}{B+C}\right) \tag{31.2}$$

The final acceptance test would probably be made by measuring two of the three quantities and using the appropriate equation. For a completely valid efficiency test the effect of hold-up (D) and loss (E) must also be taken into account.

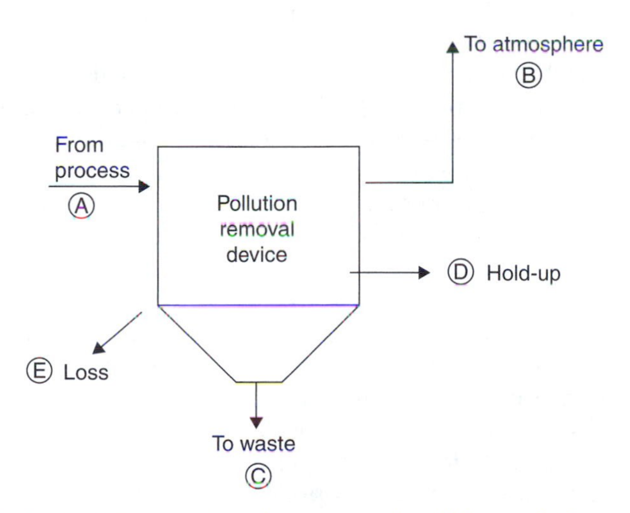

Fig. 31.4. Typical pollution control device as shown for efficiency calculations.

To remove a pollutant from the carrying stream, some property of the pollutant that is different from the carrier must be exploited. The pollutant may have different size, inertia, electrical, or absorption properties. Removal requires that the equipment be designed to apply the scientific principles necessary to perform the separation.

V. DISPOSAL OF POLLUTANTS

If a pollutant is removed from the carrying gas stream, disposal of the collected material becomes of vital concern. If the collected material is truly inert, it may be disposed of in a sanitary landfill. If it is at the other end of the scale, it is probably considered as a toxic waste and strict laws governing its disposal apply. Disposal of hazardous wastes is regulated by governmental agencies [7–13].

In the United States, the Resource Conservation and Recovery Act (RCRA) of 1976 is the major legislation covering the disposal of solid and hazardous wastes [2]. This act provides a multifaceted approach to solving the problems associated with the generation of approximately 5 billion metric tons of solid waste each year in the United States. It places particular emphasis on the regulation of hazardous wastes. This law established the Office of Solid Waste within the US Environmental Protection Agency and directed the agency to publish hazardous waste characteristics and criteria.

If a waste is designated as hazardous under the RCRA, regulations are applied to generators, transporters, and those who treat, store, or dispose of that waste. Regulations regarding hazardous wastes are enforced by the federal government, while the individual states are responsible for enforcing the provisions of the RCRA which apply to nonhazardous wastes. The act also provides for research, development, and demonstration grants for waste disposal.

The US Environmental Protection Agency, Office of Solid Waste Management Programs, defines hazardous waste as "wastes or combinations of wastes which pose a substantial present or potential hazard to human health or living organisms because they cause, or tend to cause, detrimental cumulative effects."

Hazardous wastes can be categorized in a way that shows their potential or immediate effect. A common system for categorizing substances is (1) toxic substances (acute or chronic damage to living systems), (2) flammable, (3) explosive or highly reactive, (4) irritating and/or sensitizing, (5) corrosive (strong oxidizing agents), (6) radioactive, (7) bioaccumulative and/or biomagnified substances (with toxic effects), and (8) genetically interactive substances (mutagenic, teratogenic, and carcinogenic). It is possible for a substance to be placed in any number of these categories, but placement in only one category is sufficient for it to be considered hazardous [7–13].

Table 31.1 indicates the four main types of hazardous material, with examples of substances of each type. Not presented in Table 31.1 are radioactive materials, which are considered as a separate type of hazardous waste [5].

TABLE 31.1

Hazardous Material Types

Miscellaneous inorganics	Halogens and interhalogens	Miscellaneous organics	Organic halogen compounds
Metals	Bromine pentafluoride	Acrolein	Aldrin
Antimony	Chlorine	Dinitrophenol	Chlordane
Bismuth	Chlorine pentafluoride	Tetrazene	1,1-dichloro-2,2-bis (*p*-chlorophenyl) ethane (DDD) Dichloro-diphenyl-trichloroethane (DDT)
Cadmium	Chlorine trifluoride	Nitroglycerine	Dieldrin
Chromium	Fluorine	Nitroaniline	Endrin
Cobalt	Perchloryl fluoride	Chloroacetophenone (CN tear gas)	Potassium cyanide
Copper			Heptachlor
Lead			Lindane
Mercury			Parathion
Nickel			Methyl bromide
Selenium			Polychlorinated biphenyls (PCBs)
Silver			
Tellurium			
Thallium			
Tin			
Zinc			
Nonmetallics			
Cyanide (ion)			
Hydrazine			
Fluorides			
Phosgene			

Table 31.2 lists some of the currently used pretreatments and ultimate disposal methods for hazardous wastes [6]. *Pretreatment* refers almost entirely to thickening or dewatering processes for liquids or sludges. This process not only reduces the volume of the waste but also allows easier handling and transport.

The general purpose of ultimate disposal of hazardous wastes is to prevent the contamination of susceptible environments. Surface water runoff, ground water leaching, atmospheric volatilization, and biological accumulation are processes that should be avoided during the active life of the hazardous waste. As a rule, the more persistent a hazardous waste is (i.e. the greater its resistance to breakdown), the greater the need to isolate it from the environment. If the substance cannot be neutralized by chemical treatment or incineration and still maintains its hazardous qualities, the only alternative is usually to immobilize and bury it in a secure chemical burial site.

TABLE 31.2

Ultimate Waste Disposal Methods

Process	Purpose	Wastes	Problems (remarks)
Cementation and vitrification	Fixation Immobilization Solidification	Sludges Liquids	Expensive
Centrifugation	Dewatering Consolidation	Sludges Liquids	
Filtration	Dewatering Volume reduction	Sludges Liquids	Expensive
Thickening (various methods)	Dewatering Volume reduction	Sludges Liquids	
Chemical addition (polyelectrolytes)	Precipitation Fixation Coagulation	Sludges Liquids	Can be used in conjunction with other processes
Submerged combustion	Dewatering	Liquids	Acceptable for aqueous organics
		Major Ultimate Disposal Methods	
Deep well injection	Partial removal from biosphere	Oil field brines; low toxicity, low-persistence wastes; refinery wastes	Monitoring difficulty Need for special geological formations
	Storage		Ground water contamination
Incineration	Volume reduction	Most organics	If poor process control, unwanted emissions produced
	Toxicity destruction		Can produce NO_x, SO_x, halo acids

Recovery	Reuse	Metals Solvents	Sometimes energy prohibitive
Landfill	Storage	Inert to radioactive	Volatilization
		Major Waste Disposal Methods	
Land application	Isolation		Leaching to ground water
Land burial	Dispersal		Access to biota
Ocean disposal	Dispersal Dilution Neutralization Isolation (?)	Acids, bases Explosives Chemical war agents Radioactive wastes	Contact with ocean ecosystem Containers unstable
		Minor Disposal Methods	
Biological degradation	Reduction of concentration Oxidation	Biodegradable organics	Most hazardous wastes do not now qualify
Chemical degradation (chlorination)	Conversion Oxidation	Some persistent pesticides	
Electrolytic processes	Oxidation	Organics	
Long-term sealed storage	Isolation Storage	Radioactive	How good are containers?
Salt deposit disposal	Isolation	Radioactive	Are salt deposits stable in terms of waste lifetimes?

REFERENCES

1. Stern, A. C. (ed.), *Air Pollution*, 3rd ed., Vol. 4. Academic Press, New York, 1977.
2. Arbuckle, J. G., Frick, G. W., Miller, M. L., Sullivan, T. F. P., and Vanderver, T. A., *Environmental Law Handbook*, 6th ed. Governmental Institutes, Inc., Washington, DC, 1979.
3. Strauss, W., *Industrial Gas Cleaning*, 2nd ed. Pergamon, Oxford, 1975.
4. Bvonicore, A. J., and Davis, W. T. (eds.), *Air Pollution Engineering Manual*. Van Nostrand Reinhold, New York, 1992.
5. Schieler, P., *Hazardous Materials*. Reinhold, New York, 1976.
6. Powers, P. W., *How to Dispose of Toxic Substances and Industrial Wastes*. Noyes Data Corp., Park Ridge, NJ, 1976.
7. Freeman, H., Harten, T., Springer, J., Randall, P., Curran, M. A., and Stone, K., Industrial pollution prevention: a critical review. *J. Air Waste Manage. Assoc.* **42**, 618 (1992).
8. US Environmental Protection Agency, *Pollution Prevention Directive*. US EPA, May 13, 1990.
9. Freeman, H., Hazardous waste minimization: a strategy for environmental improvement. *J. Air Waste Manage. Assoc.* **38**, 59 (1988).
10. Pollution Prevention Act of 1990, 42nd US Congress, 13101.
11. Pollution Prevention News, US EPA Office of Pollution Prevention and Toxics, June 1992.
12. *Pollution Prevention 1991: Progress on Reducing Industrial Pollutants*. US EPA, 21P-3003, October 1991.
13. Bringer, R. P., and Benforado, D. M., 3P plus: total quality environmental management, *Proceedings of the 85th Annual Meeting, Air & Waste Management Association*, Kansas City, June 1992.

SUGGESTED READING

Estimation of Permissible Concentrations of Pollutants for Continuous Exposure, Report 600/2-76-155. US Environmental Protection Agency, Research Triangle Park, NC, 1976.

Fischhoff, B., Slovic, P., and Lichtenstein, S., *Environment* **21** (4) (1979).

Henstock, M. E., and Biddulph, M. W. (eds.), *Solid Waste as a Resource*. Pergamon, Oxford, 1978.

Lowrance, W. W., *Of Acceptable Risk*. Kaufmann, Los Altos, CA, 1976.

QUESTIONS

1. For a given process at a plant, the cost of control can be related to the equation: dollars for control = 10 000 + $10e^x$, where x = percent of control/10. The material collected can be recovered and sold and the income determined from the equation: dollars recovered = (1000) (percent of control):
   (a) At what level of control will the control equipment just pay for itself?
   (b) At what level of control will the dollars recovered per dollars of control equipment be the maximum?
   (c) What would be the net cost to the process for increased control from 97.0% to 99.5%?
2. Give three examples of conversion of a pollutant to a less polluting form or substance.

3. List the advantages and disadvantages of a municipal sanitary landfill and a municipal incinerator.
4. List the advantages and disadvantages of recovering energy, in the form of steam, from a municipal incinerator.
5. Show by means of a flow diagram or sketch how you would treat and dispose of the fly ash collected from a municipal incinerator. The fly ash contains toxic and nontoxic metals, non-metallic inorganics, and organic halogen compounds.

32

Control Devices, Technologies, and Systems

I. INTRODUCTION

One of the methods of controlling air pollution mentioned in the previous chapter is pollution removal. For pollution removal to be accomplished, the polluted carrier gas must pass through a control device or system, which collects or destroys the pollutant and releases the cleaned carrier gas to the atmosphere. The control device or system selected must be specific for the pollutant of concern. If the pollutant is an aerosol, the device used will, in most cases, be different from the one used for a gaseous pollutant. If the aerosol is a dry solid, a different device must be used than for liquid droplets.

Not only the pollutant itself but also the carrier gas, the emitting process, and the operational variables of the process affect the selection of the control system. Table 32.1 illustrates the large number of variables which must be considered in controlling pollution from a source [1–4].

Once the control system is in place, its operation and maintenance become a major concern. Important reasons for an operation and maintenance (O&M) program [2] are (1) the necessity of continuously meeting emission regulations, (2) prolonging control equipment life, (3) maintaining productivity of the process served by the control device, (4) reducing operation costs, (5) promoting

TABLE 32.1

Key Characteristics of Pollution Control Devices and/or Systems

Factor considered	Characteristic of concern
General	Collection efficiency
	Legal limitations such as best available technology
	Initial cost
	Lifetime and salvage value
	Operation and maintenance costs
	Power requirement
	Space requirements and weight
	Materials of construction
	Reliability
	Reputation of manufacturer and guarantees
	Ultimate disposal/use of pollutants
Carrier gas	Temperature
	Pressure
	Humidity
	Density
	Viscosity
	Dewpoint of all condensibles
	Corrosiveness
	Inflammability
	Toxicity
Process	Gas flow rate and velocity
	Pollutant concentration
	Variability of gas and pollutant flow rates, temperature, etc.
	Allowable pressure drop
Pollutant (if gaseous)	Corrosiveness
	Inflammability
	Toxicity
	Reactivity
Pollutant (if particulate)	Size range and distribution
	Particle shape
	Agglomeration tendencies
	Corrosiveness
	Abrasiveness
	Hygroscopic tendencies
	Stickiness
	Inflammability
	Toxicity
	Electrical resistivity
	Reactivity

better public relations and avoiding community alienation, and (6) promoting better relations with regulatory officials.

The O&M program has the following minimum requirements: (1) an equipment and record system with equipment information, warranties, instruction manuals, etc.; (2) lubrication and cleaning schedules; (3) planning and scheduling of preventive maintenance; (4) a storeroom and inventory system for spare parts and supplies; (5) listing of maintenance personnel; (6) costs and budgets for O&M; and (7) storage of special tools and equipment.

A. The Vacuum Cleaner: Particulate Matter Control Device

Controlling particulate matter (PM) emissions tracks fairly closely the homeowner demands for devices to remove dust, so let us consider the ubiquitous vacuum cleaner. We can learn much about this technology by visiting the local department store. Amelia has asked us to pick out the optimal vacuum cleaner. Please join me on the shopping trip.[1]

First, she has specified a cyclone system, not a bag system. All vacuum cleaners work on the principle of pressure differential. The laws of potentiality state that flow moves from high to low pressure. So, if pressure can decrease significantly below that of the atmosphere, air will move to that pressure trough. If there is a big pressure difference between air outside and inside, the flow will be quite rapid. So, the "vacuum" (it is really a pressure differential) is created inside the vacuum cleaner using an electric pump. When the air rushes to the low pressure region it carries particles with it. Increasing velocity is proportional to increasing mass and numbers of particles that can be carried. This is the same principle as the "competence" of a stream, which is high (i.e. can carry heavier loads) in a flowing river, but the competence declines rapidly at the delta where stream velocity approaches zero. This causes the river to drop its particles in descending mass, i.e. sedimentation of heavier particles first, but colloidal matter remaining suspended for much longer times.

Two of the most common types of particle collection systems in industry are cyclones and bag systems. The vacuum cleaner is a microcosm of both technologies. The vacuum cleaner, the cyclone, and the bag (fabric filter) are all designed to remove particles. In the US, the Clean Air Act established the national ambient air quality standards (NAAQS) for PM in 1971. These standards were first directed at total suspended particulates (TSP) as measured by a high-volume sampler, i.e. a device that collected a large range of sizes of particles (aerodynamic diameters up to 50 μm). Particles aggravate bronchitis, asthma, and other respiratory diseases. Certain subpopulations of people are sensitive to PM effects, including those with asthma, cardiovascular or lung disease, as well as children and elderly people. Particles can also damage structures, harm vegetation, and reduce visibility. In 1987, the Environmental Protection Agency (EPA) changed the indicator for PM from TSP to PM_{10}, i.e. particle matter $\leq 10\,\mu m$ diameter.[2] The NAAQS for PM_{10}

[1] This is for the most part a true story. In fact, Amelia is my daughter.

[2] The diameter most often used for airborne particle measurements is the "aerodynamic diameter." The aerodynamic diameter (D_{pa}) for all particles greater than 0.5 μm can be approximated as the product of the Stokes particle diameter (D_{ps}) and the square root of the particle density (ρ_p):

$$D_{pa} = D_{ps}\sqrt{\rho_p} \tag{32.12}$$

(continued)

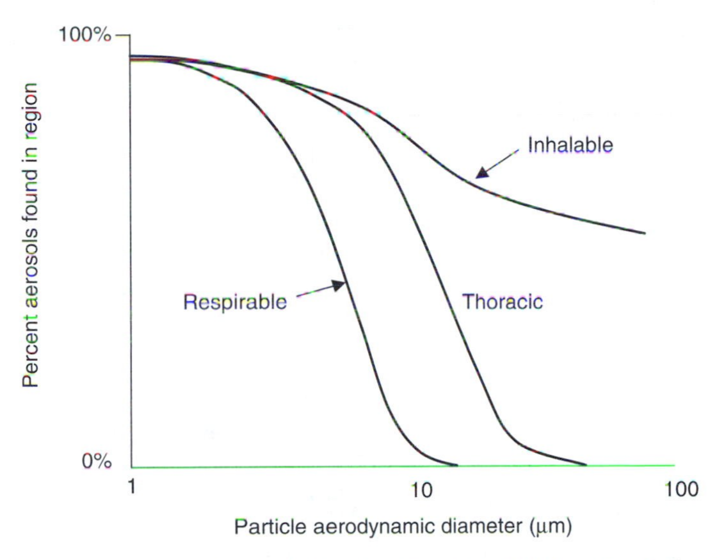

Fig. 32.1. Three regions of the respiratory system where particle matter is deposited. The inhalable fraction remains in the mouth and head area. The thoracic fraction is the mass that penetrates the airways of the respiratory system, while the smallest fraction, i.e. the respirable particulates are those that can infiltrate most deeply into the alveolar region.

was a 24-h average of 150 μg m^{-3} (not to exceed this level more than once per year), and an annual average of 50 μg m^{-3} arithmetic mean. However, even this change did not provide sufficient protection for people breathing PM contaminated air, since most of the particles that penetrate deeply into the air–blood exchange regions of the lung are quite small (see Fig. 32.1). So, in 1997, the US EPA added a new fine particle (diameters ⩽2.5) known as $PM_{2.5}$.[3]

In a cyclone (Fig. 32.2), air is rapidly circulated causing suspended particles to change directions.

Due to their inertia, the particles continue in their original direction and leave the air stream (see Fig. 32.3). This works well for larger particles because of their relatively large masses (and greater inertia), but very fine particles are more likely to remain in the air stream and stay suspended. The dusty air is

If the units of the diameters are in μm, the units of density are g cm^{-3}.

The Stokes diameter D_{ps} is the diameter of a sphere with the same density and settling velocity as the particle. The Stokes diameter is derived from the aerodynamic drag force caused by the difference in velocity of the particle and the surrounding fluid. Thus, for smooth, spherical particles, the Stokes diameter is identical to the physical or actual diameter.

Aerosol textbooks provide methods to determine the aerodynamic diameter of particles less than 0.5 μm. For larger particles gravitational settling is more important and the aerodynamic diameter is often used.

[3] For information regarding particle matter (PM) health effects and inhalable, thoracic and respirable PM mass fractions see US Environmental Protection Agency, 1996, Air Quality Criteria for Particulate Matter. Technical Report No. EPA/600/P-95/001aF, Washington, DC.

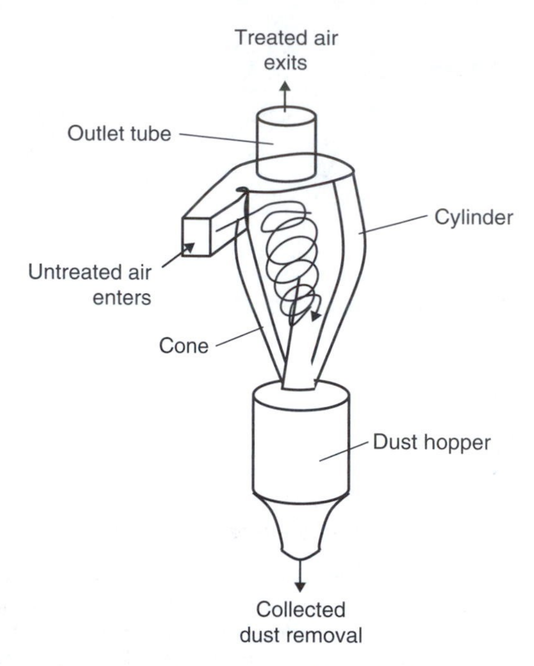

Fig. 32.2. Schematic of a simple cyclone separator. *Source*: US Environmental Protection Agency, 2004, Air Pollution Control Orientation Course, http://www.epa.gov/air/oaqps/eog/course422/ce6.html; accessed September 20, 2004.

introduced in the cyclone from the top through the inlet pipe tangential to the cylindrical portion of the cyclone. The air whirls downward to form a peripheral vortex, which creates centrifugal forces. As a result individual particles are hurled toward the cyclone wall and, after impact, fall downward where they are collected in a hopper. When the air reaches the end of the conical segment, it will change direction and move upward toward the outlet. This forms an inner vortex. The upward airflow against gravitation allows for additional separation of particles. The cyclone vacuum cleaner applies the same inertial principles, with the collected dust hitting the sides of the removable cyclone separator and falling to its bottom.

The other technology of vacuum cleaners is the bag. Actually, the engineering term for a "bag" is a fabric filter. Filtration is an important technology in every aspect of environmental engineering (i.e. air pollution, waste water treatment, drinking water, and even hazardous waste and sediment cleanup). Basically, filtration consists of four mechanical processes: (1) diffusion, (2) interception, (3) inertial impaction, and (4) electrostatics (see Fig. 32.4). Diffusion is important only for very small particles ($\leq 0.1\,\mu m$ diameter) because the Brownian motion allows them to move in a "random walk" away from the air stream. Interception works mainly for particles with diameters between 0.1 and $1\,\mu m$. The particle does not leave the air stream but comes into contact with the filter medium (e.g. a strand of fiberglass or fabric fiber). Inertial impaction, as explained in the

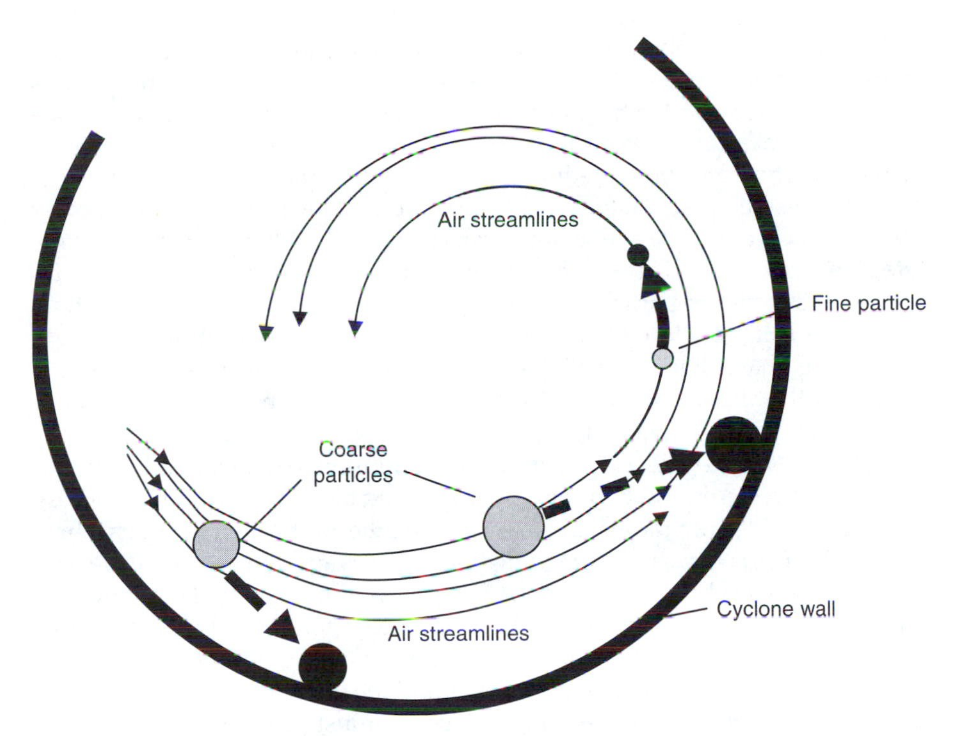

Fig. 32.3. Inertial forces in a cyclone separator. Coarse (heavier) particles' inertial forces are large enough that they leave the air stream of the vortex and collide with the cyclone wall. Fine particles have smaller mass so that their inertial forces cannot overcome the air flow, so they remain suspended in the air stream.

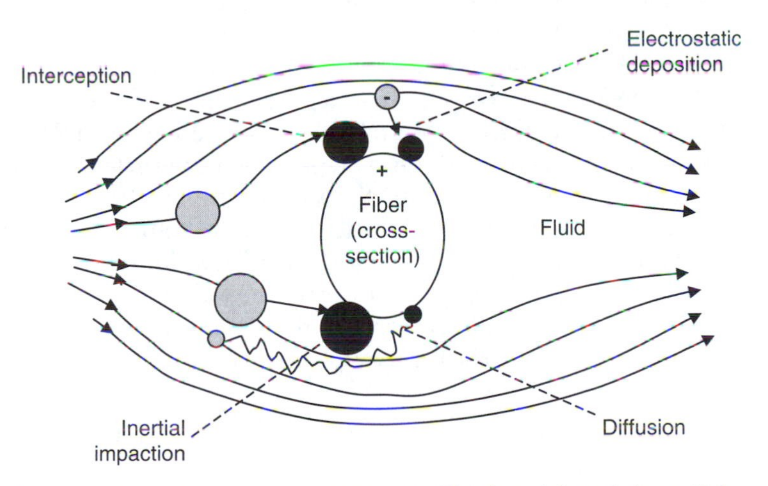

Fig. 32.4. Mechanical processes important to filtration. Adapted from: Rubow, K. L., Filtration: fundamentals and applications, in *Aerosol and Particle Measurement Short Course*. University of Minnesota, Minneapolis, MN, August 16–18, 2004.

cyclone discussion, collects particles sufficiently large to leave the air stream by inertia (diameters $\geqslant 1\,\mu m$). Electrostatics consist of electrical interactions between the atoms in the filter and those in the particle at the point of contact (Van der Waal's force), as well as electrostatic attraction (charge differences between particle and filter medium). These are the processes at work in large scale electrostatic precipitators (ESPs) that are employed in coal-fired power plant stacks around the world for particle removal. Other important factors affecting filtration efficiencies include the thickness and pore diameter or the filter, the uniformity of particle diameters and pore sizes, the solid volume fraction, the rate of particle loading onto the filter (e.g. affecting particle "bounce"), the particle phase (liquid or solid), capillarity and surface tension (if either the particle or the filter media are coated with a liquid), and characteristics of air or other carrier gases, such as velocity, temperature, pressure, and viscosity.

Environmental engineers have been using filtration to treat air and water for several decades. Air pollution controls employing fabric filters (i.e. baghouses), remove particles from the air stream by passing the air through a porous fabric. The fabric filter is efficient at removing fine particles and can exceed efficiencies of 99%. Based solely on an extrapolation of air pollution control equipment, a bag-type vacuum cleaner should be better than a cyclone-type vacuum cleaner. However, this does not take into operational efficiencies and effectiveness, which are very important to the consumer and the engineer. Changing the bag and insuring that it does not exceed its capacity must be monitored closely by the user. Also, the efficiency of the vacuum cleaner is only as good as the materials being used. Does the bag filter allow a great deal of diffusion, interception, and inertial impaction? The cyclone only requires optimization for inertia.

Selecting the correct control device is a matter of optimizing efficiencies. Which brings us to Amelia's second selection criterion, i.e. the vacuum must have a HEPA filter. I am not sure if she knows that the acronym stands for "high-efficiency particle air" filters, but she knows that it is needed to remove more of the "nasty" dust. In this case, HEPA filters are fitted to equipment to enhance "efficiency." Efficiency is often expressed as a percentage. So a 99.99% HEPA filter is efficient enough to remove 99.99% particles from the air stream. This means that if 10 000 particles enter the filter, on average only one particle would pass all the way through the filter. This is exactly the same concept that we use for incinerator efficiency, but it is known as destruction removal efficiency (DRE). For example, in the US, federal standards require that hazardous compounds can only be incinerated if the process is 99.99% efficient, and for the "nastier" compounds (i.e. "extremely hazardous wastes") the so-called "rule of six nines" applies (i.e. DRE $\geqslant$99.9999%). The HEPA and DRE calculations are quite simple:

$$\text{DRE or HEPA efficiency} = \frac{M_{\text{in}} - M_{\text{out}}}{M_{\text{in}}} \times 100 \tag{32.1}$$

Thus, if $10\,\text{mg min}^{-1}$ of a hazardous compound is fed into the incinerator, only $0.001\,\text{mg min}^{-1} = 1\,\mu\text{g min}^{-1}$ is allowed to exit the stack for a hazardous waste. If the waste is an extremely hazardous waste, only 0.00001 mg min^{-1} = 0.01 μg min^{-1} = 10 ng min^{-1} is allowed to exit the stack. Actually, this same equation is used throughout the environmental engineering discipline to calculate treatment and removal efficiencies. For example, assume that raw wastewater enters a treatment facility with 300 mg L^{-1} biochemical oxygen demand (BOD_5), 200 mg L^{-1} suspended solids (SS), and 10 mg L^{-1} phosphorous (P). The plant must meet effluent standards ≤10 mg L^{-1} BOD_5, ≤10 mg L^{-1} SS, and ≤1 mg L^{-1} P, so using the efficiency equation we know that removal rates of these contaminants must be 97% for BOD_5, 95% for SS, and 90% for P. The pure efficiency values may be misleading because the ease of removal can vary significantly with each contaminant. In this case, gravitational settling in the primary stages of the treatment plant can remove most of the SS, and the secondary treatment stage removes most of the BOD, but more complicated, tertiary treatment is needed for removing most of the nutrient P.

As I shopped for the cyclone-HEPA vacuum cleaner, I came across some additional technical terminology (i.e. "allergen removal"). I am not completely certain what this means, but I believe it is a filtration system that does not quite have the efficiency removal of a HEPA filter. However, this points out the somewhat chaotic nature of environmental engineering. Remember, the smaller particles are the ones that concern most health scientists. However, there has been a resurgence of interest in particles ranging between 2.5 and 10 μm aerodynamic diameter, referred to as coarse particles. These may consist of potentially toxic components, for example, resuspended road dust, brake lining residues, industrial byproducts, tire residues, heavy metals, and aerosols generated by organisms spores (known as "bioaerosols"), such as tree pollen and mold spores. Figure 32.1 shows that a large fraction of these coarse particles may deposit to the upper airways, causing health scientists to link them to asthma. And, since asthma appears to be increasing in children, there may be a need to reconsider the importance of larger particles. In fact, the US EPA is considering ways to develop a federal reference method (FRM)[4] for coarse particles to complement its $PM_{2.5}$ FRM.

[4] The FRM is certified reference analogous to the National Institute for Standards and Testing (NIST) standard reference material (SRM), which is standard of known quality to be used by scientists and engineers, allowing them:

(a) to help to develop and validate representative and accurate methods of analysis,
(b) to verify that tests meet validated performance criteria,
(c) to calibrate systems of measurements,
(d) to establish acceptable quality assurance (QA) programs,
(e) to provide test materials for inter-laboratory and inter-study comparisons and proficiency test programs.

Rather than materials, however, the FRM applies to environmental measurement equipment against which all other equipment is compared. For example, in ambient monitoring the US federal standard method dictates how to monitor a specific pollutant, such as aerosols or sulfur dioxide (SO_2). The FRM may be specified by technique (such as a particular physical principle, like a gravimetric or an optical technique for estimating mass) for or by design (such as a particular system for collecting and weighing particles).

Thus, equipment selection has been complicated by some fairly technical specifications. Efficiency is an important part of effectiveness, although the two terms are not synonymous. As we discussed, efficiency is simply a metric of what you get out of a system compared to what you put in. However, you can have a bunch of very efficient systems that may *en toto* be ineffective. They are all working well as designed, but they may not be working on the right things, or their overall configuration is not optimal to solve the problem at hand. So, the correct control device is the one that gives not only optimal efficiency, but one that effectively addresses the specific pollution problem at hand.

II. REMOVAL OF DRY PM

Dry aerosols, or particulate matter (PM), differ so much from the carrying gas stream that their removal should present no major difficulties. The aerosol is different physically, chemically, and electrically. It has vastly different inertial properties than the carrying gas stream and can be subjected to an electric charge. It may be soluble in a specific liquid. With such a variety of removal mechanisms that can be applied, it is not surprising that PM, such as mineral dust, can be removed by a filter, wet scrubber, or ESP with equally satisfactory results.

A. Filters

A filter removes PM from the carrying gas stream because the particulate impinges on and then adheres to the filter material. As time passes, the deposit of PM becomes greater and the deposit itself then acts as a filtering medium. When the deposit becomes so heavy that the pressure necessary to force the gas through the filter becomes excessive, or the flow reduction severely impairs the process, the filter must either be replaced or cleaned.

The filter medium can be fibrous, such as cloth; granular, such as sand; a rigid solid, such as a screen; or a mat, such as a felt pad. It can be in the shape of a tube, sheet, bed, fluidized bed, or any other desired form. The material can be natural or man-made fibers, granules, cloth, felt, paper, metal, ceramic, glass, or plastic. It is not surprising that filters are manufactured in an infinite variety of types, sizes, shapes, and materials.

The theory of filtration of aerosols from a gas stream is much more involved than the sieving action which removes particles in a liquid medium. Figure 32.4 shows three of the mechanisms of aerosol removal by a filter, as well as diffusion (which can be important for very small particles). In practice, the particles and filter elements are seldom spheres or cylinders.

Direct interception occurs when the fluid streamline carrying the particle passes within one-half of a particle diameter of the filter element. Regardless of the particle's size, mass, or inertia, it will be collected if the streamline passes sufficiently close. Inertial impaction occurs when the particle would miss the filter element if it followed the streamline, but its inertia resists the change in direction taken by the gas molecules and it continues in a direct enough course to be collected by the filter element. Electrostatic attraction occurs because the particle, the filter, or both possess sufficient electrical charge to overcome the inertial forces; the particle is then collected instead of passing the filter element. Note that size separation ("sieving") plays little or no role in filtration.

1. *Filter Efficiency*

Particles can be measured as either mass or count. Particle count is the number of particles in a given band of mass, such as particles with aerodynamic diameters greater than 10 μm (coarse fraction), those with diameters less than 10 μm, but greater than 2.5 μm (PM_{10} fraction), and those with diameters less than 2.5 μm ($PM_{2.5}$ fraction, also known as the fine fraction). However, the bands can be further subdivided. For example, there has been concern recently about the so-called "nanoparticles." These have diameters less than 100 nm. Filtration is the most common method used to measure particles in the air. So, a sample taken from a filter could show bands within the fine fraction may resemble that shown in Table 32.2.

Since filtration is important in both measuring and controlling particle matter, expressions of filter efficiency are crucial to air pollution technologies. Equation 32.1 provides the overall efficiency of any air pollution removal equipment. The efficiency (E) equation can be restated specifically for particles:

$$E = \frac{N_{\mathrm{m}} + N_{\mathrm{out}}}{N_{\mathrm{m}}} \tag{32.2}$$

$$E_{\mathrm{m}} = \frac{C_{\mathrm{m}} + C_{\mathrm{out}}}{C_{\mathrm{m}}} \tag{32.3}$$

Where N is the number of particles (count) and C is the mass concentration (m: entering; out: exiting). Note the similarity between Eqs (32.2) and (32.3) and DRE [Eq. (32.1)].

TABLE 32.2

Mass of Particles Collected on a Filter (Fictitious Data)

Size range (μm)	Count (number of particles)	Mass (μg)	Flow rate (L min^{-3})	Integration time (min)	Mass concentration (μg m^{-3})	Description
>10	2	100	16	60	96	Reentrained dust
>2.5 < 10	20	10	16	60	9.6	Tailpipe emission
>0.01 < 2.5	200	1	16	60	0.96	Suspended colloids
<0.01	20 000	0.1	16	60	0.096	Nanoparticles; mainly carbon (fullerenes = C-60)

Pollution control equipment often characterizes efficiency in terms of the fraction entering versus that exiting the filter known as particle penetration (P):

$$P = \frac{N_{\text{out}}}{N_{\text{m}}} = 1 - E \tag{32.4}$$

$$P = \frac{C_{\text{out}}}{C_{\text{m}}} = 1 - E_{\text{m}} \tag{32.5}$$

Obviously, penetration and efficiency are inversely related, so if a filter inefficient a large number or mass of particles is penetrating the filter. This is obviously not good, so the air pollution engineer needs to specify the tolerances for filtration in any design of measurement or control technologies. Inherent to penetration calculations is the velocity of air entering the system. The front (entry side) of the filter is known as the face, so face velocity (U_0) is the air's velocity just before the air enters the filter:

$$U_0 = \frac{Q}{A} \tag{32.6}$$

where Q is the volumetric flow and A is the area of the cross section through which the air is passing. However, since the flow is restricted to the void spaces of the filter, the actual velocity in the filter itself is higher than the face velocity (same air mass through less volume). This is true for flow through any porous medium, such as polyurethane traps (see Fig. 16.2) and columns of sorbant granules, like XAD resins. Thus the velocity within the filter (U_{filter}):

$$U_{\text{filter}} = \frac{Q}{A(1 - \alpha)} \tag{32.7}$$

where α is the packing density (solidity),[5] which is inverse to the filter (or trap) porosity:

$$\alpha = \frac{\text{filter fiber volume}}{\text{total filter volume}} = 1 - \text{porosity} \tag{32.8}$$

It is tempting to think of filters as sieves; however, they are really quite different. In fact, fibrous filters are more akin to numerous microscopic layers of filters, each with a specific probability of catching a particle, depending on the particle's shape and size. Therefore, the efficiency is enhanced with filter thickness.[6] Recall, that size capture is not one of the most important

[5] This is analogous to bulk density used by soil scientists.

[6] For more information on monodisperse aerosols and collection probabilities, see Chapter 9, Filtration, in *Aerosol Technology*, 2nd ed. (Hinds, W. C., ed.). Wiley-Interscience, New York).

mechanisms for collection, compared to inertial impaction, interception, and electrostatics (see Fig. 32.4).

Other lesser mechanisms that result in aerosol removal by filters are (1) gravitational settling due to the difference in mass of the aerosol and the carrying gas, (2) thermal precipitation due to the temperature gradient between a hot gas stream and the cooler filter medium which causes the particles to be bombarded more vigorously by the gas molecules on the side away from the filter element, and (3) Brownian deposition as the particles are bombarded with gas molecules that may cause enough movement to permit the aerosol to come in contact with the filter element. Brownian motion may also cause some of the particles to miss the filter element because they are moved away from it as they pass by.

Regardless of the mechanism which causes the aerosol to come in contact with the filter element, it will be removed from the gas stream only if it adheres to the surface. Aerosols arriving later at the filter element may then, in turn, adhere to the collected aerosol instead of the filter element. The result is that actual aerosol removal seldom agrees with theoretical calculations. One should also consider that certain particles do not adhere to the filter element even though they touch it. As time passes, the heavier deposits on the filter surface will be dislodged more easily than the light deposits, resulting in increased reentrainment. Because of plugging of the filter with time, the apparent size of the filter element increases, causing more interception and impaction. The general effect of all of these variables on the particle buildup and reentrainment is shown in Fig. 32.5.

The pressure drop through the filter is a function of two separate effects. The clean filter has some initial pressure drop. This is a function of filter material, depth of the filter, the superficial gas velocity, which is the gas velocity perpendicular to the filter face, and the viscosity of the gas. Added to the clean filter resistance is the resistance that occurs when the adhering particles form a cake on the filter surface. This cake increases in thickness as approximately a linear function of time, and the pressure difference necessary to cause the same gas flow also becomes a linear function with time. Usually, the pressure available at the filter is limited so that as the cake builds up the flow decreases. Filter cleaning can be based, therefore, on (1) increased pressure drop across the filter, (2) decreased volume of gas flow, or (3) time elapsed since the last cleaning.

Industrial filtration systems may be of many types. The most common type is the baghouse shown in Fig. 32.6. The filter bags are fabricated from woven material, with the material and weave selected to fit the specific application. Cotton and synthetic fabrics are used for relatively low temperatures, and glass cloth fabrics can be used for elevated temperatures, up to 290°C.

The filter ratio for baghouses, also called the *gas to cloth ratio*, varies from 0.6 to 1.5 m^3 of gas per minute per square meter of fabric. The pressure drop across the fabric is a function of the filter ratio; it ranges from about 80 mm of water for the lower filter ratios up to about 200 mm of water for the higher ratios. Before selecting any bag filter system, a thorough engineering study

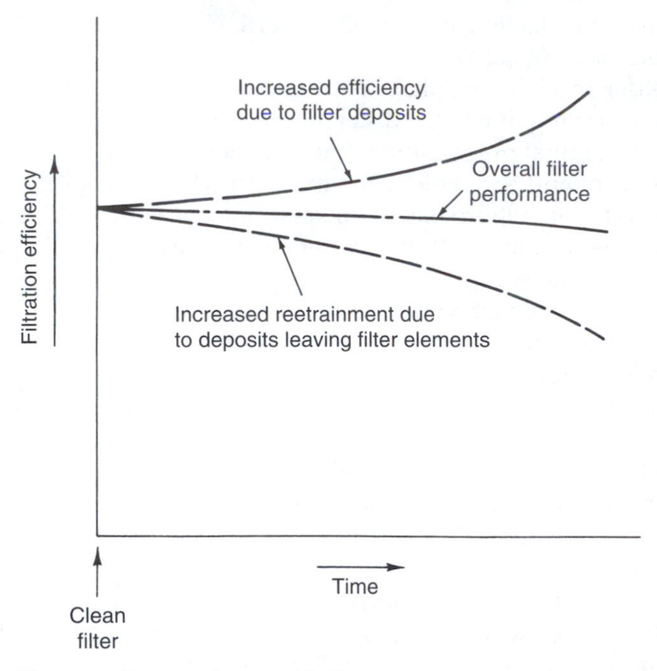

Fig. 32.5. Filtration efficiency change with time.

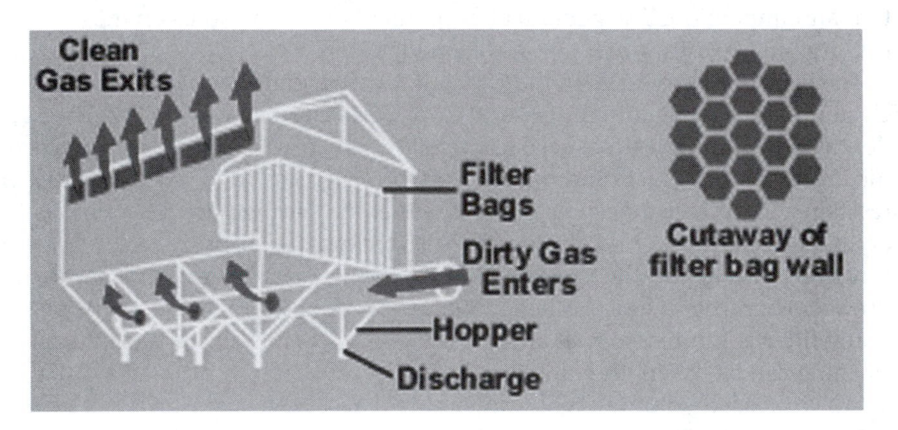

Fig. 32.6. Industrial baghouse. *Source*: U.S. Environmental Protection Agency; Air Pollution Control Orientation Course, "Source Control Technology; Control of Particulate Pollutants from Stationary Sources: Fabric Filters": www.epa.gov/apti/course422/images/baghouse.jpg; accessed on June 30, 2007.

should be made, followed by a consultation with different bag and baghouse manufacturers.

The bags must be periodically cleaned to remove the accumulated PM (Note in Fig. 32.5 the general downward trend in efficiency with time). Bag cleaning methods vary widely with the manufacturer and with baghouse style and use. Methods for cleaning include (1) mechanical shaking by agitation of the top hanger, (2) reverse flow of gas through a few of the bags at a time, (3) continuous cleaning with a reverse jet of air passing through a series of orifices on a ring as it moves up and down the clean side of the bag, and (4) collapse and pulsation cleaning methods.

The cleaning cycles are usually controlled by a timing device which deactivates the section being cleaned. The dusts removed during cleaning are collected in a hopper at the bottom of the baghouse and then removed, through an air lock or star valve, to a bin for ultimate disposal.

Other types of industrial filtration systems include (1) fixed beds or layers of granular material such as coke or sand; some of the original designs for cleaning large quantities of gases from smelters and acid plants involved passing the gases through such beds; (2) plain, treated, or charged mats or pads (common throw-away air filters used for hot air furnaces and for air conditioners are of this type); (3) paper filters of multiple plies and folds to increase filter efficiency and area (the throwaway dry air filters used on automotive engines are of this type); (4) rigid porous beds which can be made of metal, plastic, or porous ceramic (these materials are most efficient for removal of large particles such as the 30-μm particles from a wood sanding operation); and (5) fluidized beds in which the granular material of the bed is made to act as a fluid by the gas passing through it. Most fluidized beds are used for heat or mass transfer. Their use for filtration has not been extensive.

B. Electrostatic Precipitators

High-voltage ESPs have been widely used throughout the world for particulate removal since they were perfected by Fredrick Cottrell early in the twentieth century [5]. Most of the original units were used for recovery of process materials, but today gas cleaning for air pollution control is often the main reason for their installation. The ESP has distinct advantages over other aerosol collection devices: (1) it can easily handle high-temperature gases, which makes it a likely choice for boilers, steel furnaces, etc.; (2) it has an extremely small pressure drop, so that fan costs are minimized; (3) it has an extremely high-collection efficiency if operated properly on selected aerosols (many cases are on record, however, in which relatively low efficiencies were obtained because of unique or unknown dust properties); (4) it can handle a wide range of particulate sizes and dust concentrations (most precipitators work best on particles smaller than 10 μm, so that an inertial precleaner is often used to remove the large particles); and (5) if it is properly designed and constructed, its operating and maintenance costs are lower than those of any other type of particulate collection system. The ESP takes advantage of the electrostatics of particles (see Fig. 32.4). The particles move advectively with the gas stream, which is travelling horizontally through the ESP unit (into the photograph shown in Fig. 32.7). The particles become charged and then are attracted by the charge differential to either side, where they are captured.

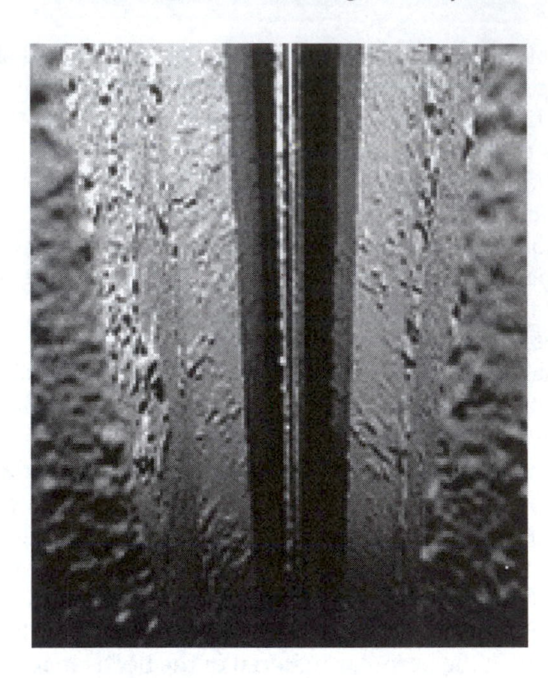

Fig. 32.7. View inward of a single gas passage through an electrostatic precipitator. The opening is about 20 cm.

Three of the disadvantages of ESPs are as follows: (1) the initial cost is the highest of any particulate collection system, (2) a large amount of space is required for the installation, and (3) ESPs are not suitable for combustible particles such as grain or wood dust.

The ESP works by charging dust with ions and then collecting the ionized particles on a surface. The collection surfaces may consist of either tubular or flat plates. For cleaning and disposal, the particles are then removed from the collection surface, usually by rapping the surface.

A high-voltage (30 kV or more) DC field is established between the central wire electrode and the grounded collecting surface. The voltage is high enough that a visible corona can be seen at the surface of the wire. The result is a cascade of negative ions in the gap between the central wire and the grounded outer surface. Any aerosol entering this gap is both bombarded and charged by these ions. The aerosols then migrate to the collecting surface because of the combined effect of this bombardment and the charge attraction. When the particle reaches the collecting surface, it loses its charge and adheres because of the attractive forces existing. It should remain there until the power is shut off and it is physically dislodged by rapping, washing, or sonic means.

In a tube-type ESP, the tubes are 8–25 cm in diameter and 1–4 m long. They are arranged vertically in banks with the central wires, about 2 mm in diameter, suspended in the center with tension weights at the bottom. Many innovations, including square, triangular, and barbed wires, are used by different manufacturers.

A plate-type ESP is similar in principle to the tubular type except that the air flows across the wires horizontally, at right angles to them. The particles are collected on vertical plates, which usually have fins or baffles to strengthen them and prevent dust reentrainment. Figure 32.8 illustrates a large plate-type precipitator. These precipitators are usually used to control and collect dry dusts.

Problems with ESPs develop because the final unit does not operate at ideal conditions. Gas channeling through the unit can result in high dust loadings in one area and light loads in another. The end result is less than optimum efficiency because of much reentrainment. The resistivity of the dust greatly affects its reentrainment in the unit. If a high-resistance dust collects on the plate surface, the effective voltage across the gap is decreased. Some power plants burning high-ash, low-sulfur coal have reported very low efficiency from the precipitator because the ash needed more SO_2 to decrease its resistivity. The suggestion that precipitator efficiency could be greatly improved by *adding* SO_2 or SO_3 to the stack gases has, not surprisingly, met with much skeptism.

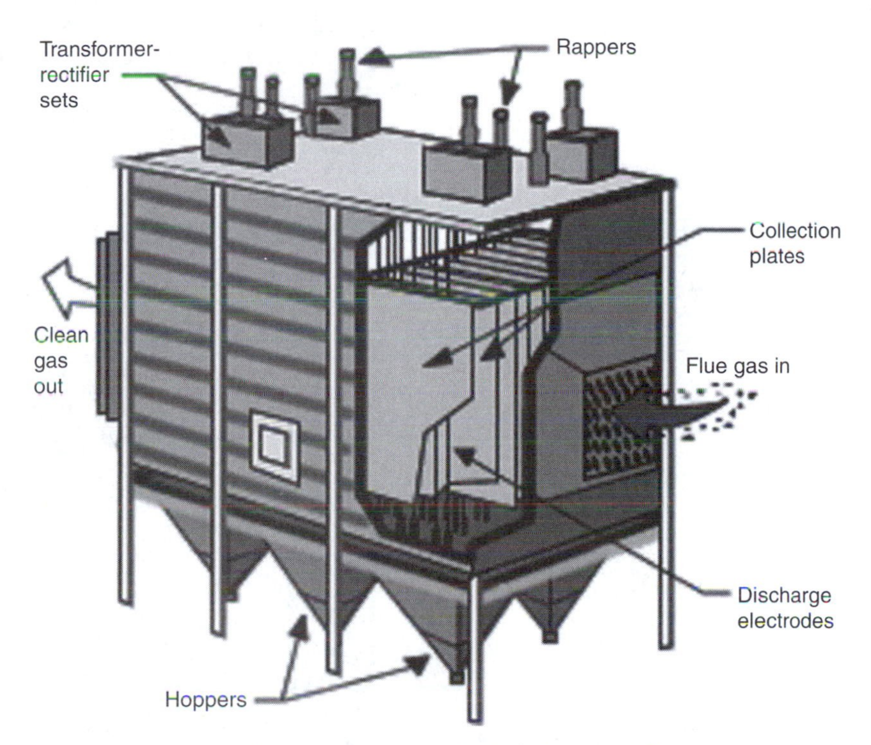

Fig. 32.8. Commercial plate-type ESP. *Source*: U.S. Environmental Protection Agency. "Basic Concepts in Environmental Sciences—Module 6: Air Pollutants and Control Techniques"; http://www.epa.gov/eogapti1/module6/matter/control/control.htm; accessed on June 30, 2007.

C. Inertial Collectors

Inertial collectors, whether cyclones, baffles, louvers, or rotating impellers, operate on the principle that the aerosol material in the carrying gas stream has a greater inertia than the gas. Since the drag forces on the particle are a function of the diameter squared and the inertial forces are a function of the diameter cubed, it follows that as the particle diameter increases, the inertial (removal) force becomes relatively greater. Inertial collectors, therefore, are most efficient for larger particles. The inertia is also a function of the mass of the particle, so that heavier particles are more efficiently removed by inertial collectors. These facts explain why an inertial collector will be highly efficient for removal of 10-μm rock dust and very inefficient for 5-μm wood particles. It would be very efficient, though, for 75-μm wood particles.

The most common inertial collector is the cyclone, which is used in two basic forms: the tangential inlet and the axial inlet. Figure 32.9 shows the two types.

In actual industrial practice, the tangential inlet type is usually a large (1–5 m in diameter) single cyclone, while the axial inlet cyclone is relatively small (about 20 cm in diameter and arranged in parallel units for the desired capacity).

For any cyclone, regardless of type, the radius of motion (curvature), the particle mass, and the particle velocity are the three factors which

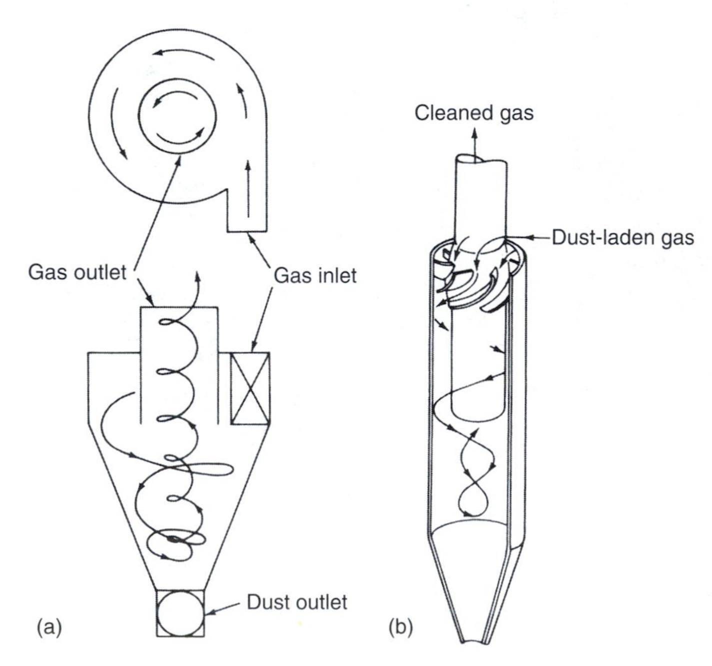

Fig. 32.9. (a) Tangential inlet cyclone. (b) Axial inlet cyclone.

determine the centrifugal force exerted on the particle. This centrifugal force may be expressed as

$$F = MA \tag{32.9}$$

where F: force (centrifugal), M: mass of the particle, and A: acceleration (centrifugal), and

$$A = \frac{V^2}{R} \tag{32.10}$$

where V: velocity of particle and R: radius of curvature. Therefore:

$$F = \frac{MV^2}{R} \tag{32.11}$$

Other types of inertial collectors which might be used for particulate separation from a carrying gas stream depend on the same theoretical principles developed for cyclones. Table 32.3 summarizes the effect of the common variables on inertial collector performance.

Although decreasing the radius of curvature and increasing the gas velocity both result in increased efficiency, the same changes cause increased pressure drop through the collector. Design of inertial collectors for maximum efficiency at minimum cost and minimum pressure drop is a problem which lends itself to computer optimization. Unfortunately, many inertial collectors, including the majority of the large single cyclones, have been designed to fit a standard-sized sheet of metal rather than a specific application and gas velocity. As tighter emission standards are adopted, inertial collectors will probably become precleaners for the more sophisticated gas cleaning devices.

D. Scrubbers

Scrubbers, or wet collectors, have been used as gas cleaning devices for many years. However, the process has two distinct mechanisms which result

TABLE 32.3

Effect of Independent Variables on Inertial Collector Efficiency

Independent variable of concern	Increase or decrease to improve efficiency
Radius of curvature	Decrease
Mass of particle	Increase
Particle diameter	Increase
Particle survace/volume ratio	Decrease
Gas velocity	Increase
Gas viscosity	Decrease

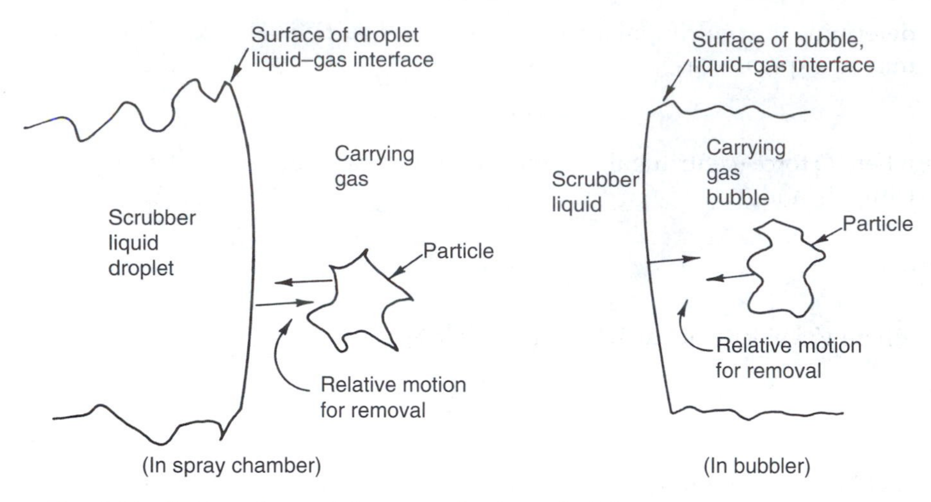

Fig. 32.10. Wetting of aerosols in a spray chamber or bubbler.

in the removal of the aerosol from the gas stream. The first mechanism involves wetting the particle by the scrubbing liquid. As shown in Fig. 32.10, this process is essentially the same whether the system uses a spray to atomize the scrubbing liquid or a diffuser to break the gas into small bubbles. In either case, it is assumed that the particle is trapped when it travels from the supporting gaseous medium across the interface to the liquid scrubbing medium. Some relative motion is necessary for the particle and liquid–gas interface to come in contact. In the spray chamber, this motion is provided by spraying the droplets through the gas so that they impinge on and make contact with the particles. In the bubbler, inertial forces and severe turbulence achieve this contact. In either case, the smaller the droplet or bubble, the greater the collection efficiency. In the scrubber, the smaller the droplet, the greater the surface area for a given weight of liquid and the greater the chance for wetting the particles. In a bubbler, smaller bubbles mean not only that more interface area is available but also that the particles have a shorter distance to travel before reaching an interface where they can be wetted.

The second mechanism important in wet collectors is removal of the wetted particles on a collecting surface, followed by their eventual removal from the device. The collecting surface can be in the form of a bed or simply a wetted surface. One common combination follows the wetting section with an inertial collector which then separates the wetted particles from the carrying gas stream.

Increasing either the gas velocity or the liquid droplet velocity in a scrubber will increase the efficiency because of the greater number of collisions per unit time. The ultimate scrubber in this respect is the venturi scrubber, which operates at extremely high gas and liquid velocities with a very high pressure drop across the venturi throat. Figure 32.11 illustrates a commercial venturi scrubber unit.

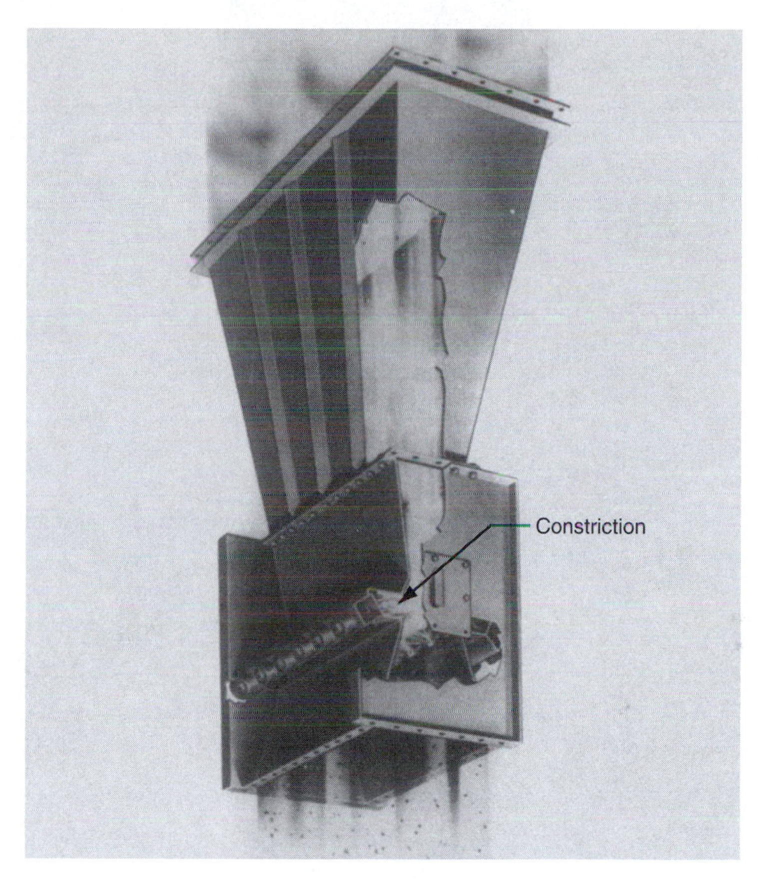

Fig. 32.11. Venturi scrubber; so-called since the velocity of the air is increased as it moves through a constricting channel. *Source*: American Air Fitter Company.

E. Dry Scrubbers

Dry scrubber is a term that has been applied to gravel bed filters that recirculate the gravel filter medium using some type of external cleaning or washing system. Some units also use an electrostatic field across the gravel bed to enhance removal of the particulate material. The dry scrubber may have to be followed by a baghouse to clean the effluent to acceptable standards. The advantage of dry scrubbers is their ability to remove large quantities of particulate pollutants, such as fly ash, from hot gas streams.

F. Comparison of Particulate Removal Systems

When selecting a system to remove particulate from a gas stream, many choices concerning equipment can be made. The selection could be made on the basis of cost, gas pressure drop, efficiency, temperature, resistance, etc. Table 32.4 summarizes these factors for comparative purposes. The tabular

TABLE 32.4

Comparison of Particulate Removal Systems

Type of collector	Particle size range (μm)	Removal efficiency	Space required	Maximum temperature (°C)	Pressure drop (cmH_2O)	Annual cost (US$ per year m^{-3})[a]
Baghouse (cotton bags)	0.1–0.1	Fair	Large	80	10	28.00
	1.0–10.0	Good	Large	80	10	28.00
	10.0–50.0	Excellent	Large	80	10	28.00
Baghouse (Dacron, nylon, Orlon)	0.1–1.0	Fair	Large	120	12	34.00
	1.0–10.0	Good	Large	120	12	34.00
	10.0–50.0	Excellent	Large	120	12	34.00
Baghouse (glass fiber)	0.1–1.0	Fair	Large	290	10	42.00
	1.0–10.0	Good	Large	290	10	42.00
	10.0–50.0	Good	Large	290	10	42.00
Baghouse (Teflon)	0.1–1.0	Fair	Large	260	20	46.00
	1.0–10.0	Good	Large	260	20	46.00
	10.0–50.0	Excellent	Large	260	20	46.00
ESP	0.1–1.0	Excellent	Large	400	1	42.00
	1.0–10.0	Excellent	Large	400	1	42.00
	10.0–50.0	Good	Large	400	1	42.00
Standard cyclone	0.1–1.0	Poor	Large	400	5	14.00
	1.0–10.0	Poor	Large	400	5	14.00
	10.0–50.0	Good	Large	400	5	14.00
High-efficiency cyclone	0.1–1.0	Poor	Moderate	400	12	22.00
	1.0–10.0	Fair	Moderate	400	12	22.00
	10.0–50.0	Good	Moderate	400	12	22.00
Spray tower	0.1–1.0	Fair	Large	540	5	50.00
	1.0–10.0	Good	Large	540	5	50.00
	10.0–50.0	Good	Large	540	5	50.00
Impingement scrubber	0.1–1.0	Fair	Moderate	540	10	46.00
	1.0–10.0	Good	Moderate	540	10	46.00
	10.0–50.0	Good	Moderate	540	10	46.00
Venturi scrubber	0.1–1.0	Good	Small	540	88	112.00
	1.0–10.0	Excellent	Small	540	88	112.00
	10.0–50.0	Excellent	Small	540	88	112.00
Dry scrubber	0.1–1.0	Fair	Large	500	10	42.00
	1.0–10.0	Good	Large	500	10	42.00
	10.0–50.0	Good	Large	500	10	42.00

[a] Includes costs for water and power, operation and maintenance, capital equipment and insurance (in 1994 US$).

values must not be considered absolute because great variations occur between types and manufacturers. No table is a substitute for a qualified consulting engineer or a reputable manufacturer's catalog.

III. REMOVAL OF LIQUID DROPLETS AND MISTS

The term *mist* generally refers to liquid droplets from submicron size to about 10 μm. If the diameter exceeds 10 μm, the aerosol is usually referred to as a *spray* or simply as *droplets*. Mists tend to be spherical because of their surface tension and are usually formed by nucleation and the condensation of vapors [6]. Larger droplets are formed by bursting of bubbles, by entrainment from surfaces, by spray nozzles, or by splash-type liquid distributors. The large droplets tend to be elongated relative to their direction of motion because of the action of drag forces on the drops.

Mist eliminators are widely used in air pollution control systems to prevent free moisture from entering the atmosphere. Usually, such mist eliminators are found downstream from wet scrubbers. The recovered mist is returned to the liquid system, resulting in lowered liquid makeup requirements.

Since mist and droplets differ significantly from the carrying gas stream, just as dry particulates do, the removal mechanisms are similar to those employed for the removal of dry particulates. Control devices developed particularly for condensing mist will be discussed separately. Mist collection is further simplified because the particles are spherical and tend to resist reentrainment, and they agglomerate after coming in contact with the surface of the collecting device.

A. Filters

Filters for mists and droplets have more open area than those used for dry particles. If a filter is made of many fine, closely spaced fibers, it will become wet due to the collected liquid. Such wetting will lead to matting of the fibers, retention of more liquid, and eventual blocking of the filter. Therefore, instead of fine, closely spaced fibers, the usual wet filtration system is composed of either knitted wire or wire mesh packed into a pad. A looser filtration medium results in a filter with a lower pressure drop than that of the filters used for dry particulates. The reported pressure drop across wire mesh mist eliminators is 1–2 cm of water at face velocities of $5\,\mathrm{m\,s^{-1}}$. The essential collection mechanisms employed for filtration of droplets and mists are inertial impaction and, to a lesser extent, direct interception.

B. Electrostatic Precipitators

ESPs for liquid droplets and mists are essentially of the wetted wall type. Figure 32.12 shows a wet-wall precipitator with tubular collection electrodes [1]. The upper ends of the tubes form weirs, and water flows over the tube ends to irrigate the collection surface.

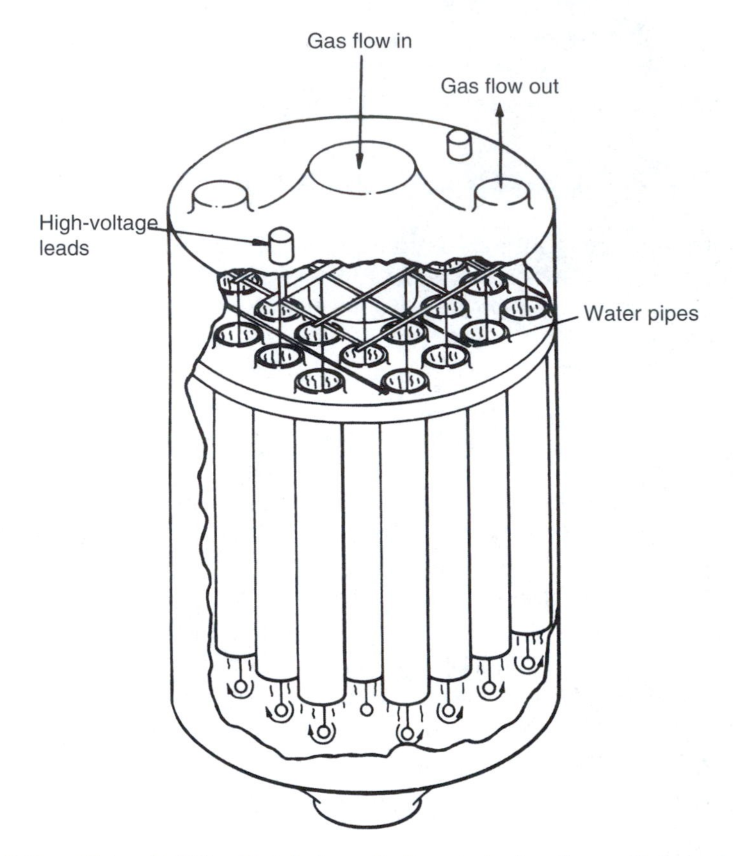

Fig. 32.12. Wet-wall ESP with tubular collection electrodes. *Source*: Oglesby Jr., S., and Nichols, G. B., Electrostatic precipitators, in *Air Pollution*, 3rd ed., Vol. IV (Stern, A. C., ed.), p. 238. Academic Press, New York, 1977.

Figure 32.13 shows an alternative type of wet precipitator with plate-type collection electrodes. In this design, sprays located in the ducts formed by adjacent collecting electrodes serve to irrigate the plates [1]. These are often supplemented by overhead sprays to ensure that the entire plate surface is irrigated. The design of such precipitators is similar to that of conventional systems except for the means of keeping insulators dry, measures to minimize corrosion, and provisions for removing the slurry.

C. Inertial Collectors

Inertial collectors for mists and droplets are widely used. They include cyclone collectors, baffle systems, and skimmers in ductwork. Inertial devices can be used as primary collection systems, precleaners for other devices, or mist eliminators. The systems are relatively inexpensive and reliable and have low pressure drops.

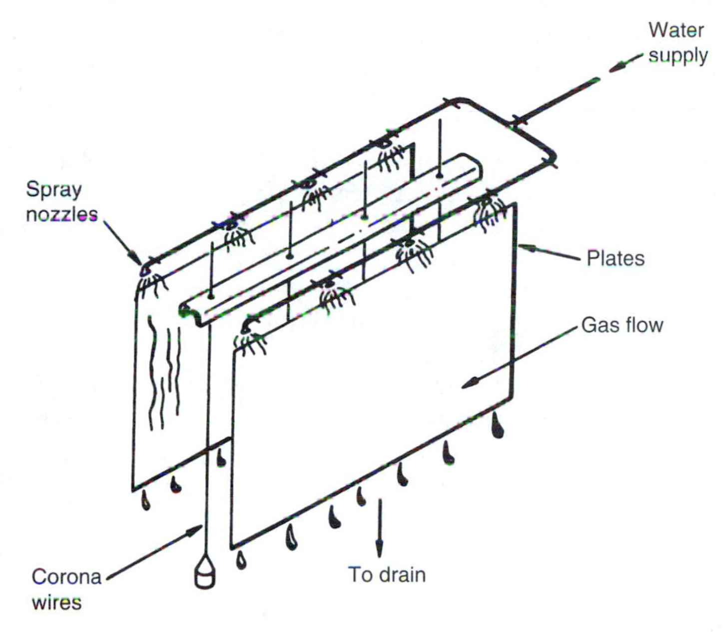

Fig. 32.13. Wet ESP with plate collection electrodes. *Source*: Oglesby Jr., S., and Nichols, G. B., Electrostatic precipitators, in *Air Pollution*, 3rd ed., Vol. IV (Stern, A. C., ed.), p. 239. Academic Press, New York, 1977.

Cyclone mist eliminators and collectors have virtually the same efficiency for both liquid aerosols and solid particles. To avoid reentrainment of the collected liquid from the walls of the cyclone, an upper limit is set to the tangential velocity that can be used. The maximum tangential velocity should be limited to the inlet velocity. Even at this speed, the liquid film may creep to the edge of the exit pipe, from which the liquid is then reentrained.

Baffle separators of the venetian blind, V, W, and wave types are widely used for spray removal. They have small space requirements and low pressure drops. They operate by diverting the gas stream and ejecting the droplets onto the collector baffles. Efficiencies of single stages may be only 40–60%, but by adding multiple stages, efficiencies approaching 100% may be obtained.

D. Scrubbers

A widely used type of scrubber for mists and droplets is the venturi scrubber. It has been used for the collection of sulfuric acid and phosphoric acid mists with very high efficiency. The scrubbing contact is made at the venturi throat, where very small droplets of the scrubbing liquid (usually water) are injected. At the throat, gas velocities as high as 130 $\mathrm{m\ s^{-1}}$ are used to increase collision efficiencies. Water, injected for acid mist control, ranges from 0.8 to

2.0 L m^{-3} of gas. Collection efficiencies approaching 100% are possible, but high efficiencies require a gas pressure drop of 60–90 cm of water across the scrubber. Normal operation, with a submicron mist, is reported to be in the 90–95% efficiency range [1].

One problem in using scrubbers to control mists and droplets is that the scrubber also acts as a condenser for volatile gases. For example, a hot plume containing volatile hydrocarbon gases, such as the exhaust from a gas turbine, may be cooled several hundred degrees by passing through a scrubber. This cooling can cause extensive condensation of the hydrocarbons, resulting in a plume with a high opacity. Teller [7] reports that cooling of exhaust gases from a jet engine in a test cell by the use of water sprays can result in droplet loadings 10–100% greater than those measured at the engine exhaust plane because of the condensation of hydrocarbons which were normally exhausted as gases.

E. Other Systems

Many unique systems have been proposed, and some used, to control the release of mists and droplets, such as:

1. Ceramic candles are thimble-shaped, porous, acid-resistant ceramic tubes. Although efficiencies exceeding 98% have been reported, the candles have high maintenance requirements because they are very fragile.
2. Electric cyclones utilize an electrode in the center of the cyclone to establish an electric field within the cyclone body. This device is more efficient than the standard cyclone. It is probably more applicable to mists and droplets than to dry particulates, due to possible fire or explosion hazards with combustible dusts.
3. Sonic agglomerators have been used experimentally for sulfuric acid mists and as mist eliminators. Commercial development is not projected at this time because the energy requirements are considerably greater than those for venturi scrubbers of similar capacity.

IV. REMOVAL OF GASEOUS POLLUTANTS

Gaseous pollutants may be easier or more difficult to remove from the carrying gas stream than aerosols, depending on the individual situation. The gases may be reactive to other chemicals, and this property can be used to collect them. Of course, any separation system relying on differences in inertial properties must be ruled out. Four general methods of separating gaseous pollutants are currently in use. These are: (1) absorption in a liquid, (2) adsorption on a solid surface, (3) condensation to a liquid, and (4) conversion into a less polluting or nonpolluting gas.

A. Absorption Devices

Absorption of pollutant gases is accomplished by using a selective liquid in a wet scrubber, packed tower, or bubble tower. Pollutant gases commonly controlled by absorption include sulfur dioxide, hydrogen sulfide, hydrogen chloride, chlorine, ammonia, oxides of nitrogen, and low-boiling hydrocarbons.

The scrubbing liquid must be chosen with specific reference to the gas being removed. The gas solubility in the liquid solvent should be high so that reasonable quantities of solvent are required. The solvent should have a low vapor pressure to reduce losses, be noncorrosive, inexpensive, nontoxic, nonflammable, chemically stable, and have a low freezing point. It is no wonder that water is the most popular solvent used in absorption devices. The water may be treated with an acid or a base to enhance removal of a specific gas. If carbon dioxide is present in the gaseous effluent and water is used as the scrubbing liquid, a solution of carbonic acid will gradually replace the water in the system.

In many cases, water is a poor scrubbing solvent. Sulfur dioxide, for example, is only slightly soluble in water, so a scrubber of very large liquid capacity would be required. SO_2 is readily soluble in an alkaline solution, so scrubbing solutions containing ammonia or amines are used in commercial applications.

Chlorine, hydrogen chloride, and hydrogen fluoride are examples of gases that are readily soluble in water, so water scrubbing is very effective for their control. For years hydrogen sulfide has been removed from refinery gases by scrubbing with diethanolamine. More recently, the light hydrocarbon vapors at petroleum refineries and loading facilities have been absorbed, under pressure, in liquid gasoline and returned to storage. All of the gases mentioned have economic importance when recovered and can be valuable raw materials or products when removed from the scrubbing solvent.

B. Adsorption Devices

Adsorption of pollutant gases occurs when certain gases are selectively retained on the surface or in the pores or interstices of prepared solids. The process may be strictly a surface phenomenon with only molecular forces involved, or it may be combined with a chemical reaction occurring at the surface once the gas and adsorber are in intimate contact. The latter type of adsorption is known as *chemisorption*.

The solid materials used as adsorbents are usually very porous, with extremely large surface-to-volume ratios. Activated carbon, alumina, and silica gel are widely used as adsorbents depending on the gases to be removed. Activated carbon, for example, is excellent for removing light hydrocarbon molecules, which may be odorous. Silica gel, being a polar material, does an excellent job of adsorbing polar gases. Its characteristics for removal of water vapor are well known.

Solid adsorbents must also be structurally capable of being packed into a tower, resistant to fracturing, and capable of being regenerated and reused after saturation with gas molecules. Although some small units use throw-away canisters or charges, the majority of industrial adsorbers regenerate the adsorbent to recover not only the adsorbent but also the adsorbate, which usually has some economic value.

The efficiency of most adsorbers is very near 100% at the beginning of operation and remains extremely high until a breakpoint occurs when the adsorbent becomes saturated with adsorbate. At this breakpoint the slope of the percentage of mass of gaseous fluid that is not sorbed increases dramatically with time. It is at the breakpoint that the adsorber should be renewed or regenerated. This is shown graphically in Fig. 32.14.

Industrial adsorption systems are engineered so that they operate in the region before the breakpoint and are continually regenerated by units. Figure 32.15 shows a schematic diagram of such a system, with steam being used to regenerate the saturated adsorbent. Figure 32.16 illustrates the actual system shown schematically in Fig. 32.15.

C. Condensers

In many situations, the most desirable control of vapor-type discharges can be accomplished by condensation. Condensers may also be used ahead of other air pollution control equipment to remove condensable components. The reasons for using condensers include (1) recovery of economically

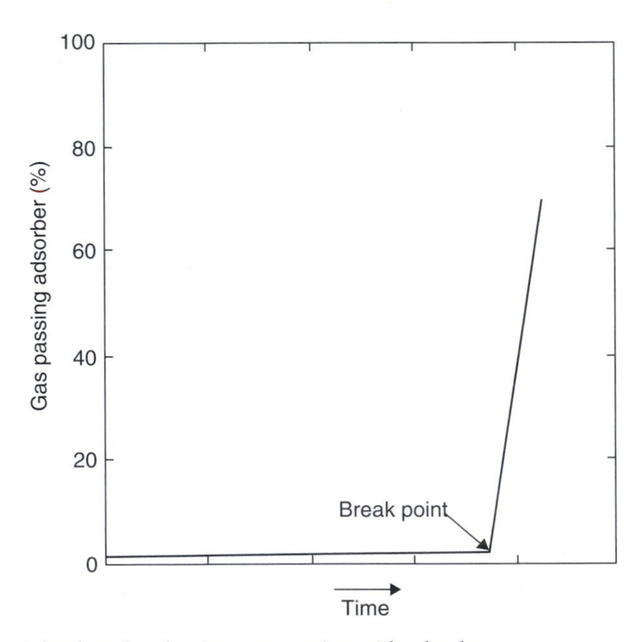

Fig. 32.14. Adsorbent breakpoint at saturation with adsorbate.

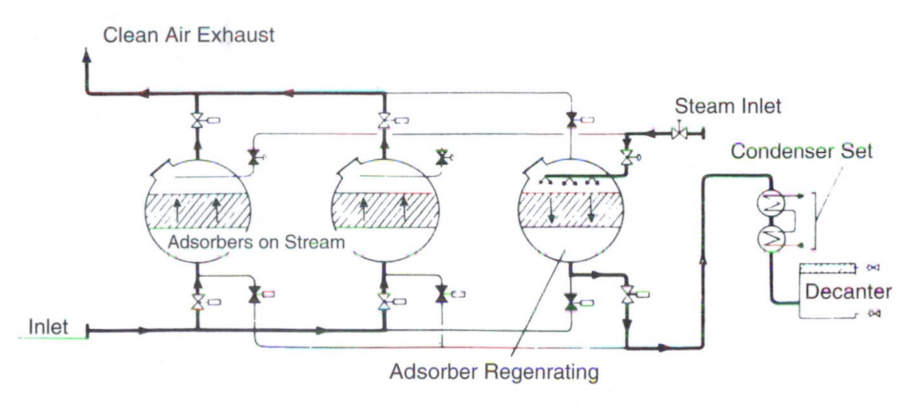

Fig. 32.15. Flow diagram for adsorber. *Source*: The British Ceca Company, Ltd.

Fig. 32.16. Pollution control facility in Milford foundry, New Hampshire includes an adsorption tower. Gaseous emissions are introduced at the base of the treatment column, where a gas diffuser ensures that they are evenly distributed through the system. The gas then contacts purification liquid in the packing of sorbents (e.g. zeolite) in the absorption zone. A second diffuser ensures that the purification liquid is evenly spread. *Source*: Environ-Access, Inc. (1996). Environmental Fact Sheet, "Treatment of Air and Gas: Treatment of Gaseous Emissions through a Wet Scrubber—MESAR Environair Inc." (F2-03-96), Sherbrooke, Quebec, Canada.

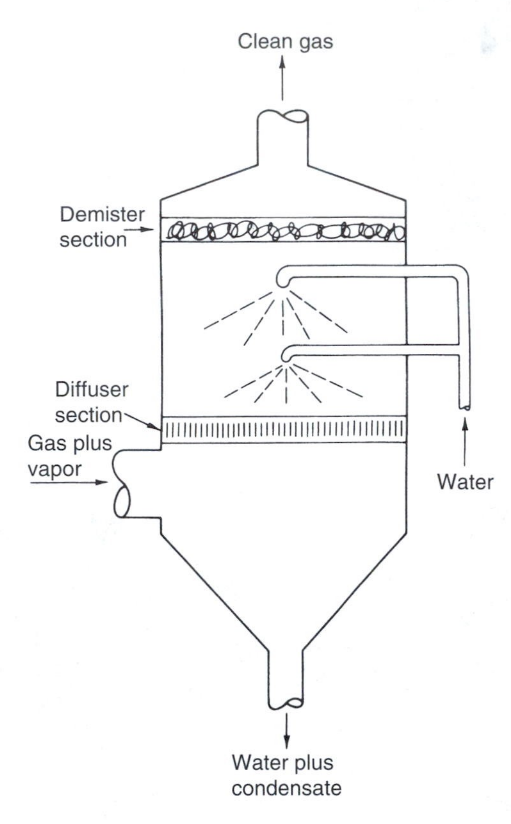

Fig. 32.17. Contact condenser.

valuable products, (2) removal of components that might be corrosive or damaging to other portions of the system, and (3) reduction of the volume of the effluent gases.

Although condensation can be accomplished either by reducing the temperature or by increasing the pressure, in gas-removal practice it is usually done by temperature reduction only.

Condensers may be of one or two general types depending on the specific application. Contact condensers operate with the coolant, vapors, and condensate intimately mixed. In surface condensers, the coolant does not come in contact with either the vapors or the condensate. The usual shell-and-tube condenser is of the surface type. Figure 32.17 illustrates a contact condenser which might be used to clean or preclean a hot corrosive gas.

Table 32.5 lists several applications of condensers currently in use. For most operations listed, air and noncondensable gases should be kept to a minimum, as they tend to reduce condenser capacity.

D. Conversion to Nonpollutant Material

A widely used system for the control of organic gaseous emissions is oxidation of the combustible components to water and carbon dioxide. Other

TABLE 32.5

Representative Applications of Condensers in Air Pollution Control

Petroleum refining	Petrochemical manufacturing	Basic chemical manufacture	Miscellaneous industries
Gasoline accumulator	Polyethylene gas vents	Ammonia	Dry cleaning
Solvents	Styrene	Chlorine solutions	Degreasers
Storage vessels	Copper napthenates		Tar dipping
Lube oil refining	Insecticides		Kraft paper
	Phthalic anhydride		
	Resin reactors		
	Solvent recover		

systems such as the oxidation of H_2S to SO_2 and H_2O are also used even though the SO_2 produced is still considered a pollutant. The trade-off occurs because the SO_2 is much less toxic and undesirable than the H_2S. The odor threshold for H_2S is about three orders of magnitude less than that for SO_2. For oxidation of H_2S to SO_2, the usual device is simply an open flare with a fuel gas pilot or auxiliary burner if the H_2S is below the stoichiometric concentration. If the SO_2 is above emission or other operation limits, it will also have to be treated (e.g. by scrubbing).

Afterburners are widely used as control devices for oxidation of undesirable combustible gases. The two general types are (1) direct-flame afterburners, in which the gases are oxidized in a combustion chamber at or above the temperature of autogenous ignition and (2) catalytic combustion systems, in which the gases are oxidized at temperatures considerably below the autogenous ignition point.

Direct-flame afterburners are the most commonly used air pollution control device in which combustible aerosols, vapors, gases, and odors are to be controlled. The components of the afterburner are shown in Fig. 32.18. They include the combustion chamber, gas burners, burner controls, and exit temperature indicator. Usual exit temperatures for the destruction of most organic materials are in the range of 650–825°C, with retention times at the elevated temperature of 0.3–0.5 s.

Direct-flame afterburners efficient and economical when properly operated. Cost to operate and maintain these systems are similar to those of the auxiliary gas fuel systems. Operating and maintenance costs are essentially those of the auxiliary gas fuel. For larger industrial applications, the overall cost of the afterburner operation may be considerably reduced by using heat recovery equipment as shown in Fig. 32.19. In fact, this is an example of the green engineering approach known as co-generation. In some innovative schemes, heat recovery can provide heat for reactors in neighboring industries. Boilers and kilns also provide efficient pollutant destruction of volatile organic compounds and other vapor phase pollutants in numerous industrial settings. Thermal processes are treated in greater detail in Chapter 33

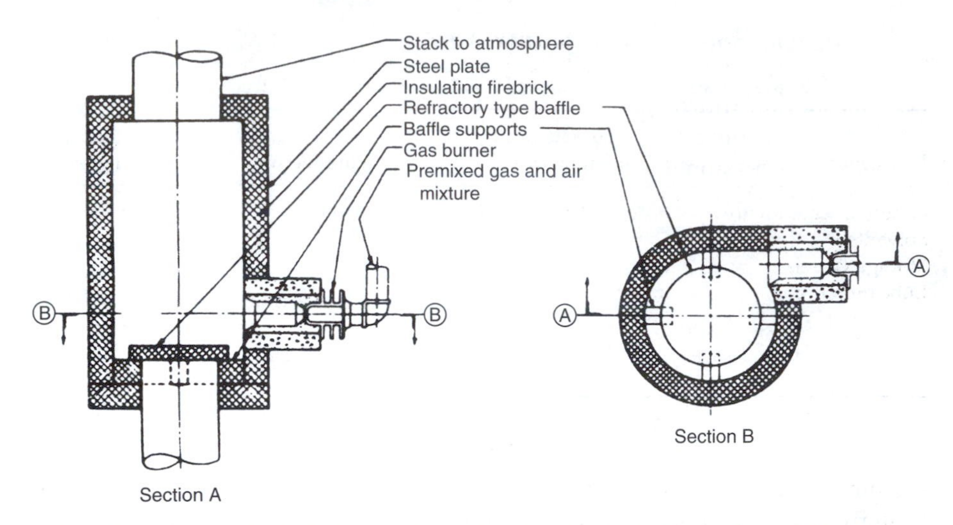

Fig. 32.18. Direct-fired afterburner. *Source*: Los Angeles Air Pollution Control District.

Fig. 32.19. Afterburner with heat recovery. (A) Fume inlet to insulated forced draft fan (310 m^3/min at 230°C). (B) Regenerative shell-and-tube heat exchanger (55% effective recovery). (C) Automatic bypass around heat exchanger for temperature control (required for excess hydrocarbons in fume steam under certain process conditions). (D) Fume inlet and burner chamber internally insulated (fume steam raised to 425°C by heat exchanger). (E) Combustion chamber, refractory lined for 815°C duty (operating at 760°C for required fume oxidation to meet local regulations). (F) Discharge stream leaving regenerative heat exchanger at 520°C enters ventilating air heat exchanger for further waste heat recovery. (G) Ventilating air fan and filter (310 m^3/min of outside air). (H) Automatic bypass with dampers for control of ventilating air temperature. (I) Heated air for winter comfort heating requirements leaves at controlled temperature. (J) Discharge stack (470°C). (K) Combustion safeguard system with dual burner manifold and controls for high turndown. (L) Remote control panel with electronic temperature controls. *Source*: Hirt Combustion Engineers.

(in particular, note the afterburner system following the rotary kiln in Fig. 33.2).

Catalytic afterburners are currently used primarily in industry for the control of solvents and organic vapor emissions from industrial ovens. They are used as emission control devices for gasoline-powered automobiles (see Chapter 35).

The main advantage of the catalytic afterburner is that the destruction of the pollutant gases can be accomplished at a temperature range of about 315–485°C, which results in considerable savings in fuel costs. However, the installed costs of the catalytic systems are higher than those of the direct-flame afterburners because of the expense of the catalyst and associated systems, so the overall annual costs tend to balance out.

In most catalytic systems there is a gradual loss of activity due to contamination or attrition of the catalyst, so the catalyst must be replaced at regular intervals. Other variables that affect the proper design and operation of catalytic systems include gas velocities through the system, amount of active catalyst surface, residence time, and preheat temperature necessary for complete oxidation of the emitted gases.

E. Biological Control Systems

Waste streams with low to moderate concentrations of volatile organic compounds (VOCs) may be treated with biological systems. These are similar to biological systems used to treat wastewater, classified as three basic types: 1. biolfilters; 2. biotrickling filters; and 3. bioscrubbers.

Biofilms of microorganisms (bacteria and fungi) are grown on porous media in biofilters and biotrickling systems. The air or other gas containing the VOCs is passed through the biologically active media, where the microbes break down the compounds to simpler compounds, eventually to carbon dioxide (if aerobic), methane (if anaerobic), and water. The major difference between biofiltration and trickling systems is how the liquid interfaces with the microbes. The liquid phase is stationary in a biofilter (see Fig. 32.20), but liquids move through the porous media of a biotrickling system (i.e. the liquid "trickles").

A particularly valuable form of biofiltration uses compost as the porous media. Compost contains numerous species of beneficial microbes already acclimated to organic wastes. Industrial compost biofilters have achieved removal rates at the 99% level. Biofilters are also the most common method for removing VOCs and odorous compounds from air streams. In addition to a wide array of volatile chain and aromatic organic compounds, biological systems have successfully removed vapor phase inorganics, such as ammonia, hydrogen sulfide and other sulfides including carbon disulfide, as well as mercaptans.

The operational key is the biofilm. The gas must interface with the film. In fact, this interface may also occur without a liquid phase (see Fig. 32.21). According to Henry's Law, the compounds partition from the gas phase

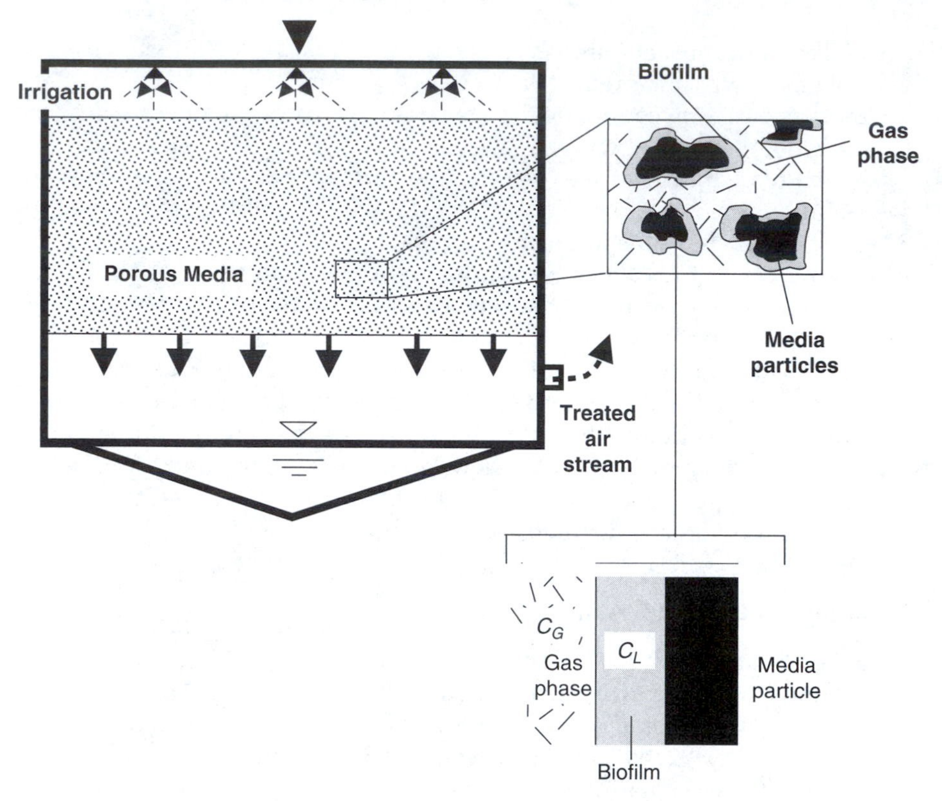

Fig. 32.20. Schematic of packed bed biological control system to treat volatile compounds. Air containing gas phase pollutants (C_G) traverse porous media. The soluble fraction of the volatilized compounds in the air stream partition into the biofilm (C_L) according to Henry's Law: $C_L = \frac{C_G}{H}$; where H is the Henry's Law constant. Adapted from Ergas, S. J. and Kinney, K. A. "Biological Control Systems" in: Air and Waste Management Association, Air Pollution Control Manual, 2nd ed, W. T. Davis (ed.), John Wiley & Sons, Inc., New York, pp. 55–65, 2000.

(in the carrier gas or air stream) to the liquid phase (biofilm). Compost has been a particularly useful medium in providing this partitioning.

The bioscrubber is a two-unit setup. The first unit is an adsorption unit (see previous discussion in this Chapter). This unit may be a spray tower, bubbling scrubber or packed column. After this unit, the air stream enters a bioreactor with a design quite similar to an activated sludge system in a wastewater treatment facility. Bioscrubbers are much less common in the US than biofiltration systems [9].

All three types of biological systems have relatively low operating costs since they are operated near ambient temperature and pressure conditons.

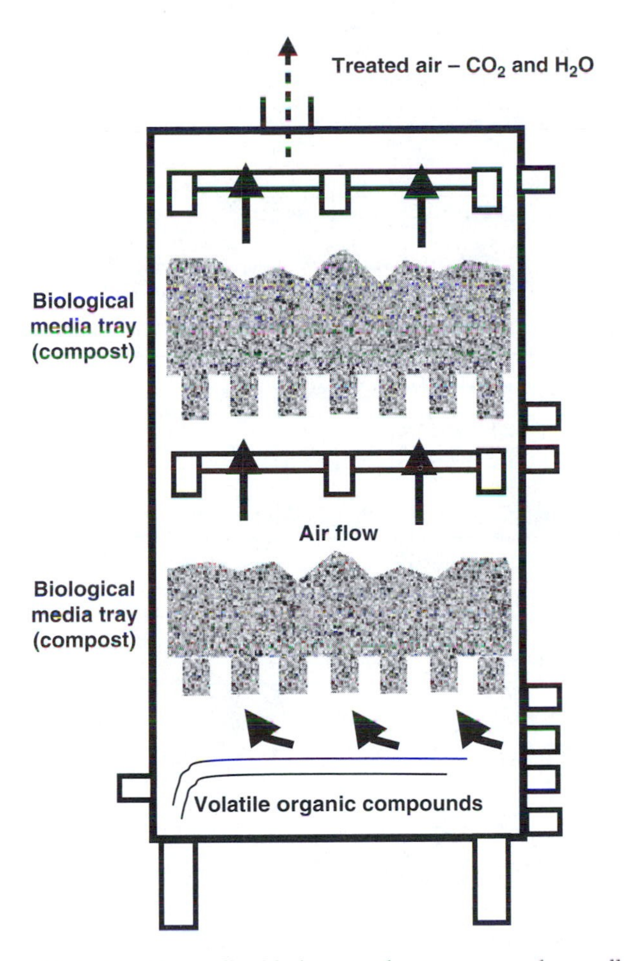

Fig. 32.21. Biofiltration without a liquid phase used to treat vapor phase pollutants. Air carrying the volatilized contaminants upward through porous media (e.g. compost) containing microbes acclimated to break down the particular contaminants. The wastes at the bottom of the system can be heated to increase the partitioning to the gas phase. Microbes in the biofilm surrounding each individual compost particle metabolize the contaminants into simpler compounds, eventually converting them into carbon dioxide and water vapor.

Power needs are generally for air movement and pressure drops are low (<10 cm $H_2O\,m^{-1}$ packed bed). Other costs include amendments (e.g. nutrients) and humidification. Another advantage is the usual small amount of toxic byproducts, as well as low rates of emissions of greenhouse gases (oxides of nitrogen and carbon dioxide), compared to thermal systems. Success is highly dependent on the degradability of the compounds present in the air stream, their fugacity and solubility needed to enter the biofilm (see Fig. 32.19), and pollutant loading rates. Care must be taken in monitoring the porous media for incomplete biodegradation, the presence of substances that may be toxic to the microbes, excessive concentrations of organic acids and alcohols, and pH. The system should also be checked for shock and the presence of dust, grease or other substances that may clog the pore spaces of the media [9].

F. Comparison of Gaseous Removal Systems

As with particulate removal systems, it is apparent that many choices are available for removal of gases from effluent streams. Table 32.6 presents some of the factors that should be considered in selecting equipment.

For the control of SO_2, several systems are currently in development and use. Table 32.7 briefly explains these systems.

TABLE 32.6

Comparison of Gaseous Pollutant Removal Systems

Type of equipment	Pressure drop (cmH_2O)	Installed cost (1990 US$ per m^3)	Annual operating cost (1990 US$ per m^3)
Scrubber	10	9.80	14.00
Absorber	10	10.40	28.00
Condenser	2.5	28.00	7.00
Direct-flame afterburner	1.2	8.20	8.40 + gas
Catalytic afterburner	2.5	11.60	28.00 + gas
Biological control systems	Low (e.g. <1 in compost)	variable (low to moderate)	variable (low to moderate)

TABLE 32.7

Possible Sulfur Dioxide Control Systems

Method	Remarks
Limestone or lime injection (dry)	Calcined limestone or lime reacts with sulfur oxides. They are then removed with a dry particulate control system.
Limestone or lime injection (wet)	Calcined limestone or lime reacts with sulfur oxides, which are then removed by wet scrubbers.
Sodium carbonate	Sodium carbonate reacts with sulfur oxides in a dry scrubber to form sodium sulfite and CO_2. Sodium sulfite is then removed with a baghouse.
Citrate process	Citrate is added to scrubbing water to enhance SO_2 solution into water. Sulfur is then removed from the citrate solution.
Copper oxide adsorption	Oxides of sulfur react with copper oxide to form copper sulfate. Removal with a dry particulate control system follows.
Caustic scrubbing	Caustic neutralizes sulfur oxides. This method is used on small processes.

V. REMOVAL OF ODORS

An odor can be described as a physiological response to activation of the sense of smell [1]. It can be caused by a chemical compound (e.g. H_2S) or a mixture of compounds (e.g. coffee roasting). Generally, if an odor is objectionable, any perceived quantity greater than the odor threshold will be cause for complaint. The control of odors, therefore, becomes a matter of reducing them to less than their odor thresholds, preventing them from entering the atmosphere, or converting them to a substance that is not odorous or has a much higher odor threshold. Odor masking is not recommended for a practical, long-term odor control system.

A. Odor Reduction by Dilution

If the odor is not a toxic substance and has no harmful effects at concentrations below its threshold, dilution may be the least expensive control technique. Dilution can be accomplished either by using tall stacks or by adding dilution air to the effluent. Tall stacks may be more costly if only capital costs are considered, but they do not require the expenditure for energy that is necessary for dilution systems. In addition, if the emission contains other pollutants, taller stacks will increase the distance travelled by the pollutant and, thus, will contribute to the long-range transport and the potential cumulative effects of these pollutants.

The odor threshold for most atmospheric pollutants may be found in the literature [1]. By properly applying the diffusion equations, one can calculate the height of a stack necessary to reduce the odor to less than its threshold at the ground or at a nearby structure. A safety factor of two orders of magnitude is suggested if the odorant is particularly objectionable.

Odor control by the addition of dilution air involves a problem associated with the breakdown of the dilution system. If a dilution fan, motor, or control system fails, the odorous material will be released to the atmosphere. If the odor is objectionable, complaints will be noted immediately. Good operation and maintenance of the dilution system become an absolute requirement, and redundant systems should be considered.

B. Odor Removal

It is sometimes possible to close an odorous system in order to prevent the release of the odor to the atmosphere. For example, a multiple-effect evaporator can be substituted for an open contact condenser on a process emitting odorous, noncondensable gases.

Another possible solution to an odor problem is to substitute a less noxious or more acceptable odor within a process. An example of this type of control is the substitution of a different resin in place of a formaldehyde-based resin in a molding or forming process.

Many gas streams can be deodorized by using solid adsorption systems to remove the odor before the stream is released to the atmosphere. Such procedures are often both effective and economical.

C. Odor Conversion

Many odorous compounds may be converted to compounds with higher odor thresholds or to nonodorous substances. An example of conversion to another compound is the oxidation of H_2S, odor threshold 0.5 ppb, to SO_2, odor threshold 0.5 ppm. The conversion results in another compound with an odor threshold three orders of magnitude greater than that of the original compound.

An example of conversion to a nonodorous substance would be the passage of a gas stream containing butyraldehyde, $CH_3CH_2CH_2CHO$, with an odor threshold of 40 ppb, through a direct-fired afterburner which converts it to CO_2 and H_2O, both nonodorous compounds. It should be noted that using a direct-fired afterburner, particularly one without heat recovery, to destroy 40 ppb is not an economical use of energy, and some other odor control system may be more desirable.

The physical and chemical characteristics of the particles and gases needing treatment determine the appropriate devices to be deployed. Engineers and planners must carefully consider many factors when selecting air pollution control systems. The next chapter provides insights into how this can be done for hazardous pollutants and other emergent contaminants.

REFERENCES

1. Stern, A. C. (ed.), *Air Pollution*, 3rd ed., Vol. 4. Academic Press, NY, 1977.
2. Theodore, L., and Buonicore, A. J., *Air Pollution Control Equipment: Selection, Design, Operation and Maintenance*. Prentice-Hall, Englewood Cliffs, NJ, 1982.
3. Strauss, W., *Industrial Gas Cleaning*, 2nd ed. Pergamon, Oxford, 1975.
4. Buonicore, A. J., and Davis, W. T. (eds.), *Air Pollution Engineering Manual*. Van Nostrand Reinhold, New York, 1992.
5. Katz, J., *Electrostatic Precipitation*. Precipitator Technology, Munhall, PA, 1979.
6. Bell, C. G., and Strauss, W., *J. Air Pollut. Control Assoc.* **23**(11), 967–969 (1973).
7. Teller, A. J., *J. Air Pollut. Control Assoc.* **27**(2), 148–149 (1977).
8. Teller, A. J., Controlling Sulfur Oxides, EPA Research Summary EPA-600/8-80-029. US Environmental Protection Agency, Research Triangle Park, NC, 1980.
9. Ergas, S. J., and Kinney, K. A., "Biological Control Systems" in *Air and Waste Management Association, Air Pollution Control Manual*, 2nd ed., Davis, W. T. (ed.), John Wiley & Sons, Inc., New York, pp. 55–65, 2000.

SUGGESTED READING

Air and Waste Management Association, Air Pollution Control Manual, 2nd Edition, W. T. Davis (Editor), John Wiley & Sons, Inc., New York, NY, 2000.

A Competitive Assessment of US Industrial Air Pollution Control Equipment Industry. US Department of Commerce, Washington, DC, 1991.

Field Operations and Enforcement Manual for Air Pollution Control, Vol. II, *Control Technology and General Source Inspection*, APTD-1101. US Environmental Protection Agency, Research Triangle Park, NC, 1972.

Operation and Maintenance of Electrostatic Precipitators. Specialty Conference Proceedings, Air Pollution Control Association, Pittsburgh, 1978.

Englund, H. M., and Beery, W. T. (eds.), *Electrostatic Precipitation of Fly Ash*, by Harry J. White. APCA Reprint Series, Air Pollution Control Association, Pittsburgh, 1977.

Kester, B. E. (ed.), *Design, Operation, and Maintenance of High Efficiency Particulate Control Equipment*. Specialty Conference Proceedings, Air Pollution Control Association, Pittsburgh, 1973.

Szabo, M. F., and Gerstle, R. W., *Operation and Maintenance of Particulate Control Devices on Coal-Fired Utility Boilers*, EPA-600/2-77-129. US Environmental Protection Agency, Research Triangle Park, NC, 1977.

QUESTIONS

1. List the similarities and differences of pollution control systems for solid PM and liquid droplets.
2. You wish to design a baghouse to clean 3000 m^3 at a filter ratio of 3 m^3 m^{-2} of cloth. The filter bags are 15 cm in diameter by 3-m long. If you design a "square" baghouse with the bags on 30-cm centers, what would be the exterior dimensions, neglecting ductwork? An alternative system uses 15-mm-diameter porous plastic tubes 1 m long on 25-mm centers. For the same filter ratio and flow, what would be the exterior dimensions for a "square" enclosure?
3. For a given cyclone collector, plot centrifugal force as a function of particle specific gravity (0.50–3.00), gas velocity (175–1750 m min^{-1}), and radius of curvature (30–250 cm).
4. List the advantages and disadvantages of using a baghouse, wet scrubber, or ESP for particulate collection from an asphalt plant drying kiln. The gases are at 250°C and contain 450 mg m^{-3} of rock dust in the 0.1–10 μm size range. Gas flow is 2000 m^3 min^{-1}. Consider initial and operation cost, space requirement, ultimate disposal, etc.
5. Suppose a gaseous process effluent of 30 m^3 min^{-1} is at 200°C and 50% relative humidity. It is cooled to 65°C by spraying with water that was initially at 20°C. What volume of saturated gas would you have to design for at 65°C? How much water per cubic meter would the system require? How much water per cubic meter would you have to remove from the system?
6. If an ESP is 90% efficient for particulate removal, what overall efficiency would you expect for two of the ESPs in series? Would the cost of the two in series be double the cost of the single ESP? List two specific cases in which you might use two ESPs in series.
7. The gaseous effluent from a process is 30 m^3 min^{-1} at 65°C. How much natural gas at 8900 kg cal m^{-3} would have to be burned per hour to raise the effluent temperature to 820°C? Natural gas requires 10 m^3 of air for every cubic meter of gas at a theoretical air/fuel ratio. Assume the air temperature is 20°C and the radiation and convection losses are 10%.

8. For Question 7, if heat recovery equipment were installed to raise the incoming effluent to 425°C, how much natural gas would have to be burned per hour?
9. Choose a representative area (a city, county, region, etc.) and prepare a table showing the change in air pollution emission if natural gas were used as a fuel instead of oil and coal.
10. Why are "oxides of nitrogen" and "oxides of sulfur" usually reported in emission inventory tables rather than the actual oxidation states?
11. For a given area estimate the yearly pollutants emitted by automobiles using the figures for gallons of gasoline sold supplied by (a) the gasoline dealers association and (b) the local taxation authorities.

33

Control of Hazardous Air Pollutants

I. AIR QUALITY AND HAZARDOUS WASTES

In Chapter 32, we generally address air pollution as a byproduct of another process, such as manufacturing. Air pollution controls are also a necessary component of directly treating other forms of wastes. The goal is to treat the waste to reduce the toxicity, to decrease exposures and, ultimately, to eliminate or at least properly the risks from hazardous substances in the waste. The type of technology applied depends on the intrinsic characteristics of the contaminants and on the substrate in which they reside. The choice must factor in all of the physical, chemical, and biological characteristics of the contaminant with respect to the matrices and substrates (if soil and sediment) or fluids (air, water, or other solvents) where the contaminants are found. Chapter 32 provides the basics of most air pollution control technologies, but when addressing particularly hazardous air pollutants, a number of additional factors must be considered. The selected approach must meet criteria for treatability (i.e. the efficiency and effectiveness of a technique in reducing the mobility and toxicity of a waste). The comprehensive remedy must consider the effects of each action taken will have on past and proceeding steps.

Eliminating or reducing pollutant concentrations in a waste stream begins with assessing the physical and chemical characteristics of each contaminant, and matching these characteristics with the appropriate treatment technology. All of the kinetics and equilibria, such as solubility, fugacity, sorption, and bioaccumulation factors, will determine the effectiveness of destruction, transformation, removal, immobilization of these contaminants. For example, Table 33.1 ranks the effectiveness of selected treatment technologies on organic and inorganic contaminants typically found in contaminated slurries, soils, sludges, and sediments. As shown, there can be synergies (e.g. innovative incineration approaches are available that not only effectively destroy organic contaminants, but in the process also destroys the inorganic cyanic compounds). Unfortunately, there are also antagonisms among certain approaches, such as the very effective incineration processes for organic contaminants that transform heavy metal species into more toxic and more mobile forms. The increased pressures and temperatures are good for breaking apart organic molecules and removing functional groups that lend them toxicity, but these same processes oxidize or in other ways transform the metals into more toxic or more bioavailable forms. So, when mixtures of organic and inorganic contaminants are targeted, more than one technology may be required to accomplish project objectives, and care must be taken not to make trade one problem (e.g. polychlorinated biphenyls, PCBs) for another (e.g. more mobile species of cadmium).

The characteristics of the substrate (e.g. solid waste, soil, sediment, or water) will affect the performance of any contaminant treatment or control. For example, sediment, sludge, slurries, and soil characteristics that will influence the efficacy of treatment technologies include particle size, solids content, and high contaminant concentration (see Table 33.2).

Particle size may be the most important limiting characteristic for application of treatment technologies to sediments. Most treatment technologies work well on sandy soils and sediments. The presence of fine-grained material adversely affects treatment system emission controls because it increases particulate generation during thermal drying, it is more difficult to dewater, and it has greater attraction to the contaminants (particularly clays). Clayey sediments that are cohesive also present materials handling problems in most processing systems. Solids content generally ranges from high (i.e. usually the *in situ* solids content (30–60% solids by weight)) to low (e.g. hydraulically dredged sediments (10–30% solids by weight)). Treatment of slurries is better at lower solids contents; but this can be achieved even for high solids contents by water addition at the time of processing. It is more difficult to change from a lower to a higher solids content, but evaporative and dewater approaches, such as those used for municipal sludges, may be employed. Also, thermal and dehalogenation processes are decreasingly efficient as solids content is reduced. More water means increased chemical costs and increased need for wastewater treatment.

Elevated levels of organic compounds or heavy metals in high concentrations must also be considered. Higher total organic carbon (TOC) content favors incineration and oxidation processes. The TOC can be the contaminant

TABLE 33.1

Effect of the Characteristics of the Contaminant on Decontamination Efficiencies

	Organic contaminants					Inorganic contaminants		
Treatment technology	PCBs	PAHs	Pesticides	Petroleum hydrocarbons	Phenolic compounds	Cyanide	Mercury	Other heavy materials
Conventional incineration	D	D	D	D	D	D	xR	pR
Innovative incineration[a]	D	D	D	D	D	D	xR	I
Pyrolysis[a]	D	D	D	D	D	D	xR	I
Vitrification[a]	D	D	D	D	D	D	xR	I
Supercritical water oxidation	D	D	D	D	D	D	U	U
Wet air oxidation	pD	D	U	D	D	D	U	U
Thermal desorption	R	R	R	R	U	U	xR	N
Immobilization	pI	pI	pI	pI	pI	pI	U	I
Solvent extraction	R	R	R	R	R	pR	N	N
Soil washing[b]	pR	pR	pR	pR	pR	pR	pR	pR
Dechlorination	D	N	pD	N	N	N	N	N
Oxidation[c]	N/D	N/D	N/D	N/D	N/D	N/D	U	xN
Bioremediation[d]	N/pd	N/D	N/D	D	D	N/D	N	N

[a] This process is assumed to produce a vitrified slag.
[b] This effectiveness of soil washing is highly dependent on the particle size of the sediment matrix, contaminant characteristics, and the type of extractive agents used.
[c] The effectiveness of oxidation depends strongly on the types of oxidant(s) involved and the target contaminants.
[d] The effectiveness of bioremediation is controlled by a large number of variables as discussed in the text.

Primary designation
D = effectively destroys contaminant
R = effectively removes contaminant
I = effectively immobilizes contaminant
N = no significant effect
N/D = effectiveness varies from no effect to highly efficient depending on the type of contaminant within each class
U = effect not known

Prefixes
p = partial
x = may cause release of non-target contaminant

Source: US Environmental Protection Agency, *Remediation Guidance Document*, Chapter 7, EPA-905-B94-003, 2003.

TABLE 33.2

Effect of Particle Size, Solids Content, and Extent of Contamination on Decontamination Efficiencies

	Predominant particle size			Solids content		High contaminant concentration	
Treatment technology	Sand	Silt	Clay	High (slurry)	Low (*in situ*)	Organic compounds	Metals
Conventional incineration	N	X	X	F	X	F	X
Innovative incineration	N	X	X	F	X	F	F
Pyrolysis	F	X	X	F	X	F	F
Vitrification	X	F	F	X	F	F	X
Supercritical water oxidation	X	F	F	X	F	F	X
Wet air oxidation	F	X	X	F	X	F	N
Thermal desorption	F	X	X	F	X	F	N
Immobilization	F	X	X	F	X	X	N
Solvent extraction	F	F	X	F	X	X	N
Soil washing	F	F	X	N	F	N	N
Dechlorination	U	U	U	F	X	X	N
Oxidation	F	X	X	N	F	X	X
Bioslurry process	N	F	N	N	F	X	X
Composting	F	N	X	F	X	F	X
Contained treatment facility	F	N	X	F	X	X	X

F: sediment characteristic favorable to the effectiveness of the process; N: sediment characteristic has no significant effect on process performance; U: effect of sediment characteristic on process is unknown; and X: sediment characteristic may impede process performance or increase cost.

Source: US Environmental Protection Agency, *Remediation Guidance Document*, Chapter 7, EPA-905-B94-003, 2003.

of concern or any organic, since they are combustibles with caloric value. Conversely, higher metal concentrations may make a technology less favorable by increasing contaminant mobility of certain metal species following application of the technology.

A number of other factors may affect selection of a treatment technology other than its effectiveness for treatment (some are listed in Table 33.3). For example, vitrification and supercritical water oxidation have only been used for relatively small projects and would require more of a proven track record before implementing them for full-scale sediment projects. Regulatory compliance and community perception are always a part of decisions regarding an incineration system. Land use considerations, including the amount of acreage needs, are commonly confronted in solidification and solid-phase bioremediation projects (as they are in sludge farming and land application). Disposing of ash and other residues following treatment must be part of any process. Treating water effluent and air emissions must be part of the decontamination decision-making process [1].

TABLE 33.3

Selected Factors on Selecting Decontamination and Treatment Approaches

Treatment technology	Implementability at full scale	Regulatory compliance	Community acceptance	Land requirements	Residuals disposal	Wastewater treatment	Air emissions control
Conventional incineration		✓	✓	✓			✓
Innovative incineration		✓	✓	✓			✓
Pyrolysis		✓					✓
Vitrification	✓	✓					✓
Supercritical water oxidation	✓						
Wet air oxidation							
Thermal desorption					✓	✓	✓
Immobilization				✓			✓
Solvent extraction					✓	✓	
Soil washing					✓	✓	
Dechlorination							✓
Oxidation	✓						
Bioslurry process	✓						✓
Composing				✓			✓
Contained treatment facility				✓		✓	✓

✓: the factor is critical in the evaluation of the technology.

Source: US Environmental Protection Agency, *Remediation Guidance Document*, Chapter 7, EPA-905-B94-003, 2003.

II. PRE-CONTROL CONSIDERATIONS

A. Estimating Contaminant Migration

Estimating potential contaminant releases (i.e. "losses" as defined by environmental regulators) from various combinations of treatment technologies is difficult due to the variability of chemical and physical characteristics of contaminated media (especially soils and sediments), the strong affinity of most contaminants for fine-grained sediment particles, and the limited track record or "scale-up" studies for many treatment technologies. Off-the-shelf models can be used for simple process operations, such as extraction or thermal vaporization applied to single contaminants in relatively pure systems. However, such models have not been appropriately evaluated for a number of other technologies because of the limited database on treatment technologies, such as for contaminated sediments or soils [1].

B. Treatability Tests

Standard engineering practice for evaluating the effectiveness of treatment technologies for any type of contaminated media (solids, liquids, or gases) requires first performing a treatability study for a sample that is representative of the contaminated material [1]. The performance data from treatability studies can aid in reliably estimating contaminant concentrations for the residues following treatment, as well as possible waste streams generated by a technology. Treatability studies may be performed at the bench-scale (in the laboratory) or at pilot-scale level (e.g. a real-world study, but limited in number of contaminants, in spatial extent, or to a specific, highly controlled form of a contaminant, e.g. one pure congener of PCBs, rather than the common mixtures). Most treatment technologies include post-treatment or controls for waste streams produced by the processing. The contaminant losses can be defined as the residual contaminant concentrations in the liquid or gaseous streams released to the environment. For technologies that extract or separate the contaminants from the bulk of the sediment, a concentrated waste stream may be produced that requires treatment offsite at a hazardous waste treatment facility, where permit requirements may require destruction and removal efficiencies greater than 99.9999% (i.e. the so-called rule of "six nines"). The other source of loss for treatment technologies is the residual contamination in the sediment after treatment. After disposal, treated wastes are subject to leaching, volatilization, and losses by other pathways. The significance of these pathways depends on the type and level of contamination that is not removed or treated by the treatment process. Various waste streams for each type of technology that should be considered in treatability evaluations are listed in Table 33.4.

TABLE 33.4

Selected Waste Streams Commonly Requiring Treatability Studies

Contaminant loss stream	Treatment technology type						
	Biological	Chemical	Extraction	Thermal desorption	Thermal destruction	Immobilization	Particle separation
Residual solids	X	X	X	X	X	X	X
Wastewater	X	X	X	X			X
Oil/organic compounds			X	X			X
Leachate						X[a]	
Stack gas				X	X		
Adsorption media			X	X			
Scrubber water					X		
Particulates (filter/cyclone)				X	X		

[a] Longe-term contaminant losses must be estimated using leaching tests and contaminant transport modeling similar to that used for sediment placed in a confined disposal facility. Leaching could be important for residual solids for other processes as well.

Source: US Environmental Protection Agency, *Remediation Guidance Document*, Chapter 7, EPA-905-B94-003, 2003.

III. CONTAMINANT TREATMENT AND CONTROL APPROACHES[1]

The life cycle viewpoint is instructive in controlling hazardous air pollutants. Five steps in sequence define an event that results in environmental contamination of the air, water, or soil pollution. These steps individually and collectively offer opportunities to intervene and to control the risks associated with hazards and thus protect public health and the environment. The steps address the presence of waste at five points in the life cycle:

SOURCE →
RELEASE →
TRANSPORT →
EXPOSURE →
RESPONSE

As a first step, the contaminant source must be identifiable. A hazardous substance must be released from the source, be transported through the air, water, or soil environment, reach a human, animal, or plant receptor in a measurable dose, and the receptor must have a quantifiable detrimental response in the form of death or illness. Intervention can occur at any one of these steps to control the risks to public health and to the environment. Of course, any intervention scheme and subsequent control by the engineer must be justified by the environmental engineer as well as the public or private client in terms of scientific evidence, sound engineering design, technological practicality, economic realities, ethical considerations, and the laws of local, state, and national governments.

A. Intervention at the Source of Contamination

A contaminant must be identifiable, either in the form of an industrial facility that generates waste byproducts, a hazardous waste processing facility, a surface or subsurface land storage/disposal facility, or an accidental spill into a water, air, or soil receiving location. The intervention must minimize or eliminate the risks to public health and the environment by utilizing technologies at this source that are economically acceptable and based on applicable scientific principles and sound engineering designs.

In the case of an industrial facility producing hazardous waste as a necessary byproduct of a profitable item, as considered here for example, the engineer can take advantage of the growing body of knowledge that has become known as life cycle analysis [4]. In the case of a hazardous waste storage facility or a spill, the engineer must take the source as a given and search for possibilities for intervention at a later step in the sequence of steps as discussed below.

[1] The principal sources for this section are collaborations with two prominent environmental engineers, Ross E. McKinney and J. Jeffrey Peirce.

Under the life cycle analysis method of intervention, the environmental manager considers the environmental impacts that could incur during the entire life cycle of (1) all of the resources that go into the product; (2) all the materials that are in the product during its use; and (3) all the materials that are available to exit from the product once it or its storage containers are no longer economically useful to society. Few simple examples exist that describe how *life cycle analysis* is conducted but consider for now any one of a number of household cleaning products. Consider that a particular cleaning product, a solvent of some sort, must be fabricated from one of several basic natural resources. Assume for now that this cleaning product currently is petroleum based. The engineer could intervene at this initial step in the life cycle of this product, as the natural resource is being selected, and consequently the engineer could preclude the formation of a source of hazardous waste by suggesting instead the production of a water-based solvent.

Similarly, intervention at the production phase of this product's life cycle and suggested fabrication techniques can preclude the formation of a source of certain contaminants from the outset. In this case the recycling of spent petroleum materials could provide for more household cleaning products with less or zero hazardous waste generation, thus controlling the risks to public health and the environment. Another example is that of "co-generation," which may allow for two manufacturing facilities to co-locate so that the "waste" of one is a "resource" for the other. An example is the location of a chemical plant near a power generation facility, so that the excess steam generated by the power plant can be piped to the nearby chemical plant, obviating the need to burn its own fuel to generate the steam needed for chemical synthesis. Another example is the use of an alcohol waste from one plant as a source for chemical processes at another.

As discussed in Chapter 30, the product under consideration must be considered long before any switches are flipped and valves turned. For example, a particular household cleaning product may result in unintended human exposure to buckets of solvent mixtures that fumigate the air in a home's kitchen or pollute the town's sewers as the bucket's liquid is flushed down a drain. In this way, life cycle analysis is type of systems engineering where a critical path is drawn, and each decision point considered.

Under the plan, the disposal of this solvent's containers must be considered from a long-term risk perspective. The challenge is that every potential and actual environmental impact of a product's fabrication, use, and ultimate disposal must be considered. This is seldom, if ever, a "straight line projection."

B. Intervention at the Point of Release

Once a contaminant source has been identified, the next step is to intervene at the point at which the waste is released into the environment. This point of release could be at the top of a stack or vent from the source of pollution to a receiving air shed, or it could be a more indirect releases, such as from the bottom most layer of a clay liner in a hazardous waste landfill connected to surrounding soil material. Similarly this point of release could be a series of points as a contaminant is released along a shoreline from a plot of land into the air, from an evaporation pond, from a landfill, near a river or through a plane of soil underlying a storage facility (i.e. a so-called "non-point source").

C. Intervention as the Contaminant Is Transported in the Environment

Wise site selection of facilities that generate, process, and store contaminants is the first step in preventing or reducing the likelihood that they will move. For example, the distance from a source to a receptor is a crucial factor in controlling the quantity and characteristics of waste as it is transported.

Meteorology is a primary determinant of the opportunities to control the atmospheric transport of contaminants. For example, manufacturing, transportation, and hazardous waste generating, processing, and storage facilities must be sited to avoid areas where specific local weather patterns are frequent and persistent. These avoidance areas include ground-based inversions, elevated inversions, valley winds, shore breezes, and city heat islands. In each of these venues, the pollutants become locked into air masses with little of no chance of moving out of the respective areas. Thus the concentrations of the pollutants can quickly and greatly pose risks to public health and the environment. In the soil environment the engineer has the opportunity to site facilities in areas of great depth-to-groundwater, as well as in soils (e.g. clays) with very slow rates of transport. In this way, engineers and scientists must work closely with city and regional planners in early in the site selection phases.[2]

D. Intervention to Control the Exposure

As mentioned in Chapter 1, the receptor of contamination can be a human, other fauna in the general scheme of living organisms, flora, or materials or constructed facilities. In the case of humans, as we discussed previously, the contaminant can be ingested, inhaled, or dermally contacted. Such exposure can be direct with human contact to, for example, particles of lead that are present in inhaled indoor air. An exposure also can be indirect as in the case of human ingestion of the cadmium and other heavy metals found in the livers of beef cattle that were raised on grasses receiving nutrition from cadmium-laced municipal wastewater treatment biosolids (commonly known as "sludge").

[2] This goes beyond zoning. Obviously, the engineer should be certain that the planned facility adheres to the zoning ordinances, land use plans, and maps of the state and local agencies. However, it behooves all of the professionals to collaborate, hopefully before any land is purchased and contractors are retained. Councils of Government (COGs) and other "A-95" organizations can be rich resources when considering options on siting. They can help avoid the need for problems long before implementation, to say nothing of contentious zoning appeal and planning commission meetings and perception problems at public hearings.

Heavy metals or chlorinated hydrocarbons similarly can be delivered to domestic animals and animals in the wild. Construction materials are also sensitive to exposure to air pollutants, from the "greening" of statutes through the de-zinc process associated with low pH rain events to the crumbling of stone bridges found in nature. Isolating potential receptors from exposure to hazardous air pollutants the engineer has an opportunity to control the risks to those receptors.

The opportunities to control exposures to contaminants are directly associated with the ability to control the amount of hazardous pollutants delivered to the receptor through source control and siting of control systems and hazardous waste management facilities. One solution to environmental contamination could be to increase their dilution in the air, water, or soil environments. We will discuss specific examples of this type of intervention later in this chapter.

E. Intervention at the Point of Response

Opportunities for intervention are grounded in basic scientific principles, engineering designs and processes, and applications of proven and developing technologies to control the risks associated with contaminants. Let us consider thermal processing as a class of hazardous control technology that is widely used in treating wastes, but which has a crucial air pollution component.

IV. THERMAL TREATMENT PROCESSES

A. Thermodynamics and Stoichiometry

Contaminants, if completely organic in structure, are, in theory, completely destructible using principles based in thermodynamics with the engineering inputs and outputs summarized as:

$$\text{Hydrocarbons} + O_2\,(+\text{ energy?}) \rightarrow CO_2 + H_2O\,(+\text{ energy?}) \quad (33.1)$$

Contaminants are mixed with oxygen, sometimes in the presence of an external energy source, and in fractions of seconds or several seconds the byproducts of gaseous carbon dioxide and water are produced to exit the top of the reaction vessel while a solid ash is produced to exit the bottom of the reaction vessel.[3] 3K Energy may also be produced during the reaction and the heat may be recovered. A derivative problem in this simple reaction could be global warming associated with the carbon dioxide.

[3] Numerous textbooks address the topic of incineration in general and hazardous waste incineration in particular. For example see Haas, C. N., and Ramos, R. J. *Hazardous and Industrial Waste Treatment*. Prentice-Hall, Inc., Englewood Cliffs, NJ, 1995; Wentz, C. A. *Hazardous Waste Management*. McGraw-Hill, Inc., New York, NY, 1989; and Peirce, J. J., Weiner, R. F., and Vesilind, P. A. *Environmental Pollution and Control*. Butterworth-Heinemann, Boston, MA, 1998.

Conversely, if the contaminant of concern to the engineer contains other chemical constituents, in particular chlorine and/or heavy metals, the original simple input and output relationship is modified to a very complex situation:

$$\begin{gathered}\text{Hydrocarbons} + O_2 (+\text{ energy?}) + \text{Cl or heavy metal(s)} + H_2O \\ + \text{ inorganic salts} + \text{nitrogen compounds} + \text{sulfur compounds} \\ + \text{ phosphorus compounds} \rightarrow CO_2 + H_2O(+\text{ energy?}) \\ + \text{ chlorinated hydrocarbons or heavy metal(s) inorganic salts} \\ + \text{ nitrogen compounds} + \text{sulfur compounds} \\ + \text{ phosphorus compounds}\end{gathered} \tag{33.2}$$

With these contaminants the potential exists for destruction of the initial contaminant, but actually exacerbating the problem by generating more hazardous off-gases containing chlorinated hydrocarbons and/or ashes containing heavy metals are produced (e.g. the improper incineration of certain chlorinated hydrocarbons can lead to the formation of the highly toxic chlorinated dioxins, furans, and hexachlorobenzene). All of the thermal systems discussed below have common attributes. All require the balancing of the three "*T*'s" of the science, engineering, and technology of incineration of any substance:

1. *T*ime of incineration
2. *T*emperature of incineration
3. *T*urbulence in the combustion chamber

The advantages of thermal systems include: (1) the potential for energy recovery; (2) volume reduction of the contaminant; (3) detoxification as selected molecules are reformulated; (4) the basic scientific principles, engineering designs, and technologies are well understood from a wide range of other applications including electric generation and municipal solid waste incineration; (5) application to most organic contaminants which compose a large percentage of the total contaminants generated worldwide; (6) the possibility to scale the technologies to handle a single gallon per pound (liter per kilogram) of waste or millions of gallon per pound (liter per kilogram) of waste: and (7) land areas that are small compared to many other facilities (e.g. landfills).

Each system design must be customized to address the specific contaminants under consideration, including the quantity of waste to be processed over the planning period as well as the physical, chemical, and microbiological characteristics of the waste also over the planning period of the project. The space required for the incinerator itself ranges from several square yards to possibly on the back of a flat bed truck to several acres to sustain a regional incinerator system. Laboratory testing and pilot studies matching a given waste to a given incinerator must be conducted prior to the design, citing, and construction of each incinerator. Generally, the same reaction applies to most thermal processes, i.e. gasification, pyrolysis, hydrolysis, and combustion [3]:

$$C_{20}H_{32}O_{10} + x_1O_2 + x_2H_2O \underset{\Delta}{\rightarrow} y_1C + y_2CO_2 + y_3CO + y_4H_2 + y_5CH_4 + y_6H_2O + y_7C_nH_m \quad (33.3)$$

The coefficients x and y balance the compounds on either side of the equation. The delta under the arrow indicates heating. In many thermal reactions, C_nH_m includes the alkanes, C_2H_2, C_2H_4, C_2H_6, C_3H_8, C_4H_{10}, C_5H_{12}, and benzene, C_6H_6. The actual reactions from test burns for commonly incinerated compounds are provided in Table 10.1.

Of all of the thermal processes, incineration is the most common process for destroying organic contaminants in industrial wastes. Incineration simply is heating wastes in the presence of oxygen to oxidize organic compounds (both toxic and non-toxic). The principal incineration steps are shown in Fig. 33.1.

B. Applying Thermal Processes for Treatment

A word of warning when choosing incineration as the recommended technology: The mere mention of "incineration" evokes controversy in communities. There have been real and perceived failures. It is also important to note that incineration alone does not "destroy" heavy metals; it simply changes the valence of the metal. In fact, incineration can increase the leachability of metals via oxidation, although processes like slagging (operating at sufficiently high temperatures to melt and remove incombustible materials) or

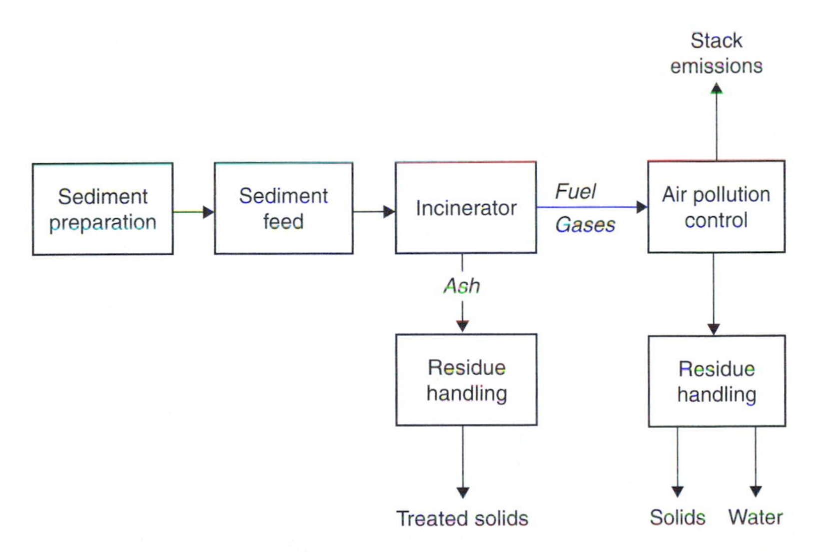

Fig. 33.1. Steps in the incineration of contaminants. *Source*: US Environmental Protection Agency, *Remediation Guidance Document*, Chapter 7, EPA-905-B94-003, 2003.

vitrification (producing non-leachable, basalt-like residue) actually reduce the mobility of many metals.

Leachability is a measure of the ease with which compounds in the waste can move into the accessible environment. The increased leachability of metals would be problematic if the ash and other residues are to be buried in landfills or stored in piles. The leachability of metals is generally measured using the toxicity characteristic leaching procedure (TCLP) test, discussed earlier. Incinerator ash that fails the TCLP must be disposed of in a waste facility approved for hazardous wastes. Enhanced leachability would be advantageous only if the residues are engineered to undergo an additional treatment step of metals. Again, the engineer must see incineration of but one component within a systematic approach for any contaminant treatment process.

There are a number of places in the incineration flow of the contaminant through the incineration process where new compounds may need to be addressed. As mentioned, ash and other residues may contain high levels of metals, at least higher than the original feed. The flue gases are likely to include both organic and inorganic compounds that have been released as a result of temperature induced volatilization and/or newly transformed products of incomplete combustion (PICs) with higher vapor pressures than the original contaminants.

The disadvantages of hazardous waste incinerators include: (1) the equipment is capital intensive, particularly the refractory material lining the inside walls of each combustion chamber that must be replaced as cracks form whenever a combustion system is cooled and/or heated; (2) the operation of the equipment requires very skilled operators and is more costly when fuel must be added to the system; (3) ultimate disposal of the ash is necessary and particularly troublesome and costly if heavy metals and/or chlorinated compounds are found during the expensive monitoring activities; and (4) air emissions may be hazardous and thus must be monitored for chemical constituents and controlled.

Given these underlying principles of incineration, seven general guidelines emerge:

1. Only liquid purely organic contaminants are true candidates for combustion.
2. Chlorine-containing organic materials deserve special consideration if in fact they are to be incinerated at all; special materials used in the construction of the incinerator, long (many seconds) of combustion time, high temperatures (>1600°C), with continuous mixing if the contaminant is in the solid or sludge form.
3. Feedstock containing heavy metals generally should not be incinerated.
4. Sulfur-containing organic material will emit sulfur oxides which must be controlled.
5. The formation of nitrogen oxides can be minimized if the combustion chamber is maintained above 1100°C.

6. Destruction depends on the interaction of a combustion chamber's temperature, dwell time, and turbulence.
7. Off-gases and ash must be monitored for chemical constituents, each residual must be treated as appropriate so the entire combustion system operates within the requirements of the local, state, and federal environmental regulators, and hazardous components of the off-gases, off-gas treatment processes, and the ash must reach ultimate disposal in a permitted facility.

V. THERMAL DESTRUCTION SYSTEMS

The types of thermal systems vary considerably. Five general categories are available to destroy contaminants: (1) rotary kiln; (2) multiple hearth; (3) liquid injection; (4) fluidized bed; and (5) multiple chamber.

A. Rotary Kiln

The combustion chamber in a rotary kiln incinerator as illustrated in Fig. 33.2 is a heated rotating cylinder that is mounted at an angle with possible baffles added to the inner face to provide the turbulence necessary for the target three "*T*'s" for the contaminant destruction process to take place. Engineering design decisions, based on the results of laboratory testing of a specific contaminant, include: (1) angle of the drum; (2) diameter and length of the drum; (3) presence and location of the baffles; (4) rotational speed of the drum; and (5) use of added fuel to increase the temperature of the combustion chamber

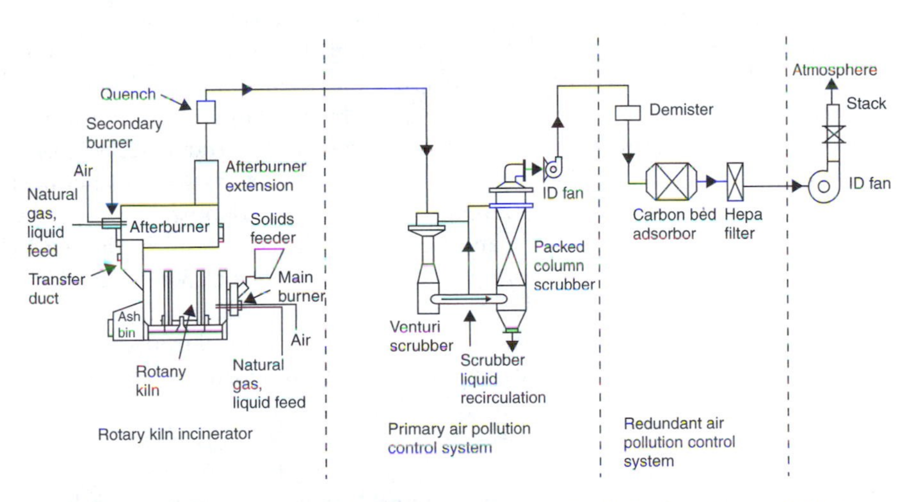

Fig. 33.2. Rotary kiln system. *Source*: US Environmental Protection Agency, 1997, Lee, J., Fournier Jr., D., King, C., Venkatesh, S., and Goldman, C. *Project Summary: Evaluation of Rotary Kiln Incinerator Operation at Low-to-Moderate Temperature Conditions.*

as the specific contaminant requires. The liquid, sludge, or solid hazardous waste is input into the upper end of the rotating cylinder, rotates with the cylinder-baffle system, and falls with gravity to the lower end of the cylinder. The heated upward moving off-gases are collected, monitored for chemical constituents, and subsequently treated as appropriate prior to release, while the ash falls with gravity to be collected, monitored for chemical constituents, and also treated as needed before ultimate disposal. The newer rotary kiln systems [4] consist of a primary combustion chamber, a transition volume, and a fired afterburner chamber. After exiting the afterburner, the flue gas is passed through a quench section followed by a primary air pollution control system (APCS). The primary APCS can be a venture scrubber followed by a packed-column scrubber. Downstream of the primary APCS, a backup secondary APCS, with a demister, an activated carbon adsorber, and a high-efficiency particulate air (HEPA) filter can collect contaminants not destroyed by the incineration.

The rotary kiln is applicable to the incineration of most organic contaminants, it is well suited for solids and sludges, and in special cases liquids and gases can be injected through auxiliary nozzles in the side of the combustion chamber. Operating temperatures generally vary from 800°C to 1650°C. Engineers use laboratory experiments to design residence times of seconds for gases and minutes or possibly hours for the incineration of solid material.

B. Multiple Hearth

In the multiple hearth illustrated in Fig. 33.3 generally contaminants in solid or sludge form are fed slowly through the top vertically stacked hearth; in special configurations hazardous gases and liquids can be injected through side nozzles. Multiple hearth incinerators, historically developed to burn municipal wastewater treatment biosolids, rely on gravity and scrapers working in the upper edges of each hearth to transport the waste through holes from upper hotter hearths to lower cooler hearths. Heated upward moving off-gases are collected, monitored for chemical constituents, and treated as appropriate prior to release; the falling ash is collected, monitored for chemical constituents, and subsequently treated prior to ultimate disposal.

Most organic wastes generally can be incinerated using a multiple hearth configuration. Operating temperatures generally vary from 300°C to 980°C. These systems are designed with residence times of seconds if gases are fed into the chambers to several hours if solid materials are placed on the top hearth and allowed to eventually drop to the bottom hearth exiting as ash.

C. Liquid Injection

Vertical or horizontal nozzles spray liquid hazardous wastes into liquid injection incinerators specially designed for the task or as a retrofit to one of the other incinerators discussed here. The wastes are atomized through the

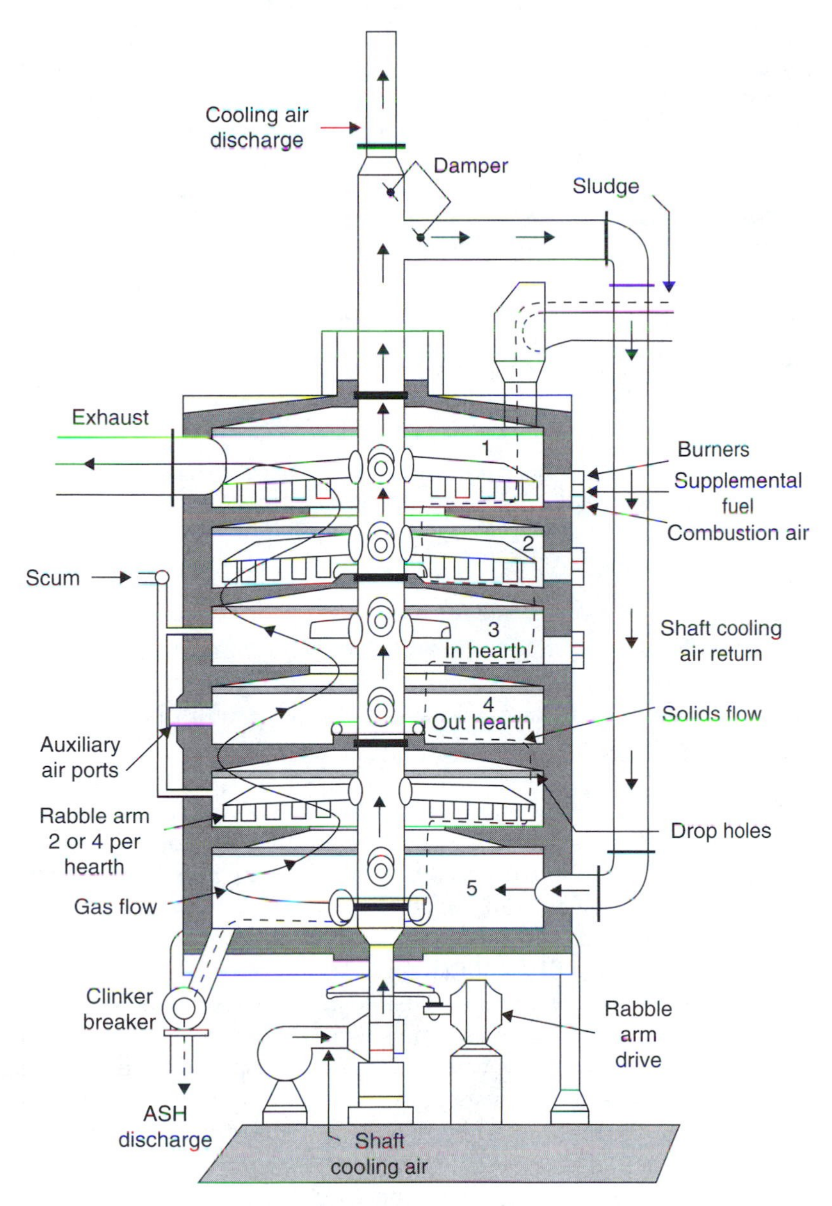

Fig. 33.3. Multiple hearth incineration system. *Source*: US Environmental Protection Agency, *Locating and Estimating Air Emissions from Sources of Benzene*, EPA-454/R-98-011, Research Triangle Park, NC, 1998.

nozzles that match the waste being handled with the combustion chamber as determined in laboratory testing. The application obviously is limited to liquids that do not clog these nozzles, though some success has been experienced with hazardous waste slurries. Operating temperatures generally vary from

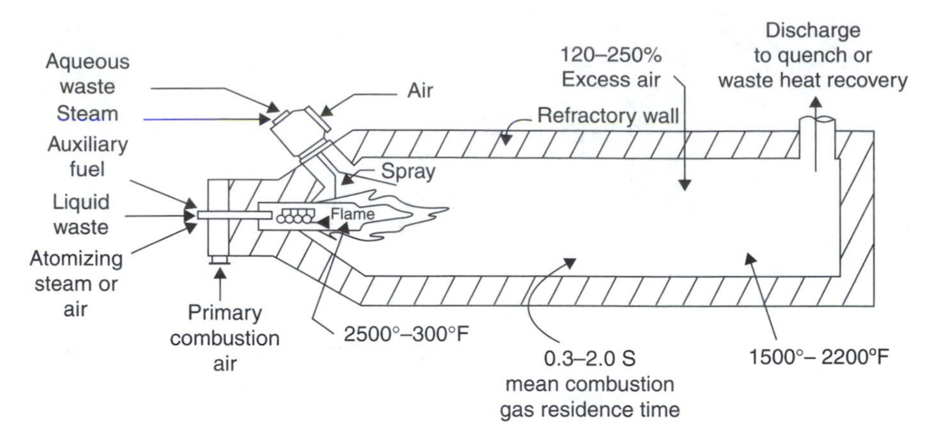

Fig. 33.4. Prototype of liquid injection system. *Source*: US Environmental Protection Agency, *Locating and Estimating Air Emissions from Sources of Benzene*, EPA-454/R-98-011, Research Triangle Park, NC, 1998.

650°C to 1650°C (1200°F to 3000°F). Liquid injection systems (Fig. 33.4) are designed with residence times of fractions of seconds as off-gases, the upward moving off-gases are collected, monitored for chemical constituents, and treated as appropriate prior to release to the lower troposphere.

D. Fluidized Bed

Contaminated feedstock is injected under pressure into a heated bed of agitated inert granular particles, usually sand, as the heat is transferred from the particles to the waste, and the combustion process proceeds as summarized in Fig. 33.5. External heat is applied to the particle bed prior to the injection of the waste and continually is applied throughout the combustion operation as the situation dictates. Heated air is forced into the bottom of the particle bed and the particles become suspended among themselves during this continuous fluidizing process. The openings created within the bed permit the introduction and transport of the waste into and through the bed. The process enables the contaminant to come into contact with particles that maintain their heat better than, for example, the gases inside a rotary kiln. The heat maintained in the particles increases the time the contaminant is in contact with a heated element and thus the combustion process could become more complete with fewer harmful byproducts. Off-gases are collected, monitored for chemical constituents, and treated as appropriate prior to release, and the falling ash is collected, monitored for chemical constituents, and subsequently treated prior to ultimate disposal.

Most organic wastes can be incinerated in a fluidized bed, while the system is best suited for liquids. Operating temperatures generally vary from 750°C to 900°C. Liquid injection systems are designed with residence times

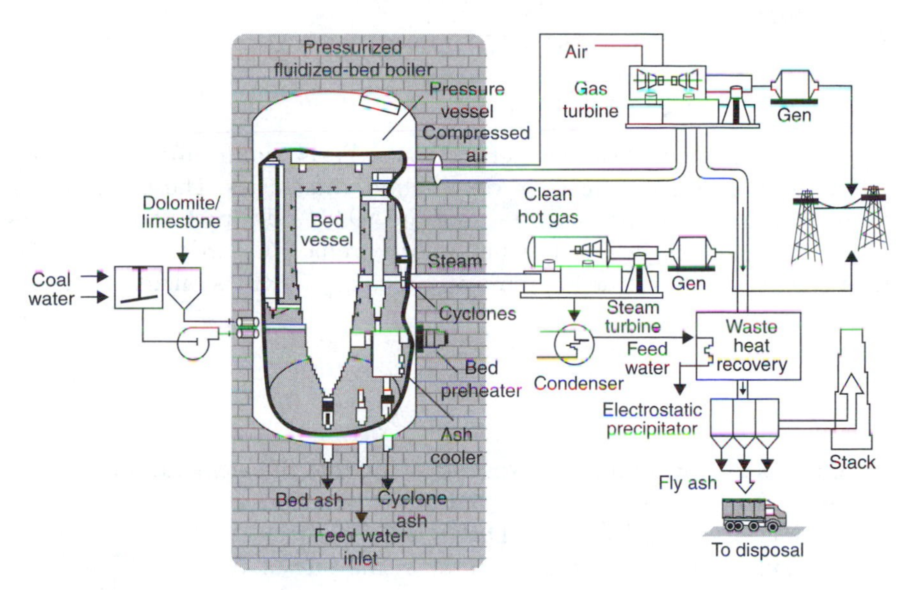

Fig. 33.5. Pressurized fluidized bed system. *Source*: US Department of Energy, 1999, TIDD PFBC Demonstration Project.

of fractions of seconds as off-gases, the upward moving off-gases are collected, monitored for chemical constituents, and treated as appropriate prior to release to the lower troposphere.

E. Multiple Chamber

Contaminants are turned to a gaseous form on a grate in the ignition chamber of a multiple chamber system. The gases created in this ignition chamber travel through baffles to a secondary chamber where the actual combustion process takes place. Often the secondary chamber is located above the ignition chamber to promote natural advection of the hot gases through the system. Heat may be added to the system in either the ignition chamber or the secondary chamber as required for specific burns.

The application of multiple chamber incinerators generally is limited to solid wastes with the waste entering the ignition chamber through an opened charging door in batch, not continuous, loading. Combustion temperatures typically hover near 540°C for most applications. These systems are designed with residence times of minutes to hours for solid hazardous wastes as off-gases are collected, monitored for chemical constituents, and treated as appropriate prior to release to the lower troposphere. At the end of each burn period the system must be cooled so that the ash can be removed prior to monitoring for chemical constituents and subsequent treatment prior to ultimate disposal.

VI. DESTRUCTION REMOVAL

A. Calculating Destruction Removal

Federal hazardous waste incineration standards require that hazardous organic compounds meet certain destruction efficiencies. These standards require that any hazardous waste undergo 99.99% destruction of all hazardous wastes and 99.9999% destruction of extremely hazardous wastes like dioxins. Recall that destruction removal efficiency (DRE) is calculated as:

$$\mathrm{DRE} = \frac{W_{in} - W_{out}}{W_{in}} \times 100 \qquad (33.4)$$

where W_{in} is the rate of mass of waste flowing into the incinerator and W_{out} is the rate of mass of waste flowing out of the incinerator.

For example, let us calculate the DRE if during a stack test, the mass of pentachlorodioxin is loaded into incinerator at the rate of $10\,\mathrm{mg\,min^{-1}}$ and the mass flow rate of the compound measured downstream in the stack is 200 picograms (pg) $\mathrm{min^{-1}}$. Is the incinerator up to code for the thermal destruction of this dioxin?

$$\mathrm{DRE} = \frac{W_{in} - W_{out}}{W_{in}} \times 100 = \frac{10\,\mathrm{mg\,min^{-1}} - 200\,\mathrm{pg\,min^{-1}}}{10\,\mathrm{mg\,min^{-1}}} \times 100$$

Since $1\,\mathrm{pg} = 10^{-12}\,\mathrm{g}$ and $1\,\mathrm{mg} = 10^{-3}$, then $1\,\mathrm{pg} = 10^{-9}\,\mathrm{mg}$. So

$$\frac{10\,\mathrm{mg\,min^{-1}} - 200 \times 10^{-9}\,\mathrm{mg\,min^{-1}}}{10\,\mathrm{mg\,min^{-1}}} \times 100 = 999999.98\%\ \text{removal}$$

Even if pentachlorodioxin is considered to be "extremely hazardous," this is better than the "rule of six nines" so the incinerator is operating up to code.

If we were to calculate the DRE during the same stack test for the mass of tetrachloromethane (CCl_4) loaded into incinerator at the rate of $100\,\mathrm{L\,min^{-1}}$ and the mass flow rate of the compound measured downstream is $1\,\mathrm{mL\,min^{-1}}$. Is the incinerator up to code for CCl_4? This is a lower removal rate since 100 L are in and 0.001 are leaving, so the DRE = 99.999. This is acceptable, i.e. better removal efficiency than 99.99% by an order of magnitude, so long as CCl_4 is not considered an extremely hazardous compound. If it were, then it would have to meet the rule of six nines (it only has five).

By the way, both of these compounds are chlorinated. As mentioned, special precautions must be taken when dealing with such halogenated compounds, since even more toxic compounds that those being treated can end up being generated. Incomplete reactions are very important sources of environmental contaminants. For example, these reactions generate PICs, such as dioxins, furans, carbon monoxide (CO), polycyclic aromatic hydrocarbons (PAHs), and hexachlorobenzene. Thus, the formation of unintended byproducts is always a possibility with thermal systems. An extensive discussion can be found in Chapter 7, for example, on the various processes that can lead to the formation of dioxins and furans. In particular, see the thermal relationships in Table 7.5 and Fig. 7.6.

VII. OTHER THERMAL PROCESSES

A. Processes Other than Incineration

Incineration is frequently used to decontaminate soils with elevated concentrations of organic hazardous constituents. High-temperature incineration, however, may not be needed to treat soils contaminated with most volatile organic compounds (VOCs). Also, in waste feeds with heavy metals, high temperature incineration will likely increase the volatilization of some of these metals into the combustion flue gas (see Tables 33.5 and 33.6). High concentrations of volatile trace metal compounds in the flue gas pose increased challenges to air pollution control. Thus, other thermal processes, i.e. thermal desorption and pyrolysis, can provide an effective alternative to incineration.

TABLE 33.5

Conservative Estimates of Heavy Metals and Metalloids Partitioning to Flue Gas As a Function of Solids Temperature and Chlorine Content[a]

	871°C		1093°C	
Metal or metalloid	Cl = 0%	Cl = 1%	Cl = 0%	Cl = 1%
Antimony	100%	100%	100%	100%
Arsenic	100%	100%	100%	100%
Barium	50%	30%	100%	100%
Beryllium	5%	5%	5%	5%
Cadmium	100%	100%	100%	100%
Chromium	5%	5%	5%	5%
Lead	100%	100%	100%	100%
Mercury	100%	100%	100%	100%
Silver	8%	100%	100%	100%
Thallium	100%	100%	100%	100%

[a] The remaining percentage of metal is contained in the bottom ash. Partitioning for liquids is estimated at 100% for all metals. The combustion gas temperature is expected to be 100–1000°F higher than the solids temperature.

Source: US Environmental Protection Agency, *Guidance on Setting Permit Conditions and Reporting Trial Burn Results: Volume II, Hazardous Waste Incineration Guidance Series*, EPA/625/6-89/019. EPA, Washington, DC, 1989.

TABLE 33.6

Metal and Metalloid Volatilization Temperatures

	Without chlorine		With 10% chlorine	
Metal or metalloid	Volatility temperature (°C)	Principal species	Volatility temperature (°C)	Principal species
Chromium	1613	CrO_2/CrO_3	1611	CrO_2/CrO_3
Nickel	1210	$Ni(OH)_2$	693	$NiCl_2$
Beryllium	1054	$Be(OH)_2$	1054	$Be(OH)_2$
Silver	904	Ag	627	AgCl
Barium	841	$Ba(OH)_2$	904	$BaCl_2$
Thallium	721	Tl_2O_3	138	TiOH
Antimony	660	Sb_2O_3	660	Sb_2O_3
Lead	627	Pb	−15	$PbCl_4$
Selenium	318	SeO_2	318	SeO_2
Cadmium	214	Cd	214	Cd
Arsenic	32	As_2O_3	32	As_2O_3
Mercury	14	Hg	14	Hg

Source: Willis, B., Howie, M., and Williams, R., *Public Health Reviews of Hazardous Waste Thermal Treatment Technologies: A Guidance Manual for Public Health Assessors*, Agency for Toxic Substances and Disease Registry, 2002.

When successful in decontaminating soils to the necessary treatment levels, thermally desorbing contaminants from soils have the additional benefits of lower fuel consumption, no formation of slag, less volatilization of metal compounds, and less complicated air pollution control demands. So, beyond monetary costs and ease of operation, a less energy (heat) intensive system can be more advantageous in terms of actual pollutant removal efficiency.

B. Pyrolysis

Pyrolysis is the process of chemical decomposition induced in organic materials by heat in the absence of oxygen. It is practicably impossible to achieve a completely oxygen-free atmosphere, so pyrolytic systems run with less stoichiometric quantities of oxygen. Because some oxygen will be present in any pyrolytic system, there will always be a small amount of oxidation. Also, desorption will occur when volatile or semivolatile compounds are present in the feed.

During pyrolysis [5] organic compounds are converted to gaseous components, along with some liquids, as coke, i.e. the solid residue of fixed carbon and ash. CO, H_2, CH_4, and other hydrocarbons are produced. It these gases cool and condense, liquids will form and leave oily tar residues and water with high concentrations of TOC. Pyrolysis generally takes place well above atmospheric pressure at temperatures exceeding 430°C. The secondary gases need their own treatment, such as by a secondary combustion chamber, by flaring, and partial condensation. Particulates must be removed by additional air pollution controls, e.g. fabric filters or wet scrubbers.

Conventional thermal treatment methods, such as rotary kiln, rotary hearth furnace, or fluidized bed furnace, are used for waste pyrolysis. Kilns or furnaces used for pyrolysis may be of the same design as those used for combustion (i.e. incineration) discussed earlier, but operate at lower temperatures and with less air than in combustion.

The target contaminant groups for pyrolysis include semivolatile organic compounds (SVOCs), including pesticides, PCBs, dioxins, and PAHs. It allows for separating organic contaminants from various wastes, including those from refineries, coal tar, wood preservatives, creosote-contaminated and hydrocarbon-contaminated soils, mixed radioactive and hazardous wastes, synthetic rubber processing, and paint and coating processes. Pyrolysis systems may be used to treat a variety of organic contaminants that chemically decompose when heated (i.e. "cracking"). Pyrolysis is not effective in either destroying or physically separating inorganic compounds that coexist with the organics in the contaminated medium. Volatile metals may be removed and transformed but their mass balance, of course, will not be changed.

C. Emerging Thermal Technologies

Other promising thermal processes include high-pressure oxidation and vitrification [6]. High-pressure oxidation combines two related technologies, i.e. wet air oxidation and supercritical water oxidation, which combine high temperature and pressure to destroy organics. Wet air oxidation can operate at pressures of about 10% of those used during supercritical water oxidation, an emerging technology that has shown some promise in the treatment of PCBs and other stable compounds that resist chemical reaction. Wet air oxidation has generally been limited to conditioning of municipal wastewater sludges, but can degrade hydrocarbons (including PAHs), certain pesticides, phenolic compounds, cyanides, and other organic compounds. Oxidation may benefit from catalysts.

Vitrification uses electricity to heat and destroy organic compounds and immobilize inert contaminants. A vitrification unit has a reaction chamber divided into two sections: the upper section to introduce the feed material containing gases and pyrolysis products, and the lower section consisting of a two-layer molten zone for the metal and siliceous components of the waste. Electrodes are inserted into the waste solids, and graphite is applied to the surface to enhance its electrical conductivity. A large current is applied, resulting in rapid heating of the solids and causing the siliceous components of the material to melt as temperatures reach about 1600°C. The end product is a solid, glass-like material that is very resistant to leaching.

VIII. INDIRECT AIR IMPACTS

In addition to direct treatment, air pollution is also a concern for other means of treating hazardous wastes, especially when these wastes are stored or treated more passively, such as in a landfill or aeration pond. Leachate collection systems (see Fig. 33.6) provide a way to collect wastes that can then be treated. However, such "pump and treat" systems can produce air pollutants. Actually, this is often intentional. For example, groundwater is treated by drilling recovery wells to pump contaminated groundwater to the surface. Commonly used groundwater treatment approaches include air stripping, filtering with granulated activated carbon (GAC), and air sparging. Air stripping transfers volatile compounds from water to air (see Fig. 33.7). Groundwater is allowed to drip downward in a tower filled with a permeable material through which a stream of air flows upward. Another method bubbles pressurized air through contaminated water in a tank. The air leaving the tank (i.e. the off-gas) is treated with methods described in Chapter 32 for removing gaseous pollutants. Filtering groundwater with GAC entails pumping the water through the GAC to trap the contaminants. In air sparging, air is pumped into the groundwater to aerate the water. Most often, a soil venting system is combined with an air sparging system for vapor extraction, with the gaseous pollutants treated, as in air stripping.

Regulatory agencies often require two or three pairs of these systems as design redundancies to protect the integrity of a hazardous waste storage or treatment facility. A primary leachate collection and treatment system must be designed like the bottom of the landfill bathtub. This leachate collection system must be graded to promote the flow of liquid within the landfill from all points in the landfill to a central collection point(s) where the liquid can

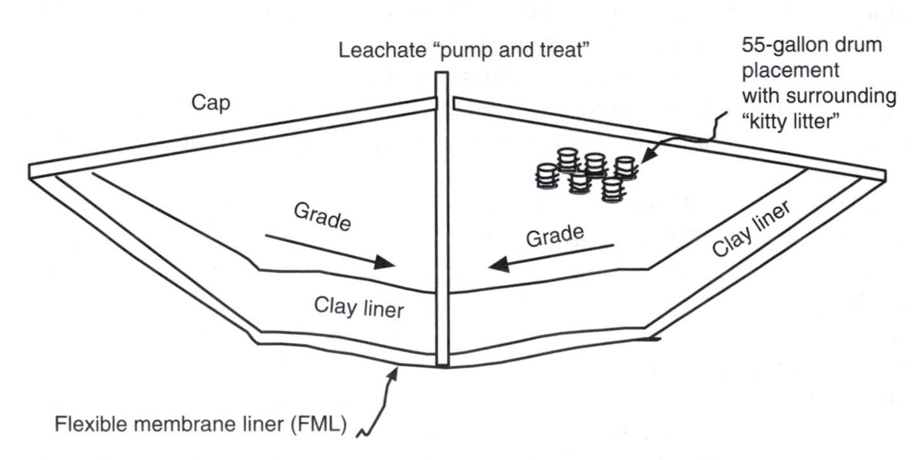

Fig. 33.6. Leachate collection system for a hazardous waste landfill. *Source*: Vallero, D., *Engineering the Risks of Hazardous Wastes*. Butterworth-Heinemann, Boston, MA, 2003.

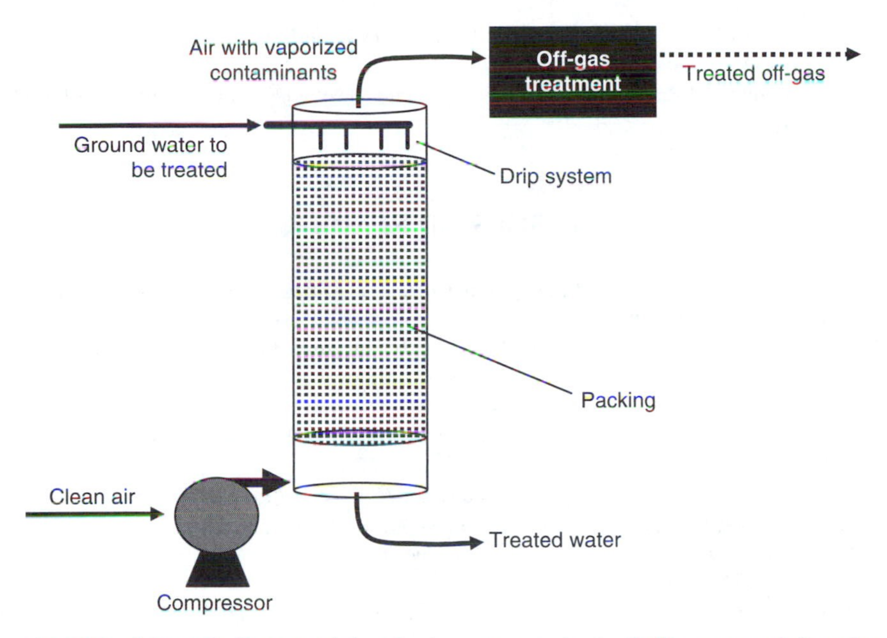

Fig. 33.7. Schematic diagram of air stripping system to treat volatile compounds in water.

be pumped to the surface for subsequent monitoring and treatment. Crushed stone and perforated pipes are used to channel the liquid along the top layer of this compacted clay liner to the pumping locations.

Thus, directly treating hazardous wastes physically and chemically, as with thermal systems and indirectly controlling air pollutants as when gases are released from pump and treat systems require a comprehensive approach. Otherwise, we are merely moving the pollutants to different locations or even making matters worse by either rendering some contaminants more toxic or exposing receptors to dangerous substances.

REFERENCES

1. US Environmental Protection Agency, *Remediation Guidance Document*, Chapter 7, EPA-905-B94-003, 2003.
2. Smith, J. K., and Peirce, J. J., Life cycle assessment standards: industrial sectors and environmental performance. *Int. J. Life Cycle Assess.* **1** (2), 115–118 (1996).
3. Biffward Programme on Sustainable Resource Use, 2003, *Thermal Methods of Municipal Waste Treatment*, http://www.biffa.co.uk/pdfs/massbalance/Thermowaste.pdf.
4. US Environmental Protection Agency, Lee, J., Fournier Jr., D., King, C., Venkatesh, S., and Goldman, C. *Project Summary: Evaluation of Rotary Kiln Incinerator Operation at Low-to-Moderate Temperature Conditions*, EPA/600/SR-96/105, Cincinnati, OH, 1997.

5. Federal Remediation Technologies Roundtable, *Remediation Technologies Screening Matrix and Reference Guide*, 4th ed, 2002.
6. US Environmental Protection Agency, *Remediation Guidance Document*, Chapter 7, EPA-905-B94-003, 2003.

SUGGESTED READING

Haas, C. N., and Ramos, R. J., *Hazardous and Industrial Waste Treatment*. Prentice-Hall, Inc., Englewood Cliffs, NJ, 1995.

Koester, C., and Hites, R., Wet and dry deposition of chlorinated dioxins and furans. *Environ. Sci. Technol.* **26**, 1375–1382 (1992); and Hites, R., Atmospheric Transport and Deposition of Polychlorinated Dibenzo-*p*-Dioxins and Dibenzofurans.

Peirce, J. J., Weiner, R. F., and Vesilind, P. A., *Environmental Pollution and Control*. Butterworth-Heinemann, Boston, MA, 1998.

Stieglitz, L., Zwick, G., Beck, J., Bautz, H., and Roth, W., *Chemosphere* **19**, 283 (1989).

US Environmental Protection Agency, *Guidance on Setting Permit Conditions and Reporting Trial Burn Results: Volume II, Hazardous Waste Incineration Guidance Series*, EPA/625/6-89/019, Washington, DC, 1989.

US Environmental Protection Agency, Lee, J., Fournier Jr., D., King, C., Venkatesh, S., and Goldman, C., *Project Summary: Evaluation of Rotary Kiln Incinerator Operation at Low-to-Moderate Temperature Conditions*, EPA/600/SR-96/105, Cincinnati, OH, 1997.

US Environmental Protection Agency, *Locating and Estimating Air Emissions from Sources of Benzene*, EPA-454/R-98-011, Research Triangle Park, NC, 1998.

Vallero, D., *Engineering the Risks of Hazardous Wastes*. Butterworth-Heinemann, Boston, MA, 2003.

Wentz, C. A., *Hazardous Waste Management*. McGraw-Hill, Inc., New York, 1989.

Willis, B., Howie, M., and Williams, R., *Public Health Reviews of Hazardous Waste Thermal Treatment Technologies: A Guidance Manual for Public Health Assessors*, Agency for Toxic Substances and Disease Registry, 2002.

QUESTIONS

1. Your company's waste stream contains a large percentage of clay-sized particles with a high content of organic compounds and chromium. Which treatment technologies are good candidates for this waste? What other assumptions do you need to make and how may these assumptions be eliminated from your decision; i.e. how can you close the "data gap"?
2. What if 20% of the organic fraction in the above waste stream is chlorinated? How would that affect your selection of treatment approaches?
3. Your city has decided to use supercritical oxidation to treat a low solid content PCB-contaminated waste. What problems, if any, do you see with this approach?
4. What characteristics of a waste increase the likelihood that it should be treated thermally? When is incineration preferable to pyrolysis and vice versa?
5. What precautions should be taken in selecting a treatment process for a halogenated waste? For a heavy metal-laden waste?
6. What are the "three T's" of thermal destruction? Give an example of problems that may ensue if each is neglected?

7. What is the best thermal destruction system for wastes containing 2,3,7,8-tetrachlorodibenzo-p-dioxin? For liquid solvents? Support your answers.
8. Is the multiple hearth system good for treating a liquid waste that breaks down at temperatures >1000°C? If the waste also contains Hg and Ni, what is the preferable treatment approach?
9. What is the difference between air stripping and sparging? Give a situation where stripping would be preferred and one where sparging is better. Explain.

34

Control of Stationary Sources

I. INTRODUCTION

Control of stationary sources of air pollution requires the application of the control concepts mentioned in Chapter 28, and usually the adaptation of the control devices mentioned in the two previous chapters. In some cases, more than one system or device must be used to achieve satisfactory control. The three general methods of control are (1) process change to a less polluting process or to a lowered emission from the existing process through modification of the operation, (2) fuel change to a fuel which will give the desired level of emissions, and (3) installation of control equipment. It is more efficient to engineer air pollution control into the source when it is first considered than to leave it until the process is operational and found to be in violation of emission standards. Most new large stationary sources of air pollution in the United States are regulated under the Clean Air Act Amendments of 1990 [1] and are legally required to comply with the Amendments.

Existing stationary sources may require modification of existing systems or installation of newer, more efficient control devices to meet more restrictive emission standards. Such changes are often required by control agencies when it can be shown that a new control technology is superior to older systems or devices being used. This is usually referred to as application of the

Best Available Control Technology (BACT). The BACT standards have been incrementally replaced with risk-based standards, meaning that sources may need to apply more stringent technologies as well as important measures to reduce emissions further (e.g. pollution prevention).

Installation of control systems may have a positive economic benefit, which will offset a portion of their cost [2]. Such benefits include (1) tax deduction provisions, (2) recovery of materials previously emitted, (3) depreciation schedules favoring the owner of the source, and (4) banking or sale of the emission offset credits if the source is in a nonattainment area.

II. ENERGY, POWER, AND INCINERATION

Thermal energy, power generation, and incineration have several factors in common. All rely on combustion, which causes the release of air pollutants; all exhaust their emissions at elevated temperatures; and all produce large quantities of ash when they consume solid or residual fuels. The ratio of the energy used to control pollution to the gross energy produced can be a deciding factor in the selection of the control system. These processes have important differences, which influence the selection of specific systems and devices for individual facilities.

A. Energy-Producing Industries

Stationary energy-producing systems are of two general types, residential and commercial space heating and industrial steam generation. The smaller systems (residential and commercial heating) are usually regulated only with respect to their smoke emission, even though they may produce appreciable amounts of other air pollutants [3]. The large industrial systems which generate steam for process use and space heating (where superheated or high-temperature saturated steam is used for a process, e.g. cogeneration, exhausted from the process at a lower energy level, and then introduced into a space heating system, where it gives up a large amount of latent energy, condensing to hot water) are required to comply with rigid standards in most countries.

Control of air pollution from energy-producing industries is a function of the fuel used and the other variables of the combustion process. The system must be thoroughly analyzed before a control system is chosen. The important variables are listed in Table 32.1. For particulate matter control, the variables of process control such as improved combustion, fuel cleaning, fuel switching, and load reduction through conservation should be considered before choosing an add-on control system. If the particulate matter emission is still found to be in excess of standards, then control devices must be used. These include inertial devices (such as multiple cyclones), baghouses, wet and dry scrubbers, electrostatic precipitators (ESPs), and some of the previously discussed novel devices. Series combinations of control devices may be necessary to achieve the required level of particulate matter emission. A commonly used system is a multiple cyclone followed by a fine-particle control system, such as a baghouse, scrubber, or ESP.

Opacity reduction is the control of fine-particulate matter (less than 1 μm). It can be accomplished through the application of the systems and devices discussed for control of particulate matter and by use of combustion control systems to reduce smoke and aerosol emission. In addition, operational practices such as continuous soot blowing and computerized fuel and air systems should be considered.

Sulfur dioxide reduction to achieve required emission levels may be accomplished by switching to lower-sulfur fuels. Use of low-sulfur coal or oil, or even biomass such as wood residue as a fuel, may be less expensive than installing an SO_2 control system after the process. This is particularly true in the wood products industry, where wood residue is often available at a relatively low cost.

If an SO_2 control device is necessary, the first decision is whether to use a wet or dry system. Many times this decision is based on the local situation regarding the disposal of the collected residue (sludge or dry material). Wet scrubbing systems, using chemical additions to the scrubbing liquid, are widely used. Various commercial systems have used lime or limestone, magnesium oxide, or sodium hydroxide slurries to remove the SO_2. Dry removal of SO_2 can be accomplished by adding the same chemicals used in wet scrubbers, but adding them in a spray drier and then removing the spent sulfates with a baghouse or ESP. This results in the collection of a dry material which may be either disposed of by landfilling or used as a raw material for other processes. The electric power generating industry has had many years of experience with SO_2 control systems [4]. These methods are discussed in more detail in the next section.

Control of oxides of nitrogen can be accomplished by catalysts or absorbants, but most control systems have concentrated on changing the combustion process to reduce the formation of NO_x. Improved burners, change in burner location, staged combustion, and low-temperature combustion utilizing fluidized-bed systems are all currently in use. These combustion improvement systems do not generate waste products, so no disposal problems exist.

B. Power Generation

In general, plants-producing electric power are much larger than those producing steam for space heating or process use. Therefore, the mass of emissions is much greater and the physical size of the control equipment larger.

The extensive control of particulate matter, opacity, SO_2, and NO_x required on new power plants is very expensive. The high-projected cost of environmental control for a new coal-fired electric generating plant has made utility companies reluctant to risk the billions of dollars necessary to use coal as a major fuel, particularly when the standards are being constantly redefined and changed [5]. It does appear that the 1990 Amendments will require more complex emission controls as an integrated part of the plant rather than the previous approach of adding control devices independently. Figure 34.1 illustrates the complexity of the current technology for the various alternative systems.

Figure 34.1(a) presents the integrated environmental control potential for maximum control of particulate matter and SO_2. Cooling tower water blowdown and treatment by-products may be used to satisfy scrubber makeup requirements. Fly ash and scrubber sludge will be produced separately. If the catalytic NO_x process is required, the integration issues will be increased significantly.

Figure 34.1(b) is similar to Fig. 34.1(a) except that an ESP is used for particulate control. This represents the most common approach for compliance when configured without a catalytic NO_x unit.

Figure 34.1(c) is distinctly different from the first two in the type of SO_2 control processes used and the sequence of the particulate matter and SO_2 controls. It is a promising approach for up to 90% SO_2 control of western United States coal, and there is a single waste product. Other features include the collection of particulate matter at temperatures below 90°C and the possibility for spray dryer cooling tower water integration. This system may or may not include a catalytic NO_x unit.

Figure 34.1(d) represents the simplest, least expensive, and lowest water consumer of all the alternatives. There is a single solid waste product. The key element is the integrated SO_2/particulate control process. Using sodium-based sorbants, compliance may be achievable for western United States coals.

Figure 34.1(e) includes a hot ESP for fly ash collection prior to a catalytic NO_x unit. Having a hot ESP dictates the use of a conventional wet scrubber and perhaps the need for a second particulate matter control device at the end of the system. Fly ash and scrubber sludge would be separate by-products, but sludge could be contaminated with NH_4 from the catalytic NO process.

Figure 34.2 illustrates a wet SO_2 desulfurization system using a spray tower absorber. Figure 34.3 illustrates a rotary atomizer injecting an alkaline slurry into a spray dryer for SO_2 control.

Selection and installation of an integrated air pollution control system do not end the concern of the utility industry. Maintenance and operational problems of the system are considered by many engineers to be the weak link in the chain of power generation equipment [6]. The reliability of the flue gas desulfurization system (FGDS) is defined as the time the system operates properly divided by the time it should have operated. For large US power plants, this has been determined by the US Environmental Protection Agency to be 83.3% in 1980 and 82.0% in 1981. These figures are cause for

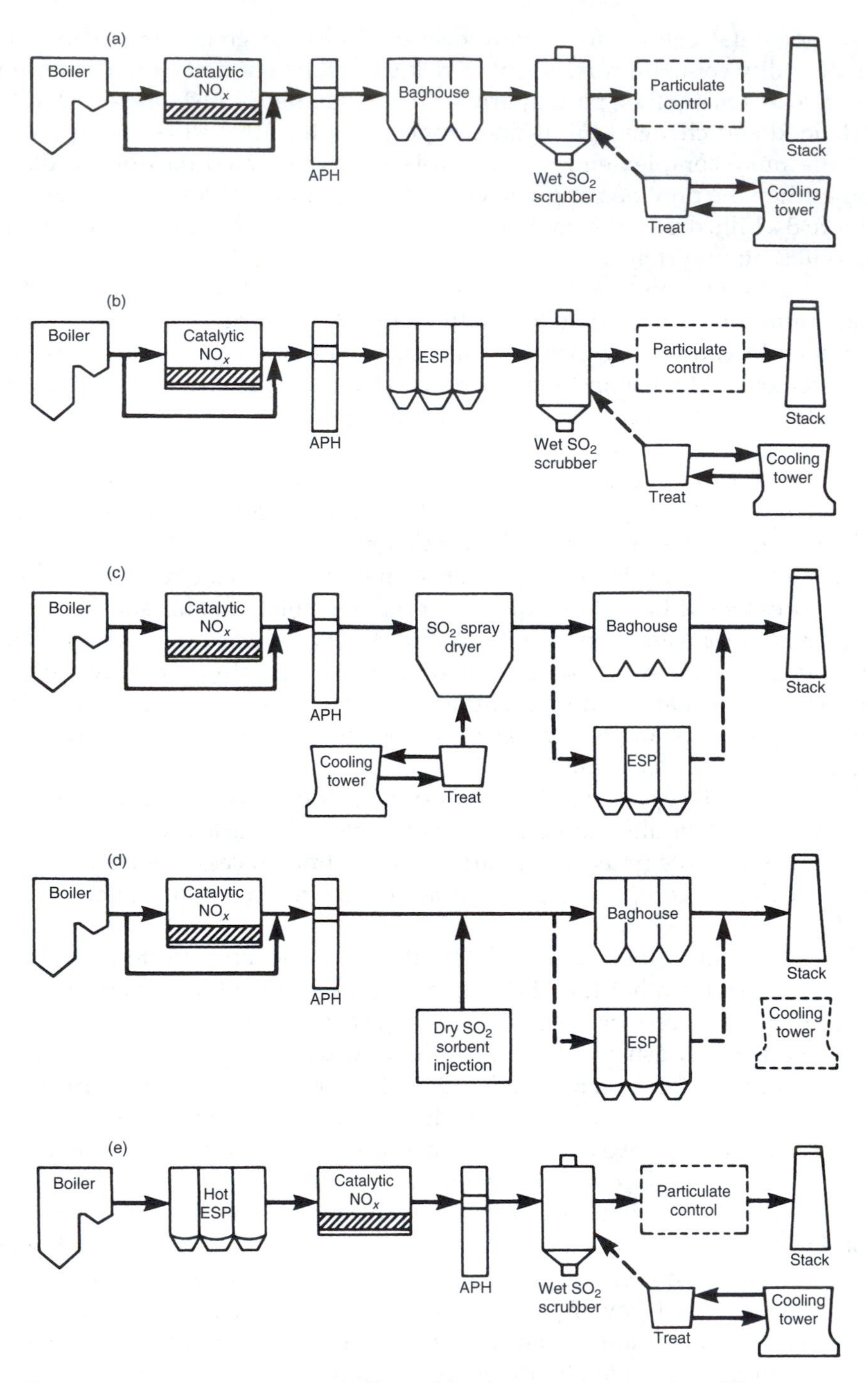

Fig. 34.1. Integrated environmental control for electric generating plants [5].

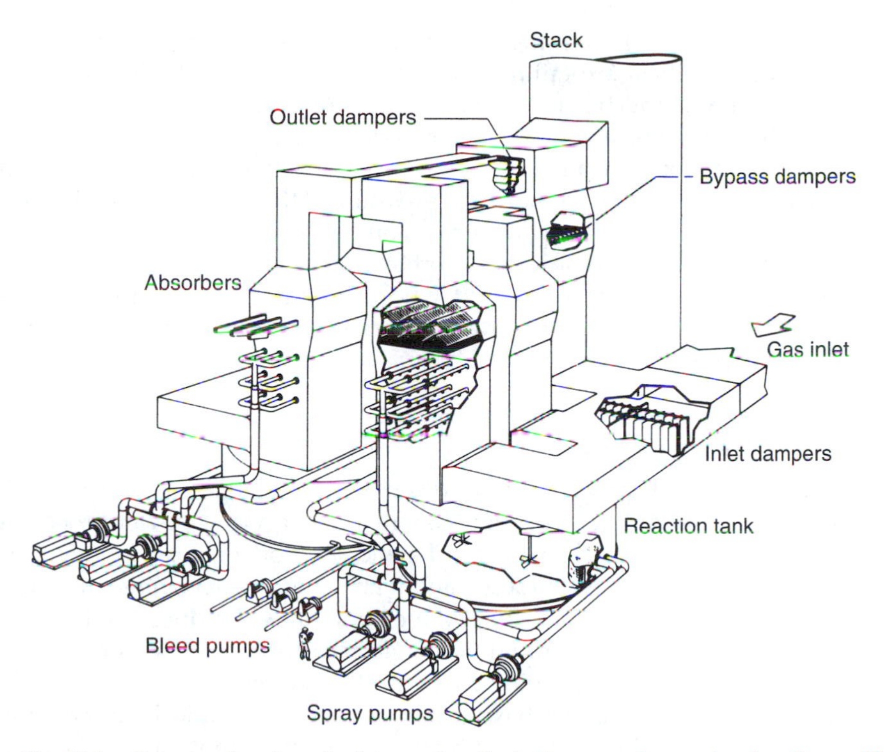

Fig. 34.2. Cutaway drawing of a flue gas desulfurization spray tower absorber. *Source*: CE Power Systems, Combustion Engineering, Inc.

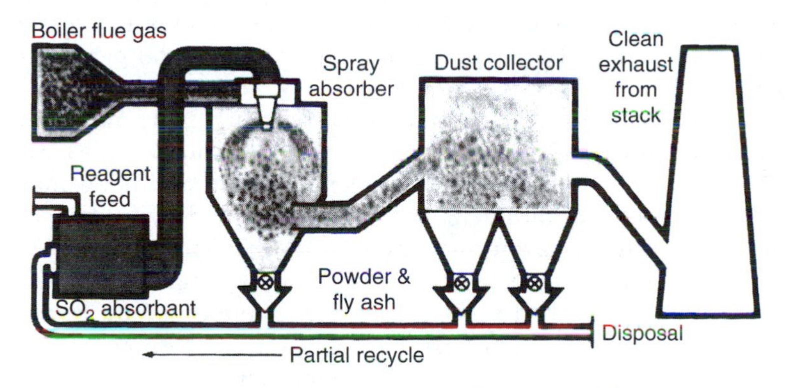

Fig. 34.3. Alkaline slurry SO_2 spray dryer. *Source:* Niro Atomizer, Inc.

concern even though the consequences to the utility may not have been extreme because the FGDS usually could be bypassed. The primary factors cited for low FGDS reliability are (1) plugging, scaling, and corrosion of scrubber internals, mist eliminators, and reheaters; (2) need for open-loop operation and blowdown of scrubber liquor to reduce the corrosive substance concentration and dissolved solids content; (3) corrosion and failure of stack liners; and (4) plugging and failure of piping, pumps, and valves. Probably more critical are failures of dampers, ducts, and baffles, because this may require plant shutdown to perform maintenance.

Other methods, which should be mentioned because they show potential benefits for pollution reduction from utility stacks include (1) coal cleaning and treatment, (2) atmospheric pressure and pressurized fluidized-bed combustors, (3) conversion of solid fuels to liquid or gaseous fuels, and (4) combustion modification through staged combustors or other systems.

Tall stacks are no longer considered to be an acceptable alternative for controlling emissions from electric power generating plants (see further discussion in Chapter 26, Section V).

C. Incineration

Thermal removal processes are discussed in detail in Chapter 33. Incineration is similar to combustion-generated energy and power processes in that fuel combines with oxygen. The incineration process, however, is designed as a waste disposal process, and if any energy is recovered, it is considered as a secondary system. Ideally, incineration will reduce combustors, [3] conversion of solid fuels to liquid or gaseous fuels, and [4] combustion modification through staged combustors or other systems.

An incinerator will usually have a fuel of varying chemical composition and physical properties, as well as varying moisture content and heating value. In addition the fuel fired in one locality may be vastly different from that fired by an incinerator of similar size and design in another locality. Refuse production in the United States has been estimated to average 2.5 g per person per day in 1970, increasing to 10 kg per person per day by the year 2000.

The air pollutants from incinerators consist of particulate matter (fly ash, carbon, metals and metal oxides, and visible smoke), combustible gases such as CO, organics, polynuclear organic material (POM), and noncombustible gases such as oxides of nitrogen, oxides of sulfur, and hydrogen chloride. The oxides of nitrogen are formed by two mechanisms: thermal NO_x, in which atmospheric nitrogen and oxygen combine at high furnace temperatures, and fuel NO_x, when nitrogen-bearing compounds are incinerated. Hydrogen chloride (HCl) emissions are causing concern because of the increased amounts of halogenated polymers, notably polyvinyl chloride (PVC), in the refuse; 1 kg of pure PVC yields about 0.6 kg of HCl.

POM emissions appear to be a function of the degree of combustion control, decreasing with increasing incinerator size (larger incinerators are more thoroughly instrumented and controlled). Table 34.1 shows the measured emission rates for POM and CO from various-sized incinerators.

Air pollution control systems using wet scrubbers will remove some water-soluble gases, but the removal of particulate matter is the primary concern for a control system. The air pollution control system, therefore, is usually a

TABLE 34.1

Generation and Emission Rates for POM and Carbon Monoxide (CO) from Incinerators

Incinerator type, size, and control	POM (g metric ton^{-1})	CO (g kg^{-1})
Municipal, 227 metric tons per day, before settling chamber	0.032	0.35
Municipal, 45 metric tons per day, before scrubber	0.258	2.00
Municipal, 45 metric tons per day, after scrubber	0.014	1.00
Commercial multiple chamber, 3 metric tons per day, no control system	1.726	12.50

Source: Brunner [7].

single device such as a wet scrubber, small-diameter multiple cyclones, fabric filters, or ESPs. The multicyclones are the least expensive system and the ESPs the most expensive.

Some novel methods of incineration offer the possibility of reduced emissions. These include "slagging" (operating at such a high temperature that incombustible materials are melted and removed as a fluid slag); fluidized beds (which are useful only on homogeneous or well-classified refuse); suspension burning in cylindrical combustion chambers, which may or may not result in slagging; and pyrolysis, which is destructive distillation in the absence of oxygen.

The emission control requirements set for municipal incinerators by the US Clean Air Act Amendments of 1990 [1] are extensive and complex. Many of the final standards have not been established as of the date of publication of this book. A thorough study of the regulations is necessary for any person dealing with incinerator technology and control.

III. CHEMICAL AND METALLURGICAL INDUSTRIES

The chemical and metallurgical industries of the world are so varied and extensive that it is impossible to cover all of the processes, emissions, and controls in a single chapter.

A. Chemical Industries

The term *chemical industry* applies to a group of industries which range from small, single owner-employee operations to huge complexes employing thousands of people. The number of environmental regulations that the chemical industry must comply with is so extensive that specialized consulting firms have been formed to aid the industry in handling them [8].

1. Inorganic Chemical Processes

Production of major inorganic chemicals in the United State exceeds 200 million metric tons per year produced in over 1300 plants [9]. These inorganic chemicals may be categorized as follows:

(a) *Acids*: The major acids produced are hydrochloric, hydrofluoric, nitric, phosphoric, and sulfuric. The emissions and usual control methods for the various acid and manufacturing processes are shown in Table 34.2.
(b) *Bases*: Major bases and caustics produced by the chemical industry are calcium oxide (lime), sodium carbonate (soda ash), and sodium hydroxide (caustic soda). The emissions and usual control methods for the various bases and their manufacturing processes are shown in Table 34.3.
(c) *Fertilizer*: Fertilizer production is dependent on the production of phosphates and nitrates. Phosphate rock preparation generates some dry particulate matter during drying, grinding, and transferring of the rock. These emissions are controlled by wet scrubbers and baghouses. The atmospheric emissions and control methods for the production processes are shown in Table 34.4.
(d) *Ammonium Nitrate Fertilizer*: It is produced by the neutralization of nitric acid with ammonia. The primary emission is the dust or fume of ammonium nitrate from the prill tower. The material is of submicron size and, therefore, highly visible. Control is usually performed by a wet scrubber followed by a mist eliminator.

TABLE 34.2

Air Pollution Emissions and Controls: Inorganic Acid Manufacture

Acid	Manufacturing process	Air pollutant emissions	Control methods in use
Hydrochloric	By-product of organic chlorination, salt process, and synthetic HCl	HCl	Absorption
Hydrofluoric	Fluorspar–sulfuric acid	SiF_4, HF	Scrubber (some with caustic)
Nitric	Pressure process and direct strong acid	NO, NO_2, N_2O_4	Catalytic reduction, adsorption, absorption
Phosphoric	Elemental phosphorus	Particulate matter, fluorides	Baghouse
	Thermal process	H_3PO_4, H_2S	Mist eliminators, alkaline scrubbers
	Wet process	SiF_4, HF	Scrubber
	Superphosphoric	Fluorides	Scrubber
Sulfuric	Contact	SO_2, acid mist	Scrubbers with mist eliminators, ESPs

TABLE 34.3

Air Pollution Emissions and Controls: Inorganic Base Manufacture

Base	Manufacturing process	Air pollutant emissions	Control methods in use
Calcium oxide (lime)	Rotary kilns, vertical and shaft kilns, fluidized-bed furnaces	Particulate matter	Cyclones plus secondary collectors (baghouse, ESP, wet scrubbers, granular bed filters, wet cyclones)
Sodium carbonate (soda ash)	Solvay (ammoniasoda)	Particulate matter	Wet scrubbers
Sodium hydroxide, caustic soda	Electrolytic	Chlorine	Alkaline scrubbers
		Mercury	Chemical scrubbing and adsorbers

TABLE 34.4

Air Pollution Emissions and Controls: Phosphate Fertilizer Plants

Process	Air pollutant emissions	Control methods in use
Normal superphosphate	SiF_4, HF	Venturi or cyclonic scrubber
	Particulate matter	Wet scrubber of baghouse
Diammonium phosphate	Gaseous F, NH_3	Venturi or cyclonic scrubber with 30% phosphoric acid
	Particulate matter	Cyclone followed by scrubber
Triple superphosphate, run of pile	SiF_4, HF	Venturi or cyclonic scrubber
Triple superphosphate, granular	SiF_4, HF, particulate matter	Venturi or packed scrubber

(e) *Chlorine*: Most of the chlorine manufactured is produced by two electrolytic methods, the diaphragm cell and the mercury cell processes. Both processes emit chlorine to the atmosphere from various streams and from handling and loading facilities. If the gas streams contain over 10% chlorine, the chlorine is recovered by absorption. If the chlorine concentration is less, the usual practice is to scrub the vent gases with an alkaline solution. Mercury is emitted from the mercury cell process from ventilation systems and by-product streams. Control techniques include (1) condensation, (2) mist elimination, (3) chemical scrubbing, (4) activated carbon adsorption, and (5) molecular sieve absorption. Several mercury cell (chloralkali) plants in Japan have been converted to diaphragm cells to eliminate the poisonous levels of methyl mercury found in fish [9].

(f) *Bromine*: All of the bromine produced in the United States is extracted from naturally occurring brines by steam extraction. The major air pollution concern is H_2S from the stripper if H_2S is present in the brine. The H_2S can either be oxidized to SO_2 in a flare or sent to a sulfur recovery plant.

2. *Organic (Petrochemical) Processes*

Most petrochemical processes are essentially enclosed and normally vent only a small amount of fugitive emissions. However, the petrochemical processes that use air-oxidation-type reactions normally vent large, continuous amounts of gaseous emissions to the atmosphere [10]. Six major petrochemical processes employ reactions using air oxidation. Table 34.5 lists the atmospheric emissions from these processes along with applicable control measures.

B. Metallurgical Industries

The metallurgical industry offers some of the most challenging air pollution control problems encountered. The gas volumes are huge, and the gas may be at a high temperature. These large, hot gas volumes may convey large quantities of dust or metal oxide fumes, some of which may be highly toxic. Also, gaseous pollutants such as SO_2 or CO may be very highly concentrated in the carrying stream. The process emissions to the atmosphere may have harmful effects on visibility, vegetation, animals, and inert materials, as well

TABLE 34.5

Air Pollution Emissions and Controls: Petrochemical Processes

Petrochemical process	Air pollutant emissions	Control methods in use
Ethylene oxide (most emissions from purge vents)	Ethane, ethylene, ethylene oxide	Catalytic afterburner
Formaldehyde (most emissions from exit gas stream of scrubber)	Formaldehyde, methanol, carbon monoxide, dimethyl ether	Wet scrubber for formaldehyde and methanol only; afterburner for organic vent gases
Phthalic anhydride (most emissions from off-gas from switch condensers)	Organic acids and anhydrides, sulfur dioxide, carbon monoxide, particulate matter	Venturi scrubber followed by cyclone separator and packed countercurrent scrubber
Acrylonitrile (most emissions from exit gas stream from product absorber)	Carbon monoxide, propylene, propane, hydrogen cyanide, acrylonitrile, acetonitrile NO_x from by-product incinerator	Thermal incinerators (gas-fired afterburners or catalytic afterburners) None
Carbon black (most emissions from exit gas stream from baghouse, some fugitive particulate)	Hydrogen, carbon monoxide, hydrogen sulfide, sulfur dioxide, methane, acetylene	Waste heat boiler or flare (no control for SO_2)
	Particulate matter (carbon black)	Baghouse
Ethylene dichloride (most emissions from exit gas stream of solvent scrubber)	Carbon monoxide, methane, ethylene, ethane, ethylene dichloride, aromatic solvent	None at present, but could use a waste heat boiler or afterburner, followed by a caustic scrubber for hydrochloric acid generated by combustion

as being detrimental to human health. It is no wonder that the metallurgical industries have spent huge sums to control emissions.

1. Nonferrous Metallurgical Operations

Nonferrous metallurgy is as varied as the ores and finished products. Almost every thermal, chemical, and physical process known to engineers is in use. The general classification scheme that follows gives an understanding of the emissions and control systems: aluminum (primary and secondary), beryllium, copper (primary and secondary), lead (primary and secondary), mercury, zinc, alloys of nonferrous metals (primary and secondary), and other nonferrous metals:

(a) *Aluminum (primary)*: The emissions from primary aluminum reduction plants may come from the primary control system, which vents the electrolytic cells through control devices, or from the secondary system, which controls the emissions from the buildings housing the cells.

Hydrogen fluoride accounts for about 90% of the gaseous fluoride emitted from the electrolytic cell. Other gaseous emissions are SO_2, CO_2, CO, NO_2, H_2S, COS, CS_2, SF_6, and various gaseous fluorocarbons. Particulate fluoride is emitted directly from the process and is also formed from condensation and solidification of the gaseous fluorides.

The fluoride removal efficiency of the control equipment at primary aluminum reduction plants is shown in Table 34.6. The removal efficiency for total fluorides is a matter of great concern.

Emission rates using BACT on the three electrolytic cell types are shown in Table 34.7.

(b) *Beryllium*: It is extracted from the ore in the form of beryllium hydroxide, which is then converted to the desired product, metal, oxide, or alloy [12]. Some of these products are extremely toxic. Table 34.8 lists

TABLE 34.6

Fluoride Removal Efficiencies: Selected Aluminum Industry Primary and Secondary Control Systems

Control system	Total fluoride removal efficiency (%)
Coated filter baghouse	94
Fluid bed dry scrubber	99
Injected alumina baghouse	98
Wet scrubber + wet ESP	99+
Dry ESP + wet scrubber	95
Floating bed	95
Spray screen (secondary)	62–77
Venturi scrubber	98
Bubbler scrubber + wet ESP	99

Source: Iverson [11].

TABLE 34.7

Emissions for Three Electrolytic Aluminum Reduction Cell Types Using Best Available Control Technology (BACT)

	Emissions with primary and secondary control	
Cell type	Fluorides (g kg^{-1} Al)	Particulate (g kg^{-1} Al)
Prebaked	0.8	3.0
Vertical stud soderberg	1.4	4.6
Horizontal stud soderberg	1.8	6.2

TABLE 34.8

Air Pollution Emissions and Controls: Beryllium Processing

Process	Air pollutant emissions	Control methods in use
Extraction	Beryllium salts	Baghouses
	Acids	Wet collectors
	Beryllium oxides	Baghouses or high-efficiency particulate air (HEPA) filters
	Beryllium dust, fume, mist	Cyclones and baghouses
Machining	Beryllium dust	Baghouses and HEPA filters
	Beryllium oxide dust	Baghouses and HEPA filters

the emissions from the various beryllium production steps, along with the control measures commonly used.

(c) *Copper*: Most copper is removed from low-grade sulfide ores using pyrometallurgical processes. The copper is first concentrated, then dewatered, and filtered. The copper smelting process consists of roasters, reverberatory furnaces, and converters. Some copper is further refined electrolytically to eliminate impurities.

Some fugitive particulate emissions occur around copper mines, concentrating, and smelting facilities, but the greatest concern is with emissions from the ore preparation, smelting, and refining processes. Table 34.9 gives the emissions of SO_2 from the smelters.

Tall stacks for SO_2 dispersion have been used in the past but are no longer acceptable as the sole means of SO_2 control. Acid plants have been installed at many smelters to convert the SO_2 to sulfuric acid, even though it may not be desirable from an economic standpoint.

The emission of volatile trace elements from roasting, smelting, and converting processes is undesirable from both air pollution and economic standpoint. Gravity collectors, cyclones, and ESPs are used to attain collection efficiencies of up to 99.7% for dust and fumes.

Treatment of slimes for economic recovery of silver, gold, selenium, tellurium, and other trace elements requires fusion and oxidation in

TABLE 34.9

Sulfur Dioxide Emission Rates from Primary Copper Smelters

Process	Emission (g of SO_2 per kg of copper)
Roasting	325–675
Reverberatory furnaces	150–475
Converters	975–1075
Reverberatory furnaces[a]	275–800
Converters[a]	850–1800

[a] Without roasting.
Source: Nelson *et al.* [13].

TABLE 34.10

Sulfur Dioxide Emission Rates from Primary Lead Smelters

Process	Emission (g of SO_2 per kg of lead)
Sintering	575–1075
Blast furnace	2.5–5
Dross reverberatory furnace	2.5–5

Source: Nelson *et al.* [13].

a furnace. The furnace gases are exhausted through a wet scrubber followed by an ESP to recover the metals.

(d) *Lead smelting and lead storage battery manufacture*: Lead ores are crushed, ground, and concentrated in a manner similar to the processing of copper ores. Fugitive emissions from these processes include dusts, fumes, and trace metals. Smelting is usually accomplished in a blast furnace after the concentrated ore is sintered. Sintering removes up to 85% of the sulfur.

Gases from the sintering process contain SO_2, dust, and metal oxide fumes. The blast furnace gases contain similar particulates plus SO_2 and CO. Table 34.10 indicates the expected SO_2 emissions.

(e) *Mercury*: Mercury is produced commercially by processing mercury sulfide (cinnabar). The mercury sulfide is thermally decomposed in a retort or roaster to produce elemental mercury and sulfur dioxide. The off-gases are cleaned by being passed through cyclonic separators, and the mercury is then condensed. The SO_2 is removed by scrubbers before the exhaust gases are released to the atmosphere. Any mercury vapors that escape are collected in refrigerated units and, usually, recovered with a baghouse or ESP. Other systems use absorption with sodium hypochlorite and sodium chloride or adsorption on activated carbon or proprietary adsorbents. The US Environmental Protection

Agency has placed a limit of 2300 g of mercury emission per 24 h on any mercury smelter or process.

(f) *Zinc*: It is processed very similarly to copper and lead. The zinc is bound in the ore as ZnS, sphalerite. Zinc is also obtained as an impurity from lead smelting, in which it is recovered from the blast furnace slag.

Dusts, fumes, and SO_2 are evolved during sintering, retorting, and roasting, as shown in Table 34.11.

Particulate emissions from zinc processing are collected in baghouses or ESPs. SO_2 in high concentrations is passed directly to an acid plant for production of sulfuric acid by the contact process. Low-concentration SO_2 streams are scrubbed with an aqueous ammonia solution. The resulting ammonium sulfate is processed to the crystalline form and marketed as fertilizer.

(g) *Other nonferrous metals and alloys*: Nonferrous metals of lesser significance include arsenic, cadmium, and refractory metals such as zirconium and titanium. Air pollution emissions from the manufacture of these metals do not constitute a major problem, although severe local problems may exist near the facility. Control of emissions is usually accomplished by a single device at the exit of the process. In many cases, the material removed by the control device has some value, either as the primary product or as a by-product. Table 34.12 shows some of the atmospheric emissions and control systems used on these metallurgical processes.

Alloys of nonferrous metals, primarily the brasses (copper and zinc) and the bronzes (copper and tin), can cause an air pollution problem during melting and casting. The type and degree of emissions depend on the furnace and the alloy. Control systems consist of hoods over the furnaces and pouring stations to collect the hot gases, ducts and fans, and baghouses or ESPs.

(h) *Secondary metals*: These metals are those recovered from scrap. Copper (including brass and bronze), lead, zinc, and aluminum are the

TABLE 34.11

Air Pollution Emissions from Primary Zinc Processing

	Emissions to the atmosphere			
Process	Dust (g dscm^{-1})	Percent of particles less than 10 μm	SO_2 (%)	SO_2 (gm kg^{-1} zinc)
Sintering	10	100	4.5–7.0	—
Horizontal retort	0.1–0.3	100	—	—
Roasting	—	—	—	825–1200

Source: Nelson *et al.* [13].

TABLE 34.12

Air Pollution Emissions from Miscellaneous Nonferrous Metallurgical Processes

Metal	Type of process	Air pollutant emissions	Control methods in use
Arsenic	By-product of copper and lead smelters	Arsenic trioxide	Baghouses or ESPs
Cadmium	By-product of zinc and lead smelters	Cadmium, cadmium oxide	Baghouses
Refractory metals Zirconium Hafnium Titanium	Kroll process, chlorination, and magnesium reduction	Chlorine, chlorides, $SiCl_4$	Wet scrubbers
Columbium Tantalum Vanadium Tungsten Molybdenum	Separation process	Ammonia	Conversion to ammonium sulfate fertilizer

principal nonferrous secondary metals. Emissions from the recovery processes are similar to those from the primary metallurgical operations except that little or no SO_2 or fluorides are evolved. Baghouses and ESPs are the commonly used control devices.

2. *Ferrous Metallurgical Operations*

Iron and steel industries are generally grouped as steel mills, which produce steel sheets or shapes, and foundries, which produce iron or steel castings. Some steel mills use electric arc furnaces with scrap steel as the raw material, but the majority are large, integrated mills with the following facilities [14]:

(a) *Coke making*: Coke is produced from blended coals by either the non-recovery beehive process or the by-product process. The by-product process produces the majority of the coke. Air pollutants from the coke-making process vary according to the point of release from the process and the time the process has been in operation. Table 34.13 shows the emissions, and their control, from the different stages of the by-product process.

Coke is produced by blending and heating bituminous coals in coke ovens to 1000–1400°C in the absence of oxygen.[1] Light weight oils and tars are distilled from the coal, generating various gases during the heating process. Every half an hour or so, the flows of gas, air, and

[1] The principal source for this section is: National Toxicology Program, 2005, Eleventh Report on Carcinogens, Coke Oven Emissions, Substance Profile; http://ntp.niehs.nih.gov/ntp/roc/eleventh/profiles/s049coke.pdf; accessed on May 11, 2005.

TABLE 34.13

Air Pollution Emissions and Controls: By-Product Coke Making

Process	Air pollutant emissions	Control methods in use
Coal and coke handling	Fugitive particulate matter	Enclose transfer points and duct to baghouses; pave and water roadways
Coke oven charging	Hydrocarbons, carbon, coal dust	Aspiration systems to draw pollutants into oven, venturi scrubbers
Coke oven discharging (pushing)	Hydrocarbons, coke dust	Hoods to fans and venturi scrubbers, low-energy scrubbers followed by ESPs (may use water spray at oven outlet)
Coke quenching	Particulate matter	Baffles and water sprays
Leaking oven doors	Hydrocarbons, carbon	Door seals with proper operation and maintenance
By-product processing	Hydrogen sulfide	Conversion to elemental sulfur or sulfuric acid by liquid absorption, wet oxidation to elemental sulfur, combustion to SO_2

waste gas are reversed to maintain uniform temperature distribution across the wall. In most modern coking systems, nearly half of the total coke oven gas produced from coking is returned to the heating flues for burning after having passed through various cleaning and co-product recovery processes. Coke oven emissions are the benzene-soluble fraction of the particulate matter generated during coke production. They are known to contain human carcinogens. These emissions comprise a highly toxic mixture of gases and aerosols.

Coke oven emissions are actually complex mixtures of gas, liquid, and solid phases, usually including a range of about 40 polycyclic aromatic hydrocarbons (PAHs), as well as other products of incomplete combustion; notably formaldehyde, acrolein, aliphatic aldehydes, ammonia, carbon monoxide, nitrogen oxides, phenol, cadmium, arsenic, and mercury. More than 60 organic compounds have been collected near coke plants. A metric ton of coal yields up to 635 kg of coke, up to 90 kg of coke breeze (large coke particulates), 7–9 kg of ammonium sulfate, 27.5–34 L of coke oven gas tar, 55–135 L of ammonia liquor, and 8–12.5 L of light oil. Up to 35% of the initial coal charge is emitted as gases and vapors. Most of these gases and vapors are collected during by-product coke production. Coke oven gas is comprised of hydrogen, methane, ethane, carbon monoxide, carbon dioxide, ethylene, propylene, butylene, acetylene, hydrogen sulfide, ammonia, oxygen, and nitrogen. Coke oven gas tar includes pyridine, tar acids, naphthalene, creosote oil, and coal-tar pitch. Benzene, xylene, toluene, and solvent naphthas may be extracted from the light oil fraction. Coke production in the US increased steadily between 1880 and the early 1950s, peaking at 65 million metric tons in 1951. In 1976, the United States was second in the world with 48 million metric tons of coke, i.e. 14.4% of the world production. By 1990, the US produced 24 million metric tons, falling to fourth in the world. A gradual decline in production has continued; production has decreased from 20 million metric tons in 1997 to 15.2 million metric tons in 2002. Demand for blast furnace coke also has declined in recent years because technological improvements have reduced the amount of coke consumed per amount of steel produced by as much as 25%.

(b) *Sintering*: It consists of mixing moist iron ore fines with a solid fuel, usually coke, and then firing the mixture to eliminate undesirable elements and produce a product of relatively uniform size, physically and chemically stable, for charging the blast furnace. Air pollutants are emitted at different points in the process, as indicated in Table 34.14.

(c) *Iron making*: It is the term used to describe how iron is produced in large, refractory-lined structures called *blast furnaces*. The iron ore, limestone, and coke are charged, heated, and then reacted to form a reducing gas, which reduces the iron oxide to metallic iron. The iron is tapped from the furnace along with the slag, which contains the impurities. A modern alternative to the blast furnace is continuous casting of iron instead of intermittent tapping. The blast furnace gas is exhausted from the top of the furnace, cooled, and cleaned of dust before it is used to fire the regenerative stoves for heating the blast furnace. The atmospheric emissions from the iron making process are listed in Table 34.15.

(d) *Steelmaking*: The open-hearth steelmaking process produced 80–90% of the steel in the United States until the 1960s, when the basic oxygen process came into wide use. By 1990 less than 5% of US steel was produced by the open-hearth process. Particulate emissions are highest

TABLE 34.14

Air Pollution Emissions and Controls: Sintering

Process	Air pollutant emissions	Control methods in use
Waste gases (main stack gases)	Particulates, CO, SO_2, chlorides, fluorides, ammonia, hydrocarbons, arsenic	Gravity separators to cyclones; then ESPs or wet scrubbers with pH adjustment
Sinter machine discharge	Particulate matter	Multiple cyclones, baghouse, or low-energy wet scrubber
Materials handling	Fugitive particulate matter	Hoods over release points to baghouse or multiple cyclones
Sinter cooler	Particulate matter (trace)	Baghouse (if required)

Source: Steiner [14].

TABLE 34.15

Air Pollution Emissions and Controls: Iron Making

Process	Air pollutant emissions	Control methods in use
Blast furnace exhaust gases	Particulate matter	Multiple cyclone plus wet scrubber or wet ESP, two-stage wet scrubber
Slag handling	H_2S, SO_2 (trace)	None
Casting	Particulate matter	Baghouse

Source: Steiner [14].

TABLE 34.16

Air Pollution Emissions and Controls: Basic Oxygen Furnace

Process	Air pollutant emissions	Control methods in use
Hot metal transfer	Graphite and iron oxide particulate matter	Multiple cyclones plus baghouses
Charging and tapping	Particulate matter	Baghouse or venturi scrubber
Furnace waste gases	Particulate matter (7–30 kg per metric ton of steel)	ESP or venturi scrubber
	Carbon monoxide	Flare

during oxygen lancing with a hot metal charge in the furnace. Particulate matter loadings are reported to be in the range of 6–11 kg per metric ton of steel. Most of the particulate matter is iron or iron oxide. Control of the open-hearth particulate matter emissions is accomplished by ESPs or high-energy scrubbers. Only small quantities of SO_2 are emitted, but if venturi scrubbers are used for particulate matter control, they will also reduce the SO_2 emissions. However, severe corrosion problems have been reported for wet scrubbers on open-hearth furnaces [14].

Basic oxygen furnaces (BOFs) have largely replaced open-hearth furnaces for steelmaking. A water-cooled oxygen lance is used to blow high-purity oxygen into the molten metal bath. This causes violent agitation and rapid oxidation of the carbon, impurities, and some of the iron. The reaction is exothermic, and an entire heat cycle requires only 30–50 min. The atmospheric emissions from the BOF process are listed in Table 34.16.

The electric arc furnace process accounted for about 25% of the 1982 US steelmaking capacity [14]. Most of the raw material used for the process is steel scrap. Pollutants generated by the electric furnace process are primarily particulate matter and CO. The furnaces are hooded, and the gas stream containing the particulate matter is collected, cooled, and passed to a bag-house for cleaning. Venturi scrubbers and ESPs are used as control devices at some mills. Charging and tapping emissions are also collected by hoods and ducted to the particulate matter control device.

After the steel is tapped from the furnace, it is poured into ingots or continuously cast into slabs or billets. Many metallurgical processes are required between the furnace and the finished product. Reheat furnaces cause no air pollution problems. Scarfing processes create a fine iron oxide fume on the steel surface and also release the same fume, which must be controlled by wet scrubbers or ESPs before it reaches the atmosphere. Pickling may result in the release of acid mists. Scrubbing with special solutions may be required for control. Galvanizing is the process of applying a zinc coating and can result in release of zinc oxide emissions to the atmosphere. Control is accomplished through local collection hoods followed by ESPs, wet scrubbers, or baghouses.

(e) *Ferrous foundry operations*: It produce castings of iron or steel. Many foundry air pollution problems are similar to those of steel mills but on a smaller scale. Potential emissions from foundry operations, along with the usual control methods, are shown in Table 34.17.

(f) *Ferroalloy production*: Ferroalloys are used to add various elements to iron or steel for specific purposes. Examples are chromium (in the form of ferrochrome) and manganese (in the form of ferromanganese) added to steel to improve its strength or hardness, or nickel and chromium added to steel to increase its corrosion resistance. In the electrolytic production of nickel, iron is not removed from nickel because nickel will be used for the production of stainless steel. The product is marketed as "ferronickel" rather than nickel. Ferrochrome, ferromanganese, and ferrosilicon are produced in high-temperature furnaces which emit copious quantities of metallic fume and particulate matter. Roasting and concentrating the ore prior to ferroalloy production produce particulate matter and oxide emissions; SO_2 and CO are released during reduction; and casting produces metal oxides and fumes.

TABLE 34.17

Air Pollution Emissions and Controls: Iron and Steel Foundries

Process	Air pollutant emissions	Control methods in use
Cupolas	Particulate matter (5–22 kg ton^{-1})	Baghouses, wet scrubbers, and ESPs
	Carbon monoxide	Afterburner
	SO_2 (25–250 ppm)	If necessary, wet scrubber
Sand conditioning, shakeout, molding	Particulate matter	Medium-energy wet scrubbers, baghouses
Core making	Hydrocarbons	Afterburners (thermal or catalytic)

Source: Steiner [14].

IV. AGRICULTURE AND FOREST PRODUCTS INDUSTRIES

The agricultural and forest products industries are dependent on renewable resources for their existence. They are also acutely aware that air pollution can damage vegetation and, therefore threaten their existence. Both industries have been exempt from many air pollution regulations in the past, but now they are finding these exemptions questioned and in some cases withdrawn [15].

A. Agriculture

The term *agriculture* refers to the operations involved in growing crops or raising animals. Dusts, smoke, gases, and odors are all emissions form various agricultural operations.

1. *Agronomy*

The preparation of soils for crops, planting, and tilling raises dust as a fugitive emission. Such operations are still exempt from air pollution regulations in most parts of the world. The application of fertilizers, pesticides, and herbicides is also exempt from air pollution regulations, but other regulations may cover the drift of these materials or runoff into surface waters. This is particularly true of the materials are hazardous or toxic.

2. *Open Burning*

A major source of particulate matter, carbon monoxide, and hydrocarbons is open burning of agricultural residue. Over 2.5 million metric tons of particulate matter per year are added to the atmosphere over the United States from burning rice, grass straw and stubble, wheat straw and stubble, weeds, prunings, and range bush. Figure 34.4 illustrates an open burn of grass straw and stubble following the harvest of the seed crop.

The major effect of such open burning is the nuisance caused by the smoke, but health effects are noticed by sensitive individuals downwind from the burn. Table 34.18 lists the pollutant emissions from grass field burning [15]. Recently, the international community has been concerned about the release of toxic compounds in addition to the more conventional pollutants in Table 34.18. For example, wood smoke is known to contain polycyclic aromatic hydrocarbons and other toxic organics, such as dioxins. The question is how similar is smoke from open burning to that of wood smoke. So-called biomass burning is currently a major area of interest for many countries, both in terms of local exposures and long-range, transcontinental and transoceanic transport.

If the open burning of agricultural residue is permitted, it should be scheduled to minimize the effect on populated areas. This requires burning when the wind is blowing away from the population centers, not burning

Fig. 34.4. Open burning of a field after a grass seed harvest.

TABLE 34.18

Air Pollution Emission Factors for Agricultural Field Burning

Pollutant	Emission (kg metric ton^{-1})
Particulate matter	8.5
Carbon monoxide	50
Hydrocarbons (as CH_4)	10
Nitrogen oxides (as NO_2)	1

Source: Faith [15].

during inversion periods, burning dry residue to establish a strong convection column rather than a smoldering fire, and burning only a certain number of acres at a time, so that the atmosphere does not become overloaded.

3. *Orchard Heating*

The practice of smudging is still carried out in many areas to protect orchards from frost. Petroleum products are burned in pots, producing both heat and smoke. Since the heat is the desirable product, smokeless heaters with return ducts to reburn the smoke are required by most air pollution control agencies. Some control agencies have passed regulations limiting the smoke to 0.5 or 1.0 g per minute per burner.

Replacement of orchard heaters by wind machines is the most desirable control measure. These large propellers force the warmer air aloft to the ground, where it mixes with the cold air, minimizing frost formation.

4. Alfalfa Dehydration and Pelletizing

Alfalfa dehydration is carried out in a direct-fired rotary dryer. The dried product is transported pneumatically to an air cooler and then to a collecting cyclone. The collected particles are ground or pelletized and then packaged for shipment. The major atmospheric emission from the process is particulate matter, which is controlled by baghouses. Odors may also be a problem, but they disperse rapidly and are no longer a problem at distances of over 1 km.

5. Animal Production

Feeding of domestic animals on a commercial basis results in large quantities of excreta, both liquid and solid. This produces obnoxious odors, which, in turn, produce complaints from citizens of the area. If the animals are concentrated in a feedlot, the odors may become so extreme that odor counteractants are necessary. However, if the feedlots are paved and regularly washed down, the odors may be kept to a satisfactory minimum with much less expense.

Manure is often recycled as a solid organic fertilizer or mixed with water and sprayed as a liquid fertilizer. If the manure is repeatedly used upwind of populated areas, complaints are sure to be filed with the air pollution control agency.

Ammonia (NH_3) is commonly released from animal operations (see Fig. 34.5). For example, in the United Kingdom, emissions from livestock and their

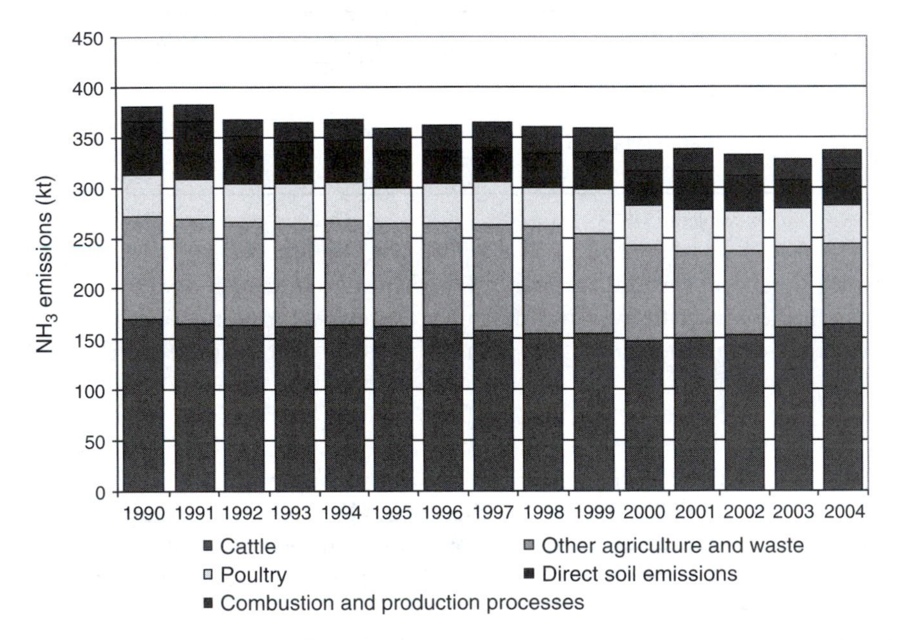

Fig. 34.5. Time series of NH_3 emissions (ktons) in the United Kingdom. *Source*: United Kingdom National Atmospheric Emissions Inventory and National Environmental Technology Centre, *UK Emissions of Air Pollutants 1970–2004*. London, UK, 2006.

wastes comprised 79% of the total ammonia emissions. The decomposition of urea in animal wastes and uric acid in poultry wastes accounted for much of these emissions.[2]

6. *Feed and Grain Milling and Handling*

All grain milling involves grinding and handling of dried grain. Air streams are used for transport of the raw material and the finished products. The result is atmospheric emissions of grain dust. Originally, the control method used was cyclones; today most systems use baghouses following the cyclones. Caution must be exercised in all phases of baghouse construction, operation, cleaning, etc., as the grain dust is explosive and can cause fires accompanied by loss of property and even of lives. Particulate emissions from uncontrolled grain-processing plants range from 0.1 to 2.0 kg of particulate matter per metric ton of grain [16].

7. *Cotton Processing*

The processing of cotton, from the field to the cloth, releases both inorganic and organic particulate matter to the atmosphere. Also, adhering pesticide residues may be emitted at the cotton gin exhaust. Table 34.19 lists the emission factors for particulate matter from cotton ginning operations.

8. *Meat and Meat Products*

The control of odors from holding pens and yards is similar to that discussed in this chapter, Section IV, A, 5. Odors can arise during rendering, cooking, smoking, and processing. Since most of the emissions from the meat products industry are odorous organics, afterburners are used successfully as control devices. Some processors have tried to use wet scrubbers or ESPs, but the emissions are often sticky and can cause severe cleaning problems. Fish processing has similar problems and solutions.

TABLE 34.19

Particulate Matter Emissions from Cotton Ginning

Process	Emissions (kg bale)[a]
Unloading fan	2.20
Cleaner	0.45
Stick and burr machine	1.35
Miscellaneous	1.35
Total	5.35

[a] One bale weighs 277 kg.
Source: Ref. [16].

[2] United Kingdom National Atmospheric Emissions Inventory and National Environmental Technology Centre, *UK Emissions of Air Pollutants 1970–2004*. London, UK, 2006.

9. Fruit and Vegetable Processing

The most severe environmental problem of fruit and vegetable processors is the potential for water pollution if the liquid wastes are not handled properly. Cooking can cause odors, which are usually controlled by using furnaces as afterburners.

One processing problem is presented by the roasting of coffee. This releases smoke, odor, and particulate matter. The particulate matter, primarily dusty and chaff, can be removed with a cyclone. The smoke and odors are usually consumed by passing the roaster exhaust gases through an afterburner. Heat recovery may be desirable if the afterburner is large enough to make it economical.

10. Miscellaneous Agricultural Industries

Other industries of interest are (1) the manufacturing of spices and flavorings, which may use activated carbon filters to remove odors from their exhaust stream; (2) the tanning industry, which uses afterburners or activated carbon for odor removal and wet scrubbers for dust removal; and (3) glue and rendering plants, which utilize sodium hypochlorite scrubbers or afterburners to control odorous emissions.

B. Forest Products

The forest products industry encompasses a broad spectrum of operations which range from the raising of trees, through cutting and removing the timber, to complete utilization of the wood residue [17].

1. Open Burning

The forest products industry (as well as governmental agencies such as the US Forest Service) practices open, prescribed, burning of logging residue (slash) as a forest management tool and as an economical means of residue disposal [18]. This burning is usually done when meteorological conditions and fuel variables, such as moisture content, can give as clean a burn as possible with a minimum effect on populated areas. On a worldwide basis, it has been estimated that approximately 90 million metric tons of particulate matter from wild and controlled forest and range fires enter the atmosphere each year. Table 34.20 lists the pollutant emissions from forest burning [16].

2. Wood-Fired Power Boilers

Wood-fired power boilers are generally found at the mills where wood products are manufactured. They are fired with waste materials from the process, such as "hogged wood," sander dust, sawdust, bark, or process trim. Little information is available on gaseous emissions from wood-fired boilers, but extensive tests of particulate matter emissions are reported [19]. The lignium component of wood is expected in the reaction: wood constituents can

TABLE 34.20

Air Pollution Emission Factors for Forest Burning

Pollutant	Emission (kg metric ton^{-1})
Particulate matter	8.5
Carbon monoxide	22
Hydrocarbons (as CH_4)	2
Nitrogen oxides (as NO_2)	1

Source: Ref. [16].

TABLE 34.21

Particulate Matter Collection Devices for Wood-Fired Boilers

Pollutant control device	Efficiency (%)
Multiple cyclone	51
Wet scrubber	67
Dry scrubber	85–97 (depending on fuel)
Baghouse	99+

Source: Boubel [19].

be quite variable, such as the cellulose, lignium, and other carbohydrate compounds. As a general rule of combustion, wood can be assumed to have the composition of $C_{6.9}H_{10.6}O_{3.5}$. However, it also contains many other elements, such as sulfur, phosphorous, and nitrogen, which are oxidized during the firing processes. These emissions range from 0.057 to 1.626 g per dry standard cubic meter, with an average of 0.343 reported for 135 tests. Collection devices for particulate matter from wood-fired boilers are shown in Table 34.21.

3. *Driers*

Driers are used in the forest products industry to lower the moisture content of the wood product being processed. Drying of dimension lumber gives it dimensional stability. This type of drying is done in steam kilns and is a batch process. No appreciable pollutants are released.

Veneer for the manufacture of plywood is dried on a continuous line in a veneer drier to assure that only dry veneer goes to the layup and gluing process. Glue will not bond if the veneer contains too much moisture. Emissions from the veneer driers are fine particulate and condensed organic material. The condensed organic material is of submicron size and appears as a blue haze coming from the stack. Control is accomplished by means of (1) a wet scrubber or (2) ducting the emissions to a wood-fired boiler, where they are burned. All of these systems must be carefully sized and operated in order to meet a 20% opacity regulation.

Wood particle and fiber driers are used to dry the raw material for particle board and similar products [20]. Just as with the veneer for plywood, the particles must be dried before being mixed with the resins and formed into board. Drying is accomplished in a gas-fired drier, a direct wood-fired drier, or steam coil driers. Many different types of driers are used in the industry. Emissions are fine particles and condensible hydrocarbons, which produce a highly visible plume. Control is accomplished with multiple cyclones or wet scrubbers. Fires in the control equipment and ductwork are quite common and must be expected periodically.

4. Kraft Process Pulp and Paper Plants

The kraft process has become the dominant process for pulp production throughout the world, primarily because of the recovery of the pulping chemicals. A schematic diagram of the kraft pulping process, with the location of atmospheric emission sources, is shown in Fig. 6.11.

Control of air pollutant emissions in modern kraft mills is accomplished by (1) proper operation of the entire mill, (2) high-efficiency ESPs on the recovery furnace (up to 99.7% efficiency for particulate matter, which is recovered and returned to the process), (3) collection of noncondensible gases from several vent points (digesters, blow tanks, washers) and ducting them to the lime kiln, where they are completely burned, and (4) high-energy wet scrubbers on the lime kiln exhaust to remove the particulate matter and sulfurous gases. A more complete analysis of the kraft process, including the emissions and their control, may be found in Ref. [17].

V. OTHER INDUSTRIAL PROCESSES

Many industries operated throughout the world do not fall into the previous categories. Some of these are universal, such as asphalt batching plants, whereas others are regional, such as bagasse-fired boilers. Each has its own emission and control problems and requires knowledgeable analysis and engineering. Some of the more widely used processes are examined in this section.

A. Mineral Products

Conversion of minerals to useful products is a major worldwide industry. Mining or quarrying of minerals can produce fugitive emissions, which may be controlled by paving work and traffic areas, wetting the materials being removed or handled, or using collection and exhaust systems at the site where the particulate matter is being generated. The usual air pollution control device is a multiple cyclone or a baghouse at the system exit. The same control techniques can be applied at other points in the process where the minerals are transported, stored, crushed and ground, concentrated, dried, and mixed [21].

1. Asphaltic Concrete Plants

Two types of asphaltic concrete plants are in common use, batch mix plants and continuous mix plants. Figure 34.6 shows a batch mix asphalt plant. Fugitive emissions occur at the handling areas and at the bin loading facility. The emissions of greatest concern, however, occur at the rotary drier, hot aggregate elevator, and hot aggregate handling systems. Each has the potential for releasing large quantities of uncontrolled particulate matter. Table 34.22 illustrates the large range of emissions from uncontrolled and controlled asphaltic concrete plants.

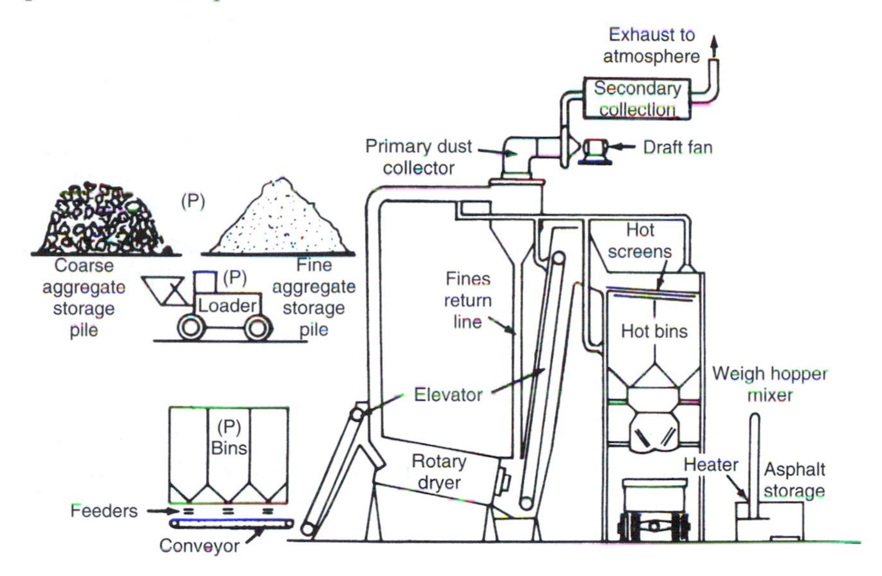

Fig. 34.6. Batch mix asphalt plant; P, denotes fugitive particulate matter emissions. *Source*: Ref. [16].

TABLE 34.22

Particulate Matter Emission Factors for Asphaltic Concrete Plants

Type of control	Emissions (kg metric ton^{-1})
Uncontrolled[a]	22.5
Cyclone precleaner	7.5
High-efficiency cyclone	0.85
Spray tower scrubber	0.20
Multiple centrifugal scrubber	0.15
Baffle spray tower scrubber	0.15
Orifice-type scrubber	0.02
Baghouse[b]	0.05

[a] Almost all plants have at least a precleaner following the rotary drier. The fines collected are returned and are an important part of the mix.
[b] Emissions from a properly designated, installed, operated, and maintained baghouse collector can be as low as 0.0025–0.010 kg per metric ton.
Source: Ref. [16].

2. Cement Plants

Portland cement manufacture accounts for about 98% of the cement production in the United States. The raw materials are crushed, processed, proportioned, ground, and blended before going to the final process, which may be either wet or dry. In the dry process, the moisture content of the raw material is reduced to less than 1% before the blending process occurs. The dry material is pulverized and fed to the rotary kiln. Further drying, decarbonating, and calcining take place as the material passes through the rotary kiln. The material leaves the kiln as clinker, which is cooled, ground, packaged, and shipped.

For the wet process, a slurry is made by adding water during the initial grinding. The homogeneous wet mixture is fed to the kiln as a wet slurry (30–40% water) or as a wet filtrate (20% water). The burning, cooling, grinding, packaging, and shipping are the same as for the dry process.

Particulate matter emissions are the primary concern with cement manufacture. Fugitive emissions and uncontrolled kiln emissions are shown in Table 34.23.

Control of particulate matter emissions from the kilns, dryers, grinders, etc. is by means of standard devices and systems: (1) multiple cyclones (80% efficiency), (2) ESPs (95% + efficiency), (3) multiple cyclones followed by ESPs (97.5% efficiency), and (4) baghouses (99.8% efficiency).

3. Glass Manufacturing Plants

Soda-lime glass accounts for about 90% of the US production. It is produced in large, direct-fired, continuous-melting furnaces in which the blended raw materials are melted at 1480°C to form glass. The emissions from soda-lime

TABLE 34.23

Air Pollution Emission Factors for Portland Cement Manufacturing without Controls

	Emissions (kg metric ton^{-1})			
	Dry process		Wet process	
Pollutant	Kilns	Dryers, grinders, etc.	Kilns	Dryers, grinders, etc.
Particulate matter	122.0	48.0	114.0	16.0
Sulfur dioxide[a]	5.1			
Mineral source	Neg	—	5.1	—
Gas combustion	2.1 × S[b]	—	Neg	—
Oil combustion	3.4 × S	—	2.1 × S[b]	—
Coal combustion		—	3.4 × S	—
Nitrogen oxides	1.3	—	1.3	—

[a] If a baghouse is used as control device, reduce SO_2 by 50% because of reactions with an alkaline filter cake.
[b] S is the percent of sulfur in the fuel.
Source: Ref. [16].

TABLE 34.24

Air Pollution Emission Factors for Fiber Glass Manufacturing without Controls

	Emissions (kg/metric ton^{-1})				
Type of process	Particulate matter	Sulfur oxides as SO2	Carbon monoxide	Nitrogen oxides as NO2	Fluorides
Textile products					
Glass furnace					
Regenerative	8.2	14.8	0.6	4.6	1.9
Recuperative	13.9	1.4	0.5	14.6	6.3
Forming	0.8	—	—	—	—
Curing ovens	0.6	—	0.8	1.3	—
Wool products					
Glass furnace					
Regenerative	10.8	5.0	0.13	2.5	0.06
Recuperative	14.2	4.8	0.13	0.9	0.06
Electric	0.3	0.02	0.03	0.14	0.01
Forming	28.8	—	—	—	—
Curing ovens[a]	1.8	—	0.9	0.6	—
Cooling[a]	0.7	—	0.1	0.1	—

[a] In addition, 0.05 kg per metric ton for phenol and 1.7 kg per metric ton for aldehyde during curing and cooling.
Source: Ref. [16].

glass melting are 1.0 kg of particulate matter per metric ton of glass and 2 × (fluoride percentage) for the fluoride emissions in kilograms per metric ton. For effective control of the emissions, baghouses are used.

Fiberglass is manufactured primarily from borosilicate glass by drawing the molten glass into fibers. Two fiberglass products are produced, textile and glass wool. The emissions from the two processes are shown in Table 34.24 [16]. Control is achieved through proper design and operation of the manufacturing operations rather than by add-on devices.

B. Petroleum Refining and Storage

A modern petroleum refinery is a complex system of chemical and physical operations. The crude oil is first separated by distillation into fractions such as gasoline, kerosene, and fuel oil. Some of the distillate fractions are converted to more valuable products by cracking, polymerization, or reforming. The products are treated to remove undesirable components, such as sulfur, and then blended to meet the final product specifications. A detailed analysis of the entire petroleum production process, including emissions and controls, is obviously well beyond the scope of this text. The reader is referred to Refs. [16, 22, 23]. Reference [16] presents an extensive tabulation of the emission sources for all processes involved in petroleum refining and production, some of which are summarized in Table 34.25.

TABLE 34.25

Sources of Emissions from Oil Refining

Type of emission	Source
Hydrocarbons	Air blowing, barometric condensers, blind changing, blowdown systems, boilers, catalyst regenerators, compressors, cooling towers, decoking operations, flares, heaters, incinerators, loading facilities, processing vessels, pumps, sampling operations, tanks, turnaround operations, vacuum jets, waste effluent handling equipment
Sulfur oxides	Boilers, catalyst regenerators, decoking operations, flares, heaters, incinerators, treaters, acid sludge disposal
Carbon monoxide	Catalyst regenerators, compressor engines, coking operations, incinerators
Nitrogen oxides	Boilers, catalyst regenerators, compressor engines, flares
Particulate matter	Boilers, catalyst regenerators, coking operations, heaters, incinerators
Odors	Air blowing, barometric condensers, drains, process vessels, steam blowing, tanks, treaters, waste effluent handling systems
Aldehydes	Catalyst regenerators, compressor engines
Ammonia	Catalyst regenerators

Source: Elkins [23].

Control of atmospheric emissions from petroleum refining can be accomplished by process change, installation of control equipment, and improved housekeeping and maintenance. In many cases, recovery of the pollutants will result in economic benefits. Table 34.26 lists some of the control measures that can be used at petroleum refineries.

C. Sewage Treatment Plants

The concern with atmospheric emissions from sewage treatment plants involves gases and odors from the plant itself, particulate matter and gaseous emissions from the sludge incinerator if one is used, and all three pollutants (gases, odors, and particulate matter) if sludge disposal is conducted at the site. The gases and odors are combustible, so afterburners or flares are used. Some plants use the sewage gas to fire small stationary boilers or fuel gas–diesel engines for plant energy. Particulate matter from sludge incinerators is usually scrubbed with treated water from the plant, and the effluent is returned to the incoming plant stream. If the odors are too persistent, masking agents are sometimes specified to lessen the objections of the public.

D. Coal Preparation Plants

Coal preparation plants are used to reduce noncombustibles and other undesirable materials in coal before it is burned.

TABLE 34.26

Control Measures for Air Pollutants from Petroleum Refining

Source	Control method
Storage vessels	Vapor recovery systems; floating roof tanks; pressure tanks; vapor balance; painting tanks white
Catalyst regenerators	Cyclones–precipitator–CO boiler; cyclones-water scrubber; multiple cyclones
Accumulator vents	Vapor recovery; vapor incineration
Blowdown systems	Smokeless flares-gas recovery
Pumps and compressors	Mechanical seals; vapor recovery; sealing glands by oil pressure; maintenance
Vacuum jets	Vapor incineration
Equipment valves	Inspection and maintenance
Pressure relief valves	Vapor recovery; vapor incinceration; rupture disks; inspection and maintenance
Effluent waste disposal	Enclosing separators; covering sewer boxes and using liquid seals; liquid seals on drains
Bulk loading facilities	Vapor collection with recovery or incineration; submerged or bottom loading
Acid treating	Continuous-type agitators with mechanical mixing; replace with catalytic hydrogenation units; incinerate all vented cases; stop sludge burning
Acid sludge storage and shipping	Caustic scrubbing; incineration, vapor return system
Spent caustic handling	Incineration; scrubbing
Doctor treating	Steam strip spent doctor solution to hydrocarbon recovery before air regeneration; replace treating unit with other, less objectionable units (Merox)
Sour water treating	Use sour water oxidizers and gas incineration; conversion to ammonium sulfate
Mercaptan disposal	Conversion to disulfides; adding to catalytic cracking charge stock; incineration; using material in organic synthesis
Asphalt blowing	Incineration; water scrubbing (nonrecirculating type)
Shutdowns, turnarounds	Depressure and purge to vapor recovery

Source: Buonicore and Davis [22].

E. Gas Turbines

Gas turbines are used as prime movers for pumps, electric generators, and large rotating machinery. Their main economic advantage is driving high-horsepower, consistent loads. Many stationary gas turbines use the same core engine as their jet engine counterpart.

REFERENCES

1. Public Law No. 101–549, 101st Congress—*Clean Air Act Amendments of 1990*, November 1990.
2. Kurtzweg, J. A., and Griffin, C. N., *J. Air Pollut. Control Assoc.* **31**, 1155–1162 (1981).

3. Cooper, J. A. (ed.), *Residential Solid Fuels: Environmental Impacts and Solutions*. Oregon Graduate Center, Beaverton, OR, 1981.
4. Preston, G. T., and Miller, M. J., *Pollut. Eng.* **14** (4), 25–29 (1982).
5. Moskowitz, J., *Mech. Eng.* **104** (4), 68–71 (1981).
6. Ellison, W., and Lefton, S. A., *Power* **126** (4), 71–76 (1982).
7. Brunner, C. R., *Handbook of Incineration Systems*. McGraw-Hill, New York, 1991.
8. Arbuckle, J. G., Frick, G. W., Miller, M. L., Sullivan, T. F. P., and Vanderver, T. A., *Environmental Law Handbook*. Government Institutes, Inc., Washington, DC, 1979.
9. *Chem. Week* **113**, 31–32 (1973).
10. Cuffe, S. T., Walsh, R. T., and Evans, L. B., Chemical industries, *in Air Pollution* (Stern, A.C., ed.), 3rd ed., Vol. 4. Academic Press, New York, 1977.
11. Iverson, R. E., *J. Met.* **25**, 19–23 (1973).
12. *Control Techniques for Beryllium Air Pollutants*, EPA Publication No. AP-116. US Environmental Protection Agency, Research Triangle Park, NC, 1973.
13. Nelson, K. W., Varner, M. O., and Smith, T. J., Nonferrous metallurgical operations *in Air Pollution* (Stern, A. C., ed.), 3rd ed., Vol. 4. Academic Press, New York, 1977.
14. Steiner, B. A., Ferrous metallurgical operations, *in Air Pollution* (Stern, A. C., ed.), 3rd ed., Vol. 4. Academic Press, New York, 1977.
15. Faith, W. L., Agriculture and agricultural-products processing, in *Air Pollution* (Stern, A. C., ed.), 3rd ed., Vol. 4. Academic Press, New York, 1977.
16. *Compilation of Air Pollutant Emission Factors*, 2nd ed. (with supplements), EPA Publication No. AP-42. US Environmental Protection Agency, Research Triangle Park, NC, 1980.
17. Hendrickson, E. R., The forest products industry, in *Air Pollution* (Stern, A. C., ed.), 3rd ed., Vol. 4. Academic Press, New York, 1977.
18. Ward, D. E., Nelson, R. M., and Adams, D. F., *Proceedings of the 70th Annual Meeting of the Air Pollution Control Association*, Pittsburgh, PA, 1977.
19. Boubel, R. W., *Control of Particulate Emissions from Wood-Fired Boilers*, Stationary Source Enforcement Series, EPA 340/1-77-026. US Environmental Protection Agency, Washington, DC, 1977.
20. Junge, D. C., and Boubel, R. W., *Analysis of Control Strategies and Compliance Schedules for Wood Particle and Fiber Dryers*, EPA Contract Report No. 68-01-3150. PEDCO Environmental Specialists, Cincinnati, OH, 1976.
21. Sussman, V. H., Mineral products industries, in *Air Pollution* (Stern, A. C., ed.), 3rd ed., Vol. 4 Academic Press, New York, 1977.
22. Buonicore, A. J., and Davis, W. T. (eds.), *Air Pollution Engineering Manual*. Van Nostrand Reinhold, New York, 1992.
23. Elkins, H. F., Petroleum refining, in *Air Pollution* (Stern, A. C., ed.), 3rd ed., Vol. 4 Academic Press, New York, 1977.

SUGGESTED READING

Power **126** (4), 375–410 (1982).

Elliot, T. C., *Power* **118**, 5.1–5.24 (1974).

Gage, S. J., *Controlling Sulfur Oxides*, Research Summary, EPA-600/8-80-029. US Environmental Protection Agency, Washington, DC, 1980.

Hesketh, H. E., *Air Pollution Control, Traditional and Hazardous Pollutants*. Technomic, Lancaster, PA, 1991.

Kokkinos, A., Cicnanowicz, J. E., Hall, R.E., and Jedman, C. B., Stationary combustion NO_x control: a summary of the 1991 symposium. *J. Air Waste Manage. Assoc.* **41** (9) (September 1991).

Lutz, S. J., McCoy, B. C., Mullingan, S. W., Christman, R. C., and Slimak, K. M., *Evaluation or Dry Sorbants and Fabric Filtration for FGD*, EPA-600/7-79-005. US Environmental Protection Agency, Research Triangle Park, NC, 1979.

Theodore, L., and Bunicore, A. J., *Air Pollution Control Equipment*. Prentice-Hall, Englewood Cliffs, NJ, 1982.

Vatauk, W. M., *Estimating Costs of Air Pollution Control*. Lewis, Chelsea, MI, 1990.

QUESTIONS

1. Rank the control systems shown in Fig. 34.1 according to their relative capital construction and operation costs.
2. Justify the statement, "Tall stacks are no longer considered as an acceptable alternative for controlling emissions from electric power generating plants."
3. For the New Source Performance Standards (NSPS) for incinerators, only particulate matter emissions are covered. Devise a standard which would also include POM, CO, and NO_x for large municipal incinerators.
4. What advantage is gained by oxidizing H_2S to SO_2? (Consider toxicity and odors.)
5. Why are total fluoride emissions from an aluminum smelter of more concern than gaseous or solid fluoride emissions?
6. What systems can be used to detect and prevent grain dust explosions?
7. Would you expect wood-fired boilers to emit more or less CO per metric ton of fuel than coal-fired boilers? More or less NO_x? More or less SO_2?
8. What would be the ultimate disposal of dry material collected by an ESP at a cement plant kiln outlet? What would be the ultimate disposal of wet sludge from a scrubber on a cement plant kiln outlet?
9. Petroleum plants may have a disposal problem with sulfur removed from their products. What are some potential uses for this sulfur?

35

Control of Mobile Sources

I. INTRODUCTION

Because mobile sources of air pollution are capable of moving from one local jurisdiction to another, they are usually regulated by the national government. In the United States, state or local agencies can have more restrictive standards, if they choose. To date, only the state of California has established extensive standards more restrictive than the US federal standards, and these only for gasoline-powered automobiles.

II. GASOLINE-POWERED VEHICLES

Gasoline-powered motor vehicles outnumber all other mobile sources combined in the number of vehicles, the amount of energy consumed, and the mass of air pollutants emitted. It is not surprising that they have received the greatest share of attention regarding emission standards and air pollution control systems. Table 27.2 shows the US federal emission control requirements for gasoline-powered passenger vehicles.

Crankcase emissions in the United States have been effectively controlled since 1963 by positive crankcase ventilation (PCV) systems which take the gases from the crankcase, through a flow control valve, and into the intake manifold. The gases then enter the combustion chamber with the fuel–air mixture, where they are burned.

Figure 35.1 shows a cross section of a gasoline engine with the PCV system.

Evaporative emissions from the fuel tank and carburetor have been controlled on all 1971 and later model automobiles sold in the United States. This has been accomplished by either a vapor recovery system which uses the crankcase of the engine for the storage of the hydrocarbon vapors or an adsorption and regeneration system using a canister of activated carbon to trap the vapors and hold them until such time as a fresh air purge through the canister carries the vapors to the induction system for burning in the combustion chamber.

The exhaust emissions from gasoline-powered vehicles are the most difficult to control. These emissions are influenced by such factors as gasoline formulation, air–fuel ratio, ignition timing, compression ratio, engine speed and load, engine deposits, engine condition, coolant temperature, and combustion chamber configuration. Consideration of control methods must be based on elimination or destruction of unburned hydrocarbons, carbon monoxide, and oxides of nitrogen. Methods used to control one pollutant may actually increase the emission of another requiring even more extensive controls.

Control of exhaust emissions for unburned hydrocarbons and carbon monoxide has followed three routes:

1. Fuel modification in terms of volatility, hydrocarbon types, or additive content. Some of the fuels currently being used are liquefied petroleum gas (LPG), liquefied natural gas (LNG), compressed natural gas (CNG), fuels with alcohol additives, and unleaded gasoline. The supply of some of these fuels is very limited. Other fuel problems involving storage, distribution, and power requirements have to be considered.
2. Minimization of pollutants from the combustion chamber. This approach consists of designing the engine with improved fuel–air distribution

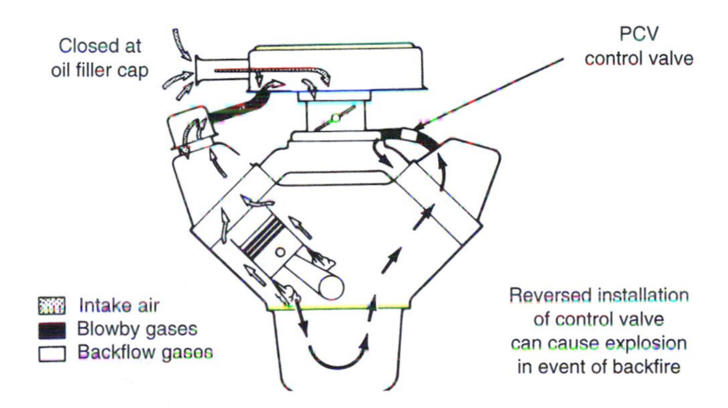

Fig. 35.1. Positive crankcase ventilation (PCV) system.

systems, ignition timing, fuel–air ratios, coolant and mixture temperatures, and engine speeds for minimum emissions. The majority of automobiles sold in the United States now use an electronic sensor/control system to adjust these variables for maximum engine performance with minimum pollutant emissions.

3. Further oxidation of the pollutants outside the combustion chamber. This oxidation may be either by normal combustion or by catalytic oxidation. These systems require the addition of air into the exhaust manifold at a point downstream from the exhaust valve. An air pump is employed to provide this air. Figure 35.2 illustrates an engine with an air pump and distribution manifold for the oxidation of CO and hydrocarbons (HC) outside the engine.

Beginning with the 1975 US automobiles, catalytic converters were added to nearly all models to meet the more restrictive emission standards. Since the lead used in gasoline is a poison to the catalyst used in the converter, a scheduled introduction of unleaded gasoline was also required. The US petroleum industry simultaneously introduced unleaded gasoline into the marketplace.

In order to lower emissions of oxides of nitrogen from gasoline engines, two general systems were developed. The first is exhaust gas recirculation (EGR), which mixes a portion of the exhaust gas with the incoming fuel–air charge, thus reducing temperatures within the combustion chamber. This recirculation is controlled by valving and associated plumbing and electronics, so that it occurs during periods of highest NO_x production, when some power reduction can be tolerated: a cruising condition at highway speed. Other alternatives are to use another catalytic converter, in series with the HC/CO converter, which decomposes the oxides of nitrogen to oxygen and nitrogen before the gases are exhausted from the tailpipe.

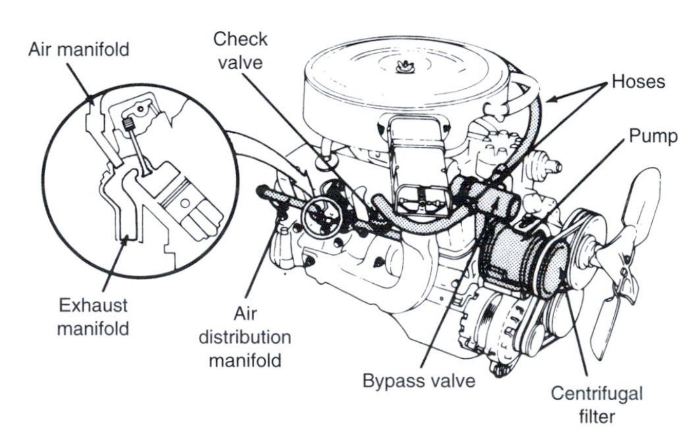

Fig. 35.2. Manifold air oxidation system.

III. DIESEL-POWERED VEHICLES

The diesel (compression ignition) cycle is regulated by fuel flow only, air flow remaining constant with engine speed. Because the diesel engine is normally operated well on the lean side of the stoichiometric mixture (40:1 or more), emission of unburned hydrocarbons and carbon monoxide is minimized. The actual emissions from a diesel engine are (1) oxides of nitrogen, as for spark ignition engines; (2) particulate matter, mainly unburned carbon, which at times can be excessive; (3) partially combusted organic compounds, many of which cause irritation to the eyes and upper respiratory system; and (4) oxides of sulfur from the use of sulfur-containing fuels. A smoking diesel engine indicates that more fuel is being injected into the cylinder than is being burned and that some of the fuel is being only partially burned, resulting in the emission of unburned carbon.

Control of diesel-powered vehicles is partially accomplished by fuel modification to obtain reduced sulfur content and cleaner burning and by proper tuning of the engine using restricted fuel settings to prevent overfueling.

Diesel fuel has begun to evolve in another, more sustainable, way. An increasing number of vehicles are running on so-called "bio-diesel" fuel. This is refined from plant-based hydrocarbons, especially from waste residues (e.g. from used cooking oils). With operational adjustments, bio-diesel is currently in use in buses as part of urban transit and college transportation systems.

Effective with the 1982 model year, particulate matter from diesel vehicles was regulated by the US Environmental Protection Agency for the first time, at a level of $0.37\,\mathrm{g\,km^{-1}}$. Diesel vehicles were allowed to meet an NO_x level of $0.93\,\mathrm{g\,km^{-1}}$ under an Environmental Protection Agency waiver. These standards were met by a combination of control systems, primarily EGR and improvements in the combustion process. For the 1985 model year, the standards decreased to 0.12 g of particulate matter per kilometer and 0.62 g of NO_x per kilometer. This required the use of much more extensive control systems [1]. The Clean Air Act Amendments of 1990 [2] have kept the emission standards at the 1985 model level with one exception: diesel-fueled heavy trucks shall be required to meet an NO_x standard of 4.0 g per brake horsepower hour.

IV. GAS TURBINES AND JET ENGINES

The modified Brayton cycle is used for both gas turbines and jet engines. The turbine is designed to produce a usable torque at the output shaft, while the jet engine allows most of the hot gases to expand into the atmosphere, producing usable thrust. Emissions from both turbines and jets are similar, as are their control methods. The emissions are primarily unburned hydrocarbons, unburned carbon which results in the visible exhaust, and oxides of nitrogen. Control of the unburned hydrocarbons and the unburned carbon may be accomplished by redesigning the fuel spray nozzles and reducing cooling air to the combustion chambers to permit more complete combustion.

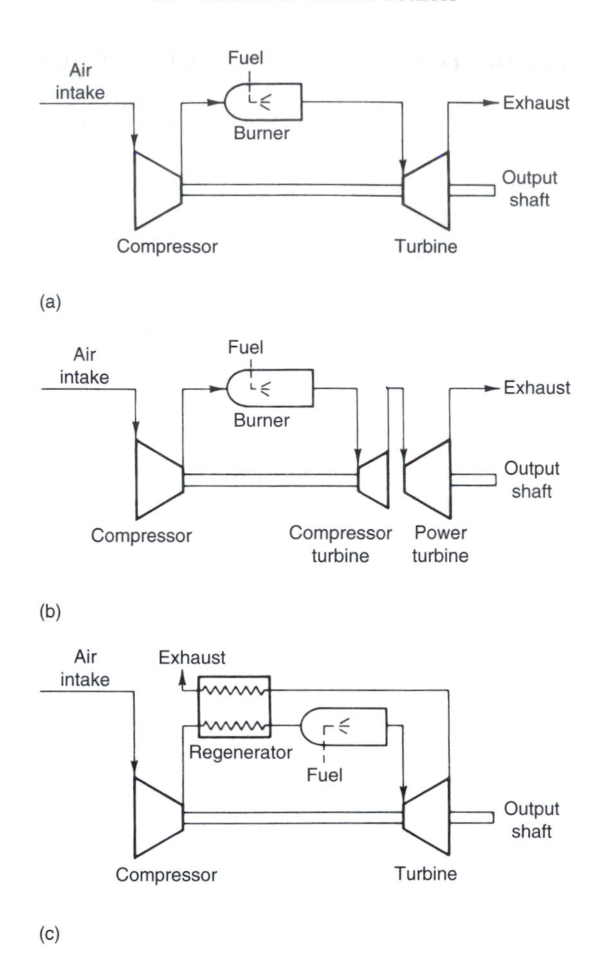

Fig. 35.3. Schematic diagrams of gas turbines: (a) simple gas turbine, (b) free-turbine engine, and (c) regenerative gas turbine.

US airlines have converted their jet fleets to lower-emission engines using these control methods. NO_x emissions may be minimized by reduction of the maximum temperature in the primary zone of the combustors.

US Environmental Protection Agency regulations for commercial, jet, and turbine-powered aircraft [3] are based on engine size (thrust) and pressure ratio (compressor outlet/compressor inlet) for the time in each mode of a standardized takeoff and landing cycle. Once the aircraft exceeds an altitude of 914 m, no regulations apply.

The gas turbine engine for automotive or truck use could be either a simple turbine, a regenerative turbine, a free turbine, or any combination. Figure 35.3 shows the basic types which have been successfully tried in automotive and truck use.

V. ALTERNATIVES TO EXISTING MOBILE SOURCES

The atmosphere of the world cannot continue to accept greater and greater amounts of emissions from mobile sources as our transportation systems expand. The present emissions from all transportation sources in the United States exceed 50 billion kg of carbon monoxide per year, 20 billion kg per year of unburned hydrocarbons, and 20 billion kg of oxides of nitrogen. If presently used power sources cannot be modified to bring their emissions to acceptable levels, we must develop alternative power sources or alternative transportation systems. All alternatives should be considered simultaneously to achieve the desired result, an acceptable transportation system with a minimum of air pollution.

One modified internal combustion engine which shows promise is the stratified-charge engine. This is a spark ignition engine using fuel injection in such a manner as to achieve selective stratification of the air/fuel ratio in the combustion chamber. The air/fuel ratio is correct for ignition at the spark plug, and the mixture is fuel lean in other portions of the combustion chamber. Only air enters the engine on the intake stroke, and the power output is controlled by the amount of fuel injected into the cylinder. Stratified-charge engines have been operated experimentally and used in some production vehicles [4]. They show promise as relatively low-emission engines. The hydrocarbon emission levels from this engine are quire variable, the CO levels low, and the NO_x levels variable but generally high.

An external combustion engine that has been widely supported as a low-emission power source is the Rankine cycle steam engine. Many different types of expanders can be used to convert the energy in the working fluid into rotary motion at a drive shaft. Expanders that have been tried or proposed are reciprocating piston engines, turbines, helical expanders, and all possible combinations of these. The advantage of the steam engine is that the combustion is continuous and takes place in a combustor with no moving parts. The result is a much lower release of air pollutants, but emissions are still not completely zero. Present technology is capable of producing a satisfactory steam-driven car, truck, or bus, but costs, operating problems, warmup time, and weight and size must be considered in the total evaluation of the system. A simple Rankine cycle steam system is shown diagrammatically in Fig. 35.4.

Electric drive systems have been tried as a means of achieving propulsion without harmful emissions. Currently, most battery-operated vehicles use lead–acid batteries, which give low power, limited range, and require frequent recharging. In power shortage areas this could be a severe additional load on the electrical system. Sulfuric acid, hydrogen, and oxygen emissions from millions of electric vehicles using lead–acid storage batteries for an energy source would be appreciable. Other types of batteries offer some promise, but their manufacture and costs present obstacles to their widespread

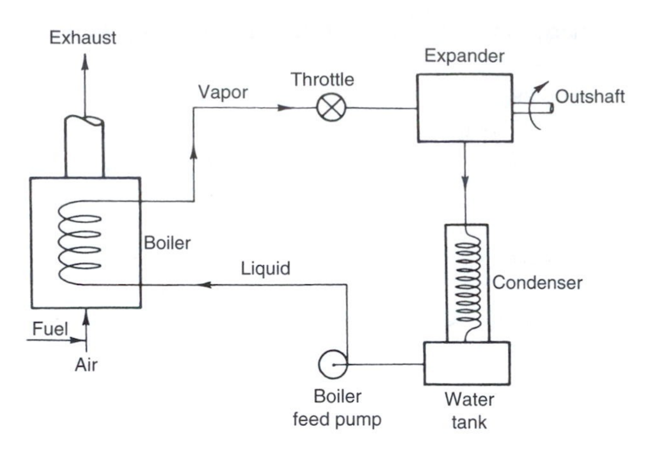

Fig. 35.4. Rankine cycle system.

automotive use. In fact, the nickel (Ni) based battery systems installed in many hybrid vehicles detract from the vehicles' overall environmental attractiveness. Viewed from a life cycle perspective, the mining of the Ni ore (e.g. from the extensive area near Sudsbury, Ontario), the processing of the ore and manufacture of Ni foam, and the shipping of the foam and other components many thousands of kilometers to vehicle assembly plants present large costs and make for a rather unsustainable network to support the batteries.

Hybrid systems consisting of two or more energy-conversion processes are now being increasingly used in very low-emission vehicles. A constant speed and load internal combustion engine driving a generator with a small battery for load surges could be made to emit less hydrocarbons, CO, and NO_x than a standard automobile engine, but the cost would be much higher. Other hybrid systems which have been proposed are steam-electric and turbine-electric. The problem associated with hybrid systems is the cost of the two engines plus the cost of the added controls and system integration [5]. However, the prices are beginning to equilibrate, and they have become more attractive to a broader consumer base since they provide a "green" alternative to ordinary internal combustion vehicles.

Fuel cells, which rely on electrochemical generation of electric power, could be used for nonpolluting sources of power for motor vehicles. Since fuel cells are not heat engines, they offer the potential for extremely low emissions with a higher thermal efficiency than internal combustion engines. Basically, a fuel cell converts chemical energy directly into electricity by combining oxygen from the air with hydrogen gas. The cell can keep producing electricity so long as hydrogen is available. The space program has used fuel cells for many decades. However, there are still obstacles to overcome before fuel cells attain widespread use. The federal government and others, especially California, have recently stated strong interests in making fuel cells more mainstream, in the interests of energy savings (and national security), as well as for improvements in air quality.

Probably the ultimate answer to the problem of emissions from millions of private automobiles is an alternative transportation system. It must be remembered, however, that even rail systems and bus systems do emit some air pollution. Rail systems are expensive and lack flexibility. A quick calculation of the number of passengers carried per minute past a single point on a freeway in private automobiles will illustrate the difficulties of a rail system in replacing the automobile. Buses offer much greater flexibility at lower cost than rail systems, but in order to operate efficiently and effectively, they would require separate roadway systems and loading stations apart from automobile traffic.

REFERENCES

1. *1982 General Motors Public Interest Report*. General Motors Corporation, Detroit, 1982.
2. Public Law No. 101–549, US 101st Congress, *Clean Air Act Amendments of 1990*, November 1990.
3. EPA proposed revisions to gaseous emissions rules for aircraft and aircraft engines. *Fed. Regist.* **43**, 12615 (1978).
4. Olson, D. R., The control of motor vehicle emissions, in *Air Pollution* (Stern, A. C., ed.), 3rd ed., Vol. IV. Academic Press, New York, 1977.
5. Harmon, R., Alternative vehicle-propulsion systems. *Mech. Eng.* **105** (4), 67–74 (March 1992).

SUGGESTED READING

Ahmed, I., *How Much Energy Does It Take to Make a Gallon of Soy Diesel?* Institute for Local Self-Reliance for the National Soy Diesel Development Board, Jefferson City, MO, 1994.

Chang, T. Y., Alternative transportation fuels and air quality. *Environ. Sci. Technol.* **25**, 1190 (1991).

Dasch, J. M., Nitrous oxide emissions from vehicles. *J. Air Waste Manage. Assoc.* **42** (1), 32–38 (January 1992).

Diesel combustion and emissions, *Proceedings P-86*. Society of Automotive Engineers, Warrendale, PA, 1980.

Diesel combustion and emissions, Part III, *SP-495*. Society of Automotive Engineers, Warrendale, PA, 1981.

Faith, W. L., and Atkisson Jr., A. A., *Air Pollution*, 2nd ed. Wiley (Interscience), New York, 1972.

Fuel and Combustion Effects on Particulate Emissions, SP-502. Society of Automotive Engineers, Warrendale, PA, 1981.

Miller, S. A., and Theis, T. L., Comparison of life-cycle inventory databases: a case study using soybean production. *J. Ind. Ecol.* **10**, 134–147, 2006.

Wolf, G. T., and Frosch, R. A., Impact of alternative fuels on vehicle emissions of greenhouse gases. *J. Air Waste Manage. Assoc.* **41** (12), 1172–1176 (December 1991).

QUESTIONS

1. Would you expect to find the same chemical composition of the hydrocarbons from the exhaust of a gasoline-powered automobile as that of gasoline in the vehicle's tank? Why?
2. What would be the effect on emissions from a gasoline-powered vehicle if it was designed to be operated on leaded fuel and an unleaded fuel was used?
3. What would be the effect on emissions from a gasoline-powered vehicle if it was designed to be operated on unleaded fuel and a leaded fuel was used?
4. Why might you expect EGR on a diesel engine to increase the particulate matter emissions?
5. Considering the wide range of sizes of automotive engines, do you feel that an emission standard in $g\,km^{-1}$ rather than $\mu g\,m^{-3}$ or ppm is equitable to all automobile manufacturers?
6. Discuss the emissions you would expect from a jet aircraft compared to those from a piston engine aircraft.
7. If a major freeway with four lanes of traffic in one direction passes four cars per second at $100\,km\,h^{-1}$ during the rush period and each car carries two people, how often would a commuter train of five cars carrying 100 passengers per car have to be operated to handle the same load? Assume the train would also operate at $100\,km\,h^{-1}$.
8. An automobile traveling $50\,km\,h^{-1}$ emits 1% CO from the exhaust. If the exhaust rate is $80\,m^3\,min^{-1}$, what is the CO emission in $g\,km^{-1}$?
9. List the following in increasing amounts from the exhaust of an idling automobile: O_2, NO_x, SO_x, N_2, unburned hydrocarbons, CO_2, and CO.
10. What are the advantages and disadvantages of bio-diesel use on a large scale? What can be done to overcome the obstacles?
11. What were problems associated with three different fuel additives? How can the lessons learned from these experiences help improve future fuel additives?

36

Source Sampling and Monitoring

I. INTRODUCTION

Air pollutants released to the atmosphere may be characterized by qualitative descriptions or quantitative analysis. For example, a plume may be characterized as brown, dense smoke, or 60% opacity. It may also be described as containing a certain concentration of particulate matter (e.g. $100 \, mg \, m^{-3}$ $PM_{2.5}$). These are qualitative descriptions made by observing the effluent as it entered the atmosphere. Arguably, of most concern are the quantitative data regarding the effluent. How many parts per million of a criteria or toxic pollutant? How many kilograms per hour? How many kilograms per year? To obtain these numbers, it becomes necessary to sample or to monitor the effluent. Sampling and monitoring, therefore, are necessary for air pollution evaluation and control. In any situation concerning atmospheric emission of pollutants, source sampling or monitoring is necessary to obtain accurate data. Figure 36.1 shows a simple source test being conducted.

II. SOURCE SAMPLING

The purpose of source sampling is to obtain as representative, precise and accurate a sample as possible of the material entering the atmosphere at a

Fig. 36.1. Source test.

minimum cost. This statement needs to be examined in light of each source test conducted. The following issues should continually be considered: (1) Is the sampling and collecting of the material representative of what is actually being released? Is this the material entering the atmosphere? Sampling at the base of a tall stack may be much easier than sampling at the top, but the fact that a pollutant exists in the breeching does not mean that it will eventually be emitted to the atmosphere. Molecules can also undergo both physical and chemical changes before leaving the stack. (2) Maximum accuracy in sampling is desirable. Is maximum accuracy attainable? Decisions regarding the total effluent will be based on what was found from a relatively small sample. Only if the sample accurately represents the total will the extrapolation to the entire effluent be valid. (3) Collecting a sample is a costly and time-consuming process. The economics of the situation must be considered and the costs minimized consistent with other objectives. It makes little sense to spend \$5000 on an extensive stack testing analysis to decide whether to purchase a \$10 000 scrubber of 95% efficiency or to try to get by with a \$7000 scrubber of 90% efficiency.

The reasons for performing a source test differ. The test might be necessary for one or more of the following reasons: (1) To obtain data concerning the emissions for an emission inventory or to identify a predominant source in the area. An example of this would be determination of the hydrocarbon release from a new type of organic solvent used in a degreasing tank. (2) To determine compliance with regulations. If authorization is obtained to construct an incinerator and the permit states that the maximum allowable particulate emission is 230 mg per standard cubic meter corrected to 12% CO_2, a source test must be made to determine compliance with the permit. (3) To gather information which will enable selection of appropriate control equipment. If a source test determines that the emission is 3000 mg of particulate per cubic meter and that it has a weight mean size of 5 μm, a control device must be chosen which will collect enough particulate to meet some required standard, such as 200 mg per cubic meter. (4) To determine the efficiency of control equipment installed to reduce emissions. If a manufacturer supplies a device guaranteed to be 95% efficient for removal of particulate with a weight mean size of 5 μm, the effluent stream must be sampled at the inlet and outlet of the device to determine if the guarantee has been met (see Chapter 32 for a discussion of how to calculate efficiency).

III. STATISTICS OF SAMPLING

Recall that most statistics are inferential. That is, we must infer the conditions of a larger population from a much smaller sample. Thus, we must be careful in how we interpret the meaning of a sample. A sample collected at the rate of $0.3\,\mathrm{L\,min^{-1}}$ from a stack discharging $2000\,\mathrm{m^3\,min^{-1}}$ to the atmosphere is likely to include substantially large error. Another term for bias is systematic error. If the sample is truly representative, it is said to be both accurate and unbiased. If the sample is not representative, it may be biased because of some consistent phenomenon (some of the hydrocarbons condense in the tubing ahead of the trap) or in error because of some uncontrolled variation (only 1.23 g of sample was collected, and the analytical technique is accurate to ±0.5 g) [1].

For practical purposes, source testing can be considered as simple random sampling [2]. The source may be considered to be composed of such a large population of samples that the population N is infinite. From this population, n units are selected in such a manner that each unit of the population has an equal chance of being chosen. For the sample, we can determine the sample mean, $\overline{y}$:

$$\overline{y} = \frac{y_1 + y_2 + \cdots + y_n}{n} \tag{36.1}$$

If the sample is unbiased we can estimate the source mean, so that:

$$\overline{\Upsilon} = \overline{y} \tag{36.2}$$

For example, if we were to take six samples of carbon monoxide from the exhaust of an idling automobile and obtain the CO percentages as shown in Table 36.1. The sample mean is:

$$\overline{y} = \frac{1.8 + 1.6 + 1.8 + 1.9 + 1.7 + 1.8}{6} = 1.767$$

TABLE 36.1

Idling Internal Combustion Engine, CO Percentages

Test number	CO (%)
1	1.8
2	1.6
3	1.8
4	1.9
5	1.7
6	1.8

The source mean is assumed to be the same if the sample is unbiased, as seen by:

$$\bar{\Upsilon} = \bar{y} = 1.767 \tag{36.3}$$

The variance of the sample and the population (source) may also be assumed equal if the sample is unbiased. The variance is S^2, defined as:

$$S^2 = \frac{\sum_1^n (y_i - \bar{y})^2}{n - 1} \tag{36.4}$$

The variance of the source is usually calculated by the formula:

$$s^2 = \frac{1}{n-1}\left[\sum y_i^2 - \frac{\left(\sum y\right)^2}{n}\right] \tag{36.5}$$

For the preceding example, this is found as follows:

$$\sum y_i^2 = 18.78, \quad \sum y_i = 10.6, \quad n = 6$$

$$s^2 = \frac{1}{6-1}\left[18.78 - \frac{(10.6)^2}{6}\right] = 0.01067$$

The standard deviation of the sample is defined as the square root of the variance. For the example, the standard deviation is:

$$s = (s^2)^{1/2} = (0.01067)^{1/2} = 0.103$$

The sample represents a population (source) which, if normally distributed, has a mean of 1.767% and a standard deviation of 0.103%. This can be illustrated as shown in Fig. 36.2.

The inference from the statistical calculations is that the true mean value of the carbon monoxide from the idling automobile has a 66.7% chance of being

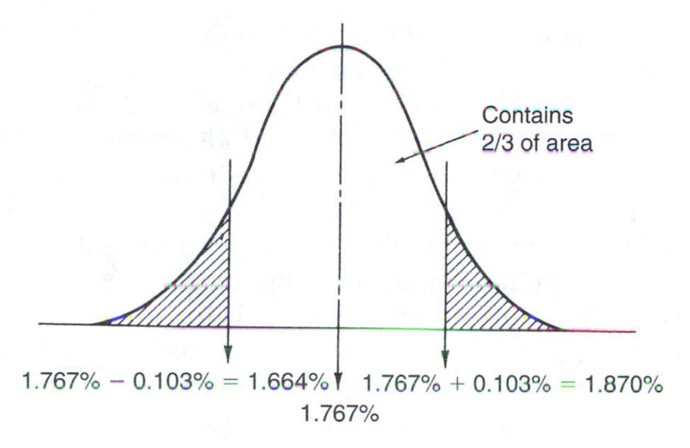

Fig. 36.2. Distribution of carbon monoxide from an automotive source.

between 1.664% and 1.870%. The best single number for the carbon monoxide emission would be 1.767% (the mean value).

Further statistical procedures can be applied to determine the confidence limits of the results. Generally, only the values for the mean and standard deviation would be reported. The reader is referred to any good statistical text to expand on the brief analysis presented here.

IV. THE SOURCE TEST

A. Test Preliminaries

The first thing that must be done for a successful source test is a complete review of all relevant background material. The test request may come in either verbal or written form. If it is verbal, it should be put into writing for the permanent record. The request may contain much or little information, but it is important to verify that it is complete and understood. Questions to ask are (1) Why should the test be made? Is it to measure a specific pollutant such as SO_2, or is it to determine less specific goal, such as identifying where a loss of a compound is occurring (e.g. between a reactor and stack) or what is causing the odor problem in the new residential area? (2) What will the test results be used for? Will it be necessary to go to court, or are the results for general information only? This may make a difference regarding the test method selected or of the necessary precision and accuracy of a given test. (3) What equipment or process is to be tested? (4) What are its operational requirements? (5) What methods would be preferred by the analytical group? (6) Are the analytical methods standard or unique? (7) Can all contaminants be sampled in a single test or will a series of test be needed (or separate tests for different target analytes)?

A literature search regarding the process and test should be conducted unless the test crew is thoroughly familiar with the source and all possible test methods. It is important to check the regulations regarding the process and specific test procedures as a part of the search [3].

When all the background material has been reviewed, it is time to inspect the source to be tested. The inspector should be accompanied by the plant manager or someone who knows the process in detail. It is also important that any technicians or mechanics be contacted at this time regarding necessary test holes, platforms, scaffolding, power requirements, etc. During this inspection, checks should be made for environmental conditions and space requirements at the sampling site. Every visitor to the site, whether an employee of the organization or a third-party inspector, should be thoroughly familiar with all safety requirements and possible hazards. In fact, safety materials should be obtained and reviewed *before* setting foot on the premises. A number of companies require any visitor to complete safety training for certain installations. Such safety training should be completed prior to entry. Testing in a noisy or dusty place at elevated temperatures is certainly uncomfortable and possibly hazardous. Rough estimates of several important factors should be made at this time. These estimates can be noted in writing during the inspection. A simple check sheet, such as the one shown in Fig. 36.3, should be a great aid.

The information obtained during the background search and from the source inspection will enable selection of the test procedure to be used. The

SOURCE TEST PRELIMINARY VISIT CHECK LIST

Plant ______

Location ______

By ______ Date ______

1. Gas flow at test point, m/min______, m^3/min ______
2. Gas temperature, °C ______
3. Gas pressure, mm of water (±) ______
4. Gas humidity, R. H., % ______
5. Pollutants of concern ______
6. Estimate of concentration ______
7. Any toxic materials? ______
8. Test crew needed ______
9. Site check:
   Electric power ______ Test holes ______
   Ambient temperature ______ Illumination ______
   Platform ______ Scaffolding ______
   Hoist ______ Ladders ______
   Test date ______
10. Environmental or safety gear ______
11. Personnel involved (names)
   Plant manager or foreman ______
   Mechanic or electrician ______

Fig. 36.3. Source test checklist.

choice will be based on the answers to several questions: (1) What are the legal requirements? For specific sources there may be only one acceptable method. (2) What range of accuracy is desirable? Should the sample be collected by a procedure that is ±5% accurate, or should a statistical technique be used on data from eight tests at ±10% accuracy? The same is true for acceptable precision. Costs of different test methods will certainly be a consideration here. (3) Which sampling and analytical methods are available that will give the required accuracy for the estimated concentration? An Orsat gas analyzer with a sensitivity limit of ±0.02% would not be chosen to sample carbon monoxide at 50–100 ppm. Conversely, an infrared gas analyzer with a full-scale deflection of 1000 ppm would not be chosen to sample CO_2 from a power boiler. (4) Is a continuous record required over many cycles of source operation, or will one or more grab samples suffice? If a source emits for only a short period of time, a method would not be selected which requires hours to gather the required sample.

The test must be scheduled well in advance for the benefit of all concerned. The plant personnel, as well as the test crew, should be given the intended date and time of the test. It is also a good idea to let the chemist or analytical service know when the testing will be conducted so that they can be ready to do their portion of the work. It may be necessary to schedule or rent equipment in advance, such as boom trucks or scaffolding. When scheduling the test, make sure that the source will be operating in its normal manner. A boiler may be operating at only one-third load on weekends because the plant steam load is off the line and only a small heating load is being carried.

B. Gas Flow Measurement

Gas flow measurement is a very important part of source testing. The volume of gaseous effluent from a source must be determined to obtain the mass loading to the atmosphere. Flow measurement through the sampling train is necessary to determine the volume of gas containing the pollutant of interest. Many of the sampling devices used for source testing have associated gas flow indicators which must be continually checked and calibrated.

Gas flows are often determined by measuring the associated pressures. Figure 36.4 illustrates several different pressure measurements commonly made on systems carrying gases. Static pressure measurements are made to adjust the absolute pressure to standard conditions specified in the test procedure.

The quantity of gaseous effluent leaving a process is usually calculated from the continuity equation, which for this use is written as

$$Q = AV \tag{36.6}$$

where Q is the flow at the specified conditions of temperature, pressure, and humidity; A is the area through which the gas flows; and V is the velocity of the effluent gas averaged over the area.

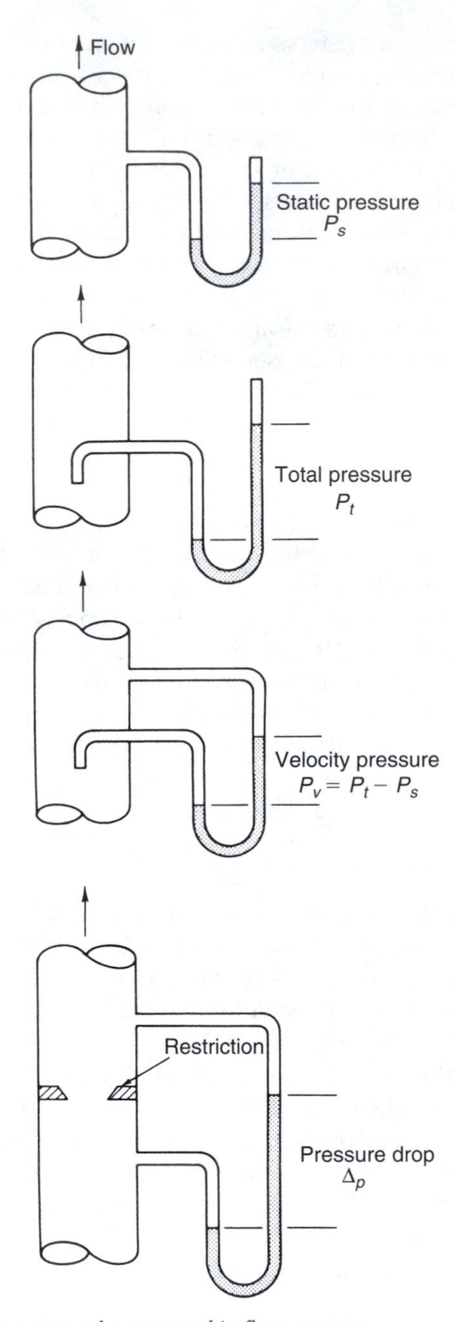

Fig. 36.4. Pressures commonly measured in flow systems.

A is commonly measured, and *V* determined, to calculate *Q*. The velocity *V* is determined at several points, in the center of equal duct areas, and averaged. Table 36.2 shows one commonly accepted method of dividing stacks or ducts into equal areas for velocity determinations.

For rectangular ducts, the area is evenly divided into the necessary number of measurement points. For circular ducts, Table 36.3 can be used to determine the location of the traverse points. In using this table, realize that traverses are made along two diameters at right angles to each other, as shown in Fig. 36.5.

In most source tests, the measurement of velocity is made with a pitotstatic tube, usually referred to simply as a *pitot tube*. Figure 36.6 illustrates the two types of pitot tubes in common use.

The standard type of pitot tube shown in this figure does not need to be calibrated, but it may be easily plugged in some high-effluent loading streams. The type S pitot tube shown in Fig. 36.6 does not plug as easily, but it does need calibration to assure its accuracy. The type S pitot tube is also more sensitive to alignment with the gas flow to obtain the correct reading. The velocity

TABLE 36.2

Number of Velocity Measurement Points

Stack diameter or (length + width)/2 (m)	Number of velocity measurement points
0.0–0.3	8
0.3–0.6	12
0.6–1.3	16
1.3–2.0	20
2.0→	24

TABLE 36.3

Velocity Sampling Locations, Diameters from Inside Wall to Traverse Point

	Number of equal areas to be sampled				
Point number	2	3	4	5	6
1	0.067	0.044	0.033	0.025	0.021
2	0.250	0.147	0.105	0.082	0.067
3	0.750	0.295	0.194	0.146	0.118
4	0.933	0.705	0.323	0.226	0.177
5		0.853	0.677	0.342	0.250
6		0.956	0.806	0.658	0.355
7			0.895	0.774	0.645
8			0.967	0.854	0.750
9				0.918	0.823
10				0.975	0.882
11					0.933
12					0.979

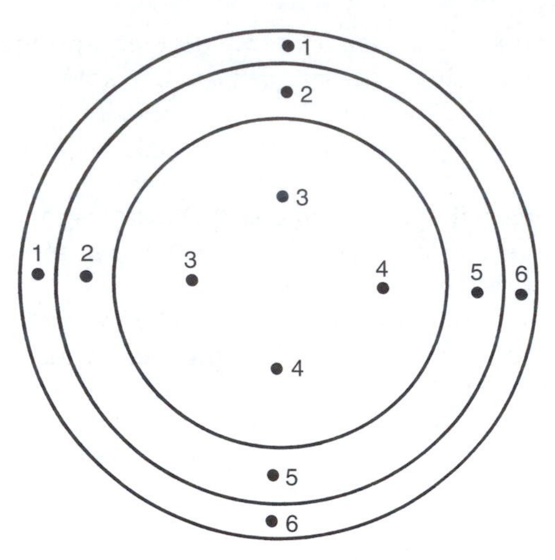

Fig. 36.5. Circular duct divided into three equal areas, as described in Table 36.3. Numbers refer to sampling points.

pressure of the flowing gas is read at each point of the traverse, and the associated gas velocity is calculated from the formula:

$$V = 420.5[(P_v/\rho)^{1/2}] \tag{36.7}$$

where V is the velocity in meters per minute, P_v is the velocity pressure in millimeters of water, and ρ is the gas density in kilograms per cubic meter. The velocities are averaged for all points of the traverse to determine the gas velocity in the duct. Velocity pressures should not be averaged, as a serious error results.

Gas velocities can also be measured with anemometers (rotating vane, hot wire, etc.), from visual observations such as the velocity of smoke puffs, or from mass balance data (knowing the fuel consumption rate, air/fuel ratio, and stack diameter).

In the sampling train itself, the gas flow must be measured to determine the sample volume. Particulates and gases are measured as micrograms per cubic meter. In either case, determination of the fraction requires that the gas volume be measured for the term in the denominator. Some sample trains contain built-in flow-indicating devices such as orifice meters, rotometers, or gas meters. These devices require calibration to assure that they read accurately at the time of the test and under test conditions.

To determine the volume through the sampling train, a positive displacement system can be used. A known volume of water is displaced by gas containing the sample. Another inexpensive procedure that works well consists of measuring the time needed for the gas to fill a plastic bag to a certain static pressure. The volume of the bag can be accurately measured under the same

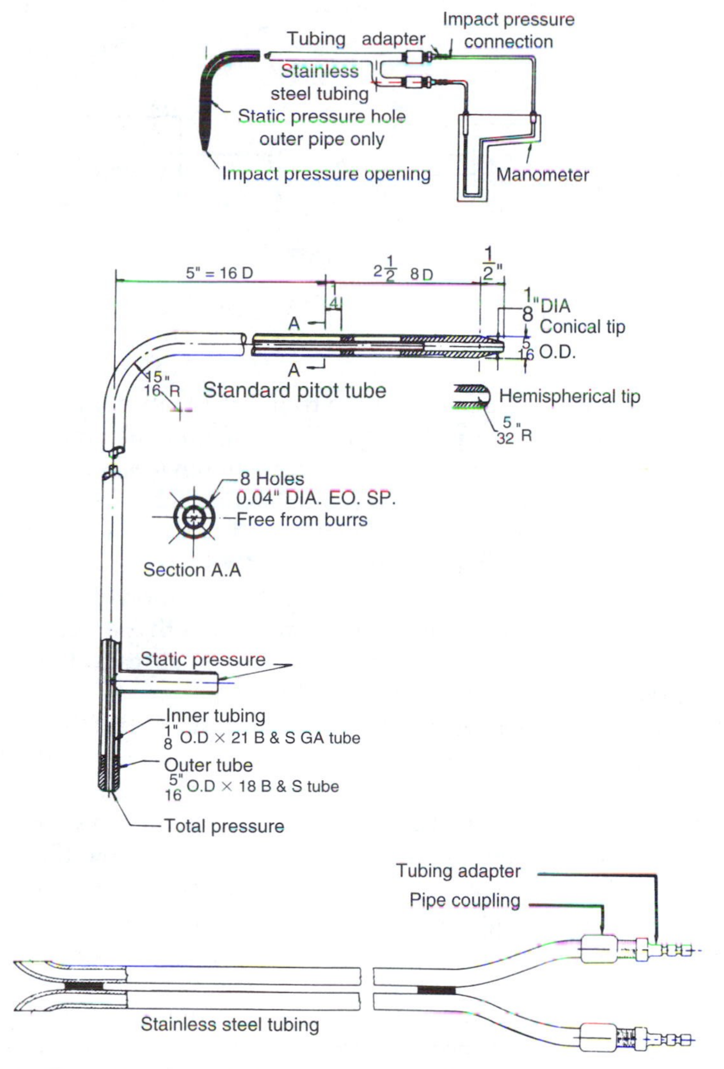

Fig. 36.6. Pitot tubes for velocity determination. *Source*: *Annual Book of Standards* [3]. *Note*: English units were used by the American Society for Testing and Materials.

conditions and hence the flow determined by dividing the bag volume by the time required to fill it.

C. Collection of the Source Sample

A typical sample train is shown in Fig. 36.7. This shows the minimum number of components, but in some systems the components may be combined. Extreme care must be exercised to assure that no leaks occur in

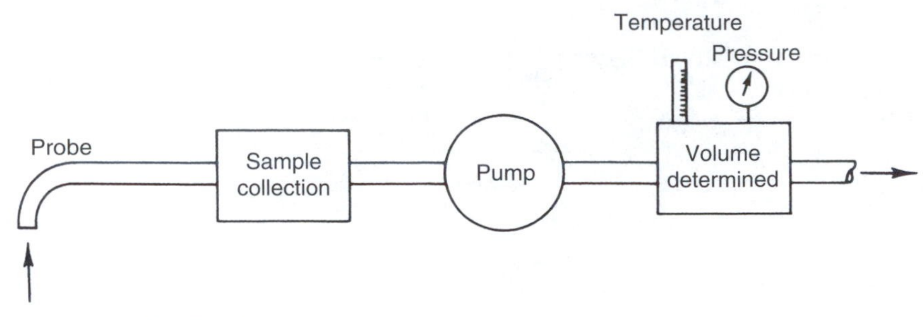

Fig. 36.7. Sampling train.

the train and that the components of the train are identical for both calibration and sampling. The pump shown in Fig. 36.7 must be both oil-less and leakproof. If the pump and volume measurement devices are interchanged, the pump no longer needs to be oil-less and leakproof, but the volume measurement will be in error unless it is adjusted for the change in static pressure. Some sampling trains become very complex as additional stages with controls and instruments are added. Many times the addition of components to a sampling train makes it so bulky and complicated that it becomes nearly impossible to use. A sampling train developed in an air-conditioned laboratory can be useless on a shaky platform in a snowstorm.

Standard sampling trains are specified for some tests. One of these standards is the system specified for large, stationary combustion sources [4]. This train was designed for sampling combustion sources and should not be selected over a simpler sampling train when sampling noncombustion sources such as low-temperature effluents from cyclones, baghouses, filters, etc. [5].

Before taking the sample train to the test site, it is wise to prepare the operating curves for the particular job. With most factory-assembled trains, these curves are a part of the package. If a sampling train is assembled from components, the curves must be developed. The type of curves will vary from source to source and from train to train. Examples of useful operating curves include (1) velocity versus velocity pressure at various temperatures [6], (2) probe tip velocity versus flowmeter readings at various temperatures, and (3) flowmeter calibration curves of flow versus pressure drop. It is much easier to take an operating point from a previously prepared curve than to take out a calculator and pad to make the calculations at the moment of the test. Remember, too, that time may be a factor and that settings must be made as rapidly as possible to obtain the necessary samples.

For sampling particulate matter, one is dealing with pollutants that have very different inertial and other characteristics from the carrying gas stream. It becomes important, therefore, to sample so that the same velocity is maintained in the probe tip as exists in the adjacent gas stream. Such sampling is called *isokinetic*. Isokinetic sampling, as well as anisokinetic sampling, is illustrated in Fig. 16.3.

If the probe velocity is less than the stack velocity, particles will be picked up by the probe, which should have been carried past it by the gas streamlines. The inertia of the particles allows them to continue on their path and be intercepted. If the probe velocity exceeds the stack velocity, the inertia of the particles carries them around the probe tip even though the carrying gases are collected. Adjustment of particulate samples taken anisokinetically to the correct stack values is possible if all of the variables of the stack gas and particulate can be accounted for in the appropriate mathematical equations.

Modern transducers and microprocessors have been used successfully to automate particulate sampling trains in order to eliminate the operating curves and manual adjustments [7]. The automated samplers adjust continuously to maintain isokinetic conditions. In addition, the microprocessor continuously calculates and displays both instantaneous sampling conditions and the total sample volume collected at any given moment. The use of the automated system with the microprocessor, therefore, eliminates both operator and calculation errors.

Several separating systems are used for particulate sampling. All rely on some principle of separating the aerosol from the gas stream. Many of the actual systems use more than one type of particulate collection device in series. If a size analysis is to be made on the collected material, it must be remembered that multiple collection devices in series will collect different size fractions. Therefore, size analyses must be made at each device and mathematically combined to obtain the size of the actual particulate in the effluent stream. In any system the probe itself removes some particulate before the carrying gas reaches the first separating device, so the probe must be cleaned and the weight of material added to that collected in the remainder of the train.

Care should be exercised when sampling for aerosols that are condensable. Some separating systems, such as wet impingers, may remove the condensables from the gas stream, whereas others, such as electrostatic precipitators, will not. Of equal concern should be possible reactions in the sampling system to form precipitates or aerosols which are not normally found when the stack gases are exhausted directly to the atmosphere. SO_3 plus other gaseous products may react in a water-filled impinger to form particulate matter not truly representative of normal SO_3 release.

When sampling particulate matter from combustion processes, it is necessary to take corresponding CO_2 readings of the effluent. Emission standards usually require combustion stack gases to be reported relative to either 12% CO_2 or 50% excess air. Adjusting to a standard CO_2 or excess air value normalizes the emission base. Also, emission standards require that the loadings be based on weight per standard cubic volume of air (usually at 20°C and 760 mmHg). In most regulations, the agency requires that the standard volume be dry, but this is not always specified.

For sampling of gases, the sample can be collected by any of several devices. Some commonly used manual methods include Orsat analyzers, absorption systems, adsorption systems, bubblers, reagent tubes, condensers, and traps.

Continuous analyzers are now more widely used than manual methods. Some types of continuous analyzers include infrared and ultraviolet instruments; flame ionization detectors; mass spectrometers; calorimetric systems; gas, liquid, and solid chromatography; coulometric and potentiometric systems; chemiluminescence; and solid-state electronic systems. Since gases undergoing analysis do not need to be sampled isokinetically, it is only necessary to insert a probe and withdraw the sample. Usually, the gas sample should be filtered to remove any accompanying particulate matter which could damage the analytical instrumentation.

For the detection and intensity of odorous substances, the nose is still the instrument usually relied upon. Since odors are gaseous, they may be sampled by simply collecting a known volume of effluent and performing some manipulation to dilute the odorous gas with known volumes of "pure" air. The odor is detected by an observer or a panel of observers. The odor-free air for dilution can be obtained by passing air through activated carbon or any other substance that removes all odors while not affecting the other gases that constitute normal air. The odor-free air is then treated by adding more and more of the odorous gas until the observer just detects the odor. The concentration is then recorded as the odor threshold as noted by that observer. The test is not truly quantitative, as much variation between observers and samples is common.

If the compound causing the odor is known and can be chemically analyzed, it may be possible to get valid quantitative data by direct gas sampling. An example would be a plant producing formaldehyde. If the effluent were sampled for formaldehyde vapor, this could be related, through proper dispersion formulas, to indicate whether the odor would cause any problems in residential neighborhoods adjacent to the plant.

Extreme care should be taken in transporting and storing the samples between the time of collection and the time of analysis. Some condensable hydrocarbon samples have been lost because the collection device was subjected to elevated temperatures during shipment. Equally disastrous is placing the sample in an oven at 105°C to drive off the moisture, only to discover that the particles of interest had a very low vapor pressure and also departed the sample. At such times, source sampling can be very frustrating.

A very important analytical tool that is overlooked by many source-testing personnel is the microscope. Microscopic analysis of a particulate sample can tell a great deal about the type of material collected as well as its size distribution. This analysis is necessary if the sample was collected to aid in the selection of a piece of control equipment. All of the efficiency curves for particulate control devices are based on fractional sizes. One would not try to remove a submicron-size aerosol with a cyclone collector, but unless a size analysis is made on the sampled material, one is merely guessing at the actual size range.

Scanning electron microscopy (SEM) may be used for analysis of particles in air (see Figures 1.4 and 36.8). Air samples are collected on filters and prepared for analysis. Filters are carbon or gold-coated and mounted so that an electron beam is directed at the sample, and emissions are measured by a detector at an angle to the electron beam. An image of the surface features on

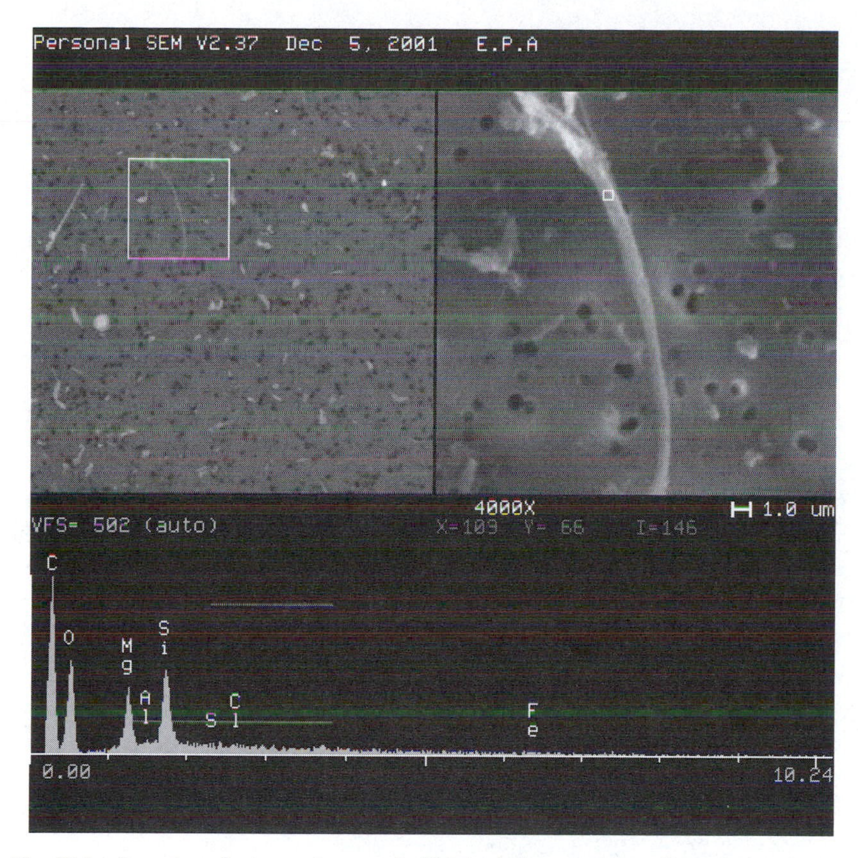

Fig. 36.8. Scanning electron micrograph (SEM) of fibers in dust collected near the World Trade Center, Manhattan, NY, in September 2001. Acquired using an Aspex Instruments, Ltd., SEM. The bottom of the micrograph represents the elemental composition of the highlighted 15-μm long fiber by energy dispersive spectroscopy (EDS). This composition (i.e. O, Si, Al, and Mg) and the morphology of the fibers indicate they are probably asbestos. The EDS carbon peak results from the dust being scanned on a polycarbonate filter. Source: US Environmental Protection Agency, 2004. Photo courtesy of T. Conner, used with permission.

the filter can then be observed at magnifications commonly up to 20 000X and higher, when needed. Many SEM systems also include energy dispersive spectrometry (EDS) to complement the morphology from SEM with the elemental composition of the particle. However, EDS will not provide individual chemical species. For example, the EDS spectrum may show that the particle consists of 20% iron and 30% zinc, but the actual compounds (e.g. iron sulfide, iron oxide, or organic iron and zinc compounds) are not known. In such cases, scientific judgment may be needed. For example, if the sample is collected near a smelter and the stack is known to generate oxidized species, metal oxides are more likely to dominate. In a reduced environment, conversely, the sulfides are likely to account for a larger part of the particle's composition.

D. Calculations and Report

Calculations that are repeatedly made can be made more accurately, and at lower cost, by using a computer. If, for example, automotive emissions are continually tested over a standardized driving cycle, a computer program to analyze the data is a necessity. Otherwise, days would be spent calculating the data obtained in hours.

For sampling a relatively small number of sources, a simplified calculation form may be used. Such forms enable the office personnel to perform the arithmetic necessary to arrive at the answers, freeing the technical staff for proposals, tests, and reports. Many of the manufacturers of source-testing equipment include example calculation forms as part of their operating manuals. Some standard sampling methods include calculation forms as a part of the method [8]. Many control agencies have developed standard forms for their own use and will supply copies on request.

The source test report is the end result of a large amount of work. It should be thorough, accurate, and written in a manner understandable to the person who intends to use it. It should state the purpose of the test, what was tested, how it was tested, the results obtained, and the conclusions reached. The actual data and calculations should be included in the appendix of the source test report so that they are available to substantiate the report if questioned.

V. SOURCE MONITORING

The monitoring of pollutant concentration or mass flow of pollutants is of interest to both plant owners and control agencies. Industry uses such measurements to keep a record of process operations and emissions for its own use and to meet regulatory requirements. Control officials use the information for compiling emission inventories, modeling of air sheds, and in some cases for enforcement.

A monitoring system is selected to meet specific needs and is tailored to the unique properties of the emissions from a particular process. It is necessary to take into account the specific process, the nature of the control devices, the peculiarities of the source, and the use of the data obtained [8].

Source monitoring can best be treated as a system concept ideally consisting of six unit operations, as shown in Table 36.4. In the United States, installation and operation of monitoring systems have been prescribed for a number of industries, as shown in Table 36.5.

A. Types of Monitors

Continuous emission monitors (CEMs) for plume opacity have been required on all utility, fossil fuel-fired, steam generators (over 264 MJ)

TABLE 36.4

System Concept of Stationary Source Measurements

Operation	Objective
Sampling site selection	Representative sampling consistent with intended interpretation of measurement
Sample transport (when applicable)	Spatial and temporal transfer of sample extract with minimum and/or known effects on sample integrity
Sample treatment (when applicable)	Physical and/or chemical conditioning of sample consistent with analytical operation, with controlled and/or known effects on sample integrity
Sample analysis	Generation of qualitative and quantitative data on pollutant or parameter of interest
Data reduction and display	Calibration and processing of analog data and display of final data in a format consistent with measurement objectives
Data interpretation	Validly relating the measurement data to the source environment within the limitations of the sampling and analytical operations

Source: Nader [9].

constructed in the United States since December 1971. These monitors are *in situ* opacity meters which measure the attenuation of a light beam projected across the stack (see Fig. 27.2). Remote-sensing monitors have been developed, but these have not yet been approved as equivalent to the *in situ* opacity monitors. CEMs for gaseous emissions are also available and required for certain facilities. Figure 36.9 illustrates the various approaches to monitoring particulate opacity and gaseous emissions.

B. Quality Assurance in Monitoring

In order to assure that the source is being accurately monitored, several requirements must be met [1]. Some of these requirements, which assure representative, noncontaminated samples, are shown in Table 36.6.

C. Monitoring of Particulate Emissions

The most common monitoring of particulate matter is for light attenuation (opacity). Less frequently used methods exist for monitoring mass concentration, size distribution, and chemical composition.

Opacity is a function of light transmission through the plume. Opacity is defined as follows:

$$\text{Opacity} = (1 - I/I_0) \times 100 \qquad (36.8)$$

where I_0 is the incident light flux and I is the light flux leaving the plume. Techniques for monitoring visible emissions (opacity) are listed in Table 36.7.

TABLE 36.5

Source Emissions Requiring Continuous Monitoring by United States New Source Performance Standards

Source	Pollutant						Scrubber pressure loss and water pressure	Flow rate
	SO_2	NO_x	CO	Opacity	H_2S	Total reduced sulfur		
Electric power plants	x	x		x				
Sulfuric acid plants	x							
Onshore natural gas processing	x							
SO_2 emissions								
Nitric acid plants		x						
Petroleum refineries	x		x	x	x	x		
Iron and steel mills (BOF)[a]							x	
Steel mills (electric arc)				x				x
Ferroalloy production				x				
Glass manufacturing plants				x				
Portland cement plants				x				x
Primary copper smelters	x			x				
Primary zinc smelters	x			x				
Primary lead smelters	x			x				
Coal preparation plants							x	
Wet process phosphoric acid plants							x	
Superphosphoric acid plants							x	
Diammonium phosphate plants							x	
Triple superphosphate plants							x	
Granular triple superphosphate plants							x	
Phosphate rock plants				x			x	
Metallic mineral processing plants							x	
Nonmetallic mineral processing plants							x	
Kraft pulp mills				x		x	x	x
Gas turbines[b]								
Lime kilns[c]				x		x	x	
Ammonium sulfate plants							x	
Lead–acid battery manufacture							x	

[a] BOF: basic oxygen furnace.
[b] Monitor sulfur and nitrogen content of fuel and water/fuel ratio.
[c] Also monitor scrubber liquid flow rate.
Source: Code of Federal Regulations [8].

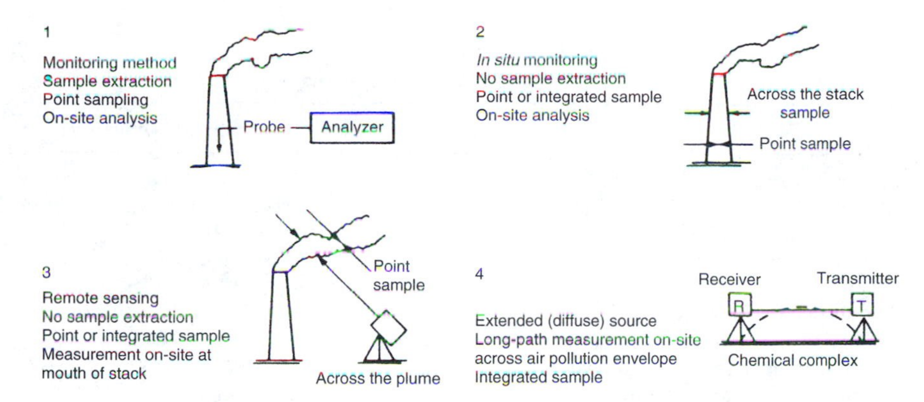

Fig. 36.9. Sampling approaches to monitoring source emissions. *Source*: Nader [9].

TABLE 36.6

Stationary Source Monitoring Requirements

Requirement	Method of attainment
Maintain gas temperature above the water or acid dew point	Heat lines or dilute with dry air
Remove water before sample enters instrument	Refrigerate or desiccate sample
Remove particulate matter before sample enters instrument	Use cyclone or filter in sample line
Dilute sample to lower the temperature to an acceptable level for the instrument	Air dilute with necessary blowers, flow measurement, and control systems
Maintain integrity of particulate sample (mass, size, and chemical composition)	Use isokinetic sampling, refrigerated sample transport, and careful handling to minimize physical or chemical changes

TABLE 36.7

Opacity Monitoring Techniques

Method of analysis	Measurement system
In-stock opacity	Optical transmissometer
Plume opacity	Lidar (light detection and ranging)
Selective opacity (for fine particles)	Extractive with light-scattering determination

D. Monitoring of Gaseous Emissions

Gas-monitoring systems are more widely used than particulate monitoring systems. They can also be used for both emission compliance monitors and process control systems. Gas monitors may be of either the *in situ* or the

TABLE 36.8

Gas Emission Monitoring Systems

Analytical scheme	Sampling approach[a]	Pollutant capability[b]
Chemielectromagnetic		
Colorimetry	E	SO_2, NO_x, H_2S, TS
Chemiluminescent	E	NO_x
Electromagnetic/electrooptical		
Flame photometry	E	SO_2, H_2S, TRS, TS
Nondispersive infrared	E	SO_2, NO, CO, HC, CO_2
Nondispersive, ultraviolet and visible	E	SO_2, NO_x, NH_3, H_2S
Dispersive, infrared and ultraviolet	I	SO_2, NO, CO, CO_2, HC, H_2S
Dispersive, infrared and ultraviolet	E	SO_2, CO, CO_2, HC, NO
Correlation, ultraviolet	I	SO_2
Correlation, ultraviolet and visible	R	SO_2, NO_x
Derivative, ultraviolet	I	SO_2, NO_2, NO, O_2, NH_3, CO
Fluorescence, ultraviolet	E	SO_2
Electrical		
Conductivity	E	SO_2, NH_3, HCl
Coulometry	E	SO_2, H_2S, TRS
Electrochemical	E, I	SO_2, NO_x, CO, H_2S, O_2
Flame ionization	E	HC
Thermal		
Oxidation	E	CO
Conductivity	E	SO_2, NH_3, CO_2
Hybrid		
Gas chromatography, flame ionization	E	CO, HC
Gas chromatography, flame photometry	E	Sulfur compounds

[a] E, extractive; I, *in situ*; R, remote.
[b] HC, hydrocarbons; TRS, total reduced sulfur; TS, total sulfur; NO_x, total oxides of nitrogen, but system may be specific for NO, NO_2, or both.
Source: Nader [9].

extractive type and use the approaches illustrated in Fig. 36.9. Table 36.8 lists the various types of gas-monitoring systems.

E. Data Reduction and Presentation

Continuous monitors usually indicate the pollutant concentration on both an indicator and a chart recording. This provides a visual indication of the instantaneous emissions, along with a permanent record of the quantitative emissions over a period of time. The monitoring system may also be equipped with an alarm device to signal the operator if the allowable emission level is being exceeded. Data-logging systems coupled with microprocessors are popular. These systems can give instantaneous values of the variables and pollutants of interest, along with the averages or totals for the period of concern.

REFERENCES

1. *Quality Assurance Handbook for Air Pollutant Measurement Systems*, Vol. III, *Stationary Source Specific Methods*, EPA-600/4-77-027b. US Environmental Protection Agency, Research Triangle Park, NC, 1977.
2. Stern, A. C. (ed.), *Air Pollution*, 3rd ed., Vol. III. Academic Press, New York, 1977.
3. *1981 Annual Book of ASTM Standards, Part 26, Gaseous Fuels; Coal and Coke; Atmospheric Analysis*. American Society for Testing and Materials, Philadelphia, PA, 1981.
4. Environmental Protection Agency—Standards of performance for new stationary sources. *Fed. Regist.* **36** (247), 2488 (1971).
5. Boubel, R. W., *J. Air. Pollut. Control Assoc.* **21**, 783–787 (1971).
6. Baumeister, T. (ed.), *Marks' Standard Handbook for Mechanical Engineers*, 8th ed. McGraw-Hill, New York, 1978.
7. Boubel, R. W., Hirsch, J. W., and Sadri, B., Particulate sampling has gone automatic, *Proceedings of the 68th Annual Meeting of the Air Pollution Control Association*, Pittsburgh, PA, 1975.
8. Code of Federal Regulations, Chapter I, Part 60, Standards of performance for new stationary sources, 1992.
9. Nader, J. S., Source monitoring, in *Air Pollution* (Stern, A. C., ed.), 3rd ed., Vol. III. Academic Press, New York, 1976.

SUGGESTED READING

Brenchley, D. L., Turley, C. D., and Yarmac, R. F., *Industrial Source Sampling*. Ann Arbor Science Publishers, Ann Arbor, MI, 1973.

Cooper, H. B. H., and Rossano, A. T., *Source Testing for Air Pollution Control*. Environmental Sciences Services Division, Wilton, CT, 1970.

Frederick, E. R. (ed.), *Proceedings, Continuous Emission Monitoring: Design, Operation, and Experience, Specialty Conference*. Air Pollution Control Association, Pittsburgh, PA, 1981.

Morrow, N. L., Brief, R. S., and Bertrand, R. R., *Chem. Eng.* **79**, 85–8 (1972).

QUESTIONS

1. During a pitot traverse of a duct, the following velocity pressures, in millimeters of water, were measured at the center of equal areas: 13.2, 29.1, 29.7, 20.6, 17.8, 30.4, 28.4, and 15.2. If the flowing fluid was air at 760 mmHg absolute and 85°C, what was the average gas velocity?
2. Would you expect errors of the same magnitude when sampling anisokinetically at 80% of stack velocity as when sampling anisokinetically at 120% of stack velocity? Explain.
3. Suppose a particulate sample from a stack is separated into two fractions by the sampling device. Both are sized microscopically and found to be lognormally distributed. One has a count mean size of 5.0 μm and a geometric deviation of 2.0. The other has a count mean size of 10.0 μm and a geometric deviation of 2.2. Two grams of the smaller-sized material were collected for each 10 g of the larger. What would be reported for the weight mean size and geometric deviation of the stack effluent?
4. A particulate sample was found to weigh 0.0216 g. The sample volume from which it was collected was 0.60 m^3 at 60°C, 760 mmHg absolute, and 90% relative humidity. What was the stack loading in milligrams per standard cubic meter?

5. A particulate sample was found to contain 350 mg m^{-3}. The CO_2 during the sampling period averaged 7.2%. If the exhaust gas flow was 2000 $m^3 min^{-1}$, what would be the particulate loading in both milligrams per cubic meter and kilograms per hour, corrected to 12% CO_2?
6. Give an example of how opacity monitoring of a coal-fired boiler could be used to improve combustion efficiency.
7. An opacity monitor is set so that the incident light is 100 units. Prepare a graph of the percentage of opacity versus the light flux leaving the plume (opacity, 0–100%; exiting light flux, 0–100 units).
8. List the advantages and disadvantages of both *in situ* and extractive gas monitors.
9. Discuss the advantages and disadvantages of one-time source testing for a specific emission versus continuous monitoring of the same emission.

37

The Future of Air Pollution

I. THE GOOD NEWS

Remarkable progress has been made in improving air quality since the 1970s. All of the criteria pollutants in many countries have been reduced dramatically. The science, engineering, and technology needed to control the release of air pollutants has steadily marched forward. New paradigms and attitudes, such as pollution prevention and green engineering, have begun to replace old ways of thinking, so that fewer pollutants have to be removed at the stack. Pollutants have been designed out of the manufacturing process.

II. STUBBORN PROBLEMS AND INNOVATIVE SOLUTIONS

The progress unfortunately been accompanied by seemingly intractable problems. Global greenhouse gas emissions continue to rise. Numerous toxic compounds continue to be released. Risks to human health persist. Sensitive ecosystems are threatened. Such problems are not readily resolved by the old "command and control" approaches. Innovations and market forces must also be part of the solution.

Hopefully, this book is one of the resources for preserving and extending what we have learned in the past few decades. Ideally, it is also a baseline for the next generation of air pollution prevention and control systems.

The problems will best be addressed by new thinking that is underpinned by sound science. The fundamentals of air pollution are based in the physical sciences, but their application to addressing present and future air pollution problems must also rely on the innovations from every aspect of contemporary society.

Index

A

Q

R

S